SO-ABA-669

# AS YOU EXAMINE *HOLT* ALGEBRA 1

**Please Notice**  HOLT ALGEBRA 1 provides a complete, balanced course designed to be mastered by all first-year algebra students. All of the concepts usually studied in a first-year algebra course are found in HOLT ALGEBRA 1.

**Format**  HOLT ALGEBRA 1's proven four-part lesson format facilitates student learning:

- Lesson objectives appear on the pupil page
- Worked-out examples demonstrate solution process
- Hints and suggestions on the pupil page make learning skills and concepts easier
- Graded exercises provide for continuous skill development

## Contents

1. The course begins with an introduction to the basic operations and properties of algebra (see pp. 1—22).

2. Worked-out examples focus on teaching the solution process without complicated explanations (see pp. 70, 193).

3. Suggestions, hints, and comments encourage students to attempt more difficult skills and concepts (see pp. 128, 212, 298).

4. Oral exercises appear before written exercise sets so that the teacher can quickly test understanding and readiness (see pp. 3, 102, 214).

5. Exercises are plentiful and are graded according to three levels of difficulty. Also, exercises within each level are graded from easiest to most challenging. Examples presented in the lesson match every type of exercise in levels Ⓐ and Ⓑ (see pp. 9, 39, 185).

6. Reading in Algebra helps students analyze and better understand the technical language of algebra (see pp. 12, 56, 197).

7. Each chapter ends with a Chapter Review and Chapter Test. The Review is diagnostic and keys items to lesson sections in which skills and concepts necessary to solve items are taught.

8. Special topics pages vary the pace and enrich the course. They include practical applications of algebra in the real world, as well as an introduction to careers that use mathematics (see pp. 449, 454).

9. Calculator and computer activities familiarize students with accepted methods of logical thought processes (see pp. 290, 395).

10. Answers to selected exercises are provided in the back of the pupil edition (see pp. 540—558). Worked-out solutions to every problem in the text are available in the separate Solutions Manual.

11. A teacher's edition with a commentary for each lesson, a complete set of answers, and daily assignments with three levels of difficulty designed to reinforce the current lesson while maintaining the previous two lessons makes teaching HOLT ALGEBRA 1 easy.

---

**Supplementary Materials on Duplicating Masters**

- **HOLT ALGEBRA 1 Skillmasters** provide two parallel sections of additional exercises for every two lessons and cumulative reviews.

- **HOLT ALGEBRA 1 Testmasters** provide an alternate form of the Chapter Tests and Cumulative and Final Tests.

- **HOLT ALGEBRA 1 Teacher's Resource Package** contains Enrichment masters, Testmasters, and Skillmasters in blackline form.

# HOLT
# ALGEBRA 1

**Teacher's Edition**

# About the Authors

**Eugene D. Nichols**
Robert O. Lawton Distinguished Professor
of Mathematics Education
Florida State University
Tallahassee, Florida

**Mervine L. Edwards**
Chairman of Mathematics Department
Shore Regional High School
West Long Branch, New Jersey

**E. Henry Garland**
Head of the Mathematics Department
Developmental Research School
DRS Professor
Florida State University
Tallahassee, Florida

**Sylvia A. Hoffman**
Resource Consultant in Mathematics
Illinois State Board of Education
State of Illinois

**Albert Mamary**
Superintendent of Schools for Instruction
Johnson City Central School District
Johnson City, New York

**William F. Palmer**
Professor of Education and Director
Center for Mathematics and Science Education
Catawba College
Salisbury, North Carolina

# FOUR-PART LESSON FORMAT

**HOLT ALGEBRA 1 uses a carefully structured four-part lesson format to involve students actively in the learning process.**

*Objectives* tell the students what they will learn in each lesson.

**Hints, Comments, and Suggestions** guide the student through the solution process.

**Examples** develop the mathematical concepts being taught in a series of carefully structured steps.

## INEQUALITIES 5.2

*Objective* **To graph solution sets of inequalities**

Recall that

$a < b$ means $a$ is less than $b$, and
$b > a$ means $b$ is greater than $a$.

When you compare the numbers 3 and 7, only one of the following relations can be true:

$$3 < 7 \qquad 3 = 7 \qquad 3 > 7$$
$$\text{true} \qquad \text{false} \qquad \text{false}$$

When you compare two numbers, you assume the following property.

**Axiom of comparison** For all real numbers $a$ and $b$, one and only one of the following is true:
$$a < b, \qquad a = b, \qquad \text{or} \qquad a > b.$$

There are many values of $x$ that make $x < 5$ true.

$$4\frac{1}{2} < 5 \qquad 4 < 5 \qquad 2 < 5 \qquad 0 < 5 \qquad -3\frac{1}{2} < 5$$

You could go on forever listing the values of $x$ that make $x < 5$ true. All numbers less than 5 will make $x < 5$ true. You can show this on a number line. Any point to the left of the point with coordinate 5 has a coordinate less than 5. Draw an arrow to show all such points.

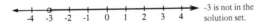

Notice that the circle at 5 is open. Since 5 is not less than 5, 5 is *not* one of the solutions. The solution set of $x < 5$ is an infinite set, {all numbers less than 5}.

*Example 1* **Graph the solution set of $a > -3$.**

The solution set of $a > -3$ is {all numbers greater than $-3$}.

-3 is not in the solution set.

The symbol for *is less than or equal to* is $\leq$. Thus, $x \leq 4$ is read as $x$ is less than or equal to 4. Similarly, $x \geq 9$ is read as *x is greater than or equal to 9*. Use of the symbol $\leq$ is illustrated in Example 2.

Inequalities

**121**

**T-8**

# BENEFITS

**HOLT ALGEBRA 1 is designed to give students a solid foundation in basic algebraic concepts and skills. By also highlighting some of the key areas of introductory geometry, basic logic, and introductory trigonometry, it provides the underlying groundwork for success in any future mathematics courses.**

| | |
|---|---|
| **FOUR-PART LESSON FORMAT** | A consistent four-part lesson format enhances the gradual development of skills and concepts. |
| **PROBLEM-SOLVING STRAND** | A clear, logical problem-solving strategy, special problem-solving application lessons, and non-routine problems provide a meaningful approach to problem-solving. |
| **ENRICHMENT** | Calculator Activities, and special topics in Logic, Geometry, Probability, and Statistics provide a wide selection of experiences for students, depending on each student's readiness and learning aptitude. |
| **COMPUTER STRAND** | Computer applications provide hands-on exercises in BASIC. References are made to further computer applications of algebraic concepts. |
| **TESTING AND REVIEW** | Extra-practice exercises, Chapter Reviews, Chapter Tests, and Cumulative Reviews based on the material taught keep students and teachers aware of progress. |
| **TEACHER'S ANNOTATED EDITION** | The Teacher's Edition includes full-size annotated pupil's pages, chapter objectives, Cumulative Reviews, Activities, additional chalkboard work, and approaches to common student errors, as well as detailed teaching suggestions. |
| **MOTIVATIONAL FEATURES** | College Prep Tests acquaint students with College Board test-taking strategies. In addition to an expanded use of calculators and computers, historical puzzles and information regarding related careers also furnish motivation. Special photographic sections pique students' interest and accentuate the relationship of algebra to our technological world. |
| **TEACHER'S RESOURCE PACKAGE** | Contains Blackline Masters for Skillmasters, Testmasters, Basic Competency Tests, and Enrichment. |
| **SUPPLEMENTARY MATERIALS** | Skillmasters (duplicating masters) for review and Extra Practice. Each Skillmaster, with two parallel forms (A and B) can be used as a pre-test or a post-test, as a diagnostic tool, as extra practice for skill-mastery, or as a regular assignment. |
| | Testmasters (duplicating masters) for review, Extra Practice, diagnosing areas of difficulty, and supplementary testing, as well as a complete testing program containing chapter tests, cumulative tests and final tests. |

# Teacher's Edition
# Contents

# Staff Credits

| | |
|---|---|
| **Editorial Development** | Everett T. Draper, Earl D. Paisley, Antoine Y. Maksoud |
| **Product Manager** | Daniel M. Loch |
| **Field Advisory Board** | Douglas A. Nash, Sam Sherwood, Robert Wolff, Wendell Anthony, Gary Crump, Jack M. Custer, Roy Eliason, Kenneth C. Scupp, Dennis Spurgeon, Jeffra Ann Nicholson |
| **Marketing Research** | Erica S. Felman, Linda A. DeLora |
| **Editorial Processing** | Margaret M. Byrne, Holly L. Massey |
| **Art and Design** | Carol Steinberg, Mary Ciuffitelli |
| **Production** | Joan McNeil, Heidi Henney |
| **Photo Resources** | Linda Sykes, Rita Longabucco |

Copyright © 1986 by Holt, Rinehart and Winston, Publishers
All rights reserved
Printed in the United States of America

ISBN: 0-03-002163-4
5678901234    032    987654321

# HOLT
# ALGEBRA 1

*Teacher's Edition*

**Eugene D. Nichols**
**Mervine L. Edwards**
**E. Henry Garland**
**Sylvia A. Hoffman**
**Albert Mamary**
**William F. Palmer**

**HOLT, RINEHART AND WINSTON, PUBLISHERS**
New York · Toronto · Mexico City · London · Sydney · Tokyo

# ENRICHMENT MATERIALS

**HOLT ALGEBRA 1** provides many resources for enriching the curriculum, depending on the students' readiness and learning aptitude.

Special topics in *Logic*, geometry, probability, and statistics develop reasoning and challenge the student.

*Geometry Review* provides algebraic application with continuous cumulative review of previously learned geometric concepts.

*Calculator* exercises expand students' practical proficiency in algebraic computation and estimation.

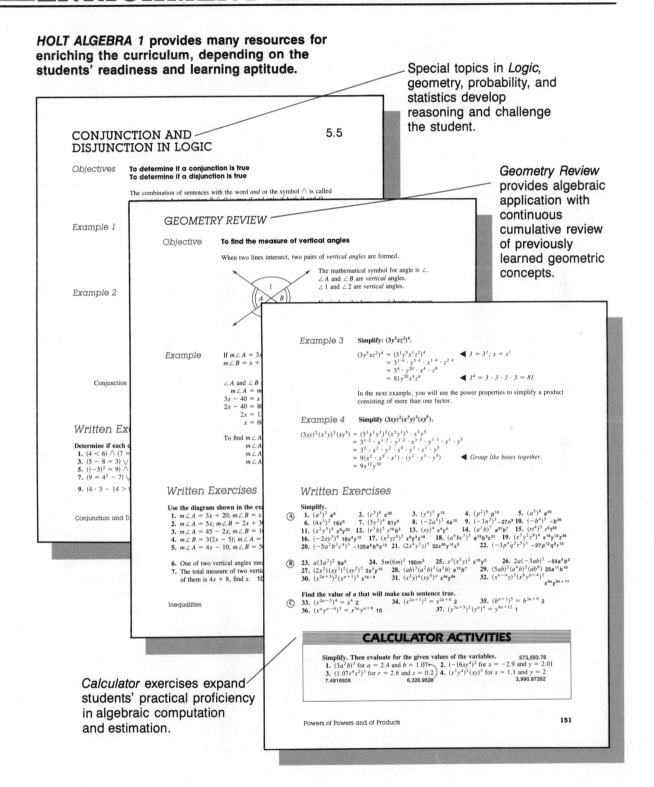

### CONJUNCTION AND DISJUNCTION IN LOGIC

5.5

Objectives  **To determine if a conjunction is true**
**To determine if a disjunction is true**

The combination of sentences with the word *and* or the symbol $\wedge$ is called

*Example 1*

*Example 2*

Conjunction

#### Written Exercises

Determine if each
1. $(4 < 6) \wedge (7 =$
3. $(5 - 8 = 3) \vee$
5. $((-3)^2 = 9) \wedge$
7. $(9 = 4^2 - 7) \wedge$
9. $(4 \cdot 3 - 14 >$

Conjunction and D

---

### GEOMETRY REVIEW

Objective  **To find the measure of vertical angles**

When two lines intersect, two pairs of *vertical angles* are formed.

The mathematical symbol for angle is $\angle$.
$\angle A$ and $\angle B$ are *vertical* angles.
$\angle 1$ and $\angle 2$ are *vertical* angles.

*Example*  If $m\angle A = 3x$
$m\angle B = x +$

$\angle A$ and $\angle B$
$m\angle A = m$
$3x - 40 = x$
$2x - 40 = 80$
$2x = 12$
$x = 60$

To find $m\angle A$
$m\angle$
$m\angle$
$m\angle$

#### Written Exercises

Use the diagram shown in the ex
1. $m\angle A = 3x + 20; m\angle B = x$
2. $m\angle A = 5x; m\angle B = 2x + 3$
3. $m\angle A = 45 - 2x; m\angle B = 1$
4. $m\angle B = 3(2x - 5); m\angle A =$
5. $m\angle A = 4x - 10; m\angle B = 5$

6. One of two vertical angles mea
7. The total measure of two vertic
of them is $4x + 8$, find $x$.  10

Inequalities

---

*Example 3*  **Simplify: $(3y^5xz^2)^4$.**

$(3y^5xz^2)^4 = (3^1y^5x^1z^2)^4$  ◀ $3 = 3^1; x = x^1$
$= 3^{1\cdot4} \cdot y^{5\cdot4} \cdot x^{1\cdot4} \cdot z^{2\cdot4}$
$= 3^4 \cdot y^{20} \cdot x^4 \cdot z^8$
$= 81y^{20}x^4z^8$  ◀ $3^4 = 3 \cdot 3 \cdot 3 \cdot 3 = 81$

In the next example, you will use the power properties to simplify a product consisting of more than one factor.

*Example 4*  **Simplify $(3xy)^2(x^3y)^3(xy^5)$.**

$(3xy)^2(x^3y)^3(xy^5) = (3^1x^1y^1)^2(x^3y^1)^3 \cdot x^1y^5$
$= 3^{1\cdot2} \cdot x^{1\cdot2} \cdot y^{1\cdot2} \cdot x^{3\cdot3} \cdot y^{1\cdot3} \cdot x^1 \cdot y^5$
$= 3^2 \cdot x^2 \cdot y^2 \cdot x^9 \cdot y^3 \cdot x^1 \cdot y^5$
$= 9(x^2 \cdot x^9 \cdot x^1) \cdot (y^2 \cdot y^3 \cdot y^5)$  ◀ *Group like bases together.*
$= 9x^{12}y^{10}$

#### Written Exercises

Simplify.
(A) 1. $(a^3)^2$  $a^6$  2. $(c^5)^6$  $c^{30}$  3. $(y^4)^3$  $y^{12}$  4. $(p^2)^6$  $p^{12}$  5. $(a^5)^8$  $a^{40}$
6. $(4x^3)^2$  $16x^6$  7. $(3y^2)^4$  $81y^8$  8. $(-2a^5)^2$  $4a^{10}$  9. $(-3n^3)^3$  $-27n^9$  10. $(-b^4)^5$  $-b^{20}$
11. $(x^2y^5)^4$  $x^8y^{20}$  12. $(r^3b)^5$  $r^{15}b^5$  13. $(xy)^4$  $x^4y^4$  14. $(a^3b)^7$  $a^{21}b^7$  15. $(rt^4)^5$  $r^5t^{20}$
16. $(-2xy^3)^4$  $16x^4y^{12}$  17. $(x^2yz^6)^3$  $x^6y^3z^{18}$  18. $(a^4bc^7)^3$  $a^{12}b^3c^{21}$  19. $(x^3y^2z^8)^6$  $x^{18}y^{12}z^{48}$
20. $(-5a^2b^3c^4)^3$  $-125a^6b^9c^{12}$  21. $(2x^4y^3z)^5$  $32x^{20}y^{15}z^5$  22. $(-3p^4q^3r^5)^3$  $-27p^{12}q^9r^{15}$

(B) 23. $a(3a^2)^2$  $9a^5$  24. $5m(6m)^2$  $180m^3$  25. $x^3(x^5y)^2$  $x^{13}y^2$  26. $2a(-3ab)^3$  $-54a^4b^3$
27. $(2x^2)(xy^3)^2(xy^2)^3$  $2x^7y^{12}$  28. $(ab)^3(a^2b)^3(a^3b)$  $a^{12}b^7$  29. $(5ab)^2(a^4b)^2(ab^6)$  $25a^{11}b^{10}$
30. $(x^{2a+3})^2(x^{a+1})^3$  $x^{7a+9}$  31. $(x^3y)^a(xy^3)^4$  $x^{4a}y^{6a}$  32. $(x^{n-4}y)^3(x^6y^{n+4})^2$  $x^{3n}y^{2n+11}$

Find the value of $a$ that will make each sentence true.
(C) 33. $(x^{2a-3})^4 = x^4$  2  34. $(x^{2a+1})^2 = x^{2a+6}$  2  35. $(b^{a+1})^5 = b^{3a+9}$  2
36. $(x^ay^{a-4})^3 = x^{3a}y^{a+8}$  10  37. $(y^{3a+5})^2(y^a)^4 = y^{8a+12}$  1

---

#### CALCULATOR ACTIVITIES

Simplify. Then evaluate for the given values of the variables.  573,593.76
1. $(3a^2b)^3$ for $a = 2.4$ and $b = 1.07$  2. $(-16xy^4)^2$ for $x = -2.9$ and $y = 2.01$
3. $(1.07r^4s^2)^3$ for $r = 2.6$ and $s = 0.2$  4. $(x^2y^4)^2(xy)^3$ for $x = 1.1$ and $y = 2$
7.4818928  6,320.9528  3,990.97262

Powers of Powers and of Products

151

Computer lessons at the end of chapters include computer applications of algebraic content. Hands-on programming exercises in BASIC encourage students with computer expertise to extend the algebra concepts further.

## Computer Section

COMPUTER SECTION

Objective

### THE INPUT STATEMENT

**To use INPUT statements in programs that evaluate formulas**

The INPUT statement allows you to enter different data into a program each time you RUN it. The following program contains a simple INPUT statement.

```
10   PRINT "ENTER A NUMBER."        Line 10 PRINTS the message within quotes.
20   INPUT X                         Line 20 displays a ? and a blinking cursor,
30   PRINT 2 * X                     the signal for requesting data.
40   END                             Line 30 computes the answer and displays it
                                     on the screen when the program is RUN.
```

To display a LISTing of your program, enter the command LIST. To debug any errors, enter the line number, correct the program line, and re-enter it. Now RUN the program. In response to the ? and blinking cursor, enter a number, say 6. The computer then stores 6 in the memory place X. The computer multiplies 6 by 2 and

...statement prints a message in addition to requesting ...INPUT is:

...UE FOR W. ''; W

...d. It is placed within '' ''.

...g quotation mark. When the program is RUN, it ...the blinking cursor.

...ssary punctuation in this form of the INPUT

...x − 2y for any values of x and y.
...ue of 3x − 2y for x = −3 and y = −2.

```
ALUE OF X. ";X          Lines 10 and 20 request values
ALUE OF Y. ";Y          for X and Y.
* Y                     Line 30 assigns the expression
F V IS "V".             3 * X − 2 * Y to V.
                        Line 40 PRINTS the result.

-3                      In response to the two INPUT messages, you enter
-2                      −3 for X and −2 for Y. Line 40 PRINTS the
                        phrase in quotes, the value of V, and the period.
```

...to translate familiar mathematical formulas into their ...following BASIC formulas compute the sum and ...and b.

Computer Section

## COMPUTER ACTIVITIES

### Solving Equations

Equations of the type $x + 5 = 8$ or $x - 9 = 12$ can be represented by the general equation $x + b = c$.

One computer program can solve any equation of this type. Each time it is given values for $b$ and $c$, it can print the solution of $x + b = c$. However, the computer must be told how to solve the equation.

The way to do this is suggested by the method used to solve the equation $x + 5 = 8$.

$$x + 5 = 8 \qquad\qquad x + b = c$$
$$x + 5 - 5 = 8 - 5 \qquad x + b - b = c - b$$
$$x = 3 \qquad\qquad x = c - b$$

The program below will solve equations of the type $x + b = c$.

```
10   FOR I = 1 TO 2              Line 10 sets up the program to solve
20   INPUT "WHAT IS B? ";B       2 equations.
30   INPUT "WHAT IS C? ";C
40   LET X = C - B               Line 40 contains the formula for
50   PRINT "THE SOLUTION OF X +  solving the equation.
     "B" = "C" IS "X"."
60   NEXT I
70   END
```

See the Computer Section beginning on page 488 for more information.

### Exercises    For Ex. 3 and 6, see TE Answer Section.

**Enter and RUN the program above to solve the following equations.**
1. $x + 5 = 8$  3
2. $x - 219.73 = -54.086$  165.644

3. Modify the program above to solve equations to the type $4x = 12$. The general equation is $ax = c$. The solution of $ax = c$ is $x = ?$ Enter and RUN the program to solve $4x = 12$ and $8x = 4$.  3, .5

**Consider the equations $3x - 4 = 16$ and $2x + 8 = 14$.**
4. Solve $3x - 4 = 16$ and $2x + 8 = 14$ without a computer.  $6\frac{2}{3}$, 3
5. Using the pattern for solving equations of this type, write the general form of this type of equation and solve it for $x$.  $ax + b = c; x = \frac{c - b}{a}$
6. Using the general solution for $x$ you found in Exercise 5, modify the program above to solve this general type of equation.
7. Enter and RUN your modified program to solve $3x - 4 = 16$ and $2x + 8 = 14$.  6.66, 3

Solving Equations

**111**

The back of the book contains a 15-page Computer Section designed to give students experiences in programming that relate to exploring ideas in algebra. Looping commands, as well as special computer functions, are employed.

Problem-solving applications in real-world careers and in mathematics usage offer abundant motivation to practice problem-solving strategies.

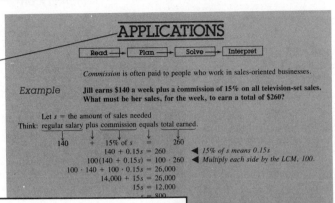

## APPLICATIONS

Read → Plan → Solve → Interpret

*Commission* is often paid to people who work in sales-oriented businesses.

**Example**    **Jill earns \$140 a week plus a commission of 15% on all television-set sales. What must be her sales, for the week, to earn a total of \$260?**

Let $s$ = the amount of sales needed
Think: regular salary plus commission equals total earned.

$$140 \quad + \quad 15\% \text{ of } s \quad = \quad 260$$
$$140 + 0.15s = 260 \quad \blacktriangleleft \; 15\% \text{ of } s \text{ means } 0.15s$$
$$100(140 + 0.15s) = 100 \cdot 260 \quad \blacktriangleleft \; \text{Multiply each side by the LCM, 100.}$$
$$100 \cdot 140 + 100 \cdot 0.15s = 26,000$$
$$14,000 + 15s = 26,000$$
$$15s = 12,000$$
$$s = 800$$

Tony is paid only a commission of 30% of all sales. What must be his total sales so that he will earn \$270? **\$900**

Marie is paid \$165 a week plus a 10% commission on all camera sales. How much must her sales be to earn a total of \$215? **\$500**

109

---

*Example 3*    **Julie's salary is 3 times Liz's salary. Together they earn \$320 a month. Find the salary of each girl.**

READ ▶    The problem asks you to find the salary of each girl.
PLAN ▶    Let $s$ = Liz's salary
$\quad\quad 3s$ = Julie's salary   ◀ *Julie's salary is 3 times Liz's salary.*
Together their sum is 320
$\quad\quad\quad\quad \downarrow$
$\quad\quad s + 3s = 320$    ◀ *Liz's salary plus Julie's salary is \$320.*
SOLVE ▶    $\quad\quad 4s = 320$
$\quad\quad\quad s = 80$

To find each girl's salary, use the representation statements:
Liz's salary, $s$, is \$80.
Julie's salary, $3s$, is $3 \cdot 80$, or \$240.

INTERPRET ▶    Check the salaries in both parts of the problem.

| Julie's salary is 3 times Liz's salary. | | | Together they earn \$320. | |
|---|---|---|---|---|
| 240 | $3 \cdot 80$ | | $80 + 240$ | 320 |
| | 240 | | 320 | |
| | $240 = 240$ | True | $320 = 320$ | True |

**Thus, Liz's salary is \$80 a month and Julie's salary is \$240 a month.**

*Example 4*    **Lola's age is 14 more than 6 times Juan's age. The sum of their ages is 35. How old is each?**

READ ▶    The problem asks you to find the age of each person.
PLAN ▶    $\quad\quad$ Let $a$ = Juan's age
$\quad 6a + 14$ = Lola's age   ◀ *14 more than 6a*
The sum of their ages is 35.
$\quad\quad\quad \downarrow \quad\quad\quad\quad \downarrow$
$\quad a + 6a + 14 \quad\quad = 35$
SOLVE ▶    $\quad\quad\quad 7a + 14 = 35$
$\quad\quad 7a + 14 - 14 = 35 - 14$
$\quad\quad\quad\quad\quad 7a = 21$
$\quad\quad\quad\quad\quad a = 3$
Juan's age, $a$, is 3.
Lola's age, $6a + 14$, is $6 \cdot 3 + 14$, or 32.

INTERPRET ▶

| Lola's age is 14 more than 6 times Juan's age. | | The sum of their ages is 35. | |
|---|---|---|---|
| 32 | $6 \cdot 3 + 14$ | $3 + 32$ | 35 |
| | $18 + 14$ | 35 | |
| | 32 | $35 = 35$ | True |
| | $32 = 32$ True | | |

**Thus, Juan is 3 years old and Lola is 32 years old.**

A four-part problem-solving strategy, continually highlighted, encourages students to approach problem solving in a logical manner.

*Summary* provides a quick review of the key concepts and procedures taught in the lesson.

**Non-Routine Problems** provide students with the opportunity to solve unusual problems using non-standard approaches or strategies.

Summary

Adding the *same number* to (or subtracting the *same number* from) each side of an inequality produces an inequality of the *same order*.

Multiplying or dividing each side of an inequality by a *positive* number produces an inequality of the *same order*.

Multiplying or dividing each side of an inequality by a *negative* number produces an inequality of *reversed order*.

*Reading in Algebra*

...ways true, or always true, or never true.
...number produces an
...y reverses the order.   **never true**
...tive number
...ue

...of the inequality be changed?
...y −2. **yes**  3. −8 < 3 Multiply by 3.  **no**
...by −1.  **6.** 2 < 4 Divide by −1.  **yes**

## 6 INTRODUCTION TO FACTORING

**Non-Routine Problem Solving**

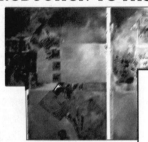

**Problem 1**

An artist has an oddly-shaped canvas. Square *A* is ... square *C* is 64 in². The total surface area is 81 in² ... into the fewest number of pieces that will fit togeth... canvas? **See TE Answer Section.**

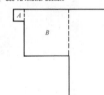

**Problem 2**

Four goats were tied, one at each corner of a squar... × 100 m. The ropes that tie them allow each goat ... 50 m radius. This leaves an ungrazed portion in th... three of the goats are moved, the rope tying the fo... allows him to graze over an area equal to the comb... four goats. How long is the rope? **100 m**

---

### COLLEGE PREP TEST

Choose the one best answer to each question or problem.

**1.** One year, a factory granted raises at three different times. If 62% of the workers
**D** received raises in the first and second periods and 10% of these did not receive a raise in the third period, what percent of the workers received raises in all three periods?
(A) 6.2%  (B) 38%  (C) 52%
(D) 55.8%  (E) 90%

**2.** If $3x − 12 = 2x − 10$, then $x + 3 = ?$
**A** (A) 5  (B) 4  (C) −3  (D) −4  (E) −5

**3.** $x + 9$ is how much more than 5?
**C** (A) 4  (B) $x + 14$  (C) $x + 4$  (D) $x$  (E) 14

**4.** $x$ is what percent of 40% of 20?
**C** (A) $\frac{x}{800}$  (B) $\frac{8}{100}x$  (C) $\frac{100}{8}x$  (D) $\frac{800}{x}$
(E) none of these

**5.** If rain is falling at the rate of 3 cm/h, how much rain will fall in 10 min?
**A** (A) 0.5 cm  (B) 18 cm  (C) 30 cm
(D) 180 cm  (E) none of these

**6.** How many 1-cm squares are needed to make a border around the shaded square, which has a 6-cm side?
**E**

6cm

(A) 24  (B) 28  (C) 36  (D) 40
(E) cannot tell from the given information

**7.** If $y = 4$, then $y^2 + 2 = ?$
**B** (A) 32  (B) 18  (C) 10  (D) 8  (E) 4

**8.** Which of the following is greater than $\frac{1}{4}$?
**D** (A) 0.04  (B) $\left(\frac{1}{4}\right)^2$  (C) $\frac{1}{8}$  (D) $\frac{1}{0.04}$
(E) none of these

**9.** What is the total maximum weight of 12 eggs if 5 of them weigh 12 to 20 oz each
**A** and the others weigh 25 to 30 oz each?
(A) 310 oz  (B) 360 oz  (C) 460 oz
(D) 500 oz  (E) none of these

**10.** A merchant paid $24 for an article. The merchant wants to place a price tag on it
**A** so that a 20% discount on the marked price can be advertised and still bring in a profit of 25% of the cost. What price should be marked on the tag?
(A) $37.50  (B) $30  (C) $27.60
(D) $25.20  (E) none of these

**11.** *PQRS* is a square.
$VR = 3$ m
**D** $PW = 7$ m
The area of the shaded portion is 39 m².
The area of *TUVS* is ?

(A) 18 m²  (B) 49 m²  (C) 60 m²
(D) 61 m²  (E) 139 m²

**College Prep Tests** give students practice and experience in strategies they will use in taking multiple-choice tests that they will encounter in the future.

116

College Prep Test

# TESTING AND REVIEW

**Tests in *HOLT ALGEBRA 1* assist the teacher in measuring the students' understanding of the major skills and concepts taught.**

*Cumulative Reviews* test the students' understanding of algebraic concepts and corresponding problem-solving skills that have been previously taught.

*Chapter Tests* are designed to measure the students' mastery of the chapter objectives. The difficulty level is indicated.

## CHAPTER FOUR TEST

Solve each equation. Check the solution.

1. $\frac{3}{5}m = 9$  **15**

2. $7 + \frac{2}{3}n = 13$  **9**

13. The length of a rectangle is 5 m more than twice the width. The perimeter is 46 m. Find the length and the width. **length, 17 m; width, 6 m**

14. ... parts such that the ... the smaller. Find each

... 12? **25%**

...angle is 2 km longer ... third side is twice as ... The perimeter is 34 ... of each side. **10 km, 8 km, 16 km**

## CHAPTER FOUR REVIEW

Ⓐ Exercises are 1–2, 5–10, 12–17, 20–22, 24–25, 29–36.  Ⓒ Exercises are starred. The rest are Ⓑ Exercises

**Vocabulary**

base [4.7]
congruent [4.7]
isosceles [4.7]
least common multiple [4.4]

percent [4.6]
sum [4.2]
selling price [4.6]

15. $8 + \frac{1}{3}x = \frac{1}{2}x$  **48**

16. $8 = \frac{2}{3}x + 4$  **6**

17. $\frac{2}{3}a + 9 = 11$  **3**

Solve each equation. Check the solution.

1. $-6 + 4a = -12 + 6a$  **3**
2. $13p + 5 = -3 + 9p$  **−2**

Solve each equation. [4.1]

3. $7m + 5 - 5m = -2 + m - 9$  **−1**
4. $6y - 4 - 7y = 8 + y - 9$  **−$\frac{3}{2}$**

Solve each equation. Check the solution.

5. $2(9 + y) = 30$  **6**
6. $5b + 3(b - 4) = 2(b + 3)$  **3**

Solve each equation. [4.2]

7. $7b + 2(1 - b) = 8b - 4$  **2**
8. $16 - (3 - 4y) = 2(y + 2)$  **−$\frac{9}{2}$**

Solve each problem. [4.3]

9. Mona's age is 4 less than twice Bill's. The sum of their ages is 38. Find each their ages. **Bill, 14 y; Mona, 24 y**

10. Separate 23 into two parts so that the part has 5 more than twice the second part. Find each part. **6, 17**

11. The second of three numbers is twice first. The third is 6 less than the seco If the second is decreased by 3 times third, the result is 50. Find the numb **−8, −1...**

Solve each equation. (Ex. 12–23)

12. $\frac{2}{7}x = 4$  [4.4]  **14**

13. $6 + \frac{3}{5}a = 12$  **10**

14. $\frac{2}{3}a = \frac{5}{7}$  **$\frac{15}{14}$**

**112**

## CUMULATIVE REVIEW

(Chapters 1–4)

Each exercise has five choices of answers. Choose the one best answer.

1. Simplify $5x + 3 + 4x$, then evaluate if $x = 3$.
   (A) $12x, 36$  (B) $20x + 3, 63$  (C) $23x, 69$
   (D) $9x + 3, 30$  (E) None of these  **D**

2. Compute $\frac{8 \cdot 2 + 4}{5 - 3}$.
   (A) 10  (B) 24  (C) 18  (D) 1  **A**
   (E) None of these

3. Solve $3x - 5 = 7$.
   (A) $\frac{4}{3}$  (B) 4  (C) 9  (D) $\frac{1}{12}$  **B**
   (E) None of these

4. Rewrite $a^5$ without exponents.
   (A) $5a$  (B) $a \div 5$  (C) $a \cdot a \cdot a \cdot a \cdot a$  **C**
   (D) $a(5)$  (E) None of these

5. Simplify $3(2x + 5) + 4(3x + 6)$.
   (A) 102  (B) $18x + 11$  (C) $18x + 21$  **D**
   (D) $18x + 39$  (E) None of these

6. Divide $(24 \div 4) \div -2$.
   (A) 12  (B) $-8$  (C) 3  (D) $\frac{1}{8}$  **E**
   (E) None of these

7. Subtract $2a - 3$ from $-4a + 2$.
   (A) $-6a - 1$  (B) $-2a - 1$  (C) $6a - 5$  **E**
   (D) $-1a$  (E) None of these

8. Write an algebraic representation for 7 less than twice a number $n$.
   (A) $7 - 2n$  (B) $2n - 7$  (C) $2n + 7$  **B**
   (D) $5n$  (E) None of these

9. Find the difference $6 - 8$. Then give its opposite.
   (A) $-2$  (B) 14  (C) 2  (D) $-14$  **C**
   (E) None of these

What words must be used to fill in the blanks to make a true statement?

10. $8 \cdot 4 = 4 \cdot 8$  **commutative, multiplication** by the _____ property of _____.
11. $6 \cdot (3 + 4) = 6 \cdot 3 + 6 \cdot 4$  **distributive** by the _____ property.
12. $7 + (-7) = 0$  **additive inverse** by the property of _____.
13. $a + 0 = a$  **addition, zero** by the _____ property of _____.
14. $(7a + 2a) + 3a = 7a + (2a + 3a)$  **associative** by the _____ property of _____.
15. $-(x - 8) = -1(x - 8)$  **multiplication, −1** by the _____ property of _____.
16. **associative, addition**

Compute.

16. $7 \cdot 3 + 4$  **25**
17. $(-8)(-4)$  **32**
18. $\frac{+24}{-8}$  **−3**
19. $-\frac{2}{3} - \frac{8}{9}$  **−$\frac{14}{9}$**
20. $-9 + 2$  **−7**
21. $-4\frac{3}{4} + 6$  **1$\frac{1}{4}$**
22. $4 - 5 - 7$  **−8**
23. $16 - 2.4$  **13.6**
24. $-7 - 4(-2)$  **1**
25. $-4 \cdot 3 + 12$  **0**
26. $\frac{7 \cdot 1 + 6}{4 \cdot 2 - 3}$  **$\frac{13}{5}$**
27. $|8 - 12|$  **4**

Evaluate for the given value(s) of the variable(s).

28. $x^2$ if $x = -\frac{4}{5}$  **$\frac{16}{25}$**
29. $3a^5$ if $a = 2$  **96**
30. $\frac{x^2 - 4x}{2x}$ if $x = 2$  **−1**
31. $4ab^2c^3$ if $a = 2, b = 3$ and $c = -1$  **−72**
32. $-8 - 4x$ if $x = -3$  **4**

**114**                Cumulative Review

*Chapter Reviews* contain exercise sets that are lesson-coded for easy reference to corresponding lessons when review is needed. Exercises are also coded to difficulty level: A, B, or C.

Clarifies process and techniques for students.

**The Teacher's Edition of *HOLT ALGEBRA 1* provides an annotated full-size reproduction of each pupil page and includes detailed chapter and lesson commentaries.**

Highlights vocabulary.

Gives ideas for presentation and individualization.

Focuses on the key concepts to be taught.

References all special feature pages.

### 11.1 Coordinates of Points in a Plane (page 293)

**Vocabulary**
abscissa, horizontal line, ordered pair, ordinate, origin, quadrant, vertical line

**Background**
The concept of "Algebraic Geometry" (Analytic Geometry) was first conceived by René Descartes (born in 1596 near Tours, France). Legend has it that the idea first came to him as he watched a fly moving about the ceiling in his room. He noted that the position of the fly could be described if one knew a relationship connecting the fly's distance from two adjacent walls. The technical words "coordinates," "abscissa," and "ordinate" are attributed to another mathematician, Leibnitz, born in 1646.

You might assign the further researching of the historical background of the "coordinate" system as a voluntary project. A good reference is "An Introduction to the History of Mathematics," by Howard Eves.

**Suggestions**
Introduce the idea of an ordered pair by asking students to describe the seat location of a person in the class. Stress the need for two numbers, seat and row. Point out that the order is important so that people are not mixed up as to whether row or seat is meant.

**Errors That Students Make**
A typical error made by students is to write the coordinates of a point in the wrong order. Stress the writing of the $x$-coordinate first.

**For Additional Chalkboard Work**
For each ordered pair, tell which is the abscissa and which is the ordinate.
1. $A(3,4)$ 2. $B(-2,5)$ 3. $C(3,-1)$ 4. $D(-2,-3)$
1st coord. is abscissa, 2nd coord. is ordinate

**Cumulative Review**
Solve. Graph the solution set on a number line.
1. $x + 4 = 3x - 10$    2. $x + 4 < 3x - 6$
7           $x > 5$

### 11.2 Point Plotting (page 296)

**Vocabulary**
graph

**Background**
Students have learned in the previous lesson how to describe the position of a point using an ordered pair of coordinates. This lesson is a simple extension of the concept to plotting points whose coordinates are given.

**Suggestions**
Stress that the first co ordered pair refers to positive, to the right,

The second coordin ordered pair refers to positive, up, negative

**Enrichment**
Give the class the fol plot them, and conne in order alphabetically they finally connect $P$ separate point $R(2,1)$ What is it?
$A(2,7)$, $B(-1,3)$, $C($
$E(-1,-5)$, $F(-1,-$
$I(-1,-11)$, $J(0,-12$
$M(12,-1)$, $N(12,4)$,
Plot $R(2,1)$.
Assign volunteers to and post the best one

**Cumulative Rev**
1. Write the ratio 2:5 as a fraction. $\frac{2}{5}$
3. Factor $2x^3 - 32x$.

Keeps previously learned skills current.

Reinforces the lesson skills.

## 11

## Chapter and Lesson Commentaries

### Chapter 11 Graphing in a Coordinate Plane

**Objectives**

To give the coordinates of a point in a plane
To identify the quadrant in which a given point is located
To plot a point, given its coordinates
To find the slope of a nonvertical line segment
To determine the slant of a line using its slope
To determine if two lines are parallel using their slopes
To determine if three points are on the same line
To write an equation of a line given the slope and one point
To write an equation of a line given two points
To write a table of values given an equation of a line
To find the slope and $y$-intercept of a line given its equation
To write an equation of a line given its slope and $y$-intercept
To graph a line given its equation
To graph an inequality in two variables

**Overview**

The first two lessons teach students how to read the coordinates of points in a plane and how to graph points whose coordinates are given. Following this is the development of the concept of slope of a line or line segment. Once students see that the slope of a line is the same for any two points on the line, they are then ready to find the equation of a line given the coordinates of any two points on the line or one point and the slope. It is easier for students to learn one general method for writing the equation of a line.

Lesson 11.5 is a direct follow-through of Lesson 11.4 on slope. Students see that since the slope of a line is the same for any two points, writing the equation merely involves setting the numerical value of the slope equal to the value of the slope obtained using one point and a general point, $P(x, y)$. This leads to a general understanding. We chose to avoid the traditional beginning of giving students an equation and having them plot points and draw a line. We first establish *why* an equation of the form $y = mx + b$ represents a line.

**Special Features**

| | |
|---|---|
| Famous Mathematicians: Descartes | p. 292 |
| Calculator Activities | p. 309 |
| Reading in Algebra | pp. 295, 316 |
| Cumulative Review | pp. 298, 301, 313, 322 |
| Non-Routine Problems | p. 322 |
| Applications | p. 323 |
| Geometry Review | p. 325 |
| Chapter Review | p. 326 |
| Chapter Test | p. 327 |
| College Prep Test | p. 328 |

**René Descartes (page 292)**

Descartes left school at the age of fourteen and shortly after devoted some time to the study of mathematics in Paris. After a varied career of military life, travel, and constructing optical instruments, he spent twenty years in Holland pursuing his study of philosophy, mathematics, and science. Descartes and Fermat conceived the ideas of modern analytic geometry, which establishes a correspondence between ordered pairs of real numbers and points in the plane. Descartes is also credited with recognizing the relationship $v - e + f = 2$, connecting the number of vertices, edges, and faces of a convex polyhedron.

Enables the teacher to anticipate difficulties that are typical for students.

Indicates mathematical content.

Provides practice in problem formulation, estimation, and application.

# SUPPLEMENTARY MATERIALS

*HOLT ALGEBRA 1*'s supplementary materials include Testmasters, Skillmasters, and a Teacher's Resource Package with Blackline Masters so that the teacher can instruct according to each student's needs and skill levels.

| | |
|---|---|
| **Skillmasters** | contain two parallel sections for every two lessons, which can be used as pre- or posttests for diagnosis, extra practice, quizzes, and assignments. Cumulative Review for every few lessons. |
| **Testmasters** | supply Chapter Tests, Cumulative Tests, and Final Tests. |
| **Teacher's Resource Package** | contains Skillmasters and Testmasters in blackline form plus Enrichment masters and Competency Tests. The Enrichment masters provide additional material that can be used to challenge all students: applications, history, and non-routine problems. The Competency Tests contain the type of material found on standard competency tests used in many states. |
| **Solutions Manual** | provides worked-out solutions to every problem and exercise in the text. |

# ABOUT HOLT ALGEBRA I

***Success Pattern*** HOLT ALGEBRA 1 guides students through the process of developing their mathematical abilities by helping them experience a pattern of successful learning that builds self-confidence. First, students review fundamental concepts: a simple version of variables, properties of operations, and integers and their opposites. Next, they build a bridge between these fundamental arithmetic concepts and simple algebraic expressions. As students become more familiar with these, they learn to solve increasingly difficult expressions. This pattern of instruction is characteristic of the approach used throughout HOLT ALGEBRA 1.

***Organization of the Course*** HOLT ALGEBRA 1 contains all of the major algebra concepts necessary to meet the instructional demands in your school successfully. In addition to feedback materials, which include tests for each unit with items keyed to lesson objectives, HOLT ALGEBRA 1 provides materials in Logic, Geometry, and Probability and Statistics to supplement basic coursework. It also provides a computer strand incorporated into every chapter, to correlate algebraic concepts with relevant and motivating BASIC-language programming skills. These pages are referenced to on-hand materials in the Computer Section at the back of the book. Lessons using calculators and practice for standardized college admission tests offer further supplementation.

***Instructional Advantage*** The instructional advantages of Holt's approach are clear and well documented. First, frequent evaluation — Chapter Reviews, tests at the end of chapters, and Cumulative Reviews every four chapters — provides continuous feedback for both you and the student. Second, a varied instructional technique provides students with the opportunity to participate actively in learning and to find an instructional method that best suits their needs. Students who learn quickly and can work independently need the challenge provided by the supplements and extensions to the basic course. Others benefit from selected problems that challenge them at their own levels of competence. HOLT ALGEBRA 1 addresses the needs of both types of students.

***Students with Handicapping Conditions*** Instructional materials are also available for helping students who need to overcome learning problems that may be due to physical disabilities (orthopedic or visual impairments), communicative disabilities (hearing loss, dyslexia), mild mental retardation, or disabling emotional disturbance. Teachers should be on the alert to such possibilities so that students with undiagnosed disabilities can be treated professionally.

Whenever possible, students with handicapping conditions should participate either partially or totally in regular school programs and classes, as part of the requirement of Federal Law PL94-142, which states that handicapped children must be educated in the least restrictive environment possible.

By using the chapter and lesson objectives in the book, the teacher can select appropriate lesson goals for each student and communicate them easily to parents or special-education teachers who may be assisting the handicapped. The Skillmasters and Testmasters in the Teacher's Resource Package are valuable tools for assessing each individual student's level of comprehension and skill in a given lesson or topic.

***Success of HOLT ALGEBRA 1*** HOLT ALGEBRA 1 has been used successfully with a variety of students. It has been found to be stimulating, highly teachable, and motivational, with a good balance of skill, theory, and application. This new edition of HOLT ALGEBRA 1, with its expanded and new features is based on suggestions from users of the previous editions.

*Example 2*  **Graph the solution set of $x \leq 4$.**

The solution set of $x \leq 4$ is {all numbers less than or equal to 4}.

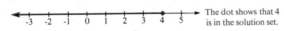 The dot shows that 4 is in the solution set.

It is easier to graph the solution set of an inequality like $-2 \leq x$ if you rewrite the inequality with the variable first. Write $-2 \leq x$ as $x \geq -2$.

*Example 3*  **Graph the solution set of $-2 \leq x$.**

Rewrite $-2 \leq x$ as $x \geq -2$.

The solution set is {all numbers greater than or equal to $-2$}.

## Written Exercises

Graph the solution set of each sentence. **Ex. 1–37, check students' graphs.**

**(A)**
1. $b < 4$
2. $x > 1$
3. $x < 8$
4. $a > -2$
5. $p < -1$
6. $x \leq -3$
7. $y \geq -5$
8. $3 \leq t$
9. $r \geq -1$
10. $-7 \leq y$
11. $6 < g$
12. $x \leq 0$
13. $a < -2$
14. $w > 0$
15. $y \leq -4$
16. $c < -6$
17. $d \leq -1$
18. $-8 \geq x$
19. $r \leq -10$
20. $d \leq -7$
21. $5 \leq b$
22. $m \leq -5$
23. $9 \geq x$
24. $u \leq -12$
25. $0 \leq x$

**(B)**  Graph the solution set of each sentence.

*Example*  **Graph the solution set of $y \neq \frac{1}{4}$.**

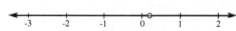

26. $x \neq -\frac{1}{2}$
27. $-\frac{1}{4} > x$
28. $y < 2\frac{1}{2}$
29. $-1\frac{1}{2} \leq p$

30. $1.5 \geq g$
31. $0 \neq x$
32. $x \geq 3.5$
33. $-1\frac{1}{4} \neq a$

**(C)**
34. $x \nleq 4$
35. $1 - 61 \ngtr y$
36. $-3.5 \nleq t$
37. $x \nless -7.1$

**122**

Chapter Five

Graded Exercises (levels A, B, and C) promote continuous development and challenge for students of all abilities.

**A variety of features guide students through activities designed to stimulate thinking.**

*Reading in Algebra* helps students develop the skills needed to read, interpret, and use the language of algebra effectively.

*Chapter Openers* provide opportunities to practice problem formulation, estimation, and application. Related careers and historical puzzles also furnish motivation.

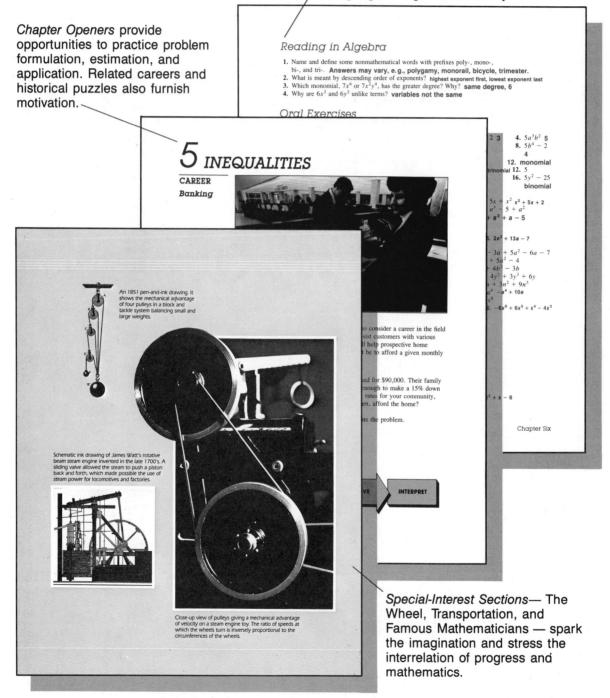

### Reading in Algebra

1. Name and define some nonmathematical words with prefixes poly-, mono-, bi-, and tri-. **Answers may vary, e.g., polygamy, monorail, bicycle, trimester.**
2. What is meant by descending order of exponents? **highest exponent first, lowest exponent last**
3. Which monomial, $7x^6$ or $7x^2y^4$, has the greater degree? Why? **same degree, 6**
4. Why are $6x^2$ and $6y^2$ unlike terms? **variables not the same**

### Oral Exercises

## 5 INEQUALITIES

**CAREER**
*Banking*

An 1851 pen-and-ink drawing. It shows the mechanical advantage of four pulleys in a block and tackle system balancing small and large weights.

Schematic ink drawing of James Watt's rotative beam steam engine invented in the late 1700's. A sliding valve allowed the steam to push a piston back and forth, which made possible the use of steam power for locomotives and factories.

Close-up view of pulleys giving a mechanical advantage of velocity on a steam engine toy. The ratio of speeds at which the wheels turn is inversely proportional to the circumferences of the wheels.

to consider a career in the field
sist customers with various
l help prospective home
t be to afford a given monthly

ed for $90,000. Their family
enough to make a 15% down
rates for your community,
en, afford the home?

te the problem.

INTERPRET

|  |  | 2 3 | **4.** $5a^3b^2$ **5** |
| --- | --- | --- | --- |
|  |  |  | **8.** $5b^4 - 2$ |
|  |  |  | **4** |
|  |  |  | **12. monomial** |
|  |  | trinomial | **12. 5** |
|  |  |  | **16.** $5y^2 - 25$ |
|  |  |  | **binomial** |

$5x + x^2$  $x^2 + 5x + 2$
$a^3 + 5 + a^2$
$+ a^2 + a - 5$

$2a^2 + 13a - 7$
$- 3a + 5a^2 - 6a - 7$
$+ 5a^2 - 4$
$+ 4b^3 - 3b$
$4y^2 + 3y^3 + 6y$
$+ 3n^2 + 9n^3$
$a^4 - a^4 + 10a$
$x^5$
$- 6x^6 + 6x^5 + x^4 - 4x^3$

$^2 + x - 6$

Chapter Six

*Special-Interest Sections—* The Wheel, Transportation, and Famous Mathematicians — spark the imagination and stress the interrelation of progress and mathematics.

T-10

# Reference Chart for Algebra I Skillmasters

| LESSON PAGES | SKILLMASTER NUMBER | LESSON PAGES | SKILLMASTER NUMBER |
|---|---|---|---|
| 1, 4 | 1 | 221, 225 | 25 |
| 7, 10, 13 | 2 | 236, 238, 242 | 26 |
| 16, 18, 22 | 3 | 245, 248, 251 | 27 |
| 28, 32, 34 | 4 | 253, 255 | 28 |
| 37, 40, 41 | 5 | 264, 269 | 29 |
| 44, 47, 49 | 6 | 273, 276 | 30 |
| 51, 53, 55, 58 | 7 | 280, 283 | 31 |
| 64, 67, 69 | 8 | 293, 296, 299 | 32 |
| 72, 74, 76, 80 | 9 | 302, 305 | 33 |
| 86, 89, 92 | 10 | 308, 310 | 34 |
| 96, 100 | 11 | 314, 317, 319 | 35 |
| 103, 106 | 12 | 330, 333, 335 | 36 |
| 118, 121, 124 | 13 | 338, 341, 344 | 37 |
| 128, 131, 132 | 14 | 347, 351 | 38 |
| 136, 138 | 15 | 365, 368 | 39 |
| 148, 150, 152 | 16 | 371, 373, 377, 382 | 40 |
| 155, 157 | 17 | 389, 391, 394 | 41 |
| 159, 161, 164 | 18 | 396, 399, 402 | 42 |
| 171, 173, 175 | 19 | 404, 406, 409 | 43 |
| 177, 180, 183 | 20 | 412, 416 | 44 |
| 186, 188 | 21 | 426, 429, 432 | 45 |
| 191, 195 | 22 | 435, 438, 441, 445 | 46 |
| 205, 209, 212 | 23 | 455, 459 | 47 |
| 215, 218 | 24 | 462, 465, 467, 470 | 48 |

# Acknowledgments

Holt is grateful to thousands of teachers and to hundreds of thousands of students who have used our algebra texts in the past and who have contributed their comments and constructive criticism to help make the teachable textbooks of the past even more teachable and enjoyable.

We are also grateful to teachers, supervisors, and administrators across the country who have participated in systematic surveys so that Holt can give America's high school students the kind of materials that best facilitate their learning.

# Chapter and Lesson Commentaries

## Chapter 1 Operations, Variables, Formulas

### Objectives

To simplify expressions by using the rules for the order of operations

To evaluate algebraic expressions and formulas for given value(s) of the variable(s)

To identify terms, variables, and numerical coefficients

To recognize the commutative, associative, and distributive properties

To simplify expressions by using the commutative, associative, and distributive properties

To simplify algebraic expressions by combining like terms

To evaluate algebraic expressions containing exponents

To recognize which numbers, if any, are solutions of an open sentence

### Overview

Students work with the operations of addition, subtraction, multiplication, and division. Special attention is given to work with fractions. This helps students maintain the proficiency with fractions that is required by many standardized and state tests. Once the order of operations and the use of grouping symbols are mastered, the students then apply these concepts to evaluating algebraic expressions with variables, including expressions containing exponents. Students are introduced to the basic commutative, associative, and distributive properties. The distributive property is used to justify combining like terms. Stress is placed upon the importance of algebra in working with formulas in a geometric setting.

### Non-Routine Problem Solving (page x)

**Non-Routine Problems** provide a strand throughout the text. They occur in chapter openers and are scattered within the chapters. They are designed to challenge students to seek original ways to attack problems. Providing students with experience in solving **Non-Routine Problems** will help prepare them for SATs as well as math league competition.

## 1.1 Rules for Order of Operations (page 1)

### Suggestions
Use the introduction to stress the need for the order of operations. A good warm-up for the lesson might be to write $5 \cdot 4 + 2$ on the chalkboard and ask the class what the correct answer is. This might lead to a discussion among students that the answer might be 22 or 30, thus demonstrating the need for a rule. Students will need to get used to using the raised dot and parentheses for multiplication. Once the introduction has been presented, simple drill to be done orally with the class might be given:

| | |
|---|---|
| $5 \cdot 4 + 2$ | Answer: $20 + 2 = 22$ |
| $6 - 2 \cdot 2$ | Answer: $6 - 4 = 2$ |
| $8 \cdot 2 + 1$ | Answer: $16 + 1 = 17$ |
| $4 + 3(2)$ | Answer: $4 + 6 = 10$ |
| $9 - 2(3)$ | Answer: $9 - 6 = 3$ |

Students sometimes have difficulty with the $\div$ symbol. They usually want to divide in any way that does not produce a fraction. For example, $2 \div 8$ will be called 4 since 2 divides into 8 evenly. Stress the reading of the symbol $\div$ as *divided by*. Try the following oral drill. Ask the students to first read the problem, then give the answer.

| | |
|---|---|
| $8 \div 2$ | Answer: 8 divided by 2 is 4. |
| $15 \div 5$ | Answer: 15 divided by 5 is 3. |
| $1 \div 2$ | Answer: 1 divided by 2 is $\frac{1}{2}$. |
| $3 \div 4$ | Answer: 3 divided by 4 is $\frac{3}{4}$. |

### Errors That Students Make
Even after the order of operations rule has been developed, students will still make errors like the following: $14 - 4 \cdot 3 = 10 \cdot 3 = 30$. They look at the 14 and 4 first, ignoring the 3. One way to help overcome this is to ask students how grouping symbols could be used to make the answer come out to be 30. Answer: $(14 - 4)3 = 10 \cdot 3 = 30$.

### Enrichment
Ask for volunteers for the **C** exercises at the chalkboard. Assign the following problem:

$$\frac{8 \cdot 2 + 4[3 \cdot 15 - 4(12 \div 6 + 4)]}{5[4 + 3(6 - 2 \cdot 1)] + 20}$$

## 1.2 Expressions with Variables (page 4)

### Vocabulary
variable, numerical, coefficient, evaluate, term

### Suggestions
The concept of variable might be easy to explain if you compare a variable with the English grammar concept of pronoun. The pronoun *he* in a sentence is like the variable $x$ in algebra. *He* is a placeholder for a name. Likewise, $x$ is a placeholder for a number.

Stress that a term can be a variable, a number, or a product of a variable and a number.

Review multiplication of fractions. For some students it might be helpful to rewrite $12 \cdot \frac{3}{4}$ as $\frac{12}{1} \cdot \frac{3}{4}$.

### Errors That Students Make
$5 + 2a$ is not equal to $7a$. This can be shown by letting $a$ be any value, except 1. For example, if $a = 2$, $5 + 2a = 5 + 2 \cdot 2$, or 9. If $5 + 2a = 7a$, the value of $7a$ if $a = 2$ is $7 \cdot 2$, or 14.

### Calculator Activities
Stress that many calculators cannot automatically handle the order of operations, and many do not have memories. For example, to compute $3 + 4 \cdot 8$, the student will have to use the commutative property to rewrite the problem as $4 \cdot 8 + 3$ so that the multiplication will come first. You might wish to provide drill with the order of operations by first giving some practice on just calculation without the need for evaluation.

**1.** $718 \cdot 619 - 249 \cdot 518$

**2.** $\dfrac{845 \cdot 612 - 48 \cdot 35}{252 - 16 \cdot 12}$

**3.** $\dfrac{122{,}540 - 1{,}165 \cdot 23}{14{,}628 - 38 \cdot 49}$

### For Additional Chalkboard Work
Name the numerical coefficient of each variable.
**1.** $3a$  **2.** $5b + 4m$  **3.** $3x + 2y$  **4.** $4 + 3k + 2g$

### Cumulative Review
Simplify.
**1.** $7 + 4 \cdot 3$           **2.** $7 - 15 \div 3$

**3.** $\dfrac{6 \cdot 3 + 4 \cdot 2}{10 - 4 \cdot 2}$

## 1.3 The Commutative and Associative Properties (page 7)

### Vocabulary
commutative, associative

### Suggestions
The primary outcome of this lesson is the conclusion that addition and multiplication can be done in any order. This will allow students later in the chapter to regroup like terms so that they may be combined. You might ask the students to look up the English dictionary definitions of ''commute'' and ''associate'' so they can see a reason for the choice of these words in the mathematical sense. For example, to *commute* means to exchange or substitute one thing for another. Most students will be familiar with the concept of commuting from home to school and vice versa. To *associate* means to join together or combine. So we can associate the first and second of three numbers to be added or multiplied, or we can associate the second and third.

### Errors That Students Make
When simplifying an expression like $2(5t)(4)$, some students tend to multiply the 2 times both the 5 and 4 to get $80t$ as an answer instead of $40t$. Have these students rewrite the problem as $2 \cdot 5 \cdot 4 \cdot t$.

### For Additional Chalkboard Work
Which property is illustrated?
**1.** $3 + 4 = 4 + 3$
**2.** $7 \cdot (5 \cdot 8) = (7 \cdot 5) \cdot 8$
**3.** $6(5m) = (5m)6$
**4.** $t + (r + y) = (t + r) + y$
Rewrite by using the associative and commutative properties so that computation will be easy. Then simplify.
**5.** $99 + 47 + 1$
**6.** $50 \cdot 19 \cdot 2$
**7.** $58 + 60 + 2$
Simplify.
**8.** $4(3m)2$ **9.** $6(3y)4$ **10.** $4(5m)3$ **11.** $6 \cdot 2(3a)$

### Cumulative Review
**1.** Simplify $18 - 9 \div 3 + 4 \cdot 5$.
**2.** Evaluate $4a + 7b$ for $a = 2$ and $b = 3$.
**3.** Simplify $9 \cdot 2 - 4 \cdot 4 \div 2$.

## 1.4 The Distributive Property (page 10)

### Vocabulary
distributive

### Suggestions
Stress both the right- and left-handed forms of the distributive property: $a(b + c) = ab + ac$ and $(b + c)a = ba + ca$. The use of the distributive property ''in reverse'' serves as the basis for justifying combining like terms in the next lesson and common monomial factoring in Chapter 6.

The distributive property provides ample opportunity to maintain fraction skills. Before doing Example 2, you may want to review products like $\frac{1}{4} \cdot 8 = \frac{1}{\overset{4}{\underset{1}{\cancel{4}}}} \cdot \overset{2}{\cancel{8}} = \frac{2}{1}$, or 2.

### Errors That Students Make
There are two types of errors that commonly appear within a few days after the distributive property has been taught.
**1.** $3(2a + 4) = 6a + 4$   The student has forgotten to also multiply 4 by 3.
**2.** $3(2a + 4) = 24a$   The student sees only multiplication, ignoring the $+$.

### For Additional Chalkboard Work
Simplify.
**1.** $5(2a + 6)$ **2.** $(7x + 2)4$

**3.** $4(9m + 1)$ **4.** $8(4t + 2)$

### Non-Routine Problems
If you have a high-level class, you might want to postpone this challenge until after the lesson on exponents. Then ask if anyone can express the value of the money in terms of exponents.
Solution: 1, or $2^0$    2, or $2^1$    $2 \cdot 2$, or $2^2$
1st day    2nd day    3rd day
$2^{30-1}$, or $2^{29}$
30th day

### Cumulative Review
Simplify.
**1.** $9 + 4 \cdot 2 - 3 \cdot 4$    **2.** $50 \cdot 37 \cdot 2$
**3.** $199 + 35 + 5 + 1$

# 1.5 Combining Like Terms (page 13)

## Vocabulary
multiplicative identity

## Suggestions
Emphasize the prose just preceding Example 1, which uses the distributive property "in reverse," to justify why like terms can be combined. Stress in the beginning that it is helpful to rewrite an expression like $4x + x$ as $4x + \underline{1}x$, using the property of the multiplicative identity.

## Errors That Students Make
If the multiplicative identity is not stressed, students will simplify $3x + 4x + x$ as $7x$, thinking of the last $x$ as "nothing" since there is no numerical coefficient written.

Another typical error is highlighted in Example 2. Some students simplify $5a + 3b + 7$ as $15ab + 7$.

Some students make the error of first adding the 5 and 6 in simplifying $5 + 6(4c + 2) + 3c$ of Example 5. Stress the order of operations here.

## For Additional Chalkboard Work
Simplify.
1. $5x + 4x + x$   $10x$
2. $7a + 3a + 2$   $10a + 2$
3. $4a + 2b$   cannot be simplified
4. $\frac{2}{3}(9x + 12) + 3x$   $9x + 8$
5. $4 + (10 + 15x)\frac{3}{5}$   $10 + 9x$

## Enrichment
All you have is a 3-minute sandglass and an 8-minute sandglass. How can you cook an egg in exactly 13 minutes?
*Solution:* Turn both upside down at the same time. As soon as the 3-minute one is empty (5 minutes left on the 8-minute one) start cooking the egg. When the 8-minute one is empty on the top, turn it over again. When it is empty again, $5 + 8$, or 13, minutes will have passed.

## Cumulative Review
Evaluate.
1. $6 + 5x$ for $x = 2$   2. $9m + 4$ for $m = 3$
       16                              31
3. $8a + 2 + 7b$ for $a = 5$, $b = 6$
       84

# 1.6 Factors and Exponents (page 16)

## Vocabulary
factor, base, exponent

## Suggestions
The following analogy is helpful: Just as repeated addition, $4 + 4 + 4$, can be interpreted as multiplication, $4 \cdot 3$, so can repeated multiplication, $4 \cdot 4 \cdot 4$, be indicated with an exponent, $4^3$. The extension of the rules for order of operations is necessary for evaluation of polynomials like $x^2 + 5x$ for $x = 3$. You may want to illustrate this example.
$$3^2 + 5 \cdot 3$$
$$9 + 15$$
$$24$$

Point out to the students that the exponent is evaluated first. Then multiply and add.

## Errors That Students Make
A typical error in the beginning is to compute $4^2$ as 8. Stress that $4^2$ means 4 multiplied as a factor twice, $4 \cdot 4$, NOT $4 \cdot 2$.

Another kind of error arises in an evaluation like that shown in Example 3.

Evaluate $5a^3$ for $a = 2$. The error is to multiply 5 times 2 first to get 10, then to compute $10^3$. This can be easily avoided if the student writes $5a^3$ as $5 \cdot a \cdot a \cdot a$. The second part of Example 3—evaluate $(5a)^3$ for $a = 2$—is given to highlight how parentheses can change the base. Stress how to determine what the base is to eliminate student errors in evaluation.

## For Additional Chalkboard Work
Compute.
1. $2^4$   2. $4^2$   3. $5^2$   4. $3^3$   5. $6^2$   6. $2^5$
   16        16        25        27        36        32
Evaluate for the given value of the variable.
7. $x^3$ for $x = 5$  8. $b^4$ for $b = 3$  9. $6a^3$ for $a = 2$
       125                    81                    48

## Cumulative Review
1. Name the coefficient of $x$ in $2 + 7x$.   7
2. Simplify $4(3a + 6)$.   $12a + 24$
3. Simplify $9x + 4x + x$.   $14x$

## 1.7 Formulas in Geometry (page 18)

### Vocabulary
area, formula, perimeter

### Suggestions
This lesson ties together several concepts of the chapter: basic properties, evaluation, and exponents. This lesson helps the students to see an application of algebra in geometry.

You might want to ask the class for an alternate formula ($A = \frac{1}{2}bh$) for the area of a triangle than the one given in Example 2.

### Enrichment
How many square yards of carpet would be needed to cover the floor of this classroom? How much would it cost? This prepares the way for the problem-solving strategy to be introduced to students in future chapters. It is important that the students learn that most real problems in life are not already written. Frequently, one must formulate on one's own what one needs in order to solve a general problem.

Encourage the class to come up with the questions that must be answered, such as the ones listed below;
1. What is the area of the room? (Use $A = lw$.)
2. If the area is found in square feet, how do you convert this to square yards? (Divide by 9 since there are 9 square feet to a square yard.)
3. How much does carpet cost per square yard?
4. What would be the formula for finding the total cost if the measurement of the room is found in terms of square feet? ($T = c \cdot \dfrac{lw}{9}$, where $c$ is the cost per square yard.)

### Cumulative Review
Simplify.
1. $4y + y + 9y$   $14y$
2. $6 + 2(3x + 8) + 2x$   $8x + 22$
3. $\dfrac{8 + 4 \div 2}{5 \cdot 3 - 5}$   $1$

## Applications (page 20)

Stress the importance of formulas in everyday life, such as problems of the consumer. Try to get students to formulate their own formulas. This will pave the way for the word problem-solving strategy that will begin in Chapter 3.

For example, discuss the meaning of time-and-a-half overtime. Ask for a formula. What would you have to know to compute a person's weekly wage?
Answer: You would have to know the person's regular hourly wage, the number of hours worked, and the number of hours considered overtime.

## 1.8 Recognizing Solutions of Open Sentences (page 21)

### Vocabulary
open sentence, solution

### Background
This lesson does not involve the formal solution of equations. Students merely SUBSTITUTE given numbers into an equation to determine if any of them are solutions of the equation.

### Suggestions
Before introducing the lesson, try a brief review that stresses both substitution and order of operations.
Evaluate for the given value(s) of the variable(s).
1. $3x + 4$; $x = 2$, $x = 5$   $10, 19$
2. $12 - 2a$; $a = 4$, $a = 6$   $4, 0$
3. $3x - 7$; $x = 5$, $x = 3$   $8, 2$
Then stress the similarity of an open sentence to an English sentence containing a pronoun. This is illustrated in the prose just preceding Example 1.

In a later chapter we shall talk about equations with no solution or many solutions. Therefore, it is essential that we not mislead students in this opening lesson. There CAN be more than one solution to an equation. This is illustrated in Exercises 19, 20, 24, and 29, which turn out to be identities.

## Computer Activities: Evaluating Algebraic Expressions (page 23)

This first lesson on programming in BASIC introduces the student to the LET, PRINT, END, RUN, and NEW commands. It provides a direct application of the rules for order of operations. Stress that RUN is not a part of the program and therefore has no line number. Explain to students that NEW is necessary to clear the computer's memory before entering a new program. Pressing the RETURN key (ENTER key on TRS-80) is like the carriage return or return key on a typewriter.

## Chapter Review (page 24)

Notice that the practice is grouped by lesson. Students can refer back to the indicated lesson for review of concepts as needed. The starred exercises are equivalent to **C** exercises of the individual lessons.

## Chapter Test (page 25)

The Chapter Test may be used as additional review if necessary. Another alternative is to use the Chapter Test for class review and then let the students do the Chapter Review, to which they have the odd-numbered answers, for more practice at home the day before the actual test.

## College Prep Test (page 26)

Stress that practicing now for college entrance tests will help the students to do well on them when they take them in the future.

# Chapter and Lesson Commentaries

## Chapter 2 Real Numbers

### Objectives

To use integers to describe real-life situations
To name the coordinates of points graphed on a number line
To graph numbers on a number line
To find the absolute value of a number
To compare integers
To identify numbers as rational or irrational
To add, subtract, multiply, and divide real numbers
To read and interpret the symbol $-a$
To identify the properties for addition of real numbers
To simplify and evaluate algebraic expressions containing real numbers
To simplify algebraic expressions by combining like terms with real number coefficients
To simplify and evaluate algebraic expressions containing parentheses
To subtract algebraic expressions

### Overview

In the beginning of this chapter, addition is indicated with the aid of parentheses. For example, $4 + (-3)$ is read as positive 4 PLUS negative 3. In other words, 4 is read as POSITIVE 4 (no symbol used) and $-3$ is read as NEGATIVE 3. The "+" is always interpreted as the operational symbol for addition. Later in the chapter it is agreed that the parentheses may be dropped, but are still understood. Thus,

$$-8 + 5 - 4$$

is read negative 8 PLUS positive 5 PLUS negative 4. This approach was adopted so students will have fewer difficulties in interpreting a problem like $-8 + 5 - 4$ as an addition problem.

## Overview (continued)

Other topics taught in this chapter are graphing integers on a number line, using integers to describe real-life situations, and absolute value. The second lesson introduces the students to real numbers as the set of rational numbers together with the set of irrational numbers. These concepts, along with the properties for addition of real numbers, help ease the students into subtracting real numbers. Multiplying and dividing real numbers are also discussed, along with simplifying and evaluating algebraic expressions with real number coefficients. Examples and exercises include work with decimal and fractional numbers to maintain students' proficiency in both.

### Special Features

### François Viète (page 27)

Students might be interested in when the symbol $\pi$ first came into use. Many people erroneously believe that the Greek mathematicians initiated the use of the symbol. The symbol was first used by early mathematicians of England such as David Gregory to represent the circumference of a circle. William Jones (1675–1749) was the first to use the symbol for the ratio of the circumference of a circle to its diameter. However, the symbol was not generally used in this context until the famous mathematician Euler adopted it in 1737.

## 2.1 Integers and the Number Line (page 28)

### Vocabulary

natural number, whole number, integer, origin, coordinate, graph, absolute value

### Suggestions

Stress that the concept of signed numbers is not new to the students. They will have encountered these numbers in weather reports that refer to temperatures *below* zero as well as on thermometers. Stress the practicality of signed numbers through the following real-life applications: temperature, distance above and below sea level, gain or loss of yardage in football, profit and loss, overdrawn checking accounts, and so on.

The concept of absolute value will be important when the rules for adding signed numbers will be given in terms of adding or subtracting absolute values.

Students are introduced to the inequality symbols $<$ and $>$ as a means of comparing numbers.

Use the **Reading in Algebra** questions to stress that if there is no sign in front of a number, it is understood to be positive.

### For Additional Chalkboard Practice

On a number line, graph the points with the given coordinates.

| **1.** $A:-7$ | **2.** $B:6$ | **3.** $C:-8$ | **4.** $D:-10$ |
|---|---|---|---|
| 7 left | 6 right | 8 left | 10 left |

Find the absolute value.

| **5.** $|-9|$ | **6.** $|35|$ | **7.** $|-45|$ | **8.** $|-19|$ |
|---|---|---|---|
| 9 | 35 | 45 | 19 |

### Cumulative Review

Simplify. Then evaluate for $x = 6$ and $y = 5$.

**1.** $3x + 2y + x + 4y$    $4x + 6y$; 54

**2.** $7y + 4x + 2 + y + x + 7$    $8y + 5x + 9$; 79

## 2.2 Real Numbers (page 32)

### Vocabulary
irrational number, rational number, real number, repeating decimal, terminating decimal

### Suggestions
Before starting the lesson, have students make up their own terminating decimals and repeating decimals. Then ask students to make up a decimal that neither terminates nor repeats. Explain that this represents an irrational number, as does $\pi$ and $\sqrt{7}$. Stress that division by 0 is not defined.

### Errors That Students Make
Some students may think that a decimal like 0.121221222... is a repeating decimal since the 1's and 2's repeat. However, since a period of digits does not repeat, 0.121221222... names an irrational number.

Another common error is to say that $\frac{4}{0} = 0$ or $\frac{4}{0} = 4$. Remind students that they can divide into 0 but not by 0.

### For Additional Chalkboard Work
Which are rational numbers, which are irrational numbers, and which are neither?

1. $0.24\overline{24}$
2. $0.383383338\ldots$
3. $-0.159$
4. $2\pi$
5. $0.263263\ldots$
6. $9$
7. $\sqrt{3}$
8. $-\frac{5}{0}$

### Cumulative Review
1. Simplify $\frac{2}{3}(6x + 9) + 5x$.
2. Evaluate $\left(\frac{2}{3}a\right)^3$ for $a = 6$.
3. Evaluate $2x - 3\frac{1}{2}$ for $x = 5$.

## 2.3 Adding Real Numbers (page 34)

### Vocabulary
closure

### Suggestions
In the beginning, stress adding numbers in terms of moves on the number line as shown in Examples 1, 3, and 4. If the class appears to be having difficulty with the concept, you might want to split the lesson and do it in two days.

    First day: Adding on the number line (*no* rules)
           Homework: Exercises 1–24
    Second day: Adding by using the rules
           Homework: Exercises 25–34

Stress the **Summary** on page 36.

Illustrate the *closure property* through concrete illustrative examples. The set of whole numbers is *closed* under addition. If you add two whole numbers, you always get a whole number; the sum is a member of the original set of whole numbers. This is *not* true for subtraction; for example, $5 - 8 = -3$, and $-3$ is *not* a member of the original set of whole numbers.

### For Additional Chalkboard Work
Assign students to show the following sums using a number line. Then for each problem ask a different student to explain how to get the same answer by using the rules.
Add.

1. $-5 + 2$
2. $7 + (-9)$
3. $-8 + (-1)$
4. $-3\frac{1}{2} + 10$
5. $-3 + (-3)$
6. $-5\frac{1}{2} + 11\frac{1}{4}$

### Enrichment
A group of friends took a bus trip. Each traveler gave the driver a tip using the same 9 coins. The total tip from the group was $8.41. How many coins did the driver receive?
*Solution:* Find the number of passengers; the only products equal to 841 are $1 \times 841$ and $29 \times 29$. Since neither 1¢ nor $8.41 can be paid in exactly 9 coins, there must have been 29 passengers paying 29¢ each. The only possible combination is 5 nickels and 4 pennies.

## 2.4 Properties for the Addition of Real Numbers (page 37)

### Vocabulary
additive identity, additive inverse, opposite

### Background
Review the commutative and associative properties already taught in Chapter 1. Stress that all the properties for addition of integers can be extended to the real numbers. The concept of the *opposite* of a number will be exceptionally important in the next lesson on subtraction: *a* subtract *b* will be interpreted as *a* PLUS the OPPOSITE of *b*.

### Suggestions
Stress that $-a$ is NOT to be interpreted as necessarily a negative number. Students can be convinced of this only by actual practice. Copy the following table on the chalkboard;

| $a$ | $-a$ |
|-----|------|
| $-4$ | |
| $5$ | |
| $-3$ | |
| $6$ | |
| $0$ | |
| $-9$ | |

Ask students to fill in the right column. Point out that it is clear from the table that $-a$ can be positive, zero, or negative, depending on the value of $a$ itself.

### For Additional Chalkboard Work
Find each sum. Then give its opposite.
1. $7 + (-4)$
2. $-9 + (-3)$
3. $-5 + 14$
4. $6 + (-6)$
5. $-15 + 8\frac{1}{2}$
6. $-9.2 + (-9.8)$

### Cumulative Review
1. Add $-19 + (-3)$.
2. Simplify $x + (8x + 1)3$.
3. Evaluate $x^3$ for $x = \frac{2}{3}$.

## 2.5 Subtracting Real Numbers (page 40)

### Background
Subtraction is taught in terms of adding the opposite. Review the *opposite* of a number from the last lesson.

### Suggestions
The following technique might help to drive home the point that subtracting a number is the same as adding its opposite. Ask how each should be read, then compute each.

$$7 - 2 \qquad\qquad 7 + (-2)$$

### For Additional Chalkboard Work
Subtract.
1. $8 - 2$
2. $-4 - 3$
3. $7 - 10$
4. $-5 - 6$
5. $4.3 - 11.1$
6. $-3\frac{3}{4} - (-9)$

### Enrichment
John and Lois have a barrel filled with 8 liters of root beer, and two containers, one that will hold 3 liters and the other, 5 liters. How can they divide the root beer equally (4 liters each) without throwing any away or using any other container?

*Solution:*

| 8 liter barrel | 5 liter barrel | 3 liter barrel |
|----------------|----------------|----------------|
| 8 | ☐ | ☐ |
| 3 | 5 | ☐ |
| 3 | 2 | 3 |
| 6 | 2 | ☐ |
| 6 | ☐ | 2 |
| 1 | 5 | 2 |
| 1 | 4 | 3 |
| 4 | 4 | ☐ |

### Cumulative Review
Add.
1. $-5 + 7$
2. $9.3 + (-12)$
3. $-6\frac{1}{2} + \left(-3\frac{1}{4}\right)$

## 2.6 Multiplying Real Numbers (page 41)

### Suggestions

The following is an alternate method for showing why a positive times a negative is a negative.

To find $4 \cdot -3$, begin with the expression $4(-3 + 3)$. $4(-3 + 3) = 4(0)$, or 0

So, $\qquad 4(-3 + 3) = 0$.

Then $4 \cdot -3 + 4 \cdot 3 = 0$ by the distributive property.

$$4 \cdot -3 + 12 = 0$$

Think, $\boxed{\phantom{xx}} + 12 = 0$

What number plus $12 = 0$?

$$\boxed{-12} + 12 = 0$$

Thus, $4 \cdot -3$ must be $-12$.

Note that it is essential to review addition while teaching this new multiplication concept. Students have a tendency to mix rules. They are now likely to say that $-3 + (-4) = 7$, thinking of the multiplication rule that two negatives make a positive.

You may want to use Examples 3 and 4 to review multiplication of fractions. It might help to rewrite $3\frac{1}{2} \cdot -8$ as $3\frac{1}{2} \cdot \frac{-8}{1}$.

### Enrichment

Use the style of the alternate method above to show that a negative times a negative is a positive.

To find $-5 \cdot -6$, begin with the expression $-5(-6 + 6)$. $-5(-6 + 6) = -5 \cdot 0$, or 0

So, $\qquad -5(-6 + 6) = 0$.

Then $-5 \cdot -6 + -5 \cdot 6 = 0$ by the distributive property.

$$-5 \cdot -6 + (-30) = 0 \text{ by (neg.)(pos.)} = \text{neg.}$$

Think, $\boxed{\phantom{xx}} + (-30) = 0$

What number plus $-30 = 0$?

$$\boxed{30} + (-30) = 0$$

Thus, $-5 \cdot -6 = 30$.

### Cumulative Review

Add.

**1.** $-3 + (-4)$ **2.** $8 + (-3.2)$ **3.** $-3\frac{5}{8} + 7$

## 2.7 Dividing Real Numbers (page 44)

### Vocabulary

multiplicative inverse, reciprocal

### Background

Review the concept that multiplication and division are inverse operations. This enables the students to easily follow the text development of the rules for division of signed numbers, which are the same as those for multiplication.

Stress the nature of the *reciprocal* of a fraction. Students can easily see that the reciprocal of $\frac{2}{3}$ is $\frac{3}{2}$. They often have trouble with whole and mixed numbers. Writing 3 as $\frac{3}{1}$ and $4\frac{2}{5}$ as $\frac{22}{5}$ is helpful.

### Suggestions

Encourage the students to write the rules for multiplication (division) and addition of signed numbers on a single sheet of paper. The following abbreviated summary will be particularly helpful to the slower students.

#### RULES FOR SIGNED NUMBERS

| Addition | Multiplication/Division |
|---|---|
| 1. Same signs:<br>Add, keep the sign. | 1. Same signs:<br>positive |
| 2. Different signs:<br>Subtract absolute values, take the sign of the larger absolute value. | 2. Different signs:<br>negative |

### Cumulative Review

Add.

**1.** $-7 + (-2)$ **2.** $7\frac{1}{2} + (-10)$ **3.** $-19 + 3.5$

## 2.8 Shortcuts in Adding and Subtracting Real Numbers (page 47)

### Suggestions

First review the rules for adding signed numbers. Stress the prose explanation written just before Example 2. The idea of interpreting an expression like $-3 + 8 - 7$ as simply the addition of three integers is helpful to students who feel overwhelmed by all the rules for addition, subtraction, multiplication, and division.

Thus, computing $-3 + 8 - 7$ involves only the rules of addition. It might help at this time to have students read some of the written exercises orally to check if they understand the correct interpretation of this new look at addition.

The **Geometry Review** will reinforce the concept of volume. Many proficiency tests use the formula $V = \frac{Bh}{3}$ for the volume of a pyramid. It would be good to explain to students that what $h$ is multiplied by represents the area of the base of the pyramid: $B$ if it is a rectangle, $s^2$ if it is a square.

### Errors That Students Make

Now that students have studied the rules for multiplication and division, a typical error is to mix them with the rules for addition. For example, students might compute $-5 - 4$ as 9, since "two negatives make a positive." You will have to remind students patiently that the rule in quotation marks is for multiplication, not for addition. So, $-5 - 4 = -9$, not 9. Stress this idea by assigning the following practice. Ask each student to give the rules used in getting the answer.

### For Additional Chalkboard Work

Compute.

1. $-6 + 4 - 12$      2. $-8 - 6 + 5$
3. $7 - 8 + 10$      4. $-2 + 6 - 4 + 1$

### Cumulative Review

Simplify.

1. $-8 \cdot \frac{3}{4}$      2. $-40 \div -\frac{1}{2}$      3. $\frac{16}{-2}$

## 2.9 Evaluating Algebraic Expressions (page 49)

### Suggestions

Students will have to use the rules for *both* addition and multiplication in many exercises in this lesson. Be sure to review these rules before beginning the lesson. You might want to try the following "getting ready" practice with the class before actually beginning the lesson.

Simplify.

1. $-5 \cdot 4$      2. $-7 + 2$
3. $-8 \cdot -3$      4. $-8 + 5 - 2 + 4$

In the beginning stress that to evaluate an expression like $-8 - 2x$ for $x = -3$ (Example 1), it is very helpful to first interpret the expression as addition with the use of parentheses. Write $-8 - 2x$ as $-8 + (-2x)$ and have students read this as negative 8 plus negative $2x$.

### For Additional Chalkboard Work

Evaluate.

1. $7 - 3x$ for $x = 2$
2. $8 - 4y$ for $y = -3$
3. $-9 - 4n$ for $n = -5$
4. $9p - 3$ for $p = -\frac{2}{3}$

### Enrichment

As a challenge, offer the following problem to the entire class, and if you see fit, give extra credit for the first one to get the correct answer.

Simplify. Then evaluate for $y = -4$ and $z = -2$.

$3(4z + 6) + 5[7 + 3(2y + 4) + 3z] + y$

### Cumulative Review

Simplify.

1. $3x + 5x + x$
2. $y + 7 + 4y + 9$
3. $2y + (10 + 15y) \frac{3}{5}$

## 2.10 Like Terms with Real Number Coefficients (page 51)

### Suggestions
Stress the grouping of like terms. The faster students may not want to be slowed down by actually writing the regrouping. Encourage the less capable students to write all the steps.

### Errors That Students Make
The slower pupils will still write $-5x - 3x = 8x$, mixing the rules for addition and multiplication. A suggestion for helping these students with the same persistent difficulty is to have them copy the rules for adding signed numbers and the rules for multiplying signed numbers side by side and use them when they do the classwork and homework. This is a good way to provide help on an individualized basis.

### For Additional Chalkboard Work
Simplify.
1. $-4x - 2x$   $-6x$
2. $-2a + 9a$   $7a$
3. $-7x + 2x$   $-5x$
4. $6r - 6r$   $0$
5. $8m - 3 - 5\frac{3}{4}m + 8$   $2\frac{1}{4}m + 5$
6. $4d - 8.1 - 6d + 9$   $-2d + 0.9$
7. $6m + 7 - 9m - 8$   $-3m - 1$
8. $7x + 3y - 2 - 5x - 5y$   $2x - 2y - 2$

### Enrichment
Provide for the brighter students with the following assignment, which serves two purposes:
1. reviews the method of simplifying algebraic expressions with parentheses and brackets, and
2. provides a challenge.
Simplify. (Then go on to homework.)
1. $3x + 2[5 + 4(7 + x)]$   $11x + 66$
2. $7a + [4 + 3(2a + 6)]3 + a + 6$   $26a + 72$

### Cumulative Review
Simplify.
1. $6a + 5(4a + 2)$   $26a + 10$
2. $9y + (7y + 2)3$   $30y + 6$
3. $-36 \div -6$   $6$

## 2.11 Multiplication Properties of 1 and −1 (page 53)

### Suggestions
As a pre-lesson warm-up, review the multiplicative property of 1 using the following problems. Simplify.
1. $5x + x + 4x$   $10x$
2. $a + 3a + 2a$   $6a$
3. $6m + 5m + m$   $12m$
4. $y + 6y + y + 8y$   $16y$
Stress, particularly for the slower students, the actual writing of the coefficients 1 and −1 where appropriate, as shown in the illustrative examples.

### For Additional Chalkboard Work
Simplify.
1. $7x - x$   $6x$
2. $3x - 4x - x$   $-2x$
3. $2m - 5m - m$   $-4m$
4. $y + 2x + 3y - x$   $x + 4y$

### Enrichment
The following enrichment problems review the use of exponents and absolute value.
1. What can you conclude about the value of $(-a)^k$ when $k$ is an odd integer and $a > 0$? The value is negative.
2. Evaluate $-a^3$ if $a = -2$.   $8$
3. Evaluate $-\dfrac{|-4x - x|}{(-x + 2x)^3}$ for $x = -1$.   $5$

### Cumulative Review
Simplify.
1. $4(2a + 6) + (3a + 2)5$   $23a + 34$
2. $2 + (8 + 10x)\frac{3}{2}$   $15x + 14$
3. $4^3 \cdot (-2)^3$   $-512$

## 2.12 Simplifying Expressions Containing Parentheses (page 55)

### Suggestions

The prose development above Example 1 aids the student in understanding the application of the distributive property to the simplification of a problem like $-2(3 + 4x)$. However, stress that this written use of the raised dot as a multiplication sign is not necessary. Students should work the exercises in the manner shown in Example 1.

### Errors That Students Make

In removing parentheses involving whole numbers, students will still tend to forget to distribute the $-4$ to BOTH the 8 and the $6a$ in simplifying $3a - 4(8 + 6a)$. This is even more likely to happen in problems like Example 3. They might write $-5 - (8 - 7b) - 6b = -5 - 8 - 7b - 6b$, forgetting to multiply the $-1$ (understood) times the $-7b$. To avoid this stress, write the $-1$ in front of the parentheses.

The **Reading in Algebra** section helps to clarify the troublesome concepts mentioned above.

### For Additional Chalkboard Work

Simplify.
**1.** $2a - 3(7 + 5a)$   $-13a - 21$
**2.** $4n - 6(1 - 2n)$   $16n - 6$
**3.** $-8 - (9 - 6b) - 5b$   $-17 + b$
Simplify. Then evaluate for $x = -3.2$.
**4.** $10x - 4(3x - 7) - 8$   $-2x + 20; 26.4$
**5.** $-2(1 - x) - (4 - x) - 4x$   $-x - 6; -2.8$

### Cumulative Review

Simplify.
**1.** $-8 \div \left(4 \div \frac{1}{3}\right)$   **2.** $|8 - 10|$   **3.** $\left(-\frac{2}{3}\right)^3$
  $-\frac{2}{3}$        $2$          $-\frac{8}{27}$

## 2.13 Subtracting Algebraic Expressions (page 58)

### Background

The students have already studied subtraction on page 40. This lesson therefore simply extends the concept to subtracting algebraic expressions.

You may wish to use the following problems at the beginning of the lesson to review the subtraction of real numbers.
Subtract.
**1.** $4 - (-3)$    **2.** $8\frac{5}{6} - 12$    **3.** $7 - 4\frac{2}{5}$
  $7$           $-3\frac{1}{6}$         $2\frac{3}{5}$
**4.** $6 - (-2)$
  $8$

### Suggestions

Once again the concept of $-a = -1(a)$ helps the student to better understand the concept of subtraction. This is stressed in all four illustrative examples.

### Errors That Students Make

Students frequently forget to write the parentheses. Thus, from $3x - 2$, subtract $4x - 6$ might mistakenly be written as $3x - 2 - 4x - 6 = -x - 8$. Stress the use of parentheses with the $-1$ technique.
From $3x - 2$, subtract $4x - 6$.
    $3x - 2 - (4x - 6)$
    $3x - 2 - 1(4x - 6)$
    $3x - 2 - 4x + 6$
      $-1x + 4$,   or   $-x + 4$

### For Additional Chalkboard Practice

Subtract.
**1.** Subtract $4x - 2$ from $x + 1$.   $-3x + 3$
**2.** From $x - 9$, subtract $-x + 1$.   $2x - 10$
**3.** Subtract $4 - 2m$ from $-m + 3$.   $m - 1$
**4.** From $5b - x$, subtract $\frac{7}{8}b - 4x$.   $4\frac{1}{8}b + 3x$

### Cumulative Review

**1.** Divide $\frac{-35}{7}$.   $-5$

**2.** Evaluate $-a + 3b$ for $a = -5$ and $b = -3$. $-4$
**3.** Simplify $7x - (-8 - 6x) + 4x$.   $17x + 8$

# Chapter and Lesson Commentaries

## Chapter 3   Equations: An Introduction

### Objectives
To solve equations with variable terms on only one
   side of the equation
To write English phrases in mathematical form
To solve word problems involving one number
To determine the truth value of a conditional

### Overview
Methods for solving equations with variables on
only one side of the equation are developed. Stress
is placed upon *undoing* operations in determining
what must be done to each side of an equation to
solve it. A logo or word problem strategy for
solving all word problems is taught. This logo will
be referred to consistently throughout the text
whenever word problems are taught. The
conditional *if-then* is an important topic in
mathematics and computer programming. The
concept of conditional including hypothesis,
conclusion, and truth value is introduced. The
chapter closes with a lesson on *proofs*. This will
help pave the way for a better understanding of the
nature of proofs in geometry.

### Special Features

### Suggestions
This page provides a good, practical introduction to
the *logo* for solving word problems. The logo is
used throughout the text as a strategy for word
problem solving. Stress the importance of the term
*interpret*. Students frequently check a word problem
by checking the solution to the equation used to
solve the problem in only the equation itself. Then,
when they solve a word problem about ages that
leads to a quadratic equation, they never question
the validity of a negative solution: "It checks in the
equation. What's wrong with it?" Stress that
checking calls for more than checking the final
equation. Is the answer reasonable? Does the
equation really interpret the intent of the problem?

## 3.1 Solving Equations: $x + b = c$
### (page 64)

**Vocabulary**

addition property for equations, check, equation, equivalent equations, subtraction property for equations

**Background**

This lesson was developed in response to feedback from teachers across the country indicating that it seems easier to begin teaching equations involving only one step. Student mastery of this and the next lesson should make it easier for students to make the transition to the more complex types of the form $ax + b = c$ and $\frac{x}{d} + b = c$.

**Suggestions**

Stress the concept of undoing an operation. Some students may be more comfortable if they can use a single technique that works for every case. Thus $x + 5 = 8$ and $x - 5 = 8$ can each be solved by basically the same method of adding the opposite (or additive inverse) of a number to each side. Thus, to solve $x + 5 = 8$, add $-5$, the opposite of 5, to each side. To solve $x - 5 = 8$, add 5, the opposite of $-5$, to each side. Experience has shown that the vertical format for showing addition or subtraction of the same number to each side of an equation seems to be easier for students to handle in their first encounters with equations. When the student gets to the lesson on equations of the type $ax + b = c$ and $\frac{x}{d} + b = c$, the transition to horizontal format is made. Then the solution of $x + 5 = 8$ will be displayed as

$$x + 5 + (-5) = 8 + (-5), \text{ or}$$
$$x + 5 - 5 = 8 - 5.$$

Notice the continual use of fractions and decimals to maintain the basic skills. You might have to review adding and subtracting fractions with unlike denominators.

**For Additional Chalkboard Work**

Solve.

1. $x - 2 = 9$  11
2. $4 + y = -3$  $-7$
3. $-8 + x = -3$  5
4. $19 = -11 + x$  30

## 3.2 Solving Equations: $ax = c; \frac{x}{a} = c$
### (page 67)

**Vocabulary**

division property for equations, multiplication property for equations

**Suggestions**

Carefully point out the use of the fraction bar to indicate division. Have students give each of the following divisions as a fraction.

1. $-8 \div 2$     2. $20 \div -4$     3. $-5 \div -7$
   $\frac{-8}{2}$            $\frac{20}{-4}$        $\frac{-5}{-7}$ or $\frac{5}{7}$

As in the last lesson, the equations in this lesson involve only one step. Emphasize that multiplication can be undone by division and division can be undone by multiplication.

**Errors That Students Make**

Since students want answers to come out to be whole numbers rather than fractions, they might solve an equation like $2 = 8x$ by dividing each side by 2: $\frac{2}{2} = \frac{8x}{2}$ and then write $x = 4$. Stress division by the coefficient of the variable.

Another common error is to ignore the negative sign in an equation like $6 = -x$ and say that the solution is 6. This can be avoided by stressing the rewriting of $6 = -x$ as $6 = -1x$.

**For Additional Chalkboard Work**

Solve.

1. $2 = 5x$    2. $\frac{a}{7} = 3$    3. $-x = 5$    4. $12 = \frac{a}{-2}$
   $\frac{2}{5}$          21         $-5$        $-24$

**Enrichment**

A phonograph record has a total diameter of 12 inches. There is an outer margin of 2 inches. The diameter of the unused center is 4 inches. There is an average of 90 grooves to the inch. How far does the needle travel when the record is played? *Solution:* The number of grooves per inch has nothing to do with the problem! The needle does not travel around the circle but moves toward the center. It travels half the diameter less half of the sum of the unused center and outer margin, or $6 - (1 + 2) = 3$.

## 3.3 Solving Equations: $ax + b = c$; $\frac{x}{d} + b = c$ (page 69)

### Background
The concept of order of operations is important in solving equations of the form $ax + b = c$ and $\frac{x}{d} + b = c$. It may be reviewed by using the following illustrative example:

Choose a number, say 2.
Multiply by 5, then add 4.   $5 \cdot 2 + 4$
How do you undo this to get back to 2?
You must reverse the order of operations of
$5 \cdot 2 + 4$.
Undo the addition of 4 by adding $-4$.
$5 \cdot 2 + 4 + (-4) = 5 \cdot 2$.
Now undo the multiplication of 5 by dividing by 5.

$\dfrac{\overset{1}{\cancel{5}} \cdot 2}{\underset{1}{\cancel{5}}} = 2$.   The result is 2, the number you started out with.

   Similarly, go through an example with division: Choose a number, say 8; divide by 4, then add 2. Now undo the operations.

### Suggestions
In this lesson the student will make the transition to the horizontal format for solving equations. Point out that this display is shorter, not necessitating the drawing of a line. You might also indicate that the writing of the parentheses is done in the text only to stress the *operation* of adding the additive inverse to each side.

### Errors That Students Make
For the equation $5x + 2 = 7$, some students might simplify it as $5x = 9$. 2 and 7 are like terms but not on the same side of the $=$ symbol. You might wish to use the following problems in addition to the **Reading in Algebra** exercises in the lesson.
1. For the equation $-5 = 7 - 3x$, you can get the variable term $-3x$ alone on the right by
   _____ .
   adding $-7$ to each side
2. The first step in solving the equation $5x - 4 = 16$ is _____ .
   to add 4 to each side

## 3.4 Conditionals in Logic (page 72)

### Vocabulary
conditional, conclusion, hypothesis, implication

### Background
The *if-then* statement is used throughout mathematics. At least a brief introduction to its nature should be made prior to a formal high school geometry course. Moreover, the computer sections on *branching* will utilize the *if-then* statement in decision making.

### Suggestions
Example 2 illustrates how a false hypothesis can still imply a true conclusion. You might want to follow through with an illustration of a false hypothesis leading to a false conclusion. For example, show that the following conditional is true.

$$\text{If } 4 = 5, \text{ then } 6 = 7.$$
$$\uparrow \qquad\qquad \uparrow$$
$$\text{false} \qquad \text{false}$$
$$\text{hypothesis} \quad \text{conclusion}$$

$$4 = 5 \qquad \blacktriangleleft \text{ False hypothesis}$$
$$4 + 2 = 5 + 2 \qquad \blacktriangleleft \text{ Add the equations.}$$
$$6 = 7 \qquad \blacktriangleleft \text{ False conclusion}$$

So, the conditional is true.

### Cumulative Review
Which property is illustrated?
1. $-7 \cdot 9 = 9 \cdot -7$       2. $\frac{3}{5} + 0 = \frac{3}{5}$
   Comm. Prop. Mult.       Add. Ident.
3. $4(5 + 8) = 4 \cdot 5 + 4 \cdot 8$
   Dist. Prop.

## 3.5 The Language of Algebra (page 74)

### Background

The phrases in this lesson are basic to all word problems that appear in the rest of the text. Because of their importance, it might be helpful for students to copy these basic translations in their notes.

### Suggestions

Stress the difference between less than and decreased by. A typical error made by students is to write 6 less than 13 as $6 - 13$. Emphasize that the "than" in "less than" tells the student to switch the order of subtraction as illustrated at the bottom of page 74. So, 6 less than 13 is $13 - 6$. Use the **Reading in Algebra** exercises to stress the use of the phrases *more than* and *less than*.

### For Additional Chalkboard Work

Write each in mathematical terms.

| | |
|---|---|
| **1.** 5 increased by 2 | $5 + 2$ |
| **2.** 5 less than 2 | $2 - 5$ |
| **3.** 5 decreased by 2 | $5 - 2$ |
| **4.** 5 more than 2 | $2 + 5$ |
| **5.** 7 increased by twice $x$ | $7 + 2x$ |
| **6.** 12 less than 4 times a number | $4x - 12$ |
| **7.** 3 times a number, increased by 1 | $3x + 1$ |
| **8.** 17 decreased by a number divided by 5 | $17 - \frac{x}{5}$ |
| **9.** Twice a number, decreased by 3 | $2x - 3$ |
| **10.** $a$ less than 7 times $b$ | $7b - a$ |

### Cumulative Review

Solve.

**1.** $2x - 5 = 7$    **2.** $13 = \frac{x}{2} + 1$    **3.** $7 = 9 - x$

     6              24             2

## 3.6 Steps for Solving Word Problems (page 76)

### Background

All word problems in this lesson involve only one number. In the next chapter students will study word problems defining relationships between more than one number. However, the logo for the four basic steps for solving word problems will not change. In the plan part of the strategy, the representation will merely involve several numbers instead of only one. Ask the students to copy the abbreviation of the logo in their notes: READ, PLAN, SOLVE, INTERPRET.

### Suggestions

Here is an alternate example you may wish to use to go over the four basic problem solving steps.

   5 more than twice a number is 15. Find the number.

*Step 1* READ carefully, What is the number you are asked to find? A typical error made by students is to say that the number is 5 or 15!

*Step 2* PLAN. You are asked to find one number. Represent that number algebraically.

Let $n$ = the number.

Now use the first sentence to write an equation.

5 more than twice a number is 15.

5 more than $2n$ is 15.

$2n + 5 = 15$ Emphasize "is" means "=".

*Step 3* SOLVE

$$2n + 5 = 15$$
$$2n + 5 + (-5) = 15 + (-5)$$
$$2n + 0 = 10$$
$$2n = 10$$
$$\frac{2n}{2} = \frac{10}{2}$$
$$n = 5$$

*Step 4* INTERPRET. Stress that checking involves more than just checking the equation. The student may have solved the equation correctly but made an error in the original interpretation of the word problem. For example, if the problem had involved the expression *less than* the pupil may have subtracted the wrong way. Thus, the student has to *check* for correct interpretation of the word problem. To check, the student must go back to the original word problem.

5 more than twice a number is 15.

| | |
|---|---|
| $2 \cdot 5 + 5$ | 15 |
| $10 + 5$ | |
| 15 | |

$15 = 15$, true

Thus, the number is 15.
You may wish to use the following problems to supplement the **Reading in Algebra** section. Use the following word problem to answer each question.
8 less than 3 times Jane's age is 24. Find her age.
1. Outline the four steps you would use to solve the problem.
   READ, PLAN, SOLVE, INTERPRET
2. Is Jane's age 24?   No
3. If you choose "$a$" as the variable for representation, what does the "$a$" represent?
   Jane's age
4. If Jane's age is represented by "$a$," which is the correct equation? $3a - 8 = 24$   or
   $8 - 3a = 24$   $3a - 8 = 24$

In order to solve **Non-Routine Problems,** you must find out that Jake pays $15\frac{1}{3}$¢ per orange and sells each for 16¢, thus making $\frac{2}{3}$¢ per orange. So, Jake must sell 150 oranges since $\frac{2}{3}$¢ $\times$ 150 = $1.00.

## Cumulative Review
Solve.
1. $7 - x = -5$
      12
2. $5 = \frac{x}{3} + 4$
      3
3. $5x - 7 - x = 17$   6

## 3.7 Proving Statements (page 80)

### Vocabulary
properties, reflexive, symmetric, transitive, substitution, theorem

### Background
Frequently, in a mathematical proof, if a statement is assumed to be true, it is called an *axiom*. We avoided this subtlety in the pupil's edition.

### Suggestions
If your class contains primarily average and below-average students, you might elect to omit this lesson. However, you may want to assign it to individual students as an extra-credit project.

The significance of the symmetric property is sometimes more easily appreciated with illustrations involving inequalities. For example, if $3 < 5$, then $5 < 3$ is not true. So the symmetric property *does not* hold for inequalities. However, the transitive property *does* hold for inequalities. If $a > b$ and $b > c$, then $a > c$. Stress the *substitution property*. It will be a vital tool later in the course for solving simultaneous equations by substitution.

### Cumulative Review
Evaluate.
1. $(2a)^3$ for $a = \frac{3}{2}$   27
2. $\frac{3}{4}a + 6$ for $a = -12$   $-3$
3. $-a + 6$ for $a = -1.8$   7.8

# 4

## Chapter
## and Lesson
## Commentaries

### Chapter 4   Equations and Word Problems

#### Objectives
To solve equations with variable terms on each side
To solve equations containing parentheses
To solve equations containing fractions
To solve equations containing decimals
To solve word problems

#### Overview
This chapter is an extension of the previous chapter. The opening lesson teaches the solution of linear equations with variable terms on each side. This lesson is followed by one on equations with parentheses. Then students have enough background in solving equations to be able to use them to solve word problems involving more than one number. This word problem lesson is followed by two more equations lessons: equations with fractions and the logical sequel, equations with decimals. An interesting application of decimal equations is that of percent, the next lesson. There is also a special lesson on perimeter problems. The next lesson is a practical-application page applying percent to commissions. The chapter closes with the first major cumulative review, which covers all the major objectives of the first four chapters.

#### Special Features

### Formulating a Problem-Solving Situation (page 85)

#### Background
The main emphasis on this page is the introduction of a new type of word problem, that of formulation. Very few real-life applications of mathematics involve prewritten word problems.

The word-problem logo introduced in the last chapter can still be used with one modification of the first step. You cannot read the problem carefully to find out what is given and what is asked. The problem is not formally written. The problem is very vague. Thus, each student has to formulate his or her own questions to solve the problem.

#### Suggestions
This kind of lesson involves a considerable amount of class discussion. Try to lead students to come up with their own questions. Some of the questions that must be asked are given on the page. This concept emphasizes the need to interpret at the end. Thus, checking is more than merely checking the arithmetic. One must be careful to check whether the original interpretation is complete.

The number of bytes available on a disk depends on how it is manufactured. A $5\frac{1}{4}$-inch, single-sided, single-density disk will contain 125,000 bytes. A double-sided, double-density disk can hold four times that amount. Many disks have 40 tracks with 16 sectors per track and 265 bytes per sector.

## 4.1 Equations with Variables on Each Side (page 86)

### Background
Point out that solving equations that have a variable term on each side is a simple extension of the types already studied in the last chapter.

### Suggestions
You may wish to use the following problems as a pre-lesson warmup.
Solve.
1. $3x + 2 = 23$  7
2. $11 - x = -4$  15
3. $9 = 5 - 4x$  −1
4. $x + 3 - 2x = 7$  −4

It is sometimes easier for students to be given a procedure that works all the time. Point out that you can begin solving an equation like $5x - 8 = 3x + 12$ (Example 1) by adding either the opposite of $3x$ to each side or the opposite of the other variable term ($5x$) to each side. This is shown in Examples 1 and 2. You could thus begin the problem in any one of four ways (adding opposite of a nonvariable to each side). A helpful hint is to add the opposite of the variable term with the smallest coefficient to each term (side) first. This eliminates having to divide by a negative coefficient.

Example 3 stresses the importance of rewriting $-n$ as $-1n$.

Stress in Example 4 that like terms on the SAME side should be combined first before adding opposites to each side.

The following problems will help students to verbalize how to solve equations with the variable on each side.
1. Tell what you can add to each side of $3x - 4 = x + 8$ to produce an equation with only one variable term.  Add either −3x or −x to each side.
2. What should be the first step in solving $3x - 4 + 2x = 8x - 3$?  Combine like terms 3x and 2x.
3. Which variable term should be rewritten in $5 - x = 2x + 10$?  Rewrite −x as −1x.

### Cumulative Review
Simplify.
1. $7x - (4 - x)$
   $8x - 4$
2. $8 - 3(2 - 3x)$
   $2 + 9x$
3. $6x - (x - 1) - 5x$  1

## 4.2 Equations Containing Parentheses (page 89)

### Vocabulary
sum

### Suggestions
You might wish to use the following problems as pre-lesson warm-up to review removing parentheses.
Simplify.
1. $4x + 3(7 - x)$
   $x + 21$
2. $-3 - (9 - 2x)$
   $-12 + 2x$
3. $x - (4 - x) - 3x + 1$  −x − 3
4. $3x - (4 - 8x)\frac{1}{2}$  7x − 2

### Errors That Students Make
After removal of parentheses in Example 2, some students forget to combine like terms.
$$2x + 12 - 6x = 2x + 6$$
Then, instead of adding $-2x$ to each side, they make the error of adding $-2x$ twice on the same side.
$$2x + (-2x) + 12 - 6x + (-2x) = 2x + 6$$
In Example 4, emphasize the meaning of the word *sum*. Many students will say that *sum* means "answer." Stress the association of + with the word *sum*.

### Geometry Review
Students should memorize the formulas for circumference and area of a circle. These appear frequently on state proficiency tests for basic skills.

### Enrichment
Show the following *magical trick*. Challenge the class by asking for an algebraic explanation as shown at the right.
1. Think of a number   $n$
2. Multiply the number by 4   $4n$
3. Add 10 to the result      $4n + 10$
4. Subtract 3 times the original number
   $4n + 10 - 3n$
5. Tell you the result   $n + 10$

You can then tell what the original number was by subtracting 10 from the result.

## 4.3 Number Problems (page 92)

### Background

The four-step logo already introduced to students in the last chapter still applies to the problems of this lesson. The only difference is that more than one number must be represented algebraically.

### Suggestions

The most difficult step for students is deciding which number to represent by the variable. It is helpful if you can drill students on recognizing what numbers are being sought in the word problem. For example, in Example 1, if you let $s$ = smaller, then the larger is 6 less than 7 times $s$.

### Errors That Students Make

A typical student error is to say that the numbers are 6 and 7, especially since "6" follows the words "numbers is." Stress that the numbers to be represented are *larger* and *smaller*. A good device to determine which number to call $s$ is to let the number following the comparison word "than" be $s$.

> larger is 6 less than 7 times smaller

Since "*smaller*" follows the word "than," let $s$ = smaller.

### For Additional Chalkboard Work

For each, what are the two numbers you must represent? Which should be called $x$? What should the other be called?

1. The larger of two numbers is 3 more than the smaller.   smaller: $x$; larger: $x + 3$
2. The first of two numbers is 2 less than 5 times the second.   second: $x$; first: $5x - 2$
3. John's age is 5 times Martha's age.
   Martha's age: $x$; John's age: $5x$

### Cumulative Review

Simplify.

1. $|7.3 - 9|$   1.7
2. $3m - (4 - m) + 9 - m$
   $3m + 5$
3. $\left(-\frac{3}{7}\right)^3$   $-\frac{27}{343}$

## 4.4 Equations with Fractions (page 96)

### Vocabulary

least common multiple

### Background

In a later chapter, students will study fractional equations, that is, equations in which the denominators contain variables. The LCM will be found by using factoring techniques that the students have not had by this current chapter. We feel, however, that there is an advantage to introducing equations with fractions as numerical coefficients of the variable, or as constants, at this early point in the course. It gives us the opportunity to introduce a wide range of practical applications without having to wait until halfway through the course. For example, the next lesson on decimal equations paves the way for all kinds of percent applications in this chapter. The denominators have been kept small so that students don't have to resort to factoring to find the LCM; this can be done by inspection.

### Suggestions

You may wish to start the lesson by reviewing multiplication of fractions in arithmetic.
Compute.

1. $\frac{3}{5} \cdot \frac{2}{7}$
   $\frac{6}{35}$
2. $4 \cdot \frac{5}{2}$ $\left(\text{Hint: Rewrite as } \frac{4}{1} \cdot \frac{5}{2}\right)$
   $10$
3. $\frac{2}{3} \cdot 6$   4
4. $\frac{1}{2} \cdot 12$   6
5. $4 \cdot \frac{3}{2}$   6
6. $8 \cdot \frac{5}{4}$   10

An equation like $\frac{3}{2}x = 5$ can be solved by multiplying each side by the reciprocal of the coefficient of $x$, $\frac{2}{3}$. This is pointed out in the first illustrative example. However, stress that it will *not* work for equations like $\frac{3}{2}x + 1 = \frac{2}{5}$. The more general method, multiplying each side by the LCM, works for *all* cases.

### Cumulative Review

Simplify.

1. $3(5 - x) - 2(4 - x)$
   $7 - x$
2. $|7 - 6 - 4|$
   3
3. $5^2 \cdot -2^3$   $-200$

## 4.5 Equations with Decimals (page 100)

### Background
This lesson provides a good opportunity to stress *why* moving the decimal point when multiplying decimals really works. To show that $(0.03)(0.2) = 0.006$, write each decimal as a fraction and multiply:

$$(0.03)(0.2)$$

$$\left(\frac{3}{100}\right)\left(\frac{2}{10}\right) = \frac{6}{1,000} = 0.006$$

Thus
$$0.03 \leftarrow 2 \text{ decimal places}$$
$$\underline{0.2} \leftarrow 1 \text{ decimal place}$$
$$0.006 \leftarrow 2 + 1, \text{ or } 3 \text{ decimal places}$$

### Suggestions
Start the lesson by having the students practice multiplying decimals.
Multiply.

**1.** $(3.45)(0.014)$     **2.** $(276)(0.024)$
     0.0483              6.624

**3.** $(2.16)(0.015)$    0.03240

Students also need practice in picking out the LCM and then in moving the decimal point when transforming the decimal equation to one with no decimals. You may wish to use the following problems to practice this concept.

Name the LCM for each equation. Then give the result of multiplying each side by the LCM.

**1.** $0.03x - 0.004 = 0.72x + 1.8$
    $1{,}000;\ 30x - 4 = 720x + 1800$

**2.** $0.2x + 0.04 = 5x - 0.36$
    $100;\ 20x + 4 = 500x - 36$

**3.** $0.5x - 1.2 = 6$    $10;\ 5x - 12 = 60$

### Errors That Students Make
When students multiply $100(2.5)$, a typical error is to count 2 places from the 2 to get 25 as an answer rather than 250.
   Stress counting from the decimal point.

### Cumulative Review
Solve.

**1.** Five more than twice a number is 35. Find the number.   15

**2.** The first of two numbers is twice the second. The sum of the numbers is 18. Find the numbers.   12, 6

## 4.6 Using Algebra in Percent Problems (page 103)

### Vocabulary
percent, selling price

### Suggestions
Review writing percents as decimals and decimals as percents by using the following exercises.
Write each as a decimal.

**1.** 25%   **2.** 4%   **3.** 40%   **4.** 120%   **5.** 115%
   0.25     0.04     0.40     1.20     1.15

Write each as a percent.

**6.** 0.32   **7.** $0.12\frac{1}{2}$   **8.** $\frac{1}{4}$   **9.** $\frac{4}{7}$   **10.** 0.03
   32%    $12\frac{1}{2}\%$    25%    $57\frac{1}{7}\%$    3%

Also review finding the percent of a number. It is the other two cases of percent that are difficult and are shown in this lesson.

   The key to solving any of the three cases of percent problems is to translate each sentence directly into an equation, reading from left to right. This is shown in the three examples below. It might be helpful to write all three on the chalkboard, spaced horizontally.

Translate each into an equation.

**1.** 5% of 48 is what number?
             $0.05 \cdot 48 = x$

**2.** 12% of what number is 36?
             $0.12 \cdot x = 36$

**3.** 6 is what percent of 8?
           $6 = x \cdot 8$

### Errors That Students Make
When solving a problem of the type "6 is what percent of 8," students will write $x = \frac{6}{8} = \frac{3}{4}$ and then forget to change the answer to the decimal 0.75, or to a percent.

### Cumulative Review

**1.** Simplify $5 - (4 - x) - 2x$.   $-x + 1$

**2.** Solve $3x - 4 = 7 - 5x$.   $\frac{11}{8}$

**3.** Evaluate $7 - 3a$ for $a = -\frac{2}{3}$.   9

## 4.7 Perimeter Problems (page 106)

### Vocabulary
base, congruent, isosceles

### Background
Students have studied perimeter in previous grades but need to be reminded of the concept. Many students frequently are also not sure of the difference between perimeter and area. Sometimes a very pragmatic, non-technical distinction helps. The perimeter of a rectangle measures the distance around the *outside* of the rectangle. The area measures the amount of space *inside* the rectangle.

Illustrate with the diagram shown. The perimeter, amount of fencing, is $6' + 6' + 8' + 8' = 28'$. The area, the number of blocks of square feet of grass inside, is $6 \cdot 8 = 48$.

GRASS YARD

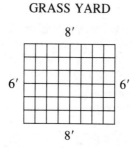

### Errors That Students Make
Many students forget to double each side's length when finding the perimeter of a rectangle; they write $p = l + w$ instead of $p = 2l + 2w$.

Another error is to think that once they have solved the equation they have found the answer and are finished. Emphasize the importance of going back to the problem and checking to see if they have fully answered the question. Does the problem call for only the length, or BOTH the length and the width?

### For Additional Chalkboard Work
The length of a rectangle is 5m more than the width. The perimeter is 38m. Find the length and width. $l = 12m, w = 7m$

### Cumulative Review
1. Simplify
   $5 - (6 - b) + 4b$. $5b - 1$
2. Solve
   $7 - (8 - m) = 3m + 9$. $-5$
3. Evaluate
   $-7 - x$ for $x = -2.3$. $-4.7$

## Applications: Commissions (page 109)

### Background
This lesson is a direct application of percent. Briefly review finding the percent of a number.

### Suggestions
It might be helpful to start the class with the following problems.
   Find the commission earned on the given item(s) at the given commission rate.
1. $500 stereo; 8%   $40
2. $6,000 car; 2%   $120
3. $75 watch; 3%   $2.25

### For Additional Chalkboard Work
Solve.
1. Joe's commission is 30% of his sales. How much must he sell to earn $330?   $1,100
2. Mary's commission is 2% of her sales. How much must she sell to earn $120?   $6,000
3. Juan earns $150 a week plus a commission of 20% on all camera sales. What must his sales be for the week to earn a total of $310?   $800
Make up a commission problem to fit the equation.
4. $120 + 0.30s = 330$
5. $40(3.50) + 0.06s = 200$
4. Jay earns $120 a week plus a commission of 30% on all sales. Find his sales to earn a total of $330.
5. Mary is paid $3.50 an hour for 40 hours a week plus a commission of 6% on all sales. Find her sales to earn a total of $200.

### Cumulative Review
1. Simplify
   $|7 - 5 - 3|$.
   1
2. Evaluate $-32x^3$
   for $x = \frac{1}{4}$.   $-\frac{1}{2}$

## Computer Activities: Solving Equations (page 111)

The discovery technique is used to lead students to discover on their own the method for using a computer to solve equations of the type $ax + b = c$. This provides a foundation for understanding the importance of literal equations and *how* solving equations really works.

# 5

## Chapter
## and Lesson
## Commentaries

### Chapter 5   Inequalities

#### Objectives

To identify and describe finite and infinite sets
To find and graph the solution sets of equations
To determine if a conjunction is true
To determine if a disjunction is true
To write a true inequality by performing a given
   operation on each side of the inequality
To solve and graph solution sets of inequalities
To solve and graph the disjunction of two
   inequalities
To solve and graph the conjunction of two
   inequalities
To solve equations with absolute value
To solve inequalities with absolute value
To solve word problems that lead to inequalities

#### Overview

In this chapter students see that open sentences can
indeed have more than one solution, or even no
solution. The concept of *set* is used to enhance the
understanding without the modern math entrapment
of awkward set-builder notation. Solving
inequalities, absolute value equations, and absolute
value inequalities provides further examples of
multiple solutions. The logic symbols of
conjunction ($\wedge$) and disjunction ($\vee$) are used to
describe the solutions of combined inequalities.

### Career: Banking (page 117)

This page really has a double purpose. Attention
is focused on the career of banking and the need
for mathematics. However, the main thrust of the
page is that of word problem formulation. Review
again the logo emphasizing that the first step
differs from that used in solving written verbal
problems. The student must now first formulate
the problem, that is, decide what questions must
be answered.

Class time schedules might prohibit complete
solution of the problem. You might, therefore,
decide to spend part of a class period exploring
possible strategies and then assign the complete
solution as a written assignment to be handed in
for extra credit within a reasonable time period.

### Alternate Project

Mr. Barton wants a loan to buy a new car that is
advertised at $8,000 with no down payment. His
income is $18,000 a year and he has no other
outstanding loans. Can he afford the car? The
format for the question is the same as that for
the home in the text with one exception: There
would be a one-time sales tax in some states, not
a yearly property tax.

## 5.1 Sets and Solution Sets (page 118)

### Vocabulary
empty set, finite, infinite, set, solution set, replacement set

### Errors That Students Make
When solving an equation like that shown in Example 6, when the student gets to the step $8x + 3 = 3 + 8x$ and then adds $-8x$ to each side to get the result $3 = 3$, the student is likely to write 3 as the answer. Similarly, in Example 5, when the student derives the equation $6 = 3$, a likely answer is 6 or 3. Students are so "brainwashed" into thinking that there is always *one* answer to every problem that it is difficult for them to accept the possibility of either *no* answer at all or *many* answers.

### For Additional Chalkboard Work
Solve.
1. $5x + 2 = 3 + 5x$      empty set
2. $4(x - 3) - 2 = -14 + 4x$    {all numbers}
3. $x - 7x + 9 = 5 - 6x + 4$    {all numbers}
4. $2(3x + 4) + x = -4 + 7x$    empty set

### Enrichment
Students can conceive of an infinite number of numbers and of an infinite number of points on a number line. However, it is difficult for them to accept the existence of an infinite number of numbers between 0 and 1.

To convince the students of this, have them plot the corresponding point on a number line for each of the following: halfway between 0 and 1, halfway between 0 and $\frac{1}{2}$, halfway between 0 and $\frac{1}{4}$, etc.

Continue in this pattern until the students complain that this could take forever. Students should now be convinced that there are an infinite number of numbers between 0 and 1 and an infinite number of points between two points on a number line.

### Cumulative Review
Solve.
1. $0.02(4 - 0.3x) = 0.15x + 3.2$   $-20$
2. $\frac{3}{5}a + \frac{3}{2} = \frac{9}{10}$   $-1$

## 5.2 Inequalities (page 121)

### Vocabulary
axiom of comparison, inequality

### Background
Students have been introduced to inequalities in the first lesson of Chapter 2 (page 28). However, they will still need reinforcement, particularly where variables are involved.

### Suggestions
Students have more trouble reading the inequality correctly when the variable is on the right of the inequality symbol. As pointed out in Example 3, it is helpful to rewrite $-2 \leq x$ as $x \geq -2$, with the variable first. Stress that the point of the arrow is always towards the *smaller* number and therefore the other number, $x$ in this case, must be the *greater* one.

### Errors That Students Make
Many students interpret $x > 5$ to be all numbers from 6 on. Once again, they tend to either ignore or forget the existence of fractions. Point out that $5\frac{1}{6}, 5\frac{1}{4}, 5\frac{1}{3}, 5\frac{1}{2}$ are all greater than 5, and that therefore the graph on the number line must begin with an open circle around the point corresponding to 5 and then continue with an arrow directed to the right.

A second typical error is to say that $-8 > -2$ since $8 > 2$. Emphasize with a physical example: "8° below 0° is colder than 2° below 0°. So, $-8 < -2$. This needs review even though it was introduced in Chapter 2.

### For Additional Chalkboard Work
Rewrite the inequality with the variable first.
1. $4 > x$     $x < 4$
2. $-2 < x$    $x > -2$
3. $6 \geq x$     $x \leq 6$

### Cumulative Review
Solve.
1. The length of a rectangle is 6m greater than the width. The perimeter is 44m. Find the length and width.   8m and 14m
2. The second of two numbers is 5 times the first. Their sum is 42. Find the numbers.   7, 35

## Geometry Review: Vertical Angles (page 123)

### Vocabulary
vertical angles

### Background
The objective of this lesson is not tested in the chapter test, which tests only the *algebra* lessons of the chapter. However, the concept of vertical angles is tested on many state proficiency tests for basic skills. Moreover, the students should see the use of algebraic skills in a geometric setting.

### Suggestions
Copy the drawing below on the chalkboard to illustrate the importance of the concept that vertical angles are formed by intersecting straight lines, not merely segments.

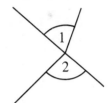

Angles 1 and 2 are not vertical angles.

If you are teaching in a state where a basic skills test is administered early in the school year, you might want to teach Lesson 16.1 on pages 455–458 now. Students study properties of angles including complementary and supplementary angles, as well as the sum of the angles of a triangle.

### Errors That Students Make
Caution students that m∠ A does not mean *m* times ∠ A. It is read as "the *measure* of angle A."

### Cumulative Review
Solve.

1. $\frac{x}{5} + 5 = -9$      2. $7x - (12 - 6x)\frac{2}{3} = 14$
   $-70$                            $2$

3. $5(x - 2) - 2x + 4 = 2x + 6$
   $12$

## 5.3 The Inequality Properties (page 124)

### Background
The students already know the equations properties for performing the same operation on each side of an equation. Point out that the same properties will hold for inequalities, with one exception.

### Suggestions
Stress the one exception rather than all the rules of inequalities: "Multiplication or division by a negative number reverses the order of the inequality." The **Reading in Algebra** questions will test if students remember the rules.

### Errors That Students Make
Even after you have emphasized the one exception, students still tend to forget to reverse the order when multiplying or dividing by a negative.

### For Additional Chalkboard Work
For each inequality perform the indicated operation on each side and write a true inequality.

1. $6 < 8$ Add $-4$.         $2 < 4$
2. $6 < 8$ Multiply by $-2$.   $-12 > -16$
3. $6 < 8$ Divide by 2.       $3 < 4$
4. $6 < 8$ Subtract 3.        $3 < 5$
5. $6 < 8$ Divide by $-2$.    $-3 > -4$

### Cumulative Review
1. What is the additive inverse of $-7$?   $7$
2. What property is illustrated? $-7(5) = 5(-7)$
   Comm. Prop. for Mult.
3. Simplify $3x - (2 - x) - 5$. Then evaluate for $x = -3$.   $4x - 7; -19$

## 5.4 Solving Inequalities (page 128)

### Suggestions

The inequality properties studied in the last lesson are now directly applied to solving inequalities.

Emphasize that the solution of an inequality should be expressed two ways: algebraically and graphically. The graphic solution will be important as a tool for solving combined inequalities in the next lesson.

### Errors That Students Make

In solving an inequality like $-2x + 14 < 6$ (Example 1), the error is usually made in the last step:

$$-2x < -8$$

Students write $x < 4$ as the solution, forgetting to reverse the order when dividing by a negative. Some students also believe that if both sides are negative, the order does not have to be reversed since two negatives make a positive and a positive does not reverse the order.

Two other errors made by students arise in solving inequalities that result in either no solution or identities. In Example 4, students are confused by the result $5 > 2$. They will want to write 5 or 2 or $x > 2$ as answers. Stress that since $5 > 2$, no matter what value $x$ has, the solution is the set of all numbers.

Likewise in Example 5, since the result $3 < 2$ is false, there is no solution. The solution is NOT 3 or 2 as many students erroneously write.

### For Additional Chalkboard Work

Solve.
1. $2x < 10$       2. $12 > -3x$       3. $-4x < -8$
   $x < 5$             $x > -4$             $x > 2$
4. $-x > 3$   $x < -3$

Solve.
5. $x + 4 < x + 7$       6. $x + 8 < x + 3$
   all numbers             no solution
7. $5(x + 2) > 5x + 3$   all numbers

### Cumulative Review

1. Evaluate $-3a - 2$       2. Simplify
   for $a = -1\frac{1}{3}$.        $-4(-1)^3$.
   2                      4
3. Simplify
   $|-4 - 8 + 2|$.   10

**T-46**

## 5.5 Conjunction and Disjunction in Logic (page 131)

### Vocabulary

conjunction, disjunction

### Background

The objectives of this second lesson on logic, unlike those of the first logic lesson in Chapter 3, *are* tested in the chapter test. The concept of *and/or* is used in the next lesson on combining inequalities. The logic concept of *and/or* is also used in writing computer programs involving decision making.

### Suggestions

Stress that the disjunction (or) statement is not the *exclusive* type used in everyday English. For example, when a parent says, "You may go to the movies Saturday or Sunday," the parent is *not* expecting *both* possibilities to occur. In logic, the disjunction (or) statement is true if *one or both* parts of $P \vee Q$ are true.

### Cumulative Review

Evaluate.
1. $(-5a)^3$ for $a = -\frac{2}{5}$      2. $|5x - 12|$ for $x = 2.3$
   8                                         0.5
3. $6 - (3 - a)$ for $a = -6$
   $-3$

## 5.6 Combining Inequalities (page 132)

### Suggestions
This lesson is a direct application of the preceding one on logic. The symbols $\wedge$ (and) and $\vee$ (or) are applied throughout.

You might want to teach this lesson to an above-average class only. Students have difficulty in distinguishing between *and/or* statements. It is sometimes helpful to illustrate the geometric interpretation of each in vertical format (one directly under the other) at the chalkboard.

$x < 6 \wedge x > 2$
It is not enough to say $x < 6$ since $1 < 6$, but 1 is NOT bigger than 2. Stress the alternate way of writing $2 < x < 6$: $x$ is between 2 and 6.

$x < 2 \vee x > 6$.

Stress that the connective AND cannot be used here. There is no number less than 2 *and* at the same time greater than 6. It is very helpful in solving combined inequalities to sketch the two given inequalities in vertical format, one directly below the other, with origins lined up with one another. For an AND statement, describe the points in common to each. A good illustration might be, in Example 1, to think of the top number line as a window shade. Use a red chalk for $x > 2$ and another color, say blue, for $x \leq 5$. If the shade $(x > 2)$ is lowered, where will the red and blue overlap one another? They will overlap between 2 and 5, including 5 but not 2.

### For Additional Chalkboard Work
Graph the solution set.
1. $x < 5 \wedge x > 1$

2. $x > 5 \vee x < 1$

3. $x > 5 \wedge x > 1$

4. $x < 5 \vee x > 1$

## 5.7 Equations with Absolute Value (page 136)

### Background
Students are already familiar with the concept of absolute value and the $\vee$ (or) symbol.

### Suggestions
This lesson deals with equations having absolute value. As a pre-lesson warmup, you might wish to review absolute value.

Once the students understand the solution of simple absolute value equations such as $|x| = 7$, it is an easy transition to the more complex types such as $|2x - 6| = 8$. A good technique is to write the equations $|2x - 6| = 8$ on the chalkboard and cover up the $2x - 6$ in the absolute value bars with your hand and then ask what number could go in here to make a true statement? Since the expression in the bars could be either 8 or $-8$, the student must write TWO equations, $2x - 6 = 8$ and $2x - 6 = -8$, then solve.

### For Additional Chalkboard Work
Solve.
1. $|x| = 9$
   9, $-9$
2. $|a| = 25$
   $-25$, 25
3. $|m| = 4$
   $-4$, 4
4. $y = |19|$
   19
5. $|2x - 6| = 10$
   8, $-2$
6. $|3x - 2| = 5$
   $\frac{7}{3}$, $-1$
7. $|5 - y| = 1$
   4, 6

### Enrichment
1. $|x - 8| = |2x - 10|$
   2, 6
2. $\left|\dfrac{x - 12}{4}\right| = 5$
   32, $-8$
3. $|x - 6| = 4 + |x|$
   1

### Cumulative Review
Solve.
1. $2x + 8 < 4x + 12$
   $x > -2$
2. $\frac{1}{2}x - 4 = \frac{1}{6}x - 5$
   $-3$
3. $0.02x = x + 9.8$
   $-10$

## 5.8 Inequalities with Absolute Value (page 138)

### Suggestions

The development of this concept is again facilitated by the use of the logic symbols for conjunction ($\wedge$) and disjunction ($\vee$).

It might help students to distinguish between the graphs of $<$ and $>$ by showing the solutions of $|x| < 3$ and $|x| > 3$ on parallel number lines with the origin of one number line directly below the other. Also stress the rewriting of each absolute value inequality as two separate inequalities.

$|x| < 3;\ x > -3$ and $x < 3$

$|x| > 3;\ x < -3$ or $x > 3$

Before going on to the more difficult ones, such as Example 3, reinforce the understanding of the basic concepts above with the exercises below. You can sketch on the chalkboard the number line graphs described by the students for each exercise.

Then the transition to the more difficult types of Examples 3 and 4 should be easier. Stress that just as $|x| \geq 3$ is solved by first rewriting two separate inequalities, $x \leq -3$ or $x \geq 3$, the solution of $|2t - 6| \geq 4$ is solved in the same manner. First rewrite it as two separate inequalities: $2t - 6 \leq -4$ or $2t - 6 \geq 4$ as shown in Example 3.

### For Additional Chalkboard Work

Graph the solution to each inequality.
1. $|x| < 7$ — points between $-7$ and 7
2. $|x| > 5$ — points to the left of $-5$ and points to the right of 5
3. $|x| \leq 4$ — points $-4$, 4, and points between $-4$ and 4
4. $|x| \geq 4$ — $-4$, 4, points to the left of $-4$ and points to the right of 4

### Cumulative Review

1. Simplify
   $6 - (1 - a)$.
   $5 + a$
2. Solve
   $5 - x = 2 - (6 - x)$.
   $\frac{9}{2}$
3. Evaluate
   $(2a)^3$ for
   $a = -3$.
   $-216$

## 5.9 Inequalities in Word Problems (page 141)

### Suggestions

Stress the meaning of *at most* ($\leq$) and *at least* ($\geq$). Many students are confused by what appears to be a contradiction in the meanings of these terms.

### Errors That Students Make

Since "most" is associated with being largest or *greatest*, students translate *at most* as $\geq$. Similarly, since "least" means *smallest*, students translate *at least* as $\leq$. Therefore, in the beginning lots of practice is needed on the correct interpretation of these phrases.

### For Additional Chalkboard Work

What is the meaning of each sentence? Tell how to write each as an inequality.
1. John's age, $x$, is at most 40.
   John's age is 40 or less. $x \leq 40$
2. Bill's mass, $x$, to qualify for wrestling, must be at least 64 kg.
   Bill's mass must be 64 kg or more. $x \geq 64$

Write an inequality for each:
3. $x$ is at most 15  $x \leq 15$
4. $y$ is at least 8  $y \geq 8$

### Cumulative Review

1. Solve
   $4x + 2 < 6x - 8$.
   $x > 5$
2. Solve
   $x - (5 - 2x) = 10$.
   $5$
3. Evaluate
   $7 - (x - 4)$ for $x = 5.2$  $5.8$

## Computer Activities: Solving Inequalities (page 143)

This lesson translates a compound inequality into a conjunction that the computer can understand. In the last exercise, students are challenged to write a program using the logic connective *or*. Stress that the inequality symbol $\leq$ cannot be written this way on a computer. It is written $<=$.

# Chapter and Lesson Commentaries

## Chapter 6  Introduction to Factoring

### Objectives

To multiply powers with the same base
To simplify the power of a power
To simplify the power of a product
To classify polynomials
To determine the degree of a polynomial
To simplify polynomials
To add and subtract polynomials
To multiply a polynomial by a monomial
To find the missing factor, given a product and
  one of its factors
To factor a polynomial as a product of a monomial
  and a polynomial

### Overview

In this chapter students work with monomials and
polynomials involving exponents. The first two
lessons deal exclusively with monomials. In the
first lesson, students multiply powers with the
same base. In the second lesson, the *power of a
power* and *power of a product* rules are taught.
The third lesson deals with classifying polynomials
by counting the number of terms, finding the
degree of a polynomial, and simplifying a
polynomial by combining like terms. The next two
lessons involve adding, subtracting, and multiplying
polynomials. Then the process of multiplying
monomials and multiplying a polynomial by a
monomial is reversed by factoring.

### Special Features

### Non-Routine Problem Solving (page 147)

This is a continuation of the strand on Non-Routine
Problems. You might want to encourage the more
mathematically talented students to explore math
contest and puzzle books for more experience in
solving such problems. Your library may have a
copy of the National Math Contest Problems books.
These are compendiums of past national exams and
solutions for several years. The National Council of
Mathematics Teachers can supply information on
ordering these books.

## 6.1 Exponent Properties (page 148)

### Background
Exponents were first introduced in Chapter 1. Review that $2^3$ means to use 2 as a base 3 times, or $2 \cdot 2 \cdot 2$ or 8.

### Errors That Students Make
Students tend to think that if a variable has no exponent written, then the exponent is "nothing," or 0. Thus, a typical student error is to interpret $a^5 \cdot a$ as $a^5$. Stress rewriting $a$ as $a^1$. Then it is easier for the students to see that $a^5 \cdot a^1$ is $a^{5+1}$ or $a^6$. Another typical student error occurs when the variable powers have numerical coefficients. For example, when students simplify $(4a^3)(3a^6)$ they first should rewrite it as $(4 \cdot 3)(a^3 \cdot a^6)$. Then, thinking of multiplication as done with 4 and 3, they write as the answer $12a^{18}$. Emphasize that only the coefficients are multiplied; the exponents of the like bases are added.

### For Additional Chalkboard Work
Simplify.

1. $x^4 \cdot x$    $x^5$    2. $a \cdot a^7$    $a^8$    3. $m \cdot m$    $m^2$
4. $(5x^2)(7x^3)$    $35x^5$    5. $(8a^6)(-3a^2)$    $-24a^8$
6. $(-2m^3)(-5m)$    $10m^4$    7. $(-a)(3a)$    $-3a^2$
8. $(6m)(5m^2)$    $30m^3$    9. $(-3a)(-3a)$    $9a^2$

### Enrichment
A used car dealer told a friend that one day the car lot sold two cars for $750 each. On one sale a profit of 25% was made and on the other, 25% was lost. The friend said, "Well, at least you broke even." The used car dealer said, "No. I lost." Who was right?

*Solution:*

Profit car:   $x + 25\%x = 750$   $x = \underline{\phantom{000}600}$
Loss car:   $x - 25\%x = 750$   $x = \underline{1{,}000}$

The total cost was thus $1,600.

Since the total sale was $1,500, $100 was lost. The used car dealer was right.

### Cumulative Review

1. $4a - (9 - 3a)\frac{2}{3} = 6a + 8$   2. $|8 - 4m| = 20$
  no solution          $-3, 7$
3. $2x - 4 > 3x - 8$   $x < 4$

## 6.2 Powers of Powers and of Products (page 150)

### Suggestions
The most difficult type of exponent problems for students to simplify is like the one occurring in Example 2. Students forget to apply the property correctly to the numerical coefficient $-3$ in simplifying $(-3x)^4$.

### Errors That Students Make
Students will simplify $(-2\,xy)^5$ as $-10x^5y^5$. To avoid this, stress the rewriting of the expression as $-2^5 \cdot x^5 \cdot y^5$ and then proceed as shown in the text.

### For Additional Chalkboard Work
Simplify.

1. $(a^4)^5$    $a^{20}$    2. $(m^2)^9$    $m^{18}$
3. $(x^2)^8$    $x^{16}$    4. $(b^3)^3$    $b^9$
5. $(x^2y^3)^7$    $x^{14}y^{21}$    6. $(a^2b^9)^5$    $a^{10}b^{45}$
7. $(mn^5)^3$    $m^3n^{15}$    8. $(p^2q)^4$    $p^8q^4$
9. $(3x^2)^3$    $27x^6$    10. $(-4a^5)^2$    $16a^{10}$
11. $(-5b^6)^2$    $25b^{12}$    12. $(6m^4)^3$    $216m^{12}$

### Enrichment
Is it true that for any positive number, the larger the power of the number the larger the number?
*Solution:* No. Consider positive fractions between 0 and 1. For example $\left(\frac{1}{2}\right)^2 = \frac{1}{4}$; $\left(\frac{1}{2}\right)^3 = \frac{1}{8}$; $\left(\frac{1}{2}\right)^5 = \frac{1}{32}$. The results are decreasing.

The larger the power of $\frac{1}{2}$, the closer the result gets to what number?
*Solution:* As $n$ increases without end, $\left(\frac{1}{2}\right)^n$ approaches the number 0, but will never actually be 0.

### Cumulative Review
1. The length of a rectangle is twice the width. The perimeter is 36. Find the length and width.   6, 12
2. A team won 8 games and lost 2 games. The number of games lost is what percent of the games played?   20%

## 6.3 Simplifying Polynomials (page 152)

### Vocabulary
binomial, degree, monomial, trinomial, polynomial

### Background
The English prefixes *mono, bi,* and *tri* are the prefixes used when classifying polynomials. Some English words that illustrate the use of these prefixes are *mono*logue, *bi*cycle, and *tri*cycle.

### Suggestions
Emphasize "descending order of terms." This concept will be important in the next chapter when we talk about solving quadratic equations that have first to be put in standard form. When there is more than one variable in a polynomial, the student is taught to write the polynomial in descending order of exponents of *one* of the named variables.

### Errors That Students Make
When combining like terms involving exponents, some students also want to add the exponents. Stress that $3x^4 + 2x^4 = 5x^4$, not $5x^8$.

Use the following additional exercises to help reinforce this idea.

### For Additional Chalkboard Work
Simplify.
1. $3x^5 + 7x^5$    2. $7x^2 + 4x^2$    3. $x^3 + 4x^3$
    $10x^5$           $11x^2$          $5x^3$
4. $-3x^5 - x^5$    5. $8a^3 + 2a^2 - 4a^3 + 6a^2$
   $-4x^5$               $4a^3 + 8a^2$
6. $3x^2 - 7x - 4x + 1$   $3x^2 - 11x + 1$

### Enrichment
Can you make up a word problem about discounts that leads to the following equation?
   $x - 0.15x = \$170$
*Solution:* A portable television is advertised at 15% discount. The new marked price is $170. Find the original selling price prior to the discount.

### Cumulative Review
Solve.
1. $3(x + 2) = 3x - 7$   no solution
2. $2(x + 4) - 1 = 2x + 7$   all numbers
3. $|x - 0.2| = 8$   $-7.8, 8.2$

## 6.4 Adding and Subtracting Polynomials (page 155)

This lesson applies familiar concepts to new situations. The concept of combining like terms is extended to adding polynomials. Also, the concept of $-a = -1(a)$ is applied to help students grasp subtraction of polynomials. In the B exercise section, students are also expected to work with fractions and decimals.

### Errors That Students Make
When students subtract polynomials, they tend to forget to write the parentheses. Thus, the error is to write "subtract $y^3 - 5x^2$ from $7y^4 - 8y^3 + 3y^2 - 4$" as $7y^4 - 8y^3 + 3y^2 - 4 - y^3 - 5x^2$.

### For Additional Chalkboard Work
Subtract.
1. $x^2 - 5x + 2$ from $-2x^2 - 4x + 1$
   $-3x^2 + x - 1$
2. $-y^3 - 4y + 6$ from $-2y^3 + y^2 - y$
   $-y^3 + y^2 + 3y - 6$
3. $-4\frac{1}{2}y^2 - y$ from $-y^3 - 5y^2 - 2y$   $-y^3 - \frac{1}{2}y^2 - y$

### Enrichment
Place the digits 1, 2, 3, 4, 5, 6, 7, 8, 9 in the figure below to complete a magic triangle in which the sum of the digits of any side of the magic triangle is always 20.

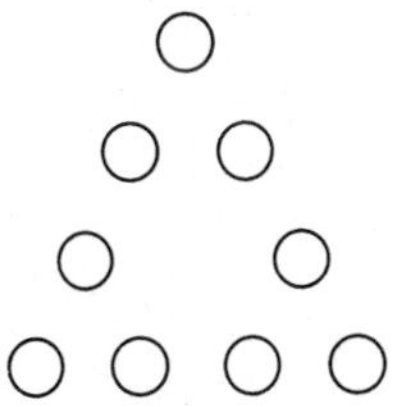

*Solution:*

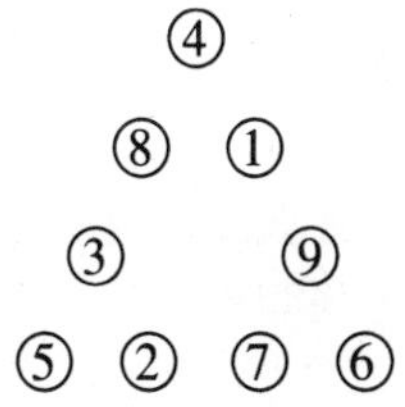

### Cumulative Review
Solve.
1. The perimeter of a triangle is 42m. The length is 3m less than twice the width. Find the length and width.   8m, 13m
2. If twice a number is decreased by 6 more than 4 times the number, the result is $-22$. Find the number.   8

## 6.5 Multiplying Polynomials (page 157)

### Suggestions

This lesson applies the distributive property to the simplification of expressions involving exponents.

Use the following practice to review the distributive property.

Simplify.

**1.** $4(3x + 2)$
$\quad 12x + 8$

**2.** $3(5x - 2)$
$\quad 15x - 6$

**3.** $7(2m - 4)$
$\quad 14m - 28$

**4.** $-2(-m + 1)$ $\quad 2m - 2$

### Alternate Example 2

For a less-than-average class, or for the benefit of several students whom you know to be slower learners, you might want to begin with simpler illustrations. The following involve only distribution of a variable to power.

Simplify.

**1.** $x^3(x^4 + x^3 + x^2)$ $\quad x^7 + x^6 + x^5$

**2.** $y^2(y^5 - y^3 + y^2 + y)$

First stress rewriting $y$ as $y^1$.

$$y^2(y^5 - y^3 + y^2 + y^1)$$
$$y^7 - y^5 + y^4 + y^3$$

### Errors That Students Make

For an exercise like $-x^3(5x^4 + 6x^2)$ in Example 2, students may misinterpret the role of the "$-$" symbol and write

$$5 - x^3 \cdot x^4 + 6 - x^3 \cdot x^2.$$

To avoid this, you might want to encourage students with this type of thinking to first rewrite $-x^3$ as $-1x^3$ and then distribute $-1x^3$.

### For Additional Chalkboard Work

Simplify.

**1.** $a^7(a^5 - a^4 + a^3)$
$\quad a^{12} - a^{11} + a^{10}$

**2.** $b^2(b^3 + b^2 - b)$
$\quad b^5 + b^4 - b^3$

**3.** $x(x^2 + x)$ $\quad x^3 + x^2$

### Cumulative Review

Solve.

**1.** $\frac{2}{3}x - 5 = 17$
$\quad 33$

**2.** $|2x - 4| = 8$
$\quad 6, -2$

**3.** $x < 4x - 9$ $\quad x > 3$

## 6.6 Finding the Missing Factor (page 159)

### Vocabulary

prime factorization, prime number

### Background

This lesson was designed to pave the way for the next lesson on factoring out a common monomial factor.

### Suggestions

You may wish to use simple exercises such as $x^3 \cdot x^4$ and $a \cdot a^3$ as a pre-lesson warmup. Following Example 2, reinforce the concept with the simple drill provided below.

Find the missing factor.

**1.** $(a^4)(?) = a^{10}$
$\quad a^6$

**2.** $(b^3)(?) = b^5$
$\quad b^2$

**3.** $(?)(n^2) = n^9$
$\quad n^7$

**4.** $(r)(?) = r^3$
$\quad r^2$

**5.** $(?)(p) = p^7$
$\quad p^6$

**6.** $(t)(?) = t^2$
$\quad t^1$ or $t$

### Errors That Students Make

As you might discover from the results of the drill above, a typical error is to write $x^3$ as the missing factor in $(x)(?) = x^3$. Once again students assume that the *absence* of a number implies that zero is the number. Stress the rewriting of $(x)(?) = x^3$ as $(x^1)(?) = x^3$; that is, $x = x^1$.

### Cumulative Review

Simplify.

**1.** $(-4.1m^3)(6m)$
$\quad -24.6m^4$

**2.** $\left(-\frac{4}{5}a^5\right)^3$
$\quad -\frac{64}{125}a^{15}$

**3.** $2m^2(m^2 - 5m + 6)$ $\quad 2m^4 - 10m^3 + 12m^2$

## 6.7 Factoring Out the Greatest Common Monomial Factor (page 161)

### Vocabulary
greatest common factor

### Suggestions
Review the application of the distributive property with exponents to pave the way for the reversal of the process, that is, factoring out the greatest common monomial factor.

Simplify.
1. $a^2(2a^2 - 5a + 4)$   $2a^4 - 5a^3 + 4a^2$
2. $3x(6x^2 + 2x - 5)$   $18x^3 + 6x^2 - 15x$
3. $4m^2(3m^3 - 7m^2 - 8m + 1)$
   $12m^5 - 28m^4 - 32m^3 + 4m^2$

This concept, for some strange reason, seems to cause more trouble for students than general trinomial factoring. Yet, many teachers would expect common monomial factoring to be easier. It might be helpful to practice separately with each of the three cases: whole number factor, variable factor, and combined whole number and variable factor.

### For Additional Chalkboard Work
Factor out the greatest common factor.
1. $3x^2 - 9x + 15$      2. $4m^2 + 12m - 20$
   $3(x^2 - 3x + 5)$         $4(m^2 + 3m - 5)$
3. $6a^2 - 18a - 24$   $6(a^2 - 3a - 4)$

Now proceed to factoring out a greatest common variable factor.

### Errors That Students Make
$x^3 + x^2 + x = x(x^2 + x)$ is the response from many students. The rationale is that "since the last term is $x$, and $x$ is being factored *out*, there is obviously no $x$ left." This might be avoided by suggesting that students rewrite
$x^3 + x^2 + x$ as $(\underline{x})(x^2) + (\underline{x})(x) + (\underline{x})(1)$
$= \underline{x}(x^2 + x + 1)$
Now reinforce with chalkboard practice using Exercise 11 on page 163.

### Cumulative Review
1. Solve $4x - (7 - 3x) = 9x - 11$.   2
2. Evaluate $-a^5b^2$ for $a = -3, b = -2$.   972
3. Simplify $\left| -8\frac{2}{3} - 4 + 6\frac{1}{6} \right|$.   $-6\frac{1}{2}$

## 6.8 Algebraic Expressions for Area (page 164)

### Background
Many of the state proficiency tests in basic skills are now requiring more than the *minimum skill* of finding the area of a circle. Finding area of shaded regions is also appearing on many tests.

### Suggestions
It might be helpful to prepare overhead transparencies to help with the visualization. For example, for Example 1 you might wish to prepare two transparencies, one of the square lightly shaded and one of the circle. By using the two transparencies, you can emphasize how to find the area of each separately. Then, by you overlaying the circle on the square it is obvious that the shaded region is the area of the square minus the area of the circle.

## Computer Activities: Computing Compound Interest (page 168)

This program demonstrates the use of the computer in calculating compound interest. An important mathematical concept is taught through the discovery approach in Exercise 3. Students find that doubling the years will not double the amount. This concept of how doubling affects results can be related to area in Section 6.8. Challenge the students to discover what doubling the side of a square does to the area and to discover algebraically why this works.

Square with side $s$: $A = s^2$

Square with side $2s$ (*double* the original side):
$A = (2s)^2 = 4s^2$, or *four times* the area of the original square.

# 7

## Chapter
## and Lesson
## Commentaries

### Chapter 7  Factoring Trinomials

#### Objectives
To multiply polynomials by polynomials
To factor polynomials completely
To solve quadratic equations by factoring
To solve polynomial equations of degree higher
than two by factoring
To solve word problems that lead to quadratic
equations
To solve word problems about consecutive integers

#### Overview
Multiplication of polynomials is developed in
general. However, the method for multiplying
two binomials is the FOIL technique. Once the
FOIL method has been established, the students
are ready for factoring. Factoring is a difficult topic
for most students. Teachers have indicated a
preference for beginning with the special cases and
gradually leading into the general method of
factoring. Therefore, the next two lessons of the
chapter deal with factoring the difference of two
squares and the perfect square trinomial. The next
lesson introduces the student to a more general type
in which the $x^2$ coefficient is.1. By this time
students are ready for the general method of
factoring involving $ax^2 + bx + c$. Factoring
trinomials of the form $ax^2 + bxy + cy^2$ is a final
basic extension of the general case.

The lesson on factoring by grouping may be
deleted for a slower class. However, the lesson is
necessary for students assigned future Ⓒ exercises.
Such students might be assigned the lesson
individually as enrichment. The chapter closes
with applications of factoring to the solution of
quadratic/polynomial equations. Quadratic
equations are then used to solve word problems.

#### Special Features

### Non-Routine Problem Solving (page 170)

This page is also part of the Non-Routine Problems
strand. Encourage students to look for different
methods of solution. Point out to the class that it is
very probable that different students will come up
with different methods of solution. Stress that
looking for a straightforward algebraic solution will
often not work. These problems do not fall into a
traditional problem type that can be solved by the
same general method or strategy. Non-Routine
Problems provide students with the opportunity to
think creatively.

## 7.1 Products of Polynomials (page 171)

### Background

The distributive property is used to justify the introduction of the FOIL method for multiplying two binomials. This in turn will pave the way for reversal of the process in subsequent lessons to factor trinomials.

### Suggestions

Have students review how to use the distributive property with the following problems.
Simplify by applying the distributive property.

**1.** $(5x + 2)3x$   **2.** $(2x - 7)2x$   **3.** $(9x - 5)2x$
    $15x^2 + 6x$      $4x^2 - 14x$      $18x^2 - 10x$

It might be beneficial to have students work out some products of two binomials by the long distributive method so that they can better appreciate the simplicity of the FOIL method. Once you have done some exercises using the distributive property, you can then illustrate the FOIL method for each one.

### For Additional Chalkboard Work

Multiply.

**1.** $(2x + 1)(x - 4)$     $2x^2 - 7x - 4$
**2.** $(3x - 2)(4x + 1)$     $12x^2 - 5x - 2$
**3.** $(5x - 3)(2x + 4)$     $10x^2 + 14x - 12$

### Errors That Students Make

Many students will state that $(3x + 5)(2x + 3) = 6x^2 + 15$. They forget to work out the middle term. This will happen again in a later lesson on squaring a binomial; they will write $(3x + 5)^2 = 9x^2 + 25$. Constant reminding of the use of the FOIL method is necessary: "Don't forget the middle term."

### Cumulative Review

**1.** Simplify $(2x^3)^5$.    $32x^{15}$
**2.** Factor out the GCF from $3x^3 - 9x^2 - 6x$.
    $3x(x^2 - 3x - 2)$
**3.** Solve $2x - 4 = -10x$.   $\frac{1}{3}$

## 7.2 A Special Product: The Difference of Two Squares (page 173)

### Vocabulary

difference of two squares

### Suggestions

Use the discovery approach. Ask the students to multiply each of the following.

**1.** $(x - 6)(x + 6)$      **2.** $(2x - 4)(2x + 4)$
    $x^2 - 36$              $4x^2 - 16$
**3.** $(3x - 2)(3x + 2)$      **4.** $(5 - 4a)(5 + 4a)$
    $9x^2 - 4$              $25 - 16a^2$

What is the pattern? Stress that theoretically there *is* a *middle term*. It is $0x$.

For the sharper students you should definitely try the shortcut $a^2 - b^2 = (a - b)(a + b)$. The less capable student might have difficulty with the pattern, particularly with exercises such as $36x^2 - 49$. Suggest that these students merely think of an expression like $36x^2 - 49$ as $36x^2 + 0x - 49$ and then factor by the FOIL method.

Use **Non-Routine Problems** to give the brighter students practice in multiplying polynomials.

### For Additional Chalkboard Work

Factor.

**1.** $y^2 - 9$           **2.** $36 - n^2$
    $(y - 3)(y + 3)$       $(6 - n)(6 + n)$
**3.** $9a^2 - 1$        **4.** $25x^2 - 64$
    $(3a - 1)(3a + 1)$    $(5x - 8)(5x + 8)$
**5.** $16 - 81y^2$      **6.** $100a^2 - 49$
    $(4 - 9y)(4 + 9y)$    $(10a - 7)(10a + 7)$

### Enrichment

Factor.

**1.** $x^4 - 625$    $(x^2 - 25)(x^2 + 25)$
           $= (x - 5)(x + 5)(x^2 + 25)$
**2.** $16a^4 - 81$    $(4a^2 - 9)(4a^2 + 9)$
           $= (2a - 3)(2a + 3)(4a^2 + 9)$

### Cumulative Review

Factor out the GCF from each.

**1.** $x^3 + 2x^2 - 7x$     **2.** $5b^2 - 10b + 35$
    $x(x^2 + 2x - 7)$       $5(b^2 - 2b + 7)$
**3.** $3y^3 - 12y^2 - 6y$
    $3y(y^2 - 4y - 2)$

## 7.3 A Special Product: A Perfect Square Trinomial (page 175)

### Vocabulary
perfect square trinomial

### Background
This lesson also lays the foundation for completing the square in a later chapter, where it leads to the development of the quadratic formula. The topic is of great importance in Algebra 2 in graphing conic sections translated off the origin as center. The significance of spiral learning is well known; the concept of perfect square trinomial occurs over and over in mathematics.

### Suggestions
Stress the *doubling* to get the middle term. Students tend to see the pattern faster through many oral exercises.

### Extra Oral Exercises
Multiply.

**1.** $(3x - 1)^2$
$9x^2 - 6x + 1$

**2.** $(2x - 3)^2$
$4x^2 - 12x + 9$

**3.** $(2a + b)^2$
$4a^2 + 4ab + b^2$

**4.** $(5m + n)^2$
$25m^2 + 10mn + n^2$

### Enrichment
Ask the class to evaluate $x^2 - 8x + 16$ for $x = 24$. Challenge the class to give you any two-digit number less than 20. You will then evaluate $x^2 - 8x + 16$ for this number almost immediately without pencil and paper. Can the class come up with the shortcut?
*Solution:* $x^2 - 8x + 16 = (x - 4)(x - 4)$ or $(x - 4)^2$.
Thus, for $x = 24$ the value of $x^2 - 8x + 16$ is $(x - 4)^2 = (24 - 4)^2 = (20)^2 = 400$.

### Cumulative Review
Solve.

**1.** $2a - 5 = 13$
$9$

**2.** $\left| x + 3\frac{1}{4} \right| = 9$
$5\frac{3}{4}, -12\frac{1}{4}$

**3.** $0.02x + 1.2 = -3.08$
$-214$

## 7.4 Factoring Trinomials: $x^2 + bx + c$ (page 177)

### Suggestions
Stress the *summary* at the end of the lesson on page 179. The pattern of signs is a significant tool in factoring. Also stress the *general* trial-and-error method of factoring. Point out that the *shortcut* method of looking for factors of $c$ whose *sum* is $b$ will *not* work when the coefficient of $x^2$ is a number other than 1.

As in previous situations throughout the text, it is helpful to write a coefficient when none is present, for example, $x^2 - x - 30 = \underline{1}x^2 - \underline{1}x - 30$.

### For Additional Chalkboard Work
Factor.

**1.** $x^2 - 10x + 16$
$(x - 8)(x - 2)$

**2.** $x^2 + 10x + 16$
$(x + 8)(x + 2)$

**3.** $x^2 - 6x - 16$
$(x - 8)(x + 2)$

**4.** $x^2 + 6x - 16$
$(x + 8)(x - 2)$

### Errors That Students Make
Students tend to forget to check the middle term. Stress the importance of checking the signs. For example, $x^2 - 5x + 6$ does *not* factor into $(x + 3)(x + 2)$. The 5 and 6 appear to check, but the 5 must be $-5$.

### Cumulative Review
**1.** Multiply $5x(x^2 - 3x + 2)$.  $5x^3 - 15x^2 + 10x$
**2.** Factor out the GCF from
$4x^2 - 8x + 24$.  $4(x^2 - 2x + 6)$
**3.** Multiply $(x - 5)(x + 5)$.  $x^2 - 25$

## 7.5 Factoring Trinomials: $ax^2 + bx + c$ (page 180)

### Suggestions

Stress that there is very little in this lesson that is new. The lesson is simply an extension of the previous lesson. However, now it is the *general method,* not the shortcut, that works. The only difference in the two lessons is that the $x^2$ coefficient is no longer always 1. Before starting this new lesson, use the following warm-up review of the previous lesson, using the general method, not the shortcut.

Factor.

1. $x^2 - 7x - 18$      2. $m^2 - m - 12$
   $(x - 9)(x + 2)$         $(m - 4)(m + 3)$
3. $a^2 - 9a + 20$   $(a - 5)(a - 4)$

Stress the reversing of factors if the first trial is unsuccessful, as shown in Example 1. Example 3 shows the situation in which a term like 12 may have more than one pair of factors, 12 and 1, 6 and 2, or 4 and 3. It might be helpful, particularly for a slower class, to begin with practice in which there is only ONE pair of factors of a term.

Use the following problems to supplement the **Reading in Algebra.** What conclusion can you draw about possible sign combinations from inspection?

1. $2x^2 + 7x + 6$   No negatives possible

2. $3x^2 - 8x + 4$   Both negative since middle term is negative, end term is positive

3. $6a^2 + 5a - 1$   Unlike signs, since last term is negative

Use a calculator to help find the factors of the last number in the Calculator Activities.

### For Additional Chalkboard Work

Factor.

1. $2x^2 + 7x - 15$      2. $6y^2 + 4y - 2$
   $(2x - 3)(x + 5)$         $(3y - 1)(2y + 2)$

### Cumulative Review

1. Solve
   $3x - 4 = 5 - x$.
   $\frac{9}{4}$

2. Solve. Sketch solution on a number line.
   $4 + x > 3x + 12$.
   $x < -4$. Points to left of $-4$.

3. Solve
   $7 - 3x = 4$.   1

## 7.6 Combined Types of Factoring (page 183)

### Suggestions

Start the lesson with the following warm-up. Factor out the GCF.

1. $5a^2 - 10a - 25$      2. $x^3 - 7x^2 - 11x$
   $5(a^2 - 2a - 5)$         $x(x^2 - 7x - 11)$
3. $2b^3 - 16b^2 + 30b$
   $2b(b^2 - 8b + 15)$

Now use the third exercise to show that $b^2 - 8b + 15$ can be factored further.

Stress that students look for three types of factors:

1. Greatest common whole number factor
2. Greatest common variable factor
3. Binomial factors.

### Errors That Students Make

After factoring out the GCF and then the resulting polynomial into two binomials, the tendency is to forget to include the GCF as a part of the final factorization. Thus, in Example 1, the final answer may be left as $(2x - 3)(x + 2)$ instead of the correct one, $3x(2x - 3)(x + 2)$.

### Enrichment

Areas of rectangles and squares can be used to show the product of two binomials.

$A = l \cdot w$
$= (a + 7)(a + 1)$

But the area of the rectangle is the sum of the areas of all of its parts.

Area of rectangle;    I       II      III     IV
                $(a)(a)$   $7(a)$   $1(a)$   $7(1)$
                  $a^2$   $+ 7a$   $+ 1a$   $+ 7$

Thus, $(a + 7)(a + 1) = a^2 + 8a + 7$.

Use the rectangle method to multiply.

1. $(a + 4)(a + 3)$      2. $(x + 7)(x + 5)$
   $a^2 + 7a + 12$          $x^2 + 12x + 35$

## 7.7 Factoring by Grouping (page 186)

### Background
This lesson is an extension of factoring out a GCF. However, the GCF might be a binomial. You might choose to skip this lesson for a below-average class. Another alternative is to assign this lesson for extra credit to the very capable students.

### Suggestions
Begin with a review of factoring out a GCF with the following exercises.
Factor out the GCF

**1.** $ax + ay$     **2.** $xm - xt$     **3.** $a^2b + a^2y$
   $a(x + y)$       $x(m - t)$       $a^2(b + y)$

### For Additional Chalkboard Work
Factor.
**1.** $b(x^2 - 36) + c(x^2 - 36)$
   $(x^2 - 36)(b + c) = (x - 6)(x + 6)(b + c)$
**2.** $y^2(m - 2) - 4(m - 2)$
   $(y^2 - 4)(m - 2) = (y - 2)(y + 2)(m - 2)$
**3.** $p^2(4y - 8) - 49(4y - 8)$    $(4y - 8)(p^2 - 49)$
   $= 4(y - 2)(p - 7)(p + 7)$

With Example 2, stress that there may be two different ways of grouping in the beginning, yet the final factorization is the same.

### Enrichment
Factoring by grouping can be used to factor all factorable trinomials of the form $ax^2 + bx + c$.
Example: Factor $2x^2 + 5x - 12$.
   Find two numbers whose product equals the product of $a \cdot c$, or $2 \cdot {}^{1}{-}12$, or $-24$, and whose sum is $b$, or $+5$. The numbers that satisfy these two conditions are $+8$ and $-3$. Now rewrite $2x^2 + 5x - 12$ as
$2x^2 + 8x - 3x - 12$
$(2x^2 + 8x) + (-3x - 12)$
$2x(x + 4) - 3(x + 4)$     So, $(2x - 3)(x + 4)$.
Factor by grouping.
**1.** $x^2 - x - 20$       **2.** $3x^2 + 10x - 8$
   $(x - 5)(x + 4)$        $(3x - 2)(x + 4)$
**3.** $6x^2 - 7x + 1$    $(6x - 1)(x - 1)$

### Cumulative Review
Solve.
**1.** $3x - 6 = 0$    **2.** $4x - 3 = 0$    **3.** $2x = 0$
    $2$            $\frac{3}{4}$         $0$

## 7.8 Solving Equations by Factoring (page 188)

### Vocabulary
quadratic equation, roots, cubic equation

### Background
Solving a quadratic equation involves setting linear factors equal to zero.

### Suggestions
Use the following exercises as a warm-up.
Solve.
**1.** $2x - 8 = 0$    **2.** $3x + 2 = 0$    **3.** $5x = 0$
    $4$            $-\frac{2}{3}$         $0$

Stress the theory that $ab = 0$ implies that $a = 0$ or $b = 0$ by having students solve quadratic equations and then *check* the solutions as shown in Examples 1, 2, and 3.

### Errors That Students Make
Quadratic equations like that shown in Example 3 trouble students. Since the factor $x$ is not a binomial, students tend to ignore it and write only the solution $x = 5$. Stress the two solutions 0 and 5. It might help to rewrite the factorization as $(x - 0)(x - 5) = 0$.
Then, $x - 0 = 0$    $x - 5 = 0$
       $x = 0$         $x = 5$
Point out that the theory can be extended to higher degree equations. Thus $x^3 - 7x^2 + 10x = 0$ becomes, after factoring;
$(x)(x^2 - 7x + 10) = 0$
$(x)(x - 5)(x - 2) = 0$
$x = 0$, or $x - 5 = 0$, or $x - 2 = 0$
$x = 0$, or $x = 5$, or $x = 2$

### For Additional Chalkboard Work
Solve.
**1.** $x^2 - 12x + 35 = 0$       **2.** $y^2 - 3y - 4 = 0$
     $5, 7$                 $4, -1$
**3.** $a^2 + 3a = 0$    $0, -3$

### Cumulative Review
Factor completely.
**1.** $2a^2 - ab - 3b^2$        **2.** $4x^2 - 36$
   $(2a - 3b)(a + b)$       $4(x - 3)(x + 3)$
**3.** $ax + ab - 2x - 2b$    $(a - 2)(x + b)$

## 7.9 Standard Form of a Quadratic Equation—Problem Solving (page 191)

### Suggestions

First review the previous lesson on quadratic equations. Use the following exercises.
Solve.

**1.** $x^2 - 7x = 0$
    0, 7

**2.** $x^2 - 11x + 30 = 0$
    5, 6

**3.** $2x^2 + 5x - 3 = 0$  $\frac{1}{2}$, $-3$

### Alternate Example 1

Try a discovery approach.

Write the following quadratic equation on the chalkboard: $a^2 - 6a = 16$. Ask the following questions:

Does this look like a quadratic equation?

What should the number on the right side of the = symbol be?

What can you add to each side of the equation to produce "0" on the RIGHT side of the = symbol?

What equation do you get if you add $-16$ to each side of $a^2 - 6a = 16$?

Then complete the solution of the equation and show the check.  8, $-2$

Now provide practice with the same type of quadratic equation.

### For Additional Chalkboard Work

Solve.

**1.** $x^2 - 4x = -3$  **2.** $x^2 = 6x$  **3.** $x^2 + 12 = 8x$
   1, 3        0, 6         2,6

**4.** The square of a number, increased by 12, is the same as 7 times the number. Find the number.
3 or 4

**5.** A number multiplied by 8 more than the number is 20. Find the number.  $-10$ or 2

### Word Problem Formulation

Can you make up a word problem about the area of a rectangle that would lead to the following quadratic equation? Recall that a quadratic equation will have two solutions. How will you have to interpret any negative solutions relative to the solution of the actual word problem itself?

$x^2 + 11x = 80$

*Solution:* The length of a rectangle is 11 more than the width. The area is 80. The solution $-5$ is a solution of the equation, but not of the word problem since lengths are not considered negative. This illustrates the importance of "interpretation." It is not enough to merely check the equation. We must also check to determine if the solution of the equation provides a meaningful interpretation of the problem itself.

### Cumulative Review

**1.** Solve
    $4x - (1 - x) = -7 - 6x.$  $-\frac{6}{11}$

**2.** Simplify
    $(-4ab^5)(7a^3b).$  $-28a^4b^6$

**3.** Factor
    $6x^2 + 26x - 20.$  $2(3x - 2)(x + 5)$

## 7.10 Problem Solving: Consecutive Integers (page 195)

### Vocabulary
consecutive integers, even integer, odd integer

### Suggestions
Students are sometimes confused by what appears to be the same representation for both odd and even integers. Stress that the difference lies in what you designate as the first integer, even or odd.
For practice, represent each.
1. 3 consecutive integers beginning with $x$.
   $x, x + 1, x + 2$
2. 4 consecutive even integers beginning with $y$.
   $y, y + 2, y + 4, y + 6$
3. 3 consecutive odd integers beginning with $x$.
   $x, x + 2, x + 4$
Point out that consecutive integers can be negative.

### Errors That Students Make
Students tend to write $-6$ as the integer consecutive to $-5$. Emphasize that 1, not $-1$, is added to get the next consecutive integer. Thus, the integer consecutive to $-5$ is $-5 + 1$ or $-4$.

### For Additional Chalkboard Work
1. Write the representation for two consecutive integers. What is the square of the second?
   $x, x + 1$; square of second is
   $(x + 1)^2 = x^2 + 2x + 1$
2. Write the representation for three consecutive odd integers. What is the square of the last?
   $x, x + 2, x + 4$; square of last is
   $(x + 4)^2 = x^2 + 8x + 16$

### Cumulative Review
Simplify.
1. $5x - (4 - 2x)\frac{1}{2} - 7x$  2. $x(x - 4) - 2x(5 - x)$
   $-x - 2$                              $3x^2 - 14x$

## Computer Activities: Finding the Sum of Consecutive Integers (page 202)

The computer is used to find the sum of consecutive integers. A thorough understanding of consecutive even and odd integers is necessary to complete the exercises correctly. The computer will find the sum for the numbers entered as INPUT.

# 8 Chapter and Lesson Commentaries

## Chapter 8   Simplifying Rational Expressions

### Objectives
To find the values, if any, for which a rational
   expression is undefined
To simplify rational expressions
To multiply rational expressions
To divide rational expressions
To simplify rational expressions that involve
   both multiplication and division of rational
   expressions
To identify the extremes and means of a
   proportion
To solve a proportion
To solve word problems involving proportions

### Overview
The chapter opens with a brief discussion of rational expressions and of why a rational expression is undefined if the denominator is zero. Then multiplication of two rational expressions is taught. Here the products are nonreducible. The property of products of fractions is then used, in reverse, to justify the technique for simplifying. When students can reduce rational expressions to lowest terms, the students then multiply rational expressions whose products can be simplified. The concepts are extended to quotients and expressions involving the operations of both multiplication and division. This is followed by a lesson on volume and a lesson on proportion which stresses the practical applications of fractions. This lesson is followed by an applications lesson using proportion to study problems about automobile gas mileage.

## Special Features

## Pythagoras (page 204)

The special numbers mentioned on this page have some interesting characteristics. *Amicable numbers* are numbers whose proper factors (all the factors of the number except the number itself) add up to the other number. A *perfect number* is a number whose proper factors add up to the number itself. An *abundant number* is a number whose proper factors add up to a number greater than the number itself. A *deficient number* is a number whose proper factors add up to a number less than the number itself.

## 8.1 Rational Expressions (page 205)

### Vocabulary
rational expression

### Background
Remind students of the basic arithmetic property for multiplying fractions $\frac{a}{b} \cdot \frac{c}{d} = \frac{a \cdot c}{b \cdot d}$. Thus, this lesson is a simple extension of the arithmetic property they have used since grade school, only now there will be variables present in the numerator or denominator, or both. The products in this lesson will be nonreducible. Reducing fractions to lowest form will be the topic of the next lesson.

### Suggestions
As a pre-lesson warmup have students multiply the following arithmetic fractions:
Multiply.

**1.** $\frac{2}{3} \cdot \frac{4}{5}$      **2.** $\frac{5}{7} \cdot \frac{1}{2}$      **3.** $\frac{2}{7} \cdot 3 \left( \text{HINT } 3 = \frac{3}{1} \right)$

  $\frac{8}{15}$              $\frac{5}{14}$            $\frac{6}{7}$

Review why a fraction is undefined if the denominator is zero. Students forget the rationale.
Example: Show that $\frac{5}{0}$ is undefined.
*Solution:* $\frac{5}{0}$ means $0\overline{)5}$. Assume that the answer is $x$. Then $0\overline{)5}^{\,x}$. Check $(0)(x) = 0$, not 5, for any value of $x$. Thus $0\overline{)5}$ or $\frac{5}{0}$ cannot have any meaning and is therefore *undefined*. Use the third Calculator Activities exercise to show that rational expressions can be undefined. Stress the rewriting of factors like 4 or $x + 5$ as $\frac{4}{1}$ and $\frac{x + 5}{1}$.

### For Additional Chalkboard Work
Multiply.

**1.** $y^3 \cdot \frac{y^2}{4}$      **2.** $\frac{a^3}{b^5} \cdot a^4$      **3.** $\frac{3}{4} \cdot (x + 2)$

  $\frac{y^5}{4}$            $\frac{a^7}{b^5}$           $\frac{3x + 6}{4}$

### Cumulative Review
Factor.

**1.** $2x^2 - 10x$          **2.** $4a^2 - 25$
    $2x(x - 5)$           $(2a - 5)(2a + 5)$
**3.** $2x^2 + 7x - 15$   $(2x - 3)(x + 5)$

## 8.2 Simplest Form (page 209)

### Vocabulary
relatively prime, simplest form

### Background
The property for multiplying fractions and the identity property provide the foundation for showing why a rational expression can be simplified.

Thus $\dfrac{ab}{bc} = \dfrac{a \cdot b}{c \cdot b} = \dfrac{a}{c} \cdot \dfrac{b}{b} = \dfrac{a}{c} \cdot 1 = \dfrac{a}{c}$ by the identity property.

Note that the procedure depends upon reversing multiplication, that is, writing a fraction as the product of two fractions. Use the following warmup to pave the way for this development.

### Suggestions
Use the following exercises as a pre-lesson warmup. Write as the product of two fractions (more than one answer is possible).

1. $\dfrac{6}{10}$     2. $\dfrac{4}{15}$     3. $\dfrac{4}{25}$

$\dfrac{3}{5} \cdot \dfrac{2}{2}$      $\dfrac{2}{5} \cdot \dfrac{2}{3}$      $\dfrac{2}{5} \cdot \dfrac{2}{5}$

Emphasize looking for common monomial factors when simplifying rational expressions. Students tend to forget to do this and look only for binomial factors.

### Errors That Students Make
Many students will simplify an expression such as $\dfrac{x-5}{2(x-5)}$ to 2. This occurs when they *"cancel."* Stress "dividing out" like factors, NOT *"crossing out"* or *"canceling."* Thus, students should write the 1's as shown:

$$\frac{\overset{1}{\cancel{x-5}}}{2\underset{1}{\cancel{(x-5)}}} = \frac{1}{2} \cdot 1 = \frac{1}{2}$$

In this way they will get the correct answer, $\frac{1}{2}$, not 2.

### Cumulative Review
Solve.

1. $9x - 5 = -2x^2$
$\frac{1}{2}, -5$

2. $4x - (7 - x) = 3$
$2$

3. $6x - 8 \le 8x + 4$    $x \ge -6$

## 8.3 The $-1$ Technique (page 212)

### Vocabulary
convenient form

### Suggestions
Prior to doing Example 1, provide the following drill on writing polynomials in "convenient form." Write in convenient form.

1. $-x^2 - 5x + 2$
$-1(x^2 + 5x - 2)$

2. $9 - x^2$
$-1(x^2 - 9)$

3. $5 - 6d - d^2$
$-1(d^2 + 6d - 5)$

4. $x^2 + 3x + 2$
Already in convenient form

### Errors That Students Make
When writing an expression such as $40 + 3a - a^2$ in convenient form, the students will usually write the terms in correct descending order, $-a^2 + 3a + 40$, but when factoring out the $-1$, will change only the sign of the first term. They will then write $-1(a^2 + 3a + 40)$. To avoid this, stress checking by applying the distributive property. They should then see that $-1(a^2 + 3a + 40) = -a^2 - 3a - 40$, which is not $-a^2 + 3a + 40$.

### For Additional Chalkboard Work
Simplify.

1. $\dfrac{x^2 - 7x + 10}{5 - x}$

$\dfrac{x - 2}{-1}$ or $-1(x - 2)$

2. $\dfrac{4 - a}{a^2 - 5a + 4}$

$\dfrac{-1}{a - 1}$

3. $\dfrac{-x^2 + 7x - 12}{x^2 - 16}$    $-\dfrac{x - 3}{x + 4}$

### Cumulative Review

1. Multiply
$(2x - 7)(x + 4)$.
$2x^2 + x - 28$

2. Factor
$2x^3 - 4x^2 + 2x$.
$2x(x - 1)(x - 1)$

3. Solve
$x^2 = 10x - 16$.    $2, 8$

## 8.4 The Quotient of Powers Property
### (page 215)

### Background
The concepts of this lesson will be important in the next chapter when students study division of polynomials by a monomial or binomial. For example, $x\overline{)6x^4 + 5x^3 + 3x^2}$ will be treated as $\frac{6x^4 + 5x^3 + 3x^2}{x} = \frac{6x^4}{x} + \frac{5x^3}{x} + \frac{3x^2}{x}$, which in turn will involve the quotient of powers property.

### Suggestions
Emphasize the difference between the two types of examples illustrated in Example 1. These show the cases of $m > n$ or $m < n$ for $\frac{x^m}{x^n}$.

### Errors That Students Make
The simplification of $\frac{a}{a^4}$ is erroneously written as $\frac{1}{a^4}$.

Stress that if there is no exponent written with the $a$, it is NOT to be construed as being NOTHING. Rewrite $\frac{a}{a^4}$ as $\frac{a^1}{a^4} = \frac{1}{a^3}$.

### For Additional Chalkboard Work
Simplify.

1. $\dfrac{a^5}{a^{11}}$   $\dfrac{1}{a^6}$

2. $\dfrac{b^9}{b^2}$   $b^7$

3. $\dfrac{m^{12}}{m^6}$   $m^6$

4. $\dfrac{p^9}{p^{12}}$   $\dfrac{1}{p^3}$

5. $\dfrac{x^3 y^4}{x^5 y}$   $\dfrac{y^3}{x^2}$

6. $\dfrac{9a^9 b^3}{12a^3 b^5}$   $\dfrac{3a^6}{4b^2}$

7. $\dfrac{16pq^4}{4p^2 q^5}$   $\dfrac{4}{pq}$

## 8.5 Simplest Form of a Product
### (page 218)

### Suggestions
Pay particular attention to the prose above Example 3 on page 219. This prose offers a review of all the techniques a student might need to simplify a product.

You may wish to use the following before you do Example 3.

For $\dfrac{2z^2 - 16z + 30}{z^3 - 25z} \cdot \dfrac{-25 - 5z}{10z - 30}$, ask the following developmental questions as you work the problem at the chalkboard.

Which polynomial must be written in convenient form? Write the result.
Ans. $-25 -- 5z = -1(5z + 25)$
$$\frac{(2z^2 - 16z + 30)(-1)(5z + 25)}{(z^3 - 25z)(10z - 30)}$$
Name any GCF's.
Factor them out.
Ans. 2 for $2z^2 - 16z + 30$
  5 for $5z + 25$
  $z$ for $z^3 - 25z$
  10 for $10z - 30$
$$\frac{(2)(z^2 - 8z + 15)(-1)(5)(z + 5)}{(z)(z^2 - 25)(10)(z - 3)}$$
Can any common factors be factored further?
Ans. yes: $10 = (5)(2)$
Can any polynomials be factored?
Ans. yes: $z^2 - 8z + 15$; $z^2 - 25$
(difference of 2 squares)
$$\frac{(2)(z - 5)(z - 3)(-1)(5)(z + 5)}{z(z - 5)(z + 5)(5)(2)(z - 3)}$$
What common factors now divide out?
Ans. 5, 2, $z - 3$,
  $z + 5$, $z - 5$
$$\frac{\overset{1}{(2)}\,\overset{1}{(z-5)}\,\overset{1}{(z-3)}\,(-1)\,\overset{1}{(5)}\,\overset{1}{(z+5)}}{(z)\,\underset{1}{(z-5)}\,\underset{1}{(z+5)}\,\underset{1}{(5)}\,\underset{1}{(2)}\,\underset{1}{(z-3)}}$$
Thus the product is $\dfrac{-1}{z}$.

### Cumulative Review
1. Find the additive inverse of $-6$.
  6

2. Simplify completely
  $7 - (3 - x) - x$.
  4

`3. Factor completely $2a^3 - 128a$.
  $2a(a - 8)(a + 8)$

## 8.6 Multiplying and Dividing (page 221)

### Background
The arithmetic property for dividing fractions,
$\frac{a}{b} \div \frac{c}{d} = \frac{a}{b} \cdot \frac{d}{c}$, also applies to dividing algebraic expressions.

### Suggestions
Before starting the lesson, you might want to try the following warm-up practice for division of arithmetic fractions. This also provides for a good review for the basic minimum competency test in basic skills now mandated in many states.
Divide.

**1.** $\frac{2}{3} \div \frac{5}{7}$     **2.** $\frac{4}{5} \div 3$    **3.** $\frac{5}{7} \div \frac{10}{21}$

$\frac{14}{15}$         $\frac{4}{15}$        $\frac{3}{2}$

Stress that when dividing algebraic fractions, once the division has been written as the product of a fraction and the *reciprocal* of the divisor, the work is merely a repeat of the type shown in the last lesson, products of fractions.

### Enrichment
You might try to prove the division property. Begin with the illustration that $16 \div 8$ is the same as $\frac{16}{8}$. So $\frac{2}{3} \div \frac{5}{7}$ can be written as $\dfrac{\frac{2}{3}}{\frac{5}{7}}$ . Now multiply numerator and denominator by $\frac{7}{5}$, the reciprocal of $\frac{5}{7}$. $\dfrac{\frac{2}{3} \cdot \frac{7}{5}}{{}^1\frac{5}{7} \cdot \frac{7}{5}{}^1} = \frac{2}{3} \cdot \frac{7}{5}$

Prove $\frac{a}{b} \div \frac{c}{d} = \frac{a}{b} \cdot \frac{d}{c}$

Solution: $\frac{a}{b} \div \frac{c}{d}$ means $\dfrac{\frac{a}{b}}{\frac{c}{d}} = \dfrac{\frac{a}{b} \cdot \frac{d}{c}}{{}^1\frac{c}{d} \cdot \frac{d}{c}{}^1} = \frac{a}{b} \cdot \frac{d}{c}$

### Cumulative Review
Solve.

**1.** $3(x + 2) = 3x + 7$    **2.** $4x - 8 = 4(x - 2)$
    no solution            all numbers

**3.** $\frac{x}{2} = \frac{3}{5}$   $\frac{6}{5}$

## Geometry Review (page 224)

### Suggestions
Many science labs have cylinders and cones that have the same height ($h$) and the same radius ($r$). You may want to borrow a set to demonstrate the experiment outlined in the lesson prior to Example 1.

Review at this time the formulas for volumes of rectangular solids and pyramids. Stress the similarity in the formulas for the volume of a cone and a pyramid. Both formulas can be written as $\frac{1}{3}hB$ where $h$ is the height and $B$ is the area of the base. This alternate form of the formula might help students to retain both formulas.

### Cumulative Review
**1.** 5 is what % of 20?   25%
**2.** 6% of what number is 30?   500
**3.** $3\frac{1}{2}$% of 40 is what number?   1.4

## 8.7 Ratio and Proportion (page 225)

### Vocabulary
extremes, means, proportion, ratio

### Background
Students need to recall that an equation containing fractions can be solved by multiplying each side by the LCM for all the denominators.

### Suggestions
Proportions in themselves are not of much value. Their main use is in solving word problems that lead to proportions. Have students write the proportions that are suggested by the short mini-problems offered below before solving them.

### For Additional Chalkboard Work
Solve.

**1.** $\dfrac{x}{7} = \dfrac{2}{3}$　**2.** $\dfrac{4}{5} = \dfrac{a}{2}$　**3.** $\dfrac{b}{5} = \dfrac{5}{9}$　**4.** $\dfrac{1}{3} = \dfrac{t}{11}$

$\dfrac{14}{3}$　　$\dfrac{8}{5}$　　$\dfrac{25}{9}$　　$\dfrac{11}{3}$

Write a proportion corresponding to each mini-word problem below.

**5.** 3 out of 5 people use a certain product. How many use the product in a city of 40,000?
$\dfrac{3}{5} = \dfrac{x}{40,000}$; 24,000

**6.** The scale on a road map is $2''{:}5$ mi. How far apart are two cities if the road-map distance is $4''$? $\dfrac{2}{5} = \dfrac{4}{x}$; 10 mi.

**7.** 3 cans of soda cost $1.20. Find the cost of 5 cans. $\dfrac{3}{1.20} = \dfrac{5}{x}$; $2.00

### Cumulative Review
Simplify.

**1.** $\dfrac{5x^3}{35x}$　**2.** $\dfrac{3-x}{9-x^2}$　**3.** $\dfrac{x^2+7x}{9b^7} \cdot \dfrac{3b^2}{49-x^2}$

$\dfrac{x^2}{7}$　　$\dfrac{1}{x+3}$　　$-\dfrac{x}{3b^5(x-7)}$

## Applications (page 228)

### Suggestions
Some students do not see how to express 40 mi/gal as a ratio. Stress reading 40 mi/gal as 40 mi:1 gal, or as the fraction $\frac{40}{1}$. In writing the proportion, also note that the terminology mi and gal need not be written.

Provide some practice in writing the correct proportion used to solve gas mileage problems.

### Cumulative Review
**1.** 6 is what % of 8?　75%
**2.** 20% of what number is 30?　150

### Word Problem Formulation
John has one week for a vacation. He lives in New York City and plans to vacation in Orlando, Florida. Would it be more expensive to fly down or drive down?
Questions for discussion:
1. How many miles is it between New York City and Orlando?
2. Assuming the car gets 40 mi/gal, what would be the cost of gas? How would the gas cost be figured?
3. What other travel costs would there be (e.g., tolls, meals on the way down and back, motels down and back)?
4. Would the motel expense be the same by plane as by car? Why?
5. How can you find the cost of flying?
6. Would he need a car to get around in Orlando? If so, what would the rental fee be?

## Computer Activities: Evaluating Rational Algebraic Expressions (page 229)

This program provides an application of the *if. . .then* statement. The main purpose of the lesson is to strengthen the student's understanding of division by zero. A good experiment is to have students RUN the program with lines 30 and 50 deleted. Students get a dramatic reinforcement of the nature of division by zero.

# 9
# Chapter
## and Lesson
## Commentaries

## Chapter 9  Combining Rational Expressions

### Objectives
To simplify rational expressions involving addition and subtraction
To divide polynomials

### Overview
Adding and subtracting rational expressions is very difficult for many students. This text adopts a philosophy that is different from most others. Instead of teaching the concept in one or two lessons with five or six illustrative examples, the text offers six lessons with 19 illustrative examples. By proceeding slowly and carefully, we believe that we can lead students to develop such confidence and expertise in working with rational expressions that by the end of the chapter they will be able to do far more difficult simplifications than most teachers previously hoped possible. The first lesson involves addition of rational expressions with the same denominator. The second lesson introduces unlike denominators that are restricted to being monomials without powers. The pace then picks up. Students progress to the following sequence of lessons.
Lesson 9.3  polynomial denominators
Lesson 9.4  mixed expressions, nonfactorable denominators
Lesson 9.5  subtraction
Lesson 9.6  simplification using the −1 technique
  Two lessons on dividing polynomials are also included in this chapter. The first involves dividing by a monomial, the second by a binomial.

## Problem Solving Formulation (page 235)

This page has two goals. The first is to show the use of mathematics in consumer planning. The second aim is to provide another experience in word-problem formulation.
  Some related skills in working on the project are as follows. How do you compute the cost of gas? Ask students to find out the advertised gas mileage of the new car. Review the formula.
Let $x$ = the number of gallons of gas needed per year. Then $15,000 = x$ (advertised mileage). Solve the equation to find $x$, the number of gallons per year. Then the students can use the current cost per gallon to find the fuel cost for the year and finally for the week. Some other expenses to be considered are license and registration, cleaning, road tolls, and depreciation (about $1,500 the first year).
  A related project is the discussion of loan interest. Explain the meaning of interest on the unpaid balance. For example, find the total interest on a $6,000 car loan at 15% a year on the unpaid balance for three years.

| Solution: | Amt. of money borrowed | | Interest |
|---|---|---|---|
| 1st yr. | $6,000 | 15% of $6,000 | $900 |
| 2nd yr. | $4,000 | 15% of $4,000 | $600 |
| 3rd yr. | $2,000 | 15% of $2,000 | $300 |
| | | Total | $1,800 |

Thus, the total interest for the loan is $1,800.

## 9.1 Addition: Like Denominators (page 236)

### Vocabulary
addition property for rational expressions

### Background
Students have used the distributive property throughout algebra to justify many processes. The distributive property is used here to justify the addition property for rational numbers.

### Suggestions
Prior to doing Example 2, which involves simplifying sums, you might need to briefly review simplification of rational expressions.
Simplify.

1. $\dfrac{x-4}{x^2-7x+12}$    2. $\dfrac{3x+5}{3x^2+5x}$    3. $\dfrac{4x-14}{2x^2+x-28}$

$\dfrac{1}{x-3}$       $\dfrac{1}{x}$       $\dfrac{2}{x+4}$

Example 3 illustrates the extension of the basic property to the sum of more than two rational expressions.

### For Additional Chalkboard Work
Add. Simplify if possible.

1. $\dfrac{4}{x}+\dfrac{7}{x}$    2. $\dfrac{m}{10}+\dfrac{3m}{10}$    3. $\dfrac{2}{15x}+\dfrac{3}{15x}$

$\dfrac{11}{x}$     $\dfrac{4m}{10}=\dfrac{2m}{5}$     $\dfrac{5}{15x}=\dfrac{1}{3x}$

4. $\dfrac{a}{a^2-4}+\dfrac{2}{a^2-4}$    5. $\dfrac{y^2}{y-3}+\dfrac{-9}{y-3}$

$\dfrac{1}{a-2}$           $y+3$

6. $\dfrac{x}{x^2+7x+10}+\dfrac{5}{x^2+7x+10}$   $\dfrac{1}{x+2}$

### Cumulative Review
Simplify.

1. $(3a^2)^3$                 2. $(3x-4)(x+5)$
$27a^6$                   $3x^2+11x-20$

3. $\dfrac{x^2-25}{5-x}$   $\dfrac{x+5}{-1}$ or $-(x+5)$

## 9.2 Addition: Monomial Denominators (page 238)

### Vocabulary
identity property for rational expressions, LCM

### Background
The entire concept of adding rational expressions with unlike denominators rests upon the property $\dfrac{a}{b}=\dfrac{a(c)}{b(c)}$.

The fundamental technique of this and subsequent lessons in this chapter will be factoring each denominator into primes, and then multiplying the numerator and denominator of each rational expression by any factors necessary to make all denominators the same. It is not enough to merely state what the LCM is. We also show in each example how the proper choice can be made.

### Alternate Example 1
To add $\dfrac{1}{5}+\dfrac{7}{10}$, first factor each denominator

into primes: $\dfrac{1}{5}+\dfrac{7}{5\cdot2}$

To be the *same,* each denominator would need 5 and 2 as factors,

$\dfrac{1}{5}+\dfrac{7}{5\cdot2}$

needs 2   has $5\cdot2$

$\dfrac{1\cdot2}{5\cdot2}+\dfrac{7}{5\cdot2}=\dfrac{2}{10}+\dfrac{7}{10}=\dfrac{9}{10}$

### Errors That Students Make
In simplifying the results of Example 3, students tend to forget to use the distributive property, and write $\dfrac{(2y+7)3}{12y}+\dfrac{(2y-1)4}{12y}$ as $\dfrac{2y+21+2y-4}{12y}$ instead of $\dfrac{6y+21+8y-4}{12y}$.

### For Additional Chalkboard Work
Add. Simplify if possible.

1. $\dfrac{3x}{14}+\dfrac{x}{7}$    2. $\dfrac{2a}{9}+\dfrac{1a}{3}$    3. $\dfrac{3n}{4}+\dfrac{n}{20}$

$\dfrac{5x}{14}$       $\dfrac{5a}{9}$       $\dfrac{16n}{20}$ or $\dfrac{4n}{5}$

## 9.3 Addition: Polynomial Denominators
### (page 242)

**Suggestions**

Before doing Example 1 at the chalkboard, you might want to provide some practice with an easier type that involves only two rational expressions.
Add.

1. $\dfrac{5}{x^2 - 9} + \dfrac{4}{x - 3}$   2. $\dfrac{7}{y - 4} + \dfrac{4}{y^2 + 2y - 24}$

$\dfrac{4x + 17}{(x - 3)(x + 3)}$   $\dfrac{7y + 46}{(y - 4)(y + 6)}$

3. $\dfrac{6}{m^2 - m} + \dfrac{4}{m}$   $\dfrac{4m + 2}{m^2 - m}$

Emphasize the importance of trying to simplify a sum. Prior to doing Example 2, it might be helpful to briefly review simplifying rational expressions.
Simplify.

1. $\dfrac{x^2 + 9x + 14}{x(x + 7)}$   2. $\dfrac{x^2 - 7x}{(x + 3)(x - 7)}$

$\dfrac{x + 2}{x}$   $\dfrac{x}{x + 3}$

3. $\dfrac{x^2 - 25}{(x - 4)(x + 5)}$   $\dfrac{x - 5}{x - 4}$

**Errors That Students Make**

Students tend to forget to apply the distributive property when simplifying after the denominators are the same. At the point in Example 1 where students have $\dfrac{-3}{(x - 5)(x + 3)} + \dfrac{4(x - 5)}{(x - 5)(x + 3)}$

$+ \dfrac{3(x + 3)}{(x - 5)(x + 3)}$, the error is to write the resulting numerator as $-3 + 4x - 5 + 3x + 3$ instead of $-3 + 4x \underline{- 20} + 3x \underline{+ 9}$.

The third example involves the need to multiply two binomials. Another typical error students make is to forget the middle term. Thus, in Example 3, students might write $(a - 4)(a + 6) = a^2 - 24$ instead of $a^2 \underline{+ 2a} - 24$.

**For Additional Chalkboard Work**

Add. Simplify if possible.

$\dfrac{5}{a - 2} + \dfrac{-20}{a^2 - 4} + \dfrac{7}{a + 2}$   $\dfrac{12}{a + 2}$

## 9.4 Some Special Cases of Addition
### (page 245)

**Suggestions**

The first example is similar to adding mixed numbers in arithmetic. Many students recall that a mixed number such as $4\frac{2}{3}$ can be written as $\dfrac{14}{3}$ by thinking $\dfrac{3 \cdot 4 + 2}{3}$ or $\dfrac{14}{3}$. Show why this works to pave the way for the algebraic application: Think, $4\frac{2}{3}$ means $4 + \dfrac{2}{3} = \dfrac{4}{1} + \dfrac{2}{3}$. Now add by getting each fraction over the same LCM.

$$\dfrac{4}{1} + \dfrac{2}{3} = \dfrac{4 \cdot 3}{1 \cdot 3} + \dfrac{2}{3} \qquad \dfrac{4 \cdot 3 + 2}{3} = \dfrac{14}{3}$$

**Errors That Students Make**

Students find it difficult to add rational expressions in which no denominator can be factored. A typical error made by students is to write $\dfrac{5}{x + 2} + \dfrac{4}{x + 4}$ as $\dfrac{9}{2x + 6}$.

Another error is to write $\dfrac{5}{(x + 2)2} + \dfrac{4}{x + 4} = \dfrac{5}{x + 4} + \dfrac{4}{x + 4}$, in which the student forgot to distribute the 2 in the denominator of the first fraction.

**For Additional Chalkboard Work**

Add.

1. $\dfrac{3}{x + 5} + \dfrac{4}{x + 10}$   2. $\dfrac{5}{x - 3} + \dfrac{2}{x - 9}$

$\dfrac{7x + 50}{(x + 5)(x + 10)}$   $\dfrac{7x - 51}{(x - 3)(x - 9)}$

**Cumulative Review**

1. For what value(s) of $x$ is $\dfrac{x - 2}{x^2 - 36}$ undefined?

$-6, 6$

2. Simplify $\dfrac{x^2 - 7x + 12}{x^2 - 25} \div \dfrac{8 - 2x}{x - 5}$.

$\dfrac{x - 3}{-2(x + 5)}$

3. Solve $5x + 24 = x^2$.   $-3, 8$

## 9.5 Subtraction (page 248)

### Background
The concept of subtracting rational expressions depends upon two previously taught concepts:

1. interpretation of $a - b$ as $a + (-b)$, and

2. $-\dfrac{a}{b} = \dfrac{-1 \cdot a}{b}$.

### Suggestions
After you have developed the concept of subtraction of rationals, $\dfrac{a}{b} - \dfrac{c}{d} = \dfrac{a}{b} + \dfrac{-1 \cdot c}{d}$, it might be helpful to the class, particularly for slower students, to offer the following as a warmup. Rename each subtraction as an addition.

1. $\dfrac{2}{x^2 - 36} - \dfrac{4}{x - 6}$

$\dfrac{2}{x^2 - 36} + \dfrac{-4}{x - 6}$

2. $\dfrac{3}{x^2 - 7x + 12} - \dfrac{-5}{x - 4}$

$\dfrac{3}{x^2 - 7x + 12} + \dfrac{5}{x - 4}$

3. $\dfrac{4}{x} - \dfrac{x - 2}{x^2 - x}$  $\dfrac{4}{x} + \dfrac{-x + 2}{x^2 - x}$

### Errors That Students Make
In Example 3, two binomials have to be multiplied. $(-1a + 2)(a - 4)$ is not $-1a^2 - 8$. Students tend to forget the middle term.

### For Additional Chalkboard Work
Subtract. Simplify if possible.

1. $\dfrac{x}{x^2 - 4x + 3} - \dfrac{2}{x - 3}$

$\dfrac{-x + 2}{(x - 3)(x - 1)}$

2. $3 - \dfrac{3x - 1}{2x}$

$\dfrac{3x + 1}{2x}$

3. $\dfrac{3}{x - 2} - \dfrac{2x - 3}{x^2 - 6x + 8}$  $\dfrac{x - 9}{(x - 2)(x - 4)}$

### Cumulative Review
Solve.

1. The sum of three numbers is 34. The first is 3 less than the second, while the third is 4 more than the second. Find the numbers.  8, 11, 15

2. Find two consecutive odd integers such that the square of the second, decreased by the first, is 44.  5 and 7

## 9.6 Simplifying: The −1 Technique (page 251)

### Suggestions
First review the portion of this technique studied in the last chapter. Students know now to write the denominator of a fraction such as $\dfrac{3}{7 - x}$ in convenient form, $7 - x = -x + 7 = -1(x - 7)$. Use the following as a warmup.

Write each fraction with its denominator in convenient form.

1. $\dfrac{a + 2}{3 - a}$

$\dfrac{a + 2}{-1(a - 3)}$

2. $\dfrac{7y - 2}{4 - 3y}$

$\dfrac{7y - 2}{-1(3y - 4)}$

3. $\dfrac{8 - m}{5m - m^2}$

$\dfrac{8 - m}{-1(m^2 - 5m)}$

It is now an easy transition to the development of the concepts in Example 1 on page 251. Once $7 - x$ is written in convenient form, the student can rewrite $\dfrac{3}{-1(x - 7)}$ so that the factor $-1$ does not appear in the denominator.

Use the property $\dfrac{a}{-1b} = \dfrac{-1 \cdot a}{b}$ to write

$\dfrac{3}{-1(x - 7)} = \dfrac{-1 \cdot 3}{x - 7}$.

### Errors that Students Make
In simplifying an expression such as exercise 6, students will rewrite it as $\dfrac{4y - 3}{5y - 10} + \dfrac{-1(3y - 4)}{y - 2}$ and then forget to distribute the $-1$ to $3y$ and $-4$.

### For Additional Chalkboard Work
Simplify.

1. $\dfrac{4x}{x^2 - 36} + \dfrac{2}{6 - x}$

$\dfrac{2}{x + 6}$

2. $\dfrac{3b + 2}{b^2 - 11b + 28} + \dfrac{2}{4 - b}$

$\dfrac{b + 16}{(b - 4)(b - 7)}$

### Cumulative Review

1. Simplify
$\dfrac{x^2 - 81}{x^3} \div \dfrac{9x - x^2}{x^2}$.

$\dfrac{x + 9}{-x^2}$

2. Solve
$\dfrac{x - 7}{4} = \dfrac{-3}{x}$.

3, 4

3. Simplify
$\dfrac{4x^7}{8x^4} \cdot \dfrac{x^3}{2}$

## 9.7 Dividing by a Monomial (page 253)

### Vocabulary
quotient

### Background
Students have learned to simplify expressions such as $\dfrac{a^3}{a^4}$ and $\dfrac{b^7}{b^2}$ by "dividing out" like factors, and then by using the "quotient of powers property" as a shortcut. This lesson shows that division of a monomial by a monomial can be interpreted as simplification of fractions with monomials for numerator and denominator. This lesson paves the way for the next lesson on division of a polynomial by a binomial.

### Suggestions
Following Example 1, try some practice on the concept with the drill below.
Divide.

1. $a^4\overline{)a^9}$  $\quad$ 2. $m\overline{)m^7}$  $\quad$ 3. $p^3\overline{)p^8}$  $\quad$ 4. $g^2\overline{)g^{10}}$
$\quad a^5 \qquad\qquad m^6 \qquad\qquad p^5 \qquad\qquad g^8$

### Errors That Students Make
Students come to believe that the quotient for the division of a monomial by a monomial must always contain a variable. Thus, a typical student error is to write the quotient for an expression such as $3x^2\overline{)21x^2}$ as $7x$, not just 7. Refer to Example 3 on page 254.

A second typical error made by students is to write the quotient for $2x^2\overline{)4x^4 + 6x^3 + 2x^2}$ as $2x^2 + 3x$. Stress that since there are three terms in the dividend, there will be three terms in the quotient. The correct quotient is therefore $2x^2 + 3x + \underline{1}$.

### For Additional Chalkboard Work
Divide.

1. $2a\overline{)4a^3 - 12a^2 + 10a}$
$\quad 2a^2 - 6a + 5$
2. $3m\overline{)6m^3 - 9m^2 + 3m}$
$\quad 2m^2 - 3m + 1$
3. $5x^2\overline{)20x^4 + 15x^3 - 30x^2}$
$\quad 4x^2 + 3x - 6$

## 9.8 Dividing by a Binomial (page 255)

### Vocabulary
dividend, divisor, remainder

### Suggestions
It might be helpful to do the illustrative prose presentation of $25\overline{)785}$, and the first example, $x + 3\overline{)x^2 + 10x + 21}$, on the chalkboard side by side in parallel steps.

You can use an alternate approach. Point out that subtraction is the same as adding the opposite. Then the subtraction can be done mentally without having to write the subtraction in the indicated horizontal format.
For example, in the first step of Example 2.

$$
\begin{array}{r}
3a \\
2a - 5\overline{)6a^2 - 7a + 5} \\
\underline{6a^2 - 15a} \\
8a
\end{array}
\quad
\begin{array}{l}
\text{Think: Add the opposite} \\
\text{of } 6a^2 - 15a \text{ to} \\
6a^2 - 7a.
\end{array}
$$

$$
\begin{array}{r}
6a^2 - \phantom{1}7a \\
-6a^2 + 15a \\
\hline
8a
\end{array}
$$
opp. of $6a^2$
opp. of $-15a$

Before going on to the more difficult types as in Example 3, provide practice on the easier ones.

### Enrichment
Exercises 35 and 36 will help the student discover a pattern for factoring the sum or difference of two cubes. For example, one might guess a factor of $x^3 - 27$ to be $x - 3$! Then division of $x^3 - 27$ by $x - 3$ yields the other factor $x^2 + 3x + 9$. Notice a pattern:
$x^3 - 27 = (x)^3 - (3)^3$. One factor is $x - 3$.
The other factor becomes $(x)^2 + x \cdot 3 + (3)^2$.
Challenge the student to derive the general pattern.
Find the factors of $a^3 - b^3$.
*Solution:* Try as a factor $a - b$. Then divide:
$(a^3 - b^3) \div (a - b) = a^2 + ab + b^2$ or
$(a)^2 + (a \cdot b) + (b)^2$. Now apply the technique to factoring $x^3 - 64$.
Use the general pattern:
$$(a)^3 - (b)^3 = (a - b)((a)^2 + (a \cdot b) + (b)^2)$$
$$(x)^3 - (4)^3 = (x - 4)((x)^2 + (x \cdot 4) + (4)^2)$$
$$64 \qquad \text{or } (x - 4)(x^2 + 4x + 16)$$

## Applications: Mixed Practice (page 258)

The following practice is designed to provide further experience with the concept of *word-problem formulation*.

**1.** Consider the equation

$$500 = x + 0.25x$$

Construct a verbal problem that would be a real-life application and would lead to this equation.

*Solution:*

A store owner lists the selling price of a TV at $500.00. The profit is 25% of the cost. Find the cost.

**2.** Consider the equation

$$\frac{7}{9} = \frac{x}{18{,}000}$$

Construct a practical verbal problem the solution of which results in this equation.

*Solution:*

Seven out of nine people in a city use a certain product. How many can be expected to use the product in a city of 18,000?

## Computer Activities: Least Common Denominator (page 259)

This program enables students to find the LCD of several fractions. Stress the meaning of the INT function and refer students to pp. 501–503. A typical student error is to think that INT *rounds* numbers. Point out that INT *truncates* numbers; for example, $\text{INT}(7.1) = \text{INT}(7.5) = \text{INT}(7.8) = 7$.

# 10

# Chapter
# and Lesson
# Commentaries

## Chapter 10   Applications of Rational Expressions

### Objectives

To solve equations containing rational expressions
To simplify complex rational expressions
To solve problems about work
To solve problems about motion
To solve literal equations and evaluate for given values of the variable

### Overview

The chapter opens with solving equations containing rational expressions. Students have previously studied equations containing fractions but not with variables in the denominator. The next lesson is a traditional application of fractional equations to work problems. Two lessons are devoted to the topic of complex fractions. The first of these involves complex fractions with only *monomial denominators*. The transition is made in the next lesson to *polynomial denominators*. The next lesson involves solving a literal equation or formula for one of its variables. Then students will evaluate these literal equations or formulas for one of the variables, given the values of the other variables. The chapter also has a traditional lesson on motion problems with some use of fractional equations. There are two problem solving application pages in the chapter.

## Special Features

## Non-Routine Problem Solving (page 263)

Point out that one of the techniques of Non-Routine Problems is sometimes to include built in "traps" to see if students are really thinking carefully. This is illustrated in the first problem. Most students will compute the first speed, 360/6 = 60. Then they will average the two speeds, 60 and 40, and get the *wrong* average, 50! The true average is the *total* distance, 720, divided by the *total* time, 6 + 9 = 15. Thus, the average is 720/15 = 48 mph.

## 10.1 Equations with Rational Expressions (page 264)

### Vocabulary
extraneous solution

### Background
Students have previously solved equations containing fractions but not with variables in the denominators and not with factoring to get the LCM. The LCM was found by *inspection*. You might want to begin with a review of this type.

Solve $\frac{x}{2} + 4 = \frac{1}{3}$. It is easy to see that the smallest number that 3 and 2 each divide evenly is 6. So the LCM is 6. Point out that if the LCM is not easily recognizable, then the students will have to use factoring, as illustrated in Example 1 on page 264 of the text.

### Suggestions
Emphasize that the LCM must be multiplied times every fraction appearing in the equation. This is an application of the distributive property.

### Errors That Students Make
A typical error occurs in the following:

$$(x - 4)(x - 1)\left(\frac{3}{x - 1} + \frac{2}{x - 4}\right) =$$

$$(x - 4)(x - 1)\frac{3x}{(x - 4)(x - 1)}$$

$$(x - 4)(3) + (2) = 3x$$

The student has not distributed the $(x - 4)(x - 1)$ to both terms on the left. To avoid this, encourage students to begin with the following as the first step:

$$(x - 4)(x - 1)\frac{3}{x - 1} + (x - 4)(x - 1)\frac{2}{x - 4} =$$

$$(x - 4)(x - 1)\frac{3x}{(x - 4)(x - 1)}$$

Emphasize checking for extraneous solutions.

### For Additional Chalkboard Work
Solve. Check for extraneous solutions.

1. $\frac{12}{x^2 - 4} = \frac{2}{x + 2} + \frac{3}{x - 2}$   no solution

2. $\frac{3x}{x^2 + 8x + 15} - \frac{2}{x + 3} = \frac{3}{x + 5}$   $x = \frac{-19}{2}$

## 10.2 Work Problems (page 269)

### Suggestions

Stress the idea that if two people work together on a job the sum of the two fractional parts of the job done by each should be the whole unit of work, or 1. Thus, if Mary does $\frac{3}{4}$ of the work and Bill completes the remaining $\frac{1}{4}$ of the work they have finished the job; $\frac{3}{4} + \frac{1}{4} = 1$, where 1 represents the whole unit of work.

For the slower student, point out in Example 2 that in the equation $\frac{x}{3} + \frac{x}{2} = 1$, 3 represents Regina's time to do the job alone, and 2 represents Carlo's time to do the job alone. Slower students might have success with work problems if you have them set up the equation by putting the time together as the numerator of each fraction and the time alone as the denominator of each fraction, then setting the sum equal to 1.

### Errors That Students Make

In Example 2, page 270, some students are careless in multiplying the fraction and write $x \cdot \frac{1}{3} = 3x$ instead of $\frac{x}{3}$. Emphasize that $x \cdot \frac{1}{3}$ means $\frac{x}{1} \cdot \frac{1}{3} = \frac{x}{3}$.

### For Additional Chalkboard Work

Solve.

1. It takes Jim 40 minutes to mow a lawn. It takes Sue 60 minutes. How long will it take them if they work together?   24 min
2. Working together, it takes Randi and Les 8 hours to paint a room. It takes Les 12 hours to do it alone. How long would it take Randi to do it alone?   24 h

## 10.3 Complex Rational Expressions (page 273)

### Vocabulary

complex rational expression

### Background

This is the first of two lessons on complex fractions. In this first lesson, all fractions contain only monomial denominators. The primary concept of the lesson is that multiplication of both the numerator and denominator of a fraction by the same nonzero number does not change the value of the fraction even though the numerator and denominator of the original fraction may themselves be fractions.

### Suggestions

Begin with illustrations involving only arithmetic fractions (no variables).

1.  $\dfrac{\frac{5}{7}}{\frac{3}{14}}$   $\dfrac{10}{3}$

2.  $\dfrac{\frac{1}{2} + \frac{1}{3}}{\frac{1}{6} - \frac{1}{3}}$   $\dfrac{5}{-1}$ or $-5$

3.  $\dfrac{\frac{2}{5} + 1}{\frac{1}{2} - \frac{3}{10}}$   $\dfrac{14}{2}$ or 7

Emphasize the importance of checking in the end whether the result can be further simplified. Example 4 illustrates this situation particularly well.

### For Additional Chalkboard Work

Simplify.

1.  $\dfrac{1 + \frac{3}{x} + \frac{2}{x^2}}{\frac{1}{x} + \frac{2}{x^2}}$   $x + 1$

2.  $\dfrac{1 - \frac{13}{x} + \frac{30}{x^2}}{\frac{1}{x} - \frac{10}{x^2}}$   $x - 3$

3.  $\dfrac{\dfrac{\frac{1}{a} + \frac{3}{a^2}}{1 - \frac{5}{a} - \frac{24}{a^2}}}{\frac{1}{a - 8}}$

### Cumulative Review

Solve.

1. $4x - 8 > 5x + 3$   $x < -11$

2. $-9c + 5 = 2c^2$   $\frac{1}{2}, -5$

## 10.4 Complex Rational Expressions with Polynomial Denominators (page 276)

### Errors That Students Make

A typical student error is pointed out below in the development of Example 1, page 276.

$$\dfrac{\dfrac{1}{(m-5)(m+2)}\left(\dfrac{4}{m-5}+\dfrac{6}{m+2}\right)}{(m-5)(m+2)\left(\dfrac{9}{(m-5)(m+2)}+\dfrac{1}{m-5}\right)}=$$

$$\dfrac{4(m+2)+6}{9+1}$$

In both the numerators, the student has forgotten to distribute the binomial to the last term: $(m-5)6$ and $(m+2)1$.

Exercises like that illustrated in Example 3, page 277, are difficult for many students. Stress the rewriting of $m+2$ as $\dfrac{m+2}{1}$. Students often forget the middle term in multiplying two binomials such as $(m-5)(m+2)$ and thus write $m^2-10$ instead of $m^2-\underline{3m}-10$.

Another typical error occurs in exercises such as Ex. 17 on page 278. When rewriting the subtraction in the numerator as addition, students might write $+\dfrac{-a-1}{a-2}$ instead of $+\dfrac{-a+1}{a-2}$. Stress that $-(a-1)=-a+1$, not $-a-1$.

### For Additional Chalkboard Work

Simplify.

1. $\dfrac{\dfrac{4}{x}-\dfrac{6}{x^2+3x}}{\dfrac{2}{x+3}+\dfrac{2}{x}}$  1

2. $\dfrac{\dfrac{4}{a^2-9}-\dfrac{2}{a+3}}{\dfrac{5}{a-3}-\dfrac{7}{a+3}}$  $\dfrac{a-5}{a-18}$

### Enrichment

Solve.  $\dfrac{\dfrac{x-7}{x^2-5x+6}}{\dfrac{5}{x-2}-\dfrac{4}{x-3}}=\dfrac{x^2-4x}{5}$

*Solution:* First simplify the complex fraction. Then solve the resulting fractional equation $1=\dfrac{x^2-4x}{5}$. The solution is 5, $-1$.

## Applications (page 279)

### Background

Take time to explain the concept of an electric circuit, covered in the prose discussion preceeding Example 1 on page 279. It might be easier for students if they first talk about simple direct current.

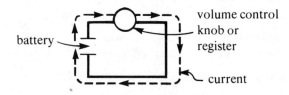

When the radio is turned on, a fixed VOLTAGE ($V$) from the battery causes CURRENT ($I$) to flow through the wire past the RESISTOR ($R$) and back to the battery. Turning the volume down is done by increasing the resistance to the current. This concept is similar to the flow of water in a garden hose. Turning the water on at the source of supply (equivalent to battery) allows the water (current) to flow. The turning of the knob acts as a resistor, increasing or decreasing the flow of water as desired. Unless there is a general change in the public availability of water, the turning of the knob does not affect the original power supply of water (similar to voltage in an electric circuit).

Point out that a famous scientist, Ohm, discovered the mathematical relationship between voltage ($V$), current ($I$), and resistance ($R$).

$$V=I\cdot R$$

This formula will be referred to later in the chapter on functions, when students study inverse variation.

### Suggestions

Challenge the students to come up with a formula for computing mentally the total resistance of two resistors. *Hint:* Find the resistance in general for two resistors $x$ and $y$.

*Solution:* $R=\dfrac{1}{\dfrac{1}{x}+\dfrac{1}{y}}$

This simplifies to $\dfrac{xy}{x+y}$. So, the total resistance for $20\,\Omega$ and $30\,\Omega$ is $\dfrac{20\cdot30}{20+30}$  $\dfrac{600}{50}=12$.

## 10.5 Formulas and Literal Equations (page 280)

### Vocabulary
literal equation

### Suggestion
This proves to be a difficult topic for many students. You might want to plan on two days for this lesson, breaking it up into special practice between illustrative examples.

First review the steps in solving a simple linear equation.

Solve.

1. $3x + 2 = 17$   5     2. $6 - 4x = -11$   $\frac{17}{4}$

3. $2 + 5x = 3$   $\frac{1}{5}$

Now show that there is a general pattern.

Alternate Example 1. Solve $px + r = t$ parallel with $5x + 4 = 8$.

| | |
|---|---|
| $5x + 4 = 8$ | $px + r = t$ |
| First add $-4$ to each side. | First add $-r$ to each side. |
| $5x + 4 - 4 = 8 - 4$ | $px + r - r = t - r$ |
| $5x = 4$ | $px = t - r$ |
| Divide each side by 5. | Divide each side by $p$. |
| $x = \dfrac{4}{5}$ | $x = \dfrac{t - r}{p}$ |

Exercises like that illustrated in Example 5 are difficult for many students. The concept involves common monomial factoring. Precede the Example with the following warmup review of the needed skill.

Factor out the GCF.

1. $mx + bx$    2. $px - 4x$    3. $7x + kx$
   $x(m + b)$     $x(p - 4)$      $x(7 + k)$

### Cumulative Review
Simplify.

1. $\dfrac{\dfrac{5}{x-4} + \dfrac{3}{x+4}}{\dfrac{7}{x^2 - 16}}$    $\dfrac{8x + 8}{7}$    2. $\dfrac{\dfrac{2}{x-3} - \dfrac{5}{x^2 - 3x}}{\dfrac{2x - 5}{x(x - 3)}}$

3. $(2x - 1)(x^2 - x + 1)$   $2x^3 - 3x^2 + 3x - 1$

## 10.6 Distance Problems (page 283)

### Suggestions
Ask students to solve the following problems as a review of the formula $d = rt$.

How far do you drive if you

1. drive 50 mph for 4 hours?    200 mi
2. drive 60 mph for 3 hours?    180 mi
3. drive 45 mph for 2 hours?     90 mi

Stress the formulation of the data in a table, as shown in the examples. It will also be helpful to students if they draw a sketch showing the direction of each car, as shown in each of the first three illustrative examples.

The following decisions should be made by a student when solving distance problems.

1. Is the distance traveled by each driver the same?
2. Is the time each driver spends driving total (or in each direction) the same?
3. If the drivers are driving away from each other, what do you know about their distances apart after so many hours?

The following problems may be used in addition to the **Reading in Algebra** for the exercises.

Two cars start at the same point and travel in opposite directions. The rate of one car is twice the rate of the slower car. After 2 h the cars are 180 mi apart.

Use this problem to answer the following questions.

1. If $x$ represents the rate of the slower car, what represents the rate of the faster car?   $2x$
2. Draw a sketch showing the distance of each car at the end of 2 hours.

faster        slower

180 mi

3. What is the relation between the distances. at the end of 2 h?   Sum is 180

## Computer Activities: Literal Equations (page 290)

The sample program is useful as a model for other programs in which the formula contains several variables. The student must be able to solve the literal equation algebraically for the requested variable before the computer can evaluate for that variable.

# 11

## Chapter and Lesson Commentaries

### Chapter 11    Graphing in a Coordinate Plane

#### Objectives

To give the coordinates of a point in a plane
To identify the quadrant in which a given point is located
To plot a point, given its coordinates
To find the slope of a nonvertical line segment
To determine the slant of a line using its slope
To determine if two lines are parallel using their slopes
To determine if three points are on the same line
To write an equation of a line given the slope and one point
To write an equation of a line given two points
To write a table of values given an equation of a line
To find the slope and $y$-intercept of a line given its equation
To write an equation of a line given its slope and $y$-intercept
To graph a line given its equation
To graph an inequality in two variables

#### Overview

The first two lessons teach students how to read the coordinates of points in a plane and how to graph points whose coordinates are given. Following this is the development of the concept of slope of a line or line segment. Once students see that the slope of a line is the same for any two points on the line, they are then ready to find the equation of a line given the coordinates of any two points on the line or one point and the slope. It is easier for students to learn one general method for writing the equation of a line.

Lesson 11.5 is a direct follow-through of Lesson 11.4 on slope. Students see that since the slope of a line is the same for *any* two points, writing the equation merely involves setting the numerical value of the slope equal to the value of the slope obtained using one point and a general point, $P(x, y)$. This leads to a general understanding. We chose to avoid the traditional beginning of giving students an equation and having them plot points and draw a line. We first establish *why* an equation of the form $y = mx + b$ represents a line.

#### Special Features

### René Descartes (page 292)

Descartes left school at the age of fourteen and shortly after devoted some time to the study of mathematics in Paris. After a varied career of military life, travel, and constructing optical instruments, he spent twenty years in Holland pursuing his study of philosophy, mathematics, and science. Descartes and Fermat conceived the ideas of modern analytic geometry, which establishes a correspondence between ordered pairs of real numbers and points in the plane. Descartes is also credited with recognizing the relationship $v - e + f = 2$, connecting the number of vertices, edges, and faces of a convex polyhedron.

## 11.1 Coordinates of Points in a Plane (page 293)

### Vocabulary

abscissa, horizontal line, ordered pair, ordinate, origin, quadrant, vertical line

### Background

The concept of "Algebraic Geometry" (Analytic Geometry) was first conceived by René Descartes (born in 1596 near Tours, France). Legend has it that the idea first came to him as he watched a fly moving about the ceiling in his room. He noted that the position of the fly could be described if one knew a relationship connecting the fly's distance from two adjacent walls. The technical words "coordinates," "abscissa," and "ordinate" are attributed to another mathematician, Leibnitz, born in 1646.

You might assign the further researching of the historical background of the "coordinate" system as a voluntary project. A good reference is "An Introduction to the History of Mathematics," by Howard Eves.

### Suggestions

Introduce the idea of an ordered pair by asking students to describe the seat location of a person in the class. Stress the need for two numbers, seat and row. Point out that the order is important so that people are not mixed up as to whether row or seat is meant.

### Errors That Students Make

A typical error made by students is to write the coordinates of a point in the wrong order. Stress the writing of the $x$-coordinate first.

### For Additional Chalkboard Work

For each ordered pair, tell which is the abscissa and which is the ordinate.
**1.** $A(3,4)$ **2.** $B(-2,5)$ **3.** $C(3,-1)$ **4.** $D(-2,-3)$
1st coord. is abscissa, 2nd coord. is ordinate

### Cumulative Review

Solve. Graph the solution set on a number line.
**1.** $x + 4 = 3x - 10$   **2.** $x + 4 < 3x - 6$
7                                    $x > 5$

## 11.2 Point Plotting (page 296)

### Vocabulary

graph

### Background

Students have learned in the previous lesson how to describe the position of a point using an ordered pair of coordinates. This lesson is a simple extension of the concept to plotting points whose coordinates are given.

### Suggestions

Stress that the first coordinate ($x$ or abscissa) of an ordered pair refers to movement along the $x$-axis: positive, to the right, negative, to the left.

The second coordinate ($y$ or ordinate) of an ordered pair refers to movement along the $y$-axis: positive, up, negative, down.

### Enrichment

Give the class the following points. They are to plot them, and connect with straight line segments in order alphabetically from letter to letter. If they finally connect $P$ to $A$, and then plot the separate point $R(2,1)$ they will get a picture. What is it?
$A(2,7)$, $B(-1,3)$, $C(-1,-1)$, $D(-3,-5)$, $E(-1,-5)$, $F(-1,-7)$, $G(2,-8)$, $H(-1,-9)$, $I(-1,-11)$, $J(0,-12)$, $K(11,-12)$, $L(11,-4)$, $M(12,-1)$, $N(12,4)$, $P(10,6)$. Join $P$ to $A$.
Plot $R(2,1)$.
Assign volunteers to make up their own pictures and post the best ones on your bulletin board.

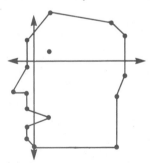

### Cumulative Review

**1.** Write the ratio 2:5 as a fraction.
$\frac{2}{5}$

**2.** Evaluate $-16a^3$ if $a = -\frac{3}{2}$.
54

**3.** Factor $2x^3 - 32x$.   $2x(x - 4)(x + 4)$

## 11.3 Slope (page 299)

### Vocabulary
slope

### Slope
The concept of slope is going to be of particular significance later in this chapter when students find the equation of a line.

### Suggestions
The idea of rise over run can sometimes be clarified with reference to the steps of a stairway. For a given run, the greater the rise the more difficult it is to climb the stairs. Note that the slope of the handrail is steeper for the stairway with the greater *rise*.

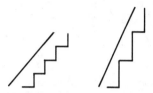

### Alternate Example 1
Arrange for use of an overhead projector for the following class-discovery approach to slope. Sketch a set of axes on each of three clear sheets of acetate so that when all three are placed in the projector glass plate simultaneously, the axes appear as one set.

On the first sheet plot the points $A(1,1)$ and $B(8,2)$. Draw $\overline{AB}$ and complete a triangle as shown below. Repeat the process for the second sheet, using the points $A(1,1)$ and $C(8,5)$. Then, on the third sheet, use the points $A(1,1)$ $D(8,10)$.

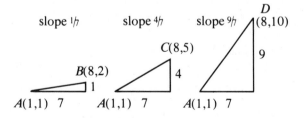

If all three overlays are placed on the projector at the same time, it is easy to see that the segment with the largest $\dfrac{\text{rise}}{\text{run}}$ is indeed the steepest line segment.

## 11.4 Slope of a Line (page 302)

### Vocabulary
parallel

### Background
The most important concept of this lesson is that the slope of a line is the SAME for ANY two points on the line.

### Suggestions
Emphasize the study of the diagram, at the bottom of page 303, that summarizes the relationship between the slant of a line and its slope. Some students are confused by the slopes of vertical and horizontal lines. You should review why division by 0 is undefined. For example, $\frac{5}{0}$ means $0\overline{)5}$. But since 0 times any number is 0, not 5, division by 0 is meaningless. However, $\frac{0}{5}$ means $5\overline{)0}$. Since $5(0) = 0$, the quotient is 0. Stress that a horizontal line has zero slope since there is no rise and a vertical line has an undefined slope since it has no run.

### Cumulative Review
Solve.

1. $\dfrac{x-4}{x+5} = \dfrac{2}{3}$    2. $\dfrac{4}{5} = \dfrac{y-4}{y-1}$    3. $\dfrac{4}{x} = \dfrac{x}{9}$

    22             16          6, −6

### For Additional Chalkboard Work
From the given points, determine whether $\overleftrightarrow{PQ}$ is parallel to $\overleftrightarrow{RS}$.

1. $P(2,7)$, $Q(3,10)$      2. $P(1,5)$, $Q(4,2)$
   $R(-3,-1)$ $S(-1,5)$     $R(4,4)$, $S(0,0)$
     $PQ \parallel RS$          $PQ \nparallel RS$

## 11.5 Equations of Lines (page 305)

### Background

The concept that the slope of a line is the same for any two points of the line can be used to find the equation of the line.

Review finding the slope of a line as a pre-lesson warmup. Find the slope of $\overleftrightarrow{AB}$ given the coordinates of two points $A$ and $B$.

**1.** $A(1,3)$, $B(3,7)$          **2.** $A(5,4)$, $B(9,5)$
$\qquad\qquad 2 \qquad\qquad\qquad\qquad \frac{1}{4}$

**3.** $A(-1,-2)$, $B(5,3)$    $\frac{5}{6}$

### Suggestions

It is important for students to learn that an equation for a line can be derived in more than one way.

### Alternate Way of Doing Example 1

Write the equation for $\overleftrightarrow{RS}$; $R(-2,-5)$, $S(1,4)$.

**(1)** Find the slope of $\overline{GR}$.

$$\frac{y - (-5)}{x - (-2)} = \frac{y + 5}{x + 2}$$

**(2)** Find the slope of $\overline{RS}$.

$$\frac{4 - (-5)}{1 - (-2)} = \frac{9}{3} \text{ or } \frac{3}{1}$$

**(3)** Set the two slopes equal to each other.

$$\frac{y + 5}{x + 2} = \frac{3}{1}$$

Solve for $y$;

$(y + 5) = 3(x + 2)$ ◀ *Solve the proportion*

$y + 5 = 3x + 6$ ◀ *Distribute 3*

$y = 3x + 1$ ◀ *Add −5 to each side to get y alone*

Now stress that the resulting equation is the SAME as when the slope of $\overline{GS}$ was used in the development of Example 1 on page 305 of the text.

Also stress the verification that each of the two given points, $R(-2,-5)$ and $S(1,4)$ do indeed satisfy the resulting equation.

Finding the equation of a line given the slope and one point is now a very simple process. Since the slope is already given, simply set it equal to the slope found by using the given point and a general point $P(x, y)$.

## 11.6 Using Equations of Lines (page 308)

### Background

In geometry the graph of a line corresponding to a given equation is thought of as a LOCUS of points satisfying a given condition. This involves two important concepts stressed below in the teaching suggestions.

**1.** All points on the line satisfy the equation.
**2.** All points satisfying the equation are on the line.

### Suggestions

After going over Example 1 ask the students to verify that points $A$ and $B$ have coordinates satisfying the equation as follows:

Verify that $(-1,-1)$ and $(9,3)$ satisfy $y = \frac{2}{5}x - \frac{3}{5}$.

Substitute $-1$ for $x$ and     Substitute 9 for $x$ and
$-1$ for $y$ in the equation.    3 for $y$ in the equation.

| $y$ | $\frac{2}{5}x - \frac{3}{5}$ |
|-----|------------------------------|
| $-1$ | $\frac{2}{5}(-1) - \frac{3}{5}$ |
| | $-\frac{2}{5} - \frac{3}{5}$ |
| | $-\frac{5}{5}$ |
| | $-1$ |

$-1 = -1$, true

| $y$ | $\frac{2}{5}x - \frac{3}{5}$ |
|-----|------------------------------|
| $3$ | $\frac{2}{5}(9) - \frac{3}{5}$ |
| | $\frac{18}{5} - \frac{3}{5}$ |
| | $\frac{15}{5}$ |
| | $3$ |

$3 = 3$, true

Thus, $A(-1,-1)$ and $B(9,3)$, two points on the line, have coordinates SATISFYING the equation.

Now use the results of Example 2 to show that a point having coordinates that SATISFY the equation must lie on the line. Ask the students to plot the two original points $A(-1,-1)$ and $B(9,3)$ on graph paper. Then draw a line through the two points. Now plot the new point of Example 2, $C(14,5)$. Stress the amazing result that indeed this point IS on the line!

### Errors That Students Make

When solving for $y$ in Example 1, students get $y = \frac{4}{10}x - 6$, forgetting to also divide the $-6$ by 10. This can be avoided by stressing that students should draw the "division bar" ALL the way across as a reminder that they must divide BOTH $4x$ and $-6$ by 10;

$$y = \frac{4x - 6}{10} = \frac{4}{10}x - \frac{6}{10}.$$

## 11.7 Slope-intercept Form of an Equation of a Line (page 310)

### Vocabulary
intercept

### Suggestions
As an alternative to the development on the first page of the lesson you might want to try the following discovery approach.
First have students do the following.
Write the equation of each line, $\overleftrightarrow{AB}$. Plot the points and draw the line through the points.
**1.** $A(0,4)$, $B(3,6)$      **2.** $A(0,-5)$, $B(2,-4)$
**3.** $A(0,1)$, $B(4,4)$
Now ask the students to write, for each problem:
**a.** the $y$-coordinate of the point where the line crosses or INTERCEPTS the $y$-axis
**b.** the slope of the line
Is there any pattern or relationship between these answers and the equation?

**1.** $y = \frac{2}{3}x + 4$      **2.** $y = \frac{1}{2}x - 5$
   slope    int.       slope    int.

**3.** $y = \frac{3}{4}x + 1$
   slope    int.

    Students tend to have difficulty graphing lines whose equations are not in the EXACT $y = mx + b$ form. Stress that for an equation like $y = 3x$ it is helpful to rewrite it so that the $m$ is a fraction and the $b$ appears as a number:
$y = \frac{3}{1}x + 0$.

### For Additional Chalkboard Work
Give the slope and the $y$-intercept of the line with the given equation.
**1.** $y = 5x + 2$      **2.** $y = -\frac{3}{4}x + 1$
    $5; 2$            $-\frac{3}{4}; 1$

### Cumulative Review
**1.** Write an equation of the line through $A(6,5)$ and $B(-3,-1)$.   $y = \frac{2}{3}x + 1$
**2.** Simplify $\dfrac{5}{x^2 - 4} - \dfrac{2}{x - 2}$.   $\dfrac{-2x + 1}{(x + 2)(x - 2)}$
**3.** Solve, sketch solution on a number line,
$-3x - 7 < 14$.   $x > -7$; all points to the right of $-7$ on the number line

## 11.8 Graphing Nonvertical Lines (page 314)

### Background
This lesson is a simple extension of the previous one. The main difference is that students must solve for $y$ in order to put the equation in the $y = mx + b$ form.

### Suggestions
Review solving an equation.
Solve.
**1.** $2x - 6 = 8$   **2.** $3x - 5 = 16$   **3.** $4 - x = 3$
    $x = 7$         $x = 7$        $x = 1$
Now show that solving an equation like $4x - 5y = 10$ for $y$ is done in much the same way as the practice above.
Solve $4x - 5y = 10$.
$$-4x + 4x - 5y = -4x + 10 \quad \blacktriangleleft \; Add \; -4x \; to \; each$$
$$-5y = -4x + 10 \qquad\qquad side$$
$$y = \frac{-4x + 10}{-5} \quad \blacktriangleleft \; Divide \; each \; side \; by \; -5$$
$$y = \frac{-4x}{-5} + \frac{10}{-5} = \frac{4}{5}x - 2$$

### For Additional Chalkboard Work
Solve for $y$. Then graph.
**1.** $3x - 7y = 21$      **2.** $4x - y = -2$
    $y = \frac{3}{7}x - 3$          $y = 4x + 2$
**3.** $-3y - x = 9$    $\frac{1}{-3}x - 3$

### Enrichment
John owns a piece of land worth $10,000. He sells it to Jane at a 10% profit based on the worth of the land. She sells the land back to him at a 10% loss. Who comes out ahead on the deal and by how much?
*Solution:* John sells the land to Jane for $11,000. She sells it back to him for $9,900. Since $11,000 - $9,900 = $1,100, John is ahead by $1,100.

### Cumulative Review
**1.** Factor       **2.** Simplify
   $2x^2 + 5x - 12$.     $5x - (4 - 3x) + 2 - x$.
   $(2x - 3)(x + 4)$     Then evaluate for
                $x = -3$.
                $7x - 2; -23$

## 11.9 Graphing Vertical Lines (page 317)

### Suggestions

Begin with the following review of the graphs of horizontal lines.

Graph the line whose equation is given.

**1.** $y = 5$    **2.** $y - 6 = -4$    **3.** $2y - 6 = -12$

Plot at least two points for each; draw a line through the two points.

$(3, 5), (1, 5)$   $(0, 2), (4, 2)$   $(0, -3), (-4, -3)$

Show that the slope of each of the three horizontal lines is zero; the $y$-coordinate is always the same.

Now students can see the similarity for graphing vertical lines. In Example 1 on page 317, the graph of $x = 2$ is a vertical line. The slope for any two points is undefined. The $x$-coordinate is always the same.

A parallel discussion of horizontal and vertical lines might help.

| To graph $y = 4$: | To graph $x = 4$: |
|---|---|
| plot two points with the same $y$-coordinate, say $(3, 4)$ and $(5, 4)$. Join with a horizontal line. | plot two points with the same $x$-coordinate, say $(4, 3)$ and $(4, 5)$. Join with a vertical line. |

### Cumulative Review

**1.** Write the equation of the line containing $A(0, 1)$ and $B(3, 3)$.

$y = \frac{2}{3}x + 1$

**2.** Graph $2y - 6 = 8$.

a horizontal line; $(1, 7), (3, 7)$

**3.** Graph $3x - 2y = 4$.

$y$-intercept $-2$, slope $\frac{3}{2}$

## 11.10 Graphing Inequalities (page 319)

### Vocabulary

open half plane

### Background

Students have already graphed solutions of linear inequalities on a number line. This lesson is an extension of that concept to two dimensions. The following chapter includes an extension of this lesson to the solution of simultaneous inequalities as well as a special lesson on business applications.

### Suggestions

Pave the way by briefly reviewing the solution of linear inequalities using a number line.

Solve each inequality. Graph each solution on a number line.

**1.** $3x + 2 < 17$
$x < 5$
points to the left of 5

**2.** $-2x - 4 \geq 8$
$x \leq -6$
$-6$ and points to the left of $-6$

**3.** $4x - 5 < 5x + 2$
$x > -7$
points to the right of $-7$

Point out that in graphing inequalities on a plane the dashed line is comparable to the open circle on the number line, and the solid line is comparable to the filled-in circle on the number line.

### Errors That Students Make

As with inequalities on the number line, students tend to forget to REVERSE the inequality sign when dividing each side by a negative number. Thus, in Example 3, students will get as far as $-2y \geq 1x - 8$, divide by $-2$, and get

$y \geq -\frac{1}{2}x + 4$ instead of the correct result,

$y \leq -\frac{1}{2}x + 4$.

### For Additional Chalkboard Work

Solve for $y$, then graph.

**1.** $3x - 4y < 20$
$y > \frac{3}{4}x - 5$

**2.** $x - y \geq 5$
$y \leq x - 5$

**3.** $2x > -2y + 8$
$-x + 4 < y$, or $y > -x + 4$

## Applications: Linear Models (page 323)

### Suggestions

Students are not used to using different scales for the axes. Point out that with large numbers it will be necessary to use a scale unless you have access to graph paper "a mile long." For example, in the table for the first practice exercise on page 324, the numbers are in the thousands. You might let each block represent 1,000 on the y-axis.

It might be better to begin with sets of points rather than word problems.

### Alternate Example

Ask the students to plot the following points: $A(0,5)$, $B(1,7)$, $C(3,10)$, $D(4,13)$, $E(5,16)$, $F(-1,3)$, and $G(-4,-3)$. Draw a line through as many points as possible. Note that the line contains all points except for $C$ and $E$. The line is then a LINEAR MODEL for the set of points $A$, $B$, $D$, $F$, and $G$. The equation of this linear model is $y = 2x + 5$.

Use the equation to predict $y$ for $x = 10$. The value of $y$ for $x = 10$ is 25. Thus, you can use this alternate example as a discovery approach to the lesson.

### For Additional Chalkboard Work

Use the following set of points to answer questions 1–3: $A(-2,11)$, $B(-1,7)$, $C(0,3)$, $D(2,-6)$, $E(3,-9)$, $F(4,-5)$, and $G(5,-17)$.
1. Draw a line through as many points as possible. Which points are not on the line? *D, F*
2. Write an equation for the linear model. $y = -4x + 3$
3. Predict $y$ if $x = 10$. $-37$

### Cumulative Review

Solve.
1. $|3x - 6| = 15$     2. $x^2 - 5x = 0$     3. $\dfrac{4}{x} = \dfrac{x}{25}$
    7, $-3$                     0, 5                     10, $-10$

## Word Problem Formulation

Mr. Hennesey has a business meeting in Chicago. His plane will land at O'Hare Airport at 10:00 AM. His meeting is at noon. He decides to rent a car when he lands. He checks on a road-map and finds that the meeting place is 100 miles from the airport. Using the math model for distance, $d = rt$, and assuming his rate to be 50 mph, he solves the equation and gets 2 hours for a solution. He will be on time! Is the solution of the equation really the solution of the problem of being on time for the meeting?

Questions for discussion:
1. Even though the road map indicates that the distance is 100 miles, will that be the exact distance he must travel?
2. Can the man get off the plane, collect his luggage, rent a car, pick up the car, and then instantly drive 50 mph?
3. Can he drive continually at a rate of 50 mph?
4. Does the meeting occur in the center of town where he will encounter heavy traffic?

## Geometry Review (page 325)

### Background

Many state proficiency tests for basic skills require students to know the properties of parallel lines relative to alternate interior and corresponding angles.

### Suggestions

Stress recognition of the two types of angles. Alternate interior angles form either a *z* or a *backward z*. For corresponding angles, think of *two* sets of axes. Angles 1 and 2 are each in the same (corresponding) third quadrant. The use of equations to find the angle measures provides a nice link between algebra and geometry.

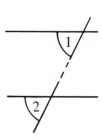

## Chapter
## and Lesson
## Commentaries

### Chapter 12   Systems of Equations

#### Objectives

To solve a system of equations in two variables
  by graphing
To determine whether a system is independent and
  consistent, inconsistent, or dependent and
  consistent
To solve a system of inequalities in two variables
  by graphing
To solve a system of equations by substitution
To solve a system of equations by the addition
  method
To solve problems involving systems of equations

#### Overview

This chapter develops three methods for solving
linear systems of equations. In the first lesson,
systems of equations are solved by graphing.
From the graph and from comparing $m$ and $b$ when
the equations are in $y = mx + b$ form, students
determine if a system is independent and consistent,
inconsistent, or consistent and dependent. The
second lesson is an extension of the first lesson, but
in this lesson systems of inequalities are solved by
graphing.

#### Overview   (continued)

In the third lesson, the systems of equations are
solved by the substitution method. The fifth and
sixth lessons use the addition and multiplication-
addition methods to solve systems of equations.
The remaining lessons involve applications of the
methods for solving systems of equations, including
coin problems, mixture problems, and digit
problems.

#### Special Features

### Career: Hotel and Motel Management (page 329)

Hotel management is a new and expanding career.
Many colleges offer programs leading to a degree
in this specialty. Some colleges require more
mathematics beyond high school, such as linear
programming, introduction to computer
programming, and statistics.

   This page also provides students with practice in
problem solving formulation. The open-endedness
of the project provides many options of approach.

## 12.1 Graphing Systems of Equations (page 330)

### Vocabulary
consistent, dependent, inconsistent, independent, parallel, system of equations

### Suggestions
Review graphing lines by the slope-intercept method before starting the lesson.

Stress that when students check the solution to a system of equations, they should substitute the $x$ and $y$ values in the original system of equations.

You should point out that if two lines are parallel, they have the same slope but different $y$-intercepts. Two lines that coincide have the same slope and the same $y$-intercept. Two lines that intersect have different slopes, and the $y$-intercept may be the same or different.

### For Additional Chalkboard Work
Solve by graphing. Check the solution.

**1.** $y = 3x + 2$
 $y = 2x - 1$
 $(-3, -7)$

**2.** $x + y = 8$
 $y = 2x + 2$
 $(2, 6)$

**3.** $3x - 2y = 4$
 $x - y = 1$
 $(2, 1)$

### Cumulative Review

**1.** Simplify $\dfrac{4}{x^2 - 36} - \dfrac{5}{6 - x}$  $\dfrac{5x + 34}{(x + 6)(x - 6)}$

**2.** Simplify $(2x^3 y)^2$.  $4x^6 y^2$

**3.** Find the value of $3x - 2y$ if $x = -1$, $y = 3$.
 $-9$

## 12.2 Graphing Systems of Inequalities (page 333)

### Suggestions
First review graphing one inequality before attempting to have students graph two inequalities on the same set of axes. The use of colored pencils might help some students to see the double-shaded region that is the solution of both inequalities.

### Errors That Students Make
A common error that students make when working with inequalities is to forget to reverse the order of the inequality when they divide or multiply by a negative number. This is why it is extremely important that when students pick a point that is in the double-shaded region, they show that its coordinates satisfy both of the original inequalities.

### For Additional Chalkboard Work
Solve by graphing.   See below.

**1.** $y \le -1$
 $x > 2$

**2.** $y \ge 3x - 1$
 $x - 2y > 3$

**3.** $x < -1$
 $y - x \ge 2$

### Cumulative Review

**1.** Simplify
 $\dfrac{x^2 - 2x - 15}{x^2 - 25} \cdot \dfrac{x + 3}{x + 5}$

**2.** Solve
 $9 - 4(2 - y) = 7y - (3 + y)$.   2

**3.** Simplify
 $\dfrac{3x^3 y^5}{9xy^7} \cdot \dfrac{x^2}{3y^2}$

1.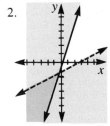

2.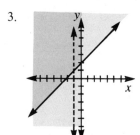

3.

## 12.3 The Substitution Method (page 335)

### Vocabulary
substitution

### Suggestions
The substitution method for solving systems of equations is most advantageous when one of the equations is already solved for either one of the variables. In Example 2, neither equation is solved for $x$ or $y$. However, point out that it is easiest to solve for $y$ in $3x - y = 5$ since the coefficient of $y$ is $-1$, and the coefficients of all the other variables are not 1 or $-1$.

Use the calculator actively to have students practice the substitution method with decimals.

### Errors That Students Make
When doing Example 2, students tend to forget to change all signs to the opposite sign when dividing by $-1$. For example, some students may write

$$3x - 1y = 5$$
$$-1y = -3x + 5$$
$$y = 3x + 5$$

### For Additional Chalkboard Work
Solve by substitution.

1. $y = x + 1$
   $x + y = 9$
   $(4, 5)$

2. $x + y = 7$
   $2x - 3y = -1$
   $(4, 3)$

3. $3x - y = 1$
   $10x - 2y = 4$
   $\left(\frac{1}{2}, \frac{1}{2}\right)$

### Cumulative Review
1. Find the slope of $\overleftrightarrow{PQ}$ for $P(3, 4)$ and $Q(7, 6)$.   $\frac{1}{2}$
2. Factor $2x^3 - 18x$ completely.
   $2x(x + 3)(x - 3)$

## 12.4 Problem Solving (page 338)

### Suggestions
Using a system of equations to solve a problem is usually helpful when a total amount is given. Once the variables have been represented, it is easy to write the two equations in terms of the two variables.

For Exercises 2, 4, and 12 on page 340, if $x$ and $y$ are used for the variable, then one of the equations will involve $x - y$. Point out that $x$ must be the greater amount for these problems to work out.

### For Additional Chalkboard Work
Solve each problem.

1. Les is 22 years older than his son Matthew. The sum of their ages is 40. Find the age of each.   9, 31

2. The length of a rectangle is 3 m less than twice the width. The perimeter is 54 m. Find the length and width of the rectangle.   17, 10

3. The difference between two numbers is 7. Twice the first number, increased by 3 times the second number, is 29. Find the two numbers. 10, 3

### Cumulative Review
1. For what value of $x$ is $\dfrac{5 + x}{x - 5}$ undefined?   5
2. Multiply $(3x - 2)(2x + 5)$.   $6x^2 + 11x - 10$
3. Solve $2x^2 + 9x - 5 = 0$.   $\frac{1}{2}, -5$

## 12.5 The Addition Method (page 341)

### Background
In this lesson the systems of equations are restricted to the type in which the coefficients of one pair of like terms are additive inverses of each other.

### Suggestions
First have students solve the system of equations in Example 1 by using the substitution method. Then carefully discuss Example 1 and show that the solutions are the same.

### For Additional Chalkboard Work
Solve by addition

1. $x + y = 7$
   $x - y = 3$
   $(5, 2)$

2. $5x - 2y = 10$
   $3x + 2y = 6$
   $(2, 0)$

3. $4x + 5y = -4$
   $-4x - 3y = 2$
   $\left(\frac{1}{4}, -1\right)$

4. $2x + 3y = 8$
   $3x - 3y = 7$
   $\left(3, \frac{2}{3}\right)$

5. $x + 2y = 15$
   $3x = 2y + 5$
   $(5, 5)$

6. $3y = -x + 2$
   $2x - 3y = 4$
   $(2, 0)$

### Cumulative Review
Graph

1. $y = -2$
   line through points
   $(1, -2), (3, -2)$

2. $y = \frac{1}{3}x - 1$
   line through points
   $(0, -1), (3, 0)$

3. $2x - 3y = 6$
   line through points
   $(3, 0), (0, -2)$

## 12.6 The Multiplication-Addition Method (page 344)

### Background
The procedures in the previous lesson motivate students to rewrite the equations in the lesson so that the coefficients in one pair of terms are additive inverses of each other.

### Suggestions
Remind students that when they multiply an equation by a number, they must multiply each term of the equation by the number. Give students practice in multiplying equations by positive and negative integers.

   In Examples 2 and 3 it is necessary to multiply both equations. To eliminate the $y$ terms in Example 2, each equation must be multiplied since 2 is not a factor of 7.

### For Additional Chalkboard Work
Solve by addition.

1. $3x + 2y = 8$
   $2x - y = 3$
   $(2, 1)$

2. $10x - 3y = 7$
   $5x + 2y = 7$
   $(1, 1)$

3. $2x + 3y = 4$
   $x - 6y = -13$
   $(-1, 2)$

4. $3x - 2y = 13$
   $5x + 4y = 29$
   $(5, 1)$

5. $2x + 5y = 1$
   $3x - 7y = 16$
   $(3, -1)$

6. $4x = -3y + 6$
   $3x + 5y = 10$
   $(0, 2)$

### Cumulative Review

1. Solve by graphing:
   $2x - 3y = 6$
   $3x + y = 9.$
   $(3, 0)$

2. Solve
   $5x = x^2 - 24.$
   $8, -3$

3. Solve
   $5x - 7 > 2x + 2.$
   $x > 3$

## 12.7 Coin and Mixture Problems (page 347)

### Suggestions

Before the lesson have students represent the total value in cents of various amounts of coins. For example, what is the total value in cents of 3 quarters and 4 nickels, 6 half-dollars and three dimes, etc. Then have students represent the total value in cents of $d$ dimes and $n$ nickels, or $q$ quarters and $h$ half-dollars. This will lead in very nicely to Examples 1 and 2.

For Example 3 students must realize that if $1.50 is 150¢ and $2.00 is 200¢, then $550 is 55,000¢.

After representing the variable and writing the equations, the solution of the problem can be done by either the substitution or addition methods.

### For Additional Chalkboard Work

Solve each problem.

1. Joshua has 1 more nickel than he has dimes. The total value is $1.85. How many coins of each type does he have?   13 nickels, 12 dimes
2. A stationery salesclerk wants to mix greeting cards costing $.40 each with cards costing $.60 each. A box of 10 of these cards will sell for $4.40. How many cards at each price should be in the box?   8 at $.40, 2 at $.60

### Cumulative Review

1. Find 3 consecutive integers whose sum is 42.   13, 14, 15
2. Find 2 positive consecutive odd integers whose product is 63.   7, 9

## 12.8 Digit Problems (page 351)

### Vocabulary
digit

### Suggestions

Start the lesson by having students identify the tens digit, units digit, sum of digits, and number formed by reversing the digits for numbers such as 83 and 25. This will help lead to the idea that if $t$ represents the tens digit and $u$ represents the units digit, then

$10t + u$ represents the number,

$10u + t$ represents the number with the digits reversed, and

$t + u$ represents the sum of the digits. Suggest that students memorize these representations.

Solving systems of equations with $t$ and $u$ as the variable might confuse some students. Give them practice on setting up and solving the following systems before starting Example 3. Solve.

1. $t + u = 12$
   $10t + u = 10u + t - 54$
   $t = 3, u = 9$
2. $3t + u = 15$
   $10u + t - 63 = 10t + u$
   $t = 2, u = 9$

### For Additional Chalkboard Work

Solve each problem.

1. The sum of the digits of a two-digit number is 12. If the digits are reversed, the new number is 36 more than the original number. Find the original number.   48
2. The units digit of a two-digit number is 5 less than the tens digit. If the digits are reversed, then new number is 45 less than the original number. Find the original number.   72

## Computer Activities: Graphing Equations of Lines (page 360)

Graphing techniques vary from one computer to another. However, discrete points suggest the shape of the graph. TAB is used to do this since it will work on any computer.

# Chapter and Lesson Commentaries

## Chapter 13    Functions and Relations

### Objectives

To determine the domain and the range of a relation

To graph a relation on a coordinate plane

To determine whether or not a relation is a function

To determine if a function is a linear function or a constant linear function

To use the standard notation, $f(x)$, for a function

To determine if a relation expresses variation, direct or inverse

To determine the constant of proportionality in a variation, direct or inverse

### Overview

The chapter opens with a discussion of relations in general and then proceeds to the special relation, the function. Students learn to pick out the domain and range of relations (functions). The special linear and constant linear functions are developed, applying the concept of the $y = mx + b$ form of the equation of a line. The all-important $f(x)$ notation is developed: given $f(x)$ and a domain, students are then asked to write the range of a relation. Two very special types of relations are then studied, inverse and direct variation. These variations give rise to many practical applications. The chapter closes with a lesson on age problems. Since this lesson is not directly related to the general objectives and lessons of the chapter, it is *not* included in the chapter review and test.

## Non-Routine Problem Solving (page 364)

The non-routine problems relate specifically to two topics covered in the chapter. Problem 1 provides a problem that can be solved using variation. Problem 2 is an age problem that provides an excellent opportunity for fraction review as well as an interesting situation for problem solving review.

## 13.1 Relations and Functions (page 365)

### Vocabulary
domain, function, range, relation

### Background
Students already know how to plot points and read coordinates of points on a rectangular coordinate system.

### Suggestions
Review graphing or plotting points.
1. Plot $R$, the following set of points.
   $R = \{(-5,4), (0,-2), (2,4), (6,7)\}$
2. Plot $T$, the following set of points.
   $T = \{(-3,-4), (-1,2), (0,0), (5,1), (8,-2)\}$

Then use the second example above as an alternate introduction to the lesson. The brackets or braces around the points indicate *set*. The set of all first coordinates of $T$, $\{-3,-1,0,5,8\}$, is called the *domain* of the relation. The set of all second coordinates of $T$, $\{-4,2,0,1,-2\}$, is called the *range* of the relation. Then discuss how a function is a relation, but a relation is not always a function.

### Errors That Students Make
Students tend to think that repetition of $y$ values is not permissible in a function. Stress that it is the first member of an ordered pair that must not repeat.

### For Additional Chalkboard Work
Which relations are NOT functions?
1. $\{(-4,5), (7,5), (4,-3)\}$
2. $\{(-6,3), (-4,-1), (-6,-4), (8,0)\}$
3. $\{(-4,2), (0,2), (3,2), (8,2)\}$
4. $\{(3,-4), (3,0), (3,-5), (3,-8)\}$
   2 and 4 are not functions

### Cumulative Review
Graph each line.
1. $y = -\frac{2}{3}x + 4$
   $y$-int. is 4, slope
   is $-\frac{2}{3}$

2. $y = 4$
   horizontal line
   through $(0,4)$

3. $x = -6$
   vertical line
   through $(-6,0)$

## 13.2 Identifying Functions (page 368)

### Vocabulary
constant linear function, linear function, vertical-line test

### Background
Students have already used the algebraic way of determining whether a relation is a function; for each first member there is only one second member. This lesson provides an easy *visual* test, the vertical-line test. However, it is also the only test possible if the student is provided only with the graph of a relation and none of the actual ordered pairs.

### Suggestions
Before proceeding to the discussion of the linear functions and constant linear functions, review the graphing of a line if given the equation in $y = mx + b$ form.
Graph each. Name the slope and $y$-intercept of each.
1. $y = -\frac{3}{4}x + 5$
   slope $-\frac{3}{4}$
   $y$-int. 5

2. $y = -4$
   Slope $\frac{0}{1}$ or 0
   $y$-int. $-4$;
   horizontal line

3. $x = 5$
   no $y$-int.
   undefined slope
   vertical line

### For Additional Chalkboard Work
Graph. Which relations are functions? Which functions are linear functions? Which functions are constant linear functions?
1. $y = \frac{3}{5}x + 1$
   linear function

2. $x = -3$
   not a function

3. $y + 2 = 4$
   constant linear function

## Enrichment

If time permits, teach the simple quadratic relations.
Graph $y = x^2$. Is this a function?
*Solution:* Find some ordered pairs. Choose the
following values for $x$: $-3, -2, -1, 0, 1, 2, 3$.
Find the $y$-coordinates. This will give the following
ordered pairs: $(-3,9), (-2,4), (-1,1), (0,0),$
$(1,1), (2,4), (3,9)$. Draw a smooth curve through
the points. The graph is called a parabola. It is a
function. Then have students graph each of the
following. Which are functions?

**1.** $y = x^2 + 2$　　**2.** $y = -x^2$　　**3.** $x = y^2$

*Solution:*

**1.** $(-3,11), (-2,6), (-1,3), (0,2), (1,3),$
$(2,6), (3,11)$; function
**2.** $(-3,-9), (-2,-4), (0,0), (2,-4), (3,-9)$;
function
**3.** $(0,0), (1,1), (1,-1), (4,2), (4,-2), (9,3),$
$(9,-3)$; this is a horizontal parabola and not
a function.

## Cumulative Review

Solve.

**1.** $x + y = 8$　　　　**2.** $3x - (4 - x) = 2x - 8$
　　$x - y = 4$　　　　　　　　　　　　　　　　$-2$
　　$(6,2)$
**3.** $x^2 - 4x = 0$　$0, 4$

## 13.3 Standard Function Notation
## (page 371)

### Suggestions

At first students have difficulty in accepting the
meaning of $f(x)$ as other than $f$ times $x$, since they
have been repeatedly told that parentheses mean
*multiplication.*

　　Calculator Activities give practice in function
notation with decimals in the domain.

### Errors That Students Make

An error sometimes arises in finding $f(-2)$ for
$f(x) = -x^2$. Students erroneously write $-(-2) = 2$
and then square the result to get 4. Stress that
$-x^2$ means $-1(x)^2$; the $-1$ is not being squared.
So $f(-2) = -1(-2)^2 = -1(4) = -4$. Use the
following exercises to stress the concept.

### For Additional Chalkboard Work

Find each indicated value.

**1.** For $f(x) = -x^3$,　　　　**2.** *For* $g(x) = -x^4$,
　　find $f(-3)$.　27　　　　　　find $g(-1)$.　$-1$
**3.** For $f(x) = -2x^5$,
　　find $f(-2)$.　64

### Enrichment

Challenge some of the sharper students in the class
with the following question.
If $f(x) = x^3$ then $f(-2) = -8$
　　　　　　　　　　　　　　　　　opposites.
　　　　　　$f(2) = 8$
It appears that in general $f(-x) = -f(x)$. Can you
come up with a counterexample of this? That is,
can you show at least one situation in which
$f(-x) \neq -f(x)$?
*Solution:* $f(x) = x^2$: $f(-2) = 4$
　　　　　　　　　　　　　　　　　NOT opposites.
　　　　　　$f(2) = 4$
Thus, in general, it is NOT true that $f(-x) = -f(x)$.

### Cumulative Review

**1.** Factor $3x^2 + 5xy - 2y^2$.　$(3x - y)(x + 2y)$
**2.** Evaluate $-3x^5y^2$ for $x = -1, y = -3$　27
**3.** Solve $x^3 - 64x = 0$.　$0, 8, -8$

## 13.4 Direct Variation (page 373)

### Vocabulary
constant of variation, direct variation

### Suggestions
You may also have students find the ratio $\frac{x}{y}$ instead of $\frac{y}{x}$ in the prose above Example 1. So, $\frac{x}{y} = \frac{1}{10}$. By using the product of the means equal to the product of the extremes, $y \cdot 1 = x \cdot 10$, or $y = 10x$, the same equation as derived on page 373. From Example 1, show that the constant of variation may be a negative number. From the Oral Exercises, show that the constant of variation may be a fraction.

### For Additional Chalkboard Work
In each of the following, $y$ varies directly as $x$.
1. $y$ is 48 when $x$ is 6. Find $y$ when $x$ is 5.   40
2. $y$ is 40 when $x$ is 8. Find $y$ when $x$ is $-3$.  $-15$
3. $y$ is 7 when $x$ is 21. Find $y$ when $x$ is 39.   13

### Cumulative Review
Solve.
1. $\frac{4}{5} + \frac{3}{x} = 2$   $2\frac{1}{2}$      2. $3x - 2y = 11$
$x - y = 3$   $(5, 2)$
3. $7(2x - 2) - 5x = 4x + 1$   3

## 13.5 Inverse Variation (page 377)

### Vocabulary
inverse variation

### Suggestions
In inverse variation, the product of two factors is always a constant. This may be a good time to review multiplying integers.
Multiply.
1. $-2 \cdot -18$   36      2. $9 \cdot 4$   36
3. $-6 \cdot 6$   $-36$      4. $12 \cdot -3$   $-36$
The multiplication problems above will help show that the product of 2 positive or 2 negative integers is a positive integer and that the product of a positive and a negative integer is a negative integer.

   The **Non-Routine Problem** can be easily solved by using a system of equations. Let $x$ = number of bicycles and $y$ = number of tricycles. Then

$$2x + 3y = 186 \text{ and } 2x + 2y = 144$$

### Cumulative Review
1. Divide $\frac{x - 8}{4} \div \frac{2x - 16}{12}$.   $\frac{3}{2}$

2. Simplify $\left| -8\frac{2}{3} \right| + \left| -5\frac{1}{2} \right|$.   $14\frac{1}{6}$

3. Simplify $\frac{2x^9 y}{16x^7 y^6}$.   $\frac{x^2}{8y^5}$

## 13.6 Age Problems (page 382)

### Suggestions

As an alternative to using two variables to solve age problems, you may use one variable, as shown below to solve Example 1.

|  | Now | 6 years from now |
|---|---|---|
| Meg's age | $4x$ | $4x + 6$ |
| Carlo's age | $x$ | $x + 6$ |

Let $x$ = Carlo's age now.
Then $4x$ = Meg's age now.
Her age 6 years from now will be twice Carlo's age in 6 years.

$4x + 6 = 2(x + 6)$
$4x + 6 = 2x + 12$
$2x + 6 = 12$ ◀ *Subtract 2x from each side.*
$2x = 6$ ◀ *Subtract 6 from each side.*
$x = 3$ ◀ *Carlo's age now*
$4x = 12$ ◀ *Meg's age now*

### For Additional Chalkboard Work

For each problem, find the age of each person now.
1. Paul is twice as old as Jane. Five years ago, he was 3 times as old as she was then.
   Paul, 20; Jane, 10
2. Mike is 8 years younger than Cheryl. Five years from now, she will be twice as old as he will be then.   Mike, 3; Cheryl, 11

### Cumulative Review

Simplify.
1. $(2x^3)(-5x^5)$
   $-10x^8$
2. $(2x^2y^3)^3$
   $8x^6y^9$
3. $\dfrac{x^2 - 16}{x^2 + 6x + 8}$   $\dfrac{x - 4}{x + 2}$

### Computer Activities: Domain and Range (page 386)

This program is actually working with ordered pairs. For advanced students, the program can be altered to accept $(X, Y)$ pairs using double subscripted arrays.

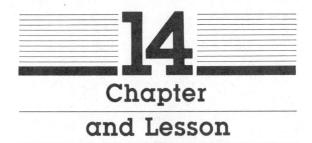

# 14
# Chapter
# and Lesson
# Commentaries

## Chapter 14   Radicals

### Objectives

To write a repeating decimal in fraction form
To find the principal square root of a perfect square
To find the approximate square root of a number in a square root table
To approximate the square root of a number using the divide-and-average method
To solve area problems that lead to the use of square roots
To simplify an expression involving square roots
To simplify square roots in which radicands contain variables
To simplify expressions containing radicals by combining like terms
To simplify products of radical expressions
To simplify quotients that involve radicals
To find the length of a side of a right triangle by using the Pythagorean property
To determine whether a triangle is a right triangle given the lengths of the three sides of the triangle

### Overview

In the first lesson of this chapter, repeating decimals are written as fractions. Then the concept of square root is introduced, and it is shown that the square roots of many whole numbers are irrational numbers.

Other lessons in the chapter involve operations with radical expressions. A lesson on the Pythagorean property is also included. A lesson on zero and negative integral exponents with an application to scientific notation is presented. The chapter closes with an applications lesson on finding the standard deviation.

## Special Features

## Leonhard Euler (page 388)

The first diagram represents the seven bridges of Königsberg joining the shores and two islands in the river of the old city of Königsberg. The mathematician Euler constructed an analog of this problem, and this is represented in the second diagram.

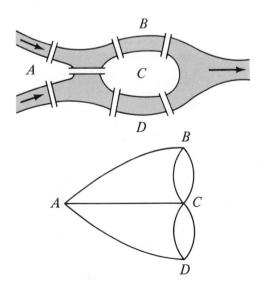

The ancient problem of the time was to try to walk across all seven bridges without crossing any bridge more than once. Euler represented each area of land by a point, which is called a vertex, and each bridge by an arc. He called his diagram a network and studied which kinds of networks can be traced without backtracking (drawing over an arc twice). A vertex is classified as odd or even based on the number of arcs emanating from it. An odd vertex has an odd number of arcs, and an even vertex has an even number of arcs. Euler discovered that a network can be traced without backtracking if and only if it has no odd vertices or exactly two odd vertices. The Königsberg bridge problem has no solution since there are four odd vertices.

Below are some additional network problems. Which of the following networks is traceable without backtracking?

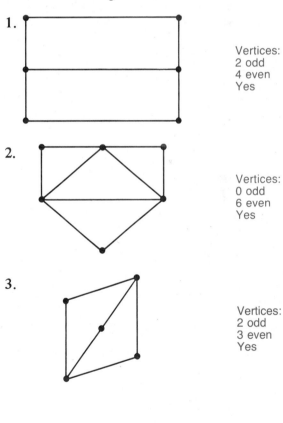

1.

Vertices:
2 odd
4 even
Yes

2.

Vertices:
0 odd
6 even
Yes

3.

Vertices:
2 odd
3 even
Yes

## 14.1 Repeating Decimals as Fractions (page 389)

### Suggestions

Review the idea that every rational number can be written in the form of a repeating or terminating decimal. Point out that in this lesson the student will learn to reverse the process and write a repeating decimal in the form of a fraction. Also review the concept that decimals that do not repeat or terminate are irrational. Students will see in the next lesson that certain square roots are also irrational.

### Errors That Students Make

When finding the fraction for a repeating decimal in which two digits repeat, some students might not remember to multiply by 100. Thus in Example 3, the student might do the following:

$$10n = 16.464646\ldots$$
$$-1n = -1.646464\ldots$$
$$\overline{\phantom{-}9n = 14.818182\ldots}$$

Then upon dividing $14.818182\ldots/9$ the student will get $1.646464\ldots$ as the quotient.

### For Additional Chalkboard Work

Find the fraction for each repeating decimal.

1. $0.\overline{4}$     2. $0.3\overline{88}$     3. $1.\overline{78}$

   $\frac{4}{9}$          $\frac{7}{18}$        $\frac{59}{33}$

### Cumulative Review

1. Solve

$$\frac{a+4}{6} - \frac{a+2}{2} = 4.$$
$$-13$$

2. Simplify

$$\frac{c-3}{c+1} - \frac{c-7}{1-c^2}.$$
$$\frac{c-4}{c-1}$$

3. Solve

$3x - 4y = 7$
$2x + 3y = 16.$    $(5,2)$

## 14.2 The Square Root of a Number (page 391)

### Vocabulary

perfect square, principal square root, radical, radicand, square root

### Suggestions

It may be helpful for future lessons if you have students make a list of the perfect squares from 1 to 400 and their principal square roots. For example:

| Perfect Square | Principal Square Root |
|:---:|:---:|
| 1 | 1 |
| 4 | 2 |
| 9 | 3 |
| 16 | 4 |
| $\vdots$ | $\vdots$ |
| 361 | 19 |
| 400 | 20 |

### For Additional Chalkboard Work

Find each of the following principal square roots.

1. $\sqrt{36}$   6     2. $\sqrt{121}$   11     3. $\sqrt{144}$   12

4. $\sqrt{9/16}$      5. $\sqrt{1/25}$      6. $\sqrt{100/81}$

  $\frac{3}{4}$           $\frac{1}{5}$         $\frac{10}{9}$

### Cumulative Review

1. Solve

$$\frac{a}{2} + \frac{a}{3} = 5.$$   6

2. For $f(x) = x^2 + 5$, find $f(-2)$.   9

3. Simplify

$$\frac{9x^5y}{12xy^3} \cdot \frac{3x^4}{4y^2}$$

## 14.3 Approximating Square Roots (page 394)

### Vocabulary
divide-and-average method

### Suggestions
As an alternate method for approximating square roots to the nearest tenth, you might try the "guess" method. The example on p. 394 might be done as follows.

*Step 1*  Guess the square root. Since $7 \cdot 7 = 49$ and $8 \cdot 8 = 64$, $\sqrt{58}$ is between 7 and 8. Guess 7.5 since 58 is about halfway between 49 and 64.

*Step 2*  Multiply 7.5 by itself.

$$
\begin{array}{r}
7.5 \\
\times\ 7.5 \\
\hline
37\ 5 \\
525\phantom{0} \\
\hline
56.25
\end{array}
$$

*Step 3*  Since $56.25 < 58$, try multiplying 7.6 by itself.

$$
\begin{array}{r}
7.6 \\
\times\ 7.6 \\
\hline
45\ 6 \\
532\phantom{0} \\
\hline
57.76
\end{array}
$$

The difference between 58 and 57.76 is 0.24.

*Step 4*  Since $57.76 < 58$, try multiplying 7.7 by itself.

$$
\begin{array}{r}
7.7 \\
\times\ 7.7 \\
\hline
53\ 9 \\
539\phantom{0} \\
\hline
59.29
\end{array}
$$

The difference between 59.29 and 58 is 1.29. Since $0.24 < 1.29$, the square root is closer to 7.6. Thus, $\sqrt{58} \doteq 7.6$.

### For Additional Chalkboard Work
Approximate each of the following to the nearest tenth.

**1.** $\sqrt{29}$  5.4   **2.** $\sqrt{68}$  8.2   **3.** $\sqrt{96}$  9.8
**4.** $\sqrt{7}$  2.6   **5.** $\sqrt{140}$  11.8  **6.** $\sqrt{375}$  19.4

## 14.4 Simplifying Square Roots (page 396)

### Suggestions
Review factoring a number into primes before starting Example 4. Students seem to have more success with the first method of Example 4 than with the second method, in which they must find the greatest perfect square factor.

### Errors That Students Make
When students try to simplify 48 by the second method in Example 4, they don't always find the greatest perfect square factor. For example students might simplify as follows:

$$
\begin{aligned}
\sqrt{48} &= \sqrt{4 \cdot 12} \\
&= \sqrt{4}\,\sqrt{12} \\
&= 2\sqrt{12}
\end{aligned}
$$

Then they think that the answer is $2\sqrt{12}$ and don't simplify the answer further.

### For Additional Chalkboard Work
Simplify each of the following.

**1.** $\sqrt{16 \cdot 9}$  12  **2.** $\sqrt{18}$  $3\sqrt{2}$  **3.** $\sqrt{32}$  $4\sqrt{2}$
**4.** $3\sqrt{12}$  $6\sqrt{3}$  **5.** $2\sqrt{48}$  $8\sqrt{3}$  **6.** $5\sqrt{50}$  $25\sqrt{2}$

### Cumulative Review
**1.** Multiply
$(3x + 5)(2x - 1)$.
$6x^2 + 7x - 5$

**2.** Solve
$y = 2x$
$x + 3y = 14$.
$(2, 4)$

**3.** Solve
$\dfrac{4}{a} = \dfrac{a}{9}$.
$6, -6$

## 14.5 Radicals with Variables (page 399)

### Background
When dealing with square roots in an elementary algebra course, it is only necessary to discuss that the square root of a negative number is not a real number. However, you may want to mention that if a number is not a real number, then it is an imaginary number. Imaginary numbers are dealt with in the Algebra 2 course.

### Suggestions
At this time it might be necessary to review the operations with inequalities, especially when multiplying or dividing by a negative number. This will be especially helpful before doing Exercise 8 on page 401.

For students who have difficulty in simplifying radicals containing variables with even exponents, try the following technique:

$$\sqrt{a^6} = \sqrt{a \cdot a \cdot a \cdot a \cdot a \cdot a}$$
$$= \underbrace{\sqrt{a} \cdot \sqrt{a}}_{a} \cdot \underbrace{\sqrt{a} \cdot \sqrt{a}}_{a} \cdot \underbrace{\sqrt{a} \cdot \sqrt{a}}_{a}$$
$$= a^3$$

### For Additional Chalkboard Work
Simplify each of the following.

1. $\sqrt{y^{12}}$   $y^6$
2. $\sqrt{25b^8}$   $5b^4$
3. $\sqrt{9m^4 n^{10}}$   $3m^2 n^5$
4. $-\sqrt{81p^{18}q^{20}r^{22}}$   $-9p^9 q^{10} r^{11}$
5. $\sqrt{27a^{16}b^4}$   $3a^8 b^2 \sqrt{3}$
6. $\sqrt{12x^6 y^{14} z^2}$   $2x^3 y^7 z \sqrt{3}$

### Cumulative Review
1. Factor
   $x^2 - 36$.
   $(x + 6)(x - 6)$
2. Solve
   $3x + 12 = 5x - 2$.
   $7$
3. Solve
   $|a| = 9$.   $9, -9$

## 14.6 Odd Powers of Variables (page 402)

### Suggestions
It may be helpful to have students write $\sqrt{x^7}$ as $\sqrt{x \cdot x \cdot x \cdot x \cdot x \cdot x \cdot x}$, then group the factors in twos as follows:

$$\sqrt{x^7} = \sqrt{x \cdot x \cdot x \cdot x \cdot x \cdot x \cdot x}$$
$$= \underbrace{\sqrt{x} \cdot \sqrt{x}}_{x} \cdot \underbrace{\sqrt{x} \cdot \sqrt{x}}_{x} \cdot \underbrace{\sqrt{x} \cdot \sqrt{x}}_{x} \cdot \sqrt{x}$$
$$= x^3 \sqrt{x}$$

### Errors That Students Make
Some students may think that the exponent of the variable must be a perfect square in order to factor it out. Therefore, they may simplify $\sqrt{x^7}$ as follows.

$$\sqrt{x^7} = \sqrt{x^4 \cdot x^3}$$
$$= \sqrt{x^4} \cdot \sqrt{x^3}$$
$$= x^2 \cdot \sqrt{x^3}$$

### For Additional Chalkboard Work
Simplify each of the following.

1. $\sqrt{g^3}$   $g\sqrt{g}$
2. $\sqrt{16b^5}$   $4b^2 \sqrt{b}$
3. $\sqrt{49a}$   $7\sqrt{a}$
4. $\sqrt{32x^7}$   $4x^3 \sqrt{2x}$
5. $-\sqrt{7n^{11}}$   $-n^5 \sqrt{7n}$
6. $\sqrt{24c^2 d^9}$   $2cd^4 \sqrt{6d}$

### Cumulative Review
1. Find the slope of $\overrightarrow{AB}$ for $A(-4, -1)$ and $B(6, 2)$.
   $\dfrac{3}{10}$
2. Solve
   $\dfrac{a + 3}{5} = \dfrac{a + 10}{10}$.
   $4$
3. Graph
   $x = 2$.
   line through $(2, 0)$, $(2, 6)$

## 14.7 Adding and Subtracting Radicals (page 404)

### Background

Combining radicals is similar to combining like terms. The distributive property can be used to show that $2\sqrt{2} + 4\sqrt{2} = 6\sqrt{2}$ as follows.

$$2\sqrt{2} + 4\sqrt{2} = (2 + 4)\sqrt{2}$$
$$= 6\sqrt{2}$$

### Suggestions

Before doing Example 3, review simplifying radicals with the following problems.
Simplify.

**1.** $5\sqrt{8}$      **2.** $4\sqrt{27}$      **3.** $\sqrt{25xy}$
  $10\sqrt{2}$         $12\sqrt{3}$          $5\sqrt{xy}$

Then stress that students should simplify any radicals before proceeding to combine like radicals.

### For Additional Chalkboard Work

Simplify each of the following.

**1.** $12\sqrt{5} - 3\sqrt{5}$      **2.** $7\sqrt{3} + 7\sqrt{2} - 2\sqrt{3}$
  $9\sqrt{5}$                      $5\sqrt{3} + 7\sqrt{2}$
**3.** $4\sqrt{27} + 3\sqrt{12} - \sqrt{48}$
  $14\sqrt{3}$

### Cumulative Review

Factor completely.
**1.** $3x^2 - 6x - 9$        **2.** $4x^2 - 16$
  $3(x - 3)(x + 1)$          $4(x + 2)(x - 2)$
**3.** $5x^3 - 7x^2 + 2x$   $x(5x - 2)(x - 1)$

## 14.8 Multiplying Radicals (page 406)

### Background

Two properties developed in earlier lessons on square roots are used to simplify products involving radicals.

$$\sqrt{a} \cdot \sqrt{b} = \sqrt{ab}$$
$$\sqrt{a} \cdot \sqrt{a} = a$$

Most products involving radicals lead to a product that contains an irrational number. However, the product of two binomials in the form $(\sqrt{a} + \sqrt{b})(\sqrt{a} - \sqrt{b})$ leads to a product that is a rational number. The two binomials are called conjugates.

### Suggestions

Review the distributive property and multiplication of two binomials before starting the lesson.
Multiply.

**1.** $3(2a + 5)$              **2.** $(2a + 3)(a - 4)$
  $6a + 15$                    $2a^2 - 5a - 12$
**3.** $(3x + 2)(3x - 2)$   $9x^2 - 4$

### Errors That Students Make

When multiplying two radical expressions some students forget to simplify the resulting radical. For example, when multiplying $3\sqrt{6} \cdot 5\sqrt{2}$, some students will leave the answer as $15\sqrt{12}$, instead of simplifying $15\sqrt{12}$ as $15\sqrt{4}\sqrt{3} = 15 \cdot 2\sqrt{3} = 30\sqrt{3}$.

### For Additional Chalkboard Work

Simplify each of the following.

**1.** $3\sqrt{10} \cdot 4\sqrt{5}$        **2.** $4\sqrt{2}(3\sqrt{6} - \sqrt{18})$
  $60\sqrt{2}$                           $24\sqrt{3} - 24$
**3.** $(3\sqrt{3} + 2\sqrt{2})(2\sqrt{3} - 5\sqrt{2})$   $-11\sqrt{6} - 2$
**4.** $(2\sqrt{5} - \sqrt{7})(2\sqrt{5} + \sqrt{7})$   $13$

### Cumulative Review

**1.** Solve                      **2.** Simplify
  $4(x + 4) = 3(x - 2)$.         $\dfrac{x^2 - 7x + 12}{x^2 - 16}$.
  $-22$                          $\dfrac{x - 3}{x + 4}$

**3.** Multiply
  $(3a^2b^4)(-5a^3b)$.   $-15a^5b^5$

## 14.9 Dividing Radicals (page 409)

### Vocabulary
rationalize

### Background
Using the property $\sqrt{a} \cdot \sqrt{b} = \sqrt{ab}$, it can be shown that $\dfrac{\sqrt{ab}}{\sqrt{a}} = \sqrt{\dfrac{ab}{a}} = \sqrt{b}$. Since multiplying both the numerator and denominator of a fraction by the same nonzero number does not change the value of the fraction, it can also be shown that $\dfrac{\sqrt{a}}{\sqrt{b}} = \dfrac{\sqrt{a} \cdot \sqrt{b}}{\sqrt{b} \cdot \sqrt{b}} = \dfrac{\sqrt{ab}}{b}$.
This is called rationalizing the denominator.

### Suggestions
The second way to rationalize the denominator, shown in Example 2, is an easier method. However, some students cannot always see the least square root needed to make the denominator a perfect square. Therefore, it may be easier for those students to always multiply by the square root in the denominator, and then simplify the resulting expression.

### For Additional Chalkboard Work
Simplify each of the following.

1. $\dfrac{\sqrt{24}}{\sqrt{6}}$  2

2. $\dfrac{\sqrt{36x^7}}{\sqrt{9x}}$  $2x^3$

3. $\dfrac{2}{\sqrt{5}}$  $\dfrac{2\sqrt{5}}{5}$

4. $\dfrac{12}{\sqrt{18}}$  $2\sqrt{2}$

5. $\sqrt{\dfrac{9a}{12a^5}}$  $\dfrac{\sqrt{3}}{2a^2}$

6. $\dfrac{5}{\sqrt{10y}}$  $\dfrac{\sqrt{10y}}{2y}$

### Cumulative Review
1. Solve
   $x^2 - 5x = 36$.
   9, −4

2. Simplify
   $2a^2(3a^3 - 2a^2 + 7a)$.
   $6a^5 - 4a^4 + 14a^3$

3. Simplify
   $(3xy^5)^3$.
   $27x^3y^{15}$

## 14.10 The Pythagorean Property (page 412)

### Vocabulary
hypotenuse, leg, Pythagorean property, right triangle

### Suggestions
Using the following illustration might be helpful in showing why the Pythagorean property works.

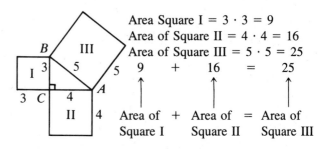

Area Square I = $3 \cdot 3 = 9$
Area of Square II = $4 \cdot 4 = 16$
Area of Square III = $5 \cdot 5 = 25$
9 + 16 = 25

Area of + Area of = Area of
Square I   Square II   Square III

Now use the letters $a$, $b$, and $c$ to represent the sides of the right triangle and show that $a^2 + b^2 = c^2$ for $a = 3$, $b = 4$, and $c = 5$.

$$a^2 + b^2 = c^2$$
$$3^2 + 4^2 = 5^2$$
$$9 + 16 = 25$$
$$25 = 25$$

### For Additional Chalkboard Work
For each right triangle, find the missing length. Give the answer in simplest radical form.

1. $a = 9$, $b = 12$
   $c = 15$

2. $a = 5$, $c = 10$
   $b = 5\sqrt{3}$

3. $b = 6$, $c = 9$  $a = 3\sqrt{5}$

### Cumulative Review
Solve.

1. $0.5x - 12 = 6.8$
   37.6

2. $\dfrac{2}{x} + \dfrac{2}{3x} = 1$
   $2\frac{2}{3}$

3. $\dfrac{2m - 5}{3} = \dfrac{m + 1}{2}$  13

## 14.11 Zero and Negative Integral Exponents (page 416)

### Background
Since this lesson is not part of the overall objectives of the chapter, the content does not appear on the Chapter Review or the Chapter Test.

### Suggestions
Rather than merely define zero and negative exponents, the intuitive foundation for the definition is laid in the text on page 416. The application of integral exponents to powers of 10 to write numbers in scientific notation is dealt with next.

### Enrichment
You might want to extend the ideas of this lesson to include rational exponents. Students have seen that $\sqrt{a}$ means $a^{\frac{1}{2}}$. Therefore, $\sqrt[3]{a}$ means $a^{\frac{1}{3}}$. In general, $\sqrt[n]{a^m} = (a^m)^{\frac{1}{n}} = \left(a^{\frac{1}{n}}\right)^m = a^{\frac{m}{n}}$. Conversely, $16^{\frac{1}{2}}$ means $\sqrt{16}$, or 4.

$$49^{-\frac{1}{2}} = \frac{1}{49^{\frac{1}{2}}} = \frac{1}{\sqrt{49}} = \frac{1}{7}$$

Practice. Simplify each.

1. $36^{\frac{1}{2}}$

   6

2. $100^{-\frac{1}{2}}$

   $\frac{1}{10}$

3. $(25a^6)^{-\frac{1}{2}}$

   $\frac{1}{5a^3}$

### Cumulative Review
Simplify.

1. $(3a - 12)\frac{2}{3} - a$  2. $3x - 4 - \frac{4}{5}x$  3. $(0.04m^2)^3$

   $-8$          $\frac{11}{5}x - 4$      $.000064m^6$

## Applications (page 419)

### Suggestions
It might be helpful to review finding the mean of a set of numbers, subtracting integers, squaring integers, and finding the square root of a number to three decimal places before having students find a standard deviation. You may also wish to discuss standard deviation in relation to the normal curve.

### For Additional Chalkboard Work
Find the standard deviation for each set of scores. scores.

1. 80, 85, 90, 85, 95

   5.099

2. 9, 10, 7, 8, 6

   1.414

### Enrichment
Have students draw a histogram for the last test scores in your class. See if the smooth curve drawn for the histogram is a normal curve. Then have the students find the standard deviation for their group of test scores.

### Cumulative Review

1. Find 2 consecutive integers whose sum is 43.  21, 22

2. Find 2 consecutive positive even integers whose product is 120.  10, 12

## Computer Activities: Pythagorean Property (page 421)
As the students become more comfortable with BASIC programming, they should begin to write their own programs without prompting. This page provides the opportunity to write two programs. The SQR function is used but the test program is provided for insertion into the Pythagorean property program.

# 15

## Chapter
## and Lesson
## Commentaries

### Chapter 15 Quadratic Functions

**Objectives**

To solve equations containing radicals

To solve word problems that lead to equations containing radicals

To solve equations of the form $(kx + b)^2 = c$ for $x$

To solve quadratic equations by completing the square

To solve quadratic equations using the quadratic formula

To solve word problems that lead to quadratic equations

To draw the graph of a quadratic equation (parabola)

To determine the coordinates of the minimum or maximum point of a parabola

**Overview**

Students learn to solve equations containing radicals and word problems that lead to equations containing radicals. Then equations of the type $(kx + b)^2 = c$ pave the way for teaching "completing the square." This in turn provides the foundation for teaching the quadratic formula. An interesting optional application of the quadratic formula is the use of it in sketching a parabola. Finding the maximum or minimum point of a parabola then provides a basis for a simple lesson on applying the maxima-minima concept to real life problems. The chapter also has word problem lessons on area and on wet mixtures (optional).

## Career: Health Care (page 425)

Point out that nursing requirements have changed over recent years. Many nursing programs now require a college education. College courses such as chemistry require extensive background in high school mathematics. If there are students in your class who are interested in nursing as a career, encourage them to find out the exact requirements for pursuing a nursing career.

Many applications involve proportion. Below are some additional practice problems.

1. A patient is to receive digoxin, 0.25 mg daily. Since the patient cannot swallow a pill, the liquid Lanoxin will be used. You have available a bottle labeled 1 mL = 0.05 mg of Lanoxin. How much medication should the patient receive?

   *Solution:* $\dfrac{1 \text{ mL}}{0.05 \text{ mg}} = \dfrac{x \text{ mL}}{0.25 \text{ mg}}$ or $\dfrac{1}{0.05} = \dfrac{x}{0.25}$

   $x = 5$. So, the patient must receive 5 mL.

2. The doctor has ordered 100 mg of sodium phenobarbital by injection for a patient. There are available ampules of phenobarbital labeled 250 mg/4 mL. How much of the medication should be given to the patient?

   *Solution:* $\dfrac{100 \text{ mg}}{x \text{ mL}} = \dfrac{250 \text{ mg}}{4 \text{ mL}}$ or $\dfrac{100}{x} = \dfrac{250}{4}$

   $x = 1.6$. So, the patient must receive 1.6 mL.

## 15.1 Equations Containing Radicals (page 426)

### Vocabulary
extraneous

### Suggestions
It is important that students see *how* extraneous roots can arise. First, briefly review the solution of quadratic equations by factoring.
Solve.

1. $x^2 = 9$    2. $x^2 = 25$    3. $x^2 = 49$
    3, −3       5, −5       7, −7

Now lead them to a discovery.

In the following practice, students begin with a linear equation that has only one solution. Upon squaring each side, they get quadratic equations that have two solutions. Yet only one solution checks in the original equation; the other is *extraneous*.

For each equation, what is the solution? Square each side. What are the two solutions? Which solution is not a solution of the original equation and therefore extraneous?

1. $x = 6$           2. $x = 8$
   6; 6, −6;          8; 8, −8;
   −6 is extraneous      −8 is extraneous

3. $x = -10$
   −10; 10, −10;
   10 is extraneous

### Errors That Students Make
In Example 2 students frequently fail to get the 7 on the other side before squaring each side, and write, in error, $4x + 12 + 49 = 1$. Stress that students should get the radical alone on one side. Another typical error arises in practice like that of Example 4. When squaring $x + 1$ the students tend to forget the middle term and write $x^2 + 1$ instead of $x^2 + 2x + 1$.

### For Additional Chalkboard Work
Solve.

1. $\sqrt{2y} = 10$   50     2. $\sqrt{x} + 5 = 9$   16
3. $\sqrt{a + 1} + 3 = 2$   $\emptyset$

## 15.2 The Solution of $x^2 = a$ (page 429)

### Suggestions
Many of the equations in this lesson will lead to simplification of square roots, so open the lesson with a brief review.
Simplify.

1. $\sqrt{20}$   2. $\sqrt{27}$   3. $\sqrt{12}$   4. $\sqrt{\frac{2}{3}}$   5. $\sqrt{\frac{3}{5}}$

   $2\sqrt{5}$     $3\sqrt{3}$     $2\sqrt{3}$    $\frac{\sqrt{6}}{3}$    $\frac{\sqrt{15}}{5}$

Stress that a quadratic equation such as $x^2 = 16$ in the opening prose on page 429 can now be solved by the shortcut:
$$x^2 = 16$$
$$x = \sqrt{16} \quad \text{or} \quad x = -\sqrt{16}$$
$$x = 4 \quad \text{or} \quad x = -4$$

### Errors That Students Make
When solving an equation like that shown in Example 4, students want to simplify the final answer as $2 + 3\sqrt{2} = 5\sqrt{2}$. Emphasize that unlike terms cannot be combined. Similarly, $2 + 3a$ is *not* $5a$.

### For Additional Chalkboard Work
Solve.

1. $x^2 = 8$     2. $y^2 = 20$     3. $m^2 = 48$
   $\pm 2\sqrt{2}$      $\pm 2\sqrt{5}$      $\pm 4\sqrt{3}$

### Enrichment
Simplify.
$$\frac{\sqrt{10}}{4} \cdot \sqrt{1 + \cfrac{1}{1 + \cfrac{1}{1 + \frac{1}{2}}}}$$

*Solution:* $1 + \cfrac{1}{1 + \cfrac{1}{1 + \frac{1}{2}}} = 1 + \cfrac{1}{1 + \cfrac{1}{\frac{3}{2}}} =$

$1 + \cfrac{1}{1 + \frac{2}{3}} = 1 + \cfrac{1}{\frac{5}{3}} = 1 + \frac{3}{5} = \frac{8}{5}; \frac{\sqrt{10}}{4} \cdot \sqrt{\frac{8}{5}} =$

$\frac{\sqrt{10}}{4} \cdot \frac{\sqrt{8}}{\sqrt{5}} = \frac{\sqrt{10}}{4} \cdot \frac{\sqrt{40}}{5} = \frac{\sqrt{400}}{4 \cdot 5} = \frac{20}{20} = 1$

### Cumulative Review

1. Factor
   $6x^2 + 22x - 8$.
   $2(3x - 1)(x + 4)$

2. Simplify
   $\frac{x + 5}{x^2 - 25} - \frac{4}{5 - x}$.
   $\frac{5}{x - 5}$

## 15.3 Completing the Square (page 432)

### Vocabulary
completing the square

### Background
Students can use the method for completing the square to solve quadratic equations. However, once the technique is understood it can then be used to develop the theory of the quadratic formula in the next lesson. The quadratic formula is a natural outcome of completing the square and provides a much easier means for solving quadratic equations that cannot be solved by factoring.

### Suggestions
Review perfect square trinomials, taught in Chapter 7. An alternate discovery approach to development of the concept of completing the square is developed below.

Square each binomial as indicated.

**1.** $(x + 4)^2$       **2.** $(x - 7)^2$
     $x^2 + 8x + 16$        $x^2 - 14x + 49$
**3.** $(x + 6)^2$    $x^2 + 12x + 36$

Now point out a pattern.

For $(x + 4)^2$, the 16 is $4^2$ or the square of $\frac{1}{2}$ of 8.

For $(x - 7)^2$, the 49 is $7^2$ or the square of $\frac{1}{2}$ of 14.

For $(x + 6)^2$, the 36 is $6^2$ or the square of $\frac{1}{2}$ of 12.

This suggests a way for finding the number that can be added to an expression like

    $y^2 - 6y$ to make a perfect square trinomial:
    $\frac{1}{2}$ of $-6$ is $-3$
      $(-3)^2 = 9$

Therefore, 9 must be added to $y^2 - 6y$ to make a perfect square trinomial.

### For Additional Chalkboard Work
Solve by completing the square.

**1.** $x^2 - 10x = 24$     **2.** $x^2 - 6x = -8$
     12, $-2$             4, 2
**3.** $x^2 + 8x - 3 = 0$ . $-4 \pm \sqrt{19}$

### Cumulative Review
**1.** Solve          **2.** Simplify
    $x - 2y = 8$
    $x + 4y = 20$.      $\dfrac{1 + \dfrac{7}{x} + \dfrac{12}{x^2}}{\dfrac{1}{x} + \dfrac{3}{x^2}}$.
      (12, 2)       $x + 4$

## 15.4 The Quadratic Formula (page 435)

### Vocabulary
quadratic formula

### Suggestions
A helpful technique is to have the students close their texts and follow the development to the concrete example as well as the formal proof (page 435) side by side on the chalkboard. Do one line at a time for each parallel development. Ask different students for answers to each step. For example: what do you get when you divide each side of $3x^2 + 1x - 1$ by 3? When you divide each side of $ax^2 + bx + c$ by $a$? Once the formula has been developed, it is important that students be really convinced that it works. Provide some easy practice as shown.

Solve each first by factoring, then by using the quadratic formula to see that the solutions are the same.

**1.** $x^2 - 4x + 3 = 0$     **2.** $x^2 - 2x - 8 = 0$
       1, 3             4, $-2$
**3.** $y^2 - 8y + 7 = 0$    7, 1

However, at this point it is necessary to show the real need for the formula, since factoring is obviously an easier method. Ask students to solve $x^2 - 1 = x$ (Example 2, page 436). After putting the equation in standard form, $1x^2 - 1x - 1 = 0$, they will quickly discover that it *cannot* be solved by factoring. The need for the formula is now seen!

### Errors That Students Make

$\dfrac{4 \pm \sqrt{2}}{4}$ is not $\dfrac{\overset{1}{\cancel{4}} \pm \sqrt{2}}{\underset{1}{\cancel{4}}}$ or $1 \pm \sqrt{2}$. It cannot be

simplified.

### For Additional Chalkboard Work
Solve by using the quadratic formula. Leave irrational roots in simplest form.

**1.** $x^2 + 4x - 21 = 0$    **2.** $x^2 - 5x + 2 = 0$
        $-7, 3$          $\dfrac{5 \pm \sqrt{17}}{2}$
**3.** $x^2 = -2x + 6$    $-1 \pm \sqrt{7}$

## 15.5 Problems about Area (page 438)

### Background

This lesson points out one of the themes of the movement on word problem formulation: interpretation. It is not enough to merely solve and then check the equation. The solution of the equation may *not* be the solution of the actual problem. The student must check to determine if the algebraic solution provides a correct interpretation of the conditions of the word problem. For example, are negative solutions of the equation acceptable interpretations for the physical conditions stipulated in the word problem?

### Suggestions

Emphasize the importance of drawing a sketch. Sometimes application of the quadratic formula in solving area problems results in square roots of numbers larger than 100, which have to be simplified before using the square root table. You may want to have students simplify problems like $3 + \sqrt{120}$ and then use a square root table to find the answer to the nearest tenth.

The calculator activity involves solving quadratic equations with larger numbers and decimals.

### For Additional Chalkboard Work

Solve.

1. The length of a rectangle is 5 m more than the width. The area is 24 m². Find the length and width.   8 m, 3 m

Solve. Use a square root table. Round the answer to the nearest tenth.

2. The base of a triangle is 4 m more than twice the height. The area is 20 m². Find the base and the height.   11.2 m, 3.6 m

### Cumulative Review

1. Graph
   $3x - 4y = 12$.
   Slope $\frac{3}{4}$
   y-intercept $-3$

2. Solve
   $\dfrac{x}{12} = \dfrac{2}{x - 2}$.
   6, $-4$

3. Simplify
   $\dfrac{x^2 - 36}{18 - 3x}$.
   $\dfrac{x + 6}{-3}$

## 15.6 Problems About Wet Mixtures (page 441)

This is a word problem lesson not directly related to the content of the rest of the chapter. Therefore, it is not tested at the end of the chapter.

## 15.7 The Parabola (page 445)

### Vocabulary

axis of symmetry, maximum, minimum, parabola, turning point

### Background

This is an optional lesson. However, it is well worth the time because of the excellent applications it affords. The next lesson deals with a variety of business and science applications of maxima-minima concepts of this lesson.

### Suggestions

The formula for the x-coordinate of the turning point $x = \dfrac{-b}{2a}$ is mandatory. The procedure illustrated in Example 2 is only motivational. It will not always work without a need for complex numbers, which are not studied until Algebra 2. Perhaps you should review the nature of a, b, and c for quadratic functions, particularly when the numerical coefficient is not actually written. For each, name a, b, and c. Then find the x-coordinate of the turning point.

1. $y = x^2 - x + 2$
   1, $-1$, 2
   $\dfrac{-(-1)}{2(1)} = \dfrac{1}{2}$

2. $y = -x^2 + x - 4$
   $-1$, 1, $-4$
   $\dfrac{-1}{2(-1)} = \dfrac{1}{2}$

3. $y = -2x^2 + 6$
   $-2$, 0, 6
   $\dfrac{-0}{2(-2)} = \dfrac{0}{-4} = 0$

## Applications (page 449)

### Background
This lesson is a direct application of the Example taught in the Ⓑ exercises of the previous lesson.

### Suggestions
First, review the method of finding the coordinates of the maximum (minimum) point of a function.
1. Maximum if $x^2$ coefficient is negative. Minimum if $x^2$ coefficient is positive.
2. Find the $x$-coordinate by using $x = \dfrac{-b}{2a}$.
3. Find the $y$-coordinate by substituting $\dfrac{-b}{2a}$ for $x$ in the function.
4. The maximum (minimum) value of the function will be the $y$-coordinate of step 3 above.

### For Additional Chalkboard Work
1. For what value of $x$ will $y = -x^2 + 6x + 8$ have a maximum? What is the maximum?
$x = 3$;
maximum 17

2. For what value of $x$ will $y = 16x^2 + 128x$ have a minimum? What is the minimum?
$x = -4$;
minimum $-256$

### Cumulative Review
1. Add
$$\frac{7}{x-2} + \frac{4}{x+3}.$$
$$\frac{11x + 13}{(x-2)(x+3)}$$

2. Solve
$$\frac{4}{5} + \frac{3}{a} = 2.$$
$$2\frac{1}{2}$$

3. Solve
$$3x + 2y = 7$$
$$2x + 3y = 8.$$
$$(1,2)$$

### Word Problem Formulation
*Project:* The Johnson family wants to build a ground-level patio deck at the back of their home. For privacy it is to be fenced in. Since the patio is up against the house, fencing will be needed for only three sides. The family has 60 feet of fencing. The family wants to have the largest area patio possible within the confines of the 60 feet of fencing. How much will it cost?

1. What are the dimensions that give maximum area? You will have to help the student here.
Let $x$ = width of fence, $60 - 2x$ = length of fence parallel to side of house.

Area: $A = x(60 - 2x)$
$\phantom{Area: }A = 60x - 2x^2$
$\phantom{Area: }A = -2x^2 + 60x$

$60 - 2x$

Now solve this as a "maximum" problem: find $x$ so that $A$ will be a maximum. *Solution:* $x = 15$. So the dimensions of the patio will be $15'$ by $30'$. The part of the problem in 1. above serves as a practical enrichment experience for the very bright student to master on his/her own with minimal help or hints from you. Once the challenge has been met, the entire class can join in on the remainder of the word problem formulation project as follows.

2. Find the cost of the patio floor.
Which is preferable: redwood timber, brick, outdoor flagstone, or concrete?
Find the cost in materials for each.
Find the cost in labor for each.
Does the labor cost depend upon the number of square feet of surface?
What other expenses have to be taken into consideration (e.g., leveling of ground, building permit)?

### Cumulative Review
Simplify.
1. $\sqrt{28a}$
$2\sqrt{7a}$

2. $(4x - 3)^2$
$16x^2 - 24x + 9$

3. $\dfrac{x^2 - 7x + 12}{16 - x^2}$
$$\frac{x - 3}{-1(x + 4)}$$

## Computer Activities: Solving Quadratic Equations (page 452)

Part of the program for the quadratic formula is provided for the student. Point out the advantage of using lines 40, 50, and 60 to find the roots X1 and X2. An attempt to do this in one line would involve unwieldy use of parentheses and a greater possibility of error.

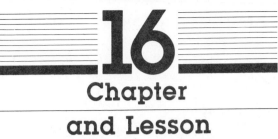

# Chapter and Lesson Commentaries

## Chapter 16   Plane Trigonometry

### Objectives

To solve problems about complementary angles
To solve problems about supplementary angles
To solve problems about the angles of a triangle
To find lengths of sides of similar triangles
To solve word problems related to similar triangles
To compute the sine, cosine, and tangent of an acute angle of a right triangle
To find lengths of sides to the nearest tenth of a unit, using trigonometric ratios
To find the measures of angles to the nearest degree using trigonometric ratios
To use trigonometric ratios to solve practical problems

### Overview

The opening lesson develops properties of supplementary and complementary angles. However, the main idea is that the "sum of the angles of a triangle is 180°" together with the resulting conclusion that the acute angles of a right triangle are complementary. These ideas pave the way for the next lesson on similar triangles. The concept of the second lesson, that corresponding sides of similar triangles are in the same ratio, paves the way for the next lesson on trigonometric ratios. Students then learn to use trig tables to find missing side lengths or missing angle measures for right triangles. Trig ratios are used to solve practical applications problems. The chapter closes with a lesson on Investment and Loan Problems.

## Special Features

## Career: Landscape Architect (page 454)

There are many facets to consider in landscaping. The suggested project can be extended to become open-ended and interdisciplinary. In addition to the mathematical considerations that were suggested, environmental considerations might include soil erosion and conservation of trees; societal considerations might include urban versus suburban settings, and how the complex might differ in appearance in different countries; linguistic considerations might include necessary signs, their translation into different languages, and the use of international symbols instead of words on signs.

## 16.1 Angle Properties (page 455)

### Background

This lesson gives students an opportunity to see the application of algebra to geometric concepts. Stress that the size of an angle is not related to the lengths of the sides of the angle. The hands of a clock at 12:15 form approximately a 90-degree angle regardless of whether you are looking at a large grandfather clock, or a tiny wristwatch. Illustrate angle formation with the aid of a circle, as pictured at the right. If $\overrightarrow{OA}$ is held fixed and $\overrightarrow{OB}$ is rotated counterclockwise, different angles are formed. A complete rotation about the circle generates 360°. Halfway around a circle would be 180°. Then $\overrightarrow{OA}$ and $\overrightarrow{OB}$ form a straight line. This motivates the basis for supplementary angles. In the diagram at the right $\overrightarrow{OA}$ and $\overrightarrow{OB}$ form a straight line. So $m\angle 1 + m\angle 2 = 180$.

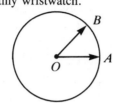

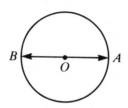

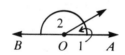

### Suggestions

Intuitively establish the property of "the sum of the measures of the three angles of any triangle is always 180." Have the students sketch a triangle on a sheet of paper. Cut out the triangle. Then cut off the three corners or angles. If the three angles are then placed adjacent to one another, as shown at the right, a straight line is formed. So $m\angle 1 + m\angle 2 + m\angle 3 = 180$.

### For Additional Chalkboard Work

Find the complement of each angle.

| **1.** 30° | **2.** 50° | **3.** 70° | **4.** $x°$ |
|---|---|---|---|
| 60° | 40° | 20° | $90° - x°$ |

## 16.2 Properties of Similar Triangles (page 459)

### Vocabulary

similar

### Suggestions

Point out some concrete everyday applications of similar figures, e.g., blueprints of houses, scale models, sewing patterns.

Have students do the following exercise to aid in teaching similar triangles. Using graph paper, plot the points $A(1,1)$, $B(5,1)$, and $C(1,4)$. Join the points to form a right triangle. Plot the points $D(7,2)$, $E(15,2)$, and $F(7,8)$. Have the students use a ruler to measure the lengths of the three sides of each triangle. Are the corresponding sides of the two triangles proportional?
Are the corresponding angles equal in measure? Students can determine this by either
1. cutting the angles out and seeing if they exactly fit one on the other, or
2. measuring with a protractor.

*Solution:*
sides of triangle *ABC:* $AB = 4$, $BC = 5$, $AC = 3$
sides of triangle *DEF:* $DE = 8$, $EF = 10$, $DF = 6$.
Ratio of proportionality is $1:2$

$m\angle A = 90$, $m\angle D = 90$
$m\angle B = 37$, $m\angle E = 37$
$m\angle C = 53$, $m\angle F = 53$

So the triangles are similar.

This experiment provides visual illustration of the reality of similar triangles plus a gentle introduction to analytic geometry techniques. A practical use of graphing is thus shown.

### For Additional Chalkboard Work

Solve.

**1.** $\dfrac{a}{2} = \dfrac{13}{4}$    **2.** $\dfrac{3}{10} = \dfrac{6}{a}$    **3.** $\dfrac{15}{x} = \dfrac{5}{3}$

$6\frac{1}{2}$          20          9

### Cumulative Review

Simplify.

**1.** $\dfrac{\sqrt{28a^3}}{2a\sqrt{7a}}$    **2.** $\dfrac{(-2x^5)^3}{-8x^{15}}$    **3.** $\dfrac{1 - \dfrac{7}{x} + \dfrac{12}{x^2}}{\dfrac{1}{x} - \dfrac{3}{x^2}}$,

$x - 4$

## 16.3 Introduction to Trigonometric Ratios (page 462)

### Vocabulary
adjacent, cosine, opposite leg, sine, tangent, trigonometry

### Background
The results of the last two lessons have paved the way for the concepts of this lesson. Similar triangle properties are used to show that the trigonometric ratios for the acute angles of the right triangle are the same regardless of the lengths of the three sides.

### For Additional Chalkboard Work
1. For right triangle $ABC$, with right angle at $C$, hypotenuse $AB = 20$, $AC = 16$, and $BC = 12$. Find $\tan A$, $\sin A$, $\cos A$, $\tan B$, $\sin B$, and $\cos B$ to 3 decimal places.
   0.75, 0.6, 0.8, 1.333, 0.8, 0.6

Simplify.

2. $\dfrac{3}{\sqrt{7}}$   3. $\dfrac{4}{\sqrt{2}}$   4. $\dfrac{5}{\sqrt{10}}$   5. $\dfrac{6}{\sqrt{5}}$

$\dfrac{3\sqrt{7}}{7}$    $2\sqrt{2}$    $\dfrac{\sqrt{10}}{2}$    $\dfrac{6\sqrt{5}}{5}$

### Enrichment
Consider any right triangle $ABC$ with right angle at $C$. Let $AC = x$, $BC = y$.
Find the length of $AB$ in terms of $x$ and $y$.
Then show that the $\tan A = \dfrac{\sin A}{\cos A}$.

*Solution:*
$$AB = \sqrt{x^2 + y^2}$$

$$\frac{\sin A}{\cos A} = \frac{\dfrac{y}{\sqrt{x^2+y^2}}}{\dfrac{x}{\sqrt{x^2+y^2}}} = \frac{y}{x} = \tan A$$

### Cumulative Review
1. Write the equation of the line containing the points $(0,1)$ and $(1,3)$.   $y = 2x + 1$
2. Graph the line with equation $y = \frac{2}{3}x - 4$.
   $y$-intercept, $-4$; slope $\frac{2}{3}$.

## 16.4 Using Trigonometric Tables (page 465)

### Suggestions
Once you have taught students to read the trigonometric tables, practice finding the angle, given the trigonometric ratio. In the beginning use only trigonometric ratios that are actually in the table. Once this is mastered, then finding the "closest value" to a trig ratio will be easier.

### For Additional Chalkboard Work
If some students have calculators with trig functions, you might want to show them how to use the inverse function to find the measure of an angle whose trig ratio is known. The technique varies with the brand of calculator. This activity should perhaps be done on an individual basis while the class is beginning a homework assignment. Then you can work with those students having the appropriate calculator.
Use the table to find $m \angle A$.
1. $\sin A = 0.515$      2. $\tan A = 5.671$
        31°               80°
3. $\cos A = 0.454$    63°

### Enrichment
An interesting pattern of the trig functions is that the sine of an angle is equal to the cosine of the complement of the angle.
Demonstrate this to the entire class through the following discovery exercise.
Find each.
1. $\sin 20°$     2. $\cos 70°$     3. $\sin 60°$
4. $\cos 30°$     5. $\sin 80°$     6. $\cos 10°$
Is there a pattern? If so, then $\sin 40° = ?$
*Solution:* $\sin 40° = \cos 50°$ since the complement of 40° is 50°.

Use the sketch at the right to show $\sin A = \cos B$. Why are $\angle A$ and $\angle B$ complementary?

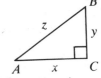

*Solution:* $A \angle$ and $\angle B$ are complementary since they are acute angles of a right triangle. (Sum of measures of all three angles is 180°.)

$$\sin A = \frac{y}{z} = \cos B = \frac{y}{z}$$

## 16.5 Right Triangle Solutions (page 467)

### Suggestions

The most difficult task for students in this lesson is determining which trig ratio to use. It might be helpful to provide practice on just writing the correct trig ratio needed to solve a problem from a picture of a right triangle.

### Enrichment

Find each answer to the nearest hundredth.
**1.** Find the area of $\triangle ABC$.    **2.** Find the area of rectangle $ABCD$.

Use the formula $A = \frac{1}{2}bh$.

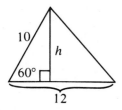
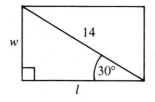

*Solution:*

$\dfrac{h}{10} = \sin 60°$

$h = 10 \sin 60°$

$h = 10(0.8660)$

Area $= \frac{1}{2}(8.66)(12) = 51.96$

$\cos 30° = \dfrac{l}{14}$

$l = 14 \cos 30°$

$l = 14(0.8660)$

$l = 12.12$

$\sin 30° = \dfrac{w}{14}$

$w = 14 \sin 30°$

$w = 14(0.5000)$

Area $= (12.12)(7) = 84.84$

## 16.6 Applying Trigonometric Ratios (page 470)

### Background

This lesson helps students see practical applications of the trigonometric ratios.

### Suggestions

The most difficult part of this lesson will be the determination, by the student, of which trig ratio to use for each application. It might help to review this process before starting the actual applications.

### For Additional Chalkboard Work

For each diagram, find the missing side or angle.
**1.**      **2.**      **3.**

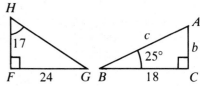

Find $p$.   9.6    Find m$\angle H$.   55°    Find $c$.   19.9

### Errors That Students Make

Many students make the following error in solving an equation like the one arising in Exercise 3 above.

$$\cos 25° = \frac{18}{c}$$

$$0.9063 = \frac{18}{c}$$

$$0.9063(18) = c$$

Urge the students to rewrite the equation as $\dfrac{0.9063}{1} = \dfrac{18}{c}$.

Then multiply each side by $c$, the LCM, to get

$$0.9063c = 18.$$

Then          $c = \dfrac{18}{0.9063}.$

### Cumulative Review

**1.** Simplify
$-(4 - x) - 6(2x + 5)$.
Then evaluate if $x = -2$.   $-11x - 34;\; -12$

**2.** Factor
$7x^3 - 21x^2 + 14x$.
$7x(x - 2)(x - 1)$

**3.** Solve
$x^2 - 7x = 0$.
$0, 7$

## 16.7 Investment and Loan Problems (page 473)

### Background

This lesson is not part of the overall objectives of the chapter. Therefore, it is not included in the Chapter Review and the Chapter Test.

Students have previously worked with the three basic "cases of percent." Stress that the three basic cases can each be approached in much the same way: translate the English sentence into an equation. Review this concept. Write an equation to solve each of the following and then solve.

**1.** Find 4% of $382.       **2.** What % of 12 is 9?
**3.** 6% of what number is 24?

    **1.** $0.04(382) = n$; $15.28
    **2.** $x(12) = 9$; $x = 0.75$; 75%
    **3.** $0.06n = 24$; 400

Point out alternative ways of doing Examples 2 and 3. In Example 2 simultaneous equations could also have been used.

    Let $x =$ amount (in $) invested at 8%
        $y =$ amount (in $) invested at 9%

Then the two equations become
$$y = 2x$$
$$x \cdot 0.08 \cdot 1 + y \cdot 0.09 \cdot 1 = 260$$
$$\text{or } 0.08x + 0.09y = 260$$
$$\text{or } 8x + 9y = 26{,}000$$

The system can now be solved by substitution. However, stress that solving Example 3 *without* simultaneous equations is somewhat harder to see.

    Let $x =$ amount borrowed at $9\frac{3}{4}\%$
  $9{,}000 - x =$ amount borrowed at $8\frac{1}{2}\%$

This concept is difficult for many students. In either case you will have to provide practice in writing fractional percents as decimals.

### For Additional Chalkboard Work

Give each percent as a decimal.

**1.** $4\frac{1}{2}\%$    **2.** $8\frac{1}{4}\%$    **3.** $4\frac{3}{4}\%$    **4.** $\frac{1}{2}\%$
  0.045       0.0825      0.0475       0.005

### Cumulative Review

**1.** Solve          **2.** Solve
  $(x - 6)^2 = 49.$       $\frac{7}{x} + 4 = 9.$
     $-1, 13$               $\frac{7}{5}$

## Computer Activities: Solving Right Triangles (page 477)

This program is not complete. The part that is given to the student demonstrates a simple procedure for labeling and filling in tables. The missing statements to compute the ratios are:

    50  LET T = D/E
    60  LET S = D/F
    70  LET C = E/F

Example 2 is more advanced and makes use of the BASIC functions SIN, COS, and TAN. The BASIC functions require that the angle measure be given in radians.

### Additional Computer Activity

This program will order a list of numbers. Since this program uses nested FOR . . . NEXT loops and arrays, it is up to the teacher to decide when its use is appropriate.

PROGRAM: ORDER

```
  5   REM THIS PROGRAM ORDERS A LIST
        OF TEN NUMBERS
 10   REM THIS LOOP INPUTS NUMBERS
 20   DIM N(10)
 30   FOR L = 1 TO 10
 40   INPUT N(L)
 50   NEXT L
 60   REM THIS LOOP ORDERS LIST
 70   FOR I = 1 TO 9
 80   FOR J = I + 1 TO 10
 90   IF N(I) < = N(J) THEN 130
100   LET T = N(I)
110   LET N(I) = N(J)
120   LET N(J) = T
130   NEXT J
140   NEXT I
150   REM PRINTS ORDERED LIST
160   FOR Y = 1 TO 10
170   PRINT N(Y)
180   NEXT Y
190   END
```

N(10) is an array to store the list of numbers.
J is an increment to test the next number.
T is a save place for switching.

## Introduction to Probability: A Simple Event

### Objectives

To determine the probability of a simple event
To determine the odds of a simple event

### Overview

The fundamentals of probability are presented in this section. Simple events are used to illustrate probability and odds.

### Vocabulary

certain event, chance, favorable outcome, impossible event, odds, probability, simple event

### Background

Review the concepts of ratio and complex fractions. Many states now have lotteries so students may be more aware of the concept of chance. Stress that probability and problems of chance are related not only to gambling devices. Illustrate this by pointing out the uses of probability in determining insurance rates. You might also ask for volunteers to research career opportunities in actuarial science.

Students may not be familiar with the terms used in connection with a deck of cards. Explain what is meant by a ''suit'' and that there are 4 suits of 13 cards in a deck of cards.

### Enrichment

Assign those students familiar with programming the computer lesson on pages 502–503 for a computer simulation of probability. Students will find applications to the ''spinner'' and ''dice'' types of problems.

## Interpreting Statistical Data

### Objective

To find the mean, mode, median, and range of a set of data

### Overview

This section will look at statistical data and methods of interpreting it. The concepts of central tendency and variability are used to show the convergence and the divergence of the data.

### Vocabulary

central tendency, mean, median, mode, range, variability

### Background

Students are familiar with the concept of average. The terminology ''mean'' may be new to some. Refer to the applications lesson on page 268 to review how to find the missing data item for a given mean. Refer also to the applications lesson on page 419 for the development of standard deviation. The measures of central tendency together with the measures of variability give a valid interpretation of the spread of a set of data. Stress how misrepresentative any one of the measures may be if taken by itself.

### Cumulative Review

Solve each equation.

1. $3x - (4 - x) = -7x - 5$
$-\frac{1}{11}$

2. $7x + 15 = -2x^2$
$1\frac{1}{2}, -5$

3. $|2x - 6| = 8$
$7, -1$

## Computer BASIC Section

### Objectives

To use INPUT statements in programs that evaluate formulas

To write programs that use BASIC algebraic expressions

To use the BASIC exponentiation symbol, $\wedge$

To use the order of operations in writing and evaluating algebraic expressions in computer programs

To write progams that use FOR. . .STEP. . .NEXT loops to perform repetitive calculations

To write programs that perform calculations on values entered in READ. . .DATA statements

To use GOTO and IF. . .THEN statements in writing programs

To write programs that use accumulators in computing sums and averages

To use the INT and RND functions to simulate chance events

### Overview

Presented in this optional unit is an introduction to computer programming in the language of BASIC. The fundamental statements of BASIC are introduced. The format of each statement should be checked against the commands given in the BASIC user's guide associated with the computer you are using in your classroom.

### Vocabulary

input, run, print, end, cursor, list, exponentiation, order of operations, for. . .next, step, loop, read. . .data, goto, if. . .then, branching, out of data, accumulator, initialize, counter, INT, RND, greatest integer, random

### Background

The computer is a tool with an increasing role in the classroom and in education. The tool you are probably using is a ''microcomputer.'' A microcomputer is defined by the fact that its central processing unit is contained on a single integrated circuit chip. This chip contains the logic and arithmetic units of the system. One of the languages your microcomputer understands is BASIC. The BASIC programming language is user-oriented and can be learned in a short amount of time. It contains the ability to develop, debug, and run programs interactively. BASIC interpreters explain the BASIC statements to the computer. The statements covered in the chapter are the simplest BASIC commands. However, many interpreters are able to understand matrix commands, format print commands, and graphic display commands. If these facilities are available to you, they will be explained in the manufacturer's user's guide. You may also need to check the user's guide for symbols, spacing techniques, and reserved words that may be different for your computer.

In addition to writing your own BASIC programs, many application programs are available for demonstration. These programs can be used to show special use of a BASIC command, a unique logic flow, or an unusual computation.

### Suggestions

The topics of programming taught in this section can serve as valuable tools in giving students a better understanding of how many mathematical principles work. It can be studied as a separate section or each statement can be introduced as needed to correlate with the programs at the end of the chapters. Most of the computer activites in the chapters lend themselves to serving as tools to further enhance the understanding of the mathematical principles of the chapters. Exercises from either the Computer Activities or the Computer Section can be adapted to provide material for special projects.

# Assignments for Reinforcement and Maintenance

Following is a suggested schedule of assignments for 170 school days. It is designed to provide a program of reinforcement and maintenance of the major skills and concepts of Algebra 1 and to allow flexibility of testing days.

Each day's assignments provide a double dimension.

In a spiral view there is provision for:

1. REINFORCEMENT of the current lesson (major part)
2. MAINTENANCE of the previous lesson
3. MAINTENANCE of the lesson taught two days ago.

In a lateral view there is a provision for:

A BASIC COURSE
B AVERAGE COURSE
C ENRICHED COURSE.

The reinforcement and maintenance program is provided as a guide for you to adopt and use in the way that you discern to be best for your class.

Included in the text are optional materials that are ancillary to the program. You may use these materials in a variety of ways:

1. Additional examples for teaching the lesson
2. Practice exercises for seat work or chalkboard work
3. Items for quizzes
4. Practice exercises for extra help outside of class time
5. Extra testing
6. Enrichment exercises.

| OPTIONAL MATERIALS | | |
|---|---|---|
| **Practice** | **Testing** | **Enrichment** |
| Cumulative Review in PE* | Cumulative Review Tests (1–4, 1–8, 1–12, and 1–16) in PE | Calculator Activities in PE |
| Chalkboard Exercises in TE** | | Non-Routine Problems in PE |
| | College Prep Tests in PE | Computer Activities in PE |
| Selected Extra Practice Section in PE | | Chapter Opener Pages in PE |
| | Cumulative Review in TE | Probability and Statistics Sections in PE |
| Geometry Review in PE | | |
| | | Enrichment in TE |

*PE: Pupil's Edition  **TE: Teacher's Edition

# Assignment Chart

| Day | Page | Level A | Page | Level B | Page | Level C |
|-----|------|---------|------|---------|------|---------|
| 1 | 3 | Ex 1–32 | 3 | Ex Odd 1–43 | 3 | Ex Even 2–20 Odd 31–43, Ex 44–47 |
| 2 | 3<br>5–6 | Ex 33, 35, 37<br>Reading Ex 1–3<br>Ex 1–4, 7–27 | 3<br>5–6 | Ex 32, 34, 36<br>Reading Ex 1–3<br>Ex Odd 1–43 | 3<br>5–6 | Ex 38, 40, 42<br>Reading Ex 1–3<br>Ex Odd 1–49 |
| 3 | 3<br>5–6<br>9 | Ex 34, 36<br>Ex 5, 28, 30<br>Ex 1–14, 17–23, 26–33 | 3<br>5–6<br>9 | Ex 38, 40<br>Ex 6, 22, 40<br>Ex Even 2–52 | 3<br>5–6<br>9 | Ex 34, 36<br>Ex 24, 40, 46<br>Ex Even 18–24, 36–58 |
| 4 | 5–6<br>9<br>12 | Ex 6, 29<br>Ex 16, 25, 35<br>Reading Ex 1, 2<br>Ex 1–7, 9–21, 24–32 | 5–6<br>9<br>12 | Ex 18, 38<br>Ex 21, 41, 49<br>Reading Ex 1, 2<br>Ex Odd 1–41 | 5–6<br>9<br>12 | Ex 44, 50<br>Ex 21, 41, 51<br>Reading Ex 1, 2<br>Ex Odd 1–43<br>Non-Routine Problems |
| 5 | 9<br>12<br>15 | Ex 15, 24<br>Ex 8, 23, 34<br>Ex 1–15, 19–28 | 9<br>12<br>15 | Ex 33, 43<br>Ex 14, 28, 38<br>Ex Even 2–42 | 9<br>12<br>15 | Ex 25, 57<br>Ex 20, 32, 38<br>Ex Even 2–46 |
| 6 | 12<br>15<br>17 | Ex 22, 33<br>Ex 16, 18, 30<br>Ex 1–4, 7–18 | 12<br>15<br>17 | Ex 20, 40<br>Ex 13, 33, 41<br>Ex 1–4, 7–19, 22, 24 | 12<br>15<br>17 | Ex 26, 44<br>Ex 17, 37, 45<br>Ex Odd 1–25 |
| 7 | 15<br>17<br>19 | Ex 17, 29<br>Ex 6, 19, 21<br>Ex Odd 1–11 | 15<br>17<br>19 | Ex 23, 43<br>Ex 6, 21, 23<br>Ex Odd 1–13 | 15<br>17<br>19<br>20 | Ex 35, 43<br>Ex 6, 18, 24<br>Ex Odd 1–13<br>Applications Even 2–6 |
| 8 | 17<br>19<br>20 | Ex 5, 20<br>Ex 4, 8, 12<br>Applications Ex 1, Even 2–6 | 17<br>19<br>20 | Ex 20, 25<br>Ex 6, 10, 14<br>Applications Ex 1, Even 2–6 | 17<br>19<br>20<br>22 | Ex 20, 22<br>Ex 6, 10, 14<br>Applications Ex 3, 5, 7<br>Ex Even 2–30 |
| 9 | 19<br>20<br>22 | Ex 6, 10<br>Applications Ex 3, 5, 7<br>Ex Even 2–14 | 19<br>20<br>22 | Ex 4, 12<br>Applications Ex 3, 5, 7<br>Ex Even 2–24 | 24 | Chapter Review |
| 10 | 24 | Chapter Review | 24 | Chapter Review | 25 | Chapter Test |
| 11 | 25 | Chapter Test | 25 | Chapter Test | 30–31 | Reading Ex 1–6<br>Ex Odd 1–43 |

| Day | Page | Level A | Page | Level B | Page | Level C |
|---|---|---|---|---|---|---|
| 12 | 30–31 | Reading Ex 1–6<br>Ex 1–3, 5–7, 9–14,<br>17–19, 21–29 | 30–31 | Reading Ex 1–6<br>Ex Odd 1–37 | 30–31<br>33 | Ex 20, 34, 42<br>Ex 1–13 |
| 13 | 30–31<br>33 | Ex 4, 15, 30<br>Ex 1–13 | 30–31<br>33 | Ex 6, 28, 38<br>Ex 1–13 | 30–31<br>36 | Ex 30, 38<br>Ex Even 2–36 |
| 14 | 30–31<br>36 | Ex 8, 20<br>Ex 1–19 | 30–31<br>36 | Ex 18, 32<br>Ex Even 2–34 | 36<br>38–39 | Ex 29, 33, 35<br>Reading Ex 1–2<br>Ex Odd 1–35 |
| 15 | 36<br>38–39 | Ex 20, 22, 23<br>Reading Ex 1, 2<br>Ex 1–3, 5–7, 10–16 | 36<br>38–39 | Ex 11, 27, 33<br>Reading Ex 1, 2<br>Ex Odd 1–31 | 36<br>38–39<br>40 | Ex 19, 27<br>Ex 16, 24, 34<br>Ex 1–19, 25 |
| 16 | 36<br>38–39<br>40 | Ex 21, 24<br>Ex 4, 9, 18<br>Ex 1–14 | 36<br>38–39<br>40 | Ex 19, 31<br>Ex 8, 14, 26<br>Ex 1–18 | 38–39<br>40<br>43 | Ex 20, 30<br>Ex 20, 22, 26<br>Ex Even 2–52 |
| 17 | 38–39<br>40<br>43 | Ex 8, 17<br>Ex 15, 17, 19<br>Ex 1–15, 17–19,<br>21–25, 27–30 | 38–39<br>40<br>43 | Ex 20, 30<br>Ex 19, 21, 23<br>Ex Even 2–46 | 40<br>43<br>46 | Ex 21, 24<br>Ex 23, 45, 51<br>Ex Odd 1–39 |
| 18 | 40<br>43<br>46 | Ex 16, 18<br>Ex 16, 20, 32<br>Ex 1–10, 13–25 | 40<br>43<br>46 | Ex 20, 22<br>Ex 23, 27, 45<br>Ex Odd 1–33 | 43<br>46<br>48 | Ex 37, 47<br>Ex 22, 30, 38<br>Ex Even 2–30 |
| 19 | 43<br>46<br>48 | Ex 26, 31<br>Ex 12, 26, 28<br>Ex 1–12 | 43<br>46<br>48 | Ex 25, 37<br>Ex 10, 24, 34<br>Ex Even 2–26 | 46<br>48<br>50 | Ex 28, 36<br>Ex 19, 27, 29<br>Ex Odd 1–39 |
| 20 | 46<br>48<br>50 | Ex 11, 27<br>Ex 13, 15, 17<br>Ex 1–16 | 46<br>48<br>50 | Ex 18, 30<br>Ex 11, 21, 25<br>Ex Odd 1–33 | 48<br>50<br>52 | Ex 17, 23<br>Ex 20, 30, 36<br>Ex Odd 1–45 |
| 21 | 48<br>50<br>52 | Ex 14, 16<br>Ex 17, 19, 21<br>Ex 1–25 | 48<br>50<br>52 | Ex 17, 23<br>Ex 14, 20, 28<br>Ex Even 2–42 | 50<br>52<br>54 | Ex 22, 38<br>Ex 20, 34, 44<br>Ex Odd 1–39 |
| 22 | 50<br>52<br>54 | Ex 18, 20<br>Ex 26, 27, 30<br>Ex 1–18 | 50<br>52<br>54 | Ex 18, 22<br>Ex 15, 29, 37<br>Ex Odd 1–35 | 52<br>54<br>56–57 | Ex 14, 38<br>Ex 14, 28, 36<br>Reading Ex 1–3<br>Ex Even 2–54 |

| Day | Page | Level A | Page | Level B | Page | Level C |
|-----|------|---------|------|---------|------|---------|
| 23 | 52 | Ex 28, 29 | 52 | Ex 17, 35 | 54 | Ex 24, 32 |
|    | 54 | Ex 19, 21, 23 | 54 | Ex 12, 28, 34 | 56–57 | Ex 25, 33, 47 |
|    | 56–57 | Reading Ex 1–3 | 56–57 | Reading Ex 1–3 | 59 | Ex Odd 1–29 |
|    |    | Ex 1–15, 19–29 |    | Ex Odd 1–47 |    |    |
| 24 | 54 | Ex 20, 22 | 54 | Ex 20, 32 | 60 | Chapter Review |
|    | 56–57 | Ex 16, 18, 30 | 56–57 | Ex 24, 36, 42 |    |    |
|    | 59 | Ex 1–18 | 59 | Ex Even 2–24 |    |    |
| 25 | 60 | Chapter Review | 60 | Chapter Review | 61 | Chapter Test |
| 26 | 61 | Chapter Test | 61 | Chapter Test | 66 | Ex Odd 1–43 |
| 27 | 66 | Ex 1–20 | 66 | Ex Odd 1–37 | 66 | Ex 8, 30, 40 |
|    |    |    |    |    | 68 | Ex Even 2–24 |
| 28 | 66 | Ex 21, 23, 25 | 66 | Ex 8, 20, 34 | 66 | Ex 34, 42 |
|    | 68 | Ex 1–15 | 68 | Ex Even 2–24 | 68 | Ex 5, 17, 23 |
|    |    |    |    |    | 71 | Reading Ex 1–4 |
|    |    |    |    |    |    | Ex Odd 1–39 |
| 29 | 66 | Ex 22, 24 | 66 | Ex 14, 30 | 68 | Ex 11, 21 |
|    | 68 | Ex 16, 18, 20 | 68 | Ex 3, 13, 21 | 71 | Ex 6, 34, 38 |
|    | 71 | Reading Ex 1–4 | 71 | Reading Ex 1–4 | 73 | Ex Even 2–8 |
|    |    | Ex 1–6, 10–27 |    | Ex Odd 1–35 |    |    |
| 30 | 68 | Ex 17, 19 | 68 | Ex 11, 19 | 71 | Ex 16, 26 |
|    | 71 | Ex 7, 9, 29 | 71 | Ex 6, 18, 34 | 73 | Ex 3, 7 |
|    | 73 | Ex Odd 1–7 | 73 | Ex Even 2–8 | 75 | Reading Ex 1–4 |
|    |    |    |    |    |    | Ex Odd 1–33 |
| 31 | 71 | Ex 8, 28 | 71 | Ex 14, 32 | 73 | Ex 1, 5 |
|    | 73 | Ex 2, 6 | 73 | Ex 3, 7 | 75 | Ex 4, 26, 32 |
|    | 75 | Reading Ex 1–4 | 75 | Reading Ex 1–4 | 78–79 | Reading Ex 1–4 |
|    |    | Ex 1–17 |    | Ex Odd 1–27 |    | Ex Even 2–20 |
|    |    |    |    |    |    | Non-Routine Problems |
| 32 | 73 | Ex 4 | 73 | Ex 1, 5 | 75 | Ex 16, 24, 30 |
|    | 75 | Ex 18, 20, 22 | 75 | Ex 4, 16, 26 | 78–79 | Ex 5, 15, 19 |
|    | 78–79 | Reading Ex 1–4 | 78–79 | Reading Ex 1–4 | 81 | Ex 1–4, 6 |
|    |    | Ex Odd 1–15 |    | Ex Odd 1–19 |    |    |
| 33 | 75 | Ex 19, 21 | 75 | Ex 10, 18 | 82 | Chapter Review |
|    | 78–79 | Ex 2, 8, 14 | 78–79 | Ex 6, 16, 18 |    |    |
|    | 81 | Ex 1, 2 | 81 | Ex 1–3, 6 |    |    |

| Day | Page | Level A | Page | Level B | Page | Level C |
|---|---|---|---|---|---|---|
| 34 | 82 | Chapter Review | 82 | Chapter Review | 83 | Chapter Test |
| 35 | 83 | Chapter Test | 83 | Chapter Test | 88 | Reading Ex 1–4<br>Ex Odd 1–45 |
| 36 | 88 | Reading Ex 1–4<br>Ex 1–22 | 88 | Reading Ex 1–4<br>Ex Odd 1–41 | 88<br>91 | Ex 16, 32, 40<br>Ex Even 2–30 |
| 37 | 88<br>91 | Ex 23, 25, 27<br>Ex Even 2–24 | 88<br>91 | Ex 10, 26, 34<br>Ex Even 2–26 | 88<br>91<br>95 | Ex 22, 38<br>Ex 17, 25, 29<br>Reading Ex 1–3<br>Ex Odd 1–17 |
| 38 | 88<br>91<br>95 | Ex 24, 26<br>Ex 5, 13, 23<br>Reading Ex 1–3<br>Ex Odd 1–11 | 88<br>91<br>95 | Ex 14, 36<br>Ex 7, 15, 25<br>Reading Ex 1–3<br>Ex Odd 1–15 | 91<br>95<br>98–99 | Ex 13, 27<br>Ex 6, 12, 16<br>Ex Even 2–36 |
| 39 | 91<br>95<br>98–99 | Ex 7, 15<br>Ex 2, 6, 10<br>Ex 1 | 91<br>95<br>98–99 | Ex 19, 23<br>Ex 4, 8, 14<br>Ex Even 2–32 | 95<br>98–99<br>102 | Ex 10, 14<br>Ex 17, 27, 33<br>Ex Odd 1–31 |
| 40 | 95<br>98–99<br>102 | Ex 4, 8<br>Ex 19, 21, 23<br>Ex 1–10 | 95<br>98–99<br>102 | Ex 6, 12<br>Ex 13, 19, 29<br>Ex Odd 1–25 | 98–99<br>102<br>105 | Ex 21, 29<br>Ex 14, 26, 30<br>Ex Even 2–20 |
| 41 | 98–99<br>102<br>105 | Ex 11, 17<br>Ex 11, 13, 15<br>Ex 1–7 | 98–99<br>102<br>105 | Ex 15, 25<br>Ex 10, 18, 24<br>Ex Even 2–18 | 102<br>105<br>108<br>109 | Ex 18, 28<br>Ex 7, 13, 19<br>Ex Odd 1–15<br>Applications Ex 1, 3, 5 |
| 42 | 102<br>105<br>108 | Ex 12, 14<br>Ex 8, 10, 12<br>Ex 1–7 | 102<br>105<br>108 | Ex 6, 16<br>Ex 5, 11, 15<br>Ex Odd 1–13 | 105<br>108<br>109<br>110 | Ex 9, 17<br>Ex 6, 12, 16<br>Applications Ex 2, 4, 6<br>Ex Odd 1–15 |
| 43 | 105<br>108<br>109 | Ex 9, 11<br>Ex 8, 10<br>Applications Ex 1–4 | 105<br>108<br>109 | Ex 7, 13<br>Ex 4, 10, 12<br>Applications Ex 1–4 | 112 | Chapter Review |
| 44 | 108<br>109<br>110 | Ex 9<br>Applications Ex 5, 6<br>Ex 1–7 | 108<br>109<br>110 | Ex 6, 14<br>Applications Ex 5, 6<br>Ex Odd 1–15 | 113 | Chapter Test |
| 45 | 112 | Chapter Review | 112 | Chapter Review | 120 | Reading Ex 1–3<br>Ex Odd 1–31 |

| Day | Page | Level A | Page | Level B | Page | Level C |
|-----|------|---------|------|---------|------|---------|
| 46 | 113 | Chapter Test | 113 | Chapter Test | 120 | Ex 8, 22, 30 |
|    |      |              |     |              | 122 | Ex Even 2–36 |
| 47 | 120 | Reading Ex 1–3 | 120 | Reading Ex 1–3 | 120 | Ex 16, 26 |
|    |      | Ex 1–4, 7, 8, 11–17 |    | Ex Odd 1–27 | 122 | Ex 10, 26, 34 |
|    |      |              |     |              | 127 | Reading Ex 1–3 |
|    |      |              |     |              |     | Ex Odd 1–27 |
| 48 | 120 | Ex 5, 9, 18 | 120 | Ex 8, 12, 22 | 122 | Ex 30, 36 |
|    | 122 | Ex 1–20 | 122 | Ex Even 2–32 | 127 | Ex 10, 16, 28 |
|    |      |              |     |              | 130 | Ex Even 2–46 |
|    |      |              |     |              |     | Non-Routine Problems |
| 49 | 120 | Ex 10, 19 | 120 | Ex 14, 24 | 127 | Ex 22, 26 |
|    | 122 | Ex 21, 23, 25 | 122 | Ex 8, 22, 26 | 130 | Ex 25, 31, 45 |
|    | 127 | Reading Ex 1–3 | 127 | Reading Ex 1–3 | 131 | Ex Odd 1–9 |
|    |      | Ex 1–18, 23, 24 |    | Ex Odd 1–27 |     |          |
| 50 | 122 | Ex 22, 24 | 122 | Ex 18, 32 | 130 | Ex 33, 41 |
|    | 127 | Ex 19, 20, 25 | 127 | Ex 8, 18, 24 | 131 | Ex 2, 6 |
|    | 130 | Ex Even 2–26 | 130 | Ex Even 2–42 | 134–135 | Reading Ex 1–3 |
|    |      |              |     |              |     | Ex Even 2–52 |
| 51 | 127 | Ex 21, 26 | 127 | Ex 14, 26 | 131 | Ex 4, 8 |
|    | 130 | Ex 5, 13, 27 | 130 | Ex 9, 23, 37 | 134–135 | Ex 29, 43, 49 |
|    | 131 | Ex Odd 1–9 | 131 | Ex Odd 1–9 | 137 | Ex Odd 1–37 |
| 52 | 130 | Ex 11, 21 | 130 | Ex 27, 33 | 134–135 | Ex 27, 45 |
|    | 131 | Ex 2, 6 | 131 | Ex 2, 6 | 137 | Ex 10, 24, 36 |
|    | 134–135 | Reading Ex 1–3 | 134–135 | Reading Ex 1–3 | 140 | Reading Ex 1, 2 |
|    |      | Ex Even 2–28 |    | Ex Even 2–44 |     | Ex Even 2–30 |
| 53 | 131 | Ex 4, 8 | 131 | Ex 4, 8 | 137 | Ex 26, 34 |
|    | 134–135 | Ex 5, 17, 25 | 134–135 | Ex 11, 27, 41 | 140 | Ex 11, 25, 29 |
|    | 137 | Ex 1–15 | 137 | Ex Odd 1–31 | 142 | Reading Ex 1–3 |
|    |      |              |     |              |     | Ex 1–8 |
| 54 | 134–135 | Ex 13, 23 | 134–135 | Ex 25, 39 | 143 | Chapter Review |
|    | 137 | Ex 16, 18, 20 | 137 | Ex 14, 26, 32 |     |          |
|    | 140 | Reading Ex 1, 2 | 140 | Reading Ex 1, 2 |     |          |
|    |      | Ex 1–6, 9–14 |    | Ex Even 2–24 |     |          |
| 55 | 137 | Ex 17, 19 | 137 | Ex 18, 30 | 144 | Chapter Test |
|    | 140 | Ex 7, 15, 17 | 140 | Ex 7, 15, 21 |     |          |
|    | 142 | Reading Ex 1–3 | 142 | Reading Ex 1–3 |     |          |
|    |      | Ex 1–6 |    | Ex 1–8 |     |          |

| Day | Page | Level A | Page | Level B | Page | Level C |
|---|---|---|---|---|---|---|
| 56 | 143 | Chapter Review | 143 | Chapter Review | 149 | Reading Ex 1–3<br>Ex 1–32, 35–38,<br>41–43 |
| 57 | 144 | Chapter Test | 144 | Chapter Test | 149<br>151 | Ex 33, 39, 44<br>Ex 1–21, 23–30,<br>33–35 |
| 58 | 149 | Reading Ex 1–3<br>Ex 1–20 | 149 | Reading Ex 1–3<br>Ex 1–30, 32, 35–38 | 149<br>151<br>154 | Ex 34, 40<br>Ex 22, 31, 36<br>Reading Ex 1–4<br>Ex Odd 1–25 |
| 59 | 149<br>152 | Ex 21, 23, 25<br>Ex 1–17 | 149<br>152 | Ex 31, 33, 39<br>Ex 1–19, 23–30 | 151<br>154<br>156 | Ex 32, 37<br>Ex 10, 20, 24<br>Ex Even 2–28 |
| 60 | 149<br>152<br>154 | Ex 22, 24<br>Ex 18, 20, 22<br>Reading Ex 1–4<br>Ex 1–10 | 149<br>152<br>154 | Ex 34, 40<br>Ex 20, 22, 31<br>Reading Ex 1–4<br>Ex Odd 1–25 | 154<br>156<br>158 | Ex 16, 22<br>Ex 7, 21, 29<br>Ex Odd 1–33 |
| 61 | 152<br>154<br>156 | Ex 19, 21<br>Ex 11, 13, 15<br>Ex 1–5, 9–15 | 152<br>154<br>156 | Ex 21, 32<br>Ex 8, 18, 22<br>Ex Even 2–26 | 156<br>158<br>160 | Ex 17, 27<br>Ex 8, 24, 32<br>Reading Ex 1–3<br>Ex Even 2–36 |
| 62 | 154<br>156<br>158 | Ex 12, 14<br>Ex 6, 8, 16<br>Ex 1–6, 9–17 | 154<br>156<br>158 | Ex 12, 24<br>Ex 7, 19, 27<br>Ex Odd 1–29 | 158<br>160<br>163 | Ex 30, 34<br>Ex 9, 31, 37<br>Ex Odd 1–35 |
| 63 | 156<br>158<br>160 | Ex 7, 17<br>Ex 7, 18, 20<br>Reading Ex 1–3<br>Ex 1–14, 17–30 | 156<br>158<br>160 | Ex 15, 21<br>Ex 6, 14, 24<br>Reading Ex 1–3<br>Ex Even 2–36 | 160<br>163<br>165 | Ex 27, 35<br>Ex 20, 28, 34<br>Ex 1–6 |
| 64 | 158<br>160<br>163 | Ex 8, 19<br>Ex 15, 31, 33<br>Ex 1–22 | 158<br>160<br>163 | Ex 8, 28<br>Ex 9, 29, 35<br>Ex Odd 1–33 | 166 | Chapter Review |
| 65 | 160<br>163<br>165 | Ex 16, 32<br>Ex 23, 25, 27<br>Ex 1–4 | 160<br>163<br>165 | Ex 27, 37<br>Ex 10, 24, 30<br>Ex 1–5 | 167 | Chapter Test |
| 66 | 166 | Chapter Review | 166 | Chapter Review | 172 | Ex Even 2–34 |

| Day | Page | Level A | Page | Level B | Page | Level C |
|-----|------|---------|------|---------|------|---------|
| 67 | 167 | Chapter Test | 167 | Chapter Test | 172 | Ex 17, 27, 33 |
| | | | | | 174 | Ex Odd 1–43 |
| | | | | | | Non-Routine Problems |
| 68 | 172 | Ex 1–19 | 172 | Ex Even 2–30 | 172 | Ex 25, 35 |
| | | | | | 174 | Ex 22, 38, 42 |
| | | | | | 176 | Ex Even 2–32 |
| 69 | 172 | Ex 20, 22, 24 | 172 | Ex 5, 17, 27 | 174 | Ex 36, 44 |
| | 174 | Ex 1–10, 13–25 | 174 | Ex Odd 1–39 | 176 | Ex 9, 23, 31 |
| | | | | | 179 | Ex Odd 1–35 |
| 70 | 172 | Ex 21, 23 | 172 | Ex 13, 25 | 176 | Ex 25, 29 |
| | 174 | Ex 11, 26, 28 | 174 | Ex 6, 24, 36 | 179 | Ex 14, 28, 34 |
| | 176 | Ex 1–10, 13–24 | 176 | Ex Even 2–32 | 181–182 | Reading Ex 1–3 |
| | | | | | | Ex Even 2–66 |
| 71 | 174 | Ex 12, 27 | 174 | Ex 26, 38 | 179 | Ex 30, 36 |
| | 176 | Ex 11, 25, 27 | 176 | Ex 5, 21, 31 | 181–182 | Ex 21, 45, 53 |
| | 179 | Ex 1–19 | 179 | Ex Odd 1–33 | 185 | Reading Ex 1–3 |
| | | | | | | Ex Odd 1–63 |
| 72 | 176 | Ex 12, 26 | 176 | Ex 23, 33 | 181–182 | Ex 37, 55 |
| | 179 | Ex 20, 22, 24 | 179 | Ex 8, 16, 28 | 185 | Ex 28, 38, 56 |
| | 181–182 | Reading Ex 1–3 | 181–182 | Reading Ex 1–3 | 187 | Reading Ex 1–3 |
| | | Ex Even 2–32 | | Ex Even 2–50 | | Ex Even 2–22 |
| 73 | 179 | Ex 21, 23 | 179 | Ex 20, 28 | 185 | Ex 42, 58 |
| | 181–182 | Ex 7, 17, 29 | 181–182 | Ex 17, 27, 45 | 187 | Ex 7, 15, 19 |
| | 185 | Reading Ex 1–3 | 185 | Reading Ex 1–3 | 190 | Reading Ex 1–4 |
| | | Ex Odd 1–33 | | Ex Odd 1–47 | | Ex Odd 1–35 |
| 74 | 181–182 | Ex 15, 25 | 181–182 | Ex 25, 37 | 187 | Ex 13, 21 |
| | 185 | Ex 8, 16, 22 | 185 | Ex 8, 32, 38 | 190 | Ex 8, 24, 32 |
| | 187 | Reading Ex 1–3 | 187 | Reading Ex 1–3 | 193–194 | Reading Ex 1–3 |
| | | Ex 1–5 | | Ex Even 2–18 | | Ex Even 2–36 |
| 75 | 185 | Ex 18, 24 | 185 | Ex 30, 46 | 190 | Ex 18, 34 |
| | 187 | Ex 6, 8 | 187 | Ex 3, 11, 15 | 193–194 | Ex 9, 25, 31 |
| | 190 | Reading Ex 1–4 | 190 | Reading Ex 1–4 | 197–198 | Reading Ex 1–5 |
| | | Ex 1–10 | | Ex Odd 1–27 | | Ex Odd 1–25 |
| 76 | 187 | Ex 7, 9 | 187 | Ex 7, 17 | 193–194 | Ex 23, 35 |
| | 190 | Ex 11, 13, 15 | 190 | Ex 8, 16, 24 | 197–198 | Ex 8, 22, 26 |
| | 193–194 | Reading Ex 1–3 | 193–194 | Reading Ex 1–3 | 199 | Ex Even 2–16 |
| | | Ex 1–13 | | Ex Even 2–28 | | |

| Day | Page | Level A | Page | Level B | Page | Level C |
|-----|------|---------|------|---------|------|---------|
| 77 | 190 | Ex 12, 14 | 190 | Ex 20, 26 | 200 | Chapter Review |
|  | 193–194 | Ex 14, 16, 18 | 193–194 | Ex 9, 19, 25 |  |  |
|  | 197–198 | Reading Ex 1–5 | 197–198 | Reading Ex 1–5 |  |  |
|  |  | Ex Odd 1–15 |  | Ex Odd 1–21 |  |  |
| 78 | 193–194 | Ex 15, 17 | 193–194 | Ex 21, 27 | 201 | Chapter Test |
|  | 197–198 | Ex 2, 8, 16 | 197–198 | Ex 10, 16, 20 |  |  |
|  | 199 | Ex Even 2–12 | 199 | Ex Even 2–14 |  |  |
| 79 | 200 | Chapter Review | 200 | Chapter Review | 207–208 | Orals Odd 1–17 |
|  |  |  |  |  |  | Ex Odd 1–43 |
| 80 | 201 | Chapter Test | 201 | Chapter Test | 207–208 | Ex 12, 36, 40 |
|  |  |  |  |  | 211 | Ex Even 2–42 |
| 81 | 207–208 | Ex 1–8, 11–23 | 207–208 | Ex Odd 1–39 | 207–208 | Ex 32, 44 |
|  |  |  |  |  | 211 | Ex 19, 33, 41 |
|  |  |  |  |  | 214 | Ex Odd 1–35 |
|  |  |  |  |  |  | Non-Routine Problems |
| 82 | 207–208 | Ex 9, 24, 26 | 207–208 | Ex 8, 22, 32 | 211 | Ex 35, 39 |
|  | 211 | Ex 1–19 | 211 | Ex Even 2–36 | 214 | Ex 12, 26, 30 |
|  |  |  |  |  | 217 | Ex Even 2–38 |
| 83 | 207–208 | Ex 10, 25 | 207–208 | Ex 18, 38 | 214 | Ex 28, 34 |
|  | 211 | Ex 20, 22, 24 | 211 | Ex 3, 19, 29 | 217 | Ex 15, 23, 35 |
|  | 214 | Ex 1–15 | 214 | Ex Odd 1–29 | 219–220 | Reading Ex 1–3 |
|  |  |  |  |  |  | Ex Odd 1–39 |
|  |  |  |  |  |  | Non-Routine Problems |
| 84 | 211 | Ex 21, 23 | 211 | Ex 21, 31 | 217 | Ex 29, 37 |
|  | 214 | Ex 16, 18, 20 | 214 | Ex 8, 18, 24 | 219–220 | Ex 14, 28, 36 |
|  | 217 | Ex 1–19 | 217 | Ex Even 2–32 | 222–223 | Ex Even 2–36 |
| 85 | 214 | Ex 17, 19 | 214 | Ex 14, 26 | 219–220 | Ex 24, 40 |
|  | 217 | Ex 20, 22, 24 | 217 | Ex 13, 21, 29 | 222–223 | Ex 13, 25, 33 |
|  | 219–220 | Reading Ex 1–3 | 219–220 | Reading Ex 1–3 | 227 | Ex Odd 1–23 |
|  |  | Ex 1–13 |  | Ex Odd 1–31 | 228 | Applications Even 2–6 |
| 86 | 217 | Ex 21, 23 | 217 | Ex 19, 31 | 230 | Chapter Review |
|  | 219–220 | Ex 14, 16, 18 | 219–220 | Ex 8, 18, 28 |  |  |
|  | 222–223 | Ex 1–15 | 222–223 | Ex Even 2–30 |  |  |
| 87 | 219–220 | Ex 15, 17 | 219–220 | Ex 16, 30 | 231 | Chapter Test |
|  | 222–223 | Ex 16, 18, 20 | 222–223 | Ex 5, 19, 25 |  |  |
|  | 227 | Ex 1–11, 13, 14 | 227 | Ex Odd 1–23 |  |  |

| Day | Page | Level A | Page | Level B | Page | Level C |
|-----|------|---------|------|---------|------|---------|
| 88 | 222–223 | Ex 17, 19 | 222–223 | Ex 17, 27 | 237 | Ex Even 2–26 |
|    | 227 | Ex 12, 15, 16 | 227 | Ex 8, 16, 22 |  |  |
|    | 228 | Applications All | 228 | Applications All |  |  |
| 89 | 230 | Chapter Review | 230 | Chapter Review | 237 | Ex 11, 21, 27 |
|    |  |  |  |  | 240–241 | Reading Ex 1–4 |
|    |  |  |  |  |  | Ex Odd 1–35 |
| 90 | 231 | Chapter Test | 231 | Chapter Test | 237 | Ex 19, 25 |
|    |  |  |  |  | 240–241 | Ex 16, 30, 34 |
|    |  |  |  |  | 244 | Reading Ex 1–4 |
|    |  |  |  |  |  | Ex Odd 1–17 |
| 91 | 237 | Ex 1–6, 9–11 | 237 | Ex Even 2–24 | 240–241 | Ex 26, 32 |
|    |  |  |  |  | 244 | Ex 8, 16, 18 |
|    |  |  |  |  | 246–247 | Reading Ex 1–4 |
|    |  |  |  |  |  | Ex Odd 1–35 |
| 92 | 237 | Ex 7, 12, 14 | 237 | Ex 7, 13, 19 | 244 | Ex 12, 14 |
|    | 240–241 | Reading Ex 1–4 | 240–241 | Reading Ex 1–4 | 246–247 | Ex 16, 24, 32 |
|    |  | Ex 1–15 |  | Ex Odd 1–29 | 249–250 | Ex Even 2–36 |
|    |  |  |  |  |  | Non-Routine Problems |
| 93 | 237 | Ex 8, 13 | 237 | Ex 11, 23 | 246–247 | Ex 28, 30 |
|    | 240–241 | Ex 16, 18, 20 | 240–241 | Ex 10, 14, 26 | 249–250 | Ex 13, 27, 33 |
|    | 244 | Reading Ex 1–4 | 244 | Reading Ex 1–4 | 252 | Ex Odd 1–25 |
|    |  | Ex Odd 1–15 |  | Ex Odd 1–15 |  |  |
| 94 | 240–241 | Ex 17, 19 | 240–241 | Ex 18, 30 | 249–250 | Ex 23, 25 |
|    | 244 | Ex 4, 8, 12 | 244 | Ex 4, 8, 12 | 252 | Ex 14, 22, 24 |
|    | 246–247 | Reading Ex 1–4 | 246–247 | Reading Ex 1–4 | 254 | Ex Even 2–32 |
|    |  | Ex 1–15 |  | Ex Odd 1–29 |  |  |
| 95 | 244 | Ex 6, 14 | 244 | Ex 6, 14 | 252 | Ex 18, 26 |
|    | 246–247 | Ex 16, 18, 20 | 246–247 | Ex 8, 18, 24 | 254 | Ex 17, 23, 31 |
|    | 249–250 | Ex 1–12, 22, 24, 26 | 249–250 | Ex Even 2–32 | 257 | Reading Ex 1–4 |
|    |  |  |  |  |  | Ex Odd 1–39 |
|    |  |  |  |  | 258 | Applications |
|    |  |  |  |  |  | Even 2–14 |
| 96 | 246–247 | Ex 17, 19 | 246–247 | Ex 16, 28 | 260 | Chapter Review |
|    | 249–250 | Ex 13, 17, 25 | 249–250 | Ex 15, 23, 27 |  |  |
|    | 252 | Ex 1–9, 13, 15 | 252 | Ex Odd 1–21 |  |  |
| 97 | 249–250 | Ex 15, 23 | 249–250 | Ex 13, 29 | 261 | Chapter Test |
|    | 252 | Ex 10, 12, 14 | 252 | Ex 4, 14, 20 |  |  |
|    | 254 | Ex 1–17, 21, 23 | 254 | Ex Even 2–30 |  |  |

| Day | Page | Level A | Page | Level B | Page | Level C |
|-----|------|---------|------|---------|------|---------|
| 98 | 252 | Ex 11, 16 | 252 | Ex 12, 22 | 267 | Reading Ex 1–3 |
|    | 254 | Ex 18, 20, 24 | 254 | Ex 11, 21, 27 |    | Ex Odd 1–23 |
|    | 257 | Reading Ex 1–4 | 257 | Reading Ex 1–4 | 268 | Applications All |
|    |     | Ex Odd 1–27 |     | Ex Odd 1–31 |    |    |
| 99 | 254 | Ex 19, 22 | 254 | Ex 19, 29 | 267 | Ex 6, 16, 24 |
|    | 257 | Ex 6, 20, 26 | 257 | Ex 12, 22, 30 | 272 | Ex Even 2–16 |
|    | 258 | Applications Ex 1–10 | 258 | Applications All |    |    |
| 100 | 260 | Chapter Review | 260 | Chapter Review | 267 | Ex 10, 22 |
|    |     |     |     |     | 272 | Ex 5, 9, 15 |
|    |     |     |     |     | 275 | Ex Odd 1–17 |
| 101 | 261 | Chapter Test | 261 | Chapter Test | 272 | Ex 7, 13 |
|    |     |     |     |     | 275 | Ex 8, 12, 16 |
|    |     |     |     |     | 278 | Ex Even 2–22 |
|    |     |     |     |     |     | Non-Routine Problems |
|    |     |     |     |     | 279 | Applications All |
| 102 | 267 | Reading Ex 1–3 | 267 | Reading Ex 1–3 | 275 | Ex 10, 14 |
|    |     | Ex 1–7, 11, 12 |     | Ex Odd 1–19 | 278 | Ex 5, 15, 21 |
|    | 268 | Applications All | 268 | Applications All | 282 | Ex Odd 1–47 |
| 103 | 267 | Ex 8, 10, 13 | 267 | Ex 4, 8, 16 | 278 | Ex 13, 19 |
|    | 272 | Ex Even 2–10 | 272 | Ex Even 2–12 | 282 | Ex 24, 34, 44 |
|    |     |     |     |     | 285–286 | Reading Ex 1–3 |
|    |     |     |     |     |     | Ex Even 2–12 |
| 104 | 267 | Ex 9, 14 | 267 | Ex 6, 18 | 282 | Ex 38, 46 |
|    | 272 | Ex 1, 5, 9 | 272 | Ex 3, 7, 11 | 285–286 | Ex 3, 7, 11 |
|    | 275 | Ex Odd 1–11 | 275 | Ex Odd 1–13 | 287 | Ex Odd 1–13, Ex 14 |
| 105 | 272 | Ex 3, 7 | 272 | Ex 5, 9 | 288 | Chapter Review |
|    | 275 | Ex 6, 10, 12 | 275 | Ex 4, 8, 14 |    |    |
|    | 278 | Ex Even 2–14 | 278 | Ex Even 2–18 |    |    |
|    | 279 | Applications All | 279 | Applications All |    |    |
| 106 | 275 | Ex 4, 8 | 275 | Ex 6, 10 | 289 | Chapter Test |
|    | 278 | Ex 5, 9, 13 | 278 | Ex 7, 11, 17 |    |    |
|    | 282 | Ex Odd 1–27 | 282 | Ex Odd 1–39 |    |    |
| 107 | 278 | Ex 7, 11 | 278 | Ex 9, 15 | 295 | Reading Ex 1–5 |
|    | 282 | Ex 10, 18, 24 | 282 | Ex 12, 24, 32 |    | Ex Odd 1–29 |
|    | 285–286 | Reading Ex 1–3 | 285–286 | Reading Ex 1–3 |    |    |
|    |     | Ex Odd 1–7, Ex 8 |     | Ex Odd 1–11 |    |    |

| Day | Page | Level A | Page | Level B | Page | Level C |
|-----|------|---------|------|---------|------|---------|
| 108 | 282 | Ex 14, 26 | 282 | Ex 26, 34 | 295 | Ex 18, 24, 26 |
|     | 285–286 | Ex 2, 4, 6 | 285–286 | Ex 4, 8, 10 | 298 | Ex Even 2–26 |
|     | 287 | Ex Odd 1–13 | 287 | Ex Odd 1–13 |     | Non-Routine Problems |
| 109 | 288 | Chapter Review | 288 | Chapter Review | 295 | Ex 22, 28 |
|     |     |     |     |     | 298 | Ex 7, 17, 25 |
|     |     |     |     |     | 301 | Ex Odd 1–27 |
| 110 | 289 | Chapter Test | 289 | Chapter Test | 298 | Ex 23, 27 |
|     |     |     |     |     | 301 | Ex 10, 22, 28 |
|     |     |     |     |     | 304 | Ex Even 2–22 |
| 111 | 295 | Reading Ex 1–5 | 295 | Reading Ex 1–5 | 301 | Ex 16, 26 |
|     |     | Ex 1–13, 17, 18, 21, 22 |     | Ex 1–14 | 304 | Ex 11, 15, 21 |
|     |     |     |     | Ex Odd 17–25 | 307 | Ex Odd 1–31 |
| 112 | 295 | Ex 14, 16, 19 | 295 | Ex 15, 18, 22 | 304 | Ex 17, 19 |
|     | 298 | Ex 1–12, 16–21 | 298 | Ex Even 2–26 | 307 | Ex 10, 24, 30 |
|     |     |     |     |     | 309 | Ex Even 2–20 |
| 113 | 295 | Ex 15, 20 | 295 | Ex 16, 20 | 307 | Ex 22, 32 |
|     | 298 | Ex 13, 15, 23 | 298 | Ex 7, 17, 25 | 309 | Ex 5, 13, 19 |
|     | 301 | Ex 1–10 | 301 | Ex Odd 1–23 | 313 | Ex Odd 1–27 |
| 114 | 298 | Ex 14, 22 | 298 | Ex 23, 27 | 309 | Ex 9, 17 |
|     | 301 | Ex 11, 13, 15 | 301 | Ex 4, 14, 20 | 313 | Ex 18, 24, 26 |
|     | 304 | Ex 1–7, 10, 13 | 304 | Ex Even 2–18 | 316 | Reading Ex 1–3 |
|     |     |     |     |     |     | Ex Even 2–22 |
| 115 | 301 | Ex 12, 14 | 301 | Ex 8, 18 | 313 | Ex 22, 28 |
|     | 304 | Ex 8, 11, 14 | 304 | Ex 7, 11, 15 | 316 | Ex 7, 17, 23 |
|     | 307 | Ex 1–10 | 307 | Ex Odd 1–25 | 318 | Ex Odd 1–21 |
| 116 | 304 | Ex 9, 12 | 304 | Ex 9, 17 | 316 | Ex 15, 21 |
|     | 307 | Ex 11, 13, 15 | 307 | Ex 6, 14, 24 | 318 | Ex 10, 16, 20 |
|     | 309 | Ex 1–4, 7–9, 11, 12 | 309 | Ex Odd 1–17 | 321–322 | Ex 2, 6, 10, 14, 18, 22, 26, 30, 34, 38, 42, 46, 50, 54, 60 |
|     |     |     |     |     |     | Non-Routine Problems |
| 117 | 307 | Ex 12, 14 | 307 | Ex 12, 26 | 318 | Ex 8, 18 |
|     | 309 | Ex 5, 10, 13 | 309 | Ex 6, 10, 14 | 321–322 | Ex 41, 47, 57 |
|     | 313 | Ex Odd 1–19 | 313 | Ex Odd 1–23 | 324 | Applications All |

| Day | Page | Level A | Page | Level B | Page | Level C |
| --- | --- | --- | --- | --- | --- | --- |
| 118 | 309 | Ex 6, 14 | 309 | Ex 4, 16 | 326 | Chapter Review |
| | 313 | Ex 4, 8, 16 | 313 | Ex 6, 16, 24 | | |
| | 316 | Reading Ex 1–3 | 316 | Reading Ex 1–3 | | |
| | | Ex 1–9, 13 | | Ex Even 2–20 | | |
| 119 | 313 | Ex 6, 18 | 313 | Ex 8, 22 | 327 | Chapter Test |
| | 316 | Ex 10, 12, 14 | 316 | Ex 9, 15, 21 | | |
| | 318 | Ex Odd 1–15 | 318 | Ex Odd 1–17 | | |
| 120 | 316 | Ex 11, 15 | 316 | Ex 3, 15 | 332 | Ex Odd 1–11 |
| | 318 | Ex 4, 12, 14 | 318 | Ex 4, 10, 18 | | |
| | 321–322 | Ex Even 2–28 | 321–322 | Ex 2, 6, 10, 14, 18, 22, 26, 30, 34, 38, 42, 46, 50 | | |
| 121 | 318 | Ex 2, 10 | 318 | Ex 12, 16 | 332 | Ex 2, 6, 12 |
| | 321–322 | Ex 11, 15, 25 | 321–322 | Ex 13, 25, 47 | 334 | Ex Even 2–14 |
| | 324 | Applications All | 324 | Applications All | | |
| 122 | 326 | Chapter Review | 326 | Chapter Review | 332 | Ex 8, 10 |
| | | | | | 334 | Ex 3, 11, 15 |
| | | | | | 337 | Ex Odd 1–27 |
| 123 | 327 | Chapter Test | 327 | Chapter Test | 334 | Ex 9, 13 |
| | | | | | 337 | Ex 8, 14, 22 |
| | | | | | 340 | Ex Even 2–16 |
| 124 | 332 | Ex Odd 1–11 | 332 | Ex Odd 1–11 | 337 | Ex 20, 24 |
| | | | | | 340 | Ex 7, 11, 15 |
| | | | | | 343 | Ex Odd 1–21 |
| 125 | 332 | Ex 2, 6, 10 | 332 | Ex 2, 6, 10 | 340 | Ex 9, 13 |
| | 334 | Ex Even 2–12 | 334 | Ex Even 2–12 | 343 | Ex 10, 18, 20 |
| | | | | | 346 | Ex Even 2–20 |
| | | | | | | Non-Routine Problems |
| 126 | 332 | Ex 4, 8 | 332 | Ex 4, 8 | 343 | Ex 12, 16 |
| | 334 | Ex 1, 5, 9 | 334 | Ex 3, 7, 11 | 346 | Ex 9, 17, 19 |
| | 337 | Ex Odd 1–15 | 337 | Ex Odd 1–21 | 349–350 | Reading Ex 1, 2 |
| | | | | | | Ex Odd 1–15 |
| 127 | 334 | Ex 3, 7 | 334 | Ex 5, 9 | 346 | Ex 11, 15 |
| | 337 | Ex 2, 8, 14 | 337 | Ex 4, 10, 18 | 349–350 | Ex 8, 12, 16 |
| | 340 | Ex Even 2–10 | 340 | Ex Even 2–14 | 353–354 | Reading Ex 1–4 |
| | | | | | | Ex Even 2–12 |

| Day | Page | Level A | Page | Level B | Page | Level C |
|---|---|---|---|---|---|---|
| 128 | 337 | Ex 4, 12 | 337 | Ex 8, 16 | 349–350 | Ex 10, 14 |
| | 340 | Ex 1, 5, 9 | 340 | Ex 3, 7, 13 | 353–354 | Ex 3, 7, 11 |
| | 343 | Ex Odd 1–15 | 343 | Ex Odd 1–17 | 355 | Ex Odd 1–15 |
| | | | | | 357 | Applications All |
| 129 | 340 | Ex 3, 7 | 340 | Ex 5, 11 | 358 | Chapter Review |
| | 343 | Ex 2, 8, 14 | 343 | Ex 6, 14, 18 | | |
| | 346 | Ex Even 2–16 | 346 | Ex Even 2–18 | | |
| 130 | 343 | Ex 4, 10 | 343 | Ex 10, 16 | 359 | Chapter Test |
| | 346 | Ex 3, 7, 13 | 346 | Ex 3, 11, 17 | | |
| | 349–350 | Reading Ex 1, 2 | 349–350 | Reading Ex 1, 2 | | |
| | | Ex Odd 1–9 | | Ex Odd 1–13 | | |
| 131 | 346 | Ex 1, 9 | 346 | Ex 5, 15 | 367 | Ex 1, 2, 4–6, 8–10 |
| | 349–350 | Ex 2, 6, 10 | 349–350 | Ex 4, 8, 12 | | Ex Odd 13–19 |
| | 353–354 | Reading Ex 1–4 | 353–354 | Reading Ex 1–4 | | |
| | | Ex Even 2–8 | | Ex Even 2–10 | | |
| 132 | 349–350 | Ex 4, 8 | 349–350 | Ex 6, 14 | 367 | Ex 7, 14, 18 |
| | 353–354 | Ex 3, 7 | 353–354 | Ex 1, 5, 9 | 370 | Ex Odd 1–29 |
| | 355 | Ex Odd 1–11 | 355 | Ex Odd 1–13 | | |
| 133 | 353–354 | Ex 1, 5 | 353–354 | Ex 3, 7 | 367 | Ex 11, 16 |
| | 355 | Ex 2, 6, 10 | 355 | Ex 4, 8, 12 | 370 | Ex 14, 22, 30 |
| | 357 | Applications All | 357 | Applications All | 372 | Ex Even 2–36 |
| 134 | 358 | Chapter Review | 358 | Chapter Review | 370 | Ex 20, 26 |
| | | | | | 372 | Ex 13, 23, 31 |
| | | | | | 375–376 | Reading Ex 1–5 |
| | | | | | | Ex Odd 1–23 |
| 135 | 359 | Chapter Test | 359 | Chapter Test | 372 | Ex 25, 33 |
| | | | | | 375–376 | Ex 8, 14, 22 |
| | | | | | 379–380 | Ex Even 2–28 |
| | | | | | | Non-Routine Problems |
| | | | | | 381 | Applications Even |
| 136 | 367 | Ex 1 | 367 | Ex 1, 2, 4–6, 8–10, | 375–376 | Ex 16, 20 |
| | | Ex Even 2–10 | | 13, 15, 17 | 379–380 | Ex 11, 21, 27 |
| | | | | | 381 | Applications Ex 3, 5, 7 |
| | | | | | 383 | Ex Odd 5–11 |
| 137 | 367 | Ex 3, 7, 11 | 367 | Ex 3, 11, 16 | 384 | Chapter Review |
| | 370 | Ex 1–15 | 370 | Ex Odd 1–29 | | |

| Day | Page | Level A | Page | Level B | Page | Level C |
|---|---|---|---|---|---|---|
| 138 | 367 | Ex 5, 9 | 367 | Ex 7, 12 | 385 | Chapter Test |
| | 370 | Ex 16, 18, 20 | 370 | Ex 8, 18, 26 | | |
| | 372 | Ex 1–4, 6–9, 11–14, 16, 18, 20 | 372 | Ex Even 2–28 | | |
| 139 | 370 | Ex 17, 19 | 370 | Ex 16, 22 | 390 | Ex Odd 1–25 |
| | 372 | Ex 5, 15, 19 | 372 | Ex 9, 19, 25 | | |
| | 375–376 | Reading Ex 1–5 Ex 1–3, 6–8, 10, 12, 14 | 375–376 | Reading Ex 1–5 Ex Odd 1–19 | | |
| 140 | 372 | Ex 10, 21 | 372 | Ex 13, 27 | 390 | Ex 10, 18, 22 |
| | 375–376 | Ex 4, 9, 13 | 375–376 | Ex 4, 8, 16 | 393 | Ex Even 2–20 |
| | 379–380 | Ex 1–3, 6, 7, 9, 10, 12, 14 | 379–380 | Ex Even 2–24 | | |
| 141 | 375–376 | Ex 5, 11 | 375–376 | Ex 6, 14 | 390 | Ex 16, 24 |
| | 379–380 | Ex 4, 8, 15 | 379–380 | Ex 9, 15, 21 | 393 | Ex 5, 11, 21 |
| | 381 | Applications Even | 381 | Applications Even | 395 | Ex Odd 1–35 |
| 142 | 379–380 | Ex 11, 13 | 379–380 | Ex 17, 23 | 393 | Ex 17, 19 |
| | 381 | Applications Ex 1, 3, 5 | 381 | Applications Ex 1, 3, 7 | 395 | Ex 20, 28, 34 |
| | 383 | Ex Odd 1–9 | 383 | Ex Odd 1–11 | 398 | Reading Ex 1–3 Ex Odd 1–41 Non-Routine Problems |
| 143 | 384 | Chapter Review | 384 | Chapter Review | 395 | Ex 26, 36 |
| | | | | | 398 | Ex 20, 32, 40 |
| | | | | | 401 | Orals All Ex Even 2–52 |
| 144 | 385 | Chapter Test | 385 | Chapter Test | 398 | Ex 28, 36 |
| | | | | | 401 | Ex 25, 41, 51 |
| | | | | | 403 | Reading Ex 1–5 Ex Odd 1–33 |
| 145 | 390 | Ex 1–8, 14 | 390 | Ex Odd 1–19 | 401 | Ex 47, 53 |
| | | | | | 403 | Ex 18, 26, 30 |
| | | | | | 405 | Ex Even 2–38 |
| 146 | 390 | Ex 9, 11, 13 | 390 | Ex 8, 14, 18 | 403 | Ex 24, 32 |
| | 393 | Ex 1–4, 7–10, 13–16 | 393 | Ex 1–4, 7–10, 13–16, 18, 20 | 405 | Ex 17, 29, 37 |
| | | | | | 407–408 | Ex 3, 7, 11 Ex Odd 15–43 |
| 147 | 390 | Ex 10, 12 | 390 | Ex 12, 16 | 405 | Ex 23, 33 |
| | 393 | Ex 6, 12, 17 | 393 | Ex 6, 12, 19 | 407–408 | Ex 18, 34, 42 |
| | 395 | Ex 1–16 | 395 | Ex Odd 1–31 | 411 | Ex Even 2–40 |

| Day | Page | Level A | Page | Level B | Page | Level C |
|-----|------|---------|------|---------|------|---------|
| 148 | 393 | Ex 5, 11 | 393 | Ex 11, 17 | 407–408 | Ex 30, 38 |
|     | 395 | Ex 17, 19, 21 | 395 | Ex 6, 20, 28 | 411 | Ex 17, 31, 37 |
|     | 398 | Reading Ex 1–3 | 398 | Reading Ex 1–3 | 414–415 | Ex Odd 5–37 |
|     |     | Ex 1–8, 11–18, 21–24 |     | Ex Odd 1–35 | 420 | Applications Ex Odd 1–7 |
| 149 | 395 | Ex 18, 20 | 395 | Ex 16, 26 | 422 | Chapter Review |
|     | 398 | Ex 10, 20, 25 | 398 | Ex 8, 24, 34 |     |     |
|     | 401 | Ex 1–7, 9–25 | 401 | Ex Even 2–44 |     |     |
| 150 | 398 | Ex 9, 19 | 398 | Ex 18, 30 | 423 | Chapter Test |
|     | 401 | Ex 8, 26, 28 | 401 | Ex 7, 21, 37 |     |     |
|     | 403 | Reading Ex 1–5 | 403 | Reading Ex 1–5 |     |     |
|     |     | Ex 1–8, 11–17 |     | Ex Odd 1–29 |     |     |
| 151 | 401 | Ex 27, 29 | 401 | Ex 23, 41 | 428 | Orals Even |
|     | 403 | Ex 10, 18, 20 | 403 | Ex 6, 16, 28 |     | Ex Odd 1–37 |
|     | 405 | Ex Even 2–26 | 405 | Ex Even 2–36 |     |     |
| 152 | 403 | Ex 9, 19 | 403 | Ex 18, 26 | 428 | Ex 16, 30, 34 |
|     | 405 | Ex 7, 15, 23 | 405 | Ex 11, 19, 31 | 430–431 | Reading Ex 1–3 |
|     | 407–408 | Ex Odd 1–29 | 407–408 | Ex Odd 1–31, 35, 37 |     | Ex Odd 1–45 |
|     |     |     |     |     |     | Non-Routine Problems |
| 153 | 405 | Ex 13, 19 | 405 | Ex 17, 33 | 428 | Ex 26, 36 |
|     | 407–408 | Ex 8, 18, 26 | 407–408 | Ex 14, 24, 36 | 430–431 | Ex 20, 32, 42 |
|     | 411 | Ex 1–12, 16–23 | 411 | Ex Even 2–34 | 434 | Ex Odd 1–41 |
| 154 | 407–408 | Ex 14, 28 | 407–408 | Ex 20, 34 | 430–431 | Ex 38, 44 |
|     | 411 | Ex 13, 15, 25 | 411 | Ex 9, 21, 33 | 434 | Ex 24, 30, 38 |
|     | 414–415 | Ex Odd 1–19 | 414–415 | Ex Odd 5–35 | 437 | Ex Even 2–30 |
| 155 | 411 | Ex 14, 24 | 411 | Ex 19, 35 | 434 | Ex 36, 40 |
|     | 414–415 | Ex 4, 12, 18 | 414–415 | Ex 10, 24, 32 | 437 | Ex 11, 21, 27 |
|     | 420 | Applications All | 420 | Applications All | 440 | Ex Odd 1–15 |
| 156 | 422 | Chapter Review | 422 | Chapter Review | 437 | Ex 25, 29 |
|     |     |     |     |     | 440 | Ex 6, 12, 16 |
|     |     |     |     |     | 443 | Ex Even 2–10 |
| 157 | 423 | Chapter Test | 423 | Chapter Test | 440 | Ex 8, 14 |
|     |     |     |     |     | 443 | Ex 3, 7, 9 |
|     |     |     |     |     | 444 | Ex Odd 1–13 |
| 158 | 428 | Ex 1–15 | 428 | Ex Odd 1–31 | 443 | Ex 1, 5 |
|     |     |     |     |     | 444 | Ex 6, 10, 14 |
|     |     |     |     |     | 448 | Ex 1–4, 7–10, 14, 16 |

| Day | Page | Level A | Page | Level B | Page | Level C |
|---|---|---|---|---|---|---|
| 159 | 428 | Ex 16, 18, 20 | 428 | Ex 14, 24, 30 | 444 | Ex 8, 12 |
| | 430–431 | Reading Ex 1–3 | 430–431 | Reading Ex 1–3 | 448 | Ex 12, 13, 15 |
| | | Ex 1–7, 10–14, 16–22 | | Ex Even 2–38 | 449 | Applications All |
| 160 | 428 | Ex 17, 19 | 428 | Ex 22, 32 | 450 | Chapter Review |
| | 430–431 | Ex 9, 15, 24 | 430–431 | Ex 11, 23, 33 | | |
| | 434 | Ex Odd 1–23 | 434 | Ex Odd 1–35 | | |
| 161 | 430–431 | Ex 8, 23 | 430–431 | Ex 21, 37 | 451 | Chapter Test |
| | 434 | Ex 8, 16, 24 | 434 | Ex 14, 22, 32 | | |
| | 437 | Ex 1–4, 7–13 | 437 | Ex Even 2–26 | | |
| 162 | 434 | Ex 14, 22 | 434 | Ex 24, 34 | 458 | Reading Ex 1–4 |
| | 437 | Ex 6, 14, 16 | 437 | Ex 5, 13, 19 | | Ex 1, 2, Odd 3–9 |
| | 440 | Ex Odd 1–9 | 440 | Ex Odd 1–13 | | |
| 163 | 437 | Ex 5, 15 | 437 | Ex 11, 25 | 458 | Ex 4, 8, 10 |
| | 440 | Ex 2, 6, 10 | 440 | Ex 4, 8, 12 | 461 | Ex 1–8, 12, 14, 16 |
| | 443 | Ex Even 2–8 | 443 | Ex Even 2–10 | | Non-Routine Problems |
| 164 | 440 | Ex 4, 8 | 440 | Ex 2, 6 | 458 | Ex 2, 6 |
| | 443 | Ex 1, 5 | 443 | Ex 3, 9 | 461 | Ex 10, 11, 15 |
| | 444 | Ex Odd 1–9 | 444 | Ex Odd 1–11 | 464 | Ex 1–5, 9, 11 |
| 165 | 443 | Ex 3, 7 | 443 | Ex 1, 7 | 461 | Ex 9, 13 |
| | 444 | Ex 2, 6, 10 | 444 | Ex 4, 8, 12 | 464 | Ex 6, 8, 12 |
| | 448 | Ex 1–4, 7–10 | 448 | Ex 1–4, 7–10, 13 | 466 | Ex 1–9, 11–21, 23–25, 27, 29 |
| 166 | 444 | Ex 4, 8 | 444 | Ex 6, 10 | 464 | Ex 7, 10 |
| | 448 | Ex 6, 12 | 448 | Ex 6, 12, 14 | 466 | Ex 10, 26, 30 |
| | 449 | Applications All | 449 | Applications All | 468–469 | Ex Odd 1–19 |
| 167 | 450 | Chapter Review | 450 | Chapter Review | 466 | Ex 22, 28 |
| | | | | | 468–469 | Ex 6, 10, 20 |
| | | | | | 472 | Ex Even 2–16 |
| 168 | 451 | Chapter Test | 451 | Chapter Test | 468–469 | Ex 8, 16 |
| | | | | | 472 | Ex 5, 9, 13 |
| | | | | | 475–476 | Ex Odd 1–13 |
| 169 | 458 | Reading Ex 1–4 | 458 | Reading Ex 1–4 | 478 | Chapter Review |
| | | Ex 1–3, 5, 7 | | Ex 1, 2, Odd 3–9 | | |
| 170 | 458 | Ex 4, 6, 8 | 458 | Ex 4, 8, 10 | 479 | Chapter Test |
| | 461 | Ex 1–12 | 461 | Ex 1–12, 14, 16 | | |

# HOLT
# ALGEBRA 1

# About the Authors

**Eugene D. Nichols**
Robert O. Lawton Distinguished Professor
of Mathematics Education
Florida State University
Tallahassee, Florida

**Mervine L. Edwards**
Chairman of Mathematics Department
Shore Regional High School
West Long Branch, New Jersey

**E. Henry Garland**
Head of the Mathematics Department
Developmental Research School
DRS Professor
Florida State University
Tallahassee, Florida

**Sylvia A. Hoffman**
Resource Consultant in Mathematics
Illinois State Board of Education
State of Illinois

**Albert Mamary**
Superintendent of Schools for Instruction
Johnson City Central School District
Johnson City, New York

**William F. Palmer**
Professor of Education and Director
Center for Mathematics and Science Education
Catawba College
Salisbury, North Carolina

# HOLT
# ALGEBRA 1

**Eugene D. Nichols**
**Mervine L. Edwards**
**E. Henry Garland**
**Sylvia A. Hoffman**
**Albert Mamary**
**William F. Palmer**

*HOLT, RINEHART AND WINSTON, PUBLISHERS*
*New York · Toronto · Mexico City · London · Sydney · Tokyo*

# Acknowledgments for Photographs

**Cover:** Windsurfer, Alastair Black/Focus on Sports.

*Page x,* Julius Baum/Bruce Coleman; *20,* S.L. Craig, Jr./Bruce Coleman; *27,* Bettmann; *63,* HRW photo by Russell Dian; *85,* HRW photo by Richard Haynes; *109,* Hazel Hankin/Stock, Boston; *117,* Alvis Upitis/Image Bank; *147,* Linda Impastato; *170,* Pat Lanza Field/Bruce Coleman; *204,* Granger; *228,* HRW photo by Richard Haynes; *235,* Tom McHugh/Photo Researchers; *258,* Paul S. Conklin/Monkmeyer; *263,* Chuck O'Rear/West Light; *268,* Sports Photo File; *292,* Culver; *329,* Lew Merrim/Monkmeyer; *364,* Lynn M. Stone/Bruce Coleman; *381,* C.C. Lockwood/Bruce Coleman; *388,* John Crerar Library, Chicago; *425,* Gabe Palmer/Image Bank; *454,* Charles Gupton/Southern Light.

**Insert 1:**
*Page 1,* Richard Pasley/Stock, Boston; *2, tl,* NYPL Picture Collection; *r,* Paul Brierley/Robert Harding Picture Library, London; *bl,* Granger; *3, tl,* Paul Brierley/Robert Harding Picture Library; *tr,* drawing by Leonardo da Vinci, National Library, Madrid; *b,* Paul Brierley/Robert Harding Picture Library; *4, tl,* Granger; *tr,* Charles Gupton/Southern Light; *b,* Bodley Head Ltd., London; *5, tl,* Chuck Place/Image Bank; *tr,* Alan Pitcairn/Grant Heilman; *b,* Werner Muller/Peter Arnold; *6, l,* Karl Hentz/Image Bank; *tr,* NYPL Picture Collection; *br,* John Burroughs Assoc., American Museum of Natural History; *7, t,* Terry A. Renna; *b,* Granger; *8, t,* Runk/Schoenberger/Grant Heilman; *bl,* drawing by Leonardo da Vinci, National Library, Madrid; *br,* drawing by Philippe de Le Hire.

**Insert 2:**
*Page 1,* Linda Dufurrena/Grant Heilman; *2, t,* William Rivelli/Image Bank; *2, b,* NYPL Picture Collection; *3, tl & b,* Bettmann; *tr,* Bohdan Hrynewych/Southern Light; *4, tl,* Yoshiomi Goto/Image Bank; *cl,* Chuck O'Rear/West Light; *bl,* Steve Allen/Peter Arnold; *r,* Jan Halaska/Photo Researchers; *5, tl,* Charles Gupton/Southern Light; *tr,* Bohdan Hrynewych/Southern Light; *b,* Victor Kennett/Robert Harding Picture Library; *6, tl,* Bradley Smith/Gemini Smith; *tr,* Lawrence Migdale/Photo Researchers; *b,* Marty Katz; *7, tl,* Tom Tracy/Stock Shop; *tr,* Bettmann; *b,* Tom Tracy/Stock Shop; *8, tl,* James A. Cook/Stock Broker; *tr,* William James Warren/West Light; *br,* Michael Melford/Peter Arnold; *bl,* Bettmann.

**Insert 3:**
*Page 1,* Charles Nicklin/Ocean Images; *2, l,* courtesy of S. Peter Stevens/Art Resource; *tr,* Dan Budnik/Woodfin Camp & Assoc.; *br,* Bettmann; *3, tl,* Le Courbusier Foundation, Paris; *tr,* Granger; *b,* Granger; *4, tl,* Culver; *tr,* Bill Pierce/Rainbow; *b,* Granger; *5, l,* Wright State University, Dayton, Ohio; *r,* Brown Brothers; *6, tl,* courtesy of Publishers Productions; *tr,* Bettmann; *br,* Granger; *7, tl,* Granger; *br,* Collection, Museum of Modern Art, N.Y., gift of Mrs. Stanley B. Resor; *bl,* Bettmann; *8, l,* Granger; *tr,* Dr. E. Miles, Jr./Miles Color Art; *br,* Scala/Art Resource.

Copyright © 1986 by Holt, Rinehart and Winston, Publishers
All rights reserved
Printed in the United States of America

ISBN: 0-03-002162-6
567890    032    987654321

# Contents

# Symbol List

| | | |
|---|---|---|
| $3^4$ | the fourth power of 3 | 16 |
| $x^3$ | the third power of $x$, or $x$ cubed | 16 |
| $\ldots$ | goes on forever | 28 |
| $<$ | is less than | 29 |
| $>$ | is greater than | 29 |
| $|x|$ | the absolute value of $x$ | 30 |
| $.3\overline{3}$ | the 3 repeats forever | 33 |
| $-a$ | the opposite of $a$ | 38 |
| LCM | least common multiple | 97 |
| $\{\ \}$ | set | 118 |
| $\varnothing$ | empty set | 119 |
| $\leq$ | is less than or equal to | 121 |
| $\geq$ | is greater than or equal to | 121 |
| GCF | greatest common factor | 161 |
| $5:7$ | the ratio 5 to 7 | 225 |
| P($x,y$) | point $P$ with coordinates $x$, $y$ | 294 |
| $\overline{AB}$ | line segment $AB$ | 299 |
| $\overleftrightarrow{MN}$ | line $MN$ | 302 |
| $f(x)$ | the value of $f$ at $x$ | 371 |
| $\sqrt{x}$ | the principal square root of $x$ | 391 |
| $\doteq$ | is approximately equal to | 392 |
| $\triangle ABC$ | triangle $ABC$ | 412 |
| $\pm$ | plus or minus | 430 |
| m$\angle A$ | degree measure of angle $A$ | 459 |
| $\sim$ | is similar to | 459 |
| $\tan A$ | tangent of m$\angle A$ | 462 |
| $\sin A$ | sine of m$\angle A$ | 462 |
| $\cos A$ | cosine of m$\angle A$ | 462 |

# 1 OPERATIONS, VARIABLES, FORMULAS

**Non-Routine Problem Solving**

**Problem 1**

A cyclist completed an 84 km course in three hours. In the first hour she covered $\frac{4}{7}$ of the distance; in the second hour she covered $\frac{1}{2}$ that distance; in the third hour she covered the remaining distance. How far did she travel in the third hour? **12 km**

**Problem 2**

The Jaspers, the Kiwis, the Koalas, and the Blips took part in a soccer competition. Each team played each of the other teams once. Two points were given for a win and one point for a tie. The Jaspers scored five points, the Koalas three points, and the Kiwis one point. The Koalas scored seven of the total thirteen goals. The Blips did not score any goals. The Koalas beat the Kiwis by a score of 4–1. What was the score in the game between the Kiwis and the Jaspers? **Jaspers 1, Kiwis 0**

# RULES FOR ORDER OF OPERATIONS

**Objectives**
**To simplify numerical expressions by using the rules for the order of operations**
**To simplify numerical expressions containing grouping symbols**

There is more than one way to show multiplication:
by using the times sign, $3 \times 6$
by using a raised dot, $3 \cdot 6$ or
by using parentheses, $3(6)$.

Each of these expressions is read as *3 times 6*.

The expression $7 \cdot 2 + 4$ involves both multiplication and addition. There could be two possible ways to simplify this expression.

Multiply first, then add.
$$7 \cdot 2 + 4$$
$$14 + 4$$
$$18$$

Add first, then multiply.
$$7 \cdot 2 + 4$$
$$7 \cdot 6$$
$$42$$

Which answer is correct, 18 or 42? To get a unique (one and only one) answer, the following agreement is made.

| Rules for order of operations | When several operations occur, <br> 1. Do all *multiplications* and *divisions* in order from left to right. <br> 2. Do all *additions* and *subtractions* in order from left to right. |
|---|---|

To simplify $7 \cdot 2 + 4$, multiply 7 and 2 first, then add 4 to the result.
$$7 \cdot 2 + 4$$
$$14 + 4$$
$$18$$

To simplify $10 \div 5 - 1$, divide 10 by 5 first, then subtract 1 from the result.
$$10 \div 5 - 1$$
$$2 - 1$$
$$1$$

**Example 1**

**Simplify.**

$10 - 2 \cdot 3$
$10 - \quad 6$ ◀ *Multiply first,*
$\quad\quad 4$ ◀ *then subtract.*

$18 + 6 \div 2$
$18 + \quad 3$ ◀ *Divide first,*
$\quad\quad 21$ ◀ *then add.*

Rules for Order of Operations

**1**

All four basic operations can be involved in an expression.

*Example 2*     **Simplify 4 · 2 + 9 − 18 ÷ 6 + 2 · 5.**

$$4 \cdot 2 + 9 - \underbrace{18 \div 6} + \underbrace{2 \cdot 5}$$

◀ *Do all multiplications and divisions from left to right.*

| | | | | |
|---|---|---|---|---|
| 8 | + 9 − | 3 | + | 10 |
| | 17 − | 3 | + | 10 |
| | | 14 | + | 10 |
| | | | | 24 |

◀ *Then do all additions and subtractions from left to right.*

To indicate a change in the order of operations, use parentheses or grouping symbols. Consider 4 + (7 + 5) · 2. This can be simplified by doing the operations within the parentheses first. Then, simplify the result by using the rules for the order of operations.

*Example 3*     **Simplify 4 + (7 + 5) · 2.**

4 + (7 + 5) · 2
4 +     12     · 2     ◀ *Do operations within the parentheses first.*
4 +     24             ◀ *Use the rules for the order of operations.*
        28

Parentheses, written ( ), are one type of grouping symbols. Brackets, written [ ], are also used. Since parentheses or brackets can be used to show multiplication as well as grouping, the raised dot is often omitted when parentheses or brackets are used. So, 4 + (7 + 5) · 2 can be written as 4 + (7 + 5)2.

*Example 4*     **Simplify 7(1 + 2) − 15 ÷ 5.**

7(1 + 2) − 15 ÷ 5
  7(3)    − 15 ÷ 5     ◀ *Simplify within the parentheses first.*
   21     −     3      ◀ *Use the rules for the order of operations.*
             18

In an expression like $\dfrac{7 + 8}{5}$, the fraction bar is a grouping symbol which indicates that the numerator and denominator are treated as single expressions. So, $\dfrac{7 + 8}{5}$ means $\dfrac{15}{5}$, or 3.

**Example 5**   **Simplify** $\dfrac{3 \cdot 4 + 8}{7 - 3}$.

$$\dfrac{3 \cdot 4 + 8}{7 - 3}$$

$$\dfrac{12 + 8}{7 - 3}$$   ◀ *Use the rules for order of operations in the numerator.*

$$\dfrac{20}{4}$$   ◀ *Simplify.*

$$5$$   ◀ *Divide.*

## Oral Exercises

1. mult.; add 2. mult.; add 3. div.; add 4. mult.; subt.

**Tell the order of operations you would use to simplify each.**

**1.** $3 + 2 \cdot 4$   **2.** $3 \cdot 2 + 1$   **3.** $4 \div 2 + 1$   **4.** $3 \cdot 2 - 4$   **5.** $7 + 2 - 3$   **6.** $1 + 8 \div 2$

**7.** $\dfrac{8 - 2}{3}$   **8.** $\dfrac{16}{5 - 3}$   **9.** $\dfrac{10 \div 5}{2}$   **10.** $\dfrac{7 + 2}{3}$   **11.** $\dfrac{4 \cdot 3}{6 - 2}$   **12.** $\dfrac{9 + 6}{4 + 1}$

5. add; subt. 6. div.; add 7. subt.; div. 8. subt.; div. 9. div.; div. 10. add; div. 11. mult.; subt.; div. 12. add; add; div.

## Written Exercises

**Simplify.**

Ⓐ  **1.** $5 \cdot 7 + 1$ **36**   **2.** $2 + 6 \cdot 5$ **32**   **3.** $9 \cdot 2 + 6$ **24**

**4.** $27 \div 3 - 5$ **4**   **5.** $14 + 28 \div 7$ **18**   **6.** $6 - 36 \div 9$ **2**

**7.** $8 + 4 \cdot 2$ **16**   **8.** $16 - 32 \div 4$ **8**   **9.** $5 + 7 \cdot 2$ **19**

**10.** $9 - 14 \div 2 + 3 \cdot 4$ **14**   **11.** $7 \cdot 8 - 4 \div 2 + 5 \cdot 6$ **84**   **12.** $14 - 16 \div 8 + 9 \cdot 5$ **57**

**13.** $48 \div 16 - 1 \cdot 2 + 7$ **8**   **14.** $19 - 60 \div 10 + 8 \div 4$ **15**   **15.** $7 \cdot 9 - 21 \div 7 + 4 \cdot 8$ **92**

**16.** $2 + 7 \cdot 4 - 15 \div 3 + 2$   **17.** $20 \div 4 + 3 \cdot 6 - 12 \div 2$   **18.** $8 \cdot 4 + 9 \div 3 - 1 \cdot 5$ **30**

**19.** $(8 - 4)3 + 1$ **13**   **27**   **20.** $6(7 - 5) + 4$ **16**   **17**   **21.** $10 - 3(5 - 2)$ **1**

**22.** $3(7 + 4) - 18 \div 9$ **31**   **23.** $25 \div 5 + (7 + 2)3$ **32**   **24.** $3(8 - 6) + 49 \div 7$ **13**

**25.** $\dfrac{18 - 2}{4}$ **4**   **26.** $\dfrac{30 + 5}{33 - 26}$ **5**   **27.** $\dfrac{27 - 12}{5 - 2}$ **5**

**28.** $\dfrac{5 \cdot 6 + 2}{12 - 4}$ **4**   **29.** $\dfrac{18 + 2 - 4}{7 - 3}$ **4**   **30.** $\dfrac{5 \cdot 4 + 2}{17 - 2 \cdot 3}$ **2**

Ⓑ  **31.** $7 + 2 \cdot 28 - 3 \cdot 9 + 39 \div 3$ **49**   **32.** $3 + 6 \cdot 9 - 5 \cdot 8 + 48 \div 6$ **25**

**33.** $5 + 2 \cdot 15 - 4 \cdot 6 + 3 + 5 \cdot 7$ **49**   **34.** $28 \cdot 5 - 4 \cdot 17 + 16 \div 8 - 7$ **67**

**35.** $(68 \div 4)3 - 3(60 - 45) + 2 \cdot 8$ **22**   **36.** $7 \cdot 5 - (18 \div 6)4 + 2(7 - 4) \div 2$ **26**

**37.** $49 - (7 + 5)2 + 2(8 + 3) + 7$ **54**   **38.** $23 \cdot 168 \div (56 - 14) - 7 \cdot 12$ **8**

**39.** $\dfrac{4 \cdot 2 + 32 \div 8}{6 - 2 - 1}$ **4**   **40.** $\dfrac{5 \cdot 9 + 18 \div 3}{2 + 3 \cdot 5}$ **3**   **41.** $\dfrac{5 \cdot 4 + 1 \cdot 15}{1 + 2 \cdot 3}$ **5**

**42.** $0.9 \cdot 0.2 + 0.6 \cdot 0.4$ **0.42**   **43.** $0.45 \div 0.5 - 0.1$ **0.8**

Ⓒ  **Simplify.** [*Hint:* **Start within the parentheses, then within the brackets.**]

**44.** $8 \div [16 - 2(3 + 4)] + 4$ **8**   **45.** $63 - 7[18 - 3(15 \div 5)]$ **0**

**46.** $\dfrac{15 + 5[5 + 3(8 \div 4 + 2)]}{7 - 45 \div [5 + 2(6 \div 3)]}$ **50**   **47.** $\dfrac{7 \cdot 2 + 3[4 \cdot 19 - 6(8 \div 4 + 2)]}{229 - 4[2 + 3(5 + 2 \cdot 6)]}$ **10**

# EXPRESSIONS WITH VARIABLES

*Objectives*
**To evaluate algebraic expressions for given value(s) of the variable(s)**
**To identify terms, variables, and numerical coefficients**

In each of the following expressions, the number by which 5 is multiplied, 2, 4, or 7, changes or varies.

$$5 \cdot 2 + 3$$
$$5 \cdot 4 + 3$$
$$5 \cdot 7 + 3$$

You can use a letter, such as $n$, to represent the number that changes. So, 5 times a number plus 3 can be written as

$$5 \cdot n + 3$$
$$5n + 3$$

Notice that the raised dot can be dropped. $5 \cdot n$ can be written as $5n$. The $n$ is called a variable. **Variables** are letters that can be replaced by numbers. The value of $5n + 3$ depends on the value substituted for $n$.

To find the value of an expression is to **evaluate** it. This is shown in Example 1.

*Example 1*   **Evaluate $7 + 2y$ for the given values of $y$.**

$$
\begin{array}{c|c|c}
y = 3 & y = 0 & y = 6 \\
7 + 2y & 7 + 2y & 7 + 2y \\
\;\;\;\downarrow & \;\;\;\downarrow & \;\;\;\downarrow \\
7 + 2 \cdot 3 & 7 + 2 \cdot 0 & 7 + 2 \cdot 6 \\
7 + 6 & 7 + 0 & 7 + 12 \\
13 & 7 & 19
\end{array}
$$

◄ *Substitute the values for the variable.*

*Example 2*   **Evaluate $12x - 2$ for $x = \dfrac{3}{4}$.**

$$12x - 2$$
$$\;\;\;\;\downarrow$$
$$12 \cdot \frac{3}{4} - 2$$
$$9 - 2 \quad ◄ \quad 12 \cdot \frac{3}{4} = \overset{3}{\cancel{12}} \cdot \frac{3}{\cancel{4}_1} = \frac{9}{1} = 9$$
$$7$$

The expression $4b + 9ac + 3$ has three terms, $4b$, $9ac$, and 3. A **term** is a number, a variable, or a product of a number and one or more variables. The term 3, with no variable, is called a constant term. In the term $4b$, 4 is the numerical coefficient of $b$. A **numerical coefficient** is the number by which the variable or variables are multiplied.

*Example 3*    **Name the terms, variables, and numerical coefficients in $5w + 7z - 3$.**

| Terms | Variables | Numerical Coefficients |
|-------|-----------|------------------------|
| $5w$  | $w$       | 5                      |
| $7z$  | $z$       | 7                      |
| 3     | none      | none                   |

*Example 4*    **Evaluate $\dfrac{3a + 6}{2b - 3c}$ for $a = 8$, $b = 6$, and $c = 2$.**

$$\frac{3a + 6}{2b - 3c}$$

$$\frac{3 \cdot 8 + 6}{2 \cdot 6 - 3 \cdot 2} \quad \blacktriangleleft \text{ Substitute 8 for } a, \text{ 6 for } b, \text{ and 2 for } c.$$

$$\frac{24 + 6}{12 - 6} \quad \blacktriangleleft \text{ Use the rules for the order of operations.}$$

$$\frac{30}{6} = 5$$

*Example 5*    **Evaluate $2xy - 4x + 5y$ for $x = 3$ and $y = 5$.**

$$2x \quad y - \quad 4x + \quad 5y$$
$$2 \cdot 3 \cdot 5 - 4 \cdot 3 + 5 \cdot 5 \quad \blacktriangleleft \text{ Substitute 3 for } x \text{ and 5 for } y.$$
$$30 - 12 \quad + 25 \quad \blacktriangleleft \text{ Use the rules for the order of operations.}$$
$$18 + 25$$
$$43$$

## Oral Exercises
**Evaluate.**
1. $2x$ for $x = 3$  **6**
2. $m + 5$ for $m = 4$  **9**
3. $6 - x$ for $x = 1$  **5**
4. $4 + g$ for $g = 5$  **9**
5. $9 - m$ for $m = 6$  **3**
6. $3a$ for $a = 5$  **15**

1. t: $3a$, $5b$, 2; v: $a$, $b$; c: 3, 5    2. t: $5x$, 2, $3y$; v: $x$, $y$; c: 5, 3
3. t: $3m$, $4b$; v: $m$, $b$; c: 3, 4
4. t: 1, $5x$, $8m$; v: $x$, $m$; c: 5, 8

## Written Exercises

Ⓐ **Name the terms, variables, and numerical coefficients of the variables.**
1. $3a + 5b + 2$
2. $5x - 2 + 3y$
3. $3m + 4b$
4. $1 + 5x + 8m$
5. $6d - 4 + 3f$
6. $7a + 2b + 3z + 4$

5. t: $6d$, 4, $3f$; v: $d$, $f$; c: 6, 3    6. t: $7a$, $2b$, $3z$, 4; v: $a$, $b$, $z$; c: 7, 2, 3

**Evaluate for the given value(s) of the variable(s).**

7. $3x + 4$ for $x = 6$ **22**

8. $10 - 4y$ for $y = 2$ **2**

9. $5 + 7a$ for $a = 3$ **26**

10. $4x - 8$ for $x = 5$ **12**

11. $9 + 5c$ for $c = 0$ **9**

12. $3m - 6$ for $m = 7$ **15**

13. $6m - 2$ for $m = \frac{2}{3}$ **2**

14. $9 + 10a$ for $a = \frac{2}{5}$ **13**

15. $15 - 8a$ for $a = \frac{3}{4}$ **9**

16. $15m - 4$ for $m = \frac{2}{5}$ **2**

17. $7 + 9g$ for $g = \frac{2}{3}$ **13**

18. $10k - 1$ for $k = \frac{3}{5}$ **5**

19. $3m - 2 + 5n$ for $m = 5$ and $n = 2$ **23**

20. $5k + 9 + 7t$ for $k = 8$ and $t = 1$ **56**

21. $6x + 3 + 4y$ for $x = 2$ and $y = 5$ **35**

22. $4a - 5b + 2$ for $a = 6$ and $b = 2$ **16**

23. $4yz + 3y + 6z$ for $y = 5$ and $z = 5$ **145**

24. $9y - 2ey + 4e$ for $y = 4$ and $e = 3$ **24**

25. $2a - 5b + ab$ for $a = 7$ and $b = 2$ **18**

26. $6c + 3d - 2cd$ for $c = 1$ and $d = 4$ **10**

27. $\dfrac{3a + 2}{b - 5}$ for $a = 5$ and $b = 6$ **17**

28. $\dfrac{2x + 8}{2y - 4}$ for $x = 2$ and $y = 3$ **6**

29. $\dfrac{4a + 6}{12 - 3b}$ for $a = 6$ and $b = 2$ **5**

30. $\dfrac{2x + 3y}{4y - 3x + 1}$ for $x = 4$ and $y = 6$ **2**

**Ⓑ Evaluate for $x = 4$, $y = 6$, and $z = 5$.**

31. $4x + 3y + 2z$ **44**

32. $7z + 8x - 4y$ **43**

33. $9y + 2z + 7x$ **92**

34. $3yz - xz$ **70**

35. $yz + 4xz$ **110**

36. $8yz - 7xz$ **100**

37. $5xyz + 4y - 3x$ **612**

38. $6xz + 4yx + 2xyz$ **456**

39. $7x + 8xyz + 4yz$ **1,108**

40. $\dfrac{xyz - 4y}{x + y}$ **$9\frac{3}{5}$**

41. $\dfrac{2xy + 5z - 1}{x + y - z}$ **$14\frac{2}{5}$**

42. $\dfrac{yz + 2z}{xy - z + 5}$ **$1\frac{2}{3}$**

43. $0.04xy + 2.9xyz + 3.8xy$ **440.16**

44. $0.04xyz + 0.04yx + 0.2xz$ **9.76**

**Ⓒ Evaluate for $w = 5$, $x = 3$, $y = 2$, and $z = 4$.**

45. $xy(5xyz + 5xy)$ **900**

46. $(3xy + 2z)(xz + y)$ **364**

47. $xz[x + y(xz + y)]$ **372**

48. $xyz[yz + x(xy + z)]$ **912**

49. $\dfrac{xzw[wz + x(yx + zw)]}{2x - y}$ **1,470**

50. $\dfrac{(wx + xz)(xy + zwx)}{x - y}$ **1,782**

## CUMULATIVE REVIEW

**Simplify.**

1. $8 \cdot 15 \div 5$ **24**

2. $9 \cdot 2 - 16 \div 4$ **14**

3. $5 \cdot 3 + 8 \cdot 2$ **31**

4. $7 \cdot 5 + 2 + 0 \cdot 4$ **37**

5. $15 + 10 \cdot 3 - 9 \div 3$ **42**

6. $27 \div 9 + 10 - 4 \cdot 2$ **5**

## CALCULATOR ACTIVITIES

Evaluate for the given value(s) of the variable(s). **3. 3,673**   **5. 0**

1. $794x + 1,148$ for $x = 62$ **50,376**

2. $3,209 - 54y$ for $y = 27$ **1,751**

3. $31a + 46b$ for $a = 71$ and $b = 32$

4. $83x - 29y$ for $x = 75$ and $y = 98$ **3,383**

5. $57n - 19kn$ for $k = 3$ and $n = 12$

6. $28m + 38rt$ for $m = 79$, $r = 4.6$, $t = 0.26$

7. $0.048y + 113.849x$ for $x = 1.3$ and $y = 21.95$ **149.0573**

**2,257.448**

# THE COMMUTATIVE AND ASSOCIATIVE PROPERTIES — 1.3

*Objectives* **To recognize the commutative and associative properties**
**To simplify expressions by using the commutative and associative properties**

You know that $7 + 3 = 10$ and $3 + 7 = 10$. So, $7 + 3 = 3 + 7$. Similarly, $5 \cdot 7 = 35$ and $7 \cdot 5 = 35$. So, $5 \cdot 7 = 7 \cdot 5$. This suggests that you can change the order when you add or multiply two numbers since the results are the same. Addition and multiplication are commutative operations.

| Commutative properties | *Addition:* For all numbers $a$ and $b$, $a + b = b + a$.<br>*Multiplication:* For all numbers $a$ and $b$, $a \cdot b = b \cdot a$. |
|---|---|

Two ways for adding $5 + 2 + 3$ are shown below.

Add 5 and 2 first, then add 3 to the result.

$$\begin{aligned} 5 + 2 + 3 &= (5 + 2) + 3 \\ &= 7 + 3 \\ &= 10 \end{aligned}$$

Add 2 and 3 first, then add the result to 5.

$$\begin{aligned} 5 + 2 + 3 &= 5 + (2 + 3) \\ &= 5 + 5 \\ &= 10 \end{aligned}$$

Since the sum is 10 in each case, $(5 + 2) + 3 = 5 + (2 + 3)$. This suggests that you may group, or associate, three numbers differently when you add them and still get the same result.

Two ways for multiplying $5 \cdot 2 \cdot 3$ are shown below.

Multiply 5 and 2 first, then multiply the result by 3.

$$\begin{aligned} 5 \cdot 2 \cdot 3 &= (5 \cdot 2) \cdot 3 \\ &= 10 \cdot 3 \\ &= 30 \end{aligned}$$

Multiply 2 and 3 first, then multiply 5 by the result.

$$\begin{aligned} 5 \cdot 2 \cdot 3 &= 5 \cdot (2 \cdot 3) \\ &= 5 \cdot 6 \\ &= 30 \end{aligned}$$

Since the product is 30 in each case, $(5 \cdot 2) \cdot 3 = 5 \cdot (2 \cdot 3)$. Thus, you may associate three numbers differently when you multiply them and still get the same result.

Addition and multiplication are associative operations.

| Associative properties | *Addition:* For all numbers $a$, $b$, and $c$, $(a + b) + c = a + (b + c)$.<br>*Multiplication:* For all numbers $a$, $b$, and $c$, $(a \cdot b) \cdot c = a \cdot (b \cdot c)$. |
|---|---|

You can use the commutative and associative properties to make computations easier. This is shown in Examples 2 and 3 on the next page.

*Example 1*    **Which property is illustrated?**

$$7b + 5a = 5a + 7b$$
$$(9 \cdot t) \cdot u = 9 \cdot (t \cdot u)$$
$$5y(3) = 3(5y)$$
$$(3a + 7b) + 4x = 3a + (7b + 4x)$$

**Answers**
Commutative property of addition
Associative property of multiplication
Commutative property of multiplication
Associative property of addition

*Example 2*    **Use the associative and commutative properties to rewrite 395 + 68 + 5 so that the computation will be easier. Then simplify.**

$$395 + 68 + 5 = (395 + 68) + 5$$
$$= 395 + (68 + 5) \quad \blacktriangleleft \textit{Associative property of addition}$$

$$= 395 + (5 + 68) \quad \blacktriangleleft \textit{Commutative property of addition}$$
$$= (395 + 5) + 68 \quad \blacktriangleleft \textit{Associative property of addition}$$
$$= 400 + 68$$
$$= 468$$

*Example 3*    **Use the associative and commutative properties to rewrite 25 · 7 · 4 so that the computation will be easier. Then simplify.**

$$25 \cdot 7 \cdot 4 = (25 \cdot 7) \cdot 4$$
$$= 25 \cdot (7 \cdot 4) \quad \blacktriangleleft \textit{Associative property of multiplication}$$

$$= 25 \cdot (4 \cdot 7) \quad \blacktriangleleft \textit{Commutative property of multiplication}$$
$$= (25 \cdot 4) \cdot 7 \quad \blacktriangleleft \textit{Associative property of multiplication}$$
$$= 100 \cdot 7$$
$$= 700$$

*Example 4*    **Simplify.**

$$4(6a)$$
$$(4 \cdot 6)a \quad \blacktriangleleft \textit{Associative property}$$
$$24a$$

$$(5b)2$$
$$2(5b) \quad \blacktriangleleft \textit{Commutative property}$$
$$(2 \cdot 5)b \quad \blacktriangleleft \textit{Associative property}$$
$$10b$$

*Summary*    When adding (or multiplying) numbers, the order in which the numbers are added (or multiplied) may be changed or the numbers may be regrouped.

## Written Exercises

**Rewrite by using the associative and commutative properties so that the computation will be easy. Then simplify.**

(A) **1.** $99 + 47 + 1$ **147**   **2.** $59 + 17 + 1$ **77**   **3.** $2 + 89 + 28$ **119**   **4.** $3 + 35 + 97$ **135**
**5.** $57 + 3 + 29$ **89**   **6.** $35 + 46 + 5$ **86**   **7.** $19 + 42 + 1$ **62**   **8.** $8 + 29 + 32$ **69**
**9.** $4 \cdot 23 \cdot 25$ **2,300**   **10.** $15 \cdot 7 \cdot 2$ **210**   **11.** $50 \cdot 14 \cdot 2$ **1,400**   **12.** $2 \cdot 47 \cdot 50$ **4,700**
**13.** $20 \cdot 43 \cdot 5$ **4,300**   **14.** $25 \cdot 66 \cdot 4$ **6,600**   **15.** $4 \cdot 5 \cdot 250$ **5,000**   **16.** $25 \cdot 17 \cdot 4$ **1,700**

**17. Comm. Prop. Add.   18. Assoc. Prop. Add   19. Assoc. Prop. Mult.**

**Which property is illustrated?   20. Comm. Prop. Mult.   21. Assoc. Prop. Mult.**
**17.** $7 + 5 = 5 + 7$      **18.** $(2 + 9) + 3 = 2 + (9 + 3)$      **19.** $(3 \cdot 7) \cdot 8 = 3 \cdot (7 \cdot 8)$
**20.** $9 \cdot 28 = 28 \cdot 9$      **21.** $a(b \cdot c) = (a \cdot b)c$      **22.** $7 + (9 + y) = (7 + 9) + y$
**23.** $5x + 3y = 3y + 5x$      **24.** $6(3b) = (6 \cdot 3)b$      **25.** $8y(3) = 3(8y)$

**22. Assoc. Prop. Add.   23. Comm. Prop. Add.**
**24. Assoc. Prop. Mult.   25. Comm. Prop. Mult.**

**Simplify.**
**26.** $7(3a)$ **21a**   **27.** $4(8b)$ **32b**   **28.** $(5b)3$ **15b**   **29.** $6(4m)$ **24m**   **30.** $(5x)2$ **10x**
**31.** $4x(3)$ **12x**   **32.** $5(9y)$ **45y**   **33.** $(6t)5$ **30t**   **34.** $4(3b)$ **12b**   **35.** $7(2x)$ **14x**

**Rewrite by using the associative and commutative properties so that the computation will be easy. Then simplify.**

(B) **36.** $199 + 56 + 4 + 1$ **260**   **37.** $2 + 165 + 148 + 5$ **320**   **38.** $299 + 74 + 6 + 1$ **380**
**39.** $788 + 189 + 12 + 11$ **1,000**   **40.** $39 + 79 + 971 + 51$ **1,140**   **41.** $22 + 355 + 178 + 145$ **700**

**42.** $2\frac{1}{3} + 5 + 4\frac{2}{3} + 7$ **19**      **43.** $7\frac{3}{8} + 6 + 3\frac{5}{8} + 2$ **19**      **44.** $\frac{1}{2} \cdot 19 \cdot 4 \cdot \frac{1}{19}$ **2**

**45.** $(2.5)(7)(4)$ **70**      **46.** $500(62.5)(2)(2)$ **125,000**   **47.** $5(74.5)(20)(2)$ **14,900**

**Simplify.**
**48.** $2(3t)(4)$ **24t**      **49.** $9y(5)(2)$ **90y**      **50.** $8(4)(5a)$ **160a**
**51.** $2(11)(50c)$ **1,100c**      **52.** $5m(7)(20)$ **700m**      **53.** $25(8b)(4)$ **800b**

(C) **54.** Is division commutative? Verify your answer.   **55.** Is subtraction associative? Verify your answer.
**No. Examples may vary.**      **No. Examples may vary.**

**Prove each of the following. Justify each step.**

*Example*   **Prove**   $(p + q) + r = (r + q) + p$   **Answers**
   $(p + q) + r = p + (q + r)$   Associative property of addition
   $\qquad\qquad = p + (r + q)$   Commutative property of addition
   $\qquad\qquad = (r + q) + p$   Commutative property of addition

**For Ex. 56–59, see TE Answer Section.**
**56.** $(a + b) + c = a + (c + b)$      **57.** $(x \cdot y) \cdot z = (z \cdot y) \cdot x$
**58.** $m \cdot (g \cdot v) = g \cdot (v \cdot m)$      **59.** $r + (t + w) = (w + r) + t$

## CUMULATIVE REVIEW

**1.** Simplify $8 - 4 \div 2 + 6 \cdot 3$. **24**      **2.** Evaluate $\dfrac{5a + 2b}{4a}$ for $a = 2$ and $b = 3$. **2**

# THE DISTRIBUTIVE PROPERTY

*Objective*  **To simplify expressions using the distributive property**

Notice what happens when the two expressions, $4(8 + 6)$ and $4 \cdot 8 + 4 \cdot 6$, are simplified.

$$4(8 + 6) \qquad \text{and} \qquad 4 \cdot 8 + 4 \cdot 6$$
$$4(14) \qquad\qquad\qquad 32 + 24$$
$$56 \qquad\qquad\qquad\quad 56$$

No matter which way you simplify the expressions, the value is the same, 56. Thus, $4(8 + 6) = 4 \cdot 8 + 4 \cdot 6$. The 4 was distributed as a multiplier over the 8 and 6. Using the commutative property, $4(8 + 6)$ can also be written as $(8 + 6)4$. So, $(8 + 6)4 = 8 \cdot 4 + 6 \cdot 4$.

Multiplication is distributive over addition.

| Distributive property of multiplication over addition | For all numbers $a$, $b$, and $c$, $a(b + c) = a \cdot b + a \cdot c$, or $ab + ac$ and $(b + c)a = b \cdot a + c \cdot a$, or $ba + ca$. |
|---|---|

*Example 1*  **Rewrite $7(5 + 3)$ by using the distributive property. Then simplify.**

$$7(5 + 3) = 7 \cdot 5 + 7 \cdot 3 \quad \blacktriangleleft \textit{ Distribute 7 over 5 and 3.}$$
$$= \quad 35 \; + \; 21 \quad \blacktriangleleft \textit{ Use the rules for order of operations.}$$
$$= \qquad 56$$

*Example 2*  **Rewrite $\frac{1}{3}(9 + 6)$ using the distributive property. Then simplify.**

$$\frac{1}{3}(9 + 6) = \frac{1}{3} \cdot 9 + \frac{1}{3} \cdot 6 \quad \blacktriangleleft \quad \frac{1}{3} \cdot 9 = \frac{1}{3} \cdot \overset{3}{\cancel{9}} = 3 \quad \vert \quad \frac{1}{3} \cdot 6 = \frac{1}{3} \cdot \overset{2}{\cancel{6}} = 2$$
$$= \quad 3 \; + 2 = 5$$

The property is also true if the multiplier is on the right of the parentheses or if you have an expression with three or more terms.

*Example 3*  **Rewrite $(7 + 6)4$ by using the distributive property. Then simplify.**

$$(7 + 6)4 = 7 \cdot 4 + 6 \cdot 4 = 28 + 24 = 52$$

*Example 4*    **Rewrite 3(8 + 5 + 2) by using the distributive property. Then simplify.**

$$3(8 + 5 + 2) = 3 \cdot 8 + 3 \cdot 5 + 3 \cdot 2 = 24 + 15 + 6 = 45$$

You can use the distributive property in reverse to rewrite $4 \cdot 9 + 4 \cdot 5$.
Notice that the 4 is distributed as a multiplier over the 9 and 5.
So, $4 \cdot 9 + 4 \cdot 5 = 4(9 + 5)$.

*Example 5*    **Rewrite by using the distributive property.**

$$7 \cdot 3 + 7 \cdot 8 \qquad 9 \cdot 5 + 1 \cdot 5 + 4 \cdot 5 + 6 \cdot 5$$
$$7(3 + 8) \qquad (9 + 1 + 4 + 6)5$$

At times, you must use the commutative property before you apply the
distributive property.

*Example 6*    **Rewrite $7 \cdot 8 + 5 \cdot 7 + 7 \cdot 9$ by using the distributive property.**

$$7 \cdot 8 + 5 \cdot 7 + 7 \cdot 9$$
$$7 \cdot 8 + 7 \cdot 5 + 7 \cdot 9 \quad \blacktriangleleft \; \textit{Commutative property.}$$
$$7(8 + 5 + 9) \qquad \blacktriangleleft \; \textit{7 is distributed over 8, 5, and 9.}$$

The distributive property can also be used to simplify expressions that contain
variables.

*Example 7*    **Rewrite by using the distributive property. Then simplify.**

$$3(4a + 2) = 3 \cdot 4a + 3 \cdot 2 \qquad \frac{1}{2}(8b + 9) = \frac{1}{2} \cdot 8b + \frac{1}{2} \cdot 9$$
$$= 12a + 6 \qquad\qquad\qquad = 4b + 4\frac{1}{2}$$

*Example 8*    **Rewrite by using the distributive property. Then simplify.**

$$3 \cdot x + 5 \cdot x \qquad 8 \cdot y + 3 \cdot y + y \cdot 2$$
$$\qquad\qquad\qquad 8 \cdot y + 3 \cdot y + 2 \cdot y$$
$$(3 + 5)x = 8x \qquad (8 + 3 + 2)y = 13y$$

The Distributive Property

# Reading in Algebra

1. Look up the meaning of the word *distributive* in a dictionary. Does its definition support the way the term is used in algebra? **Distributive–dealing a proper share to each of a group;**
2. For 6(3 + 5), what are the terms over which 6 can be distributed? **3 and 5**            **yes**

# Written Exercises
**Check students' answers for rewrites of the distributive property. Only the simplified answers are given.**

Ⓐ **Rewrite by using the distributive property. Then simplify.**
1. 7(6 + 3) **63**
2. 6(5 + 4) **54**
3. (8 + 6)2 **28**
4. (3 + 8)7 **77**
5. $\frac{1}{3}$ (12 + 6) **6**
6. $\frac{1}{4}$ (8 + 12) **5**
7. (15 + 10)$\frac{1}{5}$ **5**
8. (14 + 6)$\frac{1}{2}$ **10**

**Rewrite by using the distributive property.**
9. 5(4 + 7)
10. 8(7 + 6)
11. (9 + 1)2
12. (8 + 6)4
13. 4(7 + 1 + 6)
14. (3 + 6 + 7)3
15. 5(4 + 1 + 2)
16. (6 + 3 + 1 + 4)2
17. 5 · 7 + 5 · 3
18. 8 · 2 + 8 · 9
19. 7 · 5 + 8 · 5
20. 1 · 6 + 8 · 6
21. 6 · 3 + 6 · 2 + 6 · 8
22. 2 · 7 + 5 · 7 + 1 · 7 + 3 · 7
23. 1 · 5 + 4 · 5 + 5 · 3

**Rewrite by using the distributive property. Then simplify.**
24. 6(2x + 5) **12x + 30**
25. 3(3a + 4) **9a + 12**
26. (8k + 16)$\frac{1}{2}$ **4k + 8**
27. (14g + 7)$\frac{1}{7}$ **2g + 1**
28. 5 · x + 7 · x **12x**
29. 9 · m + 4 · m **13m**
30. 6 · t + 1 · t **7t**
31. 4 · a + 7 · a **11a**
32. 4 · p + 6 · p + 2 · p **12p**
33. q · 8 + 3 · q + 4 · q **15q**
34. 5 · b + b · 7 + b · 7 **19b**

Ⓑ **Show that each of the following is true.**
35. 9(5 + 7 + 1) = 9 · 5 + 9 · 7 + 9 · 1
    9(13) = 45 + 63 + 9   **117 = 117**
36. (7 + 6 + 3)8 = 7 · 8 + 6 · 8 + 3 · 8
    (16)8 = 56 + 48 + 24   **128 = 128**

**Compute in the easiest way.**
37. 275 · 93 + 275 · 7 **27,500**
38. 78 · 98 + 78 · 2 **7,800**
39. 33 · $\frac{1}{5}$ + 67 · $\frac{1}{5}$ **20**
40. 98(78 + 21 + 1) **9,800**
41. 47 · 92 + 52 · 92 + 1 · 92
42. 49(68 + 16 + 7 + 9) **4,900**
    **9,200**

Ⓒ **Solve these problems.**
43. Is multiplication distributive over subtraction? Does $a \cdot (b - c) = a \cdot b - a \cdot c$? Verify your answer.
    **Yes. Examples may vary.**
44. If division were distributive over addition, how would you rewrite 24 ÷ (4 + 2)? Is the resulting statement true? **No**

---

## NON-ROUTINE PROBLEMS

Would you rather have $10,000 cash or the following:
one cent on the first day, twice as much or two cents on the second day,
twice as much or four cents on the third day, and so on for 30 days?

**The second choice will yield more money.**

# COMBINING LIKE TERMS

*Objectives*  **To simplify algebraic expressions by combining like terms**
**To simplify and then evaluate algebraic expressions for given value(s) of the variable(s)**
**To simplify algebraic expressions using the property of multiplicative identity**

The expression $5y + 5y$ has two terms exactly alike. The terms of $5y + 5y$ are *like terms*.
The expression $3a + 4a + 2a$ has three terms that differ only in their coefficients, 3, 4, and 2. The terms of $3a + 4a + 2a$ are *like terms*.
The expression $3b + 5a + 2$ has three terms. The terms are *unlike terms*.

| Definition: Like terms | **Like terms** are terms that are exactly the same, or differ only in their numerical coefficients. |
|---|---|

The distributive property can be used to simplify expressions with like terms. For example, $8y + 3y = 8 \cdot y + 3 \cdot y = (8 + 3)y = 11y$. You need not go through all of these steps. Like terms can be combined. Thus, $8y + 3y = 11y$.

*Example 1*  **Simplify $9x + 5x + 2x$.**

$9x + 5x + 2x = 16x$  ◀ *Combine the like terms $9x$, $5x$, and $2x$.*
Remember, only like terms can be combined.

*Example 2*  **Simplify $5a + 3b + 7$, if possible.**

The terms $5a$, $3b$, and 7 are unlike terms. Since unlike terms cannot be combined, $5a + 3b + 7$ cannot be simplified.

The following three multiplications suggest a property.
$$1 \cdot 6 = 6 \qquad 1 \cdot 5 = 5 \qquad 8 \cdot 1 = 8$$
The product of 1 and any number is that number. Therefore, the number 1 is called the *multiplicative identity*.

| Property of multiplicative identity | For any number $a$, $1 \cdot a = a$ and $a \cdot 1 = a$. |
|---|---|

Since $x$ can be written as $1x$, the numerical coefficient of $x$ is understood to be 1.

**Example 3**   **Simplify $8x - 2x + x$.**

$$8x - 2x + x$$
$$8x - 2x + 1x \qquad \blacktriangleleft \ \textit{Rewrite x as 1x.}$$
$$7x \qquad\qquad \blacktriangleleft \ \textit{Combine the like terms.}$$

To simplify a more complex expression, first rearrange the terms so that the like terms are grouped together.

**Example 4**   **Simplify $7 + 8x + 9y + 6x - 3 - 2y$, if possible.**

$$7 + 8x + 9y + 6x - 3 - 2y$$
$$(8x + 6x) + (9y - 2y) + (7 - 3) \quad \blacktriangleleft \ \textit{Rearrange to group the like terms together.}$$
$$14x \quad + \quad 7y \quad + \quad 4 \qquad \blacktriangleleft \ \textit{Combine the like terms.}$$

To simplify an expression like $5 + 6(4c + 2) + 3c$, you will have to use both the distributive property and the rules for order of operations.

**Example 5**   **Simplify $5 + 6(4c + 2) + 3c$.**

$$5 + 6(4c + 2) + 3c$$
$$5 + 6 \cdot 4c + 6 \cdot 2 + 3c \quad \blacktriangleleft \ \textit{Distribute the 6.}$$
$$5 + 24c + 12 + 3c \qquad \blacktriangleleft \ \textit{Do the multiplications.}$$
$$(24c + 3c) + (5 + 12) \qquad \blacktriangleleft \ \textit{Group the like terms.}$$
$$27c + 17 \qquad\qquad \blacktriangleleft \ \textit{Combine the like terms.}$$

**Example 6**   **Simplify $9 + (8y + 12)\dfrac{3}{4} + y$.**

$$9 + (8y + 12)\frac{3}{4} + y$$

$$9 + 8y \cdot \frac{3}{4} + 12 \cdot \frac{3}{4} + y \quad \blacktriangleleft \ \textit{Distribute the } \frac{3}{4}.$$

$$9 + \quad 6y \quad + \quad 9 \quad + y \qquad \blacktriangleleft \ 8y \cdot \frac{3}{4} = 8 \cdot \frac{3}{4} \cdot y = \overset{2}{\cancel{8}} \cdot \frac{3}{\underset{1}{\cancel{4}}} \cdot y = \frac{6}{1} \cdot y \ \textit{or } 6y$$

$$9 + 6y + 9 + 1y \qquad\qquad \blacktriangleleft \ y = 1y$$
$$(6y + 1y) + (9 + 9) \qquad \blacktriangleleft \ \textit{Group the like terms.}$$
$$7y + 18$$

## Oral Exercises

**Simplify, if possible.**

**1.** $3x + 4x$ **7x**    **2.** $5p + 8p$ **13p**    **3.** $7x + 5$ **can't**    **4.** $8a + 7a$ **15a**    **5.** $4y + 7y$ **11y**
**6.** $5a + 3b$ **can't 7.** $7r + 9r$ **16r**    **8.** $9 + 4t$ **can't**    **9.** $3m + 9m$ **12m 10.** $8x + 7x$ **15x**

## Written Exercises

**5. 11a + 8**    **7. 5z + 22**    **15. 3g + 10**    **16. 8a + 15b**    **17. 10c + 11d + 4**

Ⓐ **Simplify, if possible.**    **18. 11x + 11y + 7**    **19. 11a + 13**    **20. 10b + 9**    **21. 13x + 9**

**1.** $7x + 4x + 5x$ **16x**    **2.** $5x + 7x + 6x$ **18x**    **3.** $6y + 10y + 3y$ **19y**    **4.** $2m + 3m + 4m$ **9m**
**5.** $4a + 7a + 8$      **6.** $3r + 9 + r$ **4r + 9 7.** $8 + 5z + 14$      **8.** $8 + 6z + 9z$ **15z + 8**
**9.** $6x + 2y + 8$ **can't 10.** $a + 3 + 5a$ **6a + 3 11.** $3m + 4y$ **can't**    **12.** $7y + 3 + 8y$ **15y + 3**
**13.** $7 + 4y - 3 - 2y$ **2y + 4**    **14.** $8 + 5m - 7 - m$ **4m + 1 15.** $5g + 4 + 2g - 4g + 6$
**16.** $5a + 6b + 9b + 3a$      **17.** $2c + 6d + 8c + 5d + 4$    **18.** $8x + 7y + 3x + 2 + 4y + 5$
**22. 6m + 40**    **23. 10z + 15**    **24. 22n + 17**    **25. 21x + 30**    **26. 33x + 28**    **27. 17z + 62**

**Simplify.**

**19.** $3 + 2(4a + 5) + 3a$      **20.** $4b + 2(3b + 1) + 7$      **21.** $3x + 5(2x + 1) + 4$
**22.** $8 + (2m + 9)3 + 5$      **23.** $4z + (2z + 5)3$      **24.** $5 + (9n + 6)2 + 4n$
**25.** $2x + 5(6 + 3x) + 4x$      **26.** $3x + 8 + (4 + 6x)5$      **27.** $3z + 7(8 + 2z) + 6$
**28.** $\frac{1}{2}(8a + 4) + 4a + 3$      **29.** $(3y + 9)\frac{2}{3} + 4y + 2$      **30.** $5p + (6 + 12p)\frac{5}{6} + 6p$
        **8a + 5**                 **6y + 8**                       **21p + 5**

Ⓑ **31.** $(7 + 5r)3 + (8 + 4r)2$    **32.** $7(5y + 4) + 8(1 + 3y)$    **33.** $6(4k + 7) + 3(8k + 9)$
**34.** $(5 + 2x)4 + (3x + 6)3$    **35.** $7(2x + 8) + (5 + 7x)4$    **36.** $(7a + 1)3 + 7(2 + 6a)$
**37.** $11 + (c + 5)8 + 2(5c + 1)$ **18c + 53**    **38.** $4(2n + 5) + 6n + 5(n + 3)$ **19n + 35**
**39.** $7 + (3a + 2)4 + 5a + 3(2a + 5)$ **23a + 30**    **31. 23r + 37**    **32. 59y + 36**
          **33. 48k + 69**    **34. 17x + 38**    **35. 42x + 76**    **36. 63a + 17**

*Example*      **Simplify $3(4a + 2) + (6 + 2a)3$. Then evaluate for $a = 2$.**
                $3(4a + 2) + (6 + 2a)3$
                   $12a + 6 + 18 + 6a$
                      $18a + 24$
                    $18 \cdot 2 + 24$     ◀ *Substitute 2 for a.*
                      $36 + 24$
                        $60$

**Simplify. Then evaluate for the given value of the variable.**

**40.** $(6 + 2x)4 + 3(3x + 5)$ for $x = 4$      **41.** $6(8 + 3k) + (2k + 1)5$ for $k = 5$
**42.** $7 + (8a + 12)\frac{1}{4} + 5a + 3(2a + 5)$ for $a = 5$ **13a + 25; 90**
**43.** $9y + \frac{1}{4}(4y + 8) + 5 + 6y + (9 + 3y)\frac{2}{3}$ for $y = 2$ **18y + 13; 49**
**40. 17x + 39; 107**      **41. 28k + 53; 193**

Ⓒ **Simplify.**

**44.** $6[5c + 8(4c + 3)]$ **222c + 144**      **45.** $2a + 7[(4a + 9)2 + 6a]$ **100a + 126**
**46.** $3[2m + 4(m + 9) + 2] + \frac{1}{2}(2m + 8)$      **47.** $3 + 5\left[2x + \frac{3}{4}(8x + 12)\right] + (5x + 4)2$
     **19m + 118**                                       **50x + 56**

# FACTORS AND EXPONENTS

*Objective*    **To evaluate algebraic expressions containing exponents**

In the product $7 \cdot 3$, 7 and 3 are *factors*.

| | |
|---|---|
| Definition: Factor | In the product $a \cdot b$, $a$ and $b$ are **factors**. |

The product $2 \cdot 2 \cdot 2$ consists of the factor 2 used three times. A convenient way of writing $2 \cdot 2 \cdot 2$ is $2^3$, where 2 is the *base* and 3 is the *exponent*. The expression $2^3$ can be read in several ways:

the third power of two        two to the third power        two cubed

A whole number exponent indicates the number of times the base is used as a factor. You can use this to find the value of the power of a number. For example, $2^5 = 2 \cdot 2 \cdot 2 \cdot 2 \cdot 2 = 32$. $2^5$ is read *the fifth power of 2*.

| | |
|---|---|
| Definition: $a^n$ | Let $a$ be any number. Then if $n$ is any natural number, $$a^n = \underbrace{a \cdot a \cdot a \cdot \ldots \cdot a}_{n \text{ factors}} = n^{\text{th}} \text{ power of } a$$ **base   exponent** |

*Example 1*    **Rewrite without exponents.**

$a^1$

$a^1 = a$

$x^4$

$x^4 = x \cdot x \cdot x \cdot x$

$y^5$

$y^5 = y \cdot y \cdot y \cdot y \cdot y$

You can evaluate an expression containing exponents.

*Example 2*    **Evaluate $a^5$ for $a = 3$.**

$a^5 = a \cdot a \cdot a \cdot a \cdot a$  ◄ *Write a as a factor five times.*

$= \underbrace{3 \cdot 3} \cdot \underbrace{3 \cdot 3} \cdot 3$  ◄ *Substitute 3 for a.*

$= \underbrace{9 \cdot 9} \cdot 3$

$= 81 \cdot 3$

$= 243$

The expression $5a^3$ means $5 \cdot a \cdot a \cdot a$. The expression $(5a)^3$ means $5a \cdot 5a \cdot 5a$.

*Example 3*  **Evaluate $5a^3$ and $(5a)^3$ for $a = 2$.**

$$5a^3 = 5 \cdot a \cdot a \cdot a \qquad\qquad (5a)^3 = (5 \cdot a) \cdot (5 \cdot a) \cdot (5 \cdot a)$$
$$\phantom{5a^3} = 5 \cdot 2 \cdot 2 \cdot 2 \qquad\qquad\phantom{(5a)^3} = (5 \cdot 2) \cdot (5 \cdot 2) \cdot (5 \cdot 2)$$
$$\phantom{5a^3} = 5 \cdot 8 = 40 \qquad\qquad\phantom{(5a)^3} = 10 \cdot 10 \cdot 10 = 1000$$

*Example 4*  **Evaluate.**

$4a^3b^2$ for $a = 2$ and $b = 3$     $(12x)^2$ for $x = \dfrac{3}{4}$

$4a^3b^2$                     $(12x)^2$

$4 \cdot a \cdot a \cdot a \cdot b \cdot b$        $\left(12 \cdot \dfrac{3}{4}\right)^2$

$4 \cdot \underline{2 \cdot 2 \cdot 2} \cdot \underline{3 \cdot 3}$       $9^2$    ◀ $12 \cdot \dfrac{3}{4} = \overset{3}{\cancel{12}} \cdot \dfrac{3}{\underset{1}{\cancel{4}}} = \dfrac{9}{1} = 9$

     $\underline{4 \cdot 8} \quad \cdot \quad 9$          $81$

       $32 \cdot 9$

        $288$

You can now extend the rules for order of operations to include exponents.
1. Raise to a power.
2. Do operations within the parentheses.
3. Multiply or divide from left to right.
4. Add or subtract from left to right.

# Written Exercises

**Rewrite without exponents.**    4. $6n \cdot 6n \cdot 6n$ or $216 \cdot n \cdot n \cdot n$     6. $a \cdot a \cdot a \cdot a \cdot b \cdot b \cdot b$

(A)   **1.** $x^5$        **2.** $x^3$        **3.** $6a^2$        **4.** $(6n)^3$      **5.** $3(a)^5$      **6.** $a^4b^3$

$x \cdot x \cdot x \cdot x \cdot x$     $x \cdot x \cdot x$       $6 \cdot a \cdot a$         **5.** $3 \cdot a \cdot a \cdot a \cdot a \cdot a$

**Evaluate for the given value of the variable(s).**    15. 48     16. 1,728     17. 4,374

  **7.** $x^3$ for $x = 4$  **64**     **8.** $b^5$ for $b = 2$  **32**     **9.** $m^4$ for $m = 3$  **81**    **10.** $m^8$ for $m = 2$  **256**

**11.** $4m^2$ for $m = 2$  **16**   **12.** $3a^5$ for $a = 2$  **96**    **13.** $2x^3$ for $x = 5$  **250**   **14.** $5a^6$ for $a = 2$  **320**

**15.** $ab^4$ for $a = 3$ and $b = 2$     **16.** $4x^3y^2$ for $x = 3$ and $y = 4$    **17.** $x^7y$ for $x = 3$ and $y = 2$

**18.** $(3a)^3$ for $a = \dfrac{2}{3}$  **8**    **19.** $(4m)^2$ for $m = \dfrac{1}{2}$  **4**   **20.** $(4a)^2$ for $a = \dfrac{1}{4}$  **1**    **21.** $(20x)^4$ for $x = \dfrac{3}{10}$

(B)   **22.** $a^3b^2c$ for $a = 4$, $b = 2$, and $c = 0$  **0**       **23.** $5xy^3z^2$ for $x = 6$, $y = 2$, and $z = 4$  **3,840**

**24.** $\dfrac{a^2 + b^2}{a - b}$ for $a = 6$ and $b = 4$  **26**        **25.** $\dfrac{a^3 - 3c}{ab}$ for $a = 3$, $b = 4$, and $c = 5$  **1**

                                                                  **21.** 1,296

# FORMULAS IN GEOMETRY

*Objective*  **To evaluate formulas for perimeter and area**

The geometric figure shown is a rectangle. The distance around it is 2 cm + 4 cm + 2 cm + 4 cm, or 12 cm. This is called the *perimeter*.

RECTANGLE

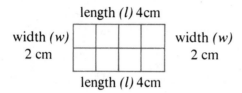

length *(l)* 4cm

width *(w)*  
2 cm

width *(w)*  
2 cm

length *(l)* 4cm

| Definition: Perimeter | The **perimeter** of a figure is the distance around it. |

The formula for the perimeter of a rectangle is
$$p = l + l + w + w$$
$$= 2l + 2w \text{ or } 2(l + w)$$
where $p$ = perimeter, $l$ = length, and $w$ = width.
In Example 1, you will find the perimeter of a triangle and the perimeter of a square.

*Example 1*  **Find the perimeter of each figure. Use the formulas and values given.**

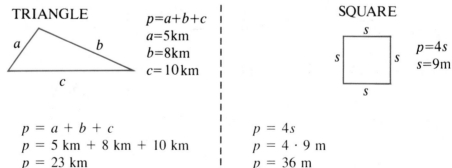

TRIANGLE

$p=a+b+c$
$a=5\text{km}$
$b=8\text{km}$
$c=10\text{km}$

SQUARE

$p=4s$
$s=9\text{m}$

$p = a + b + c$
$p = 5 \text{ km} + 8 \text{ km} + 10 \text{ km}$
$p = 23 \text{ km}$

$p = 4s$
$p = 4 \cdot 9 \text{ m}$
$p = 36 \text{ m}$

| Definition: Area | The **area** of a geometric figure is the number of unit squares it contains. |

The rectangle contains 2 rows of 4 squares each.
The formula for the area of a rectangle is $A = lw$.
The area of this rectangle is 4 cm · 2 cm, or 8 cm$^2$.
The area is read as *8 square centimeters*.

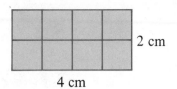

2 cm

4 cm

In Example 2, you will find the area of a triangle and the area of a trapezoid.

## *Example 2*   **Find the area of each figure. Use the formulas and values given.**

### TRIANGLE

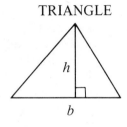

$A = \dfrac{bh}{2}$

$b = 6$ m

$h = 4$ m

$A = \dfrac{bh}{2}$

$A = \dfrac{6\text{ m} \cdot 4\text{ m}}{2}$

$A = \dfrac{24\text{ m}^2}{2}$

$A = 12$ m$^2$

### TRAPEZOID

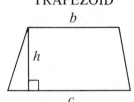

$A = \dfrac{h(b + c)}{2}$

$b = 6$ ft, $c = 8$ ft

$h = 4$ ft

$A = \dfrac{h(b + c)}{2} = \dfrac{4\text{ ft }(6\text{ ft} + 8\text{ ft})}{2}$

$A = \dfrac{4\text{ ft }(14\text{ ft})}{2}$

$A = \dfrac{56\text{ ft}^2}{2}$

$A = 28$ ft$^2$

## *Written Exercises*

**(A)**  **Use the given formulas and values to find the perimeter of each figure.**

**1.**

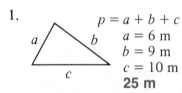

$p = a + b + c$
$a = 6$ m
$b = 9$ m
$c = 10$ m
**25 m**

**2.**

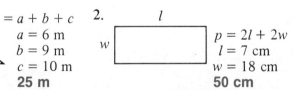

$p = 2l + 2w$
$l = 7$ cm
$w = 18$ cm
**50 cm**

**3.**

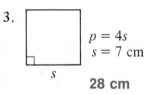

$p = 4s$
$s = 7$ cm
**28 cm**

**Use the given formulas and values to find the area of each figure.**

**4.**

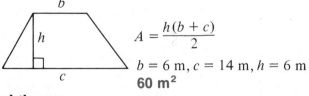

$A = \dfrac{h(b + c)}{2}$

$b = 6$ m, $c = 14$ m, $h = 6$ m
**60 m$^2$**

**5.**

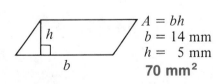

$A = bh$
$b = 14$ mm
$h = 5$ mm
**70 mm$^2$**

**(B)**  **Find the area.**
  **6.** Find the area of a triangle with height 14 in. and base 11 in.  **77 in.$^2$**
  **7.** Find the area of a trapezoid with bases 6 m and 7 m and height 3 m.  **$19\frac{1}{2}$ m$^2$**

Formulas in Geometry

# APPLICATIONS

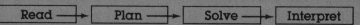

$$\boxed{\text{Read}} \longrightarrow \boxed{\text{Plan}} \longrightarrow \boxed{\text{Solve}} \longrightarrow \boxed{\text{Interpret}}$$

**1.** Some carpet installers are paid overtime for extra hours worked. A formula for the wages paid for work beyond 40 hours/week is

$$W = r\left(40 + h + \frac{h}{2}\right)$$

where $W$ is the wages, $r$ is the hourly wage, and $h$ is the number of hours over 40. Find Mrs. Crockett's weekly salary if $r$ is \$3.45 and $h$ is 8. **\$179.40**

**2.** The following formula is used by some taxi companies to figure how much to charge for a ride:

$$C = f + rd$$

where $C$ is the charge for the ride, $f$ is a fixed charge, $r$ is the rate per tenth of a kilometer, and $d$ is the distance of the ride in tenths of a kilometer. Find the charge for 2.5 km if $f$ is \$1.10 and $r$ is \$0.10. **\$3.60**

**3.** An important formula in photography is

$$F = \frac{l}{d}$$

where $F$ is the f/stop setting for a photograph, $l$ is the focal length of the lens, and $d$ is the diameter of the lens opening. Find the proper f/stop for a lens of length 100 mm with an opening of 25 mm. **4**

**4.** The formula for the approximate weekly take-home pay after all deductions is:

$$w = \frac{s - ts}{50}$$

where $w$ is the take-home pay, $s$ is the yearly salary, and $t$ is the decimal rate of deduction. Find Mark's weekly take-home pay if $s$ is \$18,500 and $t = 0.30$. **\$259**

**5.** Mr. Irvington owns and operates Appealing Apple Pies. A consultant gave him the following formula to analyze his profits,

$$p = 2x - 0.01x^2$$

where $p$ is the profit and $x$ is the number of pies baked. Find the profit on 60 pies. **\$84**

**6.** The manager of a restaurant figures the cost of making coffee is:

$$m = 0.049x$$

where $m$ is the cost of making $x$ cups. Find the cost of making 600 cups of coffee. **\$29.40**

# RECOGNIZING SOLUTIONS OF OPEN SENTENCES

*Objective*　**To recognize which numbers, if any, are solutions of an open sentence**

The English sentence "He was the first president of the United States" is neither true nor false. However, it can be changed into a true or false sentence depending upon what name is used to replace the pronoun "He." For example,

1. If we replace "He" with "Washington," the sentence would read "Washington was the first president of the United States."
   This is *true*.
2. If we replace "He" with "Jefferson," the sentence would read "Jefferson was the first president of the United States."
   This is *false*.

The pronoun "He" in the sentence above is similar to a variable in algebra. In the mathematical sentence $7x + 4 = 25$, the variable $x$ behaves like "He" in the English sentence above. So $7x + 4 = 25$ is neither true nor false. Both the English sentence and the mathematical sentence are called *open sentences*.

*Example 1*　**In the sentence $7x + 4 = 25$, replace $x$ with the indicated values. Are the resulting sentences true or false?**

| Replace $x$ with 3. | | Replace $x$ with 5. | |
|---|---|---|---|
| $7x + 4$ | 25 | $7x + 4$ | 25 |
| $7 \cdot 3 + 4$ | 25 | $7 \cdot 5 + 4$ | 25 |
| $21 + 4$ | | $35 + 4$ | |
| 25 | | 39 | |
| $25 = 25$ | | $39 \neq 25$ | |
| True | | False | |

*Write 3 in place of x.* ▶　　　◀ *Write 5 in place of x.*

**Thus,** 3 is a *solution* of $7x + 4 = 25$. When $x$ is replaced with 3 the open sentence $7x + 4 = 25$ becomes a true sentence.

Definitions:
Open sentence
Solution

An algebraic sentence containing a variable, or variables, is called an **open sentence**. Each value of the variable(s) that makes the sentence true is called a **solution** of the open sentence.

*Example 2*    **Which of the given values, 1, 6, or 5, is the solution of $4c - 2 = 3c + 3$?**

Replace $c$ with 1.

| $4c - 2$ | $3c + 3$ |
|---|---|
| $4 \cdot 1 - 2$ | $3 \cdot 1 + 3$ |
| $4 - 2$ | $3 + 3$ |
| 2 | 6 |

$2 \neq 6$  False

Replace $c$ with 6.

| $4c - 2$ | $3c + 3$ |
|---|---|
| $4 \cdot 6 - 2$ | $3 \cdot 6 + 3$ |
| $24 - 2$ | $18 + 3$ |
| 22 | 21 |

$22 \neq 21$  False

Replace $c$ with 5.

| $4c - 2$ | $3c + 3$ |
|---|---|
| $4 \cdot 5 - 2$ | $3 \cdot 5 + 3$ |
| $20 - 2$ | $15 + 3$ |
| 18 | 18 |

$18 = 18$  True

**Thus,** 5 is a solution of $4c - 2 = 3c + 3$.

# Oral Exercises

**Which are open sentences? For each sentence that is not open, indicate whether it is true or false. 3. open**

 1. He is chief of the fire company. **open**
 2. Texas is the smallest state in the U.S.A. **false**
 3. It is the fastest car made in the world.
 4. Martha Washington was the first First Lady.
 5. $7x + 3 = 24$ **open**
 6. $14 = 7 \cdot 2$ **true**
 7. $4 + 2x = 18$ **open**    **4. true**

# Written Exercises

**Which of the given values is the solution of the open sentence?**

Ⓐ  1. $5x + 14 = 24$ **2**
     1, 2
  2. $42 = 4y + 6$ **9**
     8, 9, 10
  3. $12 + 4z = 48$ **9**
     4, 2, 9
  4. $20 = 6 + 2d$ **7**
     7, 8
  5. $7r - 1 = 34$ **5**
     1, 3, 5
  6. $31 = 4g + 23$ **2**
     2, 6, 8
  7. $3a - 4 = 8$ **4**
     2, 4, 7
  8. $5a - 4 = 26$ **6**
     6, 8, 2
  9. $10r = 7r + 9$ **3**
     1, 5, 3
 10. $5x = 3x + 14$ **7**
     6, 7, 8
 11. $4a = 8 + 2a$ **4**
     0, 6, 4
 12. $15 - 2x = 3x$ **3**
     2, 3, 5
 13. $15 + y = 5y + 3$ **3**
     3, 4, 5
 14. $4x + 3 = x + 9$ **2**
     1, 2, 3
 15. $18 - 5r = 2r + 4$ **2**
     1, 2, 3, 4

Ⓑ 16. $4y - 13 = 7 - y$ **4**
     5, 4, 6
 17. $24 + 5g = 19 + 10g$ **1**
     2, 1, 6
 18. $8z - 4 = 41 - 7z$ **3**
     4, 3, 1
 19. $3(x + 4) = 3x + 12$
     0, 1, 5    **0, 1, 5**
 20. $2(6x - 3) = 3(4x - 2)$
     1, 4, 6    **1, 4, 6**
 21. $3k - 5 - k = 7 + k$ **12**
     10, 11, 12
 22. $3x + 5 - 2x = 2x + 5$
     0, 6, 8    **0**
 23. $32 - 2(x + 3) = 3(x - 3)$
     4, 5, 7    **7**
 24. $4 - (3 - x) = x + 1$
     1, 2, 3    **1, 2, 3**

Ⓒ 25. $1 + 5x = 10 - x$ **1.5**
     1.5, 2, 3.5
 26. $3n - 4 = 2(4 - n)$ **2.4**
     2.2, 2.3, 2.4
 27. $7 - [5 - (2 + x)] = 5$ **1**
     0, 1, 2
 28. $\dfrac{3x + 2}{6 - x} = 7$ **4**
     5, 4, 1
 29. $\dfrac{5m - 20}{m - 4} = 5$ **5, 6, 10**
     5, 6, 10
 30. $\dfrac{x^2 - 5x}{x} = 7$ **12**
     10, 12, 15

# COMPUTER ACTIVITIES

## Evaluating Algebraic Expressions

The computer language BASIC uses + and − as symbols for addition and subtraction. BASIC uses the symbols * for multiplication and / for division.

The rules for order of operations used in this chapter also hold for expressions written in BASIC. So, in BASIC, $S = 9 + 7 \cdot 3 - 12 \div 6$ becomes $S = 9 + 7 * 3 - 12 / 6$.

To find the value of S as the computer would, do all multiplications and divisions first. Then do all additions and subtractions.

$S = 9 + 7 * 3 - 12 / 6$
$S = 9 + 21 - 2$
$S = 28$

In performing the multiplications and divisions, $7 * 3$ means $7 \cdot 3$ or 21; $12 / 6$ means $\frac{12}{6}$ or 2.

A computer program is a set of written instructions for a computer. The program below demonstrates how the computer can be used to evaluate an expression like $9 + 7 \cdot 3 - 12 \div 6$. Notice that the lines are numbered by multiples of 10 to allow insertion of additional lines, if needed.

Enter the program into a computer. At the end of each line, press the RETURN key to get to the next line.

```
10   LET S = 9 + 7 * 3 -
     12 / 6
20   PRINT S
30   END
```

In line 10, the LET command assigns an expression to S.
Line 20 tells the computer to PRINT the value of S.
Line 30 tells the computer the program is ENDed.

See the Computer Section beginning on page 488 for more information.

After entering the program, enter the command RUN, which tells the computer to execute the program. When you RUN this program, the screen will display the value of S, 28.

Whenever you type in a new program, first type NEW.
Entering NEW erases the old program from the computer's memory.

## Exercises  For Ex. 1–3, see TE Answer Section.

**Simplify each expression. Then write, enter and RUN a program to check your results.**

1. $r = 32 - 45 \div 15 + 8 \cdot 9$ **101**

2. $b = 42 \div 2.1 + 13 \cdot 6 - 4$ **94**

3. Write a program to find the area of a triangle if the height is 42 and the base is 12. **252**

# CHAPTER ONE REVIEW

**Vocabulary**

| | |
|---|---|
| area [1.7] | multiplicative |
| associative [1.3] | identity [1.5] |
| base [1.6] | numerical |
| commutative [1.3] | coefficient [1.2] |
| distributive [1.4] | open sentence [1.8] |
| evaluate [1.2] | perimeter [1.7] |
| exponent [1.6] | solution [1.8] |
| factor [1.6] | term [1.2] |
| formula [1.7] | variable [1.2] |

Simplify. [1.1]

**1.** $8 - 3 \cdot 2$ **2**   **2.** $8 - 15 \div 3 + 6 \cdot 4$ **27**

★**3.** $22 - 3[17 - 2(15 \div 3)]$ **1**   **4.** $\dfrac{7 \cdot 3 + 5}{6 \cdot 2 + 1}$ **2**

Name the terms, variables, and numerical coefficients of the variables. [1.2]

**5.** $7x + 5y + 14$   **6.** $3n + 195$
**5. t: 7x, 5y, 14; v: x, y; nc: 7, 5**

Evaluate for the given value(s) of the variable(s). [1.2] **6. t: 3n, 195; v: n; nc: 3**

**7.** $6 + 8r$ for $r = 2$ **22**

**8.** $4abc + 3b - 2c$ for $a = \dfrac{1}{4}$, $b = \dfrac{2}{3}$, and $c = \dfrac{1}{2}$ **1$\frac{1}{3}$**

**9.** $\dfrac{2ab + 6c - abc}{a + b - 6c}$ for $a = 6$, $b = 8$, and $c = 2$ **6**

★**10.** $ab[a + b(3a + 5b)]$ for $a = 5$ and $b = 3$ **1,425**   **11. Assoc. Prop. Mult.**

Which property is illustrated? [1.3]
**11.** $3 \cdot (4 \cdot 5) = (3 \cdot 4) \cdot 5$
**12.** $24 \cdot 6 = 6 \cdot 24$ **Comm. Prop. Mult.**
**13.** $3(5 + 2) = 3 \cdot 5 + 3 \cdot 2$ **Distributive**

Simplify. [1.3]
**14.** $8(5a)$ **40a**   **15.** $(7t)3$ **21t**

Rewrite each by using the distributive property.   $4 \cdot 2 + 4 \cdot 5$
**16.** $4(3 + 2 + 5)$   **17.** $9 \cdot 4 + 5 \cdot 9 + 9 \cdot 2$
  **17. 9(4 + 5 + 2)**
Rewrite each by using the distributive property. Then simplify. [1.4]   **(4 + 2)x; 6x**
**18.** $7(3x + 4)$   **19.** $4 \cdot x + 2 \cdot x$

$7 \cdot 3x + 7 \cdot 4; 21x + 28$

Simplify, if possible. [1.5]   **22. 17y + 24**
**20.** $5x + 7 + 4x$ **9x + 7**   **23. 6x − 9**
**21.** $7a + 3b + 9 + 5b + 2$ **7a + 8b + 11**

**22.** $(4y + 8)3 + 5y$   **23.** $\dfrac{3}{4}(8x - 12)$

★**24.** $9y + 3[(4 + 2y)5 + 6y] + 9$ **57y + 69**

Simplify. Then evaluate for $x = 6$. [1.5]
**25.** $3(3x + 5) + (5x + 1)4$ **29x + 19; 193**

**26.** $6 + (3x + 2)4 + \dfrac{2}{5}(5x + 10)$ **14x + 18; 102**

Evaluate for the given value of the variable(s). [1.6]   **29. 10,240   30. 64   31. 1**

**27.** $x^3$ for $x = 7$ **343**   **28.** $2a^5$ for $a = 2$ **64**
**29.** $4a^3b^3c$ for $a = 2$, $b = 4$, and $c = 5$

**30.** $(6a)^3$ for $a = \dfrac{2}{3}$   **31.** $\dfrac{x^2 - 3x}{2x}$ for $x = 5$

Use the given formula to find each. [1.7]

**32.**

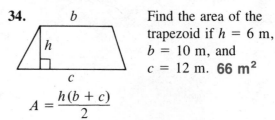

$l$
$p = 2l + 2w$
$w$ **33.**
$h$
$b$
$A = bh$

Find the perimeter if $l = 9$ cm and $w = 7$ cm. **32 cm**

Find the area if $h = 9$ in. and $b = 14$ in. **126 in.²**

**34.**
$b$
$h$
$c$
$A = \dfrac{h(b + c)}{2}$

Find the area of the trapezoid if $h = 6$ m, $b = 10$ m, and $c = 12$ m. **66 m²**

**35.** Find the area of a triangle with height 12 ft and base 18 ft. **108 ft²**

Which of the given values are solutions of the open sentence? [1.8]
**36.** $5x - 3 = 27$; 4, 6, 8 **6**
**37.** $8x - 3 = x + 2(3x + 4)$; 9, 10, 11 **11**
★**38.** $\dfrac{x^2 + 5x + 4}{x + 4} = x + 1$; 4, 7, 9 **4, 7, 9**

Name the terms, variables, and numerical coefficients of the variables.

1. $6 + 9a + 5b$
   **t: 6, 9a, 5b; v: a, b; c: 9, 5**

Which property is illustrated?

2. $(3 + 4) + 5 = 3 + (4 + 5)$
3. $6(3a + 4) = 6 \cdot 3a + 6 \cdot 4$ **Distributive**
4. $5 \cdot 9 = 9 \cdot 5$ **Comm. Prop. Mult.**
   **2. Assoc. Prop. Add.**

Simplify.

5. $9 + 4 \cdot 3$ **21**
6. $6 \cdot 1 + 7 + 5 \cdot 6$ **43**
7. $0 \cdot 9 + 6 - 10 \div 5 + 2 \cdot 4$ **12**
8. $\dfrac{7 + 4 \cdot 2}{9 - 3 \cdot 2}$ **5**

Simplify, if possible.

9. $8x + 9 + 3x$ **11x + 9**
10. $6m + 5p$ **cannot be simplified**
11. $6y + 1 + 2z + 8y + 5z$ **14y + 7z + 1**
12. $5x + x + 7x$ **13x**
13. $3(4x + 5) + (5x + 3)7$ **47x + 36**
14. $3a + \dfrac{1}{2}(6 + 4a)$ **5a + 3**

Evaluate for the given value(s) of the variables.

15. $4 + 7y$ for $y = 9$ **67**
16. $3a + 8b + 5$ for $a = 6$ and $b = 2$ **39**
17. $3x^2$ for $x = 6$ **108**
18. $2a^3 b^2 c^4$ for $a = 2$, $b = 5$, and $c = 3$    **32,400**
19. $(10a)^3$ for $a = \dfrac{2}{5}$ **64**
20. $\dfrac{y + 5x}{2x}$ for $x = 3$ and $y = 9$ **4**

Simplify. Then evaluate for $x = 4$.

21. $2(3x + 2) + 5(x + 4)$ **11x + 24; 68**
22. $7 + (4x + 2)3 + \dfrac{2}{3}(3x + 15)$

Rewrite each by using the distributive property.

23. $8(5 + 7)$ **8 · 5 + 8 · 7**
24. $(9 + 3 + 5)2$ **9 · 2 + 3 · 2 + 5 · 2**
25. $7 \cdot 6 + 4 \cdot 6 + 6 \cdot 9$ **6(7 + 4 + 9)**

Use the given formulas to find each.

26.

$A = \dfrac{bh}{2}$

Find the area if $h = 8$ m and $b = 4$ m. **16 m²**

27.

$p = 2l + 2w$

Find the perimeter if $l = 7$ ft and $w = 13$ ft. **40 ft**

28.

$A = \dfrac{h(b + c)}{2}$

Find the area if $b = 16$ cm, $c = 18$ cm, and $h = 3$ cm. **51 cm²**

29. Find the area of a triangle with base 22 in. and height 12 in. **132 in.²**

Which of the given values are solutions of the open sentence?

30. $3x - 18 = 21$; 11, 13, 15 **13**
31. $7a - 14 = 3a + 6$; 3, 4, 5 **5**
★32. Simplify $24 + 3[30 - 3(2 + 4) \div 6]$.
   **105**

★33. Simplify, if possible:

$4 + 2[3a + \dfrac{4}{5}(15a + 20)] + 6a$.

$4 + 6a + 12a + 16 + 6a$   **36a + 36**

★34. Is division commutative? Verify.

**14x + 23; 79**     **No. Examples may vary.**

# COLLEGE PREP TEST

Directions: Choose the one best answer to each question or problem.

1. Which number is 1,000 more than 999,999?
   D (A) 999,999 (B) 1,000,000 (C) 1,000,099 (D) 1,000,999 (E) 1,999,900

2. There are 6 people in a room. Each
   C shakes hands with everyone else in the room. How many handshakes is this?
   (A) 6 (B) 12 (C) 15 (D) 30 (E) 36

3. You spent exactly one dollar to purchase
   D some 5-cent stamps and 4-cent stamps. The number of 5-cent stamps you could not have bought is
   (A) 12 (B) 8 (C) 4
   (D) 1 (E) none of these

4. The sum of an odd number and an even
   A number is
   (A) always an odd number.
   (B) sometimes divisible by 2.
   (C) always a prime number.
   (D) always divisible by 2.
   (E) always an even number.

5. The total savings in buying thirty-six
   C 15-cent bags of peanuts for a class party at a special price of $1.42 per dozen is
   (A) $0.38 (B) $0.76 (C) $1.14
   (D) $4.26 (E) $5.40

6. The cost of electricity for operating a
   A 100-watt toaster, a 1,100-watt hairdryer, and four 75-watt lamps each for an hour at 8¢ per kilowatt-hour (1 kilowatt is equal to 1,000 watts) is
   (A) $0.12 (B) $1.18 (C) $11.88
   (D) $118.80 (E) $11,880

7. The greatest number of cubical blocks,
   E 4 inches on each edge, that can be used

to fill a cubical box, 2 ft on each edge, is
(A) 8 (B) 64 (C) 96 (D) 192 (E) 216

8. What part of 3 hours is 5 seconds?
   A (A) $\dfrac{5}{10,800}$ (B) $\dfrac{5}{3,600}$ (C) $\dfrac{5}{180}$
   (D) $\dfrac{5}{3}$ (E) none of these

9. Find the next number in the sequence.
   C 3, 8, 18, 33, 53, ___?___
   (A) 58 (B) 73 (C) 78 (D) 86 (E) 115

10. How many of the numbers between 100
    C and 400 begin or end with 2?
    (A) 100 (B) 110 (C) 120
    (D) 130 (E) 200

11. Shipping charges to a certain place are 15¢
    D each 30 g. The mass of a package for which the charges are $1.05 is
    (A) 7 g (B) 21 g (C) 70 g
    (D) 210 g (E) none of these

12. The missing number in the sequence
    C 3, 7, 12, 18, 25, ___?___, 42, 52 is
    (A) 35 (B) 34 (C) 33 (D) 32 (E) 29

13. Al owes Martha 40¢. Martha owes Al
    C 50¢. Al gives Martha 60¢. What is now needed to cancel the debt?
    (A) Martha must give Al 50¢.
    (B) Al must give Martha 50¢.
    (C) Martha must give Al 70¢.
    (D) Martha must give Al 10¢.
    (E) Al must give Martha 70¢.

14. If rain is falling at the rate of 1 cm/h,
    B how many centimeters of rain will fall in 12 minutes?
    (A) 0.02 cm (B) 0.2 cm (C) 2 cm
    (D) 12 cm (E) 20 cm

# 2 REAL NUMBERS

**HISTORY**

### François Viète 1540–1603

*The greatest French mathematician of the sixteenth century was François Viète. His work made possible many advances in mathematical knowledge in western Europe.*

Viète introduced the practice of representing unknown quantities with vowels and known quantities with consonants. Our $x$, $x^2$, and $x^3$ would have been $A$, $A$ quadratum, and $A$ cubum for Viète. He was also influential in the European movement to adopt a system of numeration based on tens, our present decimal system. Viète was the first one to have an actual formula for finding the value of $\pi$. In 1597 he found $\pi$ correct to nine decimal places.

Viète was an outstanding algebraist who made a clear distinction between algebra and arithmetic.

**Project**

Prepare a report on the history of $\pi$ and its evaluation.

# INTEGERS AND THE NUMBER LINE  2.1

*Objectives*

**To name the coordinates of points graphed on a number line**
**To graph numbers on a number line**
**To use integers to describe real-life situations**
**To compare integers**
**To find the absolute value of a number**

A number line may be used to show the relationships between numbers. A starting point, labeled 0, is called the *origin*. Numbers to the right of 0 are *positive,* and numbers to the left of 0 are *negative*. Zero is neither positive nor negative.

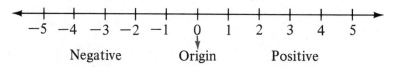

Different sets of numbers can be represented by listing their members. The three dots at the end of each list means *go on forever*.
**Natural** or **counting numbers:** N = 1, 2, 3, 4, . . .
**Whole Numbers:** W = 0, 1, 2, 3, 4, . . .
**Integers:** I = . . . −3, −2, −1, 0, 1, 2, 3, 4, . . .
The positive integers, 1, 2, 3, . . . are read as positive 1, positive 2, positive 3, and so on. The negative integers −1, −2, −3, . . . are read as negative 1, negative 2, negative 3, and so on.

These sets can be graphed on a number line. The shaded arrows mean that the graph goes on forever.

Graph of *N*: Natural numbers

Graph of *W*: Whole numbers

Graph of *I*: Integers

On a number line, the number that corresponds to a point is called the *coordinate* of the point. The point is called the *graph* of the number. In the graph of the integers above, point *A* is the graph of the number 3. The coordinate of point *A* is 3.

*Example 1*   **Name the coordinates of the points graphed.**

$-3$ is the coordinate of $C$.
0 is the coordinate of $D$.
5 is the coordinate of $E$.


*Example 2*   **Graph the points with coordinates $-4$, $-2$, and 6 on the number line. Label the points $F$, $G$, and $H$.**

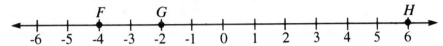

Point $F$ is the graph of $-4$.
Point $G$ is the graph of $-2$.
Point $H$ is the graph of 6.

Positive and negative numbers can be used to describe changes in direction that occur in real-life situations.


*Example 3*   **Name the integer that describes the situation.**

| 5 degrees above zero | loss of 8 yards | withdrawal of $100 |
|:---:|:---:|:---:|
| 5 | $-8$ | $-100$ |

The graphs of numbers on a number line show the *order* of the numbers. The numbers are increasing from left to right. *Symbols of inequality* can be used to show the order of two numbers.

<  means *is less than*.
>  means *is greater than*.

On a number line, $-3$ is to the left of 7.

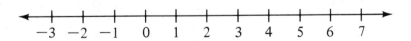

You can compare $-3$ with 7 by saying

$-3$ is less than 7.    or    7 is greater than $-3$.
$-3 \; < \; 7$              $7 \; > \; -3$

Notice that the inequality symbol points to the smaller number. On a number line, the *smaller* of two numbers is to the *left* of the larger number.

**Example 4**    **Replace each __?__ with the correct inequality symbol to make a true statement.**

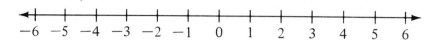

$-4 \underline{\ ?\ } -1$
$-4 < -1$ ◀ *$-4$ is to the left of $-1$.*

$3 \underline{\ ?\ } -5$
$3 > -5$ ◀ *$3$ is to the right of $-5$.*

On the number line below, notice that the integers 7 and $-7$ are on opposite sides of 0 and are the same distance from 0.

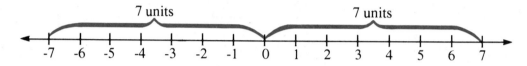

7 is 7 units to the right of 0, and $-7$ is 7 units to the left of 0. To indicate distance only, and not direction, use the *absolute value* of a number.

| Definition: Absolute value | The **absolute value** of a number is the distance of that number from 0 on a number line. The absolute value of $n$ is written $|n|$. |
| --- | --- |

The absolute value of 13 is 13. This is written $|13| = 13$.
The absolute value of $-13$ is 13. This is written $|-13| = 13$.

**Example 5**    **Find the absolute value.**

$|-2|$
$|-2| = 2$

$|5|$
$|5| = 5$

$|0|$
$|0| = 0$

# Reading in Algebra

**How do you read each of the following?**

negative 10 is less than positive 4

1. $-5$ **negative 5**     2. 7 **positive 7**     3. $-10 < 4$

4. $-6 > -8$     5. $|-4|$ **absolute value of $-4$** 6. $|12|$ **absolute value of 12**

    **negative 6 is greater than negative 8**

# Written Exercises

**(A)** **Name the number that describes each situation.**

1. loss of 3 kg  **−3**
2. gain of 4 kg  **4**
3. rise of 1,500 ft in elevation  **1,500**
4. 100 m below sea level  **−100**

**Name the coordinates of the points graphed.**

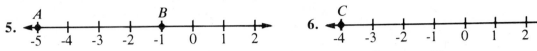

5. **A, −5; B, −1**    6. **C, −4; D, 3**    7. **E, −6; F, −2**    8. **G, −7; H, −5**

**On a number line, graph the points with the given coordinates.  Check students' graphs.**

9. $W$: $-5$
10. $X$: $1$
11. $Y$: $5$
12. $Z$: $-6$
13. $M$: $8$
14. $N$: $-1$
15. $R$: $-3$
16. $S$: $4$

**Replace each __?__ with < or > to make a true statement.**

17. $7$ __?__ $-2$  **>**
18. $-8$ __?__ $-4$  **<**
19. $-3$ __?__ $5$  **<**
20. $-2$ __?__ $-7$  **>**

**Find the absolute value.**

21. $|-8|$  **8**
22. $|4|$  **4**
23. $|-9|$  **9**
24. $|-18|$  **18**
25. $|-6|$  **6**
26. $|15|$  **15**
27. $|0|$  **0**
28. $|-5|$  **5**
29. $|1|$  **1**
30. $|-7|$  **7**

**(B)** **Compute.**

31. $|-9| + |-1|$  **10**
32. $|7| + |-8|$  **15**
33. $|-6| \cdot |-8|$  **48**
34. $|-4| + |12|$  **16**
35. $|10| + |-10|$  **20**
36. $|9| \cdot |-9|$  **81**

37. What two numbers have an absolute value of 6?  **6, −6**
38. Which number has a greater absolute value, $-8$ or 2?  **−8**

39. Name the greatest positive integer.  **none**
40. Name the least positive integer.  **1**

**Evaluate.**

41. $|x - 2|$ for $x = 2$  **0**
42. $|12 - y|$ for $y = 6$  **6**
43. $|a + 8|$ for $a = 15$  **23**

# CUMULATIVE REVIEW

**Simplify.**

1. $7m + 3m$  **10m**
2. $8t + 4t + 9t$  **21t**
3. $5c + c + 6c$  **12c**

# REAL NUMBERS

*Objective*  **To identify numbers as rational or irrational**

There are numbers on the number line that are not integers.

Numbers like $\frac{2}{5}$, $\frac{15}{4}$, $-\frac{1}{2}$, and 3 $\left(\text{or } \frac{3}{1}\right)$ are called *rational numbers*.

Recall that $\frac{16}{8} = 2$ because $8 \cdot 2 = 16$. Consider the expression $\frac{16}{0}$. $\frac{16}{0} = n$ is defined if $16 = 0 \cdot n$, but $0 \cdot n = 0$, not 16. Thus, division by 0 is undefined.

| Definition: Rational Number | A **rational number** is a number of the form $\frac{a}{b}$, where $a$ and $b$ are integers and $b \neq 0$. |
|---|---|

*Example 1*  **Verify that −14, 0, 0.7, and 3.462 are rational numbers.**

$$-14 = -\frac{14}{1} \qquad 0 = \frac{0}{1} \qquad 0.7 = \frac{7}{10} \qquad 3.462 = \frac{3,462}{1,000}$$

Every rational number can also be expressed as either a *terminating* or a *repeating decimal*. Study these examples.

$\frac{3}{8} = 3 \div 8$

$$\begin{array}{r} 0.375 \\ 8\overline{)3.000} \end{array}$$

The decimal ends, or terminates, at 5.

0.375 is a *terminating decimal*.

$\frac{1}{3} = 1 \div 3$

$$\begin{array}{r} 0.333\ldots \\ 3\overline{)1.000} \end{array}$$

The 3 repeats forever.

$\frac{2}{7} = 2 \div 7$

$$\begin{array}{r} 0.285714285714\ldots \\ 7\overline{)2.000000000000} \end{array}$$

The block of digits 285714 repeats forever.

$0.333\ldots$ and $0.285714285714\ldots$ are *repeating decimals*. The same nonzero digit, or block of digits, repeats without end.

A bar is used to indicate the digit or block of digits that repeat. Therefore, $\frac{1}{3} = 0.\overline{3}$ and $\frac{2}{7} = 0.\overline{285714}$. There are some numbers whose decimals neither terminate nor repeat. An example of such a number is 0.1231233123333123333 . . . It is an *irrational number*.

<table>
<tr><td>Definition:<br>Irrational<br>number</td><td>An **irrational number** is a number that has a decimal that neither terminates nor repeats.</td></tr>
</table>

Some examples of irrational numbers are: $-6.151551555$ . . ., $\pi = 3.14159265$ . . ., and $\sqrt{2} = 1.414213$ . . . . $\sqrt{2}$ is read *square root of 2*.

**Example 2**    **Determine whether the decimal is a rational or an irrational number.**

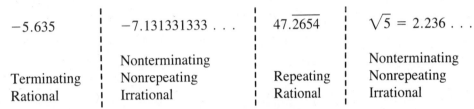

| $-5.635$ | $-7.131331333$ . . . | $47.\overline{2654}$ | $\sqrt{5} = 2.236$ . . . |
|---|---|---|---|
| Terminating<br>Rational | Nonterminating<br>Nonrepeating<br>Irrational | Repeating<br>Rational | Nonterminating<br>Nonrepeating<br>Irrational |

The number line contains points corresponding to rational numbers as well as irrational numbers.

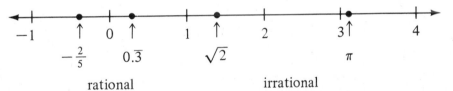

Each real number corresponds to one and only one point on the number line. Each point on the number line corresponds to a unique real number. Whenever *number* is used in this text, it will mean *real number*.

<table>
<tr><td>Definition:<br>Real numbers</td><td>The set of **real numbers** is the set of *rational numbers* together with the set of *irrational numbers*.</td></tr>
</table>

Each real number corresponds to one and only one point on the number line. Each point on the number line corresponds to a unique real number. Whenever *number* is used in this text, it will mean *real number*.

# Written Exercises

**Determine whether each is a rational number, an irrational number, or neither.**

1. $-12$ **R**    2. $0.8$ **R**    3. $\frac{7}{0}$ **N**    4. $-4.789$ **R**    5. $32.\overline{956}$ **R**    6. $\pi$ **I**    7. $0.6666$ . . . **R**

8. $\frac{4}{5}$ **R**    9. $0.\overline{35}$ **R**    10. $3.141141114$ . . . **I**    11. $-\frac{9}{0}$ **N**    12. $\frac{0}{4}$ **R**    13. $\sqrt{5}$ **I**

# ADDING REAL NUMBERS

*Objective*    **To add real numbers**

You can use the number line to add numbers. You can show the sum $4 + 3$.

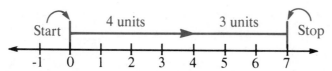

The addition is shown in terms of moves on a number line. Start at 0, move 4 units to the right (*positive direction*) to 4, then move 3 more units to the right (*positive direction*) to 7. So, $4 + 3 = 7$. This is read positive 4 plus positive 3 equals positive 7.

The same procedure can be used to find the sum of two negative numbers, as shown in Example 1. Notice that a negative number indicates a move to the left, which is the *negative direction*.

*Example 1*    **Add $-5 + (-3)$ using a number line.**

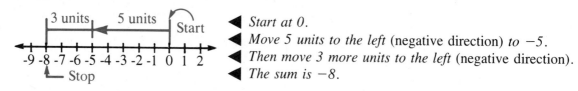

◀ *Start at 0.*
◀ *Move 5 units to the left* (negative direction) *to $-5$.*
◀ *Then move 3 more units to the left* (negative direction).
◀ *The sum is $-8$.*

**So,** $-5 + (-3) = -8$. This is read negative 5 plus negative 3 equals negative 8.

The two sums above suggest a rule that does not require a number line.

| Adding numbers with the same sign | To add two numbers with the same sign:<br>  **1.** Add their absolute values.<br>  **2.** Give the sum the same sign as the sign of each number. |
|---|---|

You can now find a sum like $-24 + (-7)$ by using the definition above. Look at Example 2.

*Example 2*    **Add $-24 + (-7)$**

$-24 + (-7)$    ◀ *The signs are the same, both negative.*
              *Add the absolute values:* $|-24| + |-7| = 24 + 7 = 31$.
    $-31$      ◀ *The sum is negative.*

A number line is used in the next two examples to help you discover a rule for finding the sum of two numbers with different signs.

*Example 3*    **Add 5 + (−7) using a number line.**

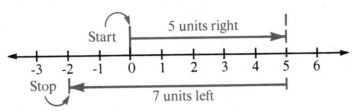

**So,** 5 + (−7) = −2. This is read positive 5 plus negative 7 equals negative 2.

*Example 4*    **Add −8 + 9$\frac{1}{2}$ using a number line.**

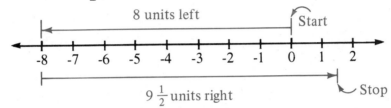

**So,** −8 + 9$\frac{1}{2}$ = 1$\frac{1}{2}$. This is read negative 8 plus positive 9$\frac{1}{2}$ equals positive 1$\frac{1}{2}$.

Examples 3 and 4 suggest a rule.

| Adding numbers with different signs | To add two numbers with different signs:<br>**1.** Subtract their absolute values, the smaller from the larger.<br>**2.** Give the result the same sign as the number with the larger absolute value. |
|---|---|

*Example 5*    **Add −26 + 5.**

−26 + 5    ◀ *The signs are different.*
*Subtract the absolute values:* $|-26| - |5| = 26 - 5 = 21.$
−21    ◀ *The sum is negative (−26 has the larger absolute value).*

*Example 6*    **Add −19 + 24.3.**

−19 + 24.3    ◀ *The signs are different, and 24.3 has the larger absolute value. Think:* $|24.3| - |-19| = 24.3 - 19 = 5.3.$
5.3    ◀ *The sum is positive (24.3 has the larger absolute value).*

Adding Real Numbers

*Example 7*    **Add $3\frac{1}{2} + \left(-5\frac{1}{4}\right)$.**

$3\frac{1}{2} + \left(-5\frac{1}{4}\right)$   ◀ *The signs are different.*    $5\frac{1}{4} = 5\frac{1}{4} = 4\frac{5}{4}$

               *Subtract:* $5\frac{1}{4} - 3\frac{1}{2}$.    $\underline{-3\frac{1}{2} = 3\frac{2}{4} = 3\frac{2}{4}}$

    $-1\frac{3}{4}$   ◀ *The sum is negative.*                    $1\frac{3}{4}$

If you add any two real numbers, the sum is also a real number. This sum is unique, that is, there is one and only one real number answer. This means that the set of real rumbers is *closed* under addition. The *closure property of addition* states that for all real numbers $a$ and $b$, $a + b$ is a unique real number.

*Summary*

To add real numbers
with the same signs:
1. Add the absolute values.
2. Keep the same sign as the sign of each number.

with different signs:
1. Subtract the absolute values.
2. Take the sign of the number with the larger absolute value.

## Written Exercises

**Add.**

Ⓐ **1.** 4 + 8   12      **2.** 8 + 5   13      **3.** 12 + 9   21      **4.** 16 + 7   23

     **5.** −5 + (−7)   −12    **6.** −12 + (−6)   −18    **7.** −8 + (−9)   −17    **8.** −6 + (−6)   −12

     **9.** 14 + (−8)   6     **10.** −26 + (−6)   −32    **11.** −15 + (−12)   −27    **12.** −18 + (−14)   −32

    **13.** 5 + (−10)   −5     **14.** −5 + 6   1        **15.** 9 + (−4)   5       **16.** −9 + (−3)   −12

    **17.** −5 + 10   5        **18.** −6 + (−6)   −12    **19.** −4 + (−12)   −16    **20.** 14 + (−3)   11

    **21.** −15 + 18.7   3.7    **22.** $-4\frac{3}{8} + \left(-5\frac{1}{8}\right)$   $-9\frac{1}{2}$   **23.** −5.7 + (7.2)   1.5    **24.** $7\frac{1}{2} + \left(-5\frac{1}{2}\right)$   2

Ⓑ **25.** $3\frac{1}{6} + \left(-8\frac{2}{3}\right)$   $-5\frac{1}{2}$   **26.** $-6 + 3\frac{1}{2}$   $-2\frac{1}{2}$    **27.** $-6\frac{2}{5} + 8\frac{7}{10}$   $2\frac{3}{10}$    **28.** $-9\frac{1}{3} + \left(-6\frac{5}{9}\right)$   $-15\frac{8}{9}$

    **29.** 0.04 + (−0.86) −0.82    **30.** −1.05 + 0.89     **31.** −2.07 + 1.49     **32.** 41 + (−8.03)

                                        −0.16                  −0.58                32.97

**Solve each problem.**

**33.** In the morning Mrs. Campbell withdrew $125 from her bank account. In the afternoon she withdrew $225. What real number describes the net change in her account? −350

**34.** A submarine was at a depth of 300 m below sea level. It rose 100 m. What real number describes the new depth of the submarine? −200

Ⓒ **35.** Is the set of even integers closed under addition? Why or why not? **Yes**

    **36.** Is the set of odd integers closed under addition? Why or why not? **No**

# PROPERTIES FOR THE ADDITION OF REAL NUMBERS

*Objectives*  **To identify the properties for addition of real numbers**
**To determine the additive inverse (or opposite) of a real number**
**To read and interpret the symbol −$a$**

All the properties of addition can be shown to be true for real numbers.
$-5 + 3 = 3 + (-5)$ is true because $-2 = -2$. The order in which two real numbers are added does not change the sum. This is the *commutative property of addition*.
You can show that the *associative property of addition* is true for real numbers.

*Example 1*  **Show that $[5 + (-7)] + (-3) = 5 + [-7 + (-3)]$.**

| $[5 + (-7)] + (-3)$ | $5 + [-7 + (-3)]$ |
| --- | --- |
| $-2 + (-3)$ | $5 + (-10)$ |
| $-5$ | $-5$ |

◀ *Add inside the brackets first.*
◀ *Answers are the same.*

**Thus, $[5 + (-7)] + (-3) = 5 + [-7 + (-3)]$.**

| Properties of addition | **Commutative:** For all real numbers $a$ and $b$, $a + b = b + a$. |
| --- | --- |
| | **Associative:** For all real numbers $a$, $b$, and $c$, $a + (b + c) = (a + b) + c$. |

What happens if you add 0 and any number? Observe some examples.

$$7.8 + 0 = 7.8 \qquad -5 + 0 = -5 \qquad 0 + \left(-9\frac{1}{2}\right) = -9\frac{1}{2}$$

From these examples you see that the sum of any number and 0 is that number.

| Property of additive identity | For any real number $a$, $a + 0 = a$ and $0 + a = a$. |
| --- | --- |
| | Zero is the **additive identity.** |

Look at 4 and −4 on the number line below.

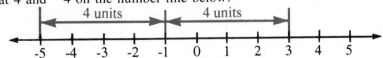

The numbers −4 and 4 are each the same distance from 0. −4 and 4 are called *opposites* or *additive inverses*. −4 is the opposite, or additive inverse, of 4.

| Symbol for opposites | The symbol $-a$ means the opposite of $a$, or the additive inverse of $a$. |
|---|---|

The opposite of a number, $-a$, can be positive, negative, or 0. For example,

$-6\frac{1}{4}$ is read as *the opposite of* $6\frac{1}{4}$,

$-(-4)$ is read as *the opposite of* $-4$, and

0 is its own opposite, that is, 0 is the opposite of 0.

*Example 2*   **For the given value of $a$, find the value of $-a$, the opposite of $a$.**

$$a = 7$$
$$-a = -7$$

The opposite of 7 is $-7$.

$$a = -5.3$$
$$-a = -(-5.3)$$
$$-a = 5.3$$

The opposite of $-5.3$ is 5.3.

What is true about the sum of a number and its opposite? Look at the following sums.

$$-8 + 8 = 0 \qquad 2\frac{1}{7} + \left(-2\frac{1}{7}\right) = 0 \qquad -5.2 + 5.2 = 0$$

The sum of any number and its opposite is 0. This is stated in the following property.

| Property of additive inverses | For any real number $a$, $a + (-a) = 0$.<br>and $(-a) + a = 0$. |
|---|---|

*Example 3*   **Which property of addition is illustrated?**

|  | **Answers** |
|---|---|
| $-9 + 9 = 0$ | Property of additive inverses |
| $0 + (-12) = -12$ | Property of additive identity |
| $7 + (-4) = -4 + 7$ | Commutative property of addition |
| $[-9 + (-3)] + 1 = -9 + [-3 + 1]$ | Associative property of addition |

# Reading in Algebra

**Indicate whether each statement is always true, sometimes true, or never true.**

**1.** The opposite of a number is a negative number. **sometimes true**

**2.** The additive identity is positive. **never true**     **3.** $x + (y + z) = (x + y) + z$ **always true**

# Written Exercises

**Show that each statement is true.  Check students' work.**

Ⓐ **1.** $[6 + (-2)] + (-4) = 6 + [-2 + (-4)]$  **2.** $-9 + [14 + (-18)] = (-9 + 14) + (-18)$
**3.** $-11 + [-8 + (-4)] = [-11 + (-8)] + (-4)$  **4.** $6 + [-3 + (-7)] = [6 + (-3)] + (-7)$

**For the given value of $a$, find the value of $-a$.**  **7.** $19\frac{3}{4}$

**5.** $a = -8$ **8**  **6.** $a = 2$ **-2**  **7.** $a = -19\frac{3}{4}$  **8.** $a = 6.3$ **-6.3**  **9.** $a = 0$ **0**

**Which property of addition is illustrated?  10. comm. prop. add.  11. prop. add. inverses**

**10.** $5 + (-2) = -2 + 5$  **11.** $-14.7 + 14.7 = 0$  **12.** $0 + \frac{4}{9} = \frac{4}{9}$ **add. ident.**

**13.** $(-8 + 3) + (-4) = -8 + [3 + (-4)]$  **14.** $-15 + 0 = -15$  **15.** $-x + x = 0$
**16.** $x + y = y + x$  **17.** $a + 0 = a$  **18.** $[-3 + (-2)] + 1 = -3 + (-2 + 1)$
**13. assoc. prop. add.  14. add. ident.  15. prop. add. inverses 16. comm. prop. add.**
**Find the value of each.  17. add. ident.  18. assoc. prop. add.**

Ⓑ **19.** $-x$ if $x = 6\frac{5}{7}$ **$-6\frac{5}{7}$**  **20.** $-b$ if $b = -4.8$ **4.8**  **21.** $a$ if $-a = -3\frac{2}{3}$ **$3\frac{2}{3}$**

*Example*  **Find each sum. Then give its opposite. $8 + (-13)$.**

$8 + (-13)$  ◄ *The signs are different. Subtract the absolute values:*
$\qquad\qquad\quad |-13| - |8| = 13 - 8 = 5$
$-5$  ◄ *The sum is negative. ($-13$ has the larger absolute value.)*
The opposite of $-5$ is 5.

**Find each sum. Then give its opposite.**

**24.** $2\frac{2}{3}, -2\frac{2}{3}$

$\quad$ **-1, 1** $\qquad\qquad$ **5, -5** $\qquad\qquad\qquad\quad$ **-5, 5** $\qquad\qquad\qquad$ **-9, 9**
**22.** $-5 + 4$  **23.** $8 + (-3)$  **24.** $-5 + 7\frac{2}{3}$  **25.** $-8 + 3$  **26.** $-7 + (-2)$

**27.** $-9 + 5$  **28.** $-6 + (-8)$  **29.** $5 + (-4)$  **30.** $-2 + (-5.7)$  **31.** $-9.8 + -2.5$
$\quad$ **-4, 4** $\qquad\qquad$ **-14, 14** $\qquad\qquad$ **1, -1** $\qquad\qquad$ **-7.7, 7.7** $\qquad$ **-12.3, 12.3**

Ⓒ *Example*  **Verify. $(-4 + 5) + (-8) = (-8 + 5) + (-4)$.**

| Expression | Reason |
|---|---|
| $(-4 + 5) + (-8)$ | Given |
| $-8 + (-4 + 5)$ | Commutative property of addition |
| $-8 + [5 + (-4)]$ | Commutative property of addition |
| $(-8 + 5) + (-4)$ | Associative property of addition |

**For Ex. 32–35, see TE Answer Section.**

**Verify.**
**32.** $(-9 + 4) + (-8) = (-8 + 4) + (-9)$  **33.** $[-8.6 + (-3)] + 8.6 = -3$
**34.** $(-x + y) + x = y$  **35.** $-m + (-r + m) = -r$

Properties for the Addition of Real Numbers  **39**

# SUBTRACTING REAL NUMBERS

*Objective*    **To subtract one real number from another**

You already know that $8 - 3 = 5$ and $8 + (-3) = 5$. Subtraction is defined as adding the opposite. So, any difference can be written as a sum.

$$8 - 3 = 8 + (-3)$$
$$= 5$$

| Definition: Subtraction | For all real numbers $a$ and $b$, $a - b = a + (-b)$. |
|---|---|

Notice that in $8 - 3$, both numbers are positive and 8 is greater than 3. In cases like this, it is more convenient to think of it as the difference of two positive numbers. However, in other cases, use the definition of subtraction.

*Example 1*    **Subtract 3 − 7.**

$3 - 7$
$3 + (-7)$  ◀ *3 − 7 means 3 plus the opposite of 7.*
$\quad -4$

*Example 2*    **Subtract.**

$6 - (-8)$  ◀ *6 − (−8) means*        $\quad -2\frac{3}{5} - (-8)$  ◀  $8 = 7\frac{5}{5}$
$6 + 8$    *6 plus the opposite*        $-2\frac{3}{5} + 8$    $-2\frac{3}{5} = -2\frac{3}{5}$
$\quad 14$    *of −8.*                $\qquad 5\frac{2}{5}$    $\overline{\qquad 5\frac{2}{5}}$

# Written Exercises

**Subtract.**

Ⓐ 
1. $9 - 4$ **5**  2. $-5 - 2$ **−7**  3. $3 - 9$ **−6**  4. $-8 - 2$ **−10**  5. $7 - 12$ **−5**
6. $6 - (-2)$ **8**  7. $-7 - (-3)$ **−4**  8. $-2 - (-1)$ **−1**  9. $2 - (-4)$ **6**  10. $8 - (-7)$ **15**
11. $-5 - 5$ **−10**  12. $7 - 8$ **−1**  13. $-2 - 1$ **−3**  14. $-3 - (-8)$ **5**  15. $5 - (-4)$ **9**

Ⓑ 
16. $-62 - (-59)$ **−3**  17. $17 - (-43)$ **60**  18. $28 - (-99)$ **127**  19. $-115 - (-99)$ **−16**
20. $-2\frac{3}{4} - 5\frac{1}{4}$ **−8**  21. $19 - 4.2$ **14.8**  22. $-7 - \left(-5\frac{1}{3}\right)$ **−1$\frac{2}{3}$**  23. $0.87 - 0.54$ **0.33**

Ⓒ 
24. Is the set of positive numbers closed under subtraction? Why or why not? **No**
25. Is subtraction commutative for the set of real numbers? Why or why not? **No**
26. Is subtraction associative for the set of real numbers? Why or why not? **No**
    **See TE Answer Section.**

# MULTIPLYING REAL NUMBERS

**Objective**

**To multiply real numbers**

Multiplying positive real numbers is like multiplying whole numbers.

$$3 \cdot 6 = 18 \qquad 1 \cdot 7\frac{2}{3} = 7\frac{2}{3} \qquad 9.2 \cdot 5.7 = 52.44$$

This suggests the following rule.

| Multiplying two positive numbers | The product of two positive numbers is a positive number.<br>positive · positive = positive |
|---|---|

The *commutative and associative properties for multiplication* are true for real numbers. Consider the following products.

$$4 \cdot 0 = 0 \qquad -2.9 \cdot 0 = 0 \qquad 0 \cdot 9\frac{1}{3} = 0 \qquad 0 \cdot 0 = 0$$

The product of every real number and 0 is 0. This is called the *property of zero for multiplication*.

| Properties of multiplication | **Commutative:** For all real numbers $a$ and $b$, $a \cdot b = b \cdot a$.<br>**Associative:** For all real numbers $a$, $b$, and $c$, $a \cdot (b \cdot c) = (a \cdot b) \cdot c$.<br>**Property of zero:** For any real number $a$, $a \cdot 0 = 0$ and $0 \cdot a = 0$. |
|---|---|

In the following patterns, the first factor remains the same.
The second factor decreases by 1 each time.

$$2 \cdot 3 \ \ = 6$$
$$2 \cdot 2 \ \ = 4 \qquad \text{The product decreases by 2 each time.}$$
$$2 \cdot 1 \ \ = 2$$
$$2 \cdot 0 \ \ = 0$$
$$2 \cdot -1 = -2$$
$$2 \cdot -2 = -4 \qquad \text{So, the pattern of decreasing by 2 continues.}$$
$$2 \cdot -3 = -6$$

This pattern suggests that the product of a positive real number and a negative real number is a negative real number.

**Example 1**

**Multiply.**

$7 \cdot 6$
$7 \cdot 6 = 42$ ◀ *pos. · pos. = pos.*

$9(-5)$
$9(-5) = -45$ ◀ *pos. · neg. = neg.*

Consider $-3(5)$. By the commutative property, $-3(5) = 5(-3)$. You have already seen that the product of a positive number and a negative number is negative. So, $-3(5) = 5(-3) = -15$. This leads to the following rule.

| Multiplying numbers with different signs | The product of a negative number and a positive number is a negative number. |
| --- | --- |
| | positive · negative = negative      negative · positive = negative |

### Example 2    **Multiply.**

$$9(-3)$$
$$9(-3) \quad \blacktriangleleft \; pos. \cdot neg. = neg.$$
$$-27$$

$$-4.1(0.2) \quad \blacktriangleleft \quad -4.1$$
$$-4.1(0.2) \qquad \times\ 0.2$$
$$-0.82 \qquad\quad -0.82$$

The following pattern suggests a rule for multiplying two negative numbers.

$$3 \cdot -2 = -6$$
$$2 \cdot -2 = -4$$
$$1 \cdot -2 = -2 \qquad \text{The product increases by 2 each time.}$$
$$0 \cdot -2 = 0$$
$$-1 \cdot -2 = 2$$
$$-2 \cdot -2 = 4 \qquad \text{So, the pattern of increasing by 2 continues.}$$
$$-3 \cdot -2 = 6$$

This pattern suggests the following rule.

| Multiplying two negative numbers | The product of two negative numbers is a positive number. |
| --- | --- |
| | negative · negative = positive |

### Example 3    **Multiply.**

$$-\frac{2}{3}\left(-\frac{4}{7}\right)$$
$$-\frac{2}{3}\left(-\frac{4}{7}\right) = \frac{8}{21}$$

$$-28\left(-\frac{1}{7}\right)$$
$$-28\left(-\frac{1}{7}\right) = 4 \quad \blacktriangleleft \; +\left(\overset{4}{\cancel{28}} \cdot \frac{1}{\cancel{7}}\right) = 4$$

If you multiply any two real numbers, the product is a unique real number. This means that the set of real numbers is *closed* under multiplication. The *closure property of multiplication* states that for all real numbers $a$ and $b$, $a \cdot b$ is a unique real number.

### Example 4    **Multiply.**

$$-4.2(0.06)$$
$$-4.2(0.06) = -0.252$$

$$3\frac{1}{2}(-8)$$
$$\frac{7}{2}(-8) = -28 \quad \blacktriangleleft \; \frac{7}{\cancel{2}} \cdot \frac{\overset{-4}{\cancel{-8}}}{1} = -28$$

## Example 5

**Multiply** $-2(8)(-5)(6)$**.**

$$-2(8)(-5)(6) = -2(-5)(8)(6) \quad \blacktriangleleft \text{ Multiplication can be done in any order.}$$
$$= 10 \cdot 48 = 480$$

## Example 6

**Multiply** $(-2)^3(3^2)$**.**

$$(-2)^3(3^2) = \underbrace{(-2)(-2)}(-2)\underbrace{(3)(3)} \quad \blacktriangleleft \; (-2)^3 = (-2)(-2)(-2) \text{ and}$$
$$= \underbrace{4(-2)} \quad \cdot \quad 9 \qquad 3^2 = 3 \cdot 3$$
$$= \quad -8 \quad \cdot \quad 9 \;=\; -72$$

## Summary

Multiplying two real numbers with like signs gives a positive product.

positive · positive = positive     negative · negative = positive

Multiplying two real numbers with different signs gives a negative product.

positive · negative = negative     negative · positive = negative

# Written Exercises

**Multiply.**

(A)
1. $4 \cdot 7$ **28**
2. $-3 \cdot 6$ **−18**
3. $6(-9)$ **−54**
4. $-3 \cdot 1$ **−3**
5. $6(-7)$ **−42**
6. $-8 \cdot 0$ **0**
7. $-3.4(0.3)$ **−1.02**
8. $-0.08(4.1)$ **−0.328**
9. $-\dfrac{3}{5}\left(\dfrac{2}{7}\right)$ $-\dfrac{6}{35}$
10. $-15\left(-\dfrac{1}{3}\right)$ **5**
11. $\dfrac{4}{5}\left(-\dfrac{2}{3}\right)$ $-\dfrac{8}{15}$
12. $-\dfrac{1}{5}(20)$ **−4**

13. $9(-7)(-3)(5)$ **945**
14. $6(0)(-7)(-9)$ **0**
15. $-5(-2)(-3)(4)$ **−120**
16. $(-4)^3(2)$ **−128**
17. $(-2)^3(-5)$ **40**
18. $(-2)(-4)^3$ **128**
19. $(-3)^2(-1)^3$ **−9**
20. $(6^2)(-1)^5$ **−36**
21. $(-2)^5(3^3)$ **−864**

(B)
22. $4\dfrac{1}{2}(-6)$ **−27**
23. $-12\left(-2\dfrac{1}{3}\right)$ **28**
24. $-3\dfrac{1}{3}\left(1\dfrac{4}{5}\right)$ **−6**
25. $4.3(-1.5)$ **−6.45**
26. $-2.3(-5.4)$ **12.42**
27. $-0.13(2.05)$ **−0.2665**
28. $(-1)^8(-3)^3(-2)$ **54**
29. $(-6)^2(-1)(-2)^3$ **288**
30. $(5^2)(-3)^2(-4)$ **−900**

(C) **Compute.** [*Hint:* Do the operations within the brackets first.]
31. $(-3 \cdot 4)[-2 + (-5)(-4)]$ **−216**
32. $(-5 + 8)[-7 + -4 \cdot 3 + (-6)]$ **−75**
33. $(-8 + 6)^2[-9 + (-4) \cdot 2]$ **−68**
34. $[-2 + (-1)]^3[7 + (-8) + (-1)]^5$ **864**

# CUMULATIVE REVIEW

**Add.**
1. $-6 + (-8)$ **−14**
2. $8 + (-7)$ **1**
3. $-12 + 18$ **6**

# DIVIDING REAL NUMBERS

*Objective*    **To divide real numbers**

Multiplication and division are inverse operations.

| *Multiplication* | *Division* |
|---|---|
| $4 \cdot 2 = 8$ | $8 \div 2 = 4$, or $\dfrac{8}{2} = 4$ |

You can use this relationship between multiplication and division to discover the rules for dividing real numbers.

| *Multiplication* | *Division* |
|---|---|
| $3 \cdot 2 = 6$ | $6 \div 2 = 3$ |
| $3 \cdot -2 = -6$ | $-6 \div -2 = 3$ |
| $-3 \cdot -2 = 6$ | $6 \div -2 = -3$ |
| $-3 \cdot 2 = -6$ | $-6 \div 2 = -3$ |

These examples suggest that the rules for dividing signed numbers are the same as the rules for multiplying signed numbers.

| Dividing numbers | The quotient of two numbers with like signs is positive. |
|---|---|

$$\frac{\text{positive}}{\text{positive}} = \text{positive} \qquad \frac{\text{negative}}{\text{negative}} = \text{positive}$$

The quotient of two numbers with different signs is negative.

$$\frac{\text{negative}}{\text{positive}} = \text{negative} \qquad \frac{\text{positive}}{\text{negative}} = \text{negative}$$

*Example 1*    **Divide.**

$$\frac{27}{3}$$

$$\frac{27}{3} = 9 \quad \blacktriangleleft \quad \frac{pos.}{pos.} = pos.$$

$$-6 \div -2$$

$$-6 \div -2 = 3 \quad \blacktriangleleft \quad neg. \div neg. = pos.$$

*Example 2*    **Divide.**

$$\frac{-36}{4}$$

$$\frac{-36}{4} = -9 \quad \blacktriangleleft \quad \frac{neg.}{pos.} = neg.$$

$$2.4 \div -0.08$$

$$2.4 \div -0.08 = -30 \quad \blacktriangleleft \quad 0.08.\overline{)2.40.} \ \ ^{30.}$$

Recall that division by zero is *undefined*. So, an expression like $\dfrac{-7}{0}$ is *undefined*. However, $\dfrac{0}{-3} = 0$. This can be checked: $-3 \cdot 0 = 0$.

Notice that each of the following products is 1.

$$\frac{4}{1} \cdot \frac{1}{4} = \frac{4}{4}, \text{ or } 1 \qquad \frac{2}{3} \cdot \frac{3}{2} = \frac{6}{6}, \text{ or } 1 \qquad \frac{3}{7} \cdot \frac{7}{3} = \frac{21}{21}, \text{ or } 1$$

Numbers whose product is 1 are called *reciprocals,* or *multiplicative inverses,* of each other. This suggests the following properties.

| | |
|---|---|
| Property of reciprocals | If $a \neq 0$ and $b \neq 0$, then $\frac{b}{a}$ is the **reciprocal (multiplicative inverse)** of $\frac{a}{b}$. $$\frac{a}{b} \cdot \frac{b}{a} = 1 \qquad \text{and} \qquad \frac{b}{a} \cdot \frac{a}{b} = 1$$ |

| | |
|---|---|
| Multiplicative inverse property | For every real nonzero number $a$, there is a unique real number $\frac{1}{a}$ such that $$a \cdot \frac{1}{a} = 1.$$ |

*Example 3*  **Find the reciprocals of $\frac{2}{3}$, 5, $1\frac{2}{7}$, and 0.**

| Number | Reciprocal | | Number | Reciprocal | | Number | Reciprocal |
|---|---|---|---|---|---|---|---|
| $\frac{2}{3}$ | $\frac{3}{2}$ | | $5 = \frac{5}{1}$ | $\frac{1}{5}$ | | $1\frac{2}{7} = \frac{9}{7}$ | $\frac{7}{9}$ |

0 has no reciprocal since $\frac{1}{0}$ is undefined.

Since multiplication and division are inverse operations, you can use the *division rule* to express division as multiplication by the reciprocal.

| | |
|---|---|
| Division rule | For all real numbers $a$ and $b$, with $b \neq 0$, $$a \div b = \frac{a}{b} = a \cdot \frac{1}{b}.$$ |

Any quotient may now be expressed as a product.

*Example 4*  **Divide.**

$$\frac{10}{9} \div -\frac{5}{3}$$

$$-\left(\frac{10}{9} \cdot \frac{3}{5}\right)$$

$$-\left(\frac{\overset{2}{\cancel{10}}}{\underset{3}{\cancel{9}}} \cdot \frac{\overset{1}{\cancel{3}}}{\underset{1}{\cancel{5}}}\right) = -\left(\frac{2}{3}\right) \text{ or } -\frac{2}{3}$$

$$-3\frac{3}{5} \div -6$$

$$+\left(\frac{18}{5} \div \frac{6}{1}\right)$$

$$+\left(\frac{\overset{3}{\cancel{18}}}{5} \cdot \frac{1}{\underset{1}{\cancel{6}}}\right) = +\frac{3}{5} \text{ or } \frac{3}{5}$$

Dividing Real Numbers

## Example 5　Compute $-48 \div (12 \div -6)$.

$-48 \div (12 \div -6)$　◀ *Divide within parentheses first.*
$-48 \div -2 = 24$

# Reading in Algebra

**Indicate whether the statement is always true, sometimes true, or never true.**

1. If two numbers are opposites, their quotient is negative. **always true**
2. The quotient of two numbers with the same sign has that sign. **sometimes true**
3. If $a \div b$ is negative and $a$ is negative, then $b$ is negative. **never true**
4. $a \div b$ is undefined. **sometimes true**

# Oral Exercises

**Give the reciprocal.**　**3. undefined**

Ⓐ 1. $\dfrac{3}{4}$ $\dfrac{4}{3}$　　2. $7$ $\dfrac{1}{7}$　　3. $0$　　4. $\dfrac{1}{3}$ 3　　5. $1$ 1　　6. $-2\dfrac{3}{4}$ $-\dfrac{4}{11}$

# Written Exercises

**Divide.**　　**12. undefined**　　　**19. −11,000**　　**20. undefined**

1. $\dfrac{8}{2}$ 4　　2. $\dfrac{-12}{3}$ −4　　3. $\dfrac{32}{-8}$ −4　　4. $\dfrac{-72}{-8}$ 9　5. $\dfrac{28}{7}$ 4　　6. $\dfrac{-40}{10}$ −4

7. $\dfrac{-45}{9}$ −5　　8. $\dfrac{9}{-9}$ −1　　9. $\dfrac{-100}{4}$ −25　10. $\dfrac{80}{-8}$ −10　11. $\dfrac{0}{5}$ 0　　12. $\dfrac{21}{0}$

13. $-60 \div 10$ **−6**　　14. $-32 \div 8$ **−4**　　15. $-0.54 \div -0.9$ **0.6** 16. $1.4 \div -0.07$ **−20**
17. $12 \div -0.6$ **−20**　　18. $3.4 \div 1.7$ **2**　　19. $-22 \div 0.002$　　20. $-32 \div 0$
21. $\dfrac{3}{7} \div -\dfrac{6}{7}$ $-\dfrac{1}{2}$　　22. $-\dfrac{4}{5} \div -\dfrac{8}{15}$ $1\dfrac{1}{2}$　23. $\dfrac{7}{8} \div -14$ $-\dfrac{1}{16}$　24. $-\dfrac{2}{3} \div \dfrac{1}{2}$ $-1\dfrac{1}{3}$
25. $-15 \div \dfrac{5}{7}$ **−21**　　26. $-\dfrac{5}{9} \div \dfrac{10}{3}$ $-\dfrac{1}{6}$　27. $1\dfrac{1}{2} \div -\dfrac{9}{8}$ $-1\dfrac{1}{3}$　28. $-\dfrac{4}{3} \div 2\dfrac{2}{3}$ $-\dfrac{1}{2}$

Ⓑ **Simplify.**
29. $-48 \div (60 \div -5)$ **4**　　　30. $(-36 \div 18) \div -2$ **1**　　　31. $72 \div (-24 \div -8)$ **24**
32. $-45 \div (-15 \div 3)$ **9**　　　33. $14 \div (28 \div -2)$ **−1**　　　34. $27 \div (-18 \div -6)$ **9**

Ⓒ 35. $(4.8 \div -6) \div (-4 \div 0.2)$ **0.04**　　　36. $[-6.4 \div (2 \cdot 0.8)] \div (144 \div -7.2)$ **0.2**
37. $(-5 + (-3.5)] \div (2.0 \div -4)$ **17**　　　38. $[-6 + (-24)] \div [-0.8 + (-0.7)]$ **20**
39. Is division associative for real numbers? Show why or why not. **No**
40. Is the set of real numbers closed under division? Why or why not? **No**
　　**Division by 0 not defined**

# SHORTCUTS IN ADDING AND SUBTRACTING REAL NUMBERS

**2.8**

*Objective*

**To add real numbers without enclosing the addends in parentheses**

In this lesson you will take a new look at addition of numbers. The associative and commutative properties of addition allow you to group numbers in any convenient way and to add in any order.
Sometimes it is easier to add numbers by grouping like signs together.

*Example 1*

**Add −3 + 8 + (−7).**

$-3 + 8 + (-7)$ ◀ *There are two negatives.*
$-3 + (-7) + 8$ ◀ *Group the negatives together.*
$\qquad -10 + 8$
$\qquad\quad -2$

In Example 1, you added the numbers −3, 8, and −7. To simplify the writing, you can omit the + ( ) notation.
Thus, $-3 + 8 + (-7)$ can be written as $-3 + 8 - 7$.

An expression like $8 - 9 + 6$ can now be interpreted as positive 8 plus negative 9 plus positive 6. This interpretation allows us to use the properties of addition to group the like signs together. This is shown in Example 2.

*Example 2*

**Simplify 8 − 9 + 6.**

$8 - 9 + 6$
$8 + 6 - 9$
$14 - 9$ ◀ *Group the positives together.*
$\quad 5$

Example 3 below shows you that it is easy to add several numbers by mentally grouping like signs.

*Example 3*

**Simplify.**

$-14 + 21 - 9$
$-14 - 9 + 21$
$\underbrace{\qquad\qquad}$
$\quad -23 \quad + 21$ ◀    $-14 - 9 = -14 + (-9)$
$\qquad -2$            $= -23$

$-5 + 8 - 6 + 4$
$-5 - 6 + 8 + 4$
$\underbrace{\qquad}\quad\underbrace{\qquad}$
$\quad -11 \quad + \quad 12$ ◀    $-5 - 6 = -5 + (-6)$
$\qquad\quad 1$            $= -11$

## Written Exercises

**Simplify.**

**(A)**
1. $5 - 11 + 3$  **−3**
2. $-8 + 4 + 2$  **−2**
3. $6 + 9 - 5$  **10**
4. $1 - 5 - 7$  **−11**
5. $-8 - 3 - 2$  **−13**
6. $-9 + 0 + 9$  **0**
7. $-5 - 12 - 2$  **−19**
8. $0 - 14 + 14$  **0**
9. $9 - 8 - 8$  **−7**
10. $6 - 15 - 3$  **−12**
11. $-8 - 8 + 16$  **0**
12. $-12 + 13 - 1$  **0**

**(B)**
13. $-7 + 5 - 3$  **−5**
14. $-6 + 2 - 8 + 6$  **−6**
15. $4 - 5 - 9 + 12$  **2**
16. $3.4 - 5 + 1.2$  **−0.4**
17. $-3.2 + 7 - 0.5$  **3.3**
18. $7.3 - 13 - 4.4$  **−10.1**
19. $-\dfrac{2}{3} + 8 - 4$  **$3\dfrac{1}{3}$**
20. $-5 + 9 - \dfrac{1}{2}$  **$3\dfrac{1}{2}$**
21. $7\dfrac{1}{3} - 8\dfrac{1}{6}$  **$-\dfrac{5}{6}$**
22. $\dfrac{-8 + 5 - 4 + 2}{-2 + 7 - 10 + 2}$  **$\dfrac{5}{3}$**
23. $\dfrac{-9 - 5 + 6 - 2}{-6 - 8 + 4 + 12}$  **−5**
24. $\dfrac{-3 - 7 + 5 - 15}{-3 - 2 + 1 + 9}$  **−4**

**(C)**
25. $-19.6 + 318.42 - 0.0098$
26. $-0.004 + 1.06 - 7.13$
27. $2.16 - 0.004 + 1.09$

25. **298.8102**  26. **−6.074**  27. **3.246**

**Simplify.** [*Hint:* Work within any innermost parentheses first.]
28. $(-8 + 3 - 7 + 4) \div (-7 - 8 + 13)$  **4**
29. $(-9 - 5 + 4 - 3) \div (6 - 8 - 4 + 3)$  **$4\frac{1}{3}$**
30. $(-7 - 4 + 2)(-5 + 3)(-8 - 4 + 12)$  **0**
31. $(-8 + 6 - 4 - 2 + 20) \div (2 \cdot 4 - 3 \cdot 6)$  **$-1\frac{1}{5}$**

## Geometry Review

**Objective**  **To find the volume of rectangular solids and pyramids**

**Example**  The formula for the volume of a pyramid with a rectangular base is $V = \dfrac{lwh}{3}$, where $V$ is the volume, $l$ is the length of the base, $w$ is the width of the base, and $h$ is the height. Find the volume when $l = 6$ ft, $w = 5$ ft, and $h = 2$ ft.

$$V = \frac{lwh}{3}$$

$$V = \frac{6 \text{ ft} \cdot 5 \text{ ft} \cdot 2 \text{ ft}}{3}$$  ◀ *Substitute 6 ft for l, 5 ft for w, and 2 ft for h.*

$$V = \frac{60 \text{ ft}^3}{3}$$  ◀ *6 ft · 5 ft · 2 ft = 6 · 5 · 2 · ft · ft · ft or 60 ft³*

$$V = 20 \text{ ft}^3$$  ◀ *Read 20 ft³ as 20 cubic feet.*

**Use the given formulas and dimensions to find the volume of each solid.**

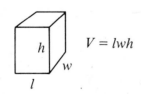

$V = lwh$

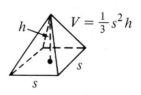

$V = \frac{1}{3}s^2 h$

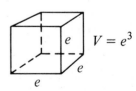

$V = e^3$

1. $l = 4$ cm, $w = 6$ cm, $h = 5$ cm
   **120 cm³**
2. $s = 7$ in., $h = 6$ in.
   **98 in.³**
3. $e = 6$ m
   **216 m³**

# EVALUATING ALGEBRAIC EXPRESSIONS 2.9

*Objective* **To simplify and evaluate algebraic expressions**

To evaluate an expression like $-8 - 2x$ for a negative value for $x$, change it to $-8 + (-2x)$ first. This will help to avoid errors with the signs.

*Example 1*  **Evaluate $-8 - 2x$ for $x = -3$.**

$$
\begin{aligned}
-8 - 2x &= -8 + (-2x) \\
&= -8 + \underbrace{(-2 \cdot -3)} \\
&= -8 + \qquad 6 \qquad \blacktriangleleft \text{ } \textit{Do the multiplication first.} \\
&= -2
\end{aligned}
$$

*Example 2*  **Evaluate $3x - 5y$ for $x = -3$ and $y = -2$.**

$$
\begin{aligned}
3x - 5y &= 3x + (-5y) \\
&= 3(-3) + (-5 \cdot -2) \\
&= -9 + 10 = 1
\end{aligned}
$$

This technique can be extended to expressions with more than two variables.

*Example 3*  **Evaluate $-4a - 7b + 6c$ for $a = -1$, $b = -3$, and $c = -0.4$.**

$$
\begin{aligned}
-4a - 7b + 6c &= -4a + (-7b) + 6c \\
&= -4(-1) + (-7 \cdot -3) + 6(-0.4) \\
&= 4 + 21 + (-2.4) \\
&= 25 - 2.4 = 22.6
\end{aligned}
$$

*Example 4*  **Evaluate.**

$18x^3$ for $x = -\dfrac{1}{3}$

$$
\begin{aligned}
18x^3 &= 18 \cdot x \cdot x \cdot x \\
&= 18 \cdot \left(-\frac{1}{3}\right) \cdot \left(-\frac{1}{3}\right) \cdot \left(-\frac{1}{3}\right) \\
&= (-6) \qquad \cdot \qquad \left(\frac{1}{9}\right) \\
&= \qquad -\frac{2}{3}
\end{aligned}
$$

$\dfrac{2}{3}x^2$ for $x = -6$

$$
\begin{aligned}
\frac{2}{3}x^2 &= \frac{2}{3} \cdot x \cdot x \\
&= \frac{2}{3} \cdot (-6) \cdot (-6) \\
&= (-4) \cdot (-6) \\
&= \qquad 24
\end{aligned}
$$

The distributive property can be extended to include real numbers.

| Distributive property | For all real numbers $a$, $b$, and $c$, $a(b + c) = a \cdot b + a \cdot c$ and $(b + c)a = b \cdot a + c \cdot a$ |
|---|---|

*Example 5*   **Simplify $(5x + 2y)3 + 8$. Then evaluate for $x = -\dfrac{1}{2}$ and $y = -\dfrac{2}{3}$.**

$$(5x + 2y)3 + 8$$
$$5x \cdot 3 + 2y \cdot 3 + 8 \qquad \blacktriangleleft \; \textit{Distribute the 3.}$$
$$15x + 6y + 8$$
$$15\left(-\frac{1}{2}\right) + 6\left(-\frac{2}{3}\right) + 8$$
$$-7\frac{1}{2} + (-4) + 8 = -3\frac{1}{2}$$

# Written Exercises

**Evaluate for the given value(s) of the variable(s).**

(A)  **1.** $5 - 2x$ for $x = 4$ **−3**   **2.** $6 - 3y$ for $y = -2$ **12**   **3.** $-3x - 12$ for $x = -2$ **−6**

**4.** $-2x - 8$ for $x = -2$ **−4**   **5.** $-2x + 5$ for $x = -3$ **11**   **6.** $-5 - 4x$ for $x = -6$ **19**

**7.** $x - 5$ for $x = -4$ **−9**   **8.** $-3x - 9$ for $x = -3$ **0**   **9.** $5 - 7m$ for $m = 0$ **5**

**10.** $4a - 3$ for $a = -6$ **−27**   **11.** $-6 - 5n$ for $n = -5$ **19**   **12.** $4 - 3b$ for $b = -6$ **22**

**13.** $4x - 2y$ for $x = -2$ and $y = 0$ **−8**   **14.** $3r - 8s$ for $r = 4$ and $s = 0.3$ **9.6**

**15.** $c - 8d$ for $c = -3$ and $d = -\dfrac{3}{4}$ **3**   **16.** $3x - 2y$ for $x = -\dfrac{2}{3}$ and $y = -2$ **2**

**17.** $-8e + 7f - 8g$ for $e = 3$, $f = -4$, and $g = -2.1$ **−35.2**

**18.** $-3p - 7q + 5r$ for $p = -3$, $q = -8$, and $r = -6$ **35**

**19.** $2y^2$ for $y = -2$ **8**   **20.** $4a^3$ for $a = -1$ **−4**   **21.** $3n^4$ for $n = -3$ **243**

(B)  **22.** $\dfrac{3}{2}p^3$ for $p = -6$ **−324**   **23.** $-\dfrac{4}{5}b^2$ for $b = 10$ **−80**   **24.** $b^5$ for $b = \dfrac{1}{2}$ **$\dfrac{1}{32}$**

25. $28x + 12y + 8$; −84   26. $15x + 6y + 9$; −39   27. $12x + 9y + 8$; −43   28. $8x + 28y + 7$; −93

**Simplify. Then evaluate for $x = -2$, $y = -3$, and $z = 8$.**   29. $27x + 6y + 7$; −65

**25.** $4(7x + 3y) + 8$   **26.** $9 + 3(5x + 2y)$   **27.** $8 + 3(4x + 3y)$

**28.** $7 + (2x + 7y)4$   **29.** $(9x + 2y)3 + 7$   **30.** $3(11x + 2y) + 8z$

**31.** $7 + \dfrac{1}{4}z + 2(3x + 2y)$   **32.** $\dfrac{1}{3}(6x + 9y + 3z) + 10$   **33.** $(2x + 18y)\dfrac{1}{2} + 4z$

30. $33x + 6y + 8z$; −20   31. $6x + 4y + \frac{1}{4}z + 7$; −15   32. $2x + 3y + z + 10$; 5

**Simplify. Then evaluate for $x = -2$, $y = -3$, and $z = -4$.**   33. $x + 9y + 4z$; 3

(C)  **34.** $3[2x + 4(3y + 8)]$ **$6x + 36y + 96$; −24**   **35.** $7x + 2[6 + 5(3y + 4)]$ **$7x + 30y + 52$; −52**

**36.** $3x + 4[2z + (5 + 3y)2]$   **37.** $2(5z + 4) + 3[4 + 2(3x + 4y)]$

**38.** $[(5x + 2y)3 - 2z + 7]\dfrac{1}{3}$   **39.** $[6x + (y + 3)9]\dfrac{2}{3} + (7z - 4)2$

36. $3x + 24y + 8z + 40$; −70   37. $18x + 24y + 10z + 20$; −128

38. $5x + 2y - \frac{2}{3}z + \frac{7}{3}$; −11   39. $4x + 6y + 14z + 10$; −72

**50**

# LIKE TERMS WITH REAL NUMBER COEFFICIENTS

*Objective*   **To simplify algebraic expressions by combining like terms with real number coefficients**

Like terms can be combined with the coefficients are real numbers because the distributive property is true for real numbers. For example,

$$-7x + 3x = -7 \cdot x + 3 \cdot x$$
$$= (-7 + 3)x$$
$$= -4x$$

However, to combine like terms you need not use the distributive property each time. Combine the coefficients of the like terms as shown in Example 1.

*Example 1*   **Simplify.**

$-7a + 2a$  ◄ *Like terms*        $-4b - 8b$  ◄ *Like terms*

$\quad -5a$  ◄ *Combine the coefficients:*      $-12b$  ◄ *Combine the coefficients:*

$\qquad -7 + 2 = -5$            $-4 - 8 = -4 + (-8) = -12$

When an expression contains both like and unlike terms, first group the like terms together. Then combine the coefficients of the like terms.

*Example 2*   **Simplify.**

$5y - 8 - 9y + 15$                $7x - 5 - 3y - 10x + 3y - 8$

$5y - 8 - 9y + 15$  ◄ *Like and unlike terms*   $7x - 5 - 3y - 10x + 3y - 8$  ◄ *Like and*

$5y - 9y - 8 + 15$  ◄ *Group the like terms*   $7x - 10x - 3y + 3y - 5 - 8$      *unlike terms*

$\quad -4y \quad + \quad 7$       *together*     $\quad -3x \quad + \quad 0y \quad - \quad 13$  ◄ $0y = 0$

$\qquad -4y + 7$                    $\qquad -3x - 13$

*Example 3*   **Simplify.**

$-2x - 6 + 5y - 3\frac{1}{2}y$                    $4x + 6y - 3.2x - 8y$

$\underbrace{-2x - 3x} + \underbrace{5y - 3\frac{1}{2}y} - 6$  ◄ $5 = 4\frac{2}{2}$     $\underbrace{4x - 3.2x} + \underbrace{6y - 8y}$  ◄ $4.0$

$\quad -5x \quad + \quad 1\frac{1}{2}y \quad - 6$   $\quad -3\frac{1}{2} = 3\frac{1}{2}$   $\quad 0.8x \quad - \quad 2y$   $\dfrac{-3.2}{0.8}$

$\qquad\qquad\qquad\qquad 1\frac{1}{2}$

In this example, you will simplify and evaluate the expression.

*Example 4*    **Simplify $-2x - 6 + 9y - 3x - 5y$. Then evaluate for $x = -3$ and $y = -8$.**

$$-2x - 6 + 9y - 3x - 5y$$
$$\underbrace{-2x - 3x} + \underbrace{9y - 5y} - 6 \quad \blacktriangleleft \text{ Group the like terms together.}$$
$$-5x \quad + \quad 4y \quad - 6$$
$$-5(-3) + \quad 4(-8) \quad - 6 \quad \blacktriangleleft \text{ Substitute } -3 \text{ for } x \text{ and } -8 \text{ for } y.$$
$$15 \quad - \quad 32 \quad - 6$$
$$15 - 38 \qquad\qquad \blacktriangleleft \; -32 - 6 = -38$$
$$-23$$

## Written Exercises

20. $-3q - 10$  27. $-13x - 10y + 7$  30. $-2d - 2e - 3.36$
31. $-8x + 4\frac{1}{4}y - 8$  32. $4\frac{3}{10}a + 2b - 11$
35. $4x + 4y - 3z; -38$

(A)  1. $-5y + 9y$ **4y**   2. $2a - 8a$ **−6a**   3. $-7b + 9b$ **2b**   4. $-3y + 8y$ **5y**
5. $4a - 7a$ **−3a**   6. $-6b + 8b$ **2b**   7. $-9x + 2x$ **−7x**   8. $-7z + 5z$ **−2z**
9. $-4c - 7c$ **−11c**   10. $-8r + 3r$ **−5r**   11. $-5y - 5y$ **−10y**   12. $2r - 2r$ **0**
13. $-7x - 3x$ **−10x**   14. $-7b + 9b$ **2b**   15. $-4k - 7k$ **−11k**   16. $15c - 4c$ **11c**
17. $6x - 7 - 3x$ **3x − 7**   18. $4z - 10 - 9z$ **−5z − 10**   19. $5c - 3.4c$ **1.6c**
20. $-5q - 2 + 2q - 8$   21. $9x - 3 - 7x + 4$ **2x + 1**   22. $9y - 8y + 3y$ **4y**
23. $2x - 3y - 5x + 4 + 8y$ **−3x + 5y + 4**   24. $4a + 2b - 3 - 6a - 5b$ **−2a + 3b − 3**
25. $-6r - 10s + 8r + 4s - 6$ **2r − 6s − 6**   26. $4p - 8q - 3 + 10q - 6p$ **−2p + 2q − 3**
27. $-9x + 7 - 2y - 4x - 8y$   28. $3d - 7.1 - 8d + 4.9$ **−5d − 2.2**
29. $-8a - 2.1y + 11a + 1.4y$ **3a − 0.7y**   30. $6d + 8e + 0.05 - 10e - 8d - 3.41$

(B)  31. $-4x - 8 + 7y - 4x - 2\frac{3}{4}y$   32. $8\frac{7}{10}a - 7b - 2 - 4\frac{2}{5}a + 9b - 9$
33. $3\frac{1}{4}m - 10p - 5$   34. $2\frac{1}{5}h - 14t - 16$
33. $-2\frac{1}{4}m - 5 - 3p + 5\frac{1}{2}m - 7p$   34. $-8 - 6t - 4\frac{4}{5}h - 8 + 7h - 8t$

36. $-3x - 6y - 6z; -15$   37. $-2x - 2y - 6; 4$   39. $-12x + 4y - 4z; 4$

**Simplify. Then evaluate for $x = -3$, $y = -2$, and $z = 6$.**
35. $-4x + 6y - 3z + 8x - 2y$   36. $-8z - 4y - 3x - 2y + 2z$
37. $-5x + 3y + 2 - 8 + 3x - 5y$   38. $-6z - 4y - 3x + 4y + 2z$ **−3x − 4z; −15**
39. $-3x + 7y - 4z - 9x - 3y$   40. $-7z - 8y - 3x - 4y + 3z$ **−3x − 12y − 4z; 9**
41. $-6x + 3 + 2y - 9 - 5x - 4y$   42. $-8y + 4z + 2y - 7 - 9z - 2y$
41. $-11x + 6y - 6; 15$   42. $-8y - 5z - 7; -21$   43. $1.2x - 2.292y - 6; -11.73$
**Simplify. Then evaluate for $x = 0$, $y = 2.5$, and $z = 0.004$.**  44. $-0.06x - 4.06z + 4.3; 4.28376$

(C)  43. $1.4x - 6 + 0.008y - 2.3y - 0.2x$   44. $4.2 - 0.06x - 8.1z + 0.1 + 4.04z$

45. $3.8z - 2.4x + 0.4y - 0.1z + 7y$   46. $2.4 - 0.08x + 0.3y - 0.1 + 3.02z$
45. $-2.4x + 7.4y + 3.7z; 18.5148$   46. $-0.08x + 0.3y + 3.02z + 2.3; 3.06208$

## CUMULATIVE REVIEW

1. Simplify $(-3)^2 \cdot (-2)^3$.   2. Evaluate $-16a^3$ for $a = -\frac{1}{2}$.   3. Simplify $\frac{2}{3}(3a + 6)$.
$-72$   $2$   $2a + 4$

# MULTIPLICATION PROPERTIES OF 1 AND −1

Objective

**To simplify algebraic expressions using the properties,
$1 \cdot a = a$ and $-1 \cdot a = -a$**

These multiplications suggest two properties.

$$1 \cdot 4 = 4 \qquad 8 \cdot 1 = 8 \qquad -1 \cdot 5 = -5 \qquad -1 \cdot -2 = 2 \qquad 7 \cdot -1 = -7$$

same      same      opposites      opposites      opposites

The product of positive one and any number is that number. The product of negative one and any number is the opposite or additive inverse of that number.

| Multiplication properties of 1 and −1 | For each real number $a$, $1 \cdot a = a$ and $a \cdot 1 = a$. <br> 1 is called the multiplicative identity. <br> For each real number $a$, $-1 \cdot a = -a$ and $a \cdot -1 = -a$. |
|---|---|

You can use these properties to simplify an expression like $-9 + 5b - 3 - b$ by first rewriting $-b$ as $-1b$ as shown in Example 1.

Example 1    **Simplify.**

$-9 + 5b + 3 - b$
$-9 + 5b + 3 - 1b$  ◀ $-b = -1b$
$5b - 1b - 9 + 3$  ◀ *Group the like terms together.*
   $4b - 6$

$5a - 4g - 6a + 5g$
$5a - 6a - 4g + 5g$  ◀ *Group the like terms together.*
  $-1a + 1g$
   $-a + g$  ◀ $-1a = -a;\ 1g = g$

Example 2    **Simplify $x - y - 2x$. Then evaluate for $x = 1.32$ and $y = -4$.**

$x - y - 2x$
$1x - 1y - 2x$  ◀ $x = 1x;\ -y = -1y$
$1x - 2x - 1y$  ◀ *Group like terms together.*
 $-1x - 1y$
$-1x + (-1 \cdot y)$
$-1 \cdot 1.32 + (-1 \cdot -4)$  ◀ *Substitute 1.32 for x and −4 for y.*
   $-1.32 + 4$
     $2.68$    ◀ $4 - 1.32 = 2.68$

*Example 3*   **Simplify** $3x - 2y - 4x + y$.
**Then evaluate for** $x = -\dfrac{1}{3}$ **and** $y = -\dfrac{1}{2}$.

$$3x - 2y - 4x + y$$
$$3x - 2y - 4x + 1y$$
$$3x - 4x - 2y + 1y$$
$$-1x - 1y$$
$$-1x + (-1 \cdot y)$$
$$-1\left(-\frac{1}{3}\right) + (-1)\left(-\frac{1}{2}\right) \blacktriangleleft \qquad \frac{1}{3} = \frac{2}{6}$$
$$\frac{1}{3} + \frac{1}{2} \qquad\qquad\qquad +\frac{1}{2} = \frac{3}{6}$$
$$\frac{5}{6} \qquad\qquad\qquad\qquad \frac{5}{6}$$

# Written Exercises

(A)  **Simplify.**    **1.** $3b - 4$   **4.** $-4x - 15$
  **1.** $-7 + 4b + 3 - b$          **2.** $-2a - 7 + a$ $-a - 7$        **3.** $9 - b - 7 - 3b$ $-4b + 2$
  **4.** $-3x - 7 - x - 8$          **5.** $-2b - 3 + b - 4$ $-b - 7$    **6.** $8y - 4 - y + 2$ $7y - 2$
  **7.** $5x - x + 4x$ **8x**       **8.** $7b - b - b$ **5b**           **9.** $3a - 7a - a$ $-5a$
 **10.** $5b - 6b - b$ **−2b**     **11.** $3b - 4 - b + 8$ **2b + 4**  **12.** $-a - 13 - a + 4$ $-2a - 9$
 **13.** $b - 4 - 2b + 8$ $-b + 4$ **14.** $6 - z - 4z + 5$ $-5z + 11$  **15.** $8c + 9 - 9c - 4$ $-c + 5$
 **16.** $-4y - b + 4y + 2b$ **b**  **17.** $-3k - 2m + 2k + m$         **18.** $10c - 7 - 11c - 3$ $-c - 10$
 **19.** $8x - 10 - 7x + 4$ **x − 6** **20.** $-4a + 5b - a - 4b$      **21.** $-7r + 8s + 2s - 9s$
 **22.** $-f + 6 - 5f - 4 + 3f - 2$ **−3f**          **23.** $-a + 5.2b + 3a + 3.8b - 10b$ **2a − b**
 **24.** $7g - 4t - 6.9g + 3t - g$ **−0.9g − t**     **17.** $-k - m$    **20.** $-5a + b$    **21.** $-7r + s$
 **25.** $x - 4y - 8; -26$   **26.** $-y + z - 5; -12$   **27.** $-3y - 8z; 12$   **28.** $x - 8y + 5z; -49$

(B)  **Simplify. Then evaluate for** $x = -2$, $y = 4$, **and** $z = -3$.
 **25.** $3x - 4y - 2x - 8$          **26.** $6y + z - 7y - 5$          **27.** $7x - y - 7x - 2y - 8z$
 **28.** $6z - 8y + x - z$           **29.** $-y + 8x - z - 2x$         **30.** $-8y - 4z + x - y - 2z + 1$
 **29.** $6x - y - z; -13$   **30.** $x - 9y - 6z + 1; -19$   **31.** $-a + b; -15.2$   **32.** $x + 2y; 8.6$
 **Simplify. Then evaluate for the given values of the variables.**
 **31.** $a - 4b - 2a + 5b; a = 7.2, b = -8$          **32.** $-x + 8y + 2x - 6y; x = -1.4, y = 5$
 **33.** $5a - 9b - 6a + 10b; a = -\dfrac{3}{4}, b = \dfrac{7}{8}$   **34.** $7x + 9 - 8x; x = 2\dfrac{3}{5}$ **−x + 9; $6\dfrac{2}{5}$**
 **35.** $-x - y - x + 2y; x = 1\dfrac{3}{5}, y = 4$   **36.** $6a - b - 7a; a = -\dfrac{1}{2}, b = 4$
   **33.** $-a + b; 1\dfrac{5}{8}$   **35.** $-2x + y; \dfrac{4}{5}$   **36.** $-a - b; -3\dfrac{1}{2}$

(C)  **Prove.**    **For Ex. 37–38, see TE Answer Section.**
 **37.** $-x \cdot -y = xy$                    **38.** $a \cdot -b = -ab$

 **39.** What can you conclude about the value of $(-a)^k$ where $k$ is an even
   integer and $a$ is negative? $(-a)^k$ **is positive**

# SIMPLIFYING EXPRESSIONS CONTAINING PARENTHESES

*Objective*

**To simplify and evaluate algebraic expressions containing parentheses**

In many cases, the distributive property can be used to simplify expressions containing parentheses. Consider the following.

$$-2(3 + 4x) = -2 \cdot 3 + (-2 \cdot 4x)$$
$$= -6 + (-8x)$$
$$= -6 - 8x$$

It is not necessary to show addition with the $+$ symbol. Here is a more compact way to show your work.

$$-2(3 + 4x) = -2 \cdot 3 - 2 \cdot 4x$$
$$= -6 - 8x$$

Below are other examples of the distributive property.

*Example 1*

**Simplify.**

$3a - 4(8 + 6a)$
$3a - 4(8 + 6a)$
$3a - 32 - 24a$
$3a - 24a - 32$
$\quad -21a - 32$

$2n - 5(6 - 3n)$
$2n - 5(6 - 3n)$ ◄ *Distribute the* $-5$.
$2n - 30 + 15n$ ◄ $-5 \cdot 6 = -30;$
$2n + 15n - 30$ $\quad -5 \cdot -3n = 15n$
$\quad 17n - 30$

In the next example you will first simplify an expression and then evaluate it for the given value of the variable.

*Example 2*

**Simplify $4x - \dfrac{1}{3}(18x - 9) - 8$. Then evaluate for $x = -3$.**

$4x - \dfrac{1}{3}(18x - 9) - 8$

$4x - 6x + 3 - 8$ ◄ $-\dfrac{1}{3} \cdot 18x = -\left(\dfrac{1}{3} \cdot \overset{6}{18} \cdot x\right) = -6x$ | $-\dfrac{1}{3}(-9) = 3$

$\quad -2x - 5$
$-2(-3) - 5$ ◄ *Substitute* $-3$ *for x.*
$\quad 6 - 5$ ◄ *Use the rules for order of operation.*
$\quad 1$

Simplifying Expressions Containing Parentheses

Sometimes you have to apply the property $-a = -1 \cdot a$ to simplify expressions containing parentheses. Consider an expression like $5x - (4 + 3x)$. It has a $-$ sign in front of the parentheses. You can use the property $-a = -1 \cdot a$ to rewrite $-(4 + 3x)$ as $-1(4 + 3x)$. Then use the distributive property to simplify and combine the like terms.

$$5x - (4 + 3x) = 5x - 1(4 + 3x)$$
$$= 5x - 4 - 3x$$
$$= 5x - 3x - 4$$
$$= 2x - 4$$

To simplify an expression like $-5 - (8 - 7b) - 6b$ you must be careful to realize that the $-$ symbol applies only to the quantity in parentheses, $8 - 7b$, not the $-6b$. The simplification is shown in Example 3.

*Example 3*    **Simplify $-5 - (8 - 7b) - 6b$.**

$-5 - (8 - 7b) - 6b$

$-5 - 1(8 - 7b) - 6b$ ◀ *Distribute the $-1$.*

$-5 - 8 + 7b - 6b$

$\quad 1b - 13$

$\quad\ b - 13$ ◀ *$1b = b$*

In Example 4 you will first simplify an expression and then evaluate it for the given value of the variable.

*Example 4*    **Simplify $-3(7 - x) - (-x + 3) - 5x$. Then evaluate for $x = -10.2$.**

$-3(7 - x) - (-x + 3) - 5x$

$-3(7 - 1x) - 1(-1x + 3) - 5x$

$-21 + 3x + 1x - 3 - 5x$

$3x + 1x - 5x - 21 - 3$

$\quad\quad -1x - 24$

$\quad\quad\quad -1(-10.2) - 24$ ◀ *Replace $x$ with $-10.2$.*

$\quad\quad\quad\quad 10.2 - 24$

$\quad\quad\quad\quad\ -13.8$

## Reading in Algebra

1. To simplify $5x - 2(3x - 4) + 3$, what number must be distributed? **$-2$**
2. What is the coefficient of $3x + 5$ in $-8x - (3x + 5)$? **$-1$**
3. In the expression $3x - 4(5x - 2)$, what number is multiplied by $-2$? **$-4$**

## Oral Exercises

**Multiply.**

                          $-8x - 20$             $-12x + 15$        $18x - 6$

**1.** $-2(5 - 4x)$    **2.** $-4(2x + 5)$    **3.** $-3(4x - 5)$    **4.** $-3(2 - 6x)$

    $8x - 10$

## Written Exercises

**Simplify.**    **4.** $-6x - 11$    **6.** $25c - 27$

Ⓐ **1.** $5x - 8(3 + 2x)$ **$-11x - 24$** **2.** $4 - 3(7 - 9z)$ **$27z - 17$**    **3.** $4y - 2(7 - 5y)$ **$14y - 14$**

**4.** $2x - 2(4x + 3) - 5$    **5.** $7c - 2(3c - 5) + 4$ **$c + 14$**  **6.** $8 - 7(5 - 3c) + 4c$

**7.** $-3(8 - 5z) + 6z$ **$21z - 24$** **8.** $5y - (y + 2)$ **$4y - 2$**        **9.** $3 - (6x - 9)$ **$-6x + 12$**

**10.** $-(e + 7) - 5e + 6$     **11.** $6 - (4 - a) - 2a$ **$-a + 2$** **12.** $-5d - (8 - d) - 9$

                                                   **10.** $-6e - 1$  **12.** $-4d - 17$

**Simplify. Then evaluate for the given value of the variable.**

**13.** $4(5y - 8) - 3y$ for $y = 8$ **$17y - 32; 104$**    **14.** $8 - 5(4 - 3c)$ for $c = -5$ **$15c - 12; -87$**

**15.** $-\dfrac{1}{3}(9x + 3) + 5x$ for $x = -4$ **$2x - 1; -9$**    **16.** $7x - \dfrac{1}{2}(4x - 6)$ for $x = -3$ **$5x + 3; -12$**

**17.** $7d - 8(4 - 3d) + 6$ for $d = 3$           **18.** $-6r + 5 - 3(2r - 1)$ for $r = -7$

**19.** $-4x - (6 - 3x)$ for $x = -6$ **$-x - 6; 0$**    **20.** $-9z - (8 + z)$ for $z = -7$ **$-10z - 8; 62$**

**21.** $5y - 9 - (7y - 6)$ for $y = 0.2$         **22.** $7r - (8 - r) + 12$ for $r = -0.8$

**17.** $31d - 26; 67$  **18.** $-12r + 8; 92$  **21.** $-2y - 3; -3.4$  **22.** $8r + 4; -2.4$

**Simplify.**

Ⓑ **23.** $-4(3x - 5) - 3(2 + 7x)$ **$-33x + 14$**    **24.** $-5(3y - 7) - 2(6 + 4y)$ **$-23y + 23$**

**25.** $6(3z - 7) - \dfrac{1}{2}(-8z - 6)$ **$22z - 39$**    **26.** $-\dfrac{1}{4}(12 - 8d) - 2(-6d + 5)$ **$14d - 13$**

**27.** $7x - 3(5x + 2) - 6(7 + 2x)$ **$-20x - 48$**    **28.** $-3(8 - 7z) + 6z - 9(4 - 3z)$ **$54z - 60$**

**Simplify. Then evaluate for the given value of the variable.**              $-a - 4; -7.4$

**29.** $-(5z - 7) - (9 - z)$ for $z = -0.2$    **30.** $-(2a - 4) - (8 - a)$ for $a = 3.4$

**31.** $-(-6r + 8) - (-7 - r)$ for $r = -2.1$    **32.** $-(-c + 5) - (-6 - c)$ for $c = -3\dfrac{1}{2}$

**33.** $-2(3 - n) - (5n + 4) - 7n$ for $n = -\dfrac{4}{5}$    **34.** $6b - 3(b + 7) - (5 - 4b)$ for $b = 0.04$

**29.** $-4z - 2; -1.2$  **31.** $7r - 1; -15.7$     **32.** $2c + 1; -6$  **33.** $-10n - 10; -2$

**Simplify.**                                     **34.** $7b - 26; -25.72$

Ⓒ **35.** $x - [3(x - 2) - (4 - 5x)]$ **$-7x + 10$**    **36.** $-2a - [-2(1 - 7a) - (5 - 3a)]$ **$-19a + 7$**

**37.** $3x - 5 - [5 - (3 - x)]$ **$2x - 7$**    **38.** $-x - [-(4 - x) - 2(3 - x)]$ **$- 5x - 2$**

                                                     $-9x + 8$

**Prove each of the following.**    **For Ex. 39–40, see TE Answer Section.**

**39.** $a - b = -(b - a)$             **40.** $-(x + y) = -y - x$

## CUMULATIVE REVIEW

**Simplify.**

**1.** $-12 \div (-16 \div 8)$ **6**        **2.** $|14 - 18|$ **4**        **3.** $(-3)^5$ **$-243$**

# SUBTRACTING ALGEBRAIC EXPRESSIONS   2.13

*Objective*    **To subtract algebraic expressions**

Recall, subtract $b$ from $a$ means

$a$ subtract $b$ or,
$a - b$

Also recall the multiplication property of $-1$, $-a = -1a$. You will use this property to subtract algebraic expressions in the examples below.

*Example 1*    **Subtract $-6a + 4$ from $2a - 5$.**

Think: $2a - 5$ subtract $-6a + 4$
$\quad\quad\; 2a - 5 \quad\; - \quad\quad (-6a + 4)$
$\quad\quad\; 2a - 5 \quad\; - \quad\quad 1(-6a + 4)$  ◄ *Property of $-1$.*
$\quad\quad\; 2a - 5 \quad\; + \quad\quad\quad 6a - 4$  ◄ *Distribute $-1$.*
$\quad\quad\quad\quad\quad 8a - 9$

Another phrase used to describe subtraction is:
$\quad\quad$ From $a$, subtract $b$.
This means $a - b$.

*Example 2*    **From $4a - 5$, subtract $-2a + 3$.**

$4a - 5$ subtract $\quad -2a + 3$
$4a - 5 \quad\quad - \quad\quad (-2a + 3)$
$4a - 5 \quad\quad - \quad\quad 1(-2a + 3)$
$4a - 5 \quad\quad + \quad\quad 2a - 3$
$\quad\quad\quad 6a - 8$

*Example 3*    **From $2a$, subtract $3b - y$.**

$2a$ subtract $\quad 3b - y$
$2a \quad\quad - \quad\quad (3b - y)$
$2a \quad\quad - \quad\quad 1(3b - 1y)$  ◄ $-y = -1y$
$2a \quad\quad - \quad\quad 3b + 1y$ , or
$2a \quad\quad - \quad\quad 3b + y$  ◄ *Simplest form*

*Example 4*   **From $3b - x$, subtract $\frac{4}{5}b - 2x$.**

$$3b - x - \left(\frac{4}{5}b - 2x\right)$$

$$3b - 1x - 1\left(\frac{4}{5}b - 2x\right)$$

$$3b - 1x - \frac{4}{5}b + 2x$$

$$3b - \frac{4}{5}b - 1x + 2x \qquad \blacktriangleleft \qquad 3 \quad = 2\frac{5}{5}$$

$$2\frac{1}{5}b + 1x \qquad\qquad\qquad -\frac{4}{5} = \frac{4}{5}$$

$$2\frac{1}{5}b + x \qquad\qquad\qquad\qquad 2\frac{1}{5}$$

# Written Exercises

**Subtract.   11.  $-20y - 5$     15.  $2a - 3b - 5$**

Ⓐ  **1.** Subtract $5x - 2$ from $7x - 6$. **$2x - 4$**     **2.** Subtract $a - 8$ from $5a + 2$. **$4a + 10$**
   **3.** Subtract $2x + 6$ from $x - 4$. **$-x - 10$**     **4.** Subtract $y - 7$ from $6y + 4$. **$5y + 11$**
   **5.** Subtract $-3b - 8$ from $b + 1$. **$4b + 9$**     **6.** Subtract $-5y - 9$ from $-3y$. **$2y + 9$**
   **7.** Subtract $4x - 5$ from $-x$. **$-5x + 5$**     **8.** Subtract $-3b$ from $-7b + 4$. **$-4b + 4$**
   **9.** Subtract $-2k - 4$ from $-7 - k$. **$k - 3$**     **10.** Subtract $2f - 3$ from $-f - 6$. **$-3f - 3$**
   **11.** From $-13y - 2$, subtract $7y + 3$.     **12.** From $18z + 3$, subtract $-6 - 3z$. **$21z + 9$**
   **13.** From $3a$, subtract $-5a + 4$. **$8a - 4$**     **14.** From $-3b - 4$, subtract $5 - b$. **$-2b - 9$**
   **15.** From $2a - 3$, subtract $3b + 2$.     **16.** From $3p - 4$, subtract $2m$. **$-2m + 3p - 4$**
   **17.** From $-c + 1$, subtract $3b + 7$.     **18.** From $5 - m$, subtract $5m + 5$. **$-6m$**

   **17.  $-3b - c - 6$**

Ⓑ  **19.** Subtract $3.04a - 2b$ from $7a - b - 5$. **$3.96a + b - 5$**
   **20.** From $3x - y - 2$, subtract $0.06x - 4y - 2$. **$2.94x + 3y$**
   **21.** From $3a - b - c$, subtract $2a + b - c$. **$a - 2b$**
   **22.** Subtract $k - 5 - d$ from $-2k - 3d + 1$. **$-3k - 2d + 6$**

   **23.** From $2a - 1$, subtract $1\frac{1}{4}a + 5$. **$\frac{3}{4}a - 6$**

   **24.** From $-3\frac{1}{6} - m$, subtract $2\frac{2}{3} - 5m$. **$4m - 5\frac{5}{6}$**

Ⓒ  **25.** Subtract $x - (-x + 3)$ from $2 - [4 - (2x + 5)]$. **6**
   **26.** From $x - [4 - (2 - x)]$, subtract $x + 1$. **$-x - 3$**
   **27.** Subtract the opposite of $[3 - (4 - x)]$ from $8x - (4 - 2x)$. **$11x - 5$**
   **28.** From $5x - 0.5[18 - (6 - 4x)]$, subtract $7x - 0.2[15 - (-50x + 10)]$. **$6x - 5$**

   **29.** Subtract the opposite of $\frac{1}{3}[8 - (-2x + 5)]$ from $\frac{2}{3}[-7x + 6]$. **$-4x + 5$**

Subtracting Algebraic Expressions                                                    **59**

Ⓒ **Exercises are starred.**

Ⓐ **Exercises are 1–7, 9–22, 24, 25, 27, 28, 31–33, 36, 39, 40, 42, 45. The rest are** Ⓑ **Exercises.**

### Vocabulary

absolute value [2.1]
additive identity [2.3]
additive inverse [2.3]
closure [2.3]
coordinate [2.1]
integer [2.1]
irrational [2.2]
multiplicative
  inverse [2.7]

origin [2.1]
opposite [2.3]
rational [2.2]
real [2.2]
reciprocal [2.7]
repeating
  decimal [2.2]
terminating
  decimal [2.2]

**1.** Replace the ? with the correct inequality symbol to make $-5$ _?_ 8 a true statement. [2.1] **<**

**2.** Name the coordinate of the point graphed. [2.1] **−2**

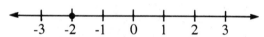

On a number line, graph the points with the given coordinates. [2.1] **Check students'**

**3.** $-3$      **4.** 11    **graphs.**

**5.** Name the number that describes a *loss* of 4 kg. [2.1] **−4**

Find the absolute value. [2.1]

**6.** $|-2.6|$ **2.6**    **7.** $|3|$ **3**   **8.** $|4| + |-4|$ **8**

Determine whether each number is rational, irrational, or neither. [2.2]

**9.** $-32$ **rational**    **10.** $\frac{5}{0}$ **neither**

**11.** $0.34345345534555\ldots$ **irrational**

Add. [2.3]

**12.** $-9 + (-8)$ **−17**    **13.** $-16 + 3\frac{1}{2}$ **$-12\frac{1}{2}$**

**14.** Find $-a$ for $a = -5$. [2.4] **5**

Which properlty of addition is illustrated? [2.4]
**15. comm. prop. add.**    **16. add. iden.**

**15.** $5 + (-2) = -2 + 5$    **16.** $b + 0 = b$
**17.** $-m + m = 0$ **add. inverses**

Subtract. [2.7]

**18.** $7 - (-2)$ **9**   **19.** $-8\frac{1}{2} - 3\frac{3}{4}$    **$-12\frac{1}{4}$**

Multiply. [2.7]

**20.** $5(-4)$ **−20**    **21.** $-\frac{3}{4}\left(-\frac{5}{7}\right)$ **$\frac{15}{28}$**

**22.** $5(-4)(-2)(3)$ **120**   **23.** $(-2)^2(-3)^3$ **−108**

Divide. [2.7]

**24.** $\frac{-12}{4}$ **−3**    **25.** $-\frac{3}{5} \div -\frac{9}{10}$ **$\frac{2}{3}$**

**26.** $(-45 \div -5) \div 3$ **3**

Simplify. [2.8]
**27.** $7 - 12$ **−5**    **28.** $-5 + 14 - 3$ **6**

**29.** $-3\frac{4}{5} + 7\frac{9}{10}$ **$4\frac{1}{10}$** **30.** $-8.1 + 12 - 2.9$ **1**

Evaluate for $x = -4$, $y = -2$, and $z = 3$. [2.9]
**31.** $6 - 2x$ **14**   **32.** $4x - 2y$ **−12**   **33.** $5y^3$
**34.** $(x + 8z)2 + 4$ **44**       **−40**
**35.** $5z + 2(5y + 3x) - 4$ **−33**

Simplify. [2.10] **36. 2a**        **2.2x − 3y**
**36.** $-7a + 9a$    **37.** $5x + 9y - 2.8x - 12y$

**38.** Simplify $-4z - y - 6 + 3\frac{1}{2}z - 3y - z$.
   Evaluate for $y = -2$ and $z = 3$. [2.10]
             **$-4y - 1\frac{1}{2}z - 6$; $-2\frac{1}{2}$**

Simplify. [2.11]
**39.** $-8 + 6b + 2 - b$ **5b − 6**
**40.** $-7g - 3a + 6g - 4a$ **−g − 7a**

**41.** Simplify $-4z - y + 3z + 2y - z$.
   Evaluate for $y = 2.4$ and $z = 3$. [2.11]
           **$y - 2z$; −3.6**

Simplify. [2.12]
**42.** $a - 6(7 + 3a)$ **−17a − 42**
**43.** $-(-6t + 5) - (1 - t)$ **7t − 6**
           **44. −6x − 1; 23** **45. 14x − 1**
**44.** Simplify $\frac{1}{3}(9 - 6x) - (4 - x) - 5x$.
   Then evaluate for $x = -4$. [2.12]

**45.** Subtract $-6x - 3$ from $8x - 4$. [2.13]

**46.** From $4x + 3y$, subtract $-6x - \frac{5}{6}y$.
          **$10x + 3\frac{5}{6}y$**

★ **47.** Prove $-b \cdot a = -ab$. [2.10]

★ **48.** Prove $-(a - b) = -a + b$. [2.11]

**For Ex. 47–48, see TE Answer Section.**

# CHAPTER TWO TEST

Ⓐ Exercises are 1–3, 7, 9–16, 18, 23–30, 32–43. Ⓒ Exercises are starred. The rest are Ⓑ Exercises.

**1.** Graph the point with the coordinate $-5$. **Check students' graphs.**

**2.** Replace the ? with the correct inequality symbol to make $8 \underline{\ ?\ } - 12$ a true statement. **>**

Simplify.

**3.** $-5 + 4$ **−1**    **4.** $-16 + 12 - 14 + 15$  **−3**

**5.** $-4\frac{2}{9} + \left(-7\frac{1}{3}\right)$  **6.** $-8.1 + 12 + (-3.4)$

$-11\frac{5}{9}$    **0.5**

Find the absolute value.

**7.** $|-6.4|$ **6.4**    **8.** $|9| + |-10|$ **19**

**9.** Name the coordinate of the point graphed.

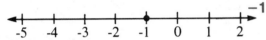

$-1$

**10.** Evaluate $-16 - 5x$ for $x = -2$. **−6**

Which property is illustrated?

**11.** $0 + a = a$ **add. identity**

**12.** $-k + k = 0$ **add. inverses**

**13.** $7 + (-8.3) = -8.3 + 7$

**14.** $(-2 + 8) + 5 = -2 + (8 + 5)$

**13. comm. prop. add. 14. asso. prop. add.**

Divide.

**15.** $\frac{-20}{4}$ **−5**    **16.** $-\frac{5}{7} \div -\frac{15}{14}$ **$\frac{2}{3}$**

**17.** $(28 \div 2) \div -7$ **−2**

**18.** If $a = -5.2$, what is the value of $-a$?

**5.2**

Simplify. Then evaluate for $a = -2$, $b = -3$, and $c = -1$.    $-3a + 6b - 12c; 0$

**19.** $3a + 4b - 5c + 2b - 6a - 7c$

**20.** $(a + 3b)2 + 5c$ **$2a + 6b + 5c; -27$**

**21.** $3(2 - a) - (5 - a) - 6a$ **$-8a + 1; 17$**

**22.** $8a + \frac{1}{5}(10a - 15b) - 5b$ **$10a - 8b; 4$**

**23.** Evaluate $4b^2$ for $b = -\frac{1}{2}$. **1**

Determine whether each is rational, irrational, or neither.

**24.** $\frac{4}{0}$ **neither**    **25.** $3.\overline{45}$ **rational**    **26.** $0.525235233\ldots$ **irrational**

Subtract.

**27.** $-8 - (-3)$ **−5**    **28.** $-5\frac{3}{4} - 9\frac{1}{8}$ **$-14\frac{7}{8}$**

**29.** From $4x - 5$, subtract $8 - x$. **$5x - 13$**

**30.** Subtract $-2a + 5.4$ from $-8a + 10$.

**$-6a + 4.6$**

**31.** What two numbers have an absolute value of 18? **18, −18**

**32.** Name the number that describes a drop of $7.3°$ in temperature. **−7.3**

Multiply.

**33.** $-3(2)$ **−6**    **34.** $7(-4)(-1)$ **28**

**35.** $-\frac{2}{3}\left(-\frac{4}{5}\right)$ **$\frac{8}{15}$**    **36.** $(-1)^3(-4)^2$ **−16**

Simplify.

**37.** $-8a + 10a - 5a$ **−3a**

**38.** $-k + 7k - k$ **5k**

**39.** $-9 + 3b + 4.5 - b$ **$2b - 4.5$**

**40.** $3p - \frac{1}{3}(6 - 3p)$ **$4p - 2$**

**41.** $-5\frac{1}{3}x + 7x$ **$1\frac{2}{3}x$**

**42.** $4a - (3 - 5a)2$ **$14a - 6$**

**43.** $2y - (5 - y)$ **$3y - 5$**

**44.** Simplify $-2a - [6(1 - 3a) - (5 - a)]$. Then evaluate for $a = -2$. **$15a - 1; -31$**
**See TE Answer Section.**

★ **45.** Prove $-(-x - y) = x + y$.

★ **46.** Simplify $[(16 - 24) \div (8 - 10)] \div -8 + 4$    **$3\frac{1}{2}$**

★ **47.** Simplify $5r - [6 - (4 - r)] - 5r - 3r$. **$-4r - 2$**

# COLLEGE PREP TEST

Directions: Choose the one best answer to each question.

**1.** A clock loses 5 minutes each day. The clock now shows the correct time. How many days will it take to reach a point where the correct time will be shown by the clock again?
**C**
(A) 288  (B) 240  (C) 144  (D) 72  (E) 36

**2.** At $a$ cents per orange, what is the price in dollars for a dozen oranges?
**D**
(A) $\dfrac{c}{100}$  (B) $\dfrac{a}{100}$  (C) $\dfrac{a}{12}$  (D) $\dfrac{12a}{100}$  (E) $\dfrac{12}{a}$

**3.** After reaching a low of $-24°F$ at 10:00 pm, the temperature began rising $2°/h$. Find the temperature at 3:00 am.
**C**
(A) $-34°F$  (B) $-16°F$  (C) $-14°F$
(D) $34°F$  (E) $60°F$

**4.** A cashier gave a man change for a dollar in dimes and quarters. How many coins did the man receive?
**D**
(A) 30  (B) 10  (C) 8  (D) 7  (E) 5

**5.** Find the value of $1^3 + 1^5$.  **D**
(A) 15  (B) 10  (C) 8  (D) 2  (E) 1

**6.** From $a - b$, subtract $b - a$.  **B**
(A) $2(b - a)$  (B) $2(a - b)$  (C) 0
(D) $2a + 2b$  (E) $a + b$

**7.** $|-5 - 6|$ is  **D**
(A) $-11$  (B) $-1$  (C) 1  (D) 11  (E) 30

**8.** $5 + \dfrac{5}{0.5} =$  **E**
(A) 5.1  (B) 5.11  (C) 5.25
(D) 6  (E) 15

**9.** How many 3-cent stamps can be purchased with $d$ cents?
**C**

(A) $\dfrac{3}{d}$  (B) $d$  (C) $\dfrac{d}{3}$  (D) $d - 3$  (E) $3d$

Use the bar graph below for questions 10 and 11.

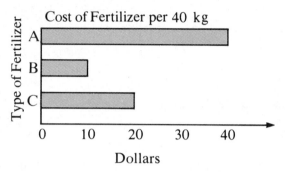

Cost of Fertilizer per 40 kg

**10.** What is the cost of 20 kg of Type C fertilizer?
**B**
(A) \$5  (B) \$10  (C) \$20  (D) \$40  (E) \$50

**11.** How many kilograms of Type B fertilizer can you get for \$5.00?
**B**
(A) 30  (B) 20  (C) 10
(D) 5  (E) none of these

**12.** The highest temperature recorded in Nevada was $112°F$. The lowest temperature was $-5°F$. What is the difference between the two temperatures?
**A**
(A) $117°$  (B) $112°$  (C) $107°$  (D) $5°$  (E) $-5°$

**13.** If a wheel rotates 12 times each minute, how many degrees does it rotate in 5 seconds? [Hint: $360°$ in one rotation]
**E**
(A) $2\tfrac{2}{5}°$  (B) $5°$  (C) $6°$  (D) $60°$  (E) $360°$

**14.** The meter on a taxicab registers \$0.70 for the first $\tfrac{1}{4}$ of a mile and \$0.20 for each additional $\tfrac{1}{4}$ of a mile. How many miles is a trip for which the meter registers \$3.10?
**D**
(A) 4  (B) $3\tfrac{3}{4}$  (C) $3\tfrac{2}{4}$  (D) $3\tfrac{1}{4}$  (E) 3

# 3 EQUATIONS: AN INTRODUCTION

## Problem Solving

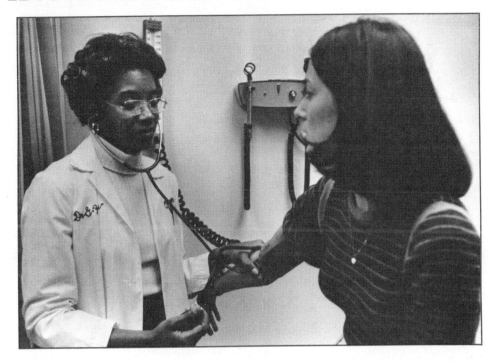

**Problem**

Blood leaves the heart's ventricles under a great amount of pressure. The contraction of the ventricles causes the blood to surge through the arteries. Arterial blood pressure is greatest at the point of contraction and is known as **systolic pressure.** The normal systolic blood pressure of a person varies with age. It can be stated by the formula

$$P = 100 + \frac{a}{2},$$

where $P$ is the systolic pressure in millimeters of mercury, and $a$ stands for age given in years.

## Four Steps in Problem Solving

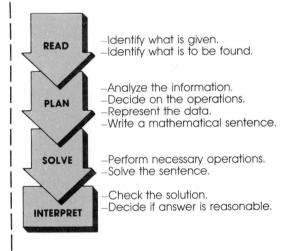

READ
—Identify what is given.
—Identify what is to be found.

PLAN
—Analyze the information.
—Decide on the operations.
—Represent the data.
—Write a mathematical sentence.

SOLVE
—Perform necessary operations.
—Solve the sentence.

INTERPRET
—Check the solution.
—Decide if answer is reasonable.

**Project**

Make a systolic blood pressure chart for yourself for the next 20 years.

# SOLVING EQUATIONS: $x + b = c$ 3.1

*Objective*   **To solve equations of the form $x + b = c$**

In Lesson 1.8, you were given an open sentence with replacement values for the variable. You were asked to find which of the values were solutions of the open sentence. However, in most of your work with open sentences you will not be given specific replacement values, one or more of which is a solution. Instead you will be asked to solve the open sentence. This means to find all of its solutions.

In this lesson and the following lessons, you will learn some properties that will help you solve equations.

| Definition: Equation | An **equation** is a mathematical sentence that contains the symbol $=$. This symbol is used between two numerical or algebraic expressions. |
|---|---|

Consider the equation $8 = 6 + 2$. It is a true statement.
If you add the same number
to each side of an equation,
another true statement is obtained.

$$8 = 6 + 2$$
$$\underline{+3 \qquad + 3}$$
$$11 = 6 + 2 + 3$$
$$11 = 11 \text{ is a true statement.}$$

Now, consider the equation $n = 4$. Its solution is 4. If you add the same number to each side of $n = 4$, you get an equation whose solution is still 4.

$$n = \quad 4$$
$$\underline{+7 \quad +7}$$
$$n + 7 = 4 + 7$$
$$n + 7 = 11$$

Equations with the same solution, such as $n = 4$ and $n + 7 = 11$, are *equivalent equations*. Adding the same number to each side of an equation will result in an equivalent equation. This suggests the following property.

| Addition property for equations | For all real numbers $a$, $b$, and $c$, if $a = b$, then $a + c = b + c$. |
|---|---|

In Example 1, the addition property for equations is used to solve $x - 7 = 6$. In this equation, notice that $x$ is not alone on the left side since $-7$ is added to $x$. To *undo* the operation, 7 (the opposite of $-7$) is added to each side.

**Example 1**

**Solve $x - 7 = 6$. Check the solution.**

$$
\begin{array}{rl}
x - 7 = & 6 \quad \blacktriangleleft \; x \text{ is not alone on the left side.} \\
\underline{\phantom{x -} 7 \phantom{=}} & \underline{7} \quad \blacktriangleleft \; 7 \text{ is the opposite of } -7. \text{ Add 7 to each side.} \\
x + 0 = & 13 \\
x = & 13 \quad \blacktriangleleft \; x \text{ is now alone.}
\end{array}
$$

Check.

$$
\begin{array}{c|c}
x - 7 & 6 \\
\hline
13 - 7 & 6 \quad \blacktriangleleft \text{ Replace } x \text{ with } 13. \\
6 & \\
\end{array}
$$

$$6 = 6 \qquad \text{True}$$

**Thus,** the solution is 13.

**Example 2**

**Solve $8\frac{1}{2} = 5\frac{1}{8} + x$. Check the solution.**

$$
\begin{array}{rl}
8\frac{1}{2} = & 5\frac{1}{8} + x \quad \blacktriangleleft \; x \text{ would be alone on the the right if there were no } 5\frac{1}{8}. \\
\underline{-5\frac{1}{8}} & \underline{-5\frac{1}{8}} \quad \blacktriangleleft \; -5\frac{1}{8} \text{ is the opposite of } 5\frac{1}{8}. \text{ Add } -5\frac{1}{8} \text{ to each side.} \\
3\frac{3}{8} = & 0 + x \quad \blacktriangleleft \quad 8\frac{1}{2} = \quad 8\frac{4}{8} \\
3\frac{3}{8} = & x \qquad\qquad \underline{-5\frac{1}{8}} \quad \underline{-5\frac{1}{8}} \\
& \qquad\qquad\qquad\qquad 3\frac{3}{8}
\end{array}
$$

Check.

$$
\begin{array}{c|c}
8\frac{1}{2} & 5\frac{1}{8} + x \\
\hline
8\frac{1}{2} & 5\frac{1}{8} + 3\frac{3}{8} \quad \blacktriangleleft \text{ Replace } x \text{ with } 3\frac{3}{8}. \\
& 8\frac{4}{8}
\end{array}
$$

$$8\frac{1}{2} = 8\frac{1}{2} \qquad \text{True} \quad \blacktriangleleft \; \frac{4}{8} = \frac{1}{2}$$

**Thus,** the solution is $3\frac{3}{8}$.

Adding $-5\frac{1}{8}$ to each side could also be interpreted as subtracting $5\frac{1}{8}$ from each side. This suggests the subtraction property for equations.

| Subtraction property for equations | For all real numbers $a$, $b$, and $c$, if $a = b$, then $a - c = b - c$. |
| --- | --- |

In the next example, you will use the subtraction property of equations. In Example 3, you subtract from, or add the opposite to, each side of the equation.

## Example 3    Solve.

$$8 + x = 14$$
$$\underline{-8 \qquad -8}$$    ◀ *Subtract 8 from each side,*
$$0 + x = 6$$       *or add the opposite of 8*
$$x = 6$$       *to each side.*

$$8.36 + n = 10.1$$ ◀ $$10.10$$
$$\underline{-8.36 \qquad -8.36}$$    $$\underline{-8.36}$$
$$0 + n = 1.74$$       $$1.74$$
$$n = 1.74$$

In order that there be no confusion between operation signs and signed numbers, you may always think *add the opposite*.

## Oral Exercises

**To solve the equation, what must you *add* to each side the equation?**
1. $x - 8 = 3$  8     2. $b + 6 = 9$  −6   3. $c - 5 = 9$  5     4. $x + 7 = 10$  −7  5. $x - 3 = 8$  3
6. $7 = 8 + f$  −8   7. $8 = 4 + c$  −4   8. $7 = 11 + n$      9. $9 = -10 + x$   10. $-5 = -4 + t$  4
                                              **8. −11   9. 10**

## Written Exercises

**Solve. Check the solution.  1. −12   7. −10   11. 30   13. −10**

(A)  1. $a + 8 = -4$     2. $4 = x + 2$  2     3. $c - 9 = 3$  12     4. $6 = 5 + x$  1     5. $b - 10 = 8$  18
   6. $7 + x = 5$  −2   7. $t + 6 = -4$     8. $9 = 10 + x$  −1  9. $4 + n = 4$  0     10. $x - 8 = -8$  0
   11. $-12 + p = 18$  12. $5 = 9 + z$  −4  13. $6 + x = -4$     14. $f - 8 = -3$  5  15. $5 + x = -1$  −6

**Solve.**
16. $n - 2 = 9$  11   17. $3 = x + 5$  −2  18. $s - 8 = 4$  12   19. $5 + x = 10$  5  20. $7 = 1 + d$  6
21. $-4 + a = -4$  22. $x - 9 = -2$  7  23. $-7 + n = -5$   24. $15 + t = 10$     25. $x - 9 = -8$  1
                                            **21. 0   23. 2   24. −5**

(B)  26. $x - 3\frac{1}{4} = 6\frac{1}{2}$  $9\frac{3}{4}$     27. $-4\frac{1}{3} + x = 3\frac{2}{9}$  $7\frac{5}{9}$   28. $4\frac{2}{5} + p = 7\frac{9}{10}$  $3\frac{1}{2}$   29. $-4\frac{1}{4} = y + 8\frac{1}{4}$  $-12\frac{1}{2}$

30. $2\frac{1}{3} + f = 9$  $6\frac{2}{3}$     31. $9 = p + 5\frac{2}{7}$  $3\frac{5}{7}$     32. $4.2 = f + 0.1$  4.1   33. $-2.4 = y + 7.6$  −10

34. $7.2 + x = 9$  1.8     35. $c - 3.81 = 9.2$     36. $3.4 = y + 8$  −4.6   37. $8.3 + g = 9.4$  1.1
                                    **35. 13.01**

(C)  **Solve.**                    **39. −9**
38. $(x + 9) - 8 = 4 - 7$  −4     39. $-15 + 4 + t = 73 - 93$     40. $-40 - 28 = -36 + z$  −32
41. $|7 - [18 \div -6]| = x + 3$  7  42. $x + 4 = |-16|$  12     43. $b - 5 = |8 - 13|$  10

## CUMULATIVE REVIEW

1. Divide $-15 \div 3$.  −5     2. Evaluate $\dfrac{a - 6}{2}$ for $a = -4$.  −5     3. Compute $\dfrac{(-3)^2}{-1}$.  −9

# SOLVING EQUATIONS: $ax = c$; $\dfrac{x}{a} = c$ 3.2

*Objectives*
### To solve equations of the form $ax = c$
### To solve equations of the form $\dfrac{x}{a} = c$

In the last lesson, you used the addition and subtraction properties of equations to solve equations such as $x + 5 = 14$ and $9 = y - 4$. Dividing each side of an equation by the same non-zero number will also result in an equivalent equation.

| | |
|---|---|
| **Division property for equations** | For all real numbers $a$, $b$, and $c$, $(c \neq 0)$, if $a = b$, then $\dfrac{a}{c} = \dfrac{b}{c}$. |

In Example 1, the division property of equations is used to solve $7x = -21$. In this equation, $x$ is not alone on the left side, since $x$ is multiplied by 7. To undo this operation, divide each side by 7.

*Example 1*    **Solve $7x = -21$. Check the solution.**

$7x = -21$    ◄ *x is not alone on the left side.*

$\dfrac{7x}{7} = \dfrac{-21}{7}$    ◄ *Divide each side by 7.*

$1x = -3$    ◄ $\dfrac{7}{7} = 1$ *and* $\dfrac{-21}{7} = -3$

$x = -3$    ◄ *x is now alone on the left.*

Check.

$$\begin{array}{c|c} 7x & -21 \\ \hline 7 \cdot -3 & -21 \\ -21 & \\ \end{array}$$    ◄ *Replace x with $-3$.*

$\qquad\qquad -21 = -21 \qquad$ True

**Thus, the solution is $-3$.**

You can also multiply each side of an equation by the same number and get an equivalent equation.

| | |
|---|---|
| **Multiplication property for equations** | For all real numbers $a$, $b$, and $c$, $(c \neq 0)$, if $a = b$, then $c \cdot a = c \cdot b$. |

In Example 2 on the next page, the multiplication property for equations is used to solve $\dfrac{x}{-3} = 5$. In this equation, $x$ is not alone on the left side, since $x$ is divided by $-3$. To undo this operation, multiply each side by $-3$.

*Example 2*     **Solve** $\dfrac{x}{-3} = 5$. **Check the solution.**

$$\frac{x}{-3} = 5$$

$$-3 \cdot \frac{x}{-3} = -3 \cdot 5 \quad \blacktriangleleft \textit{ Multiply each side by } -3.$$

$$x = -15 \quad \blacktriangleleft \;\; -3 \cdot \frac{x}{-3} = \overset{\scriptstyle 1}{\cancel{3}} \cdot \frac{x}{-\cancel{3}} = 1x \textit{ or } x$$

*Check.*

$$\begin{array}{c|c} \dfrac{x}{-3} & 5 \\ \hline \dfrac{-15}{-3} & 5 \quad \blacktriangleleft \textit{ Replace } x \textit{ by } -15. \\ 5 & \\ 5 = 5 & \textit{True} \end{array}$$

**Thus,** the solution is 15.

*Example 3*     **Solve.**

$$8 = -x$$
$$8 = -1x \quad \blacktriangleleft \;\; -x = -1x$$
$$\frac{8}{-1} = \frac{-1x}{-1} \quad \blacktriangleleft \textit{ Divide each side by } -1.$$
$$-8 = x$$

$$0.04x = 0.2$$
$$\frac{0.04x}{0.04} = \frac{0.2}{0.04} \quad \blacktriangleleft \textit{ Divide each side by } 0.04.$$
$$x = 5 \quad \blacktriangleleft \;\; 0.04. \overline{)0.20.} \overset{5.}{\phantom{)}}$$

# Written Exercises

**Solve.**

**1.** $-8x = 32$  **−4**

**2.** $20 = 5a$  **4**

**3.** $14 = 7x$  **2**

**4.** $\dfrac{x}{2} = -9$  **−18**

**5.** $15 = \dfrac{x}{-5}$  **−75**

**6.** $7 = 9t$  $\frac{7}{9}$

**7.** $-4 = 2x$  **−2**

**8.** $-3 = -5b$  $\frac{3}{5}$

**9.** $9m = -4$  $-\frac{4}{9}$

**10.** $15 = -3b$  **−5**

**11.** $\dfrac{y}{4} = -6$  **−24**

**12.** $\dfrac{w}{-2} = 8$  **−16**

**13.** $8 = \dfrac{t}{4}$  **32**

**14.** $\dfrac{m}{-3} = -7$  **21**

**15.** $12 = \dfrac{y}{4}$  **48**

**16.** $-m = 9$  **−9**

**17.** $9 = -y$  **−9**

**18.** $0.03x = 3.6$  **120**

**19.** $4.5 = -0.03x$  **−150**

**20.** $0.02x = 5.4$  **270**

**21.** $11.5 = -0.5x$  **−23**

**22.** $15.12 = 0.21x$  **72**

**23.** $\dfrac{x}{7} = 19.2$  **134.4**

**24.** $-18.3 = \dfrac{d}{-4.3}$  **78.69**

# SOLVING EQUATIONS: $ax + b = c$; $\dfrac{x}{a} + b = c$

*Objectives*    **To solve equations of the form $ax + b = c$**
**To solve equations of the form $\dfrac{x}{a} + b = c$**

Consider the equation $2x + 6 = 14$. In this equation, $x$ is multiplied by 2, then 6 is added to the result. To obtain an equivalent equation with $x$ alone on the left side will require two "*undoings*" as shown in Example 1.

*Example 1*    **Solve $2x + 6 = 14$. Check the solution.**

$$2x + 6 = 14 \quad \blacktriangleleft \; \textit{Use the addition property for equations.}$$
$$\underline{\quad -6 \quad -6} \quad \blacktriangleleft \; \textit{Add the opposite of 6 to each side.}$$
$$2x + 0 = 8$$
$$2x = 8 \quad \blacktriangleleft \; \textit{Now use the division property for equations.}$$
$$\frac{2x}{2} = \frac{8}{2} \quad \blacktriangleleft \; \textit{Divide each side by 2.}$$
$$x = 4$$

*Check.*
$$\begin{array}{c|c} 2x + 6 & 14 \\ \hline 2 \cdot 4 + 6 & 14 \\ 8 + 6 & \\ 14 & \\ 14 = 14 & \text{True} \end{array}$$

**Thus,** the solution is 4.

*Example 2*    **Solve $\dfrac{x}{5} - 7 = 23$.**

$$\frac{x}{5} - 7 = 23$$

$$\frac{x}{5} - 7 + 7 = 23 + 7 \quad \blacktriangleleft \; \textit{Add 7 to each side. Show your work horizontally.}$$

$$\frac{x}{5} + 0 = 30$$

$$\frac{x}{5} = 30$$

$$5 \cdot \frac{x}{5} = 30 \cdot 5 \quad \blacktriangleleft \; \textit{Multiply each side by 5.}$$

$$x = 150$$

**Thus,** the solution is 150.

Sometimes an equation contains like terms on the same side of the equals sign. These like terms should be combined before solving the equation. This is shown in the next example.

*Example 3*    **Solve $9 = -11 + 5a - 9a$.**

$$9 = -11 + \underbrace{5a - 9a}$$   ◀ *5a and $-9a$ are like terms on the same side of the = sign.*

$$9 = -11 \quad\;\; - 4a$$

$$11 + 9 = 11 + (-11) - 4a$$   ◀ *To get $-4a$ alone on the right, add 11 to each side.*

$$20 = 0 - 4a$$

$$20 = -4a$$

$$\frac{20}{-4} = \frac{-4a}{-4}$$   ◀ *Divide each side by $-4$.*

$$-5 = a$$

**Thus,** the solution is $-5$.

*Example 4*    **Solve $8.7 - a = 3.9$.**

$$8.7 - 1a = 3.9$$   ◀ *$-a = -1a$*

$$-8.7 + 8.7 - 1a = -8.7 + 3.9$$   ◀ *Add $-8.7$ to each side.*

$$0 - 1a = -4.8$$   ◀ *$-8.7 + 3.9 = -4.8$*

$$-1a = -4.8$$

$$\frac{-1a}{-1} = \frac{-4.8}{-1}$$   ◀ *Divide each side by $-1$.*

$$a = 4.8$$

**Thus,** the solution is 4.8.

*Example 5*    **Solve $7 = 4 + \dfrac{b}{5}$.**

$$7 = 4 + \frac{b}{5}$$

$$-4 + 7 = -4 + 4 + \frac{b}{5}$$   ◀ *Add $-4$ to each side.*

$$3 = 0 + \frac{b}{5}$$

$$3 = \frac{b}{5}$$

$$5 \cdot 3 = 5 \cdot \frac{b}{5}$$   ◀ *Multiply each side by 5.*

$$15 = b$$

**Thus,** the solution is 15.

# Reading in Algebra

**Complete each statement by writing the phrase that should be used to fill in the blanks so that the resulting statement is true.**

1. For the equation $-8 = 4x + 3$, you can get the variable term $4x$ alone on the right by _____. **adding −3 to each side**
2. For the equation $5 + 7b = 2$, you first *undo* the 5 by _____. **adding −5 to each side**
3. You can solve $-b = 4$ by _____. **dividing each side by −1**
4. The first step in solving $6 = -5r + 2$ is _____. **add −2 to each side**

# Oral Exercises

**To solve each equation, what must you *add* to each side of the equation? Then, by what number must you *multiply* or *divide* each side of the equation?**

1. $2x + 7 = 15$ **−7; 2**
2. $-7 = 2 + 3z$ **−2; 3**
3. $-8c + 40 = 48$ **−40; −8**
4. $6 + \dfrac{x}{2} = 8$ **−6; 2**
5. $-12 = 24 + \dfrac{y}{3}$ **−24; 3**
6. $\dfrac{d}{3} - 8 = 7$ **8; 3**
7. $20 - x = 15$ **−20; −1**
8. $-x + 3 = 5$ **−3; −1**
9. $-4 = 5 - y$ **−5; −1**
10. $6 + 0.4x = 30$ **−6; 0.4**
11. $3.4 + 2b = -8.6$ **−3.4; 2**
12. $7 = 5x + 4.5$ **−4.5; 5**

# Written Exercises

**Solve. Check the solution.**

(A)
1. $2x + 4 = 8$ **2**
2. $3t - 5 = -20$ **−5**
3. $4b - 2 = 14$ **4**
4. $4k - 7 = 21$ **7**
5. $9x - 5 = 4$ **1**
6. $6 + 2k = 10$ **2**
7. $-4 = -8 - 2a$ **−2**
8. $6 = -4n - 2$ **−2**
9. $4 - 3x = 13$ **−3**

**Solve.**

10. $2x + 10 = 18$ **4**
11. $-7 = 3 + 5z$ **−2**
12. $-6c + 25 = -11$ **6**
13. $17 - 3y = -10$ **9**
14. $-k + 3 = -4$ **7**
15. $8d - 14 = -22$ **−1**
16. $-3x + 7x + 16 = 20$ **1**
17. $9 - 2 - m = 7$ **0**
18. $6 = 5 - 2b + b$ **−1**
19. $76 - 20r = -24$ **5**
20. $-5x + 7x + 2 = -14$ **−8**
21. $7 - k - k = -1$ **4**
22. $\dfrac{x}{5} + 9 = 13$ **20**
23. $-7 = -9 + \dfrac{b}{3}$ **6**
24. $5 + \dfrac{y}{3} = -4$ **−27**
25. $6 + \dfrac{u}{4} = -8$ **−56**
26. $-5 = 14 + \dfrac{y}{3}$ **−57**
27. $2\dfrac{3}{4} + \dfrac{r}{6} = -5\dfrac{1}{4}$ **−48**
28. $8.2 - x = 9.7$ **−1.5**
29. $4.2 = 6.3 - y$ **2.1**
30. $18 - 0.2x = 12$ **30**

(B)
31. $p + p - 8 = -7$ **$\frac{1}{2}$**
32. $12 + 1 = -3y + 7y + 13$ **0**
33. $2x + x + 7 = 8.5$ **0.5**
34. $19 = 20 - 7d - d$ **$\frac{1}{8}$**
35. $-4 - 2m - m = -3$ **$-\frac{1}{3}$**
36. $1 - 7x - 5 - 4x = -4$ **0**

(C)
37. $x - 2x = -5 + 2$ **3**
38. $-3\dfrac{1}{8} + 5\dfrac{1}{4} = 3x - 12\dfrac{7}{8}$ **5**
39. $0.05 + \dfrac{x}{4.1} = 3.1$ **12.505**

# CONDITIONALS IN LOGIC

*Objectives*

**To recognize the hypothesis and conclusion of a conditional**
**To determine the truth value of a conditional**

In Lesson 3.1 on equations, you used the property
  *If a = b, then a + c = b + c.*
Sentences of the form *if-then* are used in mathematics.
The statement following *if* is called the *hypothesis* or *given*.
The statement following *then* is called the *conclusion*.
The sentence combining both statements is called a *conditional* or *implication*.

Conditional: If $a = b$, then $a + c = b + c$.

hypothesis          conclusion

*Example 1*

**Write the hypothesis and the conclusion of the conditional: If you live in New Jersey, then you live in the United States.**

Hypothesis: you live in New Jersey
Conclusion: you live in the United States

If $P$ represents the hypothesis and $Q$ represents the conclusion, the conditional is expressed in symbols by $P \longrightarrow Q$. This can be read if $P$, then $Q$ or $P$ implies $Q$.

A conditional is true for each of the following truth values of the hypothesis and conclusion.

| Hypothesis | Conclusion | Conditional |
|------------|------------|-------------|
| true | true | true |
| false | false | true |
| false | true | true |

The conditional is false only when:

| Hypothesis | Conclusion | Conditional |
|------------|------------|-------------|
| true | false | false |

So, a conditional is false only when a true hypothesis implies a false conclusion. Example 2 illustrates how it is possible for a false hypothesis to imply a true conclusion. Notice that the resulting conditional will still be true.

*Example 2*   **Show that the following conditional is true: If 1 = 2, then 3 = 3.**

*If* 1 = 2, *then* 3 = 3.

    false       true

    ↓         ↓

hypothesis  conclusion

**So,** the conditional is true.

$1 = 2$   ◀ *False hypothesis*
$2 = 1$   ◀ *Add the equations.*
$3 = 3$   ◀ *True conclusion*

The truth values of a conditional are summarized by the *truth table* shown, where T stands for true and F for false. A conditional is false *only* when the hypothesis is true and the conclusion is false.

| *P* | *Q* | *P* ⟶ *Q* |
|---|---|---|
| T | T | T |
| T | F | F |
| F | T | T |
| F | F | T |

*Example 3*   **Determine whether each conditional is true or false.**

$(6 + 2 = 8) \longrightarrow (7 = 5 + 4)$

$(6 + 2 = 8) \longrightarrow (7 = 5 + 4)$

    true           false

         False

$(5 + 3 = 7) \longrightarrow (9 \cdot 5 = 45)$

$(5 + 3 = 7) \longrightarrow (9 \cdot 5 = 45)$

    false           true

         True

# Written Exercises

**The hypothesis is underlined once and the conclusion is underlined twice.**

Ⓐ **Write the hypothesis and conclusion of each conditional.**

**1.** If it is raining, then the sky is cloudy.

**2.** If $3 + 4 = 7$, then $5 - 4 = 1$.

**3.** $(x + a = b) \longrightarrow (x = b - a)$

**4.** $(x - 4 = 10) \longrightarrow (x = 14)$

**Determine whether each conditional is true or false.**

**5.** $(5 + 3 \cdot 2 = 13) \longrightarrow (7 - 4 \cdot 5 = 13)$ **True**

**6.** $(6^2 - 4 \cdot 3 = 24) \longrightarrow (8 - 4 \cdot 2 = 7 - 4 \cdot 5)$ **False**

**7.** $(8 \cdot 2 - 4^3 = -48) \longrightarrow \left(7 - 2\frac{1}{2} = \frac{9}{2}\right)$ **True**

Ⓑ **8.** Determine whether the conditional $(4 = 3) \longrightarrow (8 = 6)$ is true. Show how such a conditional can be true. (HINT: Use the multiplication property for equations.) **Multiply each side of the hypothesis by 2.**

# THE LANGUAGE OF ALGEBRA

**Objective**    **To write English phrases in mathematical form**

In algebra you use mathematical symbols to indicate various operations. It is also necessary to translate English phrases into mathematical form. This will be particularly important when you solve word problems.

There are certain key phrases that often appear in written word problems. You will need to understand the meaning of these phrases and be able to write them using mathematical symbols. For example, *decreased by* means *made smaller by*. So, *decreased by* calls for *subtraction*. Similarly, *increased by* means *made greater by*. Therefore, *increased by* calls for addition.

**Example 1**    **Write in mathematical terms.**

9 decreased by 2          7 increased by 6          5 decreased by $x$

$9 - 2$                   $7 + 6$                   $5 - x$

Consider the phrase *6 increased by 3 times a number*. The number might be 4, $-2$, 0, or any number. If you let a variable represent the number, say $y$, then

6 increased by 3 times a number

becomes          $6 +$          $3 \cdot y$, or $6 + 3y$.

**Example 2**    **Write in mathematical terms 7 decreased by 5 times a number.**

7 decreased by 5 times a number

$7 -$          $5 \cdot y$    ◄ *Let y represent the number.*
$7 - 5y$

Another frequently used expression is *less than*. For example, 6 less than 13 does not mean $6 - 13$.

6 less than 13

$13 - 6$, or 7

Similarly, there is the phrase *more than*.

8 more than 3 means 3 made greater by 8.

$3 + 8$, or 11

## Example 3    Write in mathematical terms.

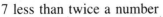
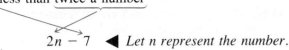

7 less than twice a number

$2n - 7$  ◀ *Let n represent the number.*

6 more than a number divided by 4

$\frac{n}{4} + 6$

## Reading in Algebra

1. How does 7 less than 3 differ from 7 decreased by 3? **$3 - 7 = -4; 7 - 3 = 4$**
2. Is 8 more than $x$ the same as $8x$ for all $x$? Why? **No: *more than* means add; 8x means**
3. Does $x$ decreased by 6 ever have the same value as $x$ less than 6? **yes; $x = 6$**     **multiply.**
4. How does 8 more than 3 differ from 8 increased by 3? **It does not since addition is**
                                                                          **commutative.**

## Written Exercises

**Write each in mathematical terms.**

Ⓐ
1. 7 decreased by 4 **$7 - 4$**
2. 8 increased by 5 **$8 + 5$**
3. 14 increased by 9 **$14 + 9$**
4. 5 decreased by 14 **$5 - 14$**
5. 16 increased by 5 times $x$ **$16 + 5x$**
6. 19 decreased by 7 times $y$ **$19 - 7y$**
7. 5 less than 8 **$8 - 5$**
8. 7 more than 6 **$6 + 7$**
9. 14 less than 8 times $x$ **$8x - 14$**
10. 3 more than twice $y$ **$2y + 3$**
11. 9 increased by 4 times a number **$9 + 4n$**
12. 2 decreased by 3 times a number **$2 - 3n$**
13. 8 less than a number divided by 7 **$\frac{n}{7} - 8$**
14. 5 more than twice a number **$2n + 5$**
15. 25 decreased by 4 times a number **$25 - 4n$**
16. 11 more than 7 times a number **$7n + 11$**
17. 5 times $x$, increased by 2 **$5x + 2$**
18. 14 less than 3 times a number **$3n - 14$**
19. 16 increased by twice a number **$16 + 2n$**
20. A number divided by 3, increased by 14
21. 9 times a number, decreased by 6 **$9n - 6$**
22. Twice a number, increased by 8 **$2n + 8$**
                                                       **20. $\frac{n}{3} + 14$**

Ⓑ
23. $x$ increased by $y$ **$x + y$**
24. $a$ more than $b$ **$b + a$**
25. 9 times $x$, decreased by twice $y$ **$9x - 2y$**
26. $m$ less than 5 times $n$ **$5n - m$**
27. 11 more than $x$ times $y$ **$xy + 11$**
28. $x$ less than $a$ divided by $b$ **$\frac{a}{b} - x$**

Ⓒ
29. 3 times a number, increased by 4 more than the same number **$3n + (n + 4)$**
30. Twice a number, decreased by 6 more than that number **$2n - (n + 6)$**
31. 2 less than 5 times a number, increased by 8 **$5n - 2 + 8$**
32. 7 more than twice a number, increased by 6 **$2n + 7 + 6$**
33. 8 less than 17, increased by 4 more than a number divided by 2 **$17 - 8 + (\frac{n}{2} + 4)$**
34. 4 less than 3 times a number, increased by 8 more than the number **$3n - 4 + (n + 8)$**

## CUMULATIVE REVIEW

**Solve.**

1. $3x - 4 = 11$ **5**
2. $15 = 2x + 9$ **3**
3. $6 = 7 - x$ **1**
4. $\frac{x}{3} + 6 = 15$ **27**

# STEPS FOR SOLVING WORD PROBLEMS 3.6

**Objectives**
**To learn the four basic steps for solving word problems**
**To solve word problems involving one number**

In this lesson you will learn a four-step approach for solving word problems. A basic skill in solving word problems is the ability to translate an English sentence into an equation. The four steps will assist you in organizing the given information so that you can write an equation that will enable you to solve the problem.

As you look at Example 1, study the four basic steps for solving word problems that are shown on the left.

*Example 1*    **9 less than 3 times a number is 33. Find the number.**

The first sentence tells what is given. The second tells what is to be found.

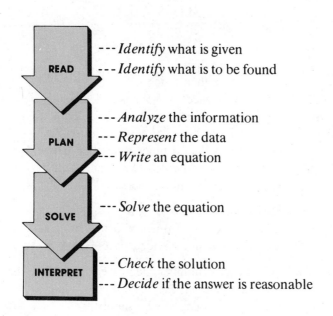

--- *Identify* what is given
--- *Identify* what is to be found

--- *Analyze* the information
--- *Represent* the data
--- *Write* an equation

--- *Solve* the equation

--- *Check* the solution
--- *Decide* if the answer is reasonable

Let $n$ = the number.
9 less than 3 times a number is 33.
9 less than $3n$ is 33.

$$3n - 9 = 33$$

$$3n - 9 = 33$$
$$3n - 9 + 9 = 33 + 9 \qquad \blacktriangleleft \;\; \text{Add 9 to each side.}$$
$$3n + 0 = 42$$
$$3n = 42$$
$$\frac{3n}{3} = \frac{42}{3} \qquad \blacktriangleleft \;\; \text{Divide each side by 3.}$$
$$n = 14$$

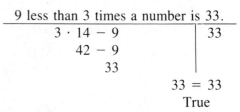

9 less than 3 times a number is 33.

| $3 \cdot 14 - 9$ | 33 |
|---|---|
| $42 - 9$ | |
| 33 | |

$$33 = 33$$
True

**Thus,** the number is 14.

Your written work may be arranged more compactly by not writing every step used in solving the equation of a word problem. This is illustrated in the next three examples of this lesson.

## Example 2

**The $500 selling price of a television is $60 more than twice the cost. Find the cost.**

The first sentence gives information about the cost.
The second sentence asks you to find the cost.

Let $c$ = cost.

500 is 60 more than twice the cost.
500 is 60 more than $2c$

$$500 = 2c + 60$$
$$500 = 2c + 60$$
$$500 - 60 = 2c + 60 - 60 \quad \blacktriangleleft \ Add\ -60\ to\ each\ side.$$
$$440 = 2c \quad \blacktriangleleft \ 60 - 60 = 0;\ 2c + 0 = 2c$$
$$220 = c \quad \blacktriangleleft \ Divide\ each\ side\ by\ 2.$$

500 is 60 more than twice the cost.

| 500 | $2 \cdot 220 + 60$ |
|-----|----------------|
|     | $440 + 60$ |
|     | 500 |

$$500 = 500$$
**Thus,** the cost of the television is $220.

## Example 3

**If Joe's age is increased by four times his age, the result is 40. How old is Joe?**

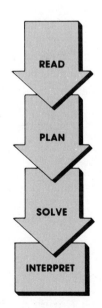

The first sentence gives information about Joe's age.
The second sentence asks you to find his age.

Let $a$ = Joe's age.

If Joe's age is increased by 4 times his age, the result is 40.

$$a \qquad + \qquad 4a \qquad = \qquad 40$$

$$1a + 4a = 40 \quad \blacktriangleleft \ a = 1a$$
$$5a = 40 \quad \blacktriangleleft \ Combine\ like\ terms\ 1a\ and\ 4a.$$
$$a = 8 \quad \blacktriangleleft \ Divide\ each\ side\ by\ 5.$$

If Joe's age is increased by 4 times his age, the result is 40.

| $8 + 4 \cdot 8$ | 40 |
|-----------------|-----|
| $8 + 32$ |  |
| 40 |  |

$$40 = 40$$

**Thus,** Joe's age is 8.

*Example 4*    **If 3 times Jill's bowling score is increased by 100 less than twice her score, the result is 900. Find her score.**

**READ**

You are given information about a bowling score.
The second sentence asks you to find that score.

Let $s$ = score.

**PLAN**

3 times Jill's score, increased by 100 less than twice her score, is 900.

$$3s \qquad + \quad 2s \quad - \quad 100 \qquad = 900$$

| | |
|---|---|
| $5s - 100 = 900$ | ◀ *Combine like terms.* |
| $5s - 100 + 100 = 900 + 100$ | ◀ *Add 100 to each side.* |
| $5s = 1{,}000$ | |
| $s = 200$ | ◀ *Divide each side by 5.* |

**SOLVE**

**INTERPRET**

3 times Jill's score, increased by 100 less than twice her score, is 900.

| | | | |
|---|---|---|---|
| $3 \cdot 200$ | $+$ | $2 \cdot 200 - 100$ | 900 |
| 600 | $+$ | $400 \quad - 100$ | |
| | | $1{,}000 - 100$ | |
| | | 900 | |

$$900 = 900$$

**Thus,** Jill's score is 200.

*Summary*    The four basic steps for solving a written word problem are:
1. **READ**    *Identify* what is given. *Identify* what is to be found.
2. **PLAN**    *Analyze* the information. *Decide* on the operations. *Represent* the data. *Write* a mathematical sentence.
3. **SOLVE**    *Perform* the necessary operations. *Solve* the mathematical sentence.
4. **INTERPRET**    *Check* the solution. *Decide* if the answer is reasonable.

# Reading in Algebra

**Use the following word problem to answer each question.**
Four less than Randy's age divided by 5 is 15. What is Randy's age?
  1. Outline the four steps you would use to solve the problem. **See above.**
  2. What does the problem ask you to find? **Randy's age**
  3. True or false? Randy's age is 15. Why? **False. It is given that 4 less than Randy's age**
  4. If Randy's age is represented by $x$, which is the correct equation? **a**    **divided by 5 is 15.**

  **a.** $\dfrac{x}{5} - 4 = 15$        **b.** $4 - \dfrac{x}{5} = 15$

## Oral Exercises

**Give an equation for each sentence.**
1. 6 more than twice $x$ is 18. **$2x + 6 = 18$**
2. 7 decreased by 4 times $n$ is 19. **$7 - 4n = 19$**
3. 9 times $n$, decreased by $n$, is 24. **$9n - n = 24$**
4. 3 more than 5 times $x$ is 33. **$5x + 3 = 33$**
5. 7 decreased by 4 times $a$ is 35. **$7 - 4a = 35$**
6. 12 is 5 more than 6 times $b$. **$12 = 6b + 5$**
7. 18 increased by twice $y$ is 40. **$18 + 2y = 40$**
8. 36 is 6 less than $b$ divided by 3. **$36 = \frac{b}{3} - 6$**

## Written Exercises

**Solve each word problem.**

Ⓐ
1. 8 more than a number is 14. Find the number. **6**
2. A number decreased by 12 is 19. Find the number. **31**
3. 20 less than Barbara's age is 50. Find Barbara's age. **70 y**
4. If the cost of a shirt is increased by $2, the result is $14. Find the cost. **$12**
5. Ten less than twice Jose's age is 64. Find Jose's age. **37 y**
6. 40 more than 3 times Jane's bowling score is 400. Find her score. **120**
7. 34 is 6 less than 5 times a number. Find the number. **8**
8. 2 more than a number divided by 3 is 4. Find the number. **6**
9. The $340 selling price of a stereo is $40 more than 3 times the cost. Find the cost. **$100**
10. Five less than 3 times the temperature is $-20°$. Find the temperature. **$-5°$**
11. If 4 times Rosa's age is decreased by 20, the result is 28. How old is she? **12 y**
12. The $60 selling price of a camera is the cost increased by twice the cost. Find the cost. **$20**
13. Two kilograms less than twice Hank's mass is 72 kg. What is Hank's mass? **37 kg**
14. Eight times Betty's age increased by 3 times her age is 110. How old is Betty? **10 y**
15. If Louis' savings were increased by 4 times his savings, the result would be $25,000. How much has he saved? **$5,000**
16. One cold day 7 times the temperature decreased by the temperature was 12°. Find the temperature. **2°**

Ⓑ
17. If 4 times Jared's age is increased by 5 less than twice his age, the result is 55. How old is he? **10 y**

18. If 5 kg more than 3 times Mac's mass is decreased by twice his mass, the result is 60 kg. What is Mac's mass? **55 kg**

19. If 5 times a number is subtracted from 8 more than 7 times the number, the result is 24. Find the number. **8**

20. If 6 is added to Marge's age divided by 3, the result is 10. How old is she? **12 y**

## NON-ROUTINE PROBLEMS

Jake buys oranges at 3 for 46¢. He sells them at 5 for 80¢. In order to make a profit of $1.00, how many oranges must he sell? **150**

# PROVING STATEMENTS

*Objective*  **To prove statements about real numbers**

Properties of real numbers are assumed to be true. Here is a summary of the properties of operation and the properties of equality for all real numbers $a$, $b$, and $c$.

| REAL NUMBER PROPERTIES | | |
|---|---|---|
| **Properties of Operation** | **Addition** | **Multiplication** |
| Commutative Property | $a + b = b + a$ | $a \cdot b = b \cdot a$ |
| Associative Property | $(a + b) + c = a + (b + c)$ | $(ab)c = a(bc)$ |
| Identity Property | $a + 0 = 0 + a = a$ | $a \cdot 1 = 1 \cdot a = a$ |
| Property of $-1$ | | $-a = -1a$ |
| Inverse Property | $a + (-a) = 0$ | $a \cdot \dfrac{1}{a} = 1$ |
| Distributive Property | $a(b + c) = ab + ac$ | |
| **Properties of Equality** | | |
| Reflexive Property | $a = a$ (A number is equal to itself.) | |
| Symmetric Property | If $a = b$, then $b = a$. | |
| Transitive Property | If $a = b$ and $b = c$, then $a = c$. | |
| Substitution Property | If $a = b$, then $a$ can be replaced by $b$ and $b$ by $a$. | |

You can use properties and definitions to prove the truth of other statements. A statement that is proved to be true is called a **theorem.** Example 1 will show you how to prove a theorem.

*Example 1*  **Prove: For all real numbers $a$ and $b$, $a + [b + (-a)] = b$.**

| **Statement** | **Reason** |
|---|---|
| 1. $a + [b + (-a)] = a + [(-a) + b]$ | Commutative property for addition |
| 2. $\quad\quad\quad\quad = [a + (-a)] + b$ | Associative property for addition |
| 3. $\quad\quad\quad\quad = 0 + b$ | Additive inverse property |
| 4. $\quad\quad\quad\quad = b$ | Identity property for addition |

You have accepted the validity of several properties for equations in this chapter. You can prove them using real number properties.

**Example 2**    **Prove: For all real numbers $a$, $b$, and $c$, if $a = b$, then $a + c = b + c$.**

$$\text{If } a = b, \text{ then } a + c = b + c.$$

$$\underbrace{\qquad\qquad}_{\substack{\text{hypothesis}\\ \text{(given)}}} \quad \underbrace{\qquad\qquad}_{\text{conclusion}}$$

| Statement | Reason |
|---|---|
| 1. $a = b$ | Given |
| 2. $a + c = a + c$ | Reflexive property |
| $\downarrow$ | |
| 3. $a + c = b + c$ | Substitution property |

**Thus,** if $a = b$, then $a + c = b + c$.

# Written Exercises

**(A)**  **Write the missing statements or reasons in each proof.**

  **1.** Prove: For all real numbers $x$, $y$, and $z$, if $x = y$, then $x - z = y - z$.

| Statement | Reason |
|---|---|
| 1. $x = y$ | __?__ **Given** |
| 2. $x - z = $ __?__ **x − z** | Reflexive property |
| 3. $x - z = $ __?__ **y − z** | __?__ **Substitution property** |

  **2.** Prove: For all real numbers $a$, $b$, and $c$, if $a + c = b + c$, then $a = b$.

| Statement | Reason |
|---|---|
| 1. $a + c = b + c$ | Given |
| 2. $(a + c) + (-c) = (b + c) + (-c)$ | __?__ property for equations **Addition** |
| 3. $a + [c + (-c)] = b + [c + (-c)]$ | Associative property |
| 4. $a + $ __? **0**$ = b + $ __? **0** | __?__ **Additive inverse property** |
| 5. $a = b$ | __?__ **Identity property for addition** |

**(B)**  **Write the proofs for each statement, including statements and reasons.**

  **3.** Prove: If $a = b$, then $ac = bc$.          **Check students' proofs.**

  **4.** Prove: If $w - y = t - y$, then $w = t$.

  **5.** Prove: If $a = b$ and $c \neq 0$, then $\dfrac{a}{c} = \dfrac{b}{c}$.

  **6.** Prove: $-(a - b) = b - a$.

  **7.** Prove: $-(-a - b) = a + b$.

# CHAPTER THREE REVIEW

**Vocabulary**

addition property for equations [3.1]

check [3.1]

conclusion [3.4]

conditional [3.4]

division property for equations [3.2]

equation [3.1]

equivalent equations [3.1]

hypothesis [3.4]

implication [3.4]

multiplication property for equations [3.2]

subtraction property for equations [3.1]

Solve. Check the solution.

1. $x - 14 = -2$ [3.1] **12**

2. $5 = m + \frac{1}{2}$ **$4\frac{1}{2}$**

3. $x - 2 = 11$ **13**

4. $-9a = -27$ [3.2] **3**

5. $4x = -24$ **$-6$**

Solve. [3.2]

6. $8 = 3t$ **$\frac{8}{3}$**

7. $36 = 0.04b$ **900**

Solve. Check the solution. [3.3]

8. $2x + 5 = 13$ **4**

9. $7 = \frac{x}{3} - 5$ **36**

10. $-5 - y = -3$ **$-2$**

Solve. [3.3]

11. $4 - 3m = -2$ **2**

12. $7x - 8x + 2 = -9$ **11**

13. $15 + 9 = 8d - 3d + 1$ **$\frac{23}{5}$**

Determine whether each conditional is true or false. [3.4]                                                    **F**

14. $(8 + 7 - 5 \cdot 3 = 0) \longrightarrow (4^2 \div 8 = 8)$

15. $(9 \cdot 2 - 6^2 = 12) \longrightarrow (7 \cdot 5 - 2 = 11 \div 3)$ **T**

Write each in mathematical terms. [3.5]

16. 5 less than 4 times a number **$4n - 5$**

17. 6 more than twice a number **$2n + 6$**

18. 9 increased by a number divided by 5 **$9 + \frac{n}{5}$**

19. 7 times a number, decreased by that number **$7n - n$**

20. $t$ less than $p$ times $q$ **$pq - t$**

★ 21. 5 increased by 9 more than 3 times a number **$5 + (3n + 9)$**

Solve each word problem. [3.6]

22. 11 more than a number is 19. Find the number. **8**

23. 6 less than Sue's age divided by 2 is 4. Find Sue's age. **20 y**

24. The $600 selling price of a couch is the cost increased by 3 times the cost. Find the cost. **$150**

25. If 5 is increased by twice a number, the result is 32. Find the number. **$\frac{27}{2}$**

26. If twice Leroy's age is increased by 4 less than 3 times his age, the result is 71. How old is Leroy? **15 y**

27. If 10 is added to 30 less than twice a number, the result is 220. Find the number. **120**

★ Which of the given values is the solution of the open sentence? [3.1]

28. $-14.2 + n + 6.2 = 84.9 - 62.9$; 9, 30, 43 **30**

29. $|9 - (2.4 \div 4)| = x + 2.9$; 3.1, 4.7, 5.5 **5.5**

★ Solve.

30. $x + 2 = |-18|$ [3.1] **16**

31. $-0.7x + 4.7x = |-8.5 - 3.9| - 4$ [3.3] **2.1**

★ 32. If 6 more than 3 times a number is increased by 5 less than the same number, the result is the absolute value of $-9$. Find the number. [3.5] **2**

Solve. Check the solution.
1. $y - 6 = -2$ **4**
2. $15 = 7 - 2m$ **−4**
3. $-4.3 - b = -9.8$ **5.5**

Write each in mathematical terms.
4. 3 less than a number divided by 7 $\frac{n}{7} - 3$
5. 6 times a number, decreased by that number **6n − n**
6. 7 increased by, 2 more than 5 times a number **7 + (5n + 2)**
7. $r$ more than $p$ times $t$ **pt + r**

Solve.
8. $a - 4 = 15$ **19**
9. $-5.4 = 9b$ **−0.6**
10. $\frac{x}{2} - 5 = 7$ **24**
11. $6 - x - 3x = 30$ **−6**
12. $3d - 5d + 4 = 12 - 6$ **−1**

Determine whether each conditional is true or false.
13. $(6^2 - 5^2 = 9 + 3) \longrightarrow (4^2 + 4 = 4 \cdot 5)$ **T**
14. $(7 + 9 \div 3 = 5 + 2) \longrightarrow (8^2 - 9 = 6 \cdot 11)$ **T**

Solve each word problem.
15. 7 less than a number is $-3$. Find the number. **4**

16. 4 more than twice Neil's age is 40. How old is Neil? **18 y**

17. If Louella's savings are increased by 3 times her savings, the result is $12,000. How much has she saved? **$3,000**

18. If Irv's mass is increased by 18 kg more than twice his mass, the result is 300 kg. Find Irv's mass. **94 kg**

19. The $500 selling price of an air conditioner is the cost increased by 4 times the cost. Find the cost. **$100**

20. If 3 times a number is increased by 5 less than twice the number, the result is 35. Find the number. **8**

21. 13 increased by a number divided by 3 is 17. Find the number. **12**

★ 22. Solve $3x + |-5 + 2| = 4 - |-2|$. **$-\frac{1}{3}$**

★ 23. Solve $|-5| - |-8 \div 2| - 3 = 2x - 7$. **$\frac{5}{2}$**

# COLLEGE PREP TEST

Directions: Choose the one best answer to each question or problem.

**1.** Find the missing number in the following
sequence: 8, 17, 26, 35, __?__

C   (A) 63    (B) 54    (C) 44
      (D) 49    (E) none of these

**2.** If the area of square $T$ is 16 and the
perimeter of square $R$ is 28, find the

C   perimeter of square $W$.

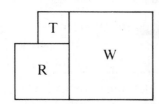

  (A) 17    (B) 28    (C) 44
  (D) 121   (E) none of these

**3.** If $3x - 6 = 9$, then $x - 2 = ?$

C   (A) 2   (B) 5   (C) 3   (D) 9   (E) 7

**4.** How much more is $m - 7$ than 7?

E   (A) $-m$    (B) 0    (C) $7 - m$
    (D) $m$      (E) $m - 14$

**5.** A train traveling 30 mi/h is stopped $2\frac{1}{2}$ mi

B   from its destination at 2:00 pm. At what
time would the train have arrived if it had
not been delayed?
(A) 2:04 pm  (B) 2:05 pm  (C) 2:065 pm
(D) 2:30 pm  (E) 2:32 pm

**6.** If golf balls sell for \$9.00/dozen and the
tax is 10% of the selling price, find the

E   charge for $2\frac{1}{3}$ dozen.

  (A) \$19.80    (B) \$21.21    (C) \$21.90
  (D) \$22       (E) \$23.10

**7.** In a group of 20 boys, 8 are on the varsity
basketball team, 9 are on the varsity

A   football, and 5 are on both varsity teams.
How many of the 20 boys are not on
either varsity team?
(A) 8  (B) 7  (C) 6  (D) 4  (E) 3

**8.** If $x - 7 = 8$ and $4 - 3y - y = 0$, then
$y - x = ?$

A   (A) $-14$ (B) $-8$ (C) 14 (D) 15 (E) 16

**9.** The symbol $\begin{vmatrix} x & y \\ z & w \end{vmatrix}$ means $xw - yz$.

C   What is the value of $\begin{vmatrix} 3 & 4 \\ 5 & 6 \end{vmatrix}$ ?

  (A) 360    (B) 38    (C) $-2$
  (D) $-18$   (E) none of these

**10.** A relationship exists between figure I and
figure II.

C

      I      II     III

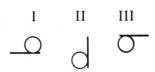

A similar relationship exists between figure
III and which of the following figures?

  (A)    (B)    (C)    (D)    (E)

# 4 EQUATIONS AND WORD PROBLEMS

## Problem Solving Formulation

When you use a word processing program, a copy of the manuscript will be stored on a diskette for later use. Most microcomputers use $5\frac{1}{4}$ in. diskettes, which are subdivided into tracks and sectors. The sectors contain bytes, which are memory locations on the disk. A byte can contain one character or a blank space.

### Project

How many diskettes would you need to store a 15-page term paper?

To answer this question you need to know the following information:

What type of diskette does your microcomputer require?
How does the diskette store information?
How many bytes are there in a sector?
How many characters per line are there?
How many lines per page are there?

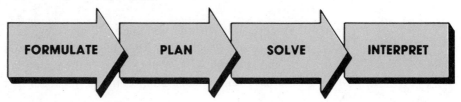

FORMULATE → PLAN → SOLVE → INTERPRET

# EQUATIONS WITH VARIABLES ON EACH SIDE

*Objectives*    **To solve equations with the variable term on each side**
**To solve problems leading to equations with the variable on each side**

You have solved equations like $2x - 8 = 12$ in which the variable is on only one side. An equation can contain the variable on each side, as in $5x - 8 = 3x + 12$. If you add $-3x$ to each side, you will have an equation with only one variable term. This is shown in Example 1.

*Example 1*    **Solve $5x - 8 = 3x + 12$. Check the solution.**

$$-3x + 5x - 8 = -3x + 3x + 12 \qquad \blacktriangleleft \textit{Add } -3x, \textit{ the opposite of } 3x, \textit{ to each side.}$$
$$2x - 8 \ = \ 0 + 12 \qquad \blacktriangleleft \textit{Combine like terms.}$$
$$2x - 8 = 12 \qquad \blacktriangleleft \textit{Now there is only one variable term.}$$
$$2x - 8 + 8 = 12 + 8 \qquad \blacktriangleleft \textit{Add 8 to each side.}$$
$$2x = 20$$
$$x = 10 \qquad \blacktriangleleft \textit{Divide each side by 2.}$$

Check.

| $5x - 8$ | $3x + 12$ |
|---|---|
| $5 \cdot 10 - 8$ | $3 \cdot 10 + 12$ |
| $50 - 8$ | $30 + 12$ |
| $42$ | $42$ |

$$42 = 42 \qquad \text{True}$$

**Thus,** the solution is 10.

*Example 2*    **Solve the equation in Example 1 by first adding $-5x$ to each side.**

$$5x - 8 = 3x + 12$$
$$-5x + 5x - 8 = -5x + 3x + 12$$
$$-8 = -2x + 12 \qquad \blacktriangleleft \textit{Only one variable term}$$
$$-8 - 12 = -2x + 12 - 12 \qquad \blacktriangleleft \textit{Add } -12 \textit{ to each side.}$$
$$-20 = -2x$$
$$10 = x \qquad \blacktriangleleft \textit{Divide each side by } -2.$$

**Thus,** the solution is 10.

To solve an equation with the variable on both sides, add the opposite of either variable term to each side. Regardless of which side of the equation you begin with, the solution will be the same. If the equation contains more than one variable term on the same side, combine these terms before solving.

*Example 3*     **Solve.**

$$6y - 19 = -14 + 7y$$
$$-7y + 6y - 19 = -14 + 7y - 7y$$
$$-1y - 19 = -14$$
$$-1y - 19 + 19 = -14 + 19$$
$$-1y = 5$$
$$y = -5 \blacktriangleleft \quad \textit{Divide each side}$$
$$\textit{by } -1.$$

**Thus,** the solution is $-5$.

$$13 - n = -7n$$
$$13 - 1n = -7n \qquad \blacktriangleleft \ -n = -1n$$
$$13 - 1n + 1n = -7n + 1n$$
$$13 = -6n$$
$$-\frac{13}{6} = n \qquad \blacktriangleleft \ \frac{pos.}{neg.} = neg.$$

**Thus,** the solution is $-\dfrac{13}{6}$.

*Example 4*     **Solve $17x - 1 = -5x + 2 + 18x$.**

$$17x - 1 = -5x + 2 + 18x \quad \blacktriangleleft \ \textit{-5x and 18x are like terms on the same side.}$$
$$17x - 1 = 13x + 2 \quad \blacktriangleleft \ \textit{Combine like terms.}$$
$$-17x + 17x - 1 = -17x + 13x + 2$$
$$-1 = -4x + 2$$
$$-1 - 2 = -4x + 2 - 2$$
$$-3 = -4x$$
$$\frac{3}{4} = x \qquad \blacktriangleleft \ \frac{neg.}{neg.} = pos.$$

**Thus,** the solution is $\dfrac{3}{4}$.

Some word problems lead to equations with the variable on both sides.

*Example 5*     **Twenty more than 4 times Jack's age is the same as 6 times his age. How old is he?**

READ ▶     The problem asks for Jack's age.

PLAN ▶     Use a variable to represent Jack's age. Let $a =$ his age.
20 more than 4 times Jack's age is the same as 6 times his age.

$$4a + 20 \qquad = \qquad 6a$$

SOLVE ▶
$$-4a + 4a + 20 = -4a + 6a$$
$$20 = 2a$$
$$10 = a$$

INTERPRET ▶     20 more than 4 times Jack's age is the same as 6 times his age.

| $4 \cdot 10 + 20$ | $6 \cdot 10$ |
|---|---|
| $40 + 20$ | $60$ |
| $60$ | |
| $60 = 60$ | True |

**Thus,** Jack is 10 years old.

Equations with Variables on Each Side

# Reading in Algebra

**True or false. If false, tell why.**

1. If $-8$ is added to each side of $3x - 4 = 7x + 8$, an equation with only one variable term is obtained. **False; −8 is not opposite of 3x or 7x.** **False; add −3x or −5x**
2. To solve $3x + 2 = 5x - 8$, you must first combine the like terms, $3x$ and $5x$. **to each side first.**
3. To solve $3x + 5x = 8 + x$, you must first combine the like terms, $3x$ and $5x$. **True**
4. To transform $3x - 4 = 7x + 8$ into an equation with only one variable term, you must begin by adding $-3x$ to each side. **False; you could add −7x to each side.**

# Written Exercises

(A) **Solve each equation. Check.**

1. $6x + 7 = 3x + 16$ **3**
2. $7y - 9 = 3y + 19$ **7**
3. $6 + 10x = 8x + 12$ **3**
4. $-4 + 5n = 18 + 3n$ **11**
5. $6p + 13 = 9p - 5$ **6**
6. $11x + 8 = -2 + 9x$ **−5**

**Solve each equation.**

7. $7y - 11 = -10 + 8y$ **−1**
8. $4m - 12 = 15 + 5m$ **−27**
9. $4 + 6x = 7x - 5$ **9**
10. $-14 + 3n = n + 14$ **14**
11. $7 - n = 5 + 3n$ $\frac{1}{2}$
12. $-9 + 8x = x - 30$ **−3**
13. $8x - 12 = 15x - 4x$ **−4**
14. $6 - a = 5 + 3a$ $\frac{1}{4}$
15. $7 - m = m + 9$ **−1**
16. $6c - 20 = 2c$ **5**
17. $9e + 14 = 11e$ **7**
18. $7d = -24 + d$ **−4**
19. $10g - 22 = 8g - 14$ **4**
20. $7z + 8 = 3z - 16$ **−6**
21. $x - 3 = 72 - 4x$ **15**
22. $-6a - 15 = -17 - 9a$ $-\frac{2}{3}$
23. $3y + 9 = -2y + 7$ $-\frac{2}{5}$
24. $7x - 2 = 6x - 2x$ $\frac{2}{3}$
25. $13x - 5 = 17x - 8$ $\frac{3}{4}$
26. $7 - 2y = 5y + 1$ $\frac{6}{7}$
27. $a + 11 = -2a + 6$ $-\frac{5}{3}$

(B) 28. $2x + 4 + 3x = 4x + 2$ **−2**
29. $3 - 4x + 7 = 4 - 2x$ **3**
30. $5x - 3 = 7x + 7 + 3x$ **−2**

34. **2**   35. $-\frac{1}{8}$

31. $5y - 5 - 4y = 7 - y$ **6**
32. $7 - b - 4 = -3b + 5$ **1**
33. $-3k - 5 + 2k = 4k - 8$ $\frac{3}{5}$
34. $7d - 6d + 5 = -1 + 4d$
35. $5l - 4 + 6 - 7l = 3 + 6l$
36. $2k - 5 - 3k = 8 - 4k + 1$ $\frac{14}{3}$
37. $8a - 5 - 9a = 4 + a - 6$ $-\frac{3}{2}$
38. $3x + x - 2 = -x + 7 - x$ $\frac{3}{2}$
39. $8a + 8 - 5a = -1 + a - 7$ **−8**

**Solve each problem.**

40. Three less than 5 times Wanda's age is the same as 3 times her age increased by 37. How old is she? **20 y**
41. Juan's salary increased by $200 is the same as twice his salary. Find his salary. **$200**
42. Ten more than twice a number is the same as 4 times the number. Find the number. **5**
43. A number decreased by 30 is the same as 14 decreased by 3 times the number. Find the number. **11**
44. Two kilograms less than 2 times the mass of an object is the same as its mass increased by 64 kg. Find the mass of the object. **66 kg**
45. The area of a square increased by 3 is the same as 35 decreased by 7 times the area. Find the area of the square. **4**

# CUMULATIVE REVIEW

**Simplify.**

$8m + 5$

1. $3 - 2(3 - 2x)$ **4x − 3**
2. $5 - (4 - a)$ **a + 1**
3. $7m - (-m - 2) + 3$

# EQUATIONS CONTAINING PARENTHESES    4.2

*Objectives*    **To solve equations that contain parentheses**
**To solve problems that lead to equations containing parentheses**

To solve an equation containing parentheses, such as $6x - 3(4 - 5x) = 30$, remove the parentheses by using the distributive property and then combine any like terms. Now you have an equation of the type you have already studied.

In Example 1, $6x - 3(4 - 5x) = 30$ becomes $21x - 12 = 30$. To solve this equation, use the addition and division properties for equations.

*Example 1*    **Solve $6x - 3(4 - 5x) = 30$.**

$$6x - 3(4 - 5x) = 30$$

| | |
|---|---|
| $6x - 12 + 15x = 30$ | ◀ *Use the distributive property.* |
| | $-3 \cdot 4 = -12; -3 \cdot -5x = 15x$ |
| $21x - 12 = 30$ | ◀ *Combine like terms.* |
| $21x - 12 + 12 = 30 + 12$ | ◀ *Addition Property for Equations* |
| $21x = 42$ | |
| $x = 2$ | ◀ *Division Property for Equations* |

**Thus,** the solution is 2.

The next example involves the removal of two sets of parentheses. After removing them and combining like terms, the result is an equation with the variable on both sides.

*Example 2*    **Solve $2x + 3(4 - 2x) = 2(x + 3)$. Check the solution.**

$$2x + 3(4 - 2x) = 2(x + 3)$$

| | |
|---|---|
| $2x + 12 - 6x = 2x + 6$ | ◀ *Use the distributive property twice.* |
| $-4x + 12 = 2x + 6$ | ◀ *Combine like terms.* |
| $-2x - 4x + 12 = -2x + 2x + 6$ | |
| $-6x + 12 = 6$ | |
| $-6x + 12 - 12 = 6 - 12$ | ◀ *Addition Property* |
| $-6x = -6$ | |
| $x = 1$ | ◀ *Division Property* |

Check.

| $2x + 3(4 - 2x)$ | $2(x + 3)$ |
|---|---|
| $2 \cdot 1 + 3(4 - 2 \cdot 1)$ | $2(1 + 3)$ |
| $2 + 3(4 - 2)$ | $2 \cdot 4$ |
| $2 + 3 \cdot 2$ | $8$ |
| $8$ | $8$ |
| $8 = 8$ | |
| True | |

**Thus,** the solution is 1.

Some equations require the use of the property of $-1$, $-a = -1a$, to remove parentheses. This is illustrated in Example 3.

*Example 3*    **Solve $-4 - (3 + 6m) = \frac{1}{2}(4 + 2m)$.**

$$-4 - 1(3 + 6m) = \frac{1}{2}(4 + 2m) \qquad \blacktriangleleft\ -(3 + 6m) = -1(3 + 6m)$$

$$-4 - 3 - 6m = 2 + 1m$$
$$-7 - 6m = 2 + 1m$$
$$-7 - 6m + 6m = 2 + 1m + 6m \qquad \blacktriangleleft\ \textit{Add 6m to each side.}$$
$$-7 = 2 + 7m$$
$$-2 - 7 = -2 + 2 + 7m \qquad \blacktriangleleft\ \textit{Add } -2 \textit{ to each side.}$$
$$-9 = 7m$$
$$\frac{-9}{7} = m$$

**Thus,** the solution is $\dfrac{-9}{7}$ or $-\dfrac{9}{7}$.

A mathematical term that appears frequently in word problems is *sum*. The sum of several numbers is the result of adding. Sum is represented by $+$.

*Example 4*    **Four times the sum of a number and 2 is the same as 10 less than the number. Find the number.**

READ ▶    Find a number when you are given information about the number.

PLAN ▶    Let $n =$ the number.
4 times the sum of $n$ and 2 is the same as 10 less than $n$.

SOLVE ▶
$$4(n + 2) = n - 10$$
$$4n + 8 = 1n - 10 \qquad \blacktriangleleft\ n = 1n$$
$$-1n + 4n + 8 = -1n + 1n - 10$$
$$3n + 8 = -10$$
$$3n + 8 - 8 = -10 - 8$$
$$3n = -18$$
$$n = -6$$

INTERPRET ▶    4 times the sum of the number and 2 is 10 less than the number.

| $4(-6 + 2)$ | $-6 - 10$ |
|---|---|
| $4 \cdot -4$ | $-16$ |
| $-16$ | |
| $-16 = -16$ | True |

**Thus,** the number is $-6$.

# Written Exercises

**Solve each equation. Check the solution.**

(A)
**1.** $3(x - 2) = 9$ **5**

**2.** $2(7 + x) = 24$ **5**

**3.** $-5(x + 2) = 20$ **−6**

**4.** $6(z - 2) = 24$ **6**

**5.** $12 - b = 2(b + 3)$ **2**

**6.** $a + 6 = 2(a - 6)$ **18**

**7.** $4y - 2(3 - 2y) = 10$ **2**

**8.** $5a + 3(2a + 1) = 25$ **2**

**9.** $2b + 4(b - 1) = 5(b - 1)$
**−1**

**Solve each equation.**

**10.** $14 - (6 + x) = 22$ **−14**

**11.** $8x - 3(2 - 5x) = 40$ **2**

**12.** $18 - (3 - 2y) = 3(y - 5)$ **30**

**13.** $7y - (4 - 2y) = 3(y + 3)$ $\frac{13}{6}$

**14.** $6p + 2(4 - 2p) = 3(p + 4)$ **−4**

**15.** $3c + 2(c + 2) = 13 - (2c - 5)$ **2**

**16.** $5 - 3(4 - 2b) = 4(b - 3) - 3$ **−4**

**17.** $-5 - (x + 8) + 6x = 4(x + 2)$ **21**

**18.** $-2(3 - 4z) + 7z = 12z - (z + 4)$ $\frac{1}{2}$

**19.** $8y - 3(4 - 2y) = 6(y + 1) - 2$ **2**

**20.** $-5y - 2(y + 4) = 6y - 9$ $\frac{1}{13}$

**21.** $8a + 5(1 - a) = 4a - 6$ **11**

**22.** $-3y - \frac{1}{2}(6y + 4) = 8y - 9$ $\frac{1}{2}$

**23.** $-\frac{1}{3}(6 - 9x) + 4x = -(2x - 8)$ $\frac{10}{9}$

**24.** $7x - (9 - 4x) = 3(x - 11)$ **−3**

**25.** $7r + 3(7 - r) = -(r + 4)$ **−5**

(B) **Solve each problem.**

**26.** Six times the sum of a number and −4 is 30. Find the number. **9**

**27.** Five times the sum of a number and 2 is 45. Find the number. **7**

(C) **Solve each equation.**

**28.** $3[5 - 3(x - 4)] = 2x + 9$ $\frac{42}{11}$

**29.** $5 - y = 6y - \left[4 + \frac{1}{7}(21y - 14)\right]$ $\frac{7}{4}$

**30.** $-9a - [2(1 + 3a) + 6] = 5a$ $-\frac{2}{5}$

# GEOMETRY REVIEW

*Objective*    **To find the circumference and area of a circle**

In circle $O$:
$\overline{OA}$ is a radius ($r$): $r = 4$
$\overline{PQ}$ is a diameter ($d$): $d = 2r = 2 \cdot 4 = 8$
The formula for the area of a circle is $A = \pi r^2$. Note that $\pi$ is approximately 3.14. The formula for the circumference of a circle is $C = 2\pi r$ or $C = \pi d$.

*Example*    **Find the area (in terms of $\pi$) of a circle with diameter 8.**

$r = \frac{1}{2}d$, so $r = \frac{1}{2} \cdot 8 = 4$    $A = \pi r^2 = \pi \cdot r \cdot r$
$\qquad\qquad\qquad\qquad\qquad\qquad\qquad = \pi \cdot 4 \cdot 4$
$\qquad\qquad\qquad\qquad\qquad\qquad\qquad = 16\pi$

**Find the circumference and area of each circle. Leave your answer in terms of $\pi$.**

**1.** radius 7
C: $14\pi$;  A: $49\pi$

**2.** diameter 10
C: $10\pi$; A: $25\pi$

**3.** radius $2\frac{2}{3}$
C: $5\frac{1}{3}\pi$ A: $7\frac{1}{9}\pi$

**4.** diameter 6.2
C: $6.2\pi$; A: $9.61\pi$

Equations Containing Parentheses

**91**

# NUMBER PROBLEMS

*Objective*    **To solve problems involving two or more numbers**

Sometimes a word problem will contain a sentence that expresses a relationship between two or more numbers. To use an equation to solve a problem of this type, you will have to decide:

**1.** which of the numbers the variable will represent and

**2.** how to represent the other numbers in terms of the variable.

For example, the second of two numbers is 8 more than twice the first. Since the second number is compared to the first, let $f$ = the first number. The second number is 8 more than twice $f$, or $2f + 8$.

You can use $x$, $y$, $t$, or any variable you choose in these problems, but you should use only one. Later in this book, you will learn how to solve equations that involve more than one variable.

*Example 1*    **The larger of two numbers is 6 less than 7 times the smaller. Represent the numbers.**

Larger is 6 less than 7 times smaller.

   Let $s$ = the smaller number          ◀ *Larger is compared to smaller.*
   $7s - 6$ = the larger number          ◀ *6 less than 7 times smaller, or 6 less than 7s*

**Thus,** the smaller number is represented by $s$, and the larger is represented by $7s - 6$.

*Example 2*    **The second of three numbers is 4 times the first. The third number is 1 more than the second. Represent the three numbers.**

Think: The second is compared to the first, and the third is compared to the second. Therefore,    let $f$ = the first number

   $4f$ = the second number    ◀ *4 times the first*

   $4f + 1$ = the third number    ◀ *1 more than the second*

**Thus,** use $f$ for the first number, $4f$ for the second, and $4f + 1$ for the third.

You can apply the four basic problem-solving steps:

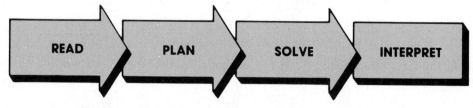

READ    PLAN    SOLVE    INTERPRET

to word problems that involve more than one number.

**Example 3**

**Julie's salary is 3 times Liz's salary. Together they earn $320 a month. Find the salary of each girl.**

READ ▶    The problem asks you to find the salary of each girl.

PLAN ▶    Let $s$ = Liz's salary

       $3s$ = Julie's salary   ◀ *Julie's salary is 3 times Liz's salary.*

       Together their sum is 320

         ↓

       $s + 3s = 320$      ◀ *Liz's salary plus Julie's salary is $320.*

SOLVE ▶         $4s = 320$

            $s = 80$

To find each girl's salary, use the representation statements:

     Liz's salary, $s$, is $80.

     Julie's salary, $3s$, is $3 \cdot 80$, or $240.

INTERPRET ▶    Check the salaries in both parts of the problem.

| Julie's salary is 3 times Liz's salary. | | | Together they earn $320. | |
|---|---|---|---|---|
| 240 | $3 \cdot 80$ | | $80 + 240$ | 320 |
| | 240 | | 320 | |
| $240 = 240$ | True | | $320 = 320$ | True |

**Thus,** Liz's salary is $80 a month and Julie's salary is $240 a month.

**Example 4**

**Lola's age is 14 more than 6 times Juan's age. The sum of their ages is 35. How old is each?**

READ ▶    The problem asks you to find the age of each person.

PLAN ▶      Let $a$ = Juan's age

     $6a + 14$ = Lola's age   ◀ *14 more than 6a*

     The sum of their ages is 35.

        ↓            ↓

     $a + 6a + 14 \quad = 35$

SOLVE ▶           $7a + 14 = 35$

       $7a + 14 - 14 = 35 - 14$

              $7a = 21$

                $a = 3$

Juan's age, $a$, is 3.

Lola's age, $6a + 14$, is $6 \cdot 3 + 14$, or 32.

INTERPRET ▶

| Lola's age is 14 more than 6 times Juan's age. | | | The sum of their ages is 35. | |
|---|---|---|---|---|
| 32 | $6 \cdot 3 + 14$ | | $3 + 32$ | 35 |
| | $18 + 14$ | | 35 | |
| | 32 | | $35 = 35$ | True |
| $32 = 32$ | True | | | |

**Thus,** Juan is 3 years old and Lola is 32 years old.

Sometimes the phrase *separate into two parts* is used in word problems. To separate 14 into two parts means to divide it into two numbers whose sum is 14. Some examples are 10 and 4, 6 and 8, 5 and 9, and so on.

*Example 5*

**Separate 89 into two parts so that the first part is 4 less than twice the second part. Find each part.**

READ ▶ You are asked to find the two parts into which 89 is separated.

PLAN ▶ *The first part is compared to the second part.* ▶ Let $p$ = the second part
*4 less than twice the second part, or 4 less than 2p* ▶ $2p - 4$ = the first part
*The sum of the parts is 89.* ▶

SOLVE ▶
$$p + (2p - 4) = 89$$
$$1p + 2p - 4 = 89$$
$$3p - 4 = 89$$
$$3p - 4 + 4 = 89 + 4$$
$$3p = 93$$
$$p = 31$$

*Use the representation to find each part.* ▶ The second part, $p$, is 31.
The first part, $2p - 4$, is 58.

INTERPRET ▶

| Separate 89 into two parts means | |
|---|---|
| Sum is 89. | |
| $31 + 58$ | $89$ |
| | $89$ |
| $89 = 89$ | True |

| First part is 4 less than twice second part. | |
|---|---|
| $58$ | $2 \cdot 31 - 4$ |
| | $62 - 4$ |
| | $58$ |
| $58 = 58$ | True |

**Thus,** the two parts are 58 and 31.

*Example 6*

**The second of three numbers is 6 times the first. The third number is 1 more than the second. If the first is decreased by twice the third, the result is −35. What are the three numbers?**

READ ▶ You are asked to find three numbers described by the first two sentences.

PLAN ▶
Let $f$ = the first number
*Second number is 6 times the first number.* ▶ $6f$ = the second number
*Third number is 1 more than the second number.* ▶ $6f + 1$ = the third number
*Write an equation.* ▶ First decreased by twice the third is −35.

$$f \quad - \quad 2(6f + 1) \quad = -35$$

SOLVE ▶
$$1f - 12f - 2 = -35$$
$$-11f - 2 = -35$$
$$-11f = -33$$
$$f = 3$$

*Find the three numbers.* ▶ The first number, $f$, is 3.
The second number, $6f$, is $6 \cdot 3$, or 18.
The third number, $6f + 1$, is $6 \cdot 3 + 1$, or 19.

INTERPRET ▶ Check on your own.
**Thus,** the three numbers are 3, 18, and 19.

# Reading in Algebra

Use the following problem to answer each question:
John's age is 3 years more than 4 times Arlene's age. Together they are
33 years old. Find their ages.

1. What two numbers must you represent? **John's age; Arlene's age**
2. What number is the basis of comparison, that is, the one you can call *x*? **Arlene's age**
3. When you write the equation, what should you use for the word *together*? **+**

# Written Exercises

**Solve each problem.**

(A)
1. Jane's salary is 4 times Mary's salary. Together they earn $150 per week. Find the salary of each girl. **Jane, $120; Mary, $30**

2. In a game, Harry's score was 3 times José's score. Together they scored 80 points. Find their scores. **Harry, 60 points; José, 20 points**

3. Noah's age is 3 years more than twice Tina's age. The sum of their ages is 24 years. Find each of their ages. **Tina, 7; Noah, 17**

4. Marie's bowling score is 5 less than 3 times Ralph's score. The sum of their scores is 215. Find the score of each. **R, 55; M, 160**

5. The second of two numbers is 8 less than twice the first. Their sum is 19. Find the two numbers. **9, 10**

6. The larger of two numbers is 7 more than 5 times the other. Their sum is 55. Find the two numbers. **8, 47**

7. Separate 90 into two parts so that the second part is 30 less than twice the first part. Find each part. **40, 50**

8. A 30-m board is separated into two pieces. The longer piece is 8 m longer than the shorter one. Find the length of each piece. **11 m, 19 m**

9. Sixty-eight students are separated into two groups. The first group is 3 times as large as the second. How many are in each group? **17, 51**

10. Separate 43 into two parts. The first part has 5 less than 3 times the second part. Find each part. **12, 31**

11. The second of three numbers is twice the first, and the third number is 6 more than the second. Their sum is 26. Find the three numbers. **4, 8, 14**

12. One month, John worked 3 hours less than Tia, and Felicia worked 4 hours more than Tia. Together they worked 196 hours. Find the number of hours worked by each.
**Tia, 65 h; John, 62 h; Felicia, 69 h**

(B)
13. The smaller of two numbers is 3 less than the greater. If the greater is decreased by twice the smaller, the result is −5. Find the two numbers. **11, 8**

14. The greater of two numbers is 3 more than the smaller. If twice the smaller is added to the greater, the result is 30. Find the numbers. **9, 12**

15. The second of three numbers is 3 times the first. The third is 5 more than the second. If the second is decreased by twice the third, the result is 5. Find the three numbers. **−5, −15, −10**

(C)
16. The second of three numbers is 7 less than 3 times the first. The third is 13 less than 6 times the first. If twice the first is decreased by the third, the result is −3. Find the three numbers. **4, 5, 11**

17. The first of two numbers is 4 less than twice the second. If 3 less than 5 times the first is subtracted from twice the second, the result is 7. Find the two numbers. **0, 2**

# EQUATIONS WITH FRACTIONS

*Objectives*  **To solve equations containing fractions**
**To solve problems leading to equations with fractions**

You have solved equations like $\frac{x}{5} = 4$ by using the multiplication property

for equations. In the equation $\frac{x}{5} = 4$, multiply each side by 5.

$$\frac{x}{5} = 4 \longleftarrow 5 \cdot \frac{x}{5} = 1x \text{ or } x$$
$$x = 20$$

Multiplying each side by 5 resulted in an equation without fractions. Similarly,

you can transform an equation like $\frac{2}{3}x = \frac{5}{6}$ into an equation without fractions

by multiplying each side by the reciprocal of $\frac{2}{3}$. Recall that the product of

a number and its reciprocal is 1. So, $\frac{2}{3} \cdot \frac{3}{2} = 1$. The use of this property

in solving the equation is shown in Example 1.

*Example 1*  **Solve $\frac{2}{3}x = \frac{5}{6}$. Check the solution.**

$$\frac{3}{2} \cdot \frac{2}{3} \cdot x = \frac{3}{2} \cdot \frac{5}{6} \quad \blacktriangleleft \text{ } Multiply \text{ } each \text{ } side \text{ } by \frac{3}{2}, \text{ } the \text{ } reciprocal \text{ } of \frac{2}{3}.$$

$$1 \cdot x = \frac{5}{4} \quad \blacktriangleleft \frac{\cancel{3}}{2} \cdot \frac{5}{\cancel{6}} = \frac{5}{4}$$

$$x = \frac{5}{4}$$

Check.

| $\frac{2}{3}x$ | $\frac{5}{6}$ |
|---|---|
| $\frac{2}{3} \cdot \frac{5}{4}$ | $\frac{5}{6}$ $\quad \blacktriangleleft \frac{\cancel{2}}{3} \cdot \frac{5}{\cancel{4}} = \frac{5}{6}$ |
| $\frac{5}{6}$ | |

$$\frac{5}{6} = \frac{5}{6} \quad \textbf{Thus,} \text{ the solution is } \frac{5}{4}.$$

The method illustrated above is most efficient when an
equation involves only *one* term on each side.

A more general method for solving equations containing
fractions is illustrated on the next page.

Sometimes an equation has several terms containing fractions, as in the equation $\frac{3}{4}x - \frac{1}{2} = \frac{5}{8}$. Multiply each side by the *least* number that is divisible by all the denominators 4, 2, and 8. The number is 8.

The number 8 is called the *least common multiple* (LCM) of 4, 2, and 8.

Now, if you multiply each side of $\frac{3}{4}x - \frac{1}{2} = \frac{5}{8}$ by 8, you will get an equation that does not contain fractions.

## Example 2    Solve $\frac{3}{4}x - \frac{1}{2} = \frac{5}{8}$.

$$8\left(\frac{3}{4}x - \frac{1}{2}\right) = 8 \cdot \frac{5}{8}$$  ◀ *Multiply each side by 8, the LCM of 4, 2, and 8.*

$$8 \cdot \frac{3}{4} \cdot x - 8 \cdot \frac{1}{2} = 8 \cdot \frac{5}{8}$$  ◀ *Distribute the 8.*

$$6x - 4 = 5$$  ◀ $\overset{2}{\cancel{8}} \cdot \frac{3}{\underset{1}{\cancel{4}}} \cdot x = 6x$  |  $\overset{4}{\cancel{8}} \cdot \frac{1}{\underset{1}{\cancel{2}}} = 4$  |  $\overset{1}{\cancel{8}} \cdot \frac{5}{\underset{1}{\cancel{8}}} = 5$

$$6x = 9$$

$$x = \frac{9}{6}$$

$$x = \frac{3}{2}$$  ◀ $\frac{9}{6} = \frac{\overset{3}{\cancel{9}}}{\underset{2}{\cancel{6}}} = \frac{3}{2}$

The solution of the equation $\frac{3}{2}x - 1 = \frac{4}{5}x + 6$ also involves the use of the distributive property, as shown in Example 3.

## Example 3    Solve $\frac{3}{2}x - 1 = \frac{4}{5} x + 6$. Check the solution.

$$\frac{3}{2}x - 1 = \frac{4}{5}x + 6$$

$$10\left(\frac{3}{2}x - 1\right) = 10\left(\frac{4}{5}x + 6\right)$$  ◀ *Multiply each side by 10, the LCM of 5 and 2.*

$$10 \cdot \frac{3}{2}x + 10 \cdot -1 = 10 \cdot \frac{4}{5}x + 10 \cdot 6$$  ◀ *Distribute the 10.*

$$\frac{30}{2}x - 10 = \frac{40}{5}x + 60$$

$$15x - 10 = 8x + 60$$

$$7x - 10 = 60$$

$$7x = 70$$

$$x = 10$$

**Thus,** the solution is 10.

Check.

| $\frac{3}{2}x - 1$ | $\frac{4}{5}x + 6$ |
|---|---|
| $\frac{3}{2} \cdot 10 - 1$ | $\frac{4}{5} \cdot 10 + 6$ |
| $\frac{30}{2} - 1$ | $\frac{40}{5} + 6$ |
| $15 - 1$ | $8 + 6$ |
| $14$ | $14$ |
| $14 = 14$ | |
| True | |

Equations with Fractions

In mathematics, the word *of* is associated with the operation of multiplication. For example, $\frac{1}{2}$ of 8 means $\frac{1}{2} \cdot 8$, or $\frac{1 \cdot 8}{2} = \frac{8}{2} = 4$. This concept is used in the word problem in Example 4. Notice that the problem leads to an equation with fractions.

*Example 4*

**Four dollars less than $\frac{1}{2}$ of Tim's weekly salary is $80. How much does Tim earn each week?**

READ ▶          You are asked to find Tim's salary.

PLAN ▶           Let $s$ = Tim's salary

*Write an equation.* ▶   4 less than $\frac{1}{2}$ of $s$ is 80.

$$\frac{1}{2}s - 4 = 80 \qquad \blacktriangleleft \text{ } \frac{1}{2} \text{ of } s \text{ means } \frac{1}{2} \cdot s, \text{ or } \frac{1}{2}s.$$

SOLVE ▶

$$2\left(\frac{1}{2}s - 4\right) = 2 \cdot 80 \qquad \blacktriangleleft \text{ } Multiply \text{ } each \text{ } side \text{ } by \text{ } 2.$$

$$2 \cdot \frac{1}{2}s + 2 \cdot -4 = 160 \qquad \blacktriangleleft \text{ } Distribute \text{ } the \text{ } 2.$$
$$1s - 8 = 160 \qquad \blacktriangleleft \text{ } Now \text{ } the \text{ } equation$$
$$s - 8 + 8 = 160 + 8 \qquad\quad contains \text{ } no \text{ } fractions.$$
$$s = 168$$

INTERPRET ▶     4 less than $\frac{1}{2}s$ is 80.

| $\frac{1}{2} \cdot 168 - 4$ | 80 |
|---|---|
| $84 - 4$ | |
| 80 | |
| $80 = 80$ | |
| True | |

**Thus,** Tim earns $168 each week.

# Oral Exercises

**For each equation, state the number by which you would multiply each side to produce an equation with no fractions.**

1. $\frac{3}{5}m = 9$  5    2. $6 = \frac{2}{3}m$  3    3. $8 = \frac{1}{2}y$  2    4. $\frac{5}{4}p = 20$  4

5. $\frac{3}{5}r = \frac{9}{10}$  10    6. $\frac{5}{4} = \frac{1}{2}x$  4    7. $\frac{2}{3}p = \frac{8}{9}$  9    8. $\frac{7}{15} = \frac{1}{3}q$  15

9. $\frac{1}{3}a + 4 = \frac{1}{2}a + 7$  6   10. $\frac{2}{3}b - 2 = \frac{1}{5}b + 3$  15   11. $\frac{4}{7}r + 2 = \frac{1}{2}r - 5$  14

12. $\frac{2}{3}a + \frac{1}{4} = \frac{5}{12}$  12    13. $\frac{4}{15} + \frac{1}{5}b = \frac{1}{3}$  15    14. $\frac{7}{3} = \frac{5}{6}b + \frac{1}{2}$  6

# Written Exercises

**Solve each equation. Check the solution.**

Ⓐ 1. $\frac{2}{5}m = 6$  **15**

2. $9 = \frac{3}{2}a$  **6**

3. $\frac{5}{3}y = 10$  **6**

4. $7 = \frac{1}{3}a$  **21**

5. $\frac{2}{7}x - \frac{1}{2} = \frac{1}{14}$  **2**

6. $\frac{5}{6} = \frac{2}{3}m - \frac{1}{2}$  **2**

7. $\frac{2}{3}a + \frac{1}{9} = \frac{1}{3}$  **$\frac{1}{3}$**

8. $\frac{1}{4} + \frac{1}{3}x = \frac{11}{12}$  **2**

**Solve.**

9. $\frac{2}{3}a = \frac{5}{9}$  **$\frac{5}{6}$**

10. $\frac{3}{7} = \frac{5}{14}x$  **$\frac{6}{5}$**

11. $\frac{3}{5}a = \frac{2}{15}$  **$\frac{2}{9}$**

12. $\frac{7}{4} = \frac{3}{2}y$  **$\frac{7}{6}$**

13. $\frac{3}{2}b + \frac{1}{4} = \frac{5}{8}$  **$\frac{1}{4}$**

14. $\frac{2}{3}y + 7 = 9$  **3**

15. $\frac{3}{10} + \frac{4}{5}x = \frac{1}{2}$  **$\frac{1}{4}$**

16. $6 + \frac{1}{2}m = 9$  **6**

17. $\frac{2}{5}a + 3 = \frac{1}{3}a$  **−45**

18. $6 + \frac{1}{2}x = \frac{1}{3}x$  **−36**

19. $\frac{3}{4}x - 6 = 3 + \frac{1}{2}x$  **36**

20. $\frac{2}{3}m + 2 = -3 + \frac{3}{2}m$  **6**

21. $\frac{1}{5}b + 7 = \frac{1}{2}b + 4$  **10**

22. $\frac{2}{3}b - 6 = \frac{1}{2}b - 5$  **6**

23. $\frac{3}{5}y + 2 = \frac{1}{2}y$  **−20**

24. $\frac{2}{3}m = \frac{4}{5}m - 2$  **15**

Ⓑ 25. $\frac{2}{3}x - 5 = 4 - \frac{1}{2}x$  **$\frac{54}{7}$**

26. $7 - \frac{2}{3}m = \frac{1}{4} + m$  **$\frac{81}{20}$**

27. $\frac{2}{3}a - \frac{1}{6} = \frac{1}{2}a$  **1**

28. $\frac{2}{5}b + \frac{1}{10} = \frac{3}{2}b$  **$\frac{1}{11}$**

29. $\frac{1}{2}x - \frac{1}{6}x = \frac{2}{3}$  **2**

30. $\frac{3}{4}m - 1 = \frac{5}{6}m$  **−12**

**Solve each problem.**

31. $20 less than $\frac{1}{3}$ of Mona's salary is $40. Find her salary. **$180**

32. Ten years more than $\frac{3}{5}$ of Jim's age is 28. How old is Jim? **30 y**

33. Rodolfo's bowling score for 3 games decreased by $\frac{1}{3}$ of his score is 220. Find his score. **330**

34. Six less than $\frac{3}{4}$ of a number is the same as the number. Find the number. **−24**

35. Myra's savings increased by $\frac{1}{2}$ of her savings is $15,000. Find her savings. **$10,000**

36. Five more than $\frac{1}{4}$ of a number is $\frac{2}{3}$ of that number. Find the number. **12**

# CUMULATIVE REVIEW

**Simplify.**

1. $2(4 - a) - 3(5 - 2a)$   **$4a - 7$**

2. $|5 - 8 - 4|$  **7**

3. $-1\frac{1}{2} + 5$  **$3\frac{1}{2}$**

## NON-ROUTINE PROBLEMS

Suppose you are able to write 5 parking tickets in 5 minutes. However, the whole police force can write 100 tickets in 1 hour and 40 minutes. How many are there on the police force? **1**

Equations with Fractions

# EQUATIONS WITH DECIMALS

*Objectives*  **To solve equations that contain decimals**
**To solve problems that lead to equations containing decimals**

Recall that multiplying a number by 10, 100, or 1,000 is equivalent to "moving" the decimal point to the right one, two, or three places, respectively. For example,

$$10 \cdot 0.4682 = 4.682 \quad 100 \cdot 0.4682 = 46.82 \quad 1,000 \cdot 0.4682 = 468.2$$

Consider an equation such as $0.7x + 0.38 = 0.645$. To solve such an equation, you can first write the decimals as fractions. Then you can use the same procedure as you used for an equation with fractions.

$0.7x + 0.38 = 0.645$ is the same as $\dfrac{7}{10}x + \dfrac{38}{100} = \dfrac{645}{1,000}$.

Multiply each side by the LCM, which is 1,000:

$$1,000\left(\frac{7}{10}x + \frac{38}{100}\right) = 1,000 \cdot \frac{645}{1,000}$$

$$1,000 \cdot \frac{7}{10}x + 1,000 \cdot \frac{38}{100} = 645$$

$$700x + 380 = 645$$

This equation now contains only integers.

As you have seen, the LCM of 0.7, 0.38, and 0.645 is 1,000. A more convenient technique for finding the LCM of several decimals follows.

In the equation $0.7x + 0.38 = 0.645$ that contains decimals,

1 digit to the    2 digits to the    3 digits to the
right of the       right of the       right of the
decimal point    decimal point    decimal point

notice that the greatest number of digits to the right of any decimal point is three. So, the LCM is 1,000.

*Example 1*    **Find the LCM for the equation $0.05m - 0.001 = 0.73m + 1.9$.**

$0.05m - 0.001 = 0.73m + 1.9$

two      three      two      one   ◀ *Number of digits to the right of the decimal point*

**Thus,** the LCM is 1,000. This is because the greatest number of digits to the right of the decimal point is three.

## Example 2

**Find the LCM for the equation $0.6b + 0.08 = 4b - 0.37$.**

$$0.6b + 0.08 = 4b - 0.37$$

one    two    none    two

**Thus,** the LCM is 100. This is because the greatest number of digits to the right of the decimal point is two.

To solve a decimal equation, first change the equation into an equation with only integers. Then solve the resulting equation. The process is illustrated below.

## Example 3

**Solve $0.007 = 0.7x - 0.21$.**

The greatest number of digits to the right of any decimal point is three, so the LCM = 1,000.

$$1{,}000 \cdot 0.007 = 1{,}000(0.7x - 0.21)$$    ◀ *Multiply each side by the LCM, 1,000.*

$$1{,}000 \cdot 0.007 = 1{,}000 \cdot 0.7x - 1{,}000 \cdot 0.21$$

◀ *To multiply by 1,000, move the decimal point three places to the right.*

$$7 = 700x - 210$$
$$7 + 210 = 700x - 210 + 210$$
$$217 = 700x$$
$$\frac{217}{700} = x$$

**Thus,** the solution is $\frac{217}{700}$.

## Example 4

**Solve $0.6a - 1 = 0.5a + 0.03$.**

Since the greatest number of digits to the right of any decimal point is two, the LCM is 100.

$$100(0.6a - 1) = 100(0.5a + 0.03)$$
$$100 \cdot 0.6a + 100 \cdot -1 = 100 \cdot 0.5a + 100 \cdot 0.03$$
$$60a - 100 = 50a + 3$$
$$10a - 100 = 3$$
$$10a = 103$$
$$a = \frac{103}{10}, \text{ or } 10\frac{3}{10}$$

**Thus,** the solution is $10\frac{3}{10}$.

If a decimal equation contains parentheses, remove the parentheses before you multiply each side by the LCM. For example, to solve $0.9x + 19.2 = 0.06(8 - 0.6x)$, first distribute 0.06 to each term in the parentheses.

*Example 5*    **Solve 0.9x + 19.2 = 0.06(8 − 0.6x) + 0.468x.**

$$0.9x + 19.2 = 0.06(8 - 0.6x) + 0.468x$$

$0.9x + 19.2 = 0.48 \quad 0.036x + 0.468x$   ◀ $0.06 \cdot 8 = 0.48 \; | \; 0.06 \cdot 0.6 = 0.036$

$1,000(0.9x + 19.2) = 1,000(0.48 - 0.036x + 0.468x)$   ◀ *The LCM is 1,000.*

$900x + 19,200 = 480 - 36x + 468x$

$900x + 19,200 = 480 + 432x$   ◀ *Combine like terms.*

$468x + 19,200 = 480$   ◀ *Add −432x to each side.*

$468x = -18,720$

$x = -40$

**Thus,** the solution is −40.

# Oral Exercises

**Name the LCM for each equation.**
1. $0.03t + 0.004 = 0.72t - 1.8$    2. $0.2y + 0.04 = 5y - 0.3$    3. $0.04 = 3.1m - 5$ **100**
4. $2.2x - 0.02 = 3.004$ **1,000**    5. $3.1g - 0.04 = 7.2$ **100**    6. $0.006 - 3x = 5.1$
7. $0.01m - 3 = 2.3 - 2.004m$    8. $1.04a - 2.5 = 0.28a$    9. $-3.1b - 0.005 = 2.6b$

**1. 1,000  7. 1,000        2. 100  8. 100        6. 1,000          1,000**

# Written Exercises

**Solve each equation.**

Ⓐ 1. $0.03x = 0.6$ **20**        2. $0.018 = 0.03t$ **0.6**        3. $0.006l - 7.3 = 0.14$ **1,240**

4. $0.5a = 1.2 + 6.4$ **15.2**        5. $0.15x - 3.2 = 1.3$ **30**        6. $0.02y - 2.6 = 0.84$ **172**

7. $0.07b = 0.7b - 0.1071$ **0.17** 8. $-4.4 = 0.08 - 0.2r$ **22.4** 9. $5.2 - 0.1x = -0.009$ **52.09**

10. $-0.6l - 0.2 = 0.5l + 0.02$ 11. $1.4q - 2 = 1.2q + 0.004$ 12. $1.5 - 0.24a = 0.09 - 5.1$

13. $0.012y - 4 = 0.112y + 2$    14. $5 - 0.03t = 0.7t - 0.11$    15. $0.8x - 0.2 = 0.23x + 119.5$

**10. −0.2       11. 10.02       12. 27.125       13. −60       14. 7       15. 210**

Ⓑ 16. $0.2(5 - 0.3x) = 0.16x + 0.208$ **3.6**       17. $0.41 - 0.02a = 0.5(0.784a - 23.9)$ **30**

18. $0.01(2 - 5x) = -0.68 + 0.3x$ **2**       19. $0.02(3 - 2y) = 0.12y - 55.14$ **345**

20. $0.45 + 9.6 = 0.06(4 - 0.3x)$ **−545**       21. $3.2 - 2.5t = -0.3(t - 480)$ **−64**

22. $0.31x - 0.1(0.02 - 0.2x) = 0.064$ **0.2**       23. $0.01 + x = 0.7x - 0.8(0.05 - x)$ **0.1**

**Solve each problem.**

24. A number decreased by 0.37 times the number is 12.6. Find the number. **20**

25. One number is 0.9 of another number. Their sum is 0.038. Find each number. **0.02; 0.018**

26. The greater of two numbers is 3.2 more than twice the smaller. If 0.3 times the smaller is added to the larger, the result is 11.48. Find the numbers. **3.6; 10.4**

**Solve each equation.**

Ⓒ 27. $0.3x - 0.02[7 - 0.1(5 - 0.04x)] = 299.79$ **1,000**    28. $7 - 0.04[6p - (2 - 0.01p)] = -43.404$ **29. 4.5**  **210**

29. $-0.02[0.4 - 0.1(2 + 3t)] = 0.004t + 0.005$    30. $0.32y - 0.2[0.5(y + 1) + 0.3] = 0.06y$ **1**

31. $-0.1[-0.07x + 0.2(0.01 - 0.02x)] = 17.2198 - 0.05x$ **300**

# USING ALGEBRA IN PERCENT PROBLEMS   4.6

*Objective*   **To solve problems involving percents**

Recall that *percent,* written %, means *per hundred.*

$$65\% = \frac{65}{100} = 0.65$$

$$4\% = \frac{4}{100} = 0.04$$

$$125\% = \frac{125}{100} = 1.25$$

In this lesson, you will learn to solve percent problems by using equations. Recall that to change a percent to a decimal, you drop the percent sign and move the decimal point two places to the left.

*Example 1*   **135% of 185 is what number?**

Let $n$ = the number
135% of 185 is $n$.

$$\begin{array}{ccc} \downarrow & \downarrow & \downarrow \\ 1.35 \cdot & 185 & = n \end{array}$$ ◀ *Write an equation.*

$$249.75 = n$$

**Thus,** 135% of 185 is 249.75.

$$\begin{array}{r} 185 \\ \times\ 1.35 \\ \hline 925 \\ 555 \\ 185 \\ \hline 249.75 \end{array}$$

To solve percent problems:
**1.** Translate the English sentence into an equation.
**2.** Then solve the equation.

*Example 2*   **7 is what percent of 9?**

Let $x$ = the percent written as a decimal.
7 is what percent of 9?

$$7 = x \cdot 9$$

$$\frac{7}{9} = x$$

$$0.77\frac{7}{9} = x$$ ◀ *Change $\frac{7}{9}$ to a decimal.*

$$0.77\frac{7}{9} \text{ means } 77\frac{7}{9}\%.$$

**Thus,** 7 is $77\frac{7}{9}\%$ of 9.

$$\begin{array}{r} 0.77 \\ 9\overline{)7.00} \\ \underline{6\ 3} \\ 70 \\ \underline{63} \\ 7 \end{array}$$

The problem in Example 3 leads to a decimal equation. You will use the technique of multiplying each side of the equation by the LCM.

## Example 3   60% of what number is 300?

Let $n$ = the number
60% of $n$ is 300?
$$\downarrow \quad \downarrow \quad \downarrow$$
| $60\% \cdot n = 300$ | ◀ *Translate the English sentence into an equation.* |
| $0.60n = 300$ | ◀ *60% of n means 0.60 · n, or 0.60n* |
| $100 \cdot 0.60n = 100 \cdot 300$ | ◀ *Multiply each side by the LCM, 100.* |
| $60n = 30{,}000$ | |
| $n = 500$ | ◀ *30,000 ÷ 60 = 500* |

**Thus,** 60% of 500 is 300.

An interesting application of percent is illustrated in Example 4. The four basic steps of problem solving help you to organize your work in finding the solution. The four steps are particularly helpful in solving more complex problems.

## Example 4   A basketball team won 6 games and lost 2 games. What percent of the games played did the team win?

READ ▶   You are told that the team won 6 games and lost 2, so the team played 6 + 2, or 8 games. You are asked to find what percent of the games played was won.

PLAN ▶   Let $x$ = the percent of the games won written as a decimal.

What percent of games played was won?
$$\downarrow \qquad \downarrow \qquad \downarrow \qquad \downarrow \quad \downarrow$$
$$x \qquad \cdot \qquad 8 \quad = \quad 6 \quad ◀ \text{ } \textit{Write an equation.}$$

SOLVE ▶
$$8x = 6$$
$$x = \frac{6}{8}$$
$$x = 0.75 \qquad\qquad ◀ \quad 8\overline{)6.00}^{\,0.75}$$

0.75 means 75%

INTERPRET ▶

| 75% of games played | games won |
|---|---|
| 75% of 8 | 6 |
| 0.75 · 8 | |
| 6 | |

$$6 = 6$$
True

**Thus,** the team won 75% of the games played.

A business application of percent involves percent of profit based on the cost of an item sold by a merchant. The next example shows an application of percent to a basic equation of business:

$$\text{cost} + \text{profit} = \text{selling price}$$
$$c + p = s$$

*Example 5*

**A store manager lists the selling price of a television set at $168. The profit is 20% of the cost. Find the cost.**

Let $c$ = the cost

$$\begin{array}{ccc} \text{cost} + & \text{profit} & = \text{selling price} \\ \downarrow & \downarrow & \end{array}$$
$$c + (20\% \text{ of } c) = 168$$
$$c + 0.20c = 168$$
$$100(1c + 0.20c) = 100 \cdot 168$$
$$100c + 20c = 16{,}800$$
$$120c = 16{,}800$$
$$c = 140$$

**Thus,** the cost of the television set is $140.

# Written Exercises

**Solve each problem.**

Ⓐ
1. 32% of 78 is what number? **24.96**
2. 43% of 6,598 is what number? **2,837.14**
3. 5 is what percent of 9? **55$\frac{5}{9}$%**
4. 8% of what number is 200? **2,500**
5. 42% of 180 is what number? **75.6**
6. 12 is what percent of 60? **20%**
7. 40% of what number is 20? **50**
8. 14% of 28 is what number? **3.92**
9. 60 is what percent of 80? **75%**
10. What percent of 12 is 8? **66$\frac{2}{3}$%**
11. 8 is 5% of what number? **160**
12. 15 is what percent of 40? **37$\frac{1}{2}$%**

Ⓑ
13. A basketball team won 5 games and lost 3. What percent of the games played did the team win? **62$\frac{1}{2}$%**
14. A parking meter contained 22 quarters and 18 dimes. The number of dimes is what percent of the number of quarters? **81$\frac{9}{11}$%**
15. The selling price of a turntable is $156. The profit is 30% of the cost. Find the cost. **$120**
16. A camera's selling price is $260. Find the cost if the profit is 30% of the cost. **$200**
17. Last year, the weekly cost of food for a family of four was $150. This year the weekly cost is $200. The increase is what percent of last year's cost? **33$\frac{1}{3}$%**
18. Last year's football team won 20 games. This season's team had 15 victories. The decrease in wins is what percent of last year's wins? **25%**

Ⓒ
19. In tryouts for a school play, 30% of the participants were eliminated the first week. Of those remaining, 20% were chosen. If 7 were chosen, how many tried out? **50**    $.20(.70p) = 7$
20. The selling price of a television set allows for a profit of 30% on the cost. Due to inflation, the selling price must be increased by 25%. If the new selling price is $195, find the cost of the television set. **$120**

# PERIMETER FORMULAS

*Objective*   **To solve problems about perimeter**

Sometimes drawing a picture will help to solve a word problem.

Recall that *perimeter* means the distance around. A formula for the perimeter of a rectangle with length *l* and width *w* is:

$$p = l + w + l + w, \text{ or}$$
$$p = 2l + 2w$$

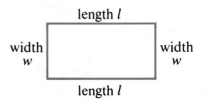

length *l*

width *w*          width *w*

length *l*

*Example 1*   **The length of a rectangle is 3 m more than 5 times the width. The perimeter is 126 m. Find the length and the width.**

READ ▶   The first sentence gives the relationship between the length and the width of the rectangle. The second sentence gives the perimeter. You are asked to find the length and the width of the rectangle.

PLAN ▶   Represent the length and the width of the rectangle. Draw a rectangle.

Let $a$ = the width   ◀ *Length is compared to width.*
$5a + 3$ = the length   ◀ *The length is 3 m more than 5a.*
Use the perimeter formula above to write an equation:

$$2l + 2w = p$$

SOLVE ▶   $2(5a + 3) + 2a = 126$   ◀ *The perimeter is 126.*

$$10a + 6 + 2a = 126$$
$$12a + 6 = 126$$
$$12a = 120$$
$$a = 10$$

5a+3

a          a

5a+3

The width, $a$, is 10.   ◀ *Use the representation to find the length and the width.*
The length, $5a + 3$, is $5 \cdot 10 + 3$, or 53.

INTERPRET ▶

| Length is 3 more than 5 times width. | |
|---|---|
| 53 | $5 \cdot 10 + 3$ |
| | $50 + 3$ |
| | 53 |
| $53 = 53$ | |
| True | |

| Perimeter is 126. | |
|---|---|
| $2 \cdot 53 + 2 \cdot 10$ | 126 |
| $106 + 20$ | |
| 126 | |
| $126 = 126$ | |
| True | |

**Thus,** the length is 53 m and the width is 10 m.

The problems in Examples 3 and 4 involve perimeters of triangles. If the lengths of the three sides of a triangle are $a$, $b$, and $c$, a formula for the perimeter of the triangle is:

$$p = a + b + c.$$

A triangle is **isosceles** if two of its sides are congruent (have the same length). These sides are called *legs,* and the third side is called the *base*. Each leg of the isosceles triangle at the right is 8 m long, and the base is 6 m long.

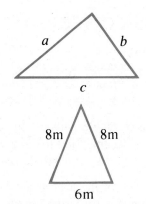

ISOSCELES TRIANGLE

*Example 2*    **The length of each leg of an isosceles triangle is 3 times the length of the base. The perimeter of the triangle is 28 m. Find the length of each side.**

Let $y$ = the length of the base    ◀ *The legs are compared to the base.*
   $3y$ = the length of each leg    ◀ *the congruent sides*
Use the perimeter formula above to write an equation:
$a + b + c = p$
$3y + 3y + y = 28$
$7y = 28$
$y = 4$
The length of the base, $y$, is 4.
The length of each leg, $3y$, is 12.
Check on your own.
**Thus,** the lengths of the sides are 4 m, 12 m, and 12 m.

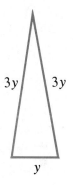

*Example 3*    **The first side of a triangle is 2 cm longer than the second side. The third side is 5 cm shorter than twice the second side. The perimeter is 49 cm. Find the length of each side.**

Let $s$ = the length of the second side    ◀ *The 1st side is compared to the 2nd.*
$s + 2$ = the length of the first side    ◀ *2 cm longer than s*
$2s - 5$ = the length of the third side    ◀ *5 cm shorter than 2s*
Equation: $s + s + 2 + 2s - 5 = 49$    ◀ *The perimeter is 49 cm.*
$4s - 3 = 49$
$4s = 52$
$s = 13$

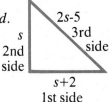

The length of the second side, $s$, is 13.
The length of the first side, $s + 2$, is $13 + 2$, or 15.
The length of the third side, $2s - 5$, is $2 \cdot 13 - 5$, or 21.
Check on your own.
**Thus,** the lengths of the sides are 15 cm, 13 cm, and 21 cm.

Perimeter Formulas

## Written Exercises

**Solve each problem.**

(A) 1. The length of a rectangle is 5 cm more than the width. The perimeter is 38 cm. Find the length and the width. **length, 12 cm; width, 7 cm**

2. The length of a rectangle is 8 m more than 6 times the width. The perimeter is 156 m. Find the length and the width. **length, 68 m; width, 10 m**

3. The length of a rectangle is twice the width. The perimeter is 42 cm. Find the length of each side. **length, 14 cm; width, 7 cm**

4. The length of a rectangle is 2 km less than 3 times the width. The perimeter is 68 km. How long is each side? **length, 25 km; width, 9 km**

5. The length of each of the legs of an isosceles triangle is 5 times the length of the base. The perimeter of the triangle is 33 m. Find the length of each side. **3 m, 15 m, 15 m**

6. The length of each of the legs of an isosceles triangle is 3 km longer than the base. The perimeter of the triangle is 24 km. Find the length of each side. **6 km, 9 km, 9 km**

7. The first side of a triangle is 3 cm longer than the second side. The third side is 4 cm shorter than twice the second side. The perimeter is 31 cm. How long is each side?

8. The perimeter of a triangle is 47 km. The first side is 5 km shorter than twice the second. The third side is 2 km longer than the first. Find the length of each side. **17 km, 11 km, 19 km**

9. One side of a triangle is twice as long as a second side. The remaining side is 3 m longer than the second side. The perimeter is 39 m. Find the length of each side.

10. One side of a triangle is 3 cm shorter than a second side. The remaining side is 5 cm longer than the second side. The perimeter is 38 cm. Find the length of each side. **9 cm, 12 cm, 17 cm**

**7. 11 cm, 8 cm, 12 cm   9. 18 m, 9 m, 12 m**

(B) 11. A square and an equilateral triangle (3 sides of equal length) have the same perimeter. Each side of the square is 15 m. Find the length of each side of the triangle. **20 m**

12. A rectangle and an equilateral triangle have the same perimeter. The length of the rectangle is twice its width. Each side of the triangle is 45 m. Find the length of each side of the rectangle. **length, 45 m; width, 22.5 m**

13. Each side of an equilateral triangle is 4 m shorter than twice the length of each side of a square. Their perimeters are the same. How long is each side of the triangle? **8 m**

14. The base of an isosceles triangle is half as long as a side of a square. Each of the legs of the triangle is twice as long as a side of the square. How long is each side of the triangle if the sum of the perimeters is 34 m? **8 m, 8 m, 2 m**

(C) 15. A rectangle's length is 3 cm less than twice its width. If the length is decreased by 2 cm and the width is decreased by 1 cm, the perimeter will be 24 cm. Find the dimensions of the original rectangle. **length, 9 cm; width, 6 cm**

16. A rectangular field is 6 times as long as it is wide. If the length is decreased by 4 m and the width is increased by 2 m, the perimeter will be 52 m. Find the dimensions of the original field. **length, 24 m; width, 4 m**

# APPLICATIONS

$$\boxed{\text{Read}} \rightarrow \boxed{\text{Plan}} \rightarrow \boxed{\text{Solve}} \rightarrow \boxed{\text{Interpret}}$$

*Commission* is often paid to people who work in sales-oriented businesses.

*Example*　**Jill earns $140 a week plus a commission of 15% on all television-set sales. What must be her sales, for the week, to earn a total of $260?**

Let $s$ = the amount of sales needed

Think: regular salary plus commission equals total earned.

$$140 \quad + \quad 15\% \text{ of } s \quad = \quad 260$$

$$140 + 0.15s = 260 \qquad \blacktriangleleft \textit{15\% of s means 0.15s}$$
$$100(140 + 0.15s) = 100 \cdot 260 \qquad \blacktriangleleft \textit{Multiply each side by the LCM, 100.}$$
$$100 \cdot 140 + 100 \cdot 0.15s = 26{,}000$$
$$14{,}000 + 15s = 26{,}000$$
$$15s = 12{,}000$$
$$s = 800$$

**Thus,** Jill must sell $800 worth of television sets.

Solve each problem.

1. Rufus works on a commission basis only. His commission is 20% of his sales. How much must he sell to earn $240? **$1,200**

2. Tony is paid only a commission of 30% of all sales. What must be his total sales so that he will earn $270? **$900**

3. Mary earns $175 a week plus a 2% commission on each car she sells. What must be her total sales to bring her total earnings to $355? **$9,000**

4. Marie is paid $165 a week plus a 10% commission on all camera sales. How much must her sales be to earn a total of $215? **$500**

Applications

# PROBLEM SOLVING: MIXED TYPES

You have studied a variety of word problems in this chapter and the previous one. You have also seen that the basic four-step strategy can be applied to each of the types. The four steps are included here. Review them and use them to solve the problems that follow.

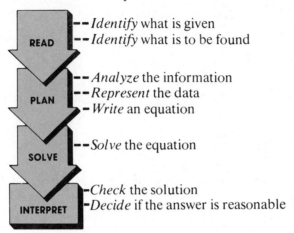

READ
--*Identify* what is given
--*Identify* what is to be found

PLAN
--*Analyze* the information
--*Represent* the data
-*Write* an equation

SOLVE
--*Solve* the equation

INTERPRET
-*Check* the solution
-*Decide* if the answer is reasonable

**Solve each problem.**

1. Jane earns $41 more a week than twice as much as Bill does. Together they earn $140. How much does each earn? **Jane, $107; Bill, $33**

2. The length of a rectangle is twice the width. The perimeter is 24 m. Find the length and the width. **8 m, 4 m**

3. Separate 25 into two parts so that the second part is 5 less than twice the first part. Find each part. **10, 15**

4. The selling price of a television set is $360. The profit is 20% of the cost. Find the cost of the television set. **$300**

5. Eight times the sum of a number and 6 is the same as 3 less than twice the number. Find the number.   $-8\frac{1}{2}$

6. The length of a rectangle is 7 yd more than 5 times the width. The perimeter is 150 yd. Find the length and the width. **$63\frac{2}{3}$ yd, $11\frac{1}{3}$ yd**

7. Two less than 6 times Harriet's age is the same as 10 more than twice her age. Find her age. **3 y**      9. **14 cm, 27 cm, 29 cm**

8. The selling price of a television set is $125 more than twice the cost. Find the cost if the selling price is $325. **$100**

9. The first side of a triangle is 40 cm shorter than twice the second. The third side is 2 cm longer than the second. The perimeter is 70 cm. Find the length of each side.

10. John, Brian, and Hank are brothers. The sum of their ages is 39 years. John's age is 4 years less than Brian's. Hank's age is 7 years more than Brian's. How old is each?

11. Tina is paid $150 a week plus a commission of 5% on all sales. Find the total sales needed to give her a salary of $230. **$1,600**

12. Ms. O'Neil can rent a car for a week for $75 plus 25¢ a mile. How far can she travel on a car-rental budget of $200? **500 mi**
**10. John, 8 y, Brian, 12 y, Hank, 19 y**

13. Each leg of an isosceles triangle is 4 m longer than the base. The perimeter of the triangle is 26 m. Find the length of each side. **6 m, 10 m, 10 m**

14. The larger of two numbers is 6 more than one half of the smaller. The sum of the two numbers is 18. Find the two numbers. **8, 10**

15. A team won 8 games and lost 2 games. The number of games won is what percent of the games played? **80%**

## Solving Equations

Equations of the type $x + 5 = 8$ or $x - 9 = 12$ can be represented by the general equation $x + b = c$.

One computer program can solve any equation of this type. Each time it is given values for $b$ and $c$, it can print the solution of $x + b = c$. However, the computer must be told how to solve the equation.

The way to do this is suggested by the method used to solve the equation $x + 5 = 8$.

$$x + 5 \qquad = 8 \qquad\qquad x + b \qquad = c$$
$$x + 5 - 5 = 8 - 5 \qquad\qquad x + b - b = c - b$$
$$x \qquad\quad = 3 \qquad\qquad x \qquad\quad = c - b$$

The program below will solve equations of the type $x + b = c$.

```
10   FOR I = 1 TO 2
20   INPUT "WHAT IS B? ";B
30   INPUT "WHAT IS C? ";C
40   LET X = C - B
50   PRINT "THE SOLUTION OF X +
     "B" = "C" IS "X"."
60   NEXT I
70   END
```

Line 10 sets up the program to solve 2 equations.

Line 40 contains the formula for solving the equation.

See the Computer Section beginning on page 488 for more information.

### Exercises    **For Ex. 3 and 6, see TE Answer Section.**

**Enter and RUN the program above to solve the following equations.**
1. $x + 5 = 8$ **3**
2. $x - 219.73 = -54.086$ **165.644**

3. Modify the program above to solve equations to the type $4x = 12$. The general equation is $ax = c$. The solution of $ax = c$ is $x = ?$ Enter and RUN the program to solve $4x = 12$ and $8x = 4$. **3, .5**

**Consider the equations $3x - 4 = 16$ and $2x + 8 = 14$.**
4. Solve $3x - 4 = 16$ and $2x + 8 = 14$ without a computer. $\mathbf{6\frac{2}{3}, 3}$
5. Using the pattern for solving equations of this type, write the general form of this type of equation and solve it for $x$. $\mathbf{ax + b = c; x = \dfrac{c - b}{a}}$

6. Using the general solution for $x$ you found in Exercise 5, modify the program above to solve this general type of equation.
7. Enter and RUN your modified program to solve $3x - 4 = 16$ and $2x + 8 = 14$. **6.66, 3**

# CHAPTER FOUR REVIEW

**Vocabulary**

base [4.7]          percent [4.6]
congruent [4.7]     sum [4.2]
isosceles [4.7]     selling price [4.6]
least common multiple [4.4]

Solve each equation. Check the solution. [4.1]
 1. $-6 + 4a = -12 + 6a$ **3**
 2. $13p + 5 = -3 + 9p$ **−2**

Solve each equation. [4.1]
 3. $7m + 5 - 5m = -2 + m - 9$ **−16**
 4. $6y - 4 - 7y = 8 + y - 9$ **$-\frac{3}{2}$**

Solve each equation. Check the solution. [4.2]
 5. $2(9 + y) = 30$ **6**
 6. $5b + 3(b - 4) = 2(b + 3)$ **3**

Solve each equation. [4.2]
 7. $7b + 2(1 - b) = 8b - 4$ **2**
 8. $16 - (3 - 4y) = 2(y + 2)$ **$-\frac{9}{2}$**

Solve each problem. [4.3]
 9. Mona's age is 4 less than twice Bill's age. The sum of their ages is 38. Find each of their ages. **Bill, 14 y; Mona, 24 y**
10. Separate 23 into two parts so that the first part has 5 more than twice the second part. Find each part. **6, 17**
11. The second of three numbers is twice the first. The third is 6 less than the second. If the second is decreased by 3 times the third, the result is 50. Find the numbers.
   **−8, −16, −22**

Solve each equation. (Ex. 12–23)
12. $\frac{2}{7}x = 4$ [4.4] **14**

13. $6 + \frac{3}{5}a = 12$ **10**

14. $\frac{2}{3}a = \frac{5}{7}$ **$\frac{15}{14}$**

15. $8 + \frac{1}{3}x = \frac{1}{2}x$ **48**

16. $8 = \frac{2}{3}x + 4$ **6**

17. $\frac{2}{3}a + 9 = 11$ **3**

18. $\frac{1}{3}x - \frac{1}{6}x = \frac{5}{2}$ **15**

19. $\frac{1}{2}b - 4 = \frac{1}{5}b + 2$ **20**

20. $0.04m = 0.8$ [4.5] **20**
21. $0.03x - 3.9 = 0.66$ **152**
22. $0.024a - 8 = 0.224a + 2$ **−50**
23. $1.25k - 3(0.25k + 6) = 12$ **60**

Solve each problem.
24. 20% of 45 is what number? [4.6] **9**
25. 6 is what percent of 30? **20%**
26. 10% of what number is 20? **200**
27. A team won 12 games and lost 8 games. What percent of the games played were won? **60%**
28. The selling price of a television set is $260. The profit is 30% of the cost. Find the cost. **$200**
29. The length of a rectangle is twice the width. The perimeter is 45 cm. Find the length and the width. [4.7]
30. The first side of a triangle is 3 km longer than the second. The third side is twice as long as the second. The perimeter is 31 km. Find the length of each side.
   **10 km, 7 km, 14 km**

Solve.
★31. $7 - m = 3m - [2 + 5(2m - 7)]$ [4.2] **$\frac{13}{3}$**
32. $-8a - [2(11 - 2a) + 4] = -9a$ **$\frac{26}{5}$**
33. The second of three numbers is 4 less than 5 times the first. The third is 11 less than 4 times the first. If twice the first is increased by the third, the result is 25. Find the three numbers. [4.3] **6, 26, 13**

**29. length, 15 cm; width, $7\frac{1}{2}$ cm.**

# CHAPTER FOUR TEST

(A) Exercises are 1–3, 5–7, 9–16.　　(C) Exercises are starred.　　The rest are (B) Exercises.

Solve each equation. Check the solution.

1. $\frac{3}{5}m = 9$ **15**

2. $7 + \frac{2}{3}p = 13$ **9**

3. $8r + 2 = 6r + 10$ **4**

4. $4b + 2(b - 4) = 8(b + 3)$ **−16**

Solve each equation.

5. $0.6 = 0.03m$ **20**

6. $\frac{1}{3}y - 2 = \frac{1}{2}y + 4$ **−36**

7. $\frac{3}{7}p = \frac{2}{5}$ **$\frac{14}{15}$**

8. $8y - 3 - 9y = 5 + y - 12$ **2**

9. $\frac{1}{5}x - \frac{1}{10}x = \frac{7}{2}$ **35**

10. $8b + 3(1 - b) = 6b - 4$ **7**

11. $7.2 - 0.5m = 32.8 - 3m$ **10.24**

Solve each problem.

12. Tom's age is 6 less than 3 times Myra's age. The sum of their ages is 26. Find each of their ages. **Myra, 8 y; Tom, 18 y**

13. The length of a rectangle is 5 m more than twice the width. The perimeter is 46 m. Find the length and the width. **length, 17 m; width, 6 m**

14. Separate 45 into two parts such that the larger part is twice the smaller. Find each part. **15, 30**

15. 3 is what percent of 12? **25%**

16. The first side of a triangle is 2 km longer than the second. The third side is twice as long as the second. The perimeter is 34 km. Find the length of each side. **10 km, 8 km, 16 km**

17. The selling price of a camera is $48. The profit is 20% of the cost. Find the cost. **$40**

18. Eight times the sum of a number and 2 is 40. Find the number. **3**

★ 19. The second of three numbers is 5 less than 4 times the first. The third is 9 less than twice the first. If 4 times the third is decreased by the first, the result is 6. Find the three numbers. **6, 19, 3**

Solve.

★ 20. $-6a - [4 - (3 - 2a)] = 5a + 6$ **$-\frac{7}{13}$**

# CUMULATIVE REVIEW

## (Chapters 1–4)

Each exercise has five choices of answers. Choose the one best answer.

1. Simplify $5x + 3 + 4x$, then evaluate if $x = 3$.
   (A) $12x$, 36    (B) $20x + 3$, 63 (C) $23x$, 69
   (D) $9x + 3$, 30 (E) None of these        D

2. Compute $\dfrac{8 \cdot 2 + 4}{5 - 3}$.
   (A) 10    (B) 24    (C) 18    (D) 1  A
   (E) None of these

3. Solve $3x - 5 = 7$.
   (A) $\frac{2}{3}$    (B) 4    (C) 9    (D) $\frac{3}{12}$  B
   (E) None of these

4. Rewrite $a^5$ without exponents.
   (A) $5a$ (B) $a + 5$ (C) $a \cdot a \cdot a \cdot a \cdot a$ C
   (D) $a(5)$    (E) None of these

5. Simplify $3(2x + 5) + 4(3x + 6)$.
   (A) 102 (B) $18x + 11$ (C) $18x + 21$ D
   (D) $18x + 39$    (E) None of these

6. Divide $(24 \div 4) \div -2$.
   (A) 12    (B) $-8$    (C) 3    (D) $\frac{1}{3}$ E
   (E) None of these

7. Subtract $2a - 3$ from $-4a + 2$.        E
   (A) $-6a - 1$ (B) $-2a - 1$ (C) $6a - 5$
   (D) $-1a$    (E) None of these

8. Write an algebraic representation for 7 less than twice a number $n$.  B
   (A) $7 - 2n$    (B) $2n - 7$    (C) $2n + 7$
   (D) $5n$    (E) None of these

9. Find the difference $6 - 8$. Then give its opposite.
   (A) $-2$    (B) 14    (C) 2    (D) $-14$  C
   (E) None of these

What words must be used to fill in the blanks to make a true statement?

10. $8 \cdot 4 = 4 \cdot 8$ commutative, multiplication
    by the _____ property of _____.

11. $6 \cdot (3 + 4) = 6 \cdot 3 + 6 \cdot 4$ distributive
    by the _____ property.

12. $7 + (-7) = 0$ additive inverse
    by the property of _____.

13. $a + 0 = a$ addition, zero
    by the _____ property of _____.

14. $(7a + 2a) + 3a = 7a + (2a + 3a)$
    by the _____ property of _____.

15. $-(x - 8) = -1(x - 8)$ multiplication, $-1$
    by the _____ property of _____.

14. associative, addition

Compute.

16. $7 \cdot 3 + 4$  25

17. $(-8)(-4)$  32

18. $\dfrac{+24}{-8}$  $-3$          19. $-\dfrac{2}{3} - \dfrac{8}{9}$  $-\dfrac{14}{9}$

20. $-9 + 2$  $-7$

21. $-4\dfrac{3}{4} + 6$  $1\dfrac{1}{4}$

22. $4 - 5 - 7$  $-8$

23. $16 - 2.4$  13.6

24. $-7 - 4(-2)$  1

25. $-4 \cdot 3 + 12$  0

26. $\dfrac{7 \cdot 1 + 6}{4 \cdot 2 - 3}$  $\dfrac{13}{5}$

27. $|8 - 12|$  4

Evaluate for the given value(s) of the variable(s).

28. $x^2$ if $x = -\dfrac{4}{5}$  $\dfrac{16}{25}$

29. $3a^5$ if $a = 2$  96

30. $\dfrac{x^2 - 4x}{2x}$ if $x = 2$  $-1$

31. $4ab^2c^3$ if $a = 2$, $b = 3$ and $c = -1$  $-72$

32. $-8 - 4x$ if $x = -3$  4

Simplify.

**33.** $-9a - 3a + 5a - 4a$  $-11a$

**34.** $-8x - 3(6 - 2x)$  $-2x - 18$

**35.** $3\frac{1}{2}y - (7 + \frac{1}{2}y)$  $3y - 7$

Simplify. Then evaluate if $a = -3$, $b = -4$, $c = -1$

**36.** $7a + 2b - c + 4b - 3a - 5c$  $4a + 66 - 6c, -30$

**37.** $4(2 - a) - (6 - a) - 5a$  $-8a + 2, 26$

**38.** $3a - (2b - a) - b$  $4a - 3b, 0$

**39.** Subtract $0.04 - y$ from $4y + 8.0$  $5y + 7.96$

**40.** Replace the ? with the correct inequality symbol to make $-5 \underline{\phantom{?}?\phantom{?}} 3$ a true statement.  $<$

Solve.

**41.** $\frac{2}{3}x = 4$  $6$

**42.** $6a + 2 = 4a - 8$  $-5$

**43.** $7y - 3 - 10y = 4 + y - 8$  $\frac{1}{4}$

**44.** $0.8 = 0.02m$  $40$

**45.** $\frac{1}{5}x - 2 = \frac{1}{3}x + 4$  $-45$

**46.** $4b - (3 - b) = 7b - 15$  $6$

**47.** $0.24y + 2.6 = 0.06(1 - 0.2y)$  $-10\frac{5}{63}$

**48.** Simplify.
$8r - [7 - (5 - r)] - 2r$  $5r - 2$

**49.** Solve.
$|-4 - |-12 \div 2|| - 2x = 8 - 4x$  $-1$

**50.** Solve.
$-5a - [3 - (5 - a)] = 4a + 6$  $-\frac{2}{5}$

**51.** Express the volume of the cylinder in terms of $k$
using $V = \pi \cdot r^2 h$  $V = 108\pi k^3$
if $r = 6k$ and $h = 3k$.

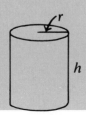

**52.** Find the area of the parallelogram using $A = hb$  $A = 12m^2$
if $h = 3m$ and $b = 4m$.

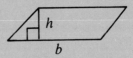

Solve each problem.

**53.** The $600 selling price of a TV is the cost increased by 3 times the cost. Find the cost.  $150

**54.** If Ron's age is increased by 5 years more than twice his age, the result is 35. Find his age.  10 yr.

**55.** Separate 25 into two parts. The larger part is 4 times the smaller part. Find each part.  20, 5

**56.** The length of a rectangle is 3 m less than twice the width. The perimeter is 30 m. Find the length and width.  9 m, 6 m

**57.** The first side of a triangle is 4 km more than the second. The third side is twice the second side. The perimeter is 28 km. Find the length of each side.
10 km, 6 km, 12 km

**58.** 4 is what percent of 20?  20%

**59.** The selling price of a watch is $65. The profit is 30% of the cost. Find the cost.
$50

**60.** The second of three numbers is 3 less than 4 times the first. The third is 8 more than twice the first. If 5 times the second is decreased by the third, the result is 67. Find the three numbers.  5, 17, 18

Determine whether each number is rational, irrational, or neither.

**61.** $\dfrac{-7}{0}$  neither  **62.** 2.78  rational

**63.** 34.676786788 . . .  irrational

Choose the one best answer to each question or problem.

**1.** One year, a factory granted raises at three different times. If 62% of the workers
**D** received raises in the first and second periods and 10% of these did not receive a raise in the third period, what percent of the workers received raises in all three periods?
(A) 6.2%   (B) 38%   (C) 52%
(D) 55.8%   (E) 90%

**2.** If $3x - 12 = 2x - 10$, then $x + 3 = $ ?
**A**   (A) 5  (B) 4  (C) $-3$  (D) $-4$  (E) $-5$

**3.** $x + 9$ is how much more than 5?
**C**   (A) 4  (B) $x + 14$  (C) $x + 4$  (D) $x$  (E) 14

**4.** $x$ is what percent of 40% of 20?
**C**   (A) $\dfrac{x}{800}$  (B) $\dfrac{8}{100}x$  (C) $\dfrac{100}{8}x$  (D) $\dfrac{800}{x}$
(E) none of these

**5.** If rain is falling at the rate of 3 cm/h, how much rain will fall in 10 min?
**A**   (A) 0.5 cm   (B) 18 cm   (C) 30 cm
(D) 180 cm   (E) none of these

**6.** How many 1-cm squares are needed to make a border around the shaded square, which has a 6-cm side?
**E**

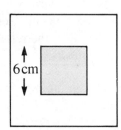

(A) 24   (B) 28   (C) 36   (D) 40
(E) cannot tell from the given information

**7.** If $y = 4$, then $y^2 + 2 = $ ?
**B**   (A) 32   (B) 18   (C) 10   (D) 8   (E) 4

**8.** Which of the following is greater than $\dfrac{1}{4}$?
**D**   (A) 0.04   (B) $\left(\dfrac{1}{4}\right)^2$   (C) $\dfrac{1}{8}$   (D) $\dfrac{1}{0.04}$
(E) none of these

**9.** What is the total maximum weight of 12 eggs if 5 of them weigh 12 to 20 oz each
**A** and the others weigh 25 to 30 oz each?
(A) 310 oz   (B) 360 oz   (C) 460 oz
(D) 500 oz   (E) none of these

**10.** A merchant paid $24 for an article. The merchant wants to place a price tag on it
**A** so that a 20% discount on the marked price can be advertised and still bring in a profit of 25% of the cost. What price should be marked on the tag?
(A) $37.50   (B) $30   (C) $27.60
(D) $25.20   (E) none of these

**11.** $PQRS$ is a square.
$VR = 3$ m
**D** $PW = 7$ m
The area of the shaded portion is 39 m$^2$.
The area of $TUVS$ is ?

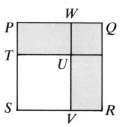

(A) 18 m$^2$   (B) 49 m$^2$   (C) 60 m$^2$
(D) 61 m$^2$   (E) 139 m$^2$

# 5 INEQUALITIES

## CAREER
## *Banking*

If you like to work with people, you may want to consider a career in the field of banking. Many banks have specialists who assist customers with various kinds of financial planning. A loan specialist will help prospective home buyers determine how much family income must be to afford a given monthly mortgage and yearly property taxes.

## *Project*

The Bordens want to buy a home that is advertised for $90,000. Their family income is $36,000 a year, and they have saved enough to make a 15% down payment. According to current mortgage and tax rates for your community, can the Bordens, who have two teen-aged children, afford the home?

Use the following procedure to help you formulate the problem.

Analyze the situation.
Focus on the question to be answered.
Identify what information is needed.
Propose the problem to be solved.
Represent the data.

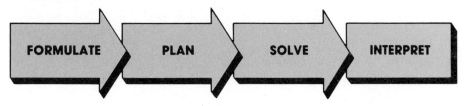

FORMULATE → PLAN → SOLVE → INTERPRET

# SETS AND SOLUTION SETS

*Objectives*
**To identify and describe finite and infinite sets**
**To find and graph the solution sets of equations**

A *set* is a collection of objects. The objects are called *members* or *elements* of the set. Braces are used to enclose the listing of the elements of a set.

The set consisting of the elements 4, 5, and 9 can be written in different ways: $A = \{4, 5, 9\}$, $A = \{9, 5, 4\}$, $A = \{5, 4, 9\}$. The order in which the elements are listed does not make a difference. When two sets have the same exact members, they are called *equal sets*. It is possible to count the members of set $A$. Such a set is called a *finite set*.

Consider the set $P$ of all positive integers. This can be written as $P = \{1, 2, 3, 4, \ldots\}$. The elements of this set continue on endlessly, so they cannot be counted. Such a set is called an *infinite set*.

*Example 1*
**Tell whether each set is finite or infinite. Then describe the set in words.**

| | |
|---|---|
| $A = \{5, 6, 7\}$ | $B = \{2, 4, 6, 8, \ldots\}$ |
| finite | infinite |
| $A$ is the set of integers between 4 and 8. It has three members. | $B$ is the set of positive even integers. |

Sometimes the members of a set may be graphed on a number line. This is shown in Example 2.

*Example 2*
$C = \{\text{integers between} -4 \text{ and } 1\}$. **List the elements of $C$. Then graph set $C$ on a number line.**

Set $C = \{-3, -2, -1, 0\}$  ◀ *−4 and 1 are not included.*

To graph $C$, place a dot at each integer between −4 and 1.

It is possible for a set to contain only one member. Such a set is shown in the next example.

*Example 3*    **Find the set of all solutions of 6x + 10 = 4x + 19 and then graph it.**

$$6x + 10 = 4x + 19$$
$$2x + 10 = 19 \qquad \blacktriangleleft \; Add -4x \; to \; each \; side.$$
$$2x = 9 \qquad \blacktriangleleft \; Add -10 \; to \; each \; side.$$
$$x = \frac{9}{2} \; or \; 4\frac{1}{2} \qquad \blacktriangleleft \; Divide \; each \; side \; by \; 2.$$

**Thus,** $\left\{4\frac{1}{2}\right\}$ is the solution set, and its graph is

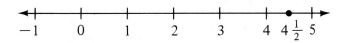

The set of numbers that the variable may represent is called the *replacement set*. Unless otherwise stated, our replacement set is the set of all real numbers.

**Definition:**
**Solution set**

The **solution set** of an open sentence is the set of all numbers that make the sentence true for the given replacement set.

Sometimes a set may contain no members. Such is the case in Example 4.

*Example 4*    **A = {integers between 7 and 8}. List the members of A.**

There are no integers between 7 and 8.

**Therefore,** A is a set with no members. A is the *empty set*.

**Definition:**
**Empty set**

The **empty set** is the set with no members. The symbol $\varnothing$ represents the empty set.

When solving an equation leads to a false statement, the solution set is the empty set, as Example 5 illustrates.

*Example 5*    **Find the solution set of x + 6 = x + 3.**

$$1x + 6 = 1x + 3 \qquad \blacktriangleleft \; 1x = x$$
$$6 = 3 \qquad \blacktriangleleft \; Add -1x \; to \; each \; side.$$

But this is false, because 6 cannot equal 3.
**Thus,** the solution set of x + 6 = x + 3 is the empty set.

Sets and Solution Sets

**119**

*Example 6*   **Find the solution set of $8x + 3 = 3 + 7x + x$. Then graph the solution set.**

$$8x + 3 = 3 + 7x + x$$
$$8x + 3 = 3 + 8x$$
$$3 = 3 \qquad \blacktriangleleft \; \text{Add} -8x \text{ to each side.}$$

This is true, since 3 does equal 3.
**Thus,** the solution set of $8x + 3 = 3 + 7x + x$ is {all real numbers}, which is an infinite set.

The graph of all real numbers is the entire number line.

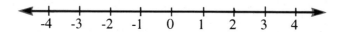

# Reading in Algebra

1. Is the number of people alive in the world today an infinite number? **no**
2. True or false? Infinite means very large. **false**
3. Is it true that every equation has a solution set? **yes (it can be the empty set)**

# Written Exercises

(A) **Tell whether each set is finite or infinite. Then describe the set in words.**   **infinite**
   1. {5, 6, 7, 8, 9} **finite**   2. {0, 1, 2, 3, 4, . . .} **infinite**   3. {1, 3, 5, 7, . . .}
   4. ∅ **finite**   5. {. . . , −3, −2, −1, 0, 1, 2, 3, . . .}   6. {−6, −5, −4, −3}
   **infinite**   **finite**

**Describe each set by listing the elements. Then graph each set.**
   7. {positive integers less than 7} **{1, 2, 3, 4, 5, 6}**   8. {integers between −6 and −3} **{−5, −4}**
   9. {integers greater than 4 and less than 6} **{5}**   10. {integers between 13 and 14} **∅**

**Find the solution set of each equation. Then graph the solution set.**   17. $\left\{\begin{matrix}\text{all}\\\text{numbers}\end{matrix}\right\}$   18. $\left\{\begin{matrix}\text{all}\\\text{numbers}\end{matrix}\right\}$
   11. $7y - 5 = 3y + 11$ **{4}**   12. $2x - 4 = 8 + 2x$ **∅**   13. $7 - 3y = -2y - 8 - y$ **∅**
   14. $7z + 18 = 9z + 4$ **{7}**   15. $5(z + 2) = 5z + 12$ **∅**   16. $3 + 4x = x + 6 + 3x$ **∅**
   17. $3a + 4 - a = 10 + 2a - 6$   18. $x - 5x + 7 = 6 - 4x + 1$   19. $5 - b = 6b + 7 - 7b$ **∅**

(B) 20. $7 - (6 - x) = -2x + 3(x + 4)$ **∅**   21. $5x - (8 - x) = 4x - 10$ **{−1}**
   22. $8p - (5 - p) = 2(p - 3) + 7p + 1$   23. $2g - (8 - g) = 7g - 10$
   24. $-3(2x + 1) + 5 = -(4x - 9) + 5x$ **{−1}**   25. $12 - 3(y + 1) = 1 - 3(y - 2)$ **∅**   $\left\{\frac{1}{2}\right\}$
   26. $7(x + 1) - 1 = 2 - (5 + 2x)$ **{−1}**   27. $6(y + 3) = 22 - 2(2 - 3y)$ **{all numbers}**
(C) 28. $7x - 3[2 - (5 - x)] = 10x - 3$ **{2}**   29. $5y - 2 - [6 - (3 - y)] = 4(y - 3) + 7$
   **{all numbers}**
   30. $1 + (3 - a) = 2 - [4 - (a + 3)]$ $\left\{\frac{3}{2}\right\}$   31. $\frac{2}{3}\left(\frac{1}{2}x - 6\right) = \frac{1}{3}x - |3 - (-1)|$ **{all numbers}**

   22. $\left\{\begin{matrix}\text{all}\\\text{numbers}\end{matrix}\right\}$   **Ex. 7–31, check students' graphs.**

# INEQUALITIES                                                    5.2

*Objective*     **To graph solution sets of inequalities**

Recall that
> $a < b$ means $a$ is less than $b$, and
> $b > a$ means $b$ is greater than $a$.

When you compare the numbers 3 and 7, only one of the following relations can be true:

$$3 < 7 \qquad 3 = 7 \qquad 3 > 7$$
$$\text{true} \qquad \text{false} \qquad \text{false}$$

When you compare two numbers, you assume the following property.

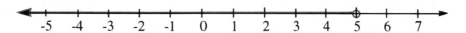

| Axiom of comparison | For all real numbers $a$ and $b$, one and only one of the following is true: |
|---|---|
| | $a < b, \qquad a = b, \qquad$ or $\qquad a > b.$ |

There are many values of $x$ that make $x < 5$ true.

$$4\tfrac{1}{2} < 5 \qquad 4 < 5 \qquad 2 < 5 \qquad 0 < 5 \qquad -3\tfrac{1}{2} < 5$$

You could go on forever listing the values of $x$ that make $x < 5$ true. All numbers less than 5 will make $x < 5$ true. You can show this on a number line. Any point to the left of the point with coordinate 5 has a coordinate less than 5. Draw an arrow to show all such points.

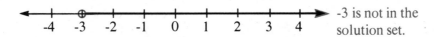

Notice that the circle at 5 is open. Since 5 is not less than 5, 5 is *not* one of the solutions. The solution set of $x < 5$ is an infinite set, {all numbers less than 5}.

*Example 1*     **Graph the solution set of $a > -3$.**

The solution set of $a > -3$ is {all numbers greater than $-3$}.

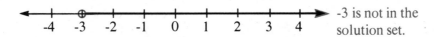     -3 is not in the
          solution set.

The symbol for *is less than or equal to* is $\leq$. Thus, $x \leq 4$ is read as $x$ is less than or equal to 4. Similarly, $x \geq 9$ is read as $x$ is greater than or equal to 9. Use of the symbol $\leq$ is illustrated in Example 2.

*Example 2*    **Graph the solution set of $x \le 4$.**

The solution set of $x \le 4$ is {all numbers less than or equal to 4}.

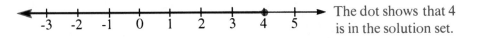

The dot shows that 4 is in the solution set.

It is easier to graph the solution set of an inequality like $-2 \le x$ if you rewrite the inequality with the variable first. Write $-2 \le x$ as $x \ge -2$.

*Example 3*    **Graph the solution set of $-2 \le x$.**

Rewrite $-2 \le x$ as $x \ge -2$.

The solution set is {all numbers greater than or equal to $-2$}.

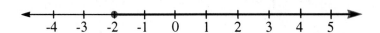

# Written Exercises

**Graph the solution set of each sentence. Ex. 1–37, check students' graphs.**

Ⓐ
1. $b < 4$
2. $x > 1$
3. $x < 8$
4. $a > -2$
5. $p < -1$
6. $x \le -3$
7. $y \ge -5$
8. $3 \le t$
9. $r \ge -1$
10. $-7 \le y$
11. $6 < g$
12. $x \le 0$
13. $a < -2$
14. $w > 0$
15. $y \le -4$
16. $c < -6$
17. $d \le -1$
18. $-8 \ge x$
19. $r \le -10$
20. $d \le -7$
21. $5 \le b$
22. $m \le -5$
23. $9 \ge x$
24. $u \le -12$
25. $0 \le x$

Ⓑ  **Graph the solution set of each sentence.**

*Example*    **Graph the solution set of $y \ne \dfrac{1}{4}$.**

26. $x \ne -\dfrac{1}{2}$
27. $-\dfrac{1}{4} > x$
28. $y < 2\dfrac{1}{2}$
29. $-1\dfrac{1}{2} \le p$

30. $1.5 \ge g$
31. $0 \ne x$
32. $x \ge 3.5$
33. $-1\dfrac{1}{4} \ne a$

Ⓒ  34. $x \not\le 4$
35. $1 - 61 \not> y$
36. $-3.5 \not\ge t$
37. $x \not< -7.1$

# GEOMETRY REVIEW

*Objective*  **To find the measure of vertical angles**

When two lines intersect, two pairs of *vertical angles* are formed.

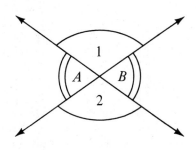

The mathematical symbol for angle is $\angle$.
$\angle A$ and $\angle B$ are *vertical* angles.
$\angle 1$ and $\angle 2$ are *vertical* angles.

Vertical angles have *equal* degree measure.
measure of $\angle A$ = measure of $\angle B$
$$m\angle A = m\angle B$$
Also, $m\angle 1 = m\angle 2$.

*Example*     If $m\angle A = 3x - 40$ and
$m\angle B = x + 80$, find $m\angle A$.

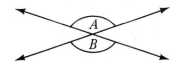

$\angle A$ and $\angle B$ are *vertical* angles.
$m\angle A = m\angle B$   ◀ *Vertical angles have equal measure.*
$3x - 40 = x + 80$   ◀ *Substitute values for $\angle A$ and $\angle B$.*
$2x - 40 = 80$        *Then solve for x.*
$\phantom{2x - 40} 2x = 120$
$\phantom{2x - 40} x = 60$

To find $m\angle A$, substitute 60 for $x$.
$$m\angle A = 3x - 40$$
$$m\angle A = 3 \cdot 60 - 40$$
$$m\angle A = 140$$

# Written Exercises

**Use the diagram shown in the example to do Exercises 1–5.**
 1. $m\angle A = 3x + 20$; $m\angle B = x + 60$   Find $m\angle A$. **80**
 2. $m\angle A = 5x$; $m\angle B = 2x + 30$   Find $m\angle B$. **50**
 3. $m\angle A = 45 - 2x$; $m\angle B = 10 + 5x$   Find $m\angle A$. **35**
 4. $m\angle B = 3(2x - 5)$; $m\angle A = x + 30$   Find $m\angle A$. **39**
 5. $m\angle A = 4x - 10$; $m\angle B = 50 - 2x$   Find $m\angle B$. **30**

 6. One of two vertical angles measures $30°$. The other is $2x + 10$. Find $x$. **10**
 7. The total measure of two vertical angles is $100°$. If the measure of one
    of them is $4x + 8$, find $x$.   **$10\frac{1}{2}$**

Inequalities

# THE INEQUALITY PROPERTIES                    5.3

*Objective*    **To write a true inequality by performing a given operation on each side of the inequality**

You can add the same number to each side of an equation and still get a true equation. Notice what happens when you add the same number to each side of an inequality. Consider the inequality $5 < 8$. Add 2 to each side.

$$5 < 8$$
$$5 + 2 \; ? \; 8 + 2$$
$$7 < 10$$

A true inequality with the same inequality symbol is obtained. The two inequalities, $5 < 8$ and $7 < 10$, are said to have the **same order.** That means that the inequality symbol in *both* inequalities is the *same*, $<$.

*Example 1*    **Write a true inequality by adding the indicated number to each side of the given inequality.**

Add 8 to each side of $6 \geq -3$.

$$6 \geq -3$$
$$6 + 8 \; ? \; -3 + 8 \qquad \textbf{same order}$$
$$14 \geq 5$$

Add $-4$ to each side of $-5 < -2$.

$$-5 < -2$$
$$-5 + (-4) \; ? \; -2 + (-4) \qquad \textbf{same order}$$
$$-9 < -6$$

These are both true inequalities.

The example above illustrates that adding the same number to each side of an inequality does *not* change the order of the inequality.

| **Addition properties of inequalities** | For all numbers $a$, $b$, and $c$, if $a < b$, then $a + c < b + c$, and if $a > b$, then $a + c > b + c$. |
|---|---|

*Example 2*    **Write a true inequality by subtracting the indicated number from each side of the given inequality.**

Subtract 6 from each side of $8 < 10$.

$$8 < 10$$
$$8 - 6 \; ? \; 10 - 6 \qquad \textbf{same order}$$
$$2 < 4$$

Subtract $-3$ from each side of $7 > -2$.

$$7 > -2$$
$$7 - (-3) \; ? \; -2 - (-3) \qquad \textbf{same order}$$
$$10 > 1$$

◀ $-2 - (-3) = -2 + 3 = 1$

Example 2 on the previous page illustrates that subtracting the same number from each side of an inequality does not change the order of the inequality.

<table>
<tr><td>Subtraction properties of inequalities</td><td>For all numbers $a$, $b$, and $c$,<br>if $a < b$, then $a - c < b - c$, and<br>if $a > b$, then $a - c > b - c$.</td></tr>
</table>

If you can *add* the *same number* to (or *subtract* the *same number* from) each side of an inequality, it would seem logical to expect that you can also multiply each side by the same number and still get a true inequality. Notice that you do get a true inequality when you multiply each side of $3 < 8$ by 4.

$$3 < 8$$
$$4 \cdot 3 \ ? \ 4 \cdot 8 \qquad \textbf{same order}$$
$$12 < 32$$

In the next example you will see what happens when you multiply each side of $3 < 8$ by a *negative* number.

*Example 3*  **Write a true inequality by multiplying each side of $3 < 8$ by $-2$.**

$$3 < 8$$
$$-2 \cdot 3 \ ? \ -2 \cdot 8 \qquad \textbf{reversed order}$$
$$-6 > -16$$

Notice that *multiplying* each side of an inequality by a *negative* number *reverses* the order of the inequality.

The multiplication properties of inequalities are stated below.

<table>
<tr><td>Multiplication properties of inequalities</td><td>If $a < b$ and $c > 0$, then $ac < bc$. If $a > b$ and $c > 0$, then $ac > bc$.<br><br>If you multiply each side of an inequality by a *positive* number, the *order* of the inequality remains the *same*.<br><br>If $a < b$ and $c < 0$, then $ac > bc$. If $a > b$ and $c < 0$, then $ac < bc$.<br><br>If you multiply each side of an inequality by a *negative* number, the *order* of the inequality is *reversed*.</td></tr>
</table>

*Example 4*    **Write a true inequality by multiplying each side of the given inequality by the indicated number.**

Multiply each side of $4 \geq -6$ by 3.     Multiply each side of $4 \geq -6$ by $-3$.

$$4 \geq -6$$
$$3 \cdot 4 \; ? \; 3 \cdot -6 \quad \textbf{same order}$$
$$12 \geq -18$$

$$4 \geq -6$$
$$-3 \cdot 4 \; ? \; -3 \cdot -6 \quad \textbf{reversed order}$$
$$-12 \leq 18$$

Since division is related to multiplication, it would seem that the division properties of inequalities would be the same as those for multiplication.

*Example 5*    **Write a true inequality by dividing each side of the given inequality by the indicated number.**

Divide each side of $-8 \leq 24$ by 4.     Divide each side of $-8 \leq 24$ by $-4$.

$$-8 \leq 24$$
$$\frac{-8}{4} \; ? \; \frac{24}{4} \quad \textbf{same order}$$
$$-2 \leq 6$$

$$-8 \leq 24$$
$$\frac{-8}{-4} \; ? \; \frac{24}{-4} \quad \textbf{reversed order}$$
$$2 \geq -6$$

Example 5 suggests the following properties.

| Division properties of inequalities | If $a < b$ and $c > 0$, then $\dfrac{a}{c} < \dfrac{b}{c}$. If $a > b$ and $c > 0$, then $\dfrac{a}{c} > \dfrac{b}{c}$. If you divide each side of an inequality by a *positive* number, the order of the inequality remains the *same*. If $a < b$ and $c < 0$, then $\dfrac{a}{c} > \dfrac{b}{c}$. If $a > b$ and $c < 0$, then $\dfrac{a}{c} < \dfrac{b}{c}$. If you divide each side of an inequality by a *negative* number, the order of the inequality is *reversed*. |
| --- | --- |

*Example 6*    **Tell what operation was performed on each side of the first inequality to obtain the second inequality.**

First inequality: $-21 < 14$
Second inequality: $3 > -2$

Each side of $-21 < 14$ was divided by $-7$ to obtain $3 > -2$, since $\dfrac{-21}{-7} = 3$ and $\dfrac{14}{-7} = -2$.

*Summary*    Adding the *same number* to (or subtracting the *same number* from) each side of an inequality produces an inequality of the *same order*.

Multiplying or dividing each side of an inequality by a *positive* number produces an inequality of the *same order*.

Multiplying or dividing each side of an inequality by a *negative* number produces an inequality of *reversed order*.

# Reading in Algebra

**Indicate whether each statement is sometimes true, always true, or never true.**
1. Multiplying each side of an inequality by the same number produces an inequality of the same order. **sometimes true**
2. Adding the same number to each side of an inequality reverses the order.    **never true**
3. Dividing each side of an inequality by the same negative number produces an inequality of reversed order. **always true**

# Oral Exercises

**If the indicated operation is performed, will the order of the inequality be changed?**
1. $7 > 2$ Add 5. **no**          2. $-3 > -6$ Divide by $-2$. **yes**  3. $-8 < 3$ Multiply by 3. **no**
4. $8 > -1$ Subtract $-4$. **no**     5. $-7 \leq -2$ Multiply by $-1$.   6. $2 < 4$ Divide by $-1$. **yes**
                                                                                                        **yes**

# Written Exercises

**For each inequality perform the indicated operation on each side and write a true inequality.**

Ⓐ  1. $5 > 4$ Add 7. **12 > 11**                    2. $9 > 3$ Subtract 2. **7 > 1**
   3. $9 < 10$ Subtract $-2$. **11 < 12**            4. $-4 < -2$ Add $-6$. **−10 < −8**
   5. $7 > -1$ Multiply by $-5$. **−35 < 5**         6. $8 > -16$ Divide by $-4$. **−2 < 4**
   7. $-12 < -2$ Divide by 2. **−6 < −1**            8. $9 > -4$ Add $-6$. **3 > −10**
   9. $-5 < 3$ Multiply by 6. **−30 < 18**          10. $6 > 0$ Multiply by 4. **24 > 0**
  11. $0 < 4$ Multiply by $-7$. **0 > −28**         12. $-12 \leq -6$ Divide by $-3$. **4 ≥ 2**
  13. $-20 < 16$ Divide by 4. **−5 < 4**            14. $-9 < 1$ Add $-3$. **−12 < −2**
  15. $11 > -2$ Multiply by $-6$. **−66 < 12**      16. $-8 \leq 2$ Divide by $-2$. **4 ≥ −1**
  17. $0 > -9$ Add $-6$. **−6 > −15**               18. $15 \geq 12$ Divide by $-3$. **−5 ≤ −4**
  19. $13 > -6$ Subtract $-2$. **15 > −4**          20. $-8 < 0$ Multiply by $-8$. **64 > 0**
  21. $-2 > -5$ Subtract 1 **−3 > −6**              22. $-8 \geq -10$ Add 4. **−4 ≥ −6**

**What operation was performed on each side of the first inequality to obtain the second inequality?**
  23. $6 > 4$; $2 > 0$ **Subtract 4.**              24. $5 \leq 7$; $15 \leq 21$ **Multiply by 3.**
  25. $-6 < 2$; $18 > -6$ **Multiply by −3.**       26. $12 > 8$; $3 > 2$ **Divide by 4.**
  27. $-2 \geq -5$; $4 \geq 1$ **Add 6.**           28. $-6 \leq 5$; $-18 \leq 15$ **Multiply by 3.**

The Inequality Properties                                                                    **127**

# SOLVING INEQUALITIES

## 5.4

**Objective**     **To solve inequalities**

The procedure for solving an inequality is similar to the procedure for solving an equation. The only difference is that when you *divide*, or *multiply*, each side of an inequality by a *negative number*, you must remember to *reverse* the order of the inequality. The examples that follow indicate the procedure for solving inequalities.

**Example 1**     **Solve $-2x + 14 < 6$. Graph the solution set.**

$$-2x + 14 < 6$$
$$-2x + 14 + (-14) < 6 + (-14)$$   ◀ *The order remains the same when adding $-14$ to each side.*
$$-2x < -8$$
$$\frac{-2x}{-2} > \frac{-8}{-2}$$   ◀ *The order is reversed when dividing each side by a negative number.*
$$x > 4$$

**Thus,** any number greater than 4 should make $-2x + 14 < 6$ true.

Check a number greater than 4.          Check a number less than 4.

Try 5.

$$
\begin{array}{c|c}
-2x + 14 & < 6 \\
\hline
-2 \cdot 5 + 14 & 6 \\
-10 + 14 & \\
4 & \\
 & 4 < 6 \quad \text{True}
\end{array}
$$

Try $-3$.

$$
\begin{array}{c|c}
-2x + 14 & < 6 \\
\hline
-2 \cdot -3 + 14 & 6 \\
\cdot \ 6 + 14 & \\
20 & \\
 & 20 < 6 \quad \text{False}
\end{array}
$$

Any number greater than 4 makes $-2x + 14 < 6$ true, and any number less than or equal to 4 makes $-2x + 14 < 6$ false. **Thus,** the solution set is {all numbers greater than 4}. Here is the graph.

4 is not a solution

**Example 2**     **Solve $5x - 8 < 6x - 6$. Graph the solution set.**

$$5x - 8 < 6x - 6$$
$$-8 < x - 6$$   ◀ *Add $-5x$ to each side.*
$$-2 < x$$   ◀ *Add 6 to each side.*
$$x > -2$$   ◀ *$-2 < x$ is the same as $x > -2$.*

**Thus,** the solution set is {all numbers greater than $-2$}.

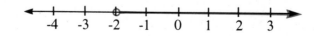

The inequality $\frac{4}{5}x + \frac{9}{10} \le \frac{1}{2}x$ can be transformed into an inequality with no fractions by multiplying each side by the LCM of 5, 10, and 2, just as you did for equations with fractions. Then solve the resulting inequality.

*Example 3*   **Solve $\frac{4}{5}x + \frac{9}{10} \le \frac{1}{2}x$. Graph the solution set.**

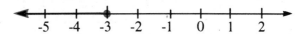

$$10\left(\frac{4}{5}x + \frac{9}{10}\right) \le 10 \cdot \frac{1}{2}x$$  ◀ *Multiply each side by 10, the LCM of 5, 10, and 2.*

$$10 \cdot \frac{4}{5}x + 10 \cdot \frac{9}{10} \le 10 \cdot \frac{1}{2}x$$  ◀ *Distribute the 10.*

$$\frac{40}{5}x + \frac{90}{10} \le \frac{10}{2}x$$

$$8x + 9 \le 5x$$

$$9 \le -3x$$  ◀ *Add $-8x$ to each side.*

$$\frac{9}{-3} \ge \frac{-3}{-3}x$$  ◀ *The order is reversed by dividing each side by a negative number.*

$$-3 \ge x$$

or $x \le -3$  ◀ $-3 \ge x$ *is the same as* $x \le -3$

The solution set is {all numbers less than or equal to $-3$}.

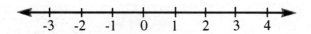

As in certain equations, the solution set of an inequality can be the set of all numbers. This is shown in Example 4.

*Example 4*   **Solve $x + 5 > 3x - 2(x - 1)$. Graph the solution set.**

$$x + 5 > 3x - 2(x - 1)$$
$$x + 5 > 3x - 2x + 2$$  ◀ *Distribute the $-2$.*
$$x + 5 > x + 2$$
$$5 > 2$$  ◀ *Add $-x$ to each side.*

This is true.

**Thus,** the solution set is {all numbers}. The graph is the entire number line.

Sometimes the solution set of an inequality can be the empty set. This is illustrated in the next example.

*Example 5*    **Solve $x + 3 < x + 2$. Graph the solution set.**

$x + 3 < x + 2$
    $3 < 2$    ◀ *Add $-x$ to each side.*
This is false.
The solution set is the empty set.
**Thus,** the solution set is the empty set.
There are *no* points on the number line for which $3 < 2$ is true.

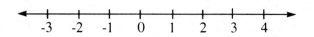

# Written Exercises

**Solve and graph the solution set of each inequality.**    5. $m \geq 7$   12. $x > 1$   23. $m \geq -7$

Ⓐ   **1.** $-3x + 12 < -15$  $x > 9$     **2.** $-5x + 15 > -20$  $x < 7$     **3.** $-17 < 9y - 8$  $y > -1$
   **4.** $7 > 3x - 8$  $x < 5$     **5.** $-3m - 27 \geq -6m - 6$     **6.** $4p + 9 \leq -19 + 11p$  $p \geq 4$
   **7.** $8a - 4 \leq 9a + 2$  $a \geq -6$     **8.** $7 - 5g \geq 9 - 4g$  $g \leq -2$     **9.** $6d - 8 < -24 - 2d$  $d < -2$
   **10.** $5r + 4 > 8r - 11$  $r < 5$     **11.** $-2k + 17 \leq 3k - 8$  $k \geq 5$     **12.** $-18 + 12x > -16 + 10x$
   **13.** $33 - 4t \geq 1 + 4t$  $t \leq 4$     **14.** $-7b + 20 \leq 2b - 7$  $b \geq 3$     **15.** $-y + 18 > 3y - 2$  $y < 5$
   **16.** $8 - 5e < -6e + 10$  $e < 2$    **17.** $9x - 7 \leq 18 + 4x$  $x \leq 5$     **18.** $7y + 8 \geq 14 + 9y$  $y \leq -3$
   **19.** $10n - 24 > 3n - 3$  $n > 3$    **20.** $12x - 36 \leq 6x - 12$  $x \leq 4$    **21.** $7a - 9 \geq 35 - 4a$  $a \geq 4$
   **22.** $2x + 8 > 2x + 14$  $\varnothing$     **23.** $5m - 4 \leq 7m + 10$     **24.** $2a - 5 > 2a + 6$  $\varnothing$
   **25.** $x + 5 < x + 10$     **26.** $3a + 4 \geq 3a + 9$  $\varnothing$     **27.** $7m + 2 < 7m - 4$  $\varnothing$
                          {all numbers}
Ⓑ   **28.** $\frac{2}{3}x + 5 < 7$  $x < 3$     **29.** $\frac{2}{3}x + 5 \geq \frac{1}{2}x + 6$  $x \geq 6$     **30.** $-2 + \frac{3}{4}x < \frac{1}{2}x$  $x < 8$

   **31.** $\frac{1}{2}m + \frac{1}{4} \geq \frac{3}{4}m$  $m \leq 1$     **32.** $\frac{2}{5}x - 7 < -8 + \frac{1}{2}x$  $x > 10$  **33.** $-4 + \frac{2}{3}y \leq \frac{1}{6}y + 5$  $y \leq 18$
   **34.** $2(5 + x) > 20$  $x > 5$     **35.** $-5(x + 4) < 25$  $x > -9$     **36.** $6(x + 4) < 6x + 5$  $\varnothing$
   **37.** $10x - 3(5 - 3x) \leq 23$  $x \leq 2$          **38.** $5 + 7g \geq -(3 - 2g) - 3g$  $g \geq -1$
   **39.** $5z + 10(-z + 14) < 95$  $z > 9$          **40.** $-7 + 5d - (d + 3) \geq d + 2$  $d \geq 4$
   **41.** $7z - (6 - z) > 10z - 10$  $z < 2$          **42.** $5x - 3 - 2(4 - x) \leq 8x - 10$  $x \geq -1$
Ⓒ   **43.** $7 - [4 - (3x - 2)] < 13 + 3x$          **44.** $2a - [8 - (3 - a)] > 2 - (5 - a)$  $\varnothing$
   **45.** $3b - [7 - (2 - b)] \geq b + 8$  $b \geq 13$     **46.** $7x - 2[4 - (5 - x)] \leq 3x - (6 - 2x)$  $\varnothing$
              43. {all numbers}          **Ex. 1–46, check student's graphs.**

# CUMULATIVE REVIEW

**1.** Evaluate $3a + 4$ for $a = -2$. $-2$    **2.** Simplify $5(-2)^3$. $-40$    **3.** Simplify $|-8 + 3|$. $5$

## NON-ROUTINE PROBLEMS

There is a frog at the bottom of a well that is 20 ft deep. During the day, he climbs 5 ft, but at night he falls asleep and slips back 4 ft. If this continues, how many days will it take the frog to escape from the well? **16 days**

# CONJUNCTION AND DISJUNCTION IN LOGIC

*Objectives*  **To determine if a conjunction is true**
**To determine if a disjunction is true**

The combination of sentences with the word *and* or the symbol $\wedge$ is called a conjunction. A conjunction $P \wedge Q$ is true if and only if *both P and Q* are true.

*Example 1*  **Which of the following conjunctions is true?**

$(8 = 7 + 1) \wedge (6 > 5)$     $(3 < -2) \wedge (-6 = -2 \cdot 3)$
True ◀ *Both parts are T.*     False ◀ *One part, 3 < -2, is F.*

The combination of sentences with *or* or the symbol $\vee$ is a disjunction. A disjunction $P \vee Q$ is true if and only if *at least P or Q* is true.

*Example 2*  **Which of the following disjunctions is true?**

$(4 > 2) \vee (14 < 8 + 2 \cdot 3)$     $(-6 > -2) \vee (-5 > 8)$
True ◀ *One part, 4 > 2, is T.*     False ◀ *Both parts are F.*

The truth tables summarize the truth values for conjunction and disjunction. A conjunction is true only when both parts are true. A disjunction is false only when both parts are false.

Conjunction

| P | Q | $P \wedge Q$ |
|---|---|---|
| T | T | T |
| T | F | F |
| F | T | F |
| F | F | F |

Disjunction

| P | Q | $P \vee Q$ |
|---|---|---|
| T | T | T |
| T | F | T |
| F | T | T |
| F | F | F |

# Written Exercises

**Determine if each conjunction or disjunction is true.**

1. $(4 < 6) \wedge (7 = 5 + 2)$ **T**
2. $(8 > 3) \vee (2 = 5)$ **T**
3. $(5 - 8 = 3) \vee (6 \neq -4)$ **T**
4. $(7 > -3) \wedge (-6 < -2)$ **T**
5. $((-3)^2 = 9) \wedge (4^2 < 2^4)$ **F**
6. $(9 = 2^3 + 1) \vee (5 > 4 \cdot 3)$ **T**
7. $(9 = 4^2 - 7) \vee (-4 > 1)$ **T**
8. $(8 - 4 \cdot 2 < 0) \wedge (6^3 \neq 216)$ **F**
9. $(4 \cdot 3 - 14 > 0) \vee \left(\left(\frac{1}{4}\right)^2 > \frac{1}{2}\right)$ **F**

# COMBINING INEQUALITIES

*Objectives*   **To solve and graph the conjunction of two inequalities**
**To solve and graph the disjunction of two inequalities**

Relationships between inequalities can be described in terms of *conjunction* or *disjunction*. A conjunction or a disjunction is a *compound inequality*. The conjunction $3 < 7 \land 9 > 7$ is true because each part is true. Rewrite $9 > 7$ as $7 < 9$; then the conjunction can also be written as $3 < 7 < 9$. This is read as 3 is less than 7 and 7 is less than 9.

To solve a conjunction of two inequalities like $x > 2$ and $x \leq 5$:

1. Graph the solution set of each inequality on a number line.
2. On a third number line, graph the set of points that are common to both inequalities.

*Example 1*   **Solve the conjunction $x > 2 \land x \leq 5$. Graph the solution set.**

Graph of $x > 2$

Graph of $x \leq 5$

Graph of points common
to both $x > 2$ and $x \leq 5$

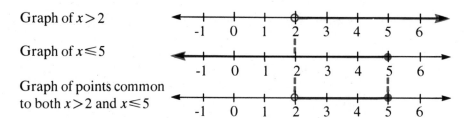

**Thus, $2 < x \leq 5$,**
or the solution set is {5, and all numbers between 2 and 5}.

*Example 2*   **Solve and graph the solution set of $2x + 8 \geq 14 \land -3x > 6$.**

First solve each inequality for $x$.

$$2x + 8 \geq 14 \qquad -3x > 6$$
$$2x \geq 6 \qquad \frac{-3x}{-3} < \frac{6}{-3}$$
$$x \geq 3 \qquad x < -2$$

◄ *The order is reversed when dividing by a negative number.*

Graph of $x \geq 3$

Graph of $x < -2$

There are no
points common
to both graphs.

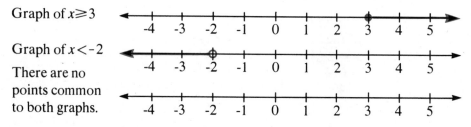

There are no points that satisfy $2x + 8 \geq 14 \land -3x > 6$. The solution set is $\varnothing$.

Recall that a disjunction is true if *either part is true* or if *both parts are true*. The disjunction $3 < 5 \lor 7 < 2$ is true because *at least one part*, $3 < 5$, is true. The procedure for graphing the solution set of a disjunction is illustrated in Example 3.

*Example 3*    **Graph the solution set of the disjunction $x < 2 \lor x > 6$.**

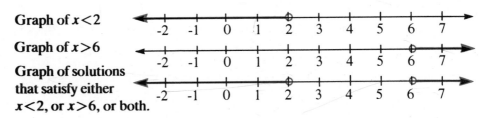

**Graph of $x < 2$**

**Graph of $x > 6$**

**Graph of solutions that satisfy either $x < 2$, or $x > 6$, or both.**

The solution set is {all numbers that are less than 2 or greater than 6}.

*Example 4*    **Graph the solution set of $x \geq -2 \lor x < 3$.**

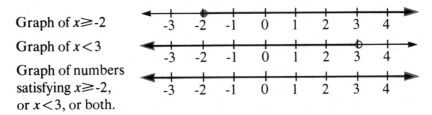

**Graph of $x \geq -2$**

**Graph of $x < 3$**

**Graph of numbers satisfying $x \geq -2$, or $x < 3$, or both.**

**Thus,** the solution set is the set of all real numbers.

Sometimes the solution set of a compound inequality is the solution set of one of the inequalities. This is shown in Example 5.

*Example 5*    **Graph and describe the solution set of $2a - 4 < 6 \lor -3a > -21$.**

First solve each inequality for $a$.

$$2a - 4 < 6 \qquad\qquad -3a > -21$$
$$2a < 10 \qquad\qquad\quad a < 7 \qquad \blacktriangleleft \textit{Dividing by } -3 \textit{ reverses the order.}$$
$$a < 5$$

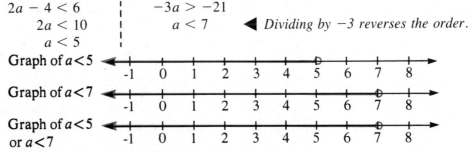

**Graph of $a < 5$**

**Graph of $a < 7$**

**Graph of $a < 5$ or $a < 7$**

Any number less than 7 satisfies $a < 5 \lor a < 7$.
**Thus,** the solution set is written as $a < 7$.

The expression $2 < x \le 5$ is read as $x$ is greater than 2 and less than or equal to 5. The next example deals with this type of combined inequality. You are asked to graph and describe the solution set of $-4 < 2x + 6 \le 8$. You should begin by rewriting $-4 < 2x + 6 \le 8$ as $-4 < 2x + 6$ and $2x + 6 \le 8$.

*Example 6*    **Graph and describe the solution set of $-4 < 2x + 6 \le 8$.**

First rewrite $-4 < 2x + 6 \le 8$ as $-4 < 2x + 6$ and $2x + 6 \le 8$.

$$-4 < 2x + 6 \qquad\qquad 2x + 6 \le 8$$
$$-10 < 2x \qquad\qquad\qquad 2x \le 2$$
$$-5 < x \qquad\qquad\qquad\quad x \le 1$$
$$x > -5$$

Graph of $x > $-5

Graph of $x \le 1$

Graph of $x > $-5 and $x \le 1$

**Thus, $-5 < x \le 1$.**
The solution set is {1, and all numbers between $-5$ and 1}.

Example 7 shows a more compact method of finding the solution set of $-4 < 2x + 6 \le 8$.

*Example 7*    **Find the solution set of $-4 < 2x + 6 \le 8$.**

$$-4 < 2x + 6 \qquad\qquad \le 8$$
$$-4 + (-6) < 2x + 6 + (-6) \le 8 + (-6) \quad \blacktriangleleft \text{ Add } -6 \text{ to each of three expressions}$$
$$-10 < 2x \qquad\qquad\qquad \le 2 \qquad\qquad\quad \text{to get } 2x \text{ alone.}$$
$$\frac{-10}{2} < \frac{2x}{2} \qquad\qquad\qquad \le \frac{2}{2} \quad \blacktriangleleft \text{ Divide each term by 2.}$$
$$-5 < x \qquad\qquad\qquad\quad \le 1$$

**Thus,** the solution set is {1, and all numbers between $-5$ and 1}.

# Reading in Algebra

**1.** True or false? $4 < x < 10$ can be rewritten as a disjunction. **False**

**Would you use a conjunction or disjunction to describe the solution sets on each number line below? 2. disjunction   3. conjunction**

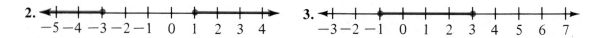

2.    $-5\ -4\ -3\ -2\ -1\ \ 0\ \ 1\ \ 2\ \ 3\ \ 4$     3.    $-3\ -2\ -1\ \ 0\ \ 1\ \ 2\ \ 3\ \ 4\ \ 5\ \ 6\ \ 7$

## Oral Exercises    See TE Answer Section.

**Tell how to sketch the graph of the solution set for each.**

1. $x > 5 \wedge x < 7$
2. $x < 4 \vee x > 8$
3. $x < -4 \vee x > 2$
4. $x > -6 \wedge x < -3$
5. $x \geq -9 \wedge x \leq 1$
6. $x \leq 3 \vee x \geq 12$
7. $x \geq -8 \wedge x < 7$
8. $x < -5 \vee x \geq 5$
9. $x \leq -5 \vee x \geq 5$
10. $3 < x < 6$
11. $-1 \leq x < 5$
12. $-8 < x \leq 2$

## Written Exercises    Ex. 1–53, check students' graphs.

**Graph the solution set of each compound inequality.**

(A)
1. $x > -3 \wedge x < 4$
2. $x > 1 \wedge x \leq 6$
3. $x > 5 \wedge x \leq 1$
4. $a \geq -3 \vee a < 4$
5. $b > 1 \vee b \leq 6$
6. $y > 5 \vee y \leq 1$
7. $p \leq -5 \wedge p > 0$
8. $x \geq -4 \wedge x \geq -2$
9. $y > -4 \vee y \leq -2$
10. $t \leq -5 \vee t > 0$
11. $b > -3 \wedge b \geq -9$
12. $g > -3 \wedge g \leq -9$
13. $r \leq -4 \wedge r > -3$
14. $x > 1 \vee x \leq -4$
15. $k \geq 0 \wedge k < 0$
16. $a > 0 \vee a \leq 0$
17. $m > -3 \vee m \leq -3$
18. $w > 5 \wedge w < 5$
19. $v < 6 \wedge v \geq 6$
20. $x \geq 8 \vee x \leq 8$
21. $x > 4 \vee x < 4$

**Solve and graph the solution set. Describe the solution set.**

22. $2x - 5 < 11 \wedge 3x + 2 \leq 20$  **$x \leq 6$**
23. $x + 5 > 8 \wedge 3x + 4 \leq 19$  **$3 < x \leq 5$**
24. $3x + 3 < 12 \vee -3x < 15$  **all numbers**
25. $2x - 8 > 6 \vee -3x > -27$  **all numbers**
26. $3x + 5 > 20 \wedge 5 + x > 6$  **$x > 5$**
27. $3x + 2 \leq 11 \wedge -2x \geq -8$  **$x \leq 3$**
28. $2x - 1 > 5 \vee -3x \leq -15$  **$x > 3$**
29. $-2 < 2x + 4 \vee -2x + 5 < 13$  **$x > -4$**

**$m > 2$**

(B)
30. $\frac{1}{2}x + 5 < 6 \wedge 3x + 5 > x + 7$  **$1 < x < 2$**
31. $\frac{2}{3}m + \frac{1}{2} > \frac{11}{6} \vee 7m + 2 > 5m + 12$

32. $x - 5 < 2x + 4 \vee -3x > 27$  **$x \neq -9$**
33. $2x + 5 < 10 \vee 4x - 1 \geq x + 11$
34. $3x - 6 < x \vee 3x - 6 \geq -x$  **all numbers**
35. $5x = 20 \wedge 3x - 5 \leq 4x - 1$  **$x = 4$**
36. $2m - 6 \geq 4m + 12 \wedge -3m \leq 27$  **$m = -9$**
37. $3x - 5 > x + 3 \wedge -5x > -20$  **$\varnothing$**
38. $-4 < 3x + 5 < 14$  **$-3 < x < 3$**
39. $3 \leq 2x - 5 < 1$  **$\varnothing$**
40. $-11 \leq 3x - 2 < 16$  **$-3 \leq x < 6$**
41. $28 > 5x - 2 \geq 18$  **$6 > x \geq 4$**
42. $-5 < -x - 2 < 4$  **$3 > x > -6$**
43. $7 \geq -x - 8 > 5$  **$-15 \leq x < -13$**
44. $\frac{1}{10}a + \frac{1}{2} \geq \frac{1}{5} \vee \frac{1}{3}a < \frac{1}{6} + \frac{1}{2}a$  **$a \geq -3$**
45. $\frac{2}{3} \geq \frac{1}{6}x + 5 > \frac{1}{2}$  **$-26 \geq x > -27$**

(C)
46. $2x - 1 < 2x + 5 < 2x + 8$  **all numbers**
47. $3x + 5 < 3x + 2 < 3x + 1$  **$\varnothing$**
48. $3x \neq 15 \wedge 2x - 4 > 6$  **$x > 5$**
49. $(3x + 4 < x + 6 \wedge x \geq 1) \vee x < 1$  **$x < 1$**
50. $8 < 2x \leq 10 \wedge 1 < x \leq 12$  **$4 < x \leq 5$**
51. $8 > 2p - 4 > 6 \vee 7 \leq 2p - 1 \leq 17$
52. $6y - (4 - y) < 17 \wedge 2 < y \leq 4$  **$2 < y < 3$**
53. $2b - (4 - b) \geq 11 \vee 9 < 2b + 1 < 15$

33. $x < \frac{5}{2} \vee x \geq 4$   51. $4 \leq p \leq 9$   53. $b > 4$

## CUMULATIVE REVIEW

1. Simplify $|-4 + 2\frac{1}{3} - 5|$  **$6\frac{2}{3}$**
2. Solve $2x - 5 = 10$.  **$\frac{15}{2}$**
3. Evaluate $3 - 8x$ for $x = -\frac{1}{2}$  **7**

# EQUATIONS WITH ABSOLUTE VALUE 5.7

*Objective*    **To solve equations with absolute value**

Recall that since 4 and $-4$ are each 4 units from the origin, the absolute value of 4 is 4, and the absolute value of $-4$ is 4.

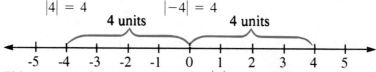

$$|4| = 4 \qquad |-4| = 4$$

This suggests that an equation like $|x| = 4$ will have *two* solutions. The solutions will be numbers whose distance from the origin is 4 units.

*Example 1*    **Solve $|x| = 4$.**

Find numbers whose absolute values are 4.

$$|x| = 4$$
$$x = 4 \lor x = -4$$

**Thus,** the solutions are 4, $-4$.

*Example 2*    **Solve $|y| = -5$.**

An absolute value can *never* be negative. **So,** there is no solution.

To solve an absolute value equation like $|x + 2| = 8$, think of $x + 2$ as some number whose distance from the origin is 8. This number could be 8 or $-8$. Therefore, you need to solve two separate equations, as shown in Example 3.

*Example 3*    **Solve $|x + 2| = 8$. Check the solutions.**

$$|x + 2| = 8$$

$$x + 2 = 8 \lor x + 2 = -8 \qquad \blacktriangleleft \textit{Write two separate equations.}$$
$$x = 6 \qquad\qquad x = -10 \qquad \blacktriangleleft \textit{Add } -2 \textit{ to each side.}$$

Check.

$$\begin{array}{c|c} |x + 2| & = 8 \\ \hline |6 + 2| & 8 \\ |8| & \\ 8 & \\ \end{array}$$
$$\qquad 8 = 8 \quad \text{True}$$

$$\begin{array}{c|c} |x + 2| & = 8 \\ \hline |-10 + 2| & 8 \\ |-8| & \\ 8 & \\ \end{array}$$
$$\qquad 8 = 8 \quad \text{True}$$

**Therefore,** the solutions are 6, $-10$.

*Example 4* **Solve $|8 - 2b| = 10$.**

$$|8 - 2b| = 10$$

| $8 - 2b = 10$ | $\lor$ | $8 - 2b = -10$ | ◄ Write two equations. |
|---|---|---|---|
| $-2b = 2$ | | $-2b = -18$ | ◄ Add $-8$ to each side. |
| $b = -1$ | | $b = 9$ | ◄ Divide each side by $-2$. |

**Thus,** the solutions are $-1$; $9$.

To solve an equation like $3|5m - 6| - 7 = 2$, first transform it into a simpler equation. Then only $|5m - 6|$ is on the left side. The complete solution follows in Example 5.

*Example 5* **Solve $3|5m - 6| - 7 = 2$.**

$$3|5m - 6| - 7 = 2$$

| $3|5m - 6| = 9$ | ◄ Add 7 to each side. |
|---|---|
| $|5m - 6| = 3$ | ◄ Divide each side by 3. |

| $5m - 6 = 3$ | $\lor$ | $5m - 6 = -3$ |
|---|---|---|
| $5m = 9$ | | $5m = 3$ |
| $m = \dfrac{9}{5}$ | | $m = \dfrac{3}{5}$ |

**Thus,** the solutions are $\dfrac{9}{5}, \dfrac{3}{5}$.

# Written Exercises

**Solve each equation if possible. Check.**

(A)
1. $|x| = 2$ **−2, 2**
2. $|b| = -1$ **∅**
3. $|r| = 4$ **−4, 4**
4. $|g| = -7$ **∅**
5. $|t - 3| = 6$ **−3, 9**
6. $|4 - m| = 8$ **−4, 12**
7. $|2m - 2| = 12$
8. $|5 - y| = 1$ **4, 6**
9. $|2x + 4| = 12$
10. $|5 - m| = -6$ **∅**
11. $|3x - 6| = 15$
12. $|6 - a| = 3$ **3, 9**

**−8, 4**
7. **−5, 7**   11. **−3, 7**

**Solve each equation if possible.**

13. $|2x - 8| = 4$ **2, 6**
14. $|4 - 3g| = 7$ **−1, $\frac{11}{3}$**
15. $|-2a + 5| = 9$ **−2, 7**
16. $|6 - 7k| = -10$ **∅**
17. $|3x - 5| = 2$ **1, $\frac{7}{3}$**
18. $|7 - p| = 13$ **−6, 20**
19. $|2m - 16| = 12$
20. $|13 - 7y| = 2$

19. **2, 14**   20. **$\frac{11}{7}, \frac{15}{7}$**

(B)
21. $3|x + 5| + 4 = 7$ **−4, −6**
22. $4|a - 3| - 3 = 9$ **0, 6**
23. $5|m - 2| - 15 = 5$ **−2, 6**
24. $8 - |2x - 5| = 1$ **−1, 6**
25. $2 - 4|3p - 1| = -6$ **$-\frac{1}{3}$, 1**
26. $6|x - 7| - 14 = 10$ **3, 11**
27. $2|3x - 1| - 4 = 12$ **$-\frac{7}{3}$, 3**
28. $-4 + 3|5 - x| = 11$ **0, 10**
29. $-8 = 6 - 2|4x - 2|$ **$-\frac{5}{4}, \frac{9}{4}$**
30. $-6 - 2|3b - 4| = 8$ **∅**
31. $5 - 2|4 - 3x| = -7$ **$-\frac{2}{3}, \frac{10}{3}$**
32. $7|5 - 4t| + 8 = 29$ **$\frac{1}{2}$, 2**

34. **−2, $\frac{14}{3}$**   35. **$-\frac{2}{3}$, 2**   38. **−26, 10**

(C)
33. $|2y - 4| = y - 6$ **∅**
34. $4 - |2m - 6| = m - 4$
35. $|[6 - 2(3 - m)]| = m + 2$
36. $|3x - 2| = |6 - x|$ **−2, 2**
37. $|4x - 2| = |6 - 2x|$ **−2, $\frac{4}{3}$**
38. $|x - 5| - |2x + 3| = -18$

# INEQUALITIES WITH ABSOLUTE VALUE     5.8

*Objective*     **To solve inequalities with absolute value**

In the inequality $|x| < 4$, at first it might appear that $x$ can be any number less than 4. Notice what happens when some values of $x$ are substituted.

For $x = 2$, $|x| = |2| = 2$.     $2 < 4$     True
For $x = -1$, $|x| = |-1| = 1$.     $1 < 4$     True
For $x = -3$, $|x| = |-3| = 3$.     $3 < 4$     True
For $x = -6$, $|x| = |-6| = 6$.     $6 < 4$     False

**Thus,** $x$ can only be a *number* whose *distance* from the origin is *less than 4 units*. This can be seen on a number line.

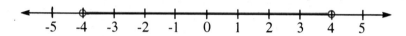

The solutions are all numbers between $-4$ and 4.
**Thus,** $-4 < x < 4$, which can also be written as $x > -4 \wedge x < 4$.

*Example 1*     **Solve $|x| \leq 6$.**

$x$ can be any number whose distance from the origin is 6 or less units.

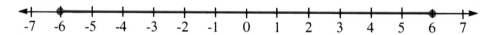

**Thus,** $-6 \leq x \leq 6$, *or* $x \geq -6 \wedge x \leq 6$.
The solution set is $\{-6, 6$, and all numbers between $-6$ and $6\}$.

The inequality $|t| > 5$ means you are looking for all numbers whose distance from the origin is *greater than 5 units*. It is clear that $t$ can be any number greater than 5. To determine whether any negative numbers satisfy the inequality, you can try some.

For $t = -3$, $|t| = |-3| = 3$     $3 > 5$     False
For $t = -4$, $|t| = |-4| = 4$     $4 > 5$     False
Now for $t = -6$, $|t| = |-6| = 6$     $6 > 5$     True
For $t = -9$, $|t| = |-9| = 9$     $9 > 5$     True

**So,** $t$ can be any number greater than 5 or less than $-5$. This is shown below.

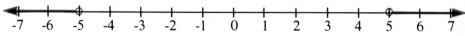

**Thus,** $t < -5 \vee t > 5$.

*Example 2*    **Solve $|x| \geq 2$.**

$x$ can be any number whose distance from the origin is 2 or more units.

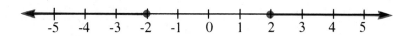

The graph includes 2 and $-2$.
**Thus, $x \leq -2 \lor x \geq 2$.**

You have seen that $|x| \geq 2$ means $x \leq -2$ or $x \geq 2$. Solving a more complex inequality works the same way. Consider $|2t - 6| \geq 4$. You want values of $t$ such that $2t - 6$ will be a distance of 4 or more units from the origin. This suggests that $2t - 6 \leq -4 \lor 2t - 6 \geq 4$. This inequality is solved in the next example.

*Example 3*    **Solve $|2t - 6| \geq 4$. Graph the solution set.**

$$|2t - 6| \geq 4$$

$2t - 6 \leq -4 \quad \lor \quad 2t - 6 \geq 4$  ◄ *Write the two inequalities.*

$\qquad 2t \leq 2 \qquad\qquad 2t \geq 10$

$\qquad\quad t \leq 1 \qquad\qquad\quad t \geq 5$

**Thus, $t \leq 1 \lor t \geq 5$.**

The graph is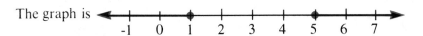

You have seen that $|y| < 4$ means $y$ is between $-4$ and 4. Similarly, $|3b + 4| < 5$ means $3b + 4$ is between $-5$ and 5. This suggests that $-5 < 3b + 4 < 5$. This can be solved as shown in Example 4.

*Example 4*    **Solve $|3b + 4| < 5$. Graph the solution set.**

$$|3b + 4| < 5$$

$-5 < \quad 3b + 4 \quad < 5$

$-5 + (-4) < 3b + 4 + (-4) < 5 + (-4)$  ◄ *Add $-4$ to each of the three expressions to get*

$\qquad -9 < \qquad 3b \qquad < 1 \qquad\qquad 3b$ *alone.*

$\qquad -3 < \qquad b \qquad < \dfrac{1}{3}$  ◄ *Divide each term by 3.*

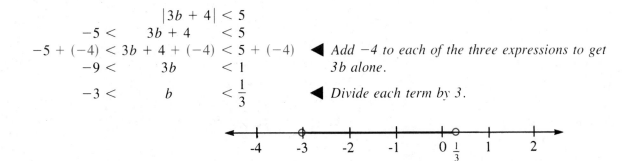

Inequalities with Absolute Value

# Reading in Algebra

**For each set of points choose the descriptions that are correct.**
1. points 5 or less units from the origin
   - **a.** $|x| \leq 5$
   - **b.** $|x| > 5$
   - **c.** $-5 \leq x \leq 5$
   - **d.** $|x| < 5$
   - **e.** $x \geq -5 \land x \leq 5$
   - **f.** $x \geq -5 \lor x \leq 5$ **a, c, e**
2. points more than 6 units from the origin
   - **a.** $-6 < x < 6$
   - **b.** $-6 < x > 6$
   - **c.** $x < -6 \lor x > 6$
   - **d.** $|x| > 6$
   - **e.** $|x| \geq 6$
   - **f.** $x > 6$ **c, d**

# Oral Exercises

**For each inequality is the graph correct? Why?**

1. $|x| < 3$ **no**

2. $|y| \geq 4$ **no**

3. $|a| > 4$ **no**

4. $|m| \leq 2$ **yes**

5. $|a| < 3$ **no**

6. $|t| \leq 1$ **no**

7. $|y| \geq 2$ **yes**

8. $|m| < 1$ **no**

# Written Exercises

**Solve each inequality.**

Ⓐ
1. $|x| \leq 8$ **$-8 \leq x \leq 8$**
2. $|x| \geq 3$ **$x \geq 3 \lor x \leq -3$**
3. $|g| < 5$ **$-5 < g < 5$**
4. $|m| > 1$ **$m > 1 \lor m < -1$**
5. $|n| < 7$ **$-7 < n < 7$**
6. $|t| \geq 4$ **$t \geq 4 \lor t \leq -4$**
7. $|b| < 2$ **$-2 < b < 2$**
8. $|k| \geq 6$ **$k \geq 6 \lor k \leq -6$**

**Solve each inequality. Graph the solution set. Ex. 9–31, check students' graphs.**
9. $|2a - 6| \leq 20$ **$-7 \leq a \leq 13$**
10. $|3x - 3| < 12$ **$-3 < x < 5$**
11. $|5y - 20| > 15$ **$y > 7 \lor y < 1$**
12. $|6 - 2a| > 14$ **$a > 10 \lor a < -4$**
13. $|3x - 6| \geq 9$ **$x \geq 5 \lor x \leq -1$**
14. $|4a - 8| \leq 16$ **$-2 \leq a \leq 6$**
15. $|2m - 5| \leq 15$ **$-5 \leq m \leq 10$**
16. $|7b - 14| > 21$ **$b > 5 \lor b < -1$**
17. $|5y - 20| < 20$ **$0 < y < 8$**

Ⓑ
18. $|2x - (x - 4)| < 1$ **$-5 < x < -3$**
19. $|7 - (5 - t)| \geq 4$ **$t \geq 2 \lor t \leq -6$**
20. $|m - (8 - m)| > 2$ **$m > 5 \lor m < 3$**
21. $|6d - (4d - 8)| < 2$ **$-5 < d < -3$**
22. $|y - 4| + 2 < 5$ **$1 < y < 7$**
23. $|x - 7| - 6 > 8$ **$x > 21 \lor x < -7$**
24. $6 - 2|x - 6| > 4$ **$5 < x < 7$**
25. $4|2p - 2| + 6 < 10$ **$\frac{1}{2} < p < \frac{3}{2}$**

Ⓒ
26. $|z - 6| > z - 2$ **$z < 4$**
27. $|x - 4| < x - 4$ **∅**
28. $|a - 8| \geq a - 8$ **all numbers**
29. $|p - 4| < p - 10$ **∅**
30. $|x| < 4 \lor |x| > 4$ **$x \neq 4, x \neq -4$**
31. $|2x - 8| < 8 \land |x| > 1$ **$1 < x < 8$**

# INEQUALITIES IN WORD PROBLEMS                    5.9

*Objective*    **To solve problems that lead to inequalities**

Many word problems involve translating an English sentence into
an inequality.

For example, Bill can be at most 82 kg to qualify for middleweight wrestling.
This means he must be 82 kg or less than 82 kg.

Therefore, *at most* is translated as $\leq$.

Similarly, if Mary's average must be at least 90 to qualify for honor roll,
her average must be 90 or greater than 90.

Therefore, *at least* is translated as $\geq$.

The basic four steps for solving word problems will have *one change*. You
will write an *inequality* instead of an equation.

*Example 1*    **The freshman class raises money by selling boxes of holiday cards at a
profit of $2 on each box. How many boxes must they sell to make a
profit of at least $500?**

Step 1
READ ▶    You are asked to find the number of boxes they must sell to make a profit
of at least $500.

Step 2
PLAN ▶    Let $n$ = the number of boxes they must sell

Write an inequality.

Think: profit on each box · number sold is at least $500

$\phantom{xxxxxxx}$ $2 $\phantom{xxxxx}$ $n$ $\phantom{xxxx}$ $\geq$ $\phantom{xx}$ $500

Step 3
SOLVE ▶    $$2n \geq 500$$
$$n \geq 250$$

Step 4
INTERPRET ▶    $2 \cdot 250 = $500

$2 times any number greater than 250 will be greater than $500.

**Thus,** the class must sell at least 250 boxes.

*Example 2*    **Tina rents a car at $150 a week plus $0.10 a kilometer. How far can she travel if she wants to spend at most $210?**

Let $d$ = the number of kilometers she can travel.

Fixed cost + cost per kilometer · number of kilometers is at most $210

$$150 + 0.10d \leq 210$$
$$100(150 + 0.10d) \leq 100 \cdot 210 \quad \blacktriangleleft \text{ \textit{Multiply each side by the LCD,}}$$
$$100 \cdot 150 + 100 \cdot 0.10d \leq 100 \cdot 210 \qquad \textit{100.}$$
$$15,000 + 10d \leq 21,000$$
$$10d \leq 6,000$$
$$d \leq 600$$

**So,** she can travel at most 600 km.

## Reading in Algebra

**For each phrase or inequality select the phrases or inequalities that do not have the same meaning.**

1. $y \leq 3$
   a. $y$ is at least 3.
   b. $y$ is at most 3.
   c. $y$ is 3 or more.
   d. $y$ is 3 or less. **a, c**

2. $x$ is at most 2.
   a. $x \leq 2$
   b. $x \geq 2$
   c. $x$ is 2 or less.
   d. $x$ is 2 or more. **b, d**

3. $a$ is at least 6.
   a. $a$ is 6 or less.
   b. $a \leq 6$
   c. $a$ is 6 or more.
   d. $a \geq 6$ **a, b**

## Written Exercises

Ⓐ **Solve each problem.**

1. A store makes a profit of $5 on each watch sold. How many watches must be sold to make a profit of at least $150? **at least 30**

2. The average cost for a used car dealer for minor repairs under a 90-day warranty is $20. How many can occur in a month if she is to lose at most $440? **at most 22**

3. Juan is paid $140 a week plus a commission of $20 on each television he sells. How many must he sell to make at least $360 a week? **at least 11**

4. Brian is paid $165 a week plus a commission of $15 on each camera he sells. How many must he sell to earn at least $315 a week? **at least 10**

5. Joan rents a car for $100 a week plus $0.10 a kilometer. How far can she travel if she wants to spend at most $175? **750 km**

6. Lovella rents a car for $56 a week plus $0.12 a kilometer. How far can she drive if she wishes to spend at most $200? **1,200 km**

Ⓑ 7. Janet earns $12,000 a year in salary and 8% commission on her sales. How much were her sales if her annual income was more than $12,800 and at most $14,000? **more than $10,000   at most $25,000**

8. A gain of 2 kg would put a boxer in the welterweight class. How much could this boxer be if the welterweight class is more than 68 kg and at most 73 kg? **more than 66 kg   at most 68 kg**

# COMPUTER ACTIVITIES

## Solving Inequalities

You have learned to solve inequalities like $8 < 2a + 4 < 16$.

You can use a computer to determine if a given number is a solution. However, most computers do not understand the meaning of an expression containing two inequality symbols. A computer can only read expressions containing one inequality symbol. It can, however, understand the conjunction of two such inequalities.

The program below uses a conjunction to determine whether a number is a solution of the inequality $8 < 2a + 4 < 16$. Note that line 35 in the program rewrites $8 < 2a + 4 < 16$ as a conjunction.

```
10   PRINT "THIS PROGRAM DETERMINES IF THE "
15   PRINT "ENTERED NUMBER IS A SOLUTION OF"
20   PRINT "8 < 2A + 4 < 16."
25   FOR I = 1 TO 9
30   INPUT "TRY A VALUE FOR A. ";A
35   IF 8 < 2 * A + 4 AND 2 * A + 4 < 16 THEN 50
40   PRINT A" IS NOT A SOLUTION."
45   GOTO 55
50   PRINT A" IS A SOLUTION."
55   PRINT : PRINT
60   NEXT I
65   END
```

See the Computer Section beginning on page 488 for more information.

## Exercises  For Ex. 2–4, see TE Answer Section.

1. Enter the program above into a computer and RUN it to determine which of the following numbers are solutions of $8 < 2a + 4 < 16$: $-4, 0, 1, 2, 3, 5, 6, 10, 16$. **3, 5**

2. Change the program above to determine which of the numbers below are solutions of $-7 \le 2x - 5 < 3$: $-4, -3, -2, 1, 4, 8, 9$. **1**

3. Write a program to determine which of the following numbers are solutions of $|3a + 6| > 18$: $-10, 0, 2, 4, 5$. (Hint: Rewrite the inequality as a disjunction.) **−10, 5**

4. Write a program to determine which of the following numbers are solutions of $|2a - 4| < 6$: $-4, -2, 0, 3, 6$. **0, 3**

# CHAPTER FIVE REVIEW

Ⓒ Exercises are starred.
The rest are Ⓑ Exercises.

## Vocabulary

axiom of comparison [5.2]
compound inequality [5.5]
conjunction [5.5]
disjunction [5.5]
empty set [5.1]
finite [5.1]

inequality [5.2]
infinite [5.1]
replacement
  set [5.1]
set [5.1]
solution set [5.1]

Determine whether each set is finite or infinite. Then describe the set in words. [5.1]

1. $\{5, 10, 15, 20\}$ **finite**
2. $\{3, 6, 9, \ldots\}$ **infinite**

Find the solution set of each equation. Then graph the solution set. [5.1]

3. $3m + 10 = -12 + 3m$ $\varnothing$
4. $8r + 3 = 7r - (-r - 3)$ **{all numbers}**
5. $7x + 6 = -7x - 22$ **–2**
6. $5y - (7 - y) = 8y + 2$ $-\frac{9}{2}$

**Ex. 3–6, check students' graphs.**

Graph the solution set of each sentence. [5.2]

7. $x < 3$
8. $-6 \le b$
9. $2 < a$
10. $y \ne -\frac{1}{2}$
11. $-\frac{3}{2} > b$

**Ex. 7–11, check students' graphs.**

For each inequality perform the indicated operation on each side and write a true inequality. [5.3]

12. $8 > -1$    Multiply by 4. **32 > –4**
13. $-8 \le 4$    Divide by –2. **4 ≥ –2**
14. $-7 < 1$    Add –3. **–10 < –2**
15. $4 > -2$    Multiply by –1. **–4 < 2**

What operation was performed on each side of the first inequality to obtain the second inequality? [5.3]

16. $-8 < 2; 24 > -6$ **Multiply by –3**
17. $-12 < 3; -4 < 1$ **Divide by 3**

Solve and graph the solution set of each inequality. [5.4]

18. $-2x + 6 < 8$ **x > –1**
19. $-36 > 9y - 9$ **y < –3**
20. $x + 2 > x + 9$ $\varnothing$
21. $3m + 5 < 3m + 1$ $\varnothing$
22. $5x - 2 \le 7x + 6$ **x ≥ –4**
23. $7 - 2(3a - 5) > -8a + 3$ **a > –7**
24. $\frac{1}{2}x - \frac{3}{10} > \frac{1}{5}x$ **x > 1**

**Ex. 18–24, check students' graphs.**

Determine if each is true. [5.5]

25. $(2 < 7) \wedge (-8 = 2 - 10)$ **True**
26. $(-3 > 5) \vee (5^2 < 20)$ **False**

Graph the solution set of each compound inequality. [5.6]

27. $x < -5 \vee x \ge 2$
28. $x \ge -2 \wedge x < 6$
29. $3a + 2 < 8 \vee -2a \le -6$ **a < 2 ∨ a ≥ 3**
30. $-14 < 3t - 5 < 7$ **–3 < t < 4**

**Ex. 27–30, check students' graphs.**

Solve each equation if possible. [5.7]

31. $|p| = 6$ **–6, 6**
32. $|6 - 2y| = -4$ $\varnothing$
33. $|2a - 6| = 8$ **–1, 7**
34. $2|y - 4| + 6 = 8$ **3, 5**

Solve each inequality. [5.8]

35. $|x| > 10$    **x > 10 ∨ x < –10**
36. $|m| \le 3$    **–3 ≤ m ≤ 3**

Solve each inequality. Graph the solution set. [5.8]

37. $|2m - 6| \le 14$ **–4 ≤ m ≤ 10**
38. $|4y - (2y - 6)| > 10$ **y > 2 ∨ y < –8**
39. $2|4p - 6| + 8 > 12$ **p > 2 ∨ p < 1**

**Ex. 37–42, check students' graphs.**

★ Solve and graph the solution set. (Ex. 41–43)

40. $4m - 3 - [9 - (2 - m)] = 3(m - 4) + 6$ [5.1] $\varnothing$
41. $4b - [6 - (2 - b)] \le 8 - (6 - b)$ [5.4] **b ≤3**
42. $|-8| \not< y$ [5.2] **y ≤ 8**

Solve each inequality.
1. $|x| > 5$  **$x > 5 \vee x < -5$**
2. $|x| < 6$  **$-6 < x < 6$**

Solve each inequality. Graph the solution set.
3. $|2a - 6| \leq 12$  **$-3 \leq a \leq 9$**
4. $2|x - 6| + 8 > 14$  **$x > 9 \vee x < 3$**
5. $|m| < 3$  **$-3 < m < 3$**

   **Ex. 3–5, check students' graphs.**

Graph the solution set of each sentence.
6. $x \geq 6$

7. $y \neq -\dfrac{1}{3}$

8. $-\dfrac{5}{2} < c$

   **Ex. 6–8, check students' graphs.**

Solve each equation if possible.
9. $|y| = -4$  **$\varnothing$**
10. $2|y + 6| - 5 = 9$  **$1, -13$**
11. $|4 - 2b| = 8$  **$-2, 6$**

Find the solution set of each equation. Then graph the solution set.
12. $2(a - 6) = a - (13 - a)$  **$\varnothing$**
13. $5x - (4 - x) = 2(3x - 2)$  **{all numbers}**

   **Ex. 12–13, check students' graphs.**

Solve.
14. 17 kg less than twice Yvonne's mass is at most 109 kg. Find her mass.  **at most 63 kg**

Solve and graph the solution set of each inequality.
15. $-3y + 7 > -8$  **$y < 5$**
16. $x + 5 < x + 1$  **$\varnothing$**
17. $-4a + 2 \geq -2a + 10$  **$a \leq -4$**

18. $\dfrac{1}{3}a - \dfrac{5}{6} < \dfrac{1}{2}a$  **$a > -5$**

   **Ex. 15–18, check students' graphs.**

For each inequality perform the indicated operation on each side and write a true inequality.
19. $7 > 2$  Multiply by $-3$.  **$-21 < -6$**
20. $8 \leq 10$  Divide by 2.  **$4 \leq 5$**

Graph the solution set of each compound inequality.
21. $x < -2 \vee x > 3$
22. $3a - 1 < 11 \wedge 2a + 5 \geq 7$  **$1 \leq a < 4$**
23. $2 < 2m - 6 < 8$  **$4 < m < 7$**

   **Ex. 21–23, check students' graphs.**

What operation was performed on each side of the first inequality to obtain the second inequality?
24. $-6 < 4; 0 < 10$  **Add 6.**
25. $-5 \geq -8; 15 \leq 24$  **Multiply by $-3$.**

Determine whether each set is finite or infinite. Then describe the set in words.
26. $\{-2, -4, -6, -8, \ldots\}$  **infinite**
27. $\{1, 3, 5, 7\}$  **finite**

Determine if each is true.
28. $(3 > -4) \wedge (-6^2 < 5^2)$  **True**
29. $\left(\dfrac{1}{3} < \dfrac{1}{2}\right) \vee (8 \cdot 2 + 5 = 21)$  **True**

**Ex. 30–32, check students' graphs.**
★ 30. Graph the solution set of $y \not> |-4|$.  **$y \leq 4$**

★ Graph the solution set of each compound inequality.
31. $4x - 2 < 6x + 8 \vee |x| > 5$  **$x \neq -5$**
32. $|2m - 2| > 2 \wedge -6 < 2m + 4 < 18$  **$\varnothing$**

# COLLEGE PREP TEST

For each item you are to compare a quantity in column 1 with a quantity in column 2. Write the letter of the correct answer from these choices:

    **A.** The quantity in column 1 is greater than the quantity in column 2.
    **B.** The quantity in column 2 is greater than the quantity in column 1.
    **C.** The quantity in column 1 is equal to the quantity in column 2.
    **D.** The relationship cannot be determined from the information given.

Notes: The information centered over both columns refers to one or both of the quantities to be compared. A symbol that appears in both columns has the same meaning in each column, and all variables represent numbers.

---

Sample question and answer.

| Column 1 | Column 2 |
|---|---|
| $x = 5$ and $y = 4$ | |
| $x^2$ | $3y + 11$ |

Answer: A, since $x^2 = 5^2 = 25$, $3y + 11 = 3 \cdot 4 + 11 = 12 + 11 = 23$ and $25 > 23$.

---

**Column 1**  **Column 2**

$x = 8$

**1.** $x + 5$      $3x + 2$
**B**

---

$y < x$

**2.** $x - y$      $y - x$
**A**

---

$a > 0, b < 0$

**3.** $ab$      $\dfrac{a}{b}$
**D**

---

$2x - 3 > 7$

**4.** $5$      $x$
**B**

---

$10 - 2x < 6$

**5.** $x$      $2$
**A**

---

$4x = 0$

**6.** $1$      $x$
**A**

---

**Column 1**  **Column 2**

$n$ is an integer

**7.** $n$      $-n$
**D**

---

Questions 8–9 refer to the number line drawn below.

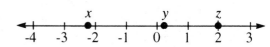

**8.** $1$      $yz$
**A**

$x < w < y$

**9.** $w$      $-y$
**D**

---

$0 < x < 10$
$0 < y < 12$

**10.** $x$      $y$
**D**

---

$3^{n+2} = 27$

**11.** $n$      $3$
**B**

---

$x > 0$ and $y > 0$

$\dfrac{x}{y} > 2$

**12.** $2y$      $x$
**B**

---

# 6 INTRODUCTION TO FACTORING

**Non-Routine
Problem
Solving**

**Problem 1**

An artist has an oddly-shaped canvas. Square $A$ is 1 in$^2$; square $B$ is 16 in$^2$; square $C$ is 64 in$^2$. The total surface area is 81 in$^2$. How can the canvas be cut into the fewest number of pieces that will fit together to make a 9 × 9 square canvas? **See TE Answer Section.**

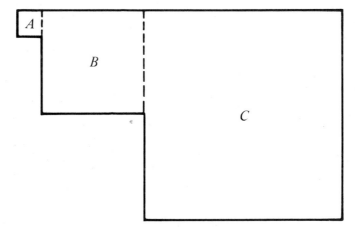

**Problem 2**

Four goats were tied, one at each corner of a square field that measures 100 m × 100 m. The ropes that tie them allow each goat to graze over an area with a 50 m radius. This leaves an ungrazed portion in the center of the field. When three of the goats are moved, the rope tying the fourth goat is lengthened. This allows him to graze over an area equal to the combined grazing area of the four goats. How long is the rope? **100 m**

# EXPONENT PROPERTIES

*Objective*

**To multiply powers with the same base**

Recall that for $a^5$, $a$ is called the *base* and $5$ is called the *exponent*. You can read $a^5$ as the *fifth power* of $a$.

Since $a^5$ means $a \cdot a \cdot a \cdot a \cdot a$ and $a^2$ means $a \cdot a$,
$a^5 \cdot a^2$ means $(a \cdot a \cdot a \cdot a \cdot a) \cdot (a \cdot a)$, or $a^7$.
**Thus, $a^5 \cdot a^2 = a^{5+2}$, or $a^7$.**

This suggests that when you multiply powers with the same base, you keep the base and add the exponents. This property of exponents is stated below.

| | |
|---|---|
| Product of powers with the same base | $x^m \cdot x^n = x^{m+n}$ |

*Example 1*

**Simplify.**

| $x^5 \cdot x^6$ ◀ *Same base* | $a^8 \cdot a$ ◀ *Same base* |
|---|---|
| $x^{5+6}$ | $a^8 \cdot a^1$ ◀ $a = a^1$ |
| $x^{11}$ | $a^{8+1}$, or $a^9$ |

*Example 2*

**Simplify $x^3 \cdot y^4$.**

$x^3 \cdot y^4 = x \cdot x \cdot x \cdot y \cdot y \cdot y \cdot y$ ◀ *Different bases*
$\qquad = x^3 \cdot y^4$
**Thus, $x^3 \cdot y^4$ cannot be further simplified.**

You can use the product of powers rule to simplify an expression like $(-6t^4)(-2t)$. The commutative and associative properties allow you to regroup the constants and the like bases so that multiplication is easier. The process is illustrated in Example 3.

*Example 3*

**Simplify.**

| $(-6t^4)(-2t)$ | $(-4a^3b^2)(-ab^5)$ |
|---|---|
| $(-6 \cdot -2)(t^4 \cdot t^1)$ ◀ $t = t^1$ | $(-4a^3b^2)(-1ab^5)$ |
| $12 \cdot t^{4+1}$ | $(-4 \cdot -1)(a^3 \cdot a^1)(b^2 \cdot b^5)$ |
| $12t^5$ | $4 \cdot a^{3+1} \cdot b^{2+5}$ |
| | $4a^4b^7$ |

**Example 4** Simplify $(-a^2)(-2a)(3a^2)$. Then evaluate for $a = -2$.

$$(-1a^2)(-2a^1)(3a^2)$$
$$(-1 \cdot -2 \cdot 3)(a^2 \cdot a^1 \cdot a^2)$$
$$6a^5 \qquad \blacktriangleleft \quad a^2 \cdot a^1 \cdot a^2 = a^{2+1+2} = a^5$$
$$6(-2)^5 \qquad \blacktriangleleft \quad \text{Substitute } -2 \text{ for } a.$$
$$6(-32), \text{ or } -192 \quad \blacktriangleleft \quad (-2)^5 = -2 \cdot -2 \cdot -2 \cdot -2 \cdot -2 = -32$$

## Reading in Algebra

**For each expression on the left, find its correct match on the right.**

1. $a^5 \cdot a$ **f**         **a.** $b^9$        **d.** $a^5$
2. $b^7 \cdot b^2$ **a**         **b.** 0        **e.** $b^{14}$
3. Exponent of $a$ in $b^3 a$ **c**         **c.** 1        **f.** $a^6$

## Written Exercises

(A) **Simplify if possible.**

1. $m^4 \cdot m^6$ **$m^{10}$**         2. $b^7 \cdot b^2$ **$b^9$**         3. $x^4 \cdot y^2$ **$x^4 y^2$**         4. $p \cdot p^2$ **$p^3$**
5. $(5a^3)(6a^7)$ **$30a^{10}$**         6. $(6n^4)(3n^3)$ **$18n^7$**         7. $(5m^4)(4m^5)$ **$20m^9$**         8. $(3b^2)(4b^3)$ **$12b^5$**
9. $(a^3)(6a^4)$ **$6a^7$**         10. $(5b^4)(b^2)$ **$5b^6$**         11. $(8a^7)(a^5)$ **$8a^{12}$**         12. $(m^6)(3m^4)$ **$3m^{10}$**
13. $(-3a^4)(5a)$ **$-15a^5$**         14. $(4b^2)(-5b)$ **$-20b^3$**         15. $(-5b^7)(3b)$ **$-15b^8$**         16. $(-3b^4)(4b)$ **$-12b^5$**
17. $(4a^2 b^4)(3a^3 b^2)$ **$12a^5 b^6$**         18. $(5x^3 y^4)(2x^4 y^3)$ **$10x^7 y^7$**         19. $(7m^3 n^2)(-5m^3 n^6)$ **$-35m^6 n^8$**
20. $(-3ab^3)(7a^2 b^5)$ **$-21a^3 b^8$**         21. $(-3a^4 b)(-5ab^2)$ **$15a^5 b^3$**         22. $(xy)(xy)$ **$x^2 y^2$**
23. $(-a^3 b)(-2a^4 b)$ **$2a^7 b^2$**         24. $(-b^3 m^5)(-2bm)$ **$2b^4 m^6$**         25. $(-4x^5 y^6)(-2xy)$ **$8x^6 y^7$**
(B) 26. $(-p^2)(4p)(-2p^5)$ **$8p^8$**         27. $(-5r^2)(-2r)(4r)$ **$40r^4$**         28. $(6a^3)(-2a)(5a)$ **$-60a^5$**
29. $(u^2 vw^3)(u^3 v^2 w^4)$ **$u^5 v^3 w^7$**         30. $(-2abc)(4a^2 bc^4)$ **$-8a^3 b^2 c^5$** 31. $(-2x^2 yz)(xy^3 z)$ **$-2x^3 y^4 z^2$**
32. $x^n \cdot x^{3n}$ **$x^{4n}$**         33. $a^{3x+1} \cdot a^{-3x+2}$ **$a^3$**         34. $a^{-3x+4} \cdot a^{-3x+5}$ **$a^{-6x+9}$**

**Simplify. Then evaluate for $a = -3$ and $b = -2$.**

35. $(-a^3)(-5a)$ **405**         36. $(-2b^2)(b)$ **16**         37. $(4a^3)(-a)$ **$-324$**
38. $(a^3)(-a)(a)$ **243**         39. $(-2b)(-3b^2)(b)$ **96**         40. $(-8b)(-3b^3)(-b^2)$ **$-1,536$**

**Find the value of $a$ that makes each sentence true.**

(C) 41. $x^{2a-1} \cdot x^{2a+10} = x^{17}$ **2**         42. $y^{3a-2} \cdot y^{-2a+5} = y^7$ **4**
43. $x^{3a-1} \cdot x^{2a-5} = x^{2a+12}$ **6**         44. $p^{2a-5} \cdot p^{3a+4} = p^{2a+1} \cdot p^{a+6}$ **4**

## CUMULATIVE REVIEW

**Solve.**

1. $3x - (4 - x) = 12$ **4**         2. $|2a - 6| = 8$ **7, $-1$**         3. $3x - 5 < 5x - 7$ **$x > 1$**

# POWERS OF POWERS AND OF PRODUCTS    6.2

*Objectives*    **To simplify the power of a power**
**To simplify the power of a product**

The expression $(x^3)^5$ is a power of a power. To simplify $(x^3)^5$, use the definition of exponents and the product of powers as shown.

$$(x^3)^5 = x^3 \cdot x^3 \cdot x^3 \cdot x^3 \cdot x^3$$
$$= x^{3+3+3+3+3}, \text{ or } x^{3 \cdot 5}$$
$$= x^{15}$$

**Thus,** $(x^3)^5 = x^{15}$

You can multiply the exponents to simplify a power of a power.

| Power of a power | $(x^m)^n = x^{m \cdot n}$ |
|---|---|

*Example 1*    **Simplify.**

$(x^4)^6$        $(x^9)^3$
$(x^4)^6 = x^{4 \cdot 6}$        $(x^9)^3 = x^{9 \cdot 3}$
      $= x^{24}$             $= x^{27}$

In Example 2 we will use the power of a product property, which is stated below.

| Power of a product | $(x \cdot y)^m = x^m \cdot y^m$ |
|---|---|

*Example 2*    **Simplify.**

$(-3x)^4 = (-3 \cdot x)^4$        $(-2xy)^5 = (-2 \cdot x \cdot y)^5$
         $= -3^4 \cdot x^4 = 81x^4$            $= -2^5 \cdot x^5 \cdot y^5 = -32x^5y^5$

Notice that a negative number raised to an *even* power is *positive*, but when raised to an *odd* power, it is *negative*.

It is easy to combine the rules for the power of a power and the power of a product.
For example, $(a^4b^5)^3 = (a^4)^3 \cdot (b^5)^3$
$$= a^{4 \cdot 3} \cdot b^{5 \cdot 3}$$
$$= a^{12} \cdot b^{15}$$
**Thus,** $(a^4b^5)^3 = a^{4 \cdot 3} \cdot b^{5 \cdot 3}$, or $a^{12}b^{15}$.

This technique is used in Example 3.

*Example 3*    **Simplify:** $(3y^5xz^2)^4$.

$$(3y^5xz^2)^4 = (3^1y^5x^1z^2)^4 \qquad \blacktriangleleft \; 3 = 3^1; \; x = x^1$$
$$= 3^{1\cdot4} \cdot y^{5\cdot4} \cdot x^{1\cdot4} \cdot z^{2\cdot4}$$
$$= 3^4 \cdot y^{20} \cdot x^4 \cdot z^8$$
$$= 81y^{20}x^4z^8 \qquad \blacktriangleleft \; 3^4 = 3 \cdot 3 \cdot 3 \cdot 3 = 81$$

In the next example, you will use the power properties to simplify a product consisting of more than one factor.

*Example 4*    **Simplify** $(3xy)^2(x^3y)^3(xy^5)$.

$$(3xy)^2(x^3y)^3(xy^5) = (3^1x^1y^1)^2(x^3y^1)^3 \cdot x^1y^5$$
$$= 3^{1\cdot2} \cdot x^{1\cdot2} \cdot y^{1\cdot2} \cdot x^{3\cdot3} \cdot y^{1\cdot3} \cdot x^1 \cdot y^5$$
$$= 3^2 \cdot x^2 \cdot y^2 \cdot x^9 \cdot y^3 \cdot x^1 \cdot y^5$$
$$= 9(x^2 \cdot x^9 \cdot x^1) \cdot (y^2 \cdot y^3 \cdot y^5) \qquad \blacktriangleleft \; \textit{Group like bases together.}$$
$$= 9x^{12}y^{10}$$

# Written Exercises

**Simplify.**

Ⓐ  **1.** $(a^3)^2$  $a^6$ **2.** $(c^5)^6$  $c^{30}$ **3.** $(y^4)^3$  $y^{12}$ **4.** $(p^2)^6$  $p^{12}$ **5.** $(a^5)^8$  $a^{40}$
**6.** $(4x^3)^2$  $16x^6$ **7.** $(3y^2)^4$  $81y^8$ **8.** $(-2a^5)^2$  $4a^{10}$ **9.** $(-3n^3)^3$  $-27n^9$ **10.** $(-b^4)^5$  $-b^{20}$
**11.** $(x^2y^5)^4$  $x^8y^{20}$ **12.** $(r^3b)^5$  $r^{15}b^5$ **13.** $(xy)^4$  $x^4y^4$ **14.** $(a^3b)^7$  $a^{21}b^7$ **15.** $(rt^4)^5$  $r^5t^{20}$
**16.** $(-2xy^3)^4$  $16x^4y^{12}$ **17.** $(x^2yz^6)^3$  $x^6y^3z^{18}$ **18.** $(a^4bc^7)^3$  $a^{12}b^3c^{21}$ **19.** $(x^3y^2z^8)^6$  $x^{18}y^{12}z^{48}$
**20.** $(-5a^2b^3c^4)^3$  $-125a^6b^9c^{12}$ **21.** $(2x^4y^3z)^5$  $32x^{20}y^{15}z^5$ **22.** $(-3p^4q^3r^5)^3$  $-27p^{12}q^9r^{15}$

Ⓑ  **23.** $a(3a^2)^2$  $9a^5$ **24.** $5m(6m)^2$  $180m^3$ **25.** $x^3(x^5y)^2$  $x^{13}y^2$ **26.** $2a(-3ab)^3$  $-54a^4b^3$
**27.** $(2x^2)(xy^3)^2(xy^2)^3$  $2x^7y^{12}$ **28.** $(ab)^3(a^2b)^3(a^3b)$  $a^{12}b^7$ **29.** $(5ab)^2(a^4b)^2(ab^6)$  $25a^{11}b^{10}$
**30.** $(x^{2a+3})^2(x^{a+1})^3$  $x^{7a+9}$ **31.** $(x^3y)^a(xy^5)^a$  $x^{4a}y^{6a}$ **32.** $(x^{n-4}y)^3(x^6y^{n+4})^2$
$$x^{3n}y^{2n+11}$$

**Find the value of $a$ that will make each sentence true.**

Ⓒ  **33.** $(x^{2a-3})^4 = x^4$  2 **34.** $(x^{2a+1})^2 = x^{2a+6}$  2 **35.** $(b^{a+1})^5 = b^{3a+9}$  2
**36.** $(x^ay^{a-4})^3 = x^{3a}y^{a+8}$  10 **37.** $(y^{3a+5})^2(y^a)^4 = y^{8a+12}$  1

# CALCULATOR ACTIVITIES

**Simplify. Then evaluate for the given values of the variables.**    573,593.76
**1.** $(3a^2b)^3$ for $a = 2.4$ and $b = 1.07$ **2.** $(-16xy^4)^2$ for $x = -2.9$ and $y = 2.01$
**3.** $(1.07r^4s^2)^3$ for $r = 2.6$ and $s = 0.2$ **4.** $(x^2y^4)^2(xy)^3$ for $x = 1.1$ and $y = 2$
7.4818928                   6,320.9528                   3,990.97262

# SIMPLIFYING POLYNOMIALS

*Objectives*
**To determine the degree of a polynomial**
**To classify polynomials**
**To simplify polynomials by combining like terms**

Each of the following is an example of a term called a monomial.
$-7$ is a constant.
$m$ is a variable.
$4y^3$ is a product of a constant and a variable.
$-3xa^5 b$ is a product of a constant and several variables.

| Definition: Monomial | A **monomial** is a term that is either a constant, a variable, or a product of a constant and one or more variables. |
|---|---|

The *degree of a monomial* is the sum of the exponents of all of its variables. A nonzero constant has degree 0, and the constant 0 has no degree.

*Example 1*  **Find the degree of each monomial.**

$7x^2$         $-8a^3 b^6$         $6xym^5$         $9$
degree: 2   degree: $3 + 6$, or 9   $6x^1 y^1 m^5$   degree: 0
                                degree: $1 + 1 + 5$, or 7

The *polynomial* $x^3 + 5x^2 - 2x + 4$ contains several terms, or monomials.

| Definition: Polynomial | A **polynomial** is a monomial, or a sum, or difference of monomials. |
|---|---|

Polynomials of one, two, or three terms have these special names:

| Terms | 1 | 2 | 3 |
|---|---|---|---|
| **Name** | *MONO*mial | *BI*nomial | *TRI*nomial |
| **Example** | $6x^5$ | $7b^2 + 1$ | $3a^2 + 5a + 2$ |

*Example 2*  **Classify each polynomial as either a monomial, binomial, or trinomial.**

$5a^2 + 6a + 8$         $5m^2 - 2$         $14x^2$
3 terms: trinomial   2 terms: binomial   one term: monomial

You can simplify a polynomial, such as $11 + 9x^3 + 3x^2 + 6x^3 - 7x^2 - 4$, by first grouping the like terms and then combining them.

$$11 + 9x^3 + 3x^2 + 6x^3 - 7x^2 - 4$$
$$(9x^3 + 6x^3) + (3x^2 - 7x^2) + (11 - 4)$$
$$15x^3 - 4x^2 + 7$$

Notice that the result, $15x^3 - 4x^2 + 7$, is written in descending order of exponents. The highest exponent, 3, comes first, then comes the next highest exponent, 2, and then the constant.

The **degree of a polynomial** is the highest degree of any of its terms after it has been simplified. For example, the degree of the polynomial $15x^3 - 4x^2 + 7$ is 3, because the highest degree of a term is 3.

*Example 3*    **Simplify: $7x^3 - 10x^5 + 4x - 3x^2 - 8x^3 - 6x + 5$. Write the result in descending order of exponents. What is the degree of the polynomial?**

$7x^3 - 10x^5 + 4x - 3x^2 - 8x^3 - 6x + 5$
$-10x^5 + (7x^3 - 8x^3) - 3x^2 + (4x - 6x) + 5$ ◀ *Group like terms in descending order of exponents.*

$-10x^5 - 1x^3 - 3x^2 - 2x + 5$ ◀ *Combine like terms.*
$-10x^5 - x^3 - 3x^2 - 2x + 5$ ◀ *$-1x^3 = -x^3$*

The result is in descending order of exponents.
The degree is 5, because the greatest degree of a term is 5.

*Example 4*    **Simplify: $-a^2 - 7 + 6a^4 - 5a + 4a^3 - 3a^4 + 7a^2 + 5a + 5$.**

$-a^2 - 7 + 6a^4 - 5a + 4a^3 - 3a^4 + 7a^2 + 5a + 5$
$(6a^4 - 3a^4) + 4a^3 + (-1a^2 + 7a^2) + (-5a + 5a) + (-7 + 5)$ ◀ *$-a^2 = -1a^2$*
$\quad 3a^4 \qquad + 4a^3 + \qquad 6a^2 \qquad + \qquad 0 \qquad - 2$ ◀ *$-5a + 5a = 0$*
$\qquad\qquad 3a^4 + 4a^3 + 6a^2 - 2$

In the polynomial $6b^5 + 8b^2$, the terms $6b^5$ and $8b^2$ are not like terms. $6b^5 + 8b^2$ cannot be simplified further.

Some of the terms of the polynomial $3x^2y^3 + 4x^3 + 8x^2y^3 - 7x^3$ contain more than one variable. In this case you can simplify by combining like terms and then write it in descending order of exponents of *one* of the variables.

*Example 5*    **Simplify $3x^2y^3 + 4x^3 + 8x^2y^3 - 7x^3$. Write in descending order of the exponents of $x$.**

$(3x^2y^3 + 8x^2y^3) + (4x^3 - 7x^3)$ ◀ *Group like terms.*
$\qquad 11x^2y^3 - 3x^3$
$\qquad -3x^3 + 11x^2y^3$ ◀ *Descending order of exponents of x*

## Reading in Algebra

1. Name and define some nonmathematical words with prefixes poly-, mono-, bi-, and tri-. **Answers may vary, e.g., polygamy, monorail, bicycle, trimester.**
2. What is meant by descending order of exponents? **highest exponent first, lowest exponent last**
3. Which monomial, $7x^6$ or $7x^2y^4$, has the greater degree? Why? **same degree, 6**
4. Why are $6x^2$ and $6y^2$ unlike terms? **variables not the same**

## Oral Exercises

**What is the degree of each polynomial?**

1. $7x^4$ **4**
2. $5x^3a^6$ **9**
3. $3x^3 + 5x^2 - 4x + 2$ **3**
4. $5a^3b^2$ **5**
5. 8 **0**
6. $5x^4 - 3x^3 + 5$ **4**
7. $x^8y^3$ **11**
8. $5b^4 - 2$ **4**

**Classify each polynomial as a monomial, binomial, or trinomial.**
**12. monomial**

9. $7x^2 - 8x$ **binomial**
10. $-6x^2$ **monomial**
11. $8m^3 - 5m^2 + 6m$ **trinomial**
12. 5
13. $x^2 - 7x + 1$ **trinomial**
14. $8x^4 - 3x^2 - 2x$ **trinomial**
15. $3x$ **monomial**
16. $5y^2 - 25$ **binomial**

**Give the polynomial in descending order of exponents.** 17. $3x^3 + 4x^2 + 2x$

17. $4x^2 + 3x^3 + 2x$
18. $-5x^2 - 6x^3 + 6x - 4$
19. $2 + 5x + x^2$  $x^2 + 5x + 2$
20. $3 - x - 2x^2 - x^3$  $-x^3 - 2x^2 - x + 3$
21. $x^3 + x^4 - 2 + x^2$  $x^4 + x^3 + x^2 - 2$
22. $a + a^3 - 5 + a^2$  $a^3 + a^2 + a - 5$

## Written Exercises

18. $-6x^3 - 5x^2 + 6x - 4$

1. $4x^2 - 3x - 9$   2. $6x^2 - 5x + 2$   3. $12a^2 - 9a - 5$   4. $13m^2 + 4m + 6$   5. $2a^2 + 13a - 7$

**Simplify. Write the result in descending order of exponents.**

Ⓐ
1. $4x^2 - 6x + 3x - 9$
2. $6x^2 - 9x + 4x + 2$
3. $2 + 7a^2 - 3a + 5a^2 - 6a - 7$
4. $7m^2 - 5m + 2 + 6m^2 + 9m + 4$
5. $9a - 3a^2 + 4a - 3 + 5a^2 - 4$
6. $-8t + 3t^2 - 6 + 4t^2 - 9t - 4$  $7t^2 - 17t - 10$
7. $3 - 5b^2 - 4b - 9 + 4b^2 - 3b$
8. $-6 + 3m^2 - 3 - 2m^2$  $m^2 - 9$
9. $-5y - 7y^2 + 6y^3 - 4y^2 + 3y^3 + 6y$
10. $5x^3 - 4x - 7x^2 + 9x - 2x^3 + 7x^2$  $3x^3 + 5x$
11. $n^3 - 8n - 4n^2 + 8n + 3n^2 + 9n^3$
12. $8a^4 - 4a^2 + 6a^4 + 5a^2$  $14a^4 + a^2$
13. $2a - 4a^4 + 8a + 3a^4$  $-a^4 + 10a$
14. $7q - 5q^2 + 2 + 4q$  $-5q^2 + 11q + 2$
15. $6x^5 - 4x^3 + x^4 - 6x^6$

7. $-b^2 - 7b - 6$   9. $9y^3 - 11y^2 + y$   11. $10n^3 - n^2$   15. $-6x^6 + 6x^5 + x^4 - 4x^3$

Ⓑ **Simplify. Write the result in descending order of exponents of $x$ or $m$.**
16. $-3m^4b + 5a^3 - 7m^4b - 6a^3$  $-10m^4b - a^3$
17. $5x^3y^2 - 4xy - 6x^3y^2 - 8xy + 5y^7$  $-x^3y^2 + 12xy + 5y^7$
18. $-6x^3y^2 + 4y^3 + 8x^3y^2 - 5y^3$  $2x^3y^2 - y^3$
19. $-7m^2 - 8m^4n^3 - 3n^2 - 2m^4n^3$  $-10m^4n^3 - 7m^2 - 3n^2$
20. $-5xy^2 - 3x^5 - 4xy^2 + 2x^5$  $-x^5 - 9xy^2$
21. $-3a^3x^2 - 4a^2 + 5a^3x^2 + 4x^4$  $4x^4 + 2a^3x^2 - 4a^2$

**Simplify. Write in descending order of exponents.**
22. $2 + 4a^3 - 5a - 3a^2 + 8a^3 - 9a + 5a - 6$  $12a^3 - 3a^2 - 9a - 4$
23. $5b^2 - 4b - 2 + b^2 - 3b^2 + 2b^3 + 7b + 4$  $2b^3 + 3b^2 + 3b + 2$
24. $x^2 + x^4 - 3x^3 - 8x - 1 + 4x^4 - 7x^3 + 9x - 5 + x^2$  $5x^4 - 10x^3 + 2x^2 + x - 6$
25. $7a - 5 + 8a^2 - 5a^3 + 3a^2 + 8a^3 - 6a^2 - a^4 - 10 + 8a$
$-a^4 + 3a^3 + 5a^2 + 15a - 15$

# ADDING AND SUBTRACTING POLYNOMIALS

*Objectives*   **To add polynomials**
**To subtract polynomials**

To add two polynomials, group like terms, in descending order of exponents if possible. Then simplify by combining like terms.

*Example 1*   **Add $(2a^2 - 2a + 1) + (3a^2 - 4a + 6)$.**

$(2a^2 - 2a + 1) + (3a^2 - 4a + 6)$
$\underbrace{2a^2 + 3a^2}\ \underbrace{-2a - 4a}\ \underbrace{+1 + 6}$   ◄ *Group like terms.*
$\qquad\quad 5a^2 - 6a + 7$   ◄ *Combine like terms.*

*Example 2*   **Add $(2a^4 - 4a + 9) + (-a^4 + a^6 + 4a - 2a^5)$.**

$(2a^4 - 4a + 9) + (-a^4 + a^6 + 4a - 2a^5)$
$(2a^4 - 4a + 9) + (-1a^4 + 1a^6 + 4a - 2a^5)$   ◄ $-a^4 = -1a^4;\ a^6 = 1a^6$
$1a^6 - 2a^5 \underbrace{+ 2a^4 - 1a^4}\ \underbrace{- 4a + 4a} + 9$
$1a^6 - 2a^5 \qquad + 1a^4 \qquad + 0 \qquad + 9$
$\qquad\quad a^6 - 2a^5 + a^4 + 9$

Recall that $-a = -1 \cdot a$. You will apply this property to simplify $-(-2x^2 + 7x - 4)$ in Example 3 and to subtract the polynomials in Example 4.

*Example 3*   **Simplify $-(-2x^2 + 7x - 4)$.**

$-(-2x^2 + 7x - 4) = -1(-2x^2 + 7x - 4)$   ◄ $-a = -1 \cdot a$
$\qquad\qquad\qquad\quad = 2x^2 - 7x + 4$   ◄ *Distribute the $-1$.*

*Example 4*   **Subtract $(y^3 + 3y^2 - 9y) - (-3y^2 - 8y + 4)$.**

$y^3 + 3y^2 - 9y - 1(-3y^2 - 8y + 4)$
$y^3 + 3y^2 - 9y + 3y^2 + 8y - 4$   ◄ *Distribute the $-1$.*
$y^3 \underbrace{+ 3y^2 + 3y^2}\ \underbrace{- 9y + 8y} - 4$   ◄ *Group like terms.*
$y^3 \qquad + 6y^2 \qquad - 1y - 4$, or $y^3 + 6y^2 - y - 4$

Recall that by the definition of subtraction, subtract $a$ from $b$ means $b - a$. You will apply this in Example 5.

*Example 5*    **Subtract $y^3 - 5y^2$ from $7y^4 - 8y^3 + 3y^2 - 4$.**

$7y^4 - 8y^3 + 3y^2 - 4 - (y^3 - 5y^2)$
$7y^4 - 8y^3 + 3y^2 - 4 - 1(y^3 - 5y^2)$   ◀ $-(y^3 - 5y^2) = -1(y^3 - 5y^2)$
$7y^4 - 8y^3 + 3y^2 - 4 - 1y^3 + 5y^2$   ◀ *Distribute the $-1$.*
$7y^4 - 8y^3 - 1y^3 + 3y^2 + 5y^2 - 4$
$7y^4 - 9y^3 + 8y^2 - 4$

## Written Exercises

**Add.**   **1. $5a^2 - 10a + 9$**                **2. $-b^2 - b + 8$  4. $-7b^2 - 9b - 5$**

Ⓐ  **1.** $(3a^2 - 7a + 4) + (2a^2 - 3a + 5)$         **2.** $(7b^2 - 3b + 2) + (-8b^2 + 2b + 6)$
**3.** $(y^2 - y + 1) + (3y^2 - y - 5)$ **$4y^2 - 2y - 4$**   **4.** $(-4b^2 - 7b + 1) + (-3b^2 - 2b - 6)$
**5.** $(a^4 - 7a + 9) + (a^5 - a^4 + 2a^3)$         **6.** $(x^3 - 2x - 4) + (x^4 - x^3 + 3x^2)$
**7.** $(x^4 - 9x + 5) + (2x^5 - x^4 + 3x^3)$         **8.** $(a^3 - 5a + 2) + (a^4 - 2a^3 + 5a - 4)$
**5. $a^5 + 2a^3 - 7a + 9$  7. $2x^5 + 3x^3 - 9x + 5$**   **6. $x^4 + 3x^2 - 2x - 4$  8. $a^4 - a^3 - 2$**

**Simplify.**                                             **11. $y^2 - 5y + 4$**
**9.** $-(5b^2 - 2b)$ **$-5b^2 + 2b$**   **10.** $-(a^2 - a - 9)$ **$-a^2 + a + 9$  11.** $-(-y^2 + 5y - 4)$
**12.** $-(-x^2 - x - 3)$ **$x^2 + x + 3$  13.** $-(6x^2 - 7x)$ **$-6x^2 + 7x$**   **14.** $-(-2m^2 - 5m + 6)$
**15.** $-(-a^2 - 5a - 2)$                **16.** $-(-7m^2 - 5m + 1)$        **17.** $-(-x^3 - x^2 + 2x - 4)$
**14. $2m^2 + 5m - 6$  15. $a^2 + 5a + 2$  16. $7m^2 + 5m - 1$  17. $x^3 + x^2 - 2x + 4$**

Ⓑ  **Subtract. (Ex. 18–21)**                **19. $a^3 + 11a^2 - 7a + 2$**
**18.** $(x^3 + 4x^2 - 6x) - (-2x^2 + 4x - 9)$   **19.** $(a^3 + 5a^2 - 4a) - (-6a^2 + 3a - 2)$
**20.** $(5m^3 - m^2 - m - 3) - (-m^2 - 2m + 8)$   **21.** $(c^3 - 4c^2 - 7c + 1)$ **21. $4c^3 + c^2 - 3c$**
       **18. $x^3 + 6x^2 - 10x + 9$  20. $5m^3 + m - 11$**       $- (-3c^3 - 5c^2 - 4c + 1)$
**22.** Subtract $x^3 - 5x^2$ from $8x^4 + x^3 + 5x - 2$.   **23.** Subtract $a^2 - 3a$ from $6a^3 - 8a^2 - 7a$.
**24.** Subtract $-a^3 - 3a$ from $a^4 + 2a^2 - 7a + 1$.   **25.** Subtract $-m^2 - 5m$ from $-m^3 - m^2 - 6m$.
**26.** Add. $(5m^3 + 2\frac{2}{3}m^2 - m) + \left(-4\frac{1}{2}m^3 + 4\frac{5}{6}m^2 + \frac{1}{3}m\right)$ **$\frac{1}{2}m^3 + 7\frac{1}{2}m^2 - \frac{2}{3}m$**
**27.** Subtract. $(3a^3 - 5.1a + 6.9) - (0.04a^3 + 8a^2 - 7a + 9)$ **$2.96a^3 - 8a^2 + 1.9a - 2.1$**
**22. $8x^4 + 5x^2 + 5x - 2$  24. $a^4 + a^3 + 2a^2 - 4a + 1$  23. $6a^3 - 9a^2 - 4a$  25. $-m^3 - m$**
Ⓒ  **28.** Subtract the sum $(-x^2 + 5x - 4) + (2x^2 - 6x - 8)$ from $-x^2 - 7x + 5$. **$-2x^2 - 6x + 17$**
**29.** From $-x^3 - x^2 + 5x - 4$, subtract the sum $(5x^3 - 4x^2 - 8x + 2) + (-6x^3 + 3x^2 + 7x - 5)$.
                                                                            **$6x - 1$**

## CUMULATIVE REVIEW

**1.** Solve $\frac{2}{3}x - \frac{1}{6} = -\frac{3}{4}$. **$-\frac{7}{8}$**    **2.** Evaluate $18a^3$ for a $= -\frac{2}{3}$. **$-\frac{16}{3}$ 3.** Solve $|2x - 4| = 8$. **6, $-2$**

# MULTIPLYING POLYNOMIALS

**Objectives**  **To simplify polynomials**
**To multiply a polynomial by a monomial**

The expression $x^2 + 5x + 6 + 3(x^2 - 4x + 5)$ can be simplified by first applying the distributive property. This is illustrated in Example 1.

*Example 1*  **Simplify $x^2 + 5x + 6 + 3(x^2 - 4x + 5)$.**

$$\begin{aligned}
x^2 + 5x + 6 + 3(x^2 - 4x + 5) &= x^2 + 5x + 6 + 3(1x^2 - 4x + 5) \\
&= x^2 + 5x + 6 + 3x^2 - 12x + 15 \quad \blacktriangleleft \textit{ Distribute the 3.} \\
&= (x^2 + 3x^2) + (5x - 12x) + (6 + 15) \quad \blacktriangleleft \textit{ Group like terms.} \\
&= 4x^2 - 7x + 21 \quad \blacktriangleleft \textit{ Combine like terms.}
\end{aligned}$$

The distributive property can also be used to simplify the product of a monomial and a polynomial when exponents are involved. The product of powers property will be used also. This is illustrated in Examples 2 and 3.

*Example 2*  **Multiply $-x^3(5x^4 + 6x^2)$.**

$$\begin{aligned}
-x^3(5x^4 + 6x^2) &= -x^3 \cdot 5x^4 + (-x^3)6x^2 \\
&= -5 \cdot x^4 \cdot x^3 - 6 \cdot x^3 \cdot x^2 \\
&= -5x^7 - 6x^5 \quad \blacktriangleleft \ x^m \cdot x^n = x^{m+n}
\end{aligned}$$

*Example 3*  **Multiply $5a^6(3a^5 - 2a^3 + 4a)$.**

$$\begin{aligned}
5a^6(3a^5 - 2a^3 + 4a) &= 5a^6(3a^5 - 2a^3 + 4a^1) \quad \blacktriangleleft \textit{ a means } a^1. \\
&= 5a^6 \cdot 3a^5 + 5a^6 \cdot -2a^3 + 5a^6 \cdot 4a^1 \\
&= 15a^{11} - 10a^9 + 20a^7
\end{aligned}$$

In the following example, there are two variables involved.

*Example 4*  **Simplify $x^4y^3(3x^4 - 5xy + y^2)$.**

$$\begin{aligned}
x^4y^3(3x^4 - 5xy + y^2) &= x^4y^3(3x^4 - 5x^1y^1 + y^2) \\
&= x^4y^3 \cdot 3x^4 + x^4y^3 \cdot -5x^1y^1 + x^4y^3 \cdot y^2 \\
&= 3x^8y^3 - 5x^5y^4 + x^4y^5
\end{aligned}$$

## Oral Exercises

**Multiply.**

1. $2(2a^2 - 6a + 7)$    $4a^2 - 12a + 14$
2. $7(3x^2 - 5x + 2)$    $21x^2 - 35x + 14$
3. $4(2a^2 - 3a + 5)$    $8a^2 - 12a + 20$
4. $a^2(a^4 + a^3)$   $a^6 + a^5$
5. $b^5(b^4 + b^2)$   $b^9 + b^7$
6. $m^2(m^3 - m^2)$   $m^5 - m^4$
7. $k^3(k^4 + k)$   $k^7 + k^4$
8. $m(m^5 + m^2)$   $m^6 + m^3$
9. $p(p^2 + p)$   $p^3 + p^2$
10. $3(4g^3 - 3g^2 + 2g + 6)$   $12g^3 - 9g^2 + 6g + 18$
11. $5(2y^3 - 4y^2 + y - 3)$   $10y^3 - 20y^2 + 5y - 15$
12. $a^2(a^3 + a^2 + a + 5)$   $a^5 + a^4 + a^3 + 5a^2$

## Written Exercises

**Ⓐ Simplify.**

1. $a^2 + 7a + 2 + 4(a^2 - 2a + 6)$   $5a^2 - a + 26$
2. $m^2 - 3m + 6 + 5(2m^2 - m - 4)$   $11m^2 - 8m - 14$
3. $-3a^2 - 4a + 8 + 2(2a^2 - a + 1)$
4. $-t^2 + 5t - 8 + 3(4t^2 - t + 5)$
5. $4c^2 - c + 3 - 2(3c^2 - c + 5)$   $-2c^2 + c - 7$
6. $-3a^2 - a + 5 - 4(a^2 - a + 1)$
7. $p^2 - p - 6 - 2(3p^2 + p - 1)$
8. $y^2 - y - 1 - 2(y^2 - y + 5)$   $-y^2 + y - 11$

3. $a^2 - 6a + 10$    7. $-5p^2 - 3p - 4$    4. $11t^2 + 2t + 7$    6. $-7a^2 + 3a + 1$

**Multiply.**

9. $f^2(3f^4 - 6f + 5)$   $3f^6 - 6f^3 + 5f^2$
10. $t^3(4t^2 - 7t + 3)$   $4t^5 - 7t^4 + 3t^3$
11. $a(3a^2 - 4a + 7)$   $3a^3 - 4a^2 + 7a$
12. $2b^2(5b + 6)$   $10b^3 + 12b^2$
13. $7a^2(a^3 - 5a^2 + 4)$
14. $2y^3(3y^4 - 7y^3 + 2y^2)$
15. $5b(3b^2 - 6b)$   $15b^3 - 30b^2$
16. $3x(4x^2 - 3x)$   $12x^3 - 9x^2$
17. $2c(6c^3 - 5c^2)$   $12c^4 - 10c^3$
18. $-2a(-3a^2 - 7a + 4)$   $6a^3 + 14a^2 - 8a$
19. $-x(3x^2 - x - 5)$   $-3x^3 + x^2 + 5x$
20. $-g(-g^2 - 4g + 2)$   $g^3 + 4g^2 - 2g$

**Ⓑ Simplify.**

21. $m^4n^3(3m^2 - 2mn + n^5)$   $3m^6n^3 - 2m^5n^4 + m^4n^8$
22. $xy(x^3 - 3xy + 5y^2)$   $x^4y - 3x^2y^2 + 5xy^3$
23. $-4ac(a^3 - 4a^2c + ac^2 - c^3)$
24. $-6x^3y(-x^3 - 2x^2y + xy^2 + y^3)$
25. $3ab(a^2 - 6ab + 7b^2)$   $3a^3b - 18a^2b^2 + 21ab^3$
26. $2xy(x^2 - xy - 3y^2)$   $2x^3y - 2x^2y^2 - 6xy^3$
27. $-x^2y^3(3x^2 - xy + 5y^2)$
28. $-tw^2(t^2 - tw^2 - w^3)$   $-t^3w^2 + t^2w^4 + tw^5$
29. $-a^3b^2(a^2 - 3ab + 5b^2)$   $-a^5b^2 + 3a^4b^3 - 5a^3b^4$
30. $-x^3y(x^2 - xy^2 + 2xy^3)$   $-x^5y + x^4y^3 - 2x^4y^4$

13. $7a^5 - 35a^4 + 28a^2$    14. $6y^7 - 14y^6 + 4y^5$

23. $-4a^4c + 16a^3c^2 - 4a^2c^3 + 4ac^4$    27. $-3x^4y^3 + x^3y^4 - 5x^2y^5$    24. $6x^6y + 12x^5y^2 - 6x^4y^3 - 6x^3y^4$

**Ⓒ**

31. $m^3n^2(5m^2 + 3mn + n^2) + m^3n^2(m^2 - 3mn + 2n^2)$   $6m^5n^2 + 3m^3n^4$
32. $3ab(a^2 - 5ab + 4b^2) - 2ab(6a^2 - 7ab - 8b^2)$   $-9a^3b - a^2b^2 + 28ab^3$
33. $-2ab^2(-a^3 - 4a^2b + ab^2 - b^3) - 5ab^2(a^3 - a^2b - ab^2 + 4b^3)$

    $-3a^4b^2 + 13a^3b^3 + 3a^2b^4 - 18ab^5$

34. $-xy^2(5x^2 - 3xy + 4y^2) + xy^2(2x^2 - xy - 5y^2)$   $-3x^3y^2 + 2x^2y^3 - 9xy^4$

## CUMULATIVE REVIEW

**Solve.**

1. The larger of two numbers is 3 less than twice the smaller. The sum of the numbers is 12. Find each number. **5, 7**

2. The perimeter of a rectangle is 52 m. The length is 2 m longer than 3 times the width. Find the length and the width. **length: 20 m, width: 6 m**

# FINDING THE MISSING FACTORS                               6.6

*Objectives*  **To factor a number into primes**
**To find the missing factor, given a product and one of its factors**

Sometimes, you can write a number as a product of factors in several ways.
For example, 40 can be factored as $20 \cdot 2$, or
$$10 \cdot 4, \text{ or}$$
$$4 \cdot 5 \cdot 2, \text{ or}$$
$$2 \cdot 2 \cdot 5 \cdot 2.$$
In the last case, neither of the factors, 2 and 5, can be factored further.
This is true because the only factors of 2 are 2 and 1, and the only factors of
5 are 5 and 1. The numbers 2 and 5 are called *prime numbers*.

| Definition: Prime number | A **prime number** is a whole number greater than 1 whose only factors are itself and 1. |
|---|---|

A number is said to be factored into primes if each of its factors is a prime
number. To find the **prime factorization** of a number, that is, to factor it
into primes, you can begin by using any two factors of the number.

*Example 1*    **Factor 24 into primes.**

You can begin in more than one way.
$$24 = 6 \cdot 4 \qquad\qquad 24 = 12 \cdot 2 \qquad\qquad 24 = 8 \cdot 3$$
$$= 3 \cdot 2 \cdot 4 \qquad\qquad = 4 \cdot 3 \cdot 2 \qquad\qquad = 4 \cdot 2 \cdot 3$$
$$= 3 \cdot 2 \cdot 2 \cdot 2 \qquad = 2 \cdot 2 \cdot 3 \cdot 2 \qquad = 2 \cdot 2 \cdot 2 \cdot 3$$
Notice that each gives the same prime factors, but in a different order. In
each case, 2 is used as a factor three times, and 3 is used as a factor once.
**Thus,** $2 \cdot 2 \cdot 2 \cdot 3$, or $2^3 \cdot 3$ is the prime factorization of 24.

According to the product of powers property, $a^5 \cdot a^3 = a^8$. You can use
this property to find a missing factor when you are given one factor and the
product.

*Example 2*    **Find the missing factor.**

$$a^7(?) = a^9$$
$$a^7 \cdot a^{\square} = a^9$$
$$a^{7+\square} = a^9 \qquad \blacktriangleleft \; x^m \cdot x^n = x^{m+n}$$
$$a^{7+2} = a^9 \qquad \textbf{Thus,} \text{ the missing factor is } a^2.$$

$$(?)(a) = a^5$$
$$a^{\square} \cdot a^1 = a^5 \qquad \blacktriangleleft \; a \text{ means } a^1.$$
$$a^{\square+1} = a^5$$
$$a^{4+1} = a^5 \qquad \textbf{Thus,} \text{ the missing factor is } a^4.$$

The procedure in Example 2 can be used when the missing factor consists of a numerical coefficient and a power. This is shown in the next example.

*Example 3*    **Find the missing factor in $(7a^5)(?) = -28a^8$.**

$$(7a^5)(?) = -28 \cdot a^8$$

$$(7a^5)(?) = 7 \cdot (?) \cdot (a^5 \cdot a^\square)$$  ◄ *You need a numerical factor and a factor of the form $a^\square$.*
$$= 7 \cdot -4 \cdot a^5 \cdot a^3$$  ◄ *$-28 = 7 \cdot (-4)$ and $a^8 = a^5 \cdot a^3$*

**Thus,** the missing factor is $-4a^3$, because $(7a^5)(-4a^3) = -28a^8$.

# Reading in Algebra

1. Give an example of a number that is not prime; then give its factors.  **6, (1, 2, 3, 6); 8, (1, 2, 4, 8);**
2. If 21 is factored into two numbers, one of which is 7, what is the **etc. Examples may vary.**
   other factor?  **3**
3. Which of the following is the prime factorization of 48?
   **a.** $2 \cdot 2 \cdot 2 \cdot 3 \cdot 3$   **b.** $12 \cdot 4$   **c.** $2^4 \cdot 3$   **d.** $2^3 \cdot 6$  **c**

# Written Exercises

Ⓐ **Factor into primes if possible.**   **1.** $2^2 \cdot 3$   **8.** $2^5$

**1.** 12   **2.** 7 **prime**  **3.** 24 $2^3 \cdot 3$  **4.** 49 $7^2$   **5.** 28 $2^2 \cdot 7$  **6.** 19 **prime**  **7.** 45 $3^2 \cdot 5$  **8.** 32
**9.** 42   **10.** 11 **prime**  **11.** 50 $2 \cdot 5^2$  **12.** 41 **prime** **13.** 48 $2^4 \cdot 3$  **14.** 60   **15.** 36 $2^2 \cdot 3^2$ **16.** 23
**9.** $2 \cdot 3 \cdot 7$   $2^2 \cdot 3 \cdot 5$   **16. prime**

**Find the missing factor.**
**17.** $(a^6)(?) = a^{11}$ $a^5$   **18.** $(a^7)(?) = a^{12}$ $a^5$   **19.** $(?)(x^6) = x^{13}$ $x^7$   **20.** $(?)(m^4) = m^8$ $m^4$
**21.** $(p)(?) = p^7$ $p^6$   **22.** $(b)(?) = b^{10}$ $b^9$   **23.** $(?)(r^6) = r^6$ $1$   **24.** $(?)(x) = x^3$ $x^2$
**25.** $(7m^5)(?) = 35m^9$ $5m^4$   **26.** $(4a^6)(?) = 28a^{10}$ $7a^4$   **27.** $(-3x^5)(?) = 30x^7$ $-10x^2$
**28.** $(?)(3x^4) = 15x^{12}$ $5x^8$   **29.** $(?)(5b^3) = 25b^7$ $5b^4$   **30.** $(?)(-8t^5) = 24t^6$ $-3t$
**31.** $(3m^4)(?) = 3m^6$ $m^2$   **32.** $(?)(5a^4) = 5a^5$ $a$   **33.** $(?)(-4a) = -16a^2$ $4a$

*Example*    $(3a^2b^4)(?) = 21a^3b^6$

$$(3a^2b^4)(?) = 21 \cdot a^3 \cdot b^6$$
$$(3a^2b^4)(?) = 3 \cdot (?) \cdot a^2 \cdot a^\square \cdot b^4 \cdot b^\square$$
$$= 3 \cdot 7 \cdot a^2 \cdot a^1 \cdot b^4 \cdot b^2$$  **Thus,** the missing factor is $7a^1b^2$, or $7ab^2$.

Ⓑ **34.** $(5a^3b^4)(?) = 30a^6b^5$ $6a^3b$   **35.** $(6a^2m^7)(?) = 24a^5m^9$ $4a^3m^2$
**36.** $(?)(3a^4m^6) = -27a^5m^8$ $-9am^2$   **37.** $(?)(2a^2b^5) = -2a^3b^6$ $-ab$
**38.** $\left(\frac{1}{3}a^2b^5\right)(?) = -5a^3b^7$ $-15ab^2$   **39.** $\left(-\frac{2}{3}m^3n^4\right)(?) = \frac{1}{2}m^3n^9$ $-\frac{3}{4}n^5$

# FACTORING OUT THE GREATEST COMMON MONOMIAL

*Objective*  **To factor a polynomial as a product of a monomial and a polynomial**

You have used the distributive property to multiply a polynomial and a monomial. For example, $3(x^2 + 7x + 5) = 3(x^2) + 3(7x) + 3(5) = 3x^2 + 21x + 15$. In this lesson, you will use the distributive property in reverse to rewrite $3x^2 + 21x + 15$ as $3(x^2 + 7x + 5)$. The 3 is called a *common monomial factor*.

The technique for factoring a polynomial as a product of a monomial and a polynomial is illustrated below. It involves finding the greatest factor common to each term of the polynomial. You then use the distributive property in reverse.

*Example 1*  **Factor the common monomial from $2a^2 + 6a + 10$.**

Step 1: Find a number that is a factor of each term, $2a^2$, $6a$, and 10. Such a number is 2. Then write each term as a product of monomials.

$$2a^2 + 6a + 10$$
$$2(a^2) + 2(3a) + 2(5)$$
$$2(a^2 + 3a + 5)$$

Step 2: Use the distributive property in reverse.

Since no number greater than 2 is a common factor of each term, $2a^2$, $6a$, and 10, 2 is the *greatest common factor*, written *GCF*, of $2a^2 + 6a + 10$.

*Example 2*  **Factor the GCF from $4b^2 - 20b - 4$.**

$$4b^2 - 20b - 4$$
$$4(b^2) + 4(-5b) + 4(-1)$$
$$4(b^2 - 5b - 1)$$

◄ *4 is the GCF of each term, because 4 is the greatest number that is a factor of each term.*
◄ *Use the distributive property in reverse.*

Sometimes, you may not recognize the GCF immediately. When this happens, you will find it helpful to factor the monomial coefficient of each term into primes.
For example, $12a^2 + 30a + 18$ can be rewritten as
$$2 \cdot 2 \cdot 3 \cdot a^2 + 2 \cdot 3 \cdot 5 \cdot a + 2 \cdot 3 \cdot 3.$$
Since there are at most one 2 and one 3 common to each term, $2 \cdot 3$, or 6, is the greatest common factor of $12a^2 + 30a + 18$.

*Example 3*     **Factor out the GCF from $12a^2 - 28a + 72$.**

$$12a^2 - 28a + 72$$
$$\underbrace{2 \cdot 2} \cdot 3 \cdot a^2 - \underbrace{2 \cdot 2} \cdot 7 \cdot a + \underbrace{2 \cdot 2} \cdot 2 \cdot 3 \cdot 3 \quad \blacktriangleleft \textit{ Factor 12, 28, and 72 into primes.}$$

Since there are at most two 2's common to each term, the GCF is $2 \cdot 2$ or 4.
$$12a^2 - 28a + 72 = 4(3a^2) + 4(-7a) + 4(18) \quad \blacktriangleleft \textit{ Use 4 as a factor of each term.}$$
$$= 4(3a^2 - 7a + 18) \quad \blacktriangleleft \textit{ Use the distributive property in reverse.}$$

The GCF of a polynomial may be a variable. The procedure in this case is very much the same as when the GCF is a whole number.

*Example 4*     **Factor the GCF from $b^3 - 2b^2 + b$.**

$$b^3 - 2b^2 + b \quad \blacktriangleleft \textit{ The coefficient of } b^3 \textit{ and } b \textit{ is 1. The GCF of the coefficients is 1.}$$
$$b^3 - 2b^2 + b^1 \quad \blacktriangleleft \textit{ } b \textit{ means } b^1.$$

$$b \cdot b \cdot b - 2 \cdot b \cdot b + b^1$$

Since there is at most one $b$ in common to each term, the GCF is $b^1$.
$$b^3 - 2b^2 + b^1 = b^1(b^2) + b^1(-2b^1) + b^1(1) \quad \blacktriangleleft \textit{ Use } b^1 \textit{ as a factor of each term.}$$
$$= b^1(b^2 - 2b^1 + 1)$$
$$= b(b^2 - 2b + 1)$$

Sometimes, the GCF consists of a greatest common whole number factor and a greatest common variable factor. This is illustrated in the next example.

*Example 5*     **Factor the GCF from $3x^5 + 9x^3 + 15x^2$.**

$$3x^5 + 9x^3 + 15x^2 \quad \blacktriangleleft \textit{ Look for the greatest common whole number factor, if any.}$$
$$3 \cdot x^5 + 3 \cdot 3 \cdot x^3 + 3 \cdot 5 \cdot x^2 \quad \blacktriangleleft \textit{ Factor the coefficients into primes.}$$
$$3(x^5 + 3x^3 + 5x^2) \quad \blacktriangleleft \textit{ Factor out 3, the greatest common whole number factor.}$$
$$3(x \cdot x \cdot x \cdot x \cdot x + 3 \cdot x \cdot x \cdot x + 5 \cdot x \cdot x) \quad \blacktriangleleft \textit{ Look for the greatest common variable factor, if any.}$$

Since there are at most two $x$'s common to each term, the GCF is $x \cdot x$ or $x^2$.
$$3[x^2(x^3) + x^2(3x^1) + x^2(5)] \quad \blacktriangleleft \textit{ Use } x^2 \textit{ as a factor of each term.}$$
$$3x^2(x^3 + 3x^1 + 5) \quad \blacktriangleleft \textit{ Use the distributive property in reverse.}$$
$$3x^2(x^3 + 3x + 5)$$

**Thus,** $3x^5 + 9x^3 + 15x^2 = 3x^2(x^3 + 3x + 5)$.

## Example 6    Factor out the GCF from $3x^7 - 10x^4 + x^3$.

$3x^7 - 10x^4 + x^3$ ◀ *Look for the greatest common whole number factor, if any.*

$3 \cdot x^7 - 2 \cdot 5 \cdot x^4 + 1 \cdot x^3$ ◀ *The GCF is 1. It is not necessary to factor out the 1.*

$\underset{\text{seven }x\text{'s}\quad\text{four }x\text{'s}\quad\text{three }x\text{'s}}{3x^7 - 10x^4 + x^3}$ ◀ *Look for the greatest common variable factor, if any.*

Since there are at most three $x$'s common to each term, the GCF is $x^3$.

$3x^7 - 10x^4 + x^3 = x^3(3x^4) + x^3(-10x^1) + x^3(1)$ ◀ *Use $x^3$ as a factor of each term.*

$\qquad = x^3(3x^4 - 10x^1 + 1)$, or $x^3(3x^4 - 10x + 1)$

## Summary    To factor out the greatest common monomial factor:

**1.** First, factor out the greatest common whole number factor, if any.

**2.** Second, factor out the greatest common variable factor, if any.

## Oral Exercises

**What is the greatest common whole number factor for each polynomial?**

**1.** $3x^2 - 3x + 9$ **3**    **2.** $2a^2 - 4a + 6$ **2**    **3.** $5m^2 - 10m + 15$ **5**    **4.** $3m^2 + m + 2$ **1**

**What is the greatest common variable factor for each polynomial?**

**5.** $m^3 + m^2 + 3m$ **m**    **6.** $p^4 - 2p^3 + 5p^2$ **$p^2$**    **7.** $a^6 + a^4 + a^3 + a^2$    **8.** $g^3 - 5g^2 + 3g$ **g**

**7.** $a^2$

## Written Exercises

**10.** $b^2(b^2 + 3b + 2)$ **13.** $b(3b^2 + 5b - 7)$ **14.** $r^2(5r^2 - 3r + 1)$ **16.** $2x^2(x^2 + 2x + 3)$ **17.** $4a(a^2 - 4a + 8)$

**Factor out the GCF, if any, from each polynomial.** **24.** $3a(a^2 - 3a + 2)$    **3.** $2(m^2 - 5m + 2)$

Ⓐ **1.** $3x^2 + 9x + 6$ $3(x^2 + 3x + 2)$    **2.** $7a^2 - 21a + 35$ $7(a^2 - 3a + 5)$ **3.** $2m^2 - 10m + 4$

**4.** $5p^2 - 15p + 10$ $5(p^2 - 3p + 2)$ **5.** $4g^2 + 20g + 32$ $4(g^2 + 5g + 8)$ **6.** $4r^2 - 12r + 28$ $4(r^2 - 3r + 7)$

**7.** $6t^2 + 18t + 30$ $6(t^2 + 3t + 5)$ **8.** $6y^2 + 12y - 48$ $6(y^2 + 2y - 8)$ **9.** $8y^2 + 6y + 4$ $2(4y^2 + 3y + 2)$

**10.** $b^4 + 3b^3 + 2b^2$    **11.** $m^3 + m^2 + m$ $m(m^2 + m + 1)$ **12.** $a^6 + a^4 + 3a^2$ $a^2(a^4 + a^2 + 3)$

**13.** $3b^3 + 5b^2 - 7b$    **14.** $5r^4 - 3r^3 + r^2$    **15.** $a^5 - a^4 + 8a^3$ $a^3(a^2 - a + 8)$

**16.** $2x^4 + 4x^3 + 6x^2$    **17.** $4a^3 - 16a^2 + 32a$    **18.** $6m^2 - 24m$ $6m(m - 4)$

**19.** $7b^3 - 14b^2 + 49b$    **20.** $4a^3 - 32a$ $4a(a^2 - 8)$    **21.** $5r^2 - 35r$ $5r(r - 7)$

**22.** $3a^3 - 5a^2 + 7a$    **23.** $5t^8 - 7t^6 + 2t^4$    **24.** $3a^3 - 9a^2 + 6a$

**25.** $7b^3 - 5b^2 + 3b$    **26.** $4y^3 - 20y^2 + 24y$    **27.** $7m^3 - 35m^2$ $7m^2(m - 5)$

**19.** $7b(b^2 - 2b + 7)$ **22.** $a(3a^2 - 5a + 7)$ **23.** $t^4(5t^4 - 7t^2 + 2)$ **25.** $b(7b^2 - 5b + 3)$ **26.** $4y(y^2 - 5y + 6)$

Ⓑ **28.** $7a^5 - 42a^4 + 21a^3 - 14a^2$    **29.** $6a^4 - 12a^3 + 24a^2 + 3a$

**30.** $3a^2 + 15ab + 39b^2$ $3(a^2 + 5ab + 13b^2)$    **31.** $78a^3 - 104a^2 + 52a$ $26a(3a^2 - 4a + 2)$

**32.** $36m^4 - 54m^3 - 90m^2 + 72m$    **33.** $6y^4 - 48y^3 + 144y^2$ $6y^2(y^2 - 8y + 24)$

Ⓒ **34.** $x^{2m+5} + x^{2m}$ $x^{2m}(x^5 + 1)$    **35.** $5y^{4m+2} + 10y^{4m+6}$    **36.** $a^a + a^{a+5}$ $a^a(1 + a^5)$

**28.** $7a^2(a^3 - 6a^2 + 3a - 2)$ **29.** $3a(2a^3 - 4a^2 + 8a + 1)$ **32.** $18m(2m^3 - 3m^2 - 5m + 4)$ **35.** $5y^{4m+2}(1 + 2y^4)$

# ALGEBRAIC EXPRESSIONS FOR AREA

*Objective*     **To write algebraic expressions for area**

The diagram shows a circle inside a square. If the circle represents a flower garden and the shaded region represents a marble border, the cost of the border depends on its *area*. To find the area of this region, recall that the formula for the area of a circle is $A = \pi r^2$ and that the formula for the area of a square is $A = s^2$.

*Example 1*     **Find the area of the shaded region in the diagram above. Write the result in factored form.**

First, by drawing two radii, you can see that the side of the square is $2r$.

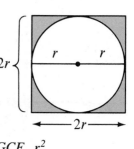

Area of shaded region = area of square − area of circle

$$A = 2r \cdot 2r - \pi r^2$$
$$= 4r^2 - \pi r^2$$
$$= r^2(4 - \pi) \quad \blacktriangleleft \ \textit{Factor out the GCF, } r^2.$$

*Example 2*     **Write a formula in factored form for the area $A$ of the region at the right.**

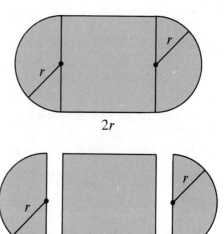

Note that the region is the combination of a square and *two* half circles. The area of the two half circles is the same as the area of the whole circle of radius $r$.

$A$ = area of square + area of circle
$A = 2r \cdot 2r + \pi r^2$
$A = 4r^2 + \pi r^2$
$A = r^2(4 + \pi) \quad \blacktriangleleft \ \textit{Factor out the GCF, } r^2.$

**Example 3**    **Write the formula for the area of the shaded region.**

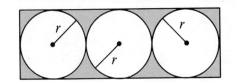

First, label the length and width of the rectangle in terms of the radius, $r$, of the circles.
length: $6r$
width: $2r$

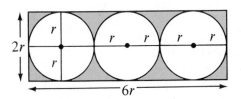

Area of shaded region = area of rectangle − area of three circles
$$A = 6r \cdot 2r - 3(\pi r^2)$$
$$= 12r^2 - 3\pi r^2$$
$$= 3r^2(4 - \pi) \quad \blacktriangleleft \ Factor\ out\ of\ the\ GCF,\ 3r^2.$$

## Written Exercises

**Write a formula in factored form for the area of the shaded region.**

**1.**    $2r^2(4 - \pi)$

**2.** 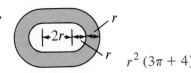   $r^2\left(8 + \dfrac{\pi}{2}\right)$

**3.**    $r^2(4 - \pi)$

**4.**   $4ab + \dfrac{\pi a^2}{2} + \dfrac{\pi b^2}{2}$

**5.**    $r^2(4 - \pi)$

**6.**    $r^2(10 - \pi)$

**7.**   $r^2(\pi - 2)$

**8.** 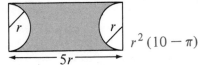   $r^2(3\pi + 4)$

**9.** 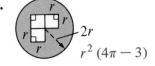   $r^2(4\pi - 3)$

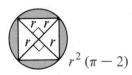

Algebraic Expressions for Area

**Vocabulary**

binomial [6.3]
degree [6.3]
greatest common factor [6.7]
monomial [6.3]
polynomial [6.3]
prime factorization [6.6]
prime number [6.6]
trinomial [6.3]

Simplify [6.1]
1. $x^4 \cdot x^8$ **$x^{12}$**   2. $(2a^2)(7a^6)$ **$14a^8$**
3. $(-p^3)(5p)(-2p^4)$ **$10p^8$**
4. $(-3a^3bc)(-a^4bc^3)$ **$3a^7b^2c^4$**
5. $(-4a^2b)(7a^3b^9)$ **$-28a^5b^{10}$**
6. $(a^{3b-4})(a^{4b+6})$ **$a^{7b+2}$**

Simplify. Then evaluate for $a = 3$ and $b = -2$. [6.1]
7. $-a^4 \cdot 3a$ **$-3a^5$, $-729$**   8. $-b^3 \cdot b^2$ **$-b^5$, $32$**

Simplify. [6.2]
9. $(x^5)^2$ **$x^{10}$**   10. $(x^3y^4)^7$ **$x^{21}y^{28}$**
11. $(4a^2b^3)^2$ **$16a^4b^6$**   12. $(x^5y^2z)^5$ **$x^{25}y^{10}z^5$**
13. $b(4b^2)^2$ **$16b^5$**   14. $(xy)^4(x^2y^3)^2$ **$x^8y^{10}$**

What is the degree of each polynomial? [6.3]
15. $6x^3 + 5x^2 - 2x + 4$ **3**   16. $3x^2y^5$ **7**

Classify each polynomial as a monomial, binomial, or trinomial. [6.3]
17. $m^2 - 4m + 2$   18. $6y$   19. $5a^3 + 4$
**17. trinomial   18. monomial   19. binomial**
Simplify. Write the result in descending order of exponents. [6.3]
20. $3x^2 + 8x + 4x + 7$ **$3x^2 + 12x + 7$**
21. $4a^2 - 7a + 2a + 6$ **$4a^2 - 5a + 6$**
22. $3 + 4a^3 - 7a - 2a^2 + 6a^3 - 8a$
$+ 4a^2 - 6a$ **$10a^3 + 2a^2 - 21a + 3$**

Add. [6.4] **23. $2a^4 + 2a^2 - 9a - 1$**
23. $(a^4 - 6a + 5) + (a^4 - 3a + 2a^2 - 6)$
24. $(-6b^2 - 8.1b + 1) + (5b^2 + 9b - 1)$
**$-b^2 + 0.9b$**

Simplify. [6.4]
25. $-(-y^3 + 4y^2 - 6y)$ **$y^3 - 4y^2 + 6y$**
26. $-(-x^3 - 8x^2 + 7x)$ **$x^3 + 8x^2 - 7x$**

27. Subtract $(x^3 + 5x^2 - 6x) - (-4x^2$
$+ 6x - 9)$.   **$x^3 + 9x^2 - 12x + 9$**
28. Subtract $6x^3 - 7x^2$ from $9x^4 - x^3$
$+ 6x^2 - 4$. **$9x^4 - 7x^3 + 13x^2 - 4$**

29. Subtract $\frac{1}{3}g^2 - 2g$ from $5g^3 + 4g^2 + g$.
**$5g^3 - 3\frac{2}{3}g^2 + 3g$**

Multiply. [6.5]
30. $x^2(3x^4 - 6x + 2)$   **$3x^6 - 6x^3 + 2x^2$**
31. $2y^3(4y^5 - 6y^3 + y^2)$ **$8y^8 - 12y^6 + 2y^5$**
32. $-3a(-4a^2 - 9a + 5)$ **$12a^3 + 27a^2 - 15a$**
33. $m^5n^2(3m^3 - 2mn^2 + n^6)$
34. $-t^2w^3(t^4 - t^2w - w^4)$
**33. $3m^8n^2 - 2m^6n^4 + m^5n^8$   34. $-t^6w^3 + t^4w^4 + t^2w^7$**
Simplify. [6.5]
35. $(x^2 + 5x + 4) + 3(-2x^2 - 7x - 3)$
36. $(5a^2 - 4a - 2) + \frac{1}{3}(-3a^2 + 6a - 9)$
**35. $-5x^2 - 16x - 5$     36. $4a^2 - 2a - 5$**
Factor into primes if possible. [6.6]
37. $54$ **$2 \cdot 3^3$**     38. $71$ **prime**
Find the missing factor. [6.6]
39. $(a^7)(?) = a^{12}$ **$a^5$**   40. $(4a^3)(?) = -8a^4$
41. $(?)(6x^3y^4) = -12x^4y^9$   **$-2xy^5$**
42. $(3a^2y^4)(?) = -15a^3y^5$   **$-5ay$**   **40. $-2a$**

Factor out the GCF, if any. [6.7]
43. $7x^2 - 14x - 28$ **$7(x^2 - 2x - 4)$**
44. $6a^2 - 13a - 11$ **no GCF**
45. $x^7 + x^5 - x^3$   **$x^3(x^4 + x^2 - 1)$**
46. $3b^3 + 15b^2 + 9b$ **$3b(b^2 + 5b + 3)$**
47. $4a^3 - 12a^2 + 16a$   **$4a(a^2 - 3a + 4)$**
48. $7a^5 - 42a^4 + 35a^3 - 21a^2$
★ 49. $3p^{4m+2} + 9p^{2m+2}$
**48. $7a^2(a^3 - 6a^2 + 5a - 3)$   49. $3p^{2m+2}(p^{2m} + 3)$**
★ 50. Find the value of $a$ that will make
$(x^{3a+1})^2 = x^{4a+8}$ true. [6.2] **3**
★ 51. Simplify: [6.5]   **51. $-x^3y - 14x^2y^2 + 16xy^3$**
$2xy(x^2 - 4xy + 2y^2) - 3xy(x^2 + 2xy - 4y^2$

# CHAPTER SIX TEST

Classify each polynomial as a monomial, binomial, or trinomial.

**1.** $3m^2 - 7m + 4$ **trinomial**

**2.** $2a$ **monomial**

**3.** $x^2 + 4x$ **binomial**

Simplify. Write the result in descending order of exponents.

**4.** $6a^2 - 3a - 3 - 5a + 7a^2$ **$13a^2 - 8a - 3$**

**5.** $5 + 4b^3 - 9b - 3b^2 + 2b^3 + 2b^2 - 6$
**$6b^3 - b^2 - 9b - 1$**

**6.** $b^5 - 4b^4 - 3b^2 - 5b^4 - 2b^5$
$+ 4b + 5b^4$ **$-b^5 - 4b^4 - 3b^2 + 4b$**

What is the degree of each polynomial?

**7.** $7a^3 - 3a^2 + 4a - 2$ **3**

**8.** $4x^3y^2$ **5**

Simplify.

**9.** $t^6 \cdot t^2$ **$t^8$**

**10.** $(a^4b^2c^3)^3$ **$a^{12}b^6c^9$**

**11.** $2m(3m)^3$ **$54m^4$**

**12.** $-(-y^3 - 4y^2 - 7y)$ **$y^3 + 4y^2 + 7y$**

**13.** $(-4ab^2c)(-ab^3c)(-2a^2b^5c)$ **$-8a^4b^{10}c^3$**

**14.** $(-p^2)(3p)(-2p^5)$ **$6p^8$**

**15.** $(3x^2y^3)^2$ **$9x^4y^6$**

**16.** $(x^3 + 4x^2 - 7x)$
$- 3(-2x^3 + 8x^2 - x + 5)$
**$7x^3 - 20x^2 - 4x - 15$**

Multiply.

**17.** $2y^3(5y^4 - 7y^3 + y^2)$ **$10y^7 - 14y^6 + 2y^5$**

**18.** $-3x^2y(x^3 - 5xy^2 + 6xy^3 - y^3)$
**$-3x^5y + 15x^3y^3 - 18x^3y^4 + 3x^2y^4$**

**19.** Add $(a^4 - 7a^2 + a) + (3a^4 - 2a^2 + 11a)$.
**$4a^4 - 9a^2 + 12a$**

**20.** Subtract $3x^2 - 5x - 2 - (4x^2 - x + 3)$.
**$-x^2 - 4x - 5$**

**21.** Subtract $x^2 + 5\frac{1}{2}x$ from $5x^3 - 6x^2 + 7\frac{1}{4}x$.
**$5x^3 - 7x^2 + 1\frac{3}{4}x$**

**22.** Factor 42 into primes. **$2 \cdot 3 \cdot 7$**

**23.** Simplify $-4a^3 \cdot 2a^2$. Then evaluate for $a = -2$. **$-8a^5$, 256**

Factor out the GCF, if any.

**24.** $5a^2 + 20a - 25$ **$5(a^2 + 4a - 5)$**

**25.** $m^4 - 3m^2 + 7m$ **$m(m^3 - 3m + 7)$**

**26.** $3x^2 + 5x - 2$ **no GCF**

**27.** $3a^3 - 18a^2 - 30a + 9$ **$3(a^3 - 6a^2 - 10a + 3)$**

**28.** $36m^4 - 27m^3 - 45m^2 + 18m$
**$9m(4m^3 - 3m^2 - 5m + 2)$**

Find the missing factor.

**29.** $(a^8)(?) = a^{13}$ **$a^5$**

**30.** $(?)(5a^2y^6) = -20a^3y^7$ **$-4ay$**

**31.** $(-3x^3y^2)(?) = 27x^5y^3$ **$-9x^2y$**

★ **32.** Factor out the GCF, if any.
$7x^{2a+5} + 35x^{3a+5}$ **$7x^{2a+5}(1 + 5x^a)$**

★Find the value of $a$ which will make each sentence true.

**33.** $x^{3a-12} \cdot x^{2a+13} = x^{9a-7}$ **2**

**34.** $(p^{4m-2})^3 = p^{18}$ **2**

# COMPUTER ACTIVITIES

## Computing Compound Interest

Many banks offer compound interest on savings accounts. When money is deposited, the bank pays interest not only on that money, the principal, but also on the interest earned. Interest payments can be made:

| annually | semiannually | quarterly |
|---|---|---|
| (once a year) | (twice a year) | (four times a year) |

The formula for finding the amount that a given principal is worth after a given number of years is $a = p\left(1 + \dfrac{r}{n}\right)^{ny}$ where

$p$ is the principal,
$n$ is the number of times a year interest is paid,
$r$ is the rate of interest written as a decimal, and
$y$ is the number of years the principal is left in the bank.

Below is a program for finding $a$ in the compound interest formula.

```
10    INPUT "WHAT IS THE PRINCIPAL? ";P
20    INPUT "HOW MANY TIMES A YEAR IS INTEREST PAID? " ;N
30    INPUT "WHAT IS THE RATE OF INTEREST AS A DECIMAL ? ";R
40    INPUT "HOW MANY YEARS IS THE PRINCIPAL LEFT IN
      THE BANK? ";Y
50    LET A = P * (1 + R / N) ^ (N * Y)
60    PRINT "THE AMOUNT AFTER "Y" YEARS IS $"A"."
70    END
```

See the Computer Section beginning on page 488 for more information.

*Example:*    John invested $2,000 at 8% compounded semiannually. To use the program above to find the amount he will have after 10 years, what must you type in for $p$, $n$, $r$, and $y$?

| | |
|---|---|
| Principal is $2,000 | $p = 2000$ |
| Interest is paid semiannually | $n = 2$ |
| Rate of interest is 8% | $r = .08$ |
| Number of years is 10 | $y = 10$ |

## Exercises

**In Exercises 1–3, round each answer to the nearest cent.**

1. RUN the program above to find $a$ for the problem in the Example. **$4,382.25**

2. Maxine inherited $10,000 which she deposited at 8% compounded semiannually. How much will she have at the end of 3 years? **$12,653.19**

3. $1,500 was invested at 6% compounded quarterly. Find the amount after 20 years. Find the amount after 40 years. Does the amount double? **$4,935.99, $16,242.69, no**

# COLLEGE PREP TEST

Directions: Choose the one best answer to each question or problem.

**1.** What is the area of the shaded region in this figure?

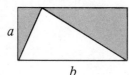

(A) $a + b$          (B) $ab$

(C) $\frac{1}{2}ab$         (D) $2a + b$

(E) cannot be determined **C**

**2.** Which value of $x$ gives the largest value of $y$ if $y = 6 - (x - 2)^2$?
(A) 1   (B) 2   (C) 5
(D) 6   (E) 10 **B**

**3.** The perimeter of a rectangle is given by the formula $p = 2l + 2w$ where $l$ is the length and $w$ is the width. If $16x + 18$ represents the perimeter and $2x + 6$ represents the width, then the length is
(A) $18x + 24$   (B) $14x + 12$   (C) $6x + 3$
(D) $12x + 6$   (E) $20x + 30$ **C**

**4.** What is the value of $1^{3a} + 1^{4a}$ for any whole number value of $a$?
(A) $1^{7a}$   (B) $12a^2$   (C) $2^{7a}$
(D) 2     (E) none of these **D**

**5.** If $0.3y = 7$, find $3.33y$.
(A) 2.1    (B) $23\frac{1}{3}$   (C) 70
(D) 77.7    (E) 777 **D**

**6.** If $3a + 12 = 27$, find the value of $2a + 4$.
(A) 15   (B) 14   (C) 11
(D) 5    (E) none of these **B**

**7.** If $x > y$ and $y > 0$, which of the following defines all possible values of $z$ if $xz < 0$?
(A) $z < 0$   (B) $z \neq 0$   (C) $z > x$
(D) $z < x$   (E) $z > 0$ **A**

**8.** How many inches are there in $k$ yards?
(A) $12k$   (B) $36k$
(C) $\frac{k}{36}$   (D) $\frac{k}{12}$   (E) $3k$ **B**

**9.** Find the next term in the pattern
1, 2, 5, 10, 17, 26, _____
(A) 33   (B) 35   (C) 37
(D) 43   (E) 52 **C**

**10.** What percentage of 1 hour is 1,800 seconds?
(A) $\frac{1}{1,800}\%$   (B) $\frac{1}{30}\%$   (C) 5%
(D) 50%     (E) 200% **D**

**11.** What is the perimeter of the figure?

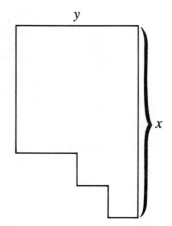

(A) $x + y$   (B) $x^2 + y^2$
(C) $xy$      (D) $2(x + y)$
(E) none of these **D**

# 7 FACTORING TRINOMIALS

**Non-Routine
Problem
Solving**

**Problem 1**  A young boy could not yet tell time but was counting the number of times the grandfather clock chimed. At 3:40 in the afternoon he told his mother that the clock had chimed 34 times. If the clock chimes the number of times of the hour on the hour and once on the half hour, what time was it when the boy began counting the chimes? **11 A.M.**

**Problem 2**  Suppose that you had a watch that had been damaged and had only the hour hand. Devise a system that will enable you to tell time with the watch. Use your system to tell the exact time it is (disregarding A.M. or P.M.) when the hour hand points to the 22-minute mark. **4:24**

# PRODUCTS OF POLYNOMIALS

*Objectives*  **To multiply two binomials**
**To multiply a polynomial and a binomial**

You used the distributive property to multiply a binomial by a monomial.
For example, $(3x + 4)2x = 3x \cdot 2x + 4 \cdot 2x$
$$= 6x^2 + 8x.$$
The distributive property can be used to simplify the product of two binomials,
such as $(3x + 4)(2x + 5)$. You can proceed as follows.
Let $a = (3x + 4)$.
Then, $(3x + 4)(2x + 5) = a(2x + 5)$.
This is the product of a monomial and a binomial
Therefore, $a(2x + 5) = a \cdot 2x + a \cdot 5$ by the distributive property.
$$\begin{aligned}
a \cdot 2x + a \cdot 5 &= (3x + 4)(2x) + (3x + 4)5 \text{ by substitution.} \\
&= 3x \cdot 2x + 4 \cdot 2x + 3x \cdot 5 + 4 \cdot 5 \\
&= 6x^2 + 8x + 15x + 20 \\
&= 6x^2 + 23x + 20
\end{aligned}$$
Notice that after $3x + 4$ was substituted for $a$, it was necessary to multiply
four times, as shown below.

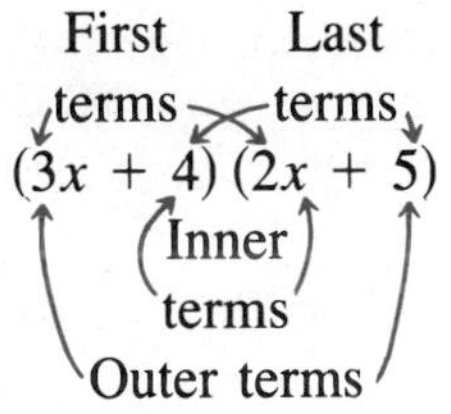

$$
\begin{array}{cc}
\text{First} & \text{Last} \\
\text{terms} & \text{terms} \\
(3x + 4) & (2x + 5) \\
& \text{Inner} \\
& \text{terms} \\
& \text{Outer terms}
\end{array}
$$

| 1. Multiply the First terms. | 2. Multiply the Outer terms. | 3. Multiply the Inner terms. | 4. Multiply the Last terms. |
|:---:|:---:|:---:|:---:|
| $3x \cdot 2x$ | $3x \cdot 5$ | $4 \cdot 2x$ | $4 \cdot 5$ |
| $6x^2$ | $15x$ | $8x$ | $20$ |
| ↑ | ↑ | ↑ | ↑ |
| F | O | I | L |

This procedure is known as the *FOIL* method.

*Example 1*  **Multiply $(3a - 5)(2a + 3)$.**

$$
\begin{array}{l}
(3a - 5)(2a + 3) \\
3a \cdot 2a + 3a \cdot 3 - 5 \cdot 2a - 5 \cdot 3 \\
\quad 6a^2 + \quad 9a - 10a \quad - 15 \\
\quad 6a^2 \qquad\quad - 1a \quad\; - 15 \\
\qquad\quad 6a^2 - a - 15 \quad \blacktriangleleft \quad -1a = -a
\end{array}
$$

## Example 2 Multiply $(4x - y)(x - 2)$.

$$(4x - 1y)(1x - 2y)$$

◀ $-y = -1y; x = 1x$

$$4x(1x) + 4x(-2y) - 1y(1x) - 1y(-2y)$$

◀ $4x \cdot -2y = 4 \cdot -2 \cdot x \cdot y = -8xy$

$$4x^2 - 8xy - 1xy + 2y^2$$

◀ $-1y \cdot 1x = -1 \cdot 1 \cdot x \cdot y = -1xy$

$$4x^2 - 9xy + 2y^2$$

The procedure for multiplying two binomials can be extended to multiplication of a binomial and a trinomial. This is illustrated in Example 3.

## Example 3 Multiply $(2x + 3)(3x^2 + 2x - 5)$.

$$\underbrace{(2x + 3)(3x^2 + 2x - 5)}$$   ◀ *Distribute 2x + 3.*

$$(2x + 3)3x^2 + (2x + 3)2x + (2x + 3) \cdot -5$$

$$6x^3 + 9x^2 + 4x^2 + 6x - 10x - 15$$

$$6x^3 + 13x^2 - 4x - 15$$   ◀ *Combine like terms: $9x^2 + 4x^2 = 13x^2$; $6x - 10x = -4x$.*

## Written Exercises   For Ex. 7–35, see TE Answer Section.

**1.** $6x^2 + 23x + 21$   **2.** $12y^2 + 31y + 20$   **3.** $6m^2 + 17m + 5$

**Multiply.**   **4.** $6a^2 + a - 1$   **5.** $8b^2 - 10b - 7$   **6.** $6r^2 - 19r + 15$

Ⓐ **1.** $(2x + 3)(3x + 7)$  **2.** $(3y + 4)(4y + 5)$  **3.** $(2m + 5)(3m + 1)$
 **4.** $(2a + 1)(3a - 1)$  **5.** $(4b - 7)(2b + 1)$  **6.** $(3r - 5)(2r - 3)$
 **7.** $(3m - 7)(m + 7)$  **8.** $(4y - 2)(y + 7)$  **9.** $(3k - 4)(2k + 5)$
 **10.** $(2k - 5)(5k - 2)$  **11.** $(4a - 1)(2a + 3)$  **12.** $(2h + 6)(h - 8)$
 **13.** $(3a - 2b)(4a + 3b)$  **14.** $(2y - m)(2y + 9m)$  **15.** $(3x + y)(2x - 3y)$
 **16.** $(3p - 2)(p + 1)$  **17.** $(2a - 3)(a - 1)$  **18.** $(2m - 5)(2m - 5)$
 **19.** $(5x - y)(2x - y)$  **20.** $(x + y)(x + y)$  **21.** $(5a + b)(2a - b)$
 **22.** $(2m + 8)(m - 1)$  **23.** $(3x - 4y)(x + y)$  **24.** $(3d - 2)(d + 3)$

Ⓑ **25.** $(3x - 1)(2x^2 + 4x + 3)$  **26.** $(2a + 7)(4a^2 - 3a + 2)$  **27.** $(x + 7)(3x^2 - x + 5)$
 **28.** $(t - 5)(t^2 - 7t + 1)$  **29.** $(3p + 2)(p^2 - 7p + 3)$  **30.** $(2g + 3)(g^2 - 8g - 4)$

Ⓒ **31.** $(3k - 2)(k^3 - 6k^2 + 5k)$  **32.** $(2x + 5)(x^3 - 2x^2 + x - 3)$  **33.** $(4 - 3a)(7 - 2a + a^2)$
 **34.** $(3x - 2)(3x + 2)(9x^2 + 4)$    **35.** $(x + y - 4)(x + y + 4)$

## CUMULATIVE REVIEW

**Simplify.**

**1.** $3\frac{1}{2} - 5$   $-1\frac{1}{2}$

**2.** $-2\frac{1}{3} \div -14$   $\frac{1}{6}$

**3.** $5x - (2 - 6x)$   $11x - 2$

# A SPECIAL PRODUCT: 7.2
# THE DIFFERENCE OF TWO SQUARES

*Objectives*  **To find products resulting in the difference of two squares**
**To factor the difference of two squares**

Sometimes the product of two binomials can result in a binomial rather than
a trinomial. This is illustrated in the example below.

*Example 1*  **Multiply $(3x + 7)(3x - 7)$.**

$$(3x + 7)(3x - 7) = 3x(3x) + 3x(-7) + 7(3x) + 7(-7)$$
$$= 9x^2 - \underbrace{21x + 21x} - 49$$
$$= 9x^2 \qquad + 0 \qquad - 49$$
$$= 9x^2 - 49$$

Notice a pattern for the product of Example 1. In the expression $9x^2 - 49$,
$9x^2$ can be written as $3x \cdot 3x$ or $(3x)^2$.
49 can be written as $7 \cdot 7$, or $(7)^2$.

Therefore, $(3x + 7)(3x - 7) = 9x^2 - 49$
$$= (3x)^2 - (7)^2, \text{ the } \textit{difference of two squares.}$$

This suggests the following formula.

| Formula for the difference of two squares | $(a + b) \quad (a - b) \quad = \quad a^2 - b^2$ |
|---|---|
| | $\qquad\uparrow \qquad\quad \uparrow \qquad\qquad\quad \uparrow$ |
| | $\text{sum} \quad \text{difference} \quad \textit{difference of two squares}$ |
| | Similarly, $(a - b)(a + b) = a^2 - b^2$. |

*Example 2*  **Use the formula for the difference of two squares to multiply.**

$(x - 5)(x + 5)$ | $(2a + 3b)(2a - 3b)$
$(x)^2 - (5)^2$ | $(2a)^2 - (3b)^2$    ◄ $(2a)^2 = 2a \cdot 2a = 4a^2;$
$x^2 - 25$ | $4a^2 - 9b^2$    $(3b)^2 = 3b \cdot 3b = 9b^2$

You can use the symmetric property of equality to write the expression
$$(a + b)(a - b) = a^2 - b^2 \text{ as}$$
$$a^2 - b^2 = (a + b)(a - b).$$

This form will be useful in *reversing* the multiplication process to find the
*factors* of an expression like $x^2 - 4$ or $25m^2 - 36$.
This is illustrated in the next example.

*Example 3*    **Factor each of the following.**

$$x^2 - 4 \qquad\qquad 25m^2 - 36 \qquad\qquad 36x^2 - y^2$$
$$(x)^2 - (2)^2 \qquad (5m)^2 - (6)^2 \qquad (6x)^2 - (y)^2$$
$$(x + 2)(x - 2) \qquad (5m + 6)(5m - 6) \qquad (6x + y)(6x - y)$$

# Oral Exercises

**Express each term in the form $(a)^2$.**

**1.** 9     **2.** 64     **3.** $16b^2$     **4.** $81x^2$     **5.** $121q^2$     **6.** $100r^2$     **7.** $49m^2$     **8.** $225f^2$
    **(3)²**       **(8)²**       **(4b)²**       **(9x)²**       **(11q)²**       **(10r)²**       **(7m)²**       **(15f)²**

# Written Exercises
**1.** $x^2 - 16$   **2.** $a^2 - 49$   **3.** $g^2 - 100$   **4.** $4x^2 - 9y^2$
**5.** $49a^2 - b^2$   **6.** $81k^2 - j^2$   **7.** $f^2 - 36h^2$   **8.** $16t^2 - 25r^2$   **9.** $4w^2 - 9y^2$
**Multiply.**   **10.** $169c^2 - 4d^2$   **11.** $64m^2 - 81n^2$   **12.** $121p^2 - 9u^2$   **13.** $(x + 4)(x - 4)$

Ⓐ **1.** $(x - 4)(x + 4)$       **2.** $(a + 7)(a - 7)$       **3.** $(g + 10)(g - 10)$
   **4.** $(2x + 3y)(2x - 3y)$     **5.** $(7a - b)(7a + b)$     **6.** $(9k + j)(9k - j)$
   **7.** $(f - 6h)(f + 6h)$       **8.** $(4t - 5r)(4t + 5r)$     **9.** $(2w + 3y)(2w - 3y)$
   **10.** $(13c - 2d)(13c + 2d)$    **11.** $(8m + 9n)(8m - 9n)$    **12.** $(11p - 3u)(11p + 3u)$
   **14.** $(m + 7)(m - 7)$   **15.** $(b + 5)(b - 5)$   **16.** $(a + 9)(a - 9)$   **17.** $(b + 1)(b - 1)$

**Factor as the difference of two squares.**    **For Ex. 21–44, see TE Answer Section.**

   **13.** $x^2 - 16$       **14.** $m^2 - 49$       **15.** $b^2 - 25$       **16.** $a^2 - 81$
   **17.** $b^2 - 1$       **18.** $m^2 - 100$      **19.** $p^2 - 64$       **20.** $y^2 - 36$
   **21.** $4b^2 - 25$      **22.** $9a^2 - 64$      **23.** $16m^2 - 25$     **24.** $49a^2 - 100$
   **25.** $1 - 64c^2$      **26.** $1 - 9x^2$       **27.** $9a^2 - 25$      **28.** $144 - 49a^2$
Ⓑ **29.** $144a^2 - 169$    **30.** $100x^2 - 121$    **31.** $121m^2 - 144$    **32.** $36m^2 - 169$
   **33.** $49 - 169a^2$     **34.** $x^2 - 196$      **35.** $169p^2 - 36$     **36.** $121m^2 - 225$

   **37.** $\dfrac{4}{25}x^2 - \dfrac{9}{49}$     **38.** $\dfrac{16}{81} - \dfrac{25}{144}a^2$     **39.** $\dfrac{121}{9}n^2 - \dfrac{25}{100}$     **40.** $\dfrac{49}{36} - \dfrac{64}{196}t^2$

Ⓒ **18.** $(m + 10)(m - 10)$    **19.** $(p + 8)(p - 8)$    **20.** $(y + 6)(y - 6)$

**Factor and then simplify.**

   **41.** $(3x - 2)^2 - (x + 5)^2$           **42.** $(5x + 3)^2 - (6x - 7)^2$
   **43.** $(k - 8)^2 - (2k - 6)^2$         **44.** $(a + b - 2c)^2 - (3a - 2b - 4c)^2$

## NON-ROUTINE PROBLEMS

At first glance, it seems impossible to factor $x^4 + 64$ because it is not the difference of two squares. Yet the correct factors are given in one of the pairs below. Which one? **d**
   **a.** $(x^2 + 8)^2$       **b.** $(x^2 + 8)(x^2 - 8)$       **c.** $(x^2 + 2x + 4)(x^2 - 8x + 16)$
   **d.** $(x^2 - 4x + 8)(x^2 + 4x + 8)$       **e.** $(x^2 - 4x + 8)(x^2 - 4x - 8)$

# A SPECIAL PRODUCT: A PERFECT SQUARE TRINOMIAL        7.3

*Objectives*   **To find the square of a binomial**
**To recognize and factor a perfect square trinomial**

*Example 1*   **Simplify.**

$(2x + 5)^2$                        $(3x - 2)^2$
$(2x + 5)(2x + 5)$                  $(3x - 2)(3x - 2)$
$4x^2 + 10x + 10x + 25$             $9x^2 - 6x - 6x + 4$
$4x^2 \qquad + 20x \quad + 25$      $9x^2 \qquad - 12x \quad + 4$

Notice a pattern for the products in Example 1.

$(2x + 5)^2 = 4x^2 + 10x + 10x + 25$        $(3x - 2)^2 = 9x^2 - 6x - 6x + 4$
$\qquad = (2x)^2 + 2(10x) + (5)^2$              $\qquad = (3x)^2 - 2(6x) + (2)^2$

This suggests a formula for the *square of a binomial*.

| | |
|---|---|
| **Formula for the square of a binomial** | $(a + b)^2 = (a)^2 + 2ab + (b)^2$ <br> $(a - b)^2 = (a)^2 - 2ab + (b)^2$ |

    **1.** *square* of the *first* term ⎯⎯⎯⎯⎯⎯⎯⎯⎯⎯
    **2.** *twice* the *product* of both terms ⎯⎯⎯⎯⎯
    **3.** *square* of the *last* term ⎯⎯⎯⎯⎯⎯⎯⎯⎯

This formula can be used to write the square of a *binomial* as a *trinomial*. The resulting trinomial is called a *perfect square trinomial*.

*Example 2*   **Simplify.**

$(4x - 3)^2$                                   $(2x + 5y)^2$
$(4x)^2 - \underbrace{2 \cdot 4x \cdot 3} + (3)^2$        $(2x)^2 + \underbrace{2 \cdot 2x \cdot 5y} + (5y)^2$

$16x^2 - \quad 24x \quad + 9$                  $4x^2 + \quad 20xy \quad + 25y^2$

You can use the *symmetric* property of equality to write the formula for squaring a binomial in a form useful for *factoring*.

$\qquad (a + b)^2 = (a)^2 + 2ab + (b)^2 \qquad (a - b)^2 = (a)^2 - 2ab + (b)^2$
$\qquad a^2 + 2ab + b^2 = (a + b)^2 \qquad a^2 - 2ab + b^2 = (a - b)^2$

This form will be useful in determining if a trinomial is a *perfect square* and then in factoring if it is one.

## Example 3

**Determine if $4x^2 - 12x + 9$ is a perfect square. If so, factor the trinomial.**

*Think*: $4x^2 - 12x + 9$ can be rewritten as
$(2x)^2 - 12x + (3)^2$ ◀ *The first and last terms are squares.*
$2 \cdot 2x \cdot 3 = 12x$ ◀ *Twice the product of 2x and 3 is the middle term (ignoring the sign).*
So, $4x^2 - 12x + 9$ is a *perfect square.*
Factor $(2x)^2 - 12x + (3)^2$.
$(2x - 3)(2x - 3)$, or $(2x - 3)^2$ ◀ *Take the sign of the middle term.*

## Example 4

**Determine if $25a^2 + 60ab + 49b^2$ is a perfect square. If so, factor the trinomial.**

*Think*: $25a^2 + 60ab + 49b^2$ can be rewritten as
$(5a)^2 + 60ab + (7b)^2$ ◀ *The first and last terms are squares.*
$2 \cdot 5a \cdot 7b = 70ab$ ◀ *This is not the middle term, 60ab.*
So, the trinomial is *not a perfect square.*
Therefore, $25a^2 + 60ab + 49b^2$ *cannot* be factored as a perfect square.

## Oral Exercises

**Which of the following are perfect squares? Answer yes or no.**

**1.** $(3x)^2 - 24x + (4)^2$ **yes**     **2.** $(2a)^2 + 20 + (5)^2$ **yes**     **3.** $(7m)^2 - 24m + (2)^2$ **no**
**4.** $25y^2 + 30y + 4$ **no**     **5.** $9a^2 - 6a + 1$ **yes**     **6.** $4y^2 - 20yx + 25x$ **yes**

## Written Exercises
1. $x^2 + 12x + 36$  2. $9a^2 - 24a + 16$  3. $25y^2 + 20y + 4$
4. $16t^2 + 8t + 1$  5. $9a^2 - 12a + 4$  6. $25y^2 - 10y + 1$  7. $36m^2 + 12m + 1$
Simplify.  8. $16g^2 - 40g + 25$  9. $4m^2 + 28mn + 49n^2$

Ⓐ
**1.** $(x + 6)^2$          **2.** $(3a - 4)^2$          **3.** $(5y + 2)^2$          **4.** $(4t + 1)^2$
**5.** $(3a - 2)^2$          **6.** $(5y - 1)^2$          **7.** $(6m + 1)^2$          **8.** $(4g - 5)^2$
**9.** $(2m + 7n)^2$          **10.** $(3y - 2x)^2$          **11.** $(b - 5c)^2$          **12.** $(3k + 5p)^2$
     10. $9y^2 - 12xy + 4x^2$  11. $b^2 - 10bc + 25c^2$  12. $9k^2 + 30kp + 25p^2$

**Determine if each is a perfect square. If so, factor the trinomial.**

**13.** $x^2 - 7x + 12$          **14.** $a^2 + 8a + 16$          **15.** $y^2 - 12y + 36$
**16.** $t^2 + 4t + 4$          **17.** $m^2 - 16m + 49$          **18.** $u^2 + 16u + 64$
**19.** $4x^2 + 6x + 1$          **20.** $9a^2 - 6a + 1$          **21.** $9a^2 - 12a + 4$
**22.** $4m^2 + 12m + 9$          **23.** $4a^2 - 30a + 25$          **24.** $16a^2 - 24a + 9$
**25.** $9x^2 + 30xy + 25y^2$          **26.** $25a^2 - 20ab + 4b^2$          **27.** $4z^2 + 8zw + w^2$
Ⓑ **28.** $49x^2 - 56xy + 16y^2$          **29.** $25x^2 - 60xy + 36y^2$          **30.** $64x^2 + 48xy + 9y^2$
**31.** $121m^2 + 44mn + 4n^2$          **32.** $25k^2 + 90kj + 81j^2$          **33.** $25a^2 + 70ab + 49b^2$
     **For Ex. 13–33, see TE Answer Section.**

# FACTORING TRINOMIALS: $x^2 + bx + c$     7.4

*Objective*     **To factor a trinomial with 1 as the coefficient of $x^2$ into two binomials**

You have seen that the product of two factors like $(x - 5)(x + 7)$ is a *trinomial*, $x^2 + 2x - 35$. Notice the following relationship.

$$\text{Trinomial: } x^2 + 2x - 35$$
$$\text{Factors: } (x - 5)(x + 7)$$

$x$ and $x$ are factors of $x^2$.     $-5$ and $7$ are factors of $-35$.

Given a trinomial, such as $x^2 - x - 30$, you can *factor* it into two binomials. Factoring is a *trial-and-error process*. You will have to try different combinations until you find the pair of binomial factors whose product is the trinomial $x^2 - x - 30$.

*Example 1*     **Factor $x^2 - x - 30$.**

$$x^2 - x - 30 = x^2 - 1x - 30 \qquad \blacktriangleleft -x = -1x$$

Trial 1.   $x^2 - 1x - 30 \stackrel{?}{=} (x - 10)(x + 3)$     ◀ *Try $x$ and $x$ as factors of $x^2$,*
Check by multiplying the factors.                              *$-10$ and $3$ as factors of $-30$.*
$(x - 10)(x + 3) = x^2 + 3x - 10x - 30$
$\qquad\qquad\qquad = x^2 - 7x - 30$
This combination *does not work*.     ◀ $x^2 - 7x - 30 \neq x^2 - 1x - 30$

Trial 2.   $x^2 - 1x - 30 \stackrel{?}{=} (x - 5)(x + 6)$     ◀ *No other factors for $x^2$.*
Check by multiplying the factors.                              *Try $-5$ and $6$ as factors of $-30$.*
$(x - 5)(x + 6) = x^2 + 6x - 5x - 30$
$\qquad\qquad\quad = x^2 + 1x - 30$
This combination *does not work*.     ◀ $x^2 + 1x - 30 \neq x^2 - 1x - 30$

Trial 3.   $x^2 - 1x - 30 \stackrel{?}{=} (x + 5)(x - 6)$     ◀ *Constant term is negative. Try*
Check by multiplying the factors.                              *reversing the signs of the factors.*
$(x + 5)(x - 6) = x^2 - 6x + 5x - 30$
$\qquad\qquad\quad = x^2 - 1x - 30$
This combination *works*.     ◀ *The trinomial is $x^2 - 1x - 30$.*

**So,** $x^2 - x - 30 = (x + 5)(x - 6)$.

Notice a pattern from the example above: $x^2 - x - 30 = (x + 5)(x - 6)$.
If the *constant* term of a trinomial is *negative*, the *factors* will have *opposite* signs.

To factor a trinomial like $x^2 - 14x + 48$, use the sign of the constant term, 48, as a clue to the signs of the factors. Since 48 is *positive*, any pair of trial factors must have the *same* sign. Since the coefficient of the middle term, $-14$, is negative, both factors must be negative.

## Example 2    Factor $x^2 - 14x + 48$.

Trial 1.    $x^2 - 14x + 48 \overset{?}{=} (x - 12)(x - 4)$        ◄ *Try x and x as factors of $x^2$,*
Check by multiplying the factors.                                                        *−12 and −4 as factors of 48.*
$(x - 12)(x - 4) = x^2 - 4x - 12x + 48$
$\qquad\qquad\qquad = x^2 - 16x + 48$

This combination *does not work*.        ◄ $x^2 - 16x + 48 \neq x^2 - 14x + 48$

Trial 2.    $x^2 - 14x + 48 \overset{?}{=} (x - 6)(x - 8)$        ◄ *Try −6 and −8 as factors of 48.*
Check by multiplying the factors.
$(x - 6)(x - 8) = x^2 - 8x - 6x + 48$
$\qquad\qquad\quad = x^2 - 14x + 48$

This combination *works*.

**So,** $x^2 - 14x + 48 = (x - 6)(x - 8)$.

The method of factoring used in Examples 1 and 2 is a *general method* that will work for all trinomials of the form $x^2 + bx + c$. By now you may have discovered an easier method for finding the correct combinations of factors of the constant term. Observe the patterns below.

$x^2 + 11x + 30 = (x + 5)(x + 6)$    ◄ *c is positive; b is positive.*
$x^2 - 11x + 30 = (x - 5)(x - 6)$    ◄ *c is positive; b is negative.*
$x^2 + 1x - 30 = (x - 5)(x + 6)$    ◄ *c is negative.*

When $c$ is positive, the sign of $b$ is the sign of the factors of $c$.

When $c$ is negative, the factors of $c$ will have opposite signs.

## Example 3    Factor $x^2 + 19x + 60$.

$x^2 + 19x + 60$

Try factors of 60. Check to see if the *sum* of the factors is 19.

$\qquad\qquad 60 = 10 \cdot 6 \qquad 10 + 6 = 16.$    ◄ *Sum ≠ 19.*
$\qquad\qquad 60 = 5 \cdot 12 \qquad 5 + 12 = 17.$    ◄ *Sum ≠ 19.*
$\qquad\qquad 60 = 4 \cdot 15 \qquad 4 + 15 = 19.$    ◄ *Sum = 19.*

**So,** $x^2 + 19x + 60 = (x + 4)(x + 15)$.

*Example 4*    **Factor $x^2 - 3x - 40$.**

$x^2 - 3x - 40$

Try factors of $-40$. Check to see if the *sum* of the factors is $-3$.

$$4(-10) = -40 \qquad 4 + (-10) = -6 \quad \blacktriangleleft \; Sum \neq -3.$$
$$-5(8) = -40 \qquad -5 + 8 = 3 \quad \blacktriangleleft \; Sum \neq -3.$$
$$5(-8) = -40 \qquad 5 + (-8) = -3 \quad \blacktriangleleft \; Sum = -3.$$

**So, $x^2 - 3x - 40 = (x + 5)(x - 8)$.**

The first two examples of this lesson introduced you to a *general method* of factoring a trinomial. This general method will be needed in the rest of the lessons of this chapter for trinomials such as $2x^2 + 7x - 4$ in which the coefficient of $x^2$ is *not* 1.

The method of Examples 3 and 4 works *only* for trinomials with 1 as the coefficient of the $x^2$ term.

*Summary*    To factor a trinomial of the form $1x^2 + bx + c$, look for factors of $c$ whose **sum** is $b$.

| If $c$ is *negative*, the factors will have *opposite* signs. | If $c$ is *positive* and $b$ is *negative*, both factors will be *negative*. | If $c$ is *positive* and $b$ is *positive*, both factors will be *positive*. |
| --- | --- | --- |

# Written Exercises

**Factor each trinomial into two binomials.**  **4.** $(m - 5)(m - 4)$  **5.** $(k + 2)(k - 6)$

Ⓐ
**1.** $x^2 + 8x + 12$ $(x + 6)(x + 2)$  **2.** $a^2 + 10a + 25$ $(a + 5)(a + 5)$  **3.** $b^2 + 9b + 14$ $(b + 7)(b + 2)$
**4.** $m^2 - 9m + 20$   **5.** $k^2 - 4k - 12$   **6.** $g^2 - 4g + 3$ $(g - 1)(g - 3)$
**7.** $x^2 - x - 20$ $(x - 5)(x + 4)$   **8.** $x^2 + x - 12$ $(x - 3)(x + 4)$   **9.** $r^2 - r - 6$ $(r - 3)(r + 2)$
**10.** $a^2 + 5a - 24$ $(a + 8)(a - 3)$  **11.** $p^2 - 9p + 18$ $(p - 6)(p - 3)$  **12.** $x^2 - 2x - 24$ $(x - 6)(x + 4)$
**13.** $t^2 + 3t + 2$ $(t + 2)(t + 1)$   **14.** $m^2 - 11m + 28$   **15.** $x^2 + 4x - 21$ $(x + 7)(x - 3)$
**16.** $b^2 - 11b + 18$ $(b - 9)(b - 2)$ **17.** $y^2 + 12y + 20$ $(y + 10)(y + 2)$ **18.** $a^2 + 2a - 8$ $(a + 4)(a - 2)$
**19.** $v^2 - 10v + 9$ $(v - 9)(v - 1)$  **20.** $c^2 - 3c - 28$ $(c - 7)(c + 4)$  **21.** $y^2 - 8y + 16$ $(y - 4)(y - 4)$
**22.** $w^2 - 8w - 20$   **23.** $m^2 + 11m + 24$   **24.** $a^2 - 6a - 27$ $(a - 9)(a + 3)$
**14.** $(m - 7)(m - 4)$  **22.** $(w - 10)(w + 2)$  **23.** $(m + 8)(m + 3)$  **29.** $(t + 6)(t - 8)$
Ⓑ
**25.** $r^2 - 16r + 48$ $(r - 12)(r - 4)$ **26.** $a^2 - 6a - 40$ $(a + 4)(a - 10)$ **27.** $w^2 - 3w - 54$ $(w - 9)(w + 6)$
**28.** $x^2 - x - 56$ $(x - 8)(x + 7)$   **29.** $t^2 - 2t - 48$   **30.** $x^2 - 7x - 60$ $(x + 12)(x - 5)$
**31.** $b^2 + 7b - 44$ $(b + 11)(b - 4)$ **32.** $y^2 - y - 72$ $(y - 9)(y + 8)$  **33.** $g^2 - 12g - 64$ $(g - 16)(g + 4)$
Ⓒ
**34.** $x^{2m} - 24x^m - 112$   **35.** $x^{4a} - 11x^{2a} - 126$   **36.** $-144 - 18a^{2p} + a^{4p}$
   **34.** $(x^m - 28)(x^m + 4)$    **35.** $(x^{2a} - 18)(x^{2a} + 7)$    **36.** $(a^{2p} - 24)(a^{2p} + 6)$

Factoring Trinomials: $x^2 + bx + c$

# FACTORING TRINOMIALS: $ax^2 + bx + c$     7.5

*Objective*     **To factor trinomials in which the coefficient of $x^2$ is not 1**

Consider the trinomial $ax^2 + bx + c$ in which $a$ is not 1. The signs of $b$ and $c$ give you clues about the signs of the trial factors. In the trinomial $2x^2 - 11x + 15$, $c$, 15, is positive and $b$, $-11$, is negative. So, the trial factors of 15 must both be negative, as shown in Example 1.

*Example 1*     **Factor $2x^2 - 11x + 15$.**

Trial 1.  $2x^2 - 11x + 15 \stackrel{?}{=} (2x - 3)(x - 5)$        ◀ *Try $2x$ and $x$ as factors of $2x^2$,*
Check by multiplying the factors.                          *$-3$ and $-5$ as factors of 15.*
$(2x - 3)(x - 5) = 2x^2 - 10x - 3x + 15$
$= 2x^2 - 13x + 15$
This combination does *not* work.                          ◀ *The result is not $2x^2 - 11x + 15$.*

Trial 2.  $2x^2 - 11x + 15 \stackrel{?}{=} (2x - 5)(x - 3)$        ◀ *Reverse the factors $-3$ and $-5$.*
Check by multiplying the factors.
$(2x - 5)(x - 3) = 2x^2 - 6x - 5x + 15$
$= 2x^2 - 11x + 15$                          ◀ *This combination works.*
**So,** $2x^2 - 11x + 15 = (2x - 5)(x - 3)$.

In the trinomial $6a^2 + 7a - 5$, the constant term, $-5$, is negative. So, the trial factors of $-5$ must have opposite signs, as shown in Example 2. Note that since the coefficient of $a^2$, 6, is not prime, you may have to try different factor combinations for the first term.

*Example 2*     **Factor $6a^2 + 7a - 5$.**

1. $6a^2 + 7a - 5 \stackrel{?}{=} (6a + 5)(1a - 1)$        ◀ *Try $6a$ and $1a$ as factors of $6a^2$,*
Check by multiplying the factors.                          *5 and $-1$ as factors of $-5$.*
$(6a + 5)(1a - 1) = 6a^2 - 6a + 5a - 5$
$= 6a^2 - 1a - 5$
This combination does *not* work.                          ◀ *The result is not $6a^2 + 7a - 5$.*

2. $6a^2 + 7a - 5 \stackrel{?}{=} (3a + 5)(2a - 1)$        ◀ *Try $3a$ and $2a$ as factors of $6a^2$.*
Check by multiplying the factors.
$(3a + 5)(2a - 1) = 6a^2 - 3a + 10a - 5$
$= 6a^2 + 7a - 5$                          ◀ *This combination works.*
**So,** $6a^2 + 7a - 5 = (3a + 5)(2a - 1)$.

There are times when you might have to try several combinations before you find the correct pair of binomial factors. For example, in $2k^2 - 5k - 12$, the possible factors of $2k^2$ are $2k$ and $k$ or, if reversed, $k$ and $2k$. Possible factors of $-12$ are 4 and $-3$, $-4$ and 3, 6 and $-2$, $-6$ and 2, 12 and $-1$, and $-12$ and 1. Similarly, each pair of factors can be reversed. As you gain experience you will discover that you do not have to go through as many trials before you find the factors of a trinomial.

## Example 3    Factor $2k^2 - 5k - 12$.

1. $2k^2 - 5k - 12 \overset{?}{=} (2k - 6)(k + 2)$    ◀ *Try $2k$ and $k$ as factors of $2k^2$,*
   Check by multiplying the factors.        *$-6$ and $+2$ as factors of $-12$.*
   $(2k - 6)(k + 2) = 2k^2 + 4k - 6k - 12$
   $\phantom{(2k - 6)(k + 2)} = 2k^2 - 2k - 12$
   This combination does *not* work.        ◀ *The result is not $2k^2 - 5k - 12$.*

2. $2k^2 - 5k - 12 \overset{?}{=} (2k + 2)(k - 6)$    ◀ *Reverse the factors $-6$ and 2.*
   Check by multiplying the factors.
   $(2k + 2)(k - 6) = 2k^2 - 12k + 2k - 12$
   $\phantom{(2k + 2)(k - 6)} = 2k^2 - 10k - 12$
   This combination also does *not* work.

3. $2k^2 - 5k - 12 \overset{?}{=} (2k + 3)(k - 4)$    ◀ *Try $+3$ and $-4$ as factors of $-12$.*
   Check by multiplying the factors.
   $(2k + 3)(k - 4) = 2k^2 - 8k + 3k - 12$
   $\phantom{(2k + 3)(k - 4)} = 2k^2 - 5k - 12$    ◀ *This combination works.*

   **So,** $2k^2 - 5k - 12 = (2k + 3)(k - 4)$.

# Reading in Algebra    2. **Check students' steps.**
$$2a^2 + 3a - 20 = (2a - 5)(a + 4)$$

1. Which pair of factors, $(2x + 3)$ and $(x + 5)$ or $(2x - 3)$ and $(x - 5)$, does not give the product for $2x^2 - 13x + 15$? **$(2x + 3)(x + 5)$**
2. Outline the steps you would use in factoring $2a^2 + 3a - 20$.
3. A student factored $6x^2 + 5x - 1$ as $(3x + 1)(2x + 1)$. In checking, the middle term was correct. Where was the error? **The constant term is negative. The signs of the factors should be opposites.**

1. **2x and x; 6, 1 or 3, 2**

# Oral Exercises    2. **3y and y; 10, 1 or 5, 2**
3. **3x and x; −4, −1 or −2, −2**

For each of the following, list the possible trial factors you would use in factoring the trinomial.    4. **2a and a; 14, −1 or −14, 1 or 7, −2 or −7, 2**

1. $2x^2 + 7x + 6$        2. $3y^2 + 11y + 10$        3. $3x^2 - 8x + 4$
4. $2a^2 - 3a - 14$       5. $3x^2 - x - 4$          6. $2b^2 - 7b - 4$
7. $6m^2 - 17m + 5$       8. $10y^2 - 19y + 7$       9. $8b^2 - 10b + 3$

5. **3x and x; 4, −1, or −4, 1 or −2, 2 or 2, −2**
6. **2b and b; −4, 1 or 4, −1 or −2, 2 or 2, −2    For Ex. 7–9, see TE Answer Section.**

Factoring Trinomials: $ax^2 + bx + c$

## Written Exercises

For Ex. 1–57 see TE Answer Section.

**Factor each trinomial into two binomials.**

(A)
1. $2x^2 + 7x + 6$
2. $3y^2 + 11y + 10$
3. $3x^2 - 8x + 4$
4. $2a^2 - 3a - 14$
5. $3x^2 - x - 4$
6. $2b^2 - 7b - 4$
7. $6m^2 - 17m + 5$
8. $10y^2 - 19y + 7$
9. $8b^2 - 10b + 3$
10. $6a^2 + 5a - 1$
11. $4n^2 + 9n + 5$
12. $3a^2 - 5a - 8$
13. $4x^2 - x - 5$
14. $2a^2 - a - 3$
15. $2y^2 - 7y + 6$
16. $2x^2 + 11x - 21$
17. $2x^2 + 3x - 20$
18. $2y^2 - y - 10$
19. $2g^2 - 7g - 15$
20. $2a^2 + 13a + 15$
21. $3b^2 + 11b + 6$
22. $3x^2 + 10x - 25$
23. $36k^2 + 12k + 1$
24. $2p^2 - 15p + 25$
25. $6a^2 - a - 1$
26. $3y^2 - 14y - 24$
27. $2m^2 - 7m + 5$
28. $2b^2 + 17b + 30$
29. $2x^2 + 5x + 2$
30. $12y^2 + 7y + 1$
31. $28y^2 - 18y + 2$
32. $4a^2 - 20a + 25$
33. $3x^2 - 13x - 30$

(B)
34. $2x^2 - 11x - 40$
35. $2a^2 + 23a + 45$
36. $3b^2 + 10b - 48$
37. $2a^2 - 25a + 50$
38. $3b^2 + 8b - 35$
39. $2m^2 - 29m + 60$
40. $3m^2 + 29m + 40$
41. $5x^2 - 42x - 27$
42. $3a^2 - 31a + 56$
43. $6a^2 - 13a - 15$
44. $8r^2 - 46r + 45$
45. $6p^2 + p - 35$
46. $15q^2 - 19q + 6$
47. $15y^2 + 17y - 18$
48. $14b^2 - 15b - 9$
49. $16y^2 + 14y - 15$
50. $18a^2 - 9a - 35$
51. $14p^2 + 31p - 10$

(C)
52. $10 + 3z - z^2$
53. $18 - 7m - m^2$
54. $15 - 2a - a^2$
55. $3 + 2a - 8a^2$
56. $5 - 8a - 4a^2$
57. $8 + 3x - 5x^2$

**Factor each trinomial. Simplify each factor of the result.**

58. $2(x - 1)^2 + 7(x - 1) - 4$ **(2x − 3)(x + 3)**
59. $6(m - 1)^2 - 7(m - 1) - 20$ **(3m + 1)(2m − 7)**
60. $12(y - 4)^2 - 31(y - 4) + 20$
61. $15(a - 1)^2 + 14(a - 1) - 8$ **(5a − 7)(3a + 1)**
62. $4(a + 2)^2 + 11(a + 2) - 20$ **(4a + 3)(a + 6)**
63. $10(g - 1)^2 - 39(g - 1) + 14$ **(5g − 7)(2g − 9)**
64. $20 + 30(x - 2) + 10(x - 2)^2$ **10x(x − 1)**
65. $12 + 29(t - 4) + 15(t - 4)^2$ **(3t − 8)(5t − 17)**
60. **(4y − 21)(3y − 16)**
66. If $2x^2 + (3b - 2)x - 15$ is factored into $(2x - 3)(x + 5)$, what is the value of $b$? **3**
67. Factor $12a^{4m} - a^{2m} - 35$ into two binomials. **(4a^{2m} − 7)(3a^{2m} + 5)**

## CUMULATIVE REVIEW

**Multiply.**
1. $(a - 4)(a + 4)$ **a² − 16**
2. $(3x - 2)(3x + 2)$ **9x² − 4**
3. $(-5a^3)(4a)$ **−20a⁴**

## CALCULATOR ACTIVITIES

**Factor each trinomial into two binomials.**
1. $x^2 + 88x + 336$
**(x + 84)(x + 4)**
2. $x^2 - 110x - 575$
**(x − 115)(x + 5)**
3. $2a^2 + 145a - 375$
**(2a − 5)(a + 75)**

# THE WHEEL

A wheel operates as an infinite series of levers. It revolutionized land transportation, increased labor's efficiency, and substituted power for human muscles.

A modern-day paddlewheeler carries tourists along the Mississippi. The paddles are connected to the spokes that in turn act as levers to move the boat through the water.

An 1851 pen-and-ink drawing. It shows the mechanical advantage of four pulleys in a block and tackle system balancing small and large weights.

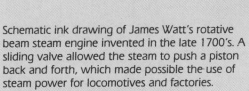

Schematic ink drawing of James Watt's rotative beam steam engine invented in the late 1700's. A sliding valve allowed the steam to push a piston back and forth, which made possible the use of steam power for locomotives and factories.

Close-up view of pulleys giving a mechanical advantage of velocity on a steam engine toy. The ratio of speeds at which the wheels turn is inversely proportional to the circumferences of the wheels.

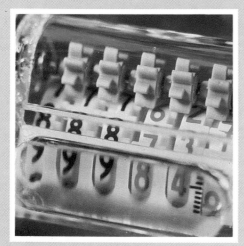

Plastic gearing in a digital counter system. Gears, toothed wheels, can transfer energy and change speeds or force.

A Leonardo da Vinci drawing showing conversion of rotary motion into reciprocating motion. As the crank turns the shaft holding the partly toothed gears, the wheel moves first clockwise then counterclockwise.

Close-up of the reduction unit of worm and wheel gearing. This reduction gearing of a power transmission slows the speed through gear wheels of different diameters.

A 100-year-old lithograph advertisement from Scotland. Sprocket wheels on bicycles and tricycles connected on separate shafts by chains give increased mechanical advantages.

Second-place winner of the Special Olympics maneuverability race at Raleigh, NC, 1984. Today such chairs are electrically as well as hand-propelled.

All-wooden cart preserved for 3000 years in the frozen solid of the Altai Mountains. The cart is now in the Hermitage Museum, USSR.

Pile of old metal-rimmed, metal-hubbed, wooden-spoked wagon wheels in Old Tucson, Arizona. These wheels are similar to those of the wagon trains that settled the West.

A spare wheel with cover of a 1935 Packard 8. Specially spoked or decorated automobile wheels are still considered to be status symbols.

Drive wheel of a steam locomotive. Combinations of wheels, gear wheels, or other connected mechanical parts for transmitting motion are usually called trains.

A pen-and-ink illustration of a bucket and well with a hand turned crank. The effort is lessened by the lengthened handle, which serves as a lever.

Canal-control wheel and gears on a canal at Lawrence, Massachusetts. Raising and lowering gates by compound wheel-and-gear machines made canal travel possible.

(Left to right) Thomas Edison, inventor; John Burroughs, naturalist; Henry Ford, carmaker; and Harvey Firestone, tiremaker; on a large water wheel.

Wheeled irrigation in California. Relatively simple wheel-and-pipe machines bring additional water supply to crops in semiarid regions.

A modern sakieh on the plain of Thebes, Egypt. Water pumps use gears and wheels to deliver adequate water for plant growth and animal sustenance in arid regions.

A blueprint drawing of a pulley. The flat front view cross-section and vertical longitudinal cross-section give the dimensions for accurate manufacture.

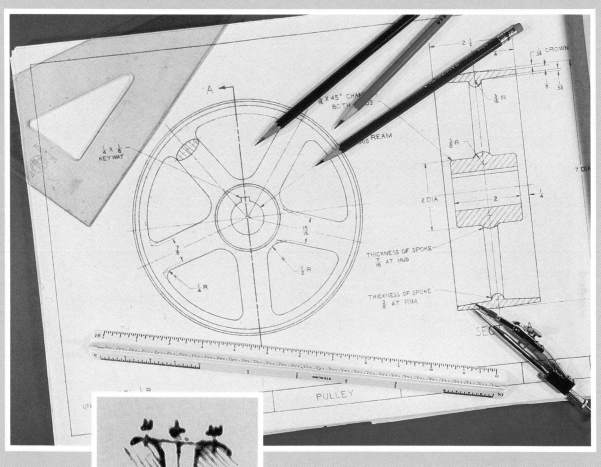

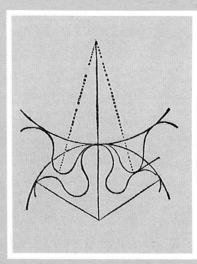

Leonardo da Vinci's designs were calculated to result in a minimum of frictional resistance in gear teeth. The shape is epicycloidal generated by one circle rolling around the outside of another.

Philippe de Le Hire's sketch of ideally shaped gear teeth to decrease friction. These were sketched 200 years after Leonardo's but are strikingly similar.

# COMBINED TYPES OF FACTORING 7.6

*Objective*    **To factor polynomials completely**

You have factored out the GCF from a polynomial. You have also factored a trinomial or binomial (difference of two squares) into two binomials. Sometimes it is necessary to combine more than one type of factoring to factor a polynomial completely.

*Example 1*    **Factor $6x^3 + 3x^2 - 18x$ completely.**

1. First find the GCF, if any, and factor it out.

$3(2)x^3 + 3(1)x^2 + 3(-6)x$         $6 \cdot x \cdot x \cdot x + 3 \cdot x \cdot x - 18 \cdot x$
The greatest common                       The greatest common
numerical factor is 3.                        variable factor is $x$.

**So,** the GCF is $3x$.

$6x^3 + 3x^2 - 18x = (3x)(2x^2) + (3x)(1x) + (3x)(-6)$    ◀ *Use 3x as a factor of*
$\qquad\qquad\qquad = 3x(2x^2 + 1x - 6)$         *each term.*

2. Try to factor $2x^2 + 1x - 6$ into two binomials.

$3x(2x^2 + 1x - 6) \overset{?}{=} 3x(2x - 3)(x + 2)$    ◀ *Try 2x and x as factors of $2x^2$,*
$\qquad\qquad\qquad\qquad\qquad\qquad$ *−3 and 2 as factors of −6.*

Check by multiplying the factors.

$3x(2x - 3)(x + 2) = 3x(2x^2 + \underline{4x - 3x} - 6)$

$\qquad\qquad\qquad = 3x(2x^2 \quad + 1x \quad - 6)$    ◀ *This combination works.*
$\qquad\qquad\qquad = 6x^3 + 3x^2 - 18x$    ◀ *The complete product is the*
$\qquad\qquad\qquad\qquad\qquad\qquad\qquad$ *original polynomial.*

**Thus,** the complete factorization of $6x^3 + 3x^2 - 18x$ is $3x(2x - 3)(x + 2)$.

Sometimes, after you have factored out the GCF, the other polynomial might be the difference of two squares, as shown in Example 2.

*Example 2*    **Factor $20x^3 - 5x$.**

$20x^3 - 5x = (5x)(4x^2) + (5x)(-1)$    ◀ *Factor out the GCF, 5x.*
$\qquad\qquad = 5x(4x^2 - 1)$
$\qquad\qquad = 5x(2x + 1)(2x - 1)$    ◀ *$4x^2 - 1 = (2x)^2 - (1)^2$*
$\qquad\qquad\qquad\qquad\qquad\qquad$ *$= (2x + 1)(2x - 1)$*

**Thus,** $20x^3 - 5x = 5x(2x + 1)(2x - 1)$.

*Example 3*    **Factor $x^2 - 9x + 14$.**

1. First look for the GCF.

   $1(x^2) + 3(-3x) + 2(7)$              $x \cdot x - 9 \cdot x + 14$
   The greatest common                  There is no variable factor
   numerical factor is 1.                common to all three terms.
   **So,** the GCF is 1.

2. Factor $x^2 - 9x + 14$ into two binomials.
   $x^2 - 9x + 14 \stackrel{?}{=} (x - 7)(x - 2)$      ◄ *Try x and x as factors of $x^2$,*
                                                         *$-7$ and $-2$ as factors of 14.*
   Check by multiplying the factors.
   $(x - 7)(x - 2) = x^2 - 2x - 7x + 14$
   $\qquad\qquad\qquad = x^2 - 9x + 14$      ◄ *This combination works.*
   **Thus,** $x^2 - 9x + 14 = (x - 7)(x - 2)$.

The trinomial $4m^2 - 2mb - 20b^2$ has two variables, $m$ and $b$. You can factor a trinomial with two variables in the same general way you factor a trinomial with one variable. The first step is to look for a GCF. There is no *variable* common to all three terms of $4m^2 - 2mb - 20b^2$. But there is a greatest common numerical factor, 2. You can factor out the 2, then factor the remaining trinomial as shown in Example 4.

*Example 4*    **Factor $4m^2 - 2mb - 20b^2$.**

1. Since the GCF is 2, $4m^2 - 2mb - 20b^2 = 2(2m^2 - 1mb - 10b^2)$.

2. Factor $2m^2 - 1mb - 10b^2$ into two binomials.
   $2(2m^2 - 1mb - 10b^2) \stackrel{?}{=} 2(2m - 5b)(m + 2b)$    ◄ *Try 2m and m as factors of $2m^2$,*
                                                                    *$-5b$ and $2b$ as factors of $-10b^2$.*
   Check by multiplying the factors.

   $2(2m - 5b)(m + 2b) = 2(2m^2 + \underbrace{4mb - 5mb} - 10b^2)$
   $\qquad\qquad\qquad\qquad = 2(2m^2 \quad - 1mb \quad - 10b^2)$    ◄ *This combination works.*
   $\qquad\qquad\qquad\qquad = 4m^2 - 2mb - 20b^2$    ◄ *The complete product is*
                                                         *the original polynomial.*
   **Thus,** $4m^2 - 2mb - 20b^2 = 2(2m - 5b)(m + 2b)$.

*Summary*    To factor a polynomial completely:
1. Factor out the GCF, if there is any.
2. Factor the remaining polynomial if possible.
   Look for the difference of two squares.
   Look for perfect square trinomials.
3. Check that each factor cannot be factored further.
4. Check that the complete product is the original polynomial.

# Reading in Algebra

1. What is the first step in factoring a polynomial completely? **Factor out the GCF, if any.**

2. Is $x$ a common variable factor of $x^2 + 5xy + 6y^2$? Why or why not? **No; $x$ does not appear in the last term.**

3. Find the error in the factorization $a^2 - 5ab + 6b^2 = (a - 3)(a - 2)$.
**The variable $b$ does not appear in the factored form.**

**For Ex. 16–21, 25–30, 34–45, 49–60, see TE Answer Section.**

# Written Exercises
1. $3(a - 1)(a - 4)$   2. $2(b + 2)(b - 5)$   3. $2(a - 2)(a + 7)$
4. $2r(r + 1)(r - 3)$   5. $3x(x + 4)(x + 1)$   6. $2m(m + 4)(m - 5)$   9. $4m(m - 5)(m + 5)$
**Factor completely.**   10. $(x - 10)(x - 3)$   11. $4(a - 1)(a - 5)$   12. $(2a - 1)(a + 7)$

(A)
1. $3a^2 - 15a + 12$
2. $2b^2 - 6b - 20$
3. $2a^2 + 10a - 28$
4. $2r^3 - 4r^2 - 6r$
5. $3x^3 + 15x^2 + 12x$
6. $2m^3 - 2m^2 - 40m$
7. $3x^3 - 3x$ $3x(x - 1)(x + 1)$
8. $3a^3 - 27a$ $3a(a + 3)(a - 3)$
9. $4m^3 - 100m$
10. $x^2 - 13x + 30$
11. $4a^2 - 24a + 20$
12. $2a^2 + 13a - 7$
13. $5n^2 - 10n$ $5n(n - 2)$
14. $6p^2 - 12p$ $6p(p - 2)$
15. $2b^2 + 4b$ $2b(b + 2)$
16. $2t^3 - 14t^2 + 24t$
17. $6g^3 - 28g^2 - 10g$
18. $6r^3 - 3r^2 - 30r$
19. $5a^2 - 35a + 50$
20. $2a^3 + 16a^2 + 30a$
21. $12w^2 + 33w - 9$
22. $2z^2 - 7z - 15$ $(2z + 3)(z - 5)$
23. $4y^2 + 7y - 2$ $(4y - 1)(y + 2)$
24. $5x^3 - 20x$ $5x(x - 2)(x + 2)$
25. $2y^2 + 20yz + 18z^2$
26. $2m^2 - 2mn - 40n^2$
27. $3x^2 - 9xy - 30y^2$
28. $3x^2 - 75y^2$
29. $4a^2 - 4b^2$
30. $12q^2 - 75k^2$
31. $8r^2 + 4rt - 30t^2$
32. $6m^2 - 9mn - 15n^2$
33. $8m^2 - 24mn + 18n^2$

31. $2(4r^2 + 2rt - 15t^2)$   32. $3(2m - 5n)(m + n)$   33. $2(2m - 3n)(2m - 3n)$

(B)
34. $6x^2 - 3x - 63$
35. $121a - 49a^3$
36. $3p^4 - 5p^3 - 28p^2$
37. $30e^3 + 22e^2 - 28e$
38. $30r^3 - 9r^2 - 12r$
39. $10b^3 + 34b^2 - 24b$
40. $a^2b^3 - 25b$
41. $r^3 - 5r^2t + 4rt^2$
42. $x^3y - xy^3$
43. $x^3 - 3x^2y + 2xy^2$
44. $3a^3b + 13a^2b^2 - 3ab^3$
45. $12m^3y - 75my^3$
46. $10x^3y + 21x^2y^2 + 9xy^3$
47. $2a^3b + 11a^2b^2 - 21ab^3$
48. $6x^3y + 7x^2y^2 - 20xy^3$

46. $xy(5x + 3y)(2x + 3y)$   47. $ab(2a - 3b)(a + 7b)$   48. $xy(3x - 4y)(2x + 5y)$

(C)
49. $2x^4 - 512$
50. $243 - 3y^4$
51. $625m^4 - 16$
52. $x^5 - 32x^3 + 256x$
53. $2a^4 - 26a^2 + 72$
54. $4r^4 - 29r^2 + 25$
55. $3a^4 - 75a^2 + 432$
56. $3t^5 - 60t^3 + 192t$
57. $8y^6 - 50y^4 + 72y^2$
58. $x^4 - 13x^2y^2 + 36y^4$
59. $16a^4 - 625b^4$
60. $2x^5y - 20x^3y^3 + 18xy^5$
61. $x^5y - 18x^3y^3 + 81xy^5$
62. $2x^5y + 34x^3y^3 + 32xy^5$
63. $x^5y - 53x^3y^3 + 196xy^5$

61. $xy(x + 3y)(x - 3y)(x + 3y)(x - 3y)$   62. $2xy(x^2 + 16y^2)(x^2 + y^2)$

63. $xy(x + 7y)(x - 7y)(x + 2y)(x - 2y)$

# CUMULATIVE REVIEW

**Solve.**

1. $\dfrac{2}{3}x - \dfrac{1}{4} = \dfrac{7}{12}$ $1\dfrac{1}{4}$

2. $-4.4 = 0.08 - 0.2r$ **22.4**

3. $5 + 7g \geq -(3 - 2g) - 3g$
$g \geq -1$

Combined Types of Factoring

# FACTORING BY GROUPING 7.7

*Objective* **To factor polynomials by grouping**

Many polynomials contain a common monomial factor and can be factored as the product of a monomial and a polynomial.

For example, since $x$ is a common factor of $3bx + 5mx$,

$$\begin{aligned} 3bx + 5mx &= (3b)x + (5m)x \\ &= x(3b + 5m). \end{aligned}$$

In the polynomial $x(t + w) + y(t + w)$, notice that the binomial $(t + w)$ is common to each of the terms, $x(t + w)$ and $y(t + w)$. Therefore, you can factor out the common binomial factor, as shown below.

$$x(t + w) + y(t + w) = (t + w)(x + y)$$

Sometimes the common factor is not obvious, as in this polynomial:

$$am + bm + an + bn.$$

However, notice that $m$ is common to the first two terms, $am$ and $bm$. Similarly, $n$ is common to the last two terms, $an$ and $bn$.

Therefore,

$$am + bm + an + bn = m(a + b) + n(a + b).$$

Now, the common binomial factor, $(a + b)$, is obvious and can be factored out, as shown below.

$$\begin{aligned} am + bm + an + bn &= m(a + b) + n(a + b) \\ &= (a + b)(m + n) \end{aligned}$$

This method of factoring is called *factoring by grouping*.

*Example 1* **Factor $6a - 6b + ax - bx$.**

Group the terms. You can begin in either of the two ways shown.

| | | |
|---|---|---|
| $(6a - 6b) + (ax - bx)$ | or | $(6a + ax) + (-6b - bx)$ |
| $6(a - b) + x(a - b)$ | | $a(6 + x) - b(6 + x)$ ◀ *a is common to the first two terms, and $-b$ is common to the last two terms.* |
| $(a - b)(6 + x)$ | | $(6 + x)(a - b)$ |

Notice that the same factors are obtained although the terms are grouped differently.

**Thus,** $6a - 6b + ax - bx = (a - b)(6 + x)$, or $(6 + x)(a - b)$.

You should always check to see if a polynomial factor can be factored further, as is the case in the next example.

## Example 2    Factor $x^2y + x^2 - 9y - 9$ completely.

$x^2y + x^2 - 9y - 9$       or $x^2y + x^2 - 9y - 9$
$(x^2y + x^2) + (-9y - 9)$       $(x^2y - 9y) + (x^2 - 9)$ ◀ *Group the y-terms*
$x^2(y + 1) - 9(y + 1)$       $y(x^2 - 9) + 1(x^2 - 9)$    *together.*
$(y + 1)(x^2 - 9)$ ◀ $x^2 - 9$ *is the difference*    $(x^2 - 9)(y + 1)$
$(y + 1)(x - 3)(x + 3)$    *of two squares.*    $(x - 3)(x + 3)(y + 1)$

**Thus,** $x^2y + x^2 - 9y - 9 = (y + 1)(x - 3)(x + 3)$ when factored completely. The factors may be listed in any order.

# Reading in Algebra

1. True or false? The correct answer for factoring by grouping depends on the way you group the terms. **False**
2. For the polynomial $ax + ay + bx + by$, what factor is common to the first two terms? to the last two terms? **a; b**
3. Show some ways to regroup the terms of $ax - 2b + 2a - xb$ so that you can factor by grouping. **by a's and b's: $ax + 2a + (-2b) + (-xb) = a(x + 2) - b(x + 2)$**
   **by x's and 2's: $ax - xb + (-2b) + 2a = x(a - b) + 2(a - b)$**

# Oral Exercises

1. $(a + b)(m + p)$   2. $(x - y)(5 + d)$   3. $(r + 5)(m^2 + 7)$

**Factor out the common binomial factor.**

1. $(a + b)m + (a + b)p$      2. $5(x - y) + d(x - y)$      3. $m^2(r + 5) + (r + 5)7$
4. $(x^2 - 25)4 - a^2(x^2 - 25)$      5. $p(36 - t^2) + (36 - t^2)q$      6. $k(g + r) - t(g + r)$
   $(x^2 - 25)(4 - a^2)$            $(36 - t^2)(p + q)$           $(g + r)(k - t)$

# Written Exercises

1. $(m + b)(x + t)$   2. $(x + y)(7 + r)$   3. $(b - c)(a + 5)$   4. $(a - b)(x + 2)$   5. $(z + 2)(m + y)$

**Factor each polynomial completely.** 6. $(r + 7)(p + t)$   7. $(a + 5)(a + x)$   8. $(m - 7)(m + y)$

(A)   1. $xm + xb + tm + bt$      2. $7x + 7y + rx + ry$      3. $ab - ac + 5b - 5c$
    4. $xa - xb + 2a - 2b$      5. $mz + 2m + yz + 2y$      6. $pr + 7p + tr + 7t$
    7. $a^2 + 5a + xa + 5x$      8. $m^2 - 7m + my - 7y$      9. $t^2 - 5t + 3t - 15$

(B)   10. $7ax + 3ay + 7bx + 3by$      11. $ta + 6t + a^2 + 6a$      12. $g^2 + 3g - ag - 3a$

9. $(t - 5)(t + 3)$   10. $(7x + 3y)(a + b)$   11. $(a + 6)(a + t)$   12. $(g + 3)(g - a)$   13. $(m + 1)(x - 2)(x + 2)$

13. $x^2m + x^2 - 4m - 4$      14. $x^2a + x^2b - 16a - 16b$      15. $p^2y + p^2 - 25y - 25$
16. $16a + 16 - y^2a - y^2$      17. $x^3 - 4y^2x + 2x^2 - 8y^2$      18. $y^2x + 2y^2 - 100x - 200$

14. $(a + b)(x - 4)(x + 4)$   15. $(y + 1)(p - 5)(p + 5)$   16. $(a + 1)(4 - y)(4 + y)$   17. $(x + 2)(x - 2y)(x + 2y)$

(C)   19. $y^4 - 81x^2y^2 - 4y^2 + 324x^2$          20. $y^2x^2 - 7xy^2 + 12y^2 - 4x^2 + 28x - 48$
    21. $r^2x^2 - 3r^2x + 2r^2 - 9x^2 + 27x - 18$      22. $y^4a^4 + 16 - y^4 - 16a^4$

18. $(x + 2)(y - 10)(y + 10)$   19. $(y - 2)(y + 2)(y - 9x)(y + 9x)$   20. $(y - 2)(y + 2)(x - 4)(x - 3)$
21. $(r - 3)(r + 3)(x - 1)(x - 2)$   22. $(a - 1)(a + 1)(a^2 + 1)(y - 2)(y + 2)(y^2 + 4)$

Factoring by Grouping                                     **187**

# SOLVING EQUATIONS BY FACTORING 7.8

*Objectives* **To solve quadratic equations by factoring**
**To solve cubic equations by factoring**

The following examples suggest a property when a product of two factors is 0.

$$4 \cdot 0 = 0 \qquad 0 \cdot -7 = 0 \qquad 0 \cdot 0 = 0$$

If a product is 0, then at least one of the factors is 0.

| Product property of zero | If $a \cdot b = 0$, then either $a = 0$ or $b = 0$, or both $a$ and $b$ are 0. |
|---|---|

You can use this property to find the value of $x$ that will make an equation such as $(x - 5)(x - 4) = 0$ true. Since the product $(x - 5)(x - 4)$ is 0, at least one of the factors, $x - 5$ or $x - 4$ must be 0.

That is, $x - 5 = 0$, so $x = 5$ or $x - 4 = 0$, so $x = 4$

Substituting 5 for $x$ in $(x - 5)(x - 4)$ gives $(5 - 5)(5 - 4) = 0 \cdot 1 = 0$.
Substituting 4 for $x$ in $(x - 5)(x - 4)$ gives $(4 - 5)(4 - 4) = -1 \cdot 0 = 0$.
**Thus,** the solutions of $(x - 5)(x - 4) = 0$ are 5 and 4.

The left side of the equation $x^2 - 9x + 20 = 0$ is a polynomial whose degree is 2. The right side of the equation is 0. This equation is called a *quadratic equation*. A quadratic equation has *degree* 2.

| Definition: Quadratic equation | An equation of the form $ax^2 + bx + c = 0$ where $a$, $b$, and $c$ are real numbers and $a \neq 0$ is called a **quadratic equation.** |
|---|---|

Some quadratic equations can be solved by factoring the polynomial and then applying the Product Property of Zero. This process is illustrated below.

*Example 1* **Solve $x^2 - 9x + 14 = 0$.**

1. Factor.
$$x^2 - 9x + 14 = 0$$
$$(x - 7)(x - 2) = 0$$

2. Set each factor equal to 0. $\quad x - 7 = 0$ or $x - 2 = 0$
3. Solve for $x$. $\qquad\qquad\qquad x = 7 \qquad x = 2$
4. *Check.*

Let $x = 7$. ▶

$$\begin{array}{c|c} x^2 - 9x + 14 = 0 & \\ \hline 7^2 - 9 \cdot 7 + 14 & 0 \\ 49 - 63 + 14 & \\ 0 & \\ 0 = 0, \text{ true} & \end{array}$$

Let $x = 2$. ▶

$$\begin{array}{c|c} x^2 - 9x + 14 = 0 & \\ \hline 2^2 - 9 \cdot 2 + 14 & 0 \\ 4 - 18 + 14 & \\ 0 & \\ 0 = 0, \text{ true} & \end{array}$$

**Thus,** the solutions of $x^2 - 9x + 14 = 0$ are 7 and 2.

The quadratic equation $x^2 - 9x + 14 = 0$ has two solutions, 7 and 2. The solutions of an equation are also called **roots**. A *second degree equation* has *two* roots.

## Example 2    Solve $3m^2 + 8m - 3 = 0$.

1. Factor.

$$3m^2 + 8m - 3 = 0$$
$$(3m - 1)(m + 3) = 0$$

2. Set each factor equal to 0.

$$3m - 1 = 0 \quad \text{or} \quad m + 3 = 0$$
$$3m = 1 \qquad\qquad m = -3$$

3. Solve for $m$.

$$m = \frac{1}{3}$$

4. *Check.*

*Let $m = \frac{1}{3}$.* ▶

$$
\begin{array}{c|c}
3m^2 + 8m - 3 = 0 & \\
3(\frac{1}{3})^2 + 8(\frac{1}{3}) - 3 & 0 \\
3(\frac{1}{9}) + \frac{8}{3} - 3 & \\
\frac{1}{3} + \frac{8}{3} - 3 & \\
\frac{9}{3} - 3 & \\
3 - 3 & \\
0 & \\
\end{array}
$$

$0 = 0$, true

*Let $m = -3$.* ▶

$$
\begin{array}{c|c}
3m^2 + 8m - 3 = 0 & \\
3(-3)^2 + 8(-3) - 3 & 0 \\
3(9) - 24 - 3 & \\
27 - 24 - 3 & \\
0 & \\
\end{array}
$$

$0 = 0$, true

**Thus,** the solutions, or roots, are $\frac{1}{3}$ and $-3$.

## Example 3    Solve $x^2 - 5x = 0$.

1. Factor.

$x^2 - 5x = 0$ ◀ *Look for a GCF.*
$x(x - 5) = 0$ ◀ *The GCF is $x$.*

2. Set each factor equal to 0.

$x = 0 \quad \text{or} \quad x - 5 = 0$

3. Solve for $x$.

$x = 5$

4. *Check.*

*Let $x = 0$.* ▶

$$
\begin{array}{c|c}
x^2 - 5x = 0 & \\
0^2 - 5 \cdot 0 & 0 \\
0 - 0 & \\
0 & \\
\end{array}
$$

$0 = 0$, true

*Let $x = 5$.* ▶

$$
\begin{array}{c|c}
x^2 - 5x = 0 & \\
5^2 - 5 \cdot 5 & 0 \\
25 - 25 & \\
0 & \\
\end{array}
$$

$0 = 0$, true

**Thus,** the roots are 0 and 5.

The equation $x^3 - 25x = 0$ is a *third degree equation,* or a **cubic equation.** You can solve $x^3 - 25x$ by extending the Product Property of Zero to more than two factors: If $a \cdot b \cdot c = 0$, then $a = 0$, or $b = 0$, or $c = 0$.

*Example 4*   **Solve $x^3 - 25x = 0$.**

$$x^3 - 25x = 0$$
$$x(x^2 - 25) = 0 \quad \blacktriangleleft \textit{Factor out the GCF, } x.$$
$$x(x - 5)(x + 5) = 0 \quad \blacktriangleleft x^2 - 25 \textit{ is the difference of two squares.}$$

So, $x = 0$ or $x - 5 = 0$ or $x + 5 = 0$.   $\blacktriangleleft$ *Set each factor equal to 0 and solve for x.*
$$x = 5 \qquad\qquad x = -5$$

**Thus,** there are three roots: 0, 5, and $-5$.   $\blacktriangleleft$ *Check.*

Notice that the degree of an equation determines the maximum number of roots.

## Reading in Algebra

1. How many solutions would you expect for the equation $x^3 - 5x^2 + 10x = 0$? **3**
2. What is another name for the *solutions* of an equation? **roots**
3. How does a quadratic equation differ from a cubic equation?
4. What is the degree of the equation $x^5 - 4x = 0$? **5**
3. **The quadratic equation has 2 roots; the cubic has 3.**

## Oral Exercises

**For what values of $x$ will each equation be true?**   1. 6, 3   2. −7, −1   3. −5, 2   4. 0, 8   5. 9, −9
1. $(x - 6)(x - 3) = 0$   2. $(x + 7)(x + 1) = 0$   3. $(x + 5)(x - 2) = 0$   4. $3x(x - 8) = 0$
5. $(x - 9)(x + 9) = 0$       6. $(x - 3)(x + 5)(x - 4) = 0$       7. $x(x - 1)(x + 2) = 0$
                                             3, −5, 4                                      0, 1, −2

## Written Exercises

**Solve each equation.**

Ⓐ
1. $x^2 - 8x + 12 = 0$  2, 6
2. $p^2 + 9p + 20 = 0$  −4, −5
3. $r^2 + 8r - 9 = 0$  1, −9
4. $2t^2 + 5t - 3 = 0$  $\frac{1}{2}$, −3
5. $x^2 - 36 = 0$  **6, −6**
6. $3m^2 - 13m - 10 = 0$  $-\frac{2}{3}$, **5**
7. $g^2 - 7g = 0$  0, 7
8. $2a^2 + 10a = 0$  **0, −5**
9. $3a^2 - 22a + 7 = 0$  $\frac{1}{3}$, 7
10. $2y^2 - 11y + 5 = 0$  $\frac{1}{2}$, 5
11. $3d^2 + 14d - 5 = 0$  $\frac{1}{3}$, −5
12. $2t^2 - 7t - 15 = 0$  $-\frac{3}{2}$, 5
13. $x^2 - 100 = 0$  **10, −10**
14. $m^2 + 13m = 0$  **0, −13**
15. $3a^2 - 7a = 0$  $\frac{7}{3}$, 0

Ⓑ
16. $2x^2 + 13x - 24 = 0$  $\frac{3}{2}$, −8
17. $3y^2 + 16y - 35 = 0$  $\frac{5}{3}$, −7
18. $5a^2 + 34a - 7 = 0$  $\frac{1}{5}$, −7
19. $2a^2 + 5a - 42 = 0$  $\frac{7}{2}$, −6
20. $5x^2 - 22x + 21 = 0$  $\frac{7}{5}$, 3
21. $2a^2 + 23a + 56 = 0$  $-\frac{7}{2}$, −8
22. $x^3 - 36x = 0$  **0, 6, −6**
23. $4b^3 - 16b = 0$  **−2, 0, 2**
24. $5x^3 - 245x = 0$  **−7, 0, 7**
25. $x^3 - 9x^2 + 20x = 0$  **0, 4, 5**
26. $a^3 - a^2 - 20a = 0$  **−4, 0, 5**
27. $4p^3 - 16p = 0$  **−2, 0, 2**

Ⓒ
28. **−5, −2, 0, 2, 5**
28. $x^5 - 29x^3 + 100x = 0$
29. **−3, −1, 0, 1, 3**
29. $x^5 - 10x^3 + 9x = 0$
30. **−4, −2, 0, 2, 4**
30. $t^5 - 20t^3 + 64t = 0$
31. $4a^5 - 13a^3 + 9a = 0$
32. $36m^4 - 85m^2 + 9 = 0$
33. $2x^6 - 100x^4 + 98x^2 = 0$
34. $8x^4 - 18x^2 = 0$  $-\frac{3}{2}, 0, 0, \frac{3}{2}$
35. $3x^5 - 39x^3 + 108x = 0$
36. $8y^5 - 58y^3 + 50y = 0$

31. $0, \frac{3}{2}, -\frac{3}{2}, 1, -1$   32. $\frac{1}{3}, -\frac{1}{3}, \frac{3}{2}, -\frac{3}{2}$   33. $7, -7, 0, 0, -1, 1$    35. $-3, -2, 0, 2, 3$   36. $-\frac{5}{2}, -1, 1, 0, \frac{5}{2}$

# STANDARD FORM OF A QUADRATIC EQUATION—PROBLEM SOLVING

**Objectives**

**To solve quadratic equations after rewriting them in standard form with a positive coefficient of the square term**
**To solve problems that lead to quadratic equations**

A quadratic equation is in *standard form* if
1. the polynomial is equal to zero, and
2. the terms are arranged in descending order of exponents.

For example, $2x^2 - 7x - 15 = 0$ is in standard form because
1. the polynomial, $2x^2 - 7x - 15$, is equal to 0, and
2. the terms $2x^2$, $-7x$, and $-15$ are in descending order of exponents.

You will find it is easier to solve quadratic equations if they are in *standard form*. It will also simplify factoring if the coefficient of the squared term is *positive*.

**Example 1**

**Write $n^2 - 12 = 4n$ in standard form.**

Get the polynomial equal to 0.

$$n^2 - 12 = 4n$$
$$n^2 - 4n - 12 = 0 \quad \blacktriangleleft \textit{ Subtract } 4n \textit{ from each side. Put the terms in}$$
$$\textit{descending order of exponents.}$$

**Example 2**

**Solve $8x = -x^2 - 15$. Check the solution.**

The coefficient of the $x^2$ term, $-1$, is not positive. If you add $1x^2$ to each side, the new equation will have a positive $x^2$ term.

1. Get the coefficient of the $x^2$ term positive. Then get the equation in standard form.
2. Factor.
3. Set each factor equal to 0.
4. Solve for $x$.
5. *Check.*

$$8x = -1x^2 - 15$$
$$1x^2 + 8x = -15 \quad \blacktriangleleft \textit{ Add } 1x^2 \textit{ to each side.}$$
$$1x^2 + 8x + 15 = 0 \quad \blacktriangleleft \textit{ Add } 15 \textit{ to each side.}$$
$$(x + 5)(x + 3) = 0$$
$$x + 5 = 0 \text{ or } x + 3 = 0$$
$$x = -5 \text{ or } x = -3$$

*Let $x = -5$.* $\blacktriangleright$

| $8x = -x^2 - 15$ | |
|---|---|
| $8 \cdot -5$ | $-(-5)^2 - 15$ |
| $-40$ | $-25 - 15$ |
| | $-40$ |

$-40 = -40$, true

*Let $x = -3$.* $\blacktriangleright$

| $8x = -x^2 - 15$ | |
|---|---|
| $8 \cdot -3$ | $-(-3)^2 - 15$ |
| $-24$ | $-9 - 15$ |
| | $-24$ |

$-24 = -24$, true

**Thus,** the roots are $-5$ and $-3$.

Sometimes it is more convenient to get the polynomial on the right side of the equation and 0 on the left side. Examples 3 and 4 show two such situations.

## Example 3    **Solve $-x^2 = 5x$. Check.**

$-1x^2 = 5x$    ◀ *The coefficient of the $x^2$ term, $-1$, is not positive.*
$\quad 0 = 1x^2 + 5x$    ◀ *Add $1x^2$ to each side. The equation is now in standard form.*
$\quad 0 = x(x + 5)$    ◀ *Factor out the GCF, $x$.*
$\quad x = 0 \text{ or } x + 5 = 0$    ◀ *Set each factor equal to 0.*
$\qquad\qquad\quad x = -5$    ◀ *Solve for $x$.*

Check.
Let $x = 0$. ▶
$$
\begin{array}{c|c}
\multicolumn{2}{c}{-x^2 = 5x} \\
\hline
-(0)^2 & 5 \cdot 0 \\
0 & 0 \\
\multicolumn{2}{c}{0 = 0, \text{ true}}
\end{array}
$$

Let $x = -5$. ▶
$$
\begin{array}{c|c}
\multicolumn{2}{c}{-x^2 = 5x} \\
\hline
-(-5)^2 & 5 \cdot -5 \\
-25 & -25 \\
\multicolumn{2}{c}{-25 = -25, \text{ true}}
\end{array}
$$

**Thus,** the roots are 0 and $-5$.

## Example 4    **Solve $4 - 7x = 2x^2$. Check.**

Notice that the $x^2$ term is on the right and positive.
Write the equation in the standard form $0 = ax^2 + bx + c$.

$\qquad 4 - 7x = 2x^2$
$\qquad\qquad 4 = 2x^2 + 7x$    ◀ *Add $7x$ to each side.*
$\qquad\qquad 0 = 2x^2 + 7x - 4$    ◀ *Subtract 4 from each side.*
$\qquad\qquad\qquad\qquad\qquad$ *The equation is now in standard form.*
$\qquad\qquad 0 = (2x - 1)(x + 4)$    ◀ *Factor.*
$2x - 1 = 0 \quad \text{or} \quad x + 4 = 0$    ◀ *Set each factor equal to 0.*
$\quad 2x = 1 \qquad\qquad x = -4$    ◀ *Solve each equation for $x$.*
$\quad\; x = \tfrac{1}{2}$

Check.
Let $x = \tfrac{1}{2}$. ▶
$$
\begin{array}{c|c}
\multicolumn{2}{c}{4 - 7x = 2x^2} \\
\hline
4 - 7(\tfrac{1}{2}) & 2(\tfrac{1}{2})^2 \\
4 - 3\tfrac{1}{2} & 2(\tfrac{1}{4}) \\
\tfrac{1}{2} & \tfrac{1}{2} \\
\multicolumn{2}{c}{\tfrac{1}{2} = \tfrac{1}{2}, \text{ true}}
\end{array}
$$

Let $x = -4$. ▶
$$
\begin{array}{c|c}
\multicolumn{2}{c}{4 - 7x = 2x^2} \\
\hline
4 - 7(-4) & 2(-4)^2 \\
4 + 28 & 2(16) \\
32 & 32 \\
\multicolumn{2}{c}{32 = 32, \text{ true}}
\end{array}
$$

**Thus,** the roots are $\tfrac{1}{2}$ and $-4$.

Sometimes a word problem leads to a quadratic equation. In this case, you still use the basic four steps for solving word problems. Such a word problem is illustrated in the next example.

*Example 5*    **The square of a number is 6 less than five times the number. Find the number.**

READ ▶    You are asked to find a number.
The first sentence gives information about the number to be found.

PLAN ▶    Represent the data.
Let $n$ = the number.
Write an equation.
Square of $n$ is 6 less than 5 times $n$.

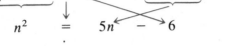

$$n^2 \quad = \quad 5n \quad - \quad 6$$

SOLVE ▶    To solve for $n$, put the quadratic equation into standard form.
$$n^2 = 5n - 6$$
$$n^2 - 5n = -6 \qquad \blacktriangleleft \text{ } \textit{Subtract 5n from each side.}$$
$$n^2 - 5n + 6 = 0 \qquad \blacktriangleleft \text{ } \textit{Add 6 to each side.}$$
$$(n - 3)(n - 2) = 0 \qquad \blacktriangleleft \text{ } \textit{Factor.}$$
$$n - 3 = 0 \text{ or } n - 2 = 0 \qquad \blacktriangleleft \text{ } \textit{Set each factor equal to 0 and solve.}$$
$$n = 3 \qquad\qquad n = 2$$

INTERPRET ▶    There seem to be two roots. Check each one.

SUBSTITUTE 3 ▶    Square of number is 6 less than 5 times the number.

| $3^2$ | $5 \cdot 3 - 6$ |
|---|---|
| 9 | $15 - 6$ |
|  | 9 |

$$9 = 9, \text{ true}$$

SUBSTITUTE 2 ▶    Square of number is 6 less than 5 times the number.

| $2^2$ | $5 \cdot 2 - 6$ |
|---|---|
| 4 | $10 - 6$ |
|  | 4 |

$$4 = 4, \text{ true}$$

**Thus,** there are two numbers, 3 and 2, that meet the conditions.

1. coefficient of $x^2$ term is positive, the polynomial is equal to zero, terms are arranged in descending order of exponents.

## Reading in Algebra

1. What are the three conditions necessary for a quadratic equation to be in standard form?
2. If the $x^2$ coefficient is negative, what should be the first step in rewriting the equation in standard form? **Use addition property to make it positive.**
3. Which of the conditions for standard form does $x^2 + 4 + 5x = 0$ not meet? **terms are not arranged in descending order of exponents.**

## Oral Exercises

**What would you add to each side of these equations to put the equation into standard form with a positive coefficient for the square term?**

**1.** $4x + 3 = -x^2$ $\;x^2$

**2.** $-x^2 = 5x$ $\;x^2$

**3.** $-12 = m^2 - 8m$ $\;12$

**4.** $-5t = t^2 + 6$ $\;5t$

**5.** $a^2 = 4a - 12$ $\;-4a + 12$

**6.** $-g^2 - 5g = -6$ $\;g^2 + 5g$

## Written Exercises

(A) **Solve each equation. Check.**

**1.** $3z = -z^2$ $\;0, -3$

**2.** $12r = -r^2 - 32$ $\;-4, -8$

**3.** $-x^2 = -7x$ $\;0, 7$

**4.** $4x = -x^2 - 4$ $\;-2, -2$

**5.** $12 - 4m = m^2$ $\;2, -6$

**6.** $6 - 5p = p^2$ $\;1, -6$

**Solve each equation.**

**7.** $x^2 + 4x = 5$ $\;1, -5$

**8.** $a^2 - 10 = 3a$ $\;5, -2$

**9.** $12g = -g^2$ $\;0, -12$

**10.** $49 = b^2$ $\;7, -7$

**11.** $9m^2 = 4$ $\;\frac{2}{3}, -\frac{2}{3}$

**12.** $5b - 2 = 3b^2$ $\;\frac{2}{3}, 1$

**13.** $n^2 - 7n = 18$ $\;9, -2$

**14.** $2b^2 + 7 = 15b$ $\;\frac{1}{2}, 7$

**15.** $3k^2 = -2k + 1$ $\;\frac{1}{3}, -1$

**16.** $3x - 20 = -2x^2$ $\;\frac{5}{2}, -4$

**17.** $3a^2 - 22a = 16$ $\;-\frac{2}{3}, 8$

**18.** $25 = 9b^2$ $\;\frac{5}{3}, -\frac{5}{3}$

(B) **19.** $12x^2 = 35 - x$ $\;-\frac{7}{4}, \frac{5}{3}$

**20.** $4m^2 - 42 = 17m$ $\;-\frac{7}{4}, 6$

**21.** $11b - 10 = -6b^2$ $\;\frac{2}{3}, -\frac{5}{2}$

**22.** $12y^2 + 15 = 29y$ $\;\frac{3}{4}, \frac{5}{3}$

**23.** $6x^2 = 21 - 11x$ $\;\frac{7}{6}, -3$

**24.** $17x - 14 = -6x^2$ $\;\frac{2}{3}, -\frac{7}{2}$

**Solve each problem.**

**25.** The square of a number is 10 less than 7 times the number. Find the number. $\;2, 5$

**26.** The square of a number is the same as twice the number. Find the number. $\;0, 2$

**27.** If 6 times a number is added to the square of the number, the result is 16. Find the number. $\;-8, 2$

**28.** Eight less than 9 times a number is the same as the square of the number. Find the number. $\;1, 8$

(C) **Solve each equation.**

**29.** $(3x - 2)(2x + 5) = 5x^2 + 5x - 18$ $\;-4, -2$

**30.** $(x - 3)(x + 3) = 2x^2 - 18$ $\;3, -3$

**31.** $4a^2 + 10a - 24 = (3a - 5)(a + 4)$ $\;1, -4$

**32.** $t^2 + 9t - 43 = (2t + 3)(t - 5)$ $\;14, 2$

**33.** $3(a - 2)^2 = (a - 2)$ $\;2, \frac{7}{3}$

**34.** $(2a - 6)^2 - 10(2a - 6) + 25 = 0$ $\;\frac{11}{2}, \frac{11}{2}$

**Solve each problem.**

**35.** If a number is multiplied by 7 more than 3 times that number, the result is 20. Find the number. $\;\frac{5}{3}, -4$

**36.** If 3 times the square of a number is increased by 20 less than 3 times the number, the result is the number squared. Find the number. $\;\frac{5}{2}, -4$

## CUMULATIVE REVIEW

**1.** Solve
$2x - (x + 1) = -8 - 6x.$ $\;-1$

**2.** Simplify
$(-3x^4y)(5xy^6).$ $\;-15x^5y^7$

**3.** Factor
$6x^2 + 27x - 15.$ $\;3(2x - 1)(x + 5)$

# PROBLEM SOLVING: CONSECUTIVE INTEGERS

Objective — **To solve problems about consecutive integers.**

If you start with any given integer and count by ones, you will get **consecutive integers.** Starting with 5, the four consecutive integers are 5, 6, 7, and 8.

If $x$ is the first integer, then four consecutive integers are $x$, $x + 1$, $x + 2$, and $x + 3$.

You can now use the four basic steps for solving problems to solve problems about consecutive integers.

Example 1 — **Find three consecutive integers whose sum is 126.**

READ ▶ You are asked to find three consecutive integers.
You are given that their sum is 126.

PLAN ▶ Represent the data.
Let $x$ = the first integer.
Let $x + 1$ = the second integer.
Let $x + 2$ = the third integer.
Use *their sum is 126* to write an equation.
$x + (x + 1) + (x + 2) = 126$ ◀ *Sum means add.*

SOLVE ▶
$$3x + 3 = 126$$
$$3x = 123 \quad ◀ \text{ } Subtract \text{ } 3 \text{ } from \text{ } each \text{ } side.$$
$$x = 41 \quad ◀ \text{ } Divide \text{ } each \text{ } side \text{ } by \text{ } 3.$$
The first integer, $x$, is 41.  ◀ *Use the representation to find the integers.*
The second integer, $x + 1$, is $41 + 1$, or 42.
The third integer, $x + 2$, is $41 + 2$, or 43.

INTERPRET ▶ The integers, 41, 42, and 43, are consecutive integers.
The sum, $41 + 42 + 43$, is 126.
**Thus,** the integers are 41, 42, and 43.

**Even integers** are integers that are divisible by 2. Some examples of even integers are $-8$, $-2$, 0, 14, and 18.

If you start with an even integer and count by twos, you will get *consecutive even integers*. For example, starting with 4, three consecutive even integers are 4, 6, and 8.

If $x$ is the first even integer, then three consecutive even integers are $x$, $x + 2$, and $x + 4$.

The next example involves consecutive even integers. The word problem also leads to a quadratic equation that may have two sets of solutions.

*Example 2*    **Find two consecutive even integers whose product is 48.**

READ ▶    You are asked to find two consecutive even integers.
You are given that their product is 48.

PLAN ▶    Represent the given data.
Let $x$ = the first even integer.
Let $x + 2$ = the second even integer.
Use *their product is 48* to write an equation.
$x(x + 2) = 48$    ◀ *Product means multiply.*

SOLVE ▶

$$x^2 + 2x = 48 \quad ◀ x(x + 2) = x \cdot x + x \cdot 2 = x^2 + 2x$$
$$x^2 + 2x - 48 = 0 \quad ◀ \text{Get the equation into standard form.}$$
$$\text{Subtract 48 from each side.}$$
$$(x + 8)(x - 6) = 0 \quad ◀ \text{Factor.}$$
$$x + 8 = 0 \quad \text{or} \quad x - 6 = 0 \quad ◀ \text{Set each factor equal to 0 and solve.}$$
$$x = -8 \qquad\qquad x = 6$$

There appear to be two solutions. Use the representation to find the integers.
First integer: $x = -8$                  or First integer: $x = 6$
Second integer: $x + 2 = -8 + 2$, or $-6$    Second integer: $x + 2 = 6 + 2$, or 8

INTERPRET ▶    $-8$ and $-6$ are consecutive even integers and $-8 \cdot -6 = 48$.
6 and 8 are consecutive even integers and $6 \cdot 8 = 48$.
There are two pairs of solutions: $-8$ and $-6$, and 6 and 8.

**Thus,** the integers are $-8$ and $-6$, and 6 and 8.

**Odd integers** are integers *not* divisible by 2. Some examples of odd integers are $-9$, $-5$, 1, 3, 11, and 13.

If you start with any odd integer and count by twos, you will get *consecutive odd integers*. For example, starting with 3, four consecutive odd integers are 3, 5, 7, and 9.

If $x$ is the first odd integer, then three consecutive odd integers are $x$, $x + 2$, and $x + 4$.

A quadratic equation may have two roots. However, sometimes only one of the roots will be a solution of a word problem, as in the next example.

*Example 3*   **Find three consecutive odd integers such that the square of the third, decreased by the first, is 46.**

Represent the three consecutive odd integers.
Let $x$ = the first odd integer.
Let $x + 2$ = the second odd integer.   ◀ *Add 2 to get the next odd integer.*
Let $x + 4$ = the third odd integer.   ◀ *$(x + 2) + 2 = x + 4$, the 3rd odd integer.*

Write an equation and solve it.
Square of third, decreased by first, is 46.

$$(x + 4)^2 \quad - \quad x = 46$$

$$x^2 + 8x + 16 - x = 46 \quad ◀ (x + 4)^2 = (x + 4)(x + 4)$$
$$x^2 + 7x + 16 = 46 \quad ◀ \text{Combine like terms: } 8x - x = 7x.$$
$$x^2 + 7x - 30 = 0 \quad ◀ \text{Subtract 46 from each side to get the equation}$$
*into standard form.*

$$(x + 10)(x - 3) = 0 \quad ◀ \text{Factor.}$$
$$x + 10 = 0 \quad \text{or} \quad x - 3 = 0 \quad ◀ \text{Set each factor equal to 0.}$$
$$x = -10 \quad\quad\quad x = 3 \quad ◀ \text{Solve each equation for } x.$$

But, only one of the roots, 3, is an *odd* integer.
So, only 3 will be a solution relevant to the word problem.

Use the representation to write the three integers.
The first integer, $x$, is 3.
The second integer, $x + 2$, is $3 + 2$, or 5.
The third integer, $x + 4$, is $3 + 4$, or 7.

Check in the original problem.

Square of third, decreased by first, is 46.

| $(7)^2 - 3$ | 46 |
| $49 - 3$ | |
| 46 | |

$$46 = 46, \text{ true}$$

**So,** the three consecutive odd integers are 3, 5, and 7.

# Reading in Algebra

**Indicate whether each statement is sometimes true, always true, or never true.**
1. Adding 2 to an integer produces an even integer. **S**
2. If a word problem leads to a quadratic equation, then there will be two solutions for the *word* problem. **S**
3. Adding 1 to an integer results in an odd integer. **S**
4. Two consecutive odd integers differ by 2. **A**
5. The sum of two odd integers is an odd integer. **N**

Problem Solving: Consecutive Integers

# Written Exercises

**Solve each problem.**

**(A)**

**1.** Find two consecutive integers whose sum is 39. **19, 20**

**2.** Find two consecutive odd integers whose sum is 36. **17, 19**

**3.** Find five consecutive integers whose sum is 45. **7, 8, 9, 10, 11**

**4.** Find three consecutive even integers whose sum is 126. **40, 42, 44**

**5.** Find three consecutive integers such that three times the first, added to the last, is 22. **5, 6, 7**

**6.** Find four consecutive odd integers such that twice the second, added to the last, is 25. **5, 7, 9, 11**

**7.** Find two consecutive even integers whose product is 80. **−10, −8 or 8, 10**

**8.** Find two consecutive integers whose product is 56. **7, 8; −8, −7**

**9.** Find two consecutive odd integers whose product is 63. **7, 9; −9, −7**

**10.** Find two consecutive integers whose product is 20. **4, 5; −5, −4**

**11.** Find two consecutive odd integers such that the square of the first, increased by the second, is 32. **5, 7**

**12.** Find two consecutive integers such that the sum of the first and the square of the second is −1. **−1, 0; −2, −1**

**13.** Find two consecutive odd integers such that the sum of their squares is 74. **5, 7; −7, −5**

**14.** Find two consecutive even integers such that the sum of their squares is 52.

**15.** Find three consecutive integers such that the square of the first, added to the last, is 8. **2, 3, 4; −3, −2, −1**

**16.** Find two consecutive odd integers such that the square of the second, decreased by the first, is 14. **−5, −3**

14. **4, 6; −6, −4**

**(B)**

**17.** Find four consecutive integers such that the sum of the squares of the second and third is 61. **4, 5, 6, 7; −7, −6, −5, −4**

**18.** Find five consecutive integers such that the square of the third, decreased by the square of the second, is 3 times the first.

18. **3, 4, 5, 6, 7**

**19.** Find three consecutive odd integers such that the first times the second is 1 less than 4 times the third. **5, 7, 9; −3, −1, 1**

**20.** Find three consecutive even integers such that the product of the first and the third is 20 less than 10 times the second. **0, 2, 4; 6, 8, 10**

**21.** Find three consecutive integers such that the square of the third is 12 more than the square of the second. **None**

**22.** Find three consecutive even integers such that twice the square of the second, decreased by the square of the third, is 6 less than the first. **2, 4, 6**

**(C)**

**23.** Find three consecutive integers such that the product of the second and the third, decreased by 4 times the second, is 5 more than 5 times the first. **−1, 0, 1; 7, 8, 9**

**24.** Find three consecutive integers such that the product of all three, decreased by the cube of the first, is 33. **3, 4, 5**

**25.** The product of the second and third of three consecutive integers, decreased by the square of the second, is 1 more than the first. Find the integers. **any 3 consecutive integers**

**26.** If the product of all three consecutive odd integers is decreased by 35 times the first, the result is 0. Find the three integers. **3, 5, 7; −9, −7, −5**

# PROBLEM SOLVING

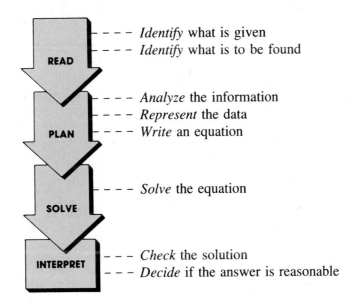

READ — — — *Identify* what is given
— — — *Identify* what is to be found

PLAN — — — *Analyze* the information
— — — *Represent* the data
— — — *Write* an equation

SOLVE — — — *Solve* the equation

INTERPRET — — — *Check* the solution
— — — *Decide* if the answer is reasonable

## Solve each word problem.

1. Juan rents a car for a week at the rate of $75 plus 30¢/km. How far can he go on a budget of $195? **400 km**

2. The length of a rectangle is 8 km more than 3 times the width. The perimeter is 80 km. Find the length and the width. **32 km, 8 km**

3. Six less than 4 times Tina's age is the same as 4 more than twice her age. How old is she? **5 y**

4. Find three consecutive even integers such that the sum of the second and third is 3 times the first integer. **6, 8, 10**

5. Five more than three times a number is at least 32. Find the number. **at least 9**

6. The selling price of a television set is $80 more than twice the cost. Find the cost if the selling price is $380. **$150**

7. Sixteen less than the square of a number is the same as 6 times the number. Find the number. **−2, 8**

8. Eight less than twice a number is almost 20. Find the number. **almost 14**

9. Joan is paid $180 a week plus 5% commission on all sales. Find the sales needed to give her a total salary of $240. **$1200**

10. Eight times the sum of a number and 6 is the same as 8 more than 3 times the number. Find the number. **−8**

11. Separate 66 into two parts so that the larger part is twice the smaller one. Find each part.

12. Find two consecutive even integers such that their product is 120. **10, 12; −12, −10**

13. Maurice's mass is 9 kg more than 3 times Julep's mass. Together their masses are 89 kg. Find the mass of each. **20 kg, 69 kg**

14. The sum of the ages of three sisters is 35. Mona's age is 6 less than Paula's. Jane's age is 5 more than Mona's. How old is each? **14. Paula, 14; Jane, 13; Mona, 8**

15. The first side of a triangle is 4 cm shorter than 3 times the second side. The third side is 6 cm longer than the second side. The perimeter is 42 cm. Find the length of each side. **20, 8, 14**

16. Last year, Hank's weekly salary was $250. This year his weekly salary is $300. His raise in salary is what percentage of last year's salary? **20%**

11. **44, 22**

# CHAPTER SEVEN REVIEW © Exercises are starred.

Ⓐ Exercises are 1–3, 6–11, 13–16, 19–20, 23–24, 27–31, 38–40.    The rest are Ⓑ Exercises.

**Vocabulary**

consecutive integers [7.10]
cubic equation [7.8]
difference of two squares [7.2]
even integer [7.10]
odd integer [7.10]
product property of zero [7.8]
perfect square trinomial [7.3]
quadratic equation [7.8]
roots [7.8]

Multiply. [7.1]

1. $(r + 6)(r + 4)$ $r^2 + 10r + 24$
2. $(2x + 5y)(x + 3y)$ $2x^2 + 11xy + 15y^2$
3. $(3p - 7)(2p + 7)$ $6p^2 + 7p - 49$
4. $(g - 5)(g^2 + 7g + 3)$ $g^3 + 2g^2 - 32g - 15$
5. $(3k - 1)(k^2 + 2k - 4)$ $3k^3 + 5k^2 - 14k + 4$

6. Multiply $(2a + 7y)(2a - 7y)$. [7.2]
$4a^2 - 49y^2$

Factor. [7.2]

7. $4y^2 - 9$ $(2y - 3)(2y + 3)$
8. $b^2 - 81r^2$ $(b - 9r)(b + 9r)$

9. Simplify $(3a - 4)^2$. [7.3] $9a^2 - 24a + 16$

Determine if each is a perfect square. If so, factor the trinomial. [7.3]

10. $25x^2 - 20x + 4$ $(5x - 2)^2$
11. $49a^2 + 14ab + 4b^2$ not a perfect square

Factor each trinomial into two binomials.

12. $y^2 + y - 42$ [7.4] $(y + 7)(y - 6)$
13. $x^2 + 8x + 12$ $(x + 6)(x + 2)$
14. $x^2 - 18x - 88$ $(x - 22)(x + 4)$
15. $2x^2 - 9x - 5$ [7.5] $(2x + 1)(x - 5)$
16. $12p^2 - 19p - 10$ $(12p + 5)(p - 2)$
17. $6p^2 + p - 1$ $(3p - 1)(2p + 1)$

Factor completely. [7.6]

18. $20x^3 + 42x^2 - 20x$ $2x(5x - 2)(2x + 5)$
19. $2b^3 - 32b$ $2b(b - 4)(b + 4)$
20. $2m^3 + 2m^2 - 40m$ $2m(m + 5)(m - 4)$
21. $2p^2 + 13pq + 15q^2$ $(2p + 3q)(p + 5q)$
22. $3x^3 - 15x^2y + 12xy^2$ $3x(x - y)(x - 4y)$

Factor each polynomial completely. [7.7]

23. $at + az + bt + bz$ $(t + z)(a + b)$
24. $m^2 + 5m - km - 5k$ $(m + 5)(m - k)$
25. $x^3 - 9xy^2 + 2x^2 - 18y^2$
     25. $(x + 2)(x - 3y)(x + 3y)$

Solve each equation. [7.8]

26. $4b^3 - 100b = 0$ $5, 0, -5$
27. $x^2 - 4x - 12 = 0$ $6, -2$
28. $m^2 + 7m = 0$ $0, -7$
29. $3y^2 + 8y - 35 = 0$ $\frac{7}{3}, -5$

Solve each equation. [7.9]

30. $-12r + 35 = -r^2$ $5, 7$
31. $25 = t^2$ $5, -5$
32. $-33a - 10 = -7a^2$ $-\frac{2}{7}, 5$
33. $11x = -6x^2 + 35$ $\frac{5}{3}, -\frac{7}{2}$
34. $2a^2 = -15a - 28$ $-\frac{7}{2}, -4$
35. $9m - 18 = -2m^2$ $\frac{3}{2}, -6$

Solve each problem. [7.9]

36. The square of a number is 8 more than −1, 8
    7 times the number. Find the number.
37. Fourteen more than 5 times a number is the same as the square of the number. Find the number. $-2, 7$

Solve each word problem. [7.10]

38. Find two consecutive integers whose sum is 43. $21, 22$
39. Find two consecutive even integers whose product is 8. $2, 4; -4, -2$
40. Find two consecutive odd integers such that the square of the second, increased by 4 times the first, is 37. $3, 5; -11, -9$
     43. $(m - 5y)(m + 5y)(m - y)(m + y)$

★ Factor each polynomial completely.

41. $-80p - 16a^2p + a^4p$ [7.2]
42. $3b^5 - 1,875b$ [7.5] $3b(b - 5)(b + 5)(b^2 + 25)$
43. $m^4 - 26m^2y^2 + 25y^4$ [7.6]
44. $a^4 - 16a^2b^2 - 9a^2 + 144b^2$ [7.7]
     44. $(a + 3)(a - 3)(a - 4b)(a + 4b)$

★ Solve each equation. [7.9] 41. $p(a^2 - 20)(a^2 + 4)$

45. $5(x - 3)^2 = (x - 3)$ $3, \frac{16}{5}$
46. $(3b - 2)^2 - 5(3b - 2) + 4 = 0$ $1, 2$
47. $(3x - 2)^2 - 7x(x - 2) = -9x + 44$ $\frac{5}{2}, -8$

Solve each equation.

**1.** $x^2 + 9x + 20 = 0$ **−5, −4**

**2.** $16m = 35 - 3m^2$ **$\frac{5}{3}$, −7**

**3.** $a^2 + 3a = 0$ **0, −3**

**4.** $2x^3 + 5x^2 - 12x = 0$ **0, $\frac{3}{2}$, −4**

**5.** $3p + 28 = p^2$ **−4, 7**

**6.** $2y^2 = -3y + 35$ **$\frac{7}{2}$, −5**

**7.** $-19x - 10 = -15x^2$ **$-\frac{2}{5}$, $\frac{5}{3}$**

**8.** $28 = 4y^2 + 9y$ **$\frac{7}{4}$, −4**

**9.** Simplify $(2a - 5)^2$. **$4a^2 - 20a + 25$**

Multiply.

**10.** $(4x - 3)(4x + 3)$ **$16x^2 - 9$**

**11.** $(2b + 5c)(4b - 3c)$ **$8b^2 + 14bc - 15c^2$**

**12.** $(2a + 3)(a^2 - 6a + 4)$
$2a^3 - 9a^2 - 10a + 12$

Factor each polynomial completely.

**13.** $x^2 - 7x - 18$ **$(x - 9)(x + 2)$**

**14.** $m^2 + 4m - 21$ **$(m + 7)(m - 3)$**

**15.** $3x^2 + 4xy - 4y^2$ **$(3x - 2y)(x + 2y)$**

**16.** $8b^2 + 8bc - 6c^2$ **$2(2b - c)(2b + 3c)$**

**17.** $169t^2 - 25$ **$(13t - 5)(13t + 5)$**

**18.** $18y^3 - 15y^2 - 63y$ **$3y(3y - 7)(2y + 3)$**

**19.** $a^3b - 49ab^3$ **$ab(a - 7b)(a + 7b)$**

Determine if each trinomial is a perfect square. If so, factor the trinomial.

**20.** $25m^2 - 10m + 4$ **not a perfect square**

**21.** $9a^2 - 12ab + 4b^2$ **$(3a - 2b)^2$**

Solve each problem.

**22.** The square of a number is 12 less than 8 times the number. Find the number. **2, 6**

**23.** If 5 times a number is added to the square of the number, the result is 36. Find the number. **−9, 4**

**24.** Find two consecutive odd integers whose product is 35. **5, 7 or −5, −7**

**25.** Find four consecutive integers whose sum is 82. **19, 20, 21, 22**

**26.** Find two consecutive even integers such that the square of the first, increased by the second, is 44. **6, 8**

★ **27.** Solve $a^5 - 17a^3 + 16a = 0$.
**−4, −1, 0, 1, 4**

★ **28.** Multiply $(3x + 2)(x - 4)(x + 1)$.
**$3x^3 - 7x^2 - 18x - 8$**

★ **29.** Factor $x^2a^2 - 25x^2 - y^2a^2 + 25y^2$ completely. **$(x - y)(x + y)(a - 5)(a + 5)$**

# COMPUTER ACTIVITIES

## Finding Sums of Consecutive Integers

Even integers have a factor of 2. Odd integers do not have a factor of 2.

The computer can perform many different operations on long lists of numbers and can be programmed to print such lists. It can also be programmed to select specific numbers from a list, arrange them in a particular order, or perform other operations upon them as specified by the programmer. It is often useful for a computer to add the numbers in a list (for example, to compute a company's total payroll). The program in this lesson illustrates a method for selecting particular numbers and computing their sum.

This program will find the sum of three consecutive odd or even integers. If the sum is for odd integers, the starting number must be odd, and if the sum is for even integers, the starting number must be even.

```
10   PRINT "THIS PROGRAM FINDS THE SUM OF 3 CONSECUTI
     VE ODD OR EVEN INTEGERS."
20   PRINT "TYPE IN THE FIRST NUMBER."
30   INPUT X
40   LET S = X + (X + 2) + (X + 4)
50   PRINT "THE THREE INTEGERS ARE "X", "X + 2", "X +
     4"."
60   PRINT "THE SUM IS "S"."
70   END
```

See the Computer Section beginning on page 488 for more information.

## Exercises  For Ex. 5–8, see TE Answer Section.

**Enter and RUN the program above for the following numbers:**

1. 4 **18**

2. 6 **24**

3. 7 **27**

4. 41 **129**

5. Change the program to find the sum of four consecutive odd or even integers.

6. Write a program to calculate the product of three consecutive odd or even integers.

7. Write a program to find the three consecutive integers for a sum if the sum is the INPUT.

8. Write a program to find the three consecutive even or odd integers for a sum if the sum is the INPUT.

# COLLEGE PREP TEST

For items 1–8, in each item you are to compare a quantity in Column 1 with a quantity in Column 2. Write the letter of the correct answer from these choices:

(A) The quantity in Column 1 is greater than the quantity in Column 2.
(B) The quantity in Column 2 is greater than the quantity in Column 1.
(C) The quantity in Column 1 is equal to the quantity in Column 2.
(D) The relationship cannot be determined from the given information.

Notes: Information centered over both columns refers to one or both of the quantities to be compared. A symbol that appears in both columns has the same meaning in each column, and all variables represent real numbers.

| | Column 1 | Column 2 | |
|---|---|---|---|
| **1.** | $5^2$ | $4^3$ | B |
| **2.** | $B = 0, A > 2, C > 2$ | | B |
| | $B(A + C)$ | $A(B + C)$ | |
| **3.** | $(x + y)^2$ | $x^2 + y^2$ | D |
| **4.** | $(x - 7)^2$ | $x^2 - 14x + 49$ | C |

| | Column 1 | Column 2 | |
|---|---|---|---|
| **5.** | $5 \cdot 144 \cdot 6$ | $12^2 \cdot 5^2$ | A |
| **6.** | The solution of $x - (6 - x) = 0$ | The solution of $x^2 - 6x + 9 = 0$ | C |
| **7.** | | $2^{n+1} = 8$ | B |
| | $n$ | $5$ | |
| **8.** | $a^3$ | $a^2$ | D |

For items 9–16, choose the letter of the correct answer.

**9.** If $3a + 3a + 3a = 36$, then $4a - 1 =$
(A) 47    (B) 15    (C) 12    **B**
(D) 7    (E) 3

**10.** What value of $x$ gives $y$ its greatest value if $y = 6 - (x - 1)^2$?
(A) 0    (B) 1    (C) 4    **B**
(D) 6    (E) 8

**11.** Find $x$ if $18 \cdot 18 = 6 \cdot 6 \cdot x$.
(A) 729    (B) 36    (C) 12    **D**
(D) 9    (E) $\frac{1}{2}$

**12.** If $x + 6$ is an even integer, which of the following is not an even integer?
(A) $2x + 6$   (B) $4x + 2$   (C) $x - 6$
(D) $x$    (E) $x + 1$    **E**

**13.** Six consecutive even integers are given. The sum of the first three is 12. What is the sum of the last three?    **C**
(A) 12    (B) 18    (C) 30
(D) 42    (E) none of these

**14.** Suppose $y = x + 2$. What is the value of $(x - y)^3$?    **A**
(A) $-8$   (B) 6   (C) 5   (D) 2   (E) 0

**15.** How many 7-cent stamps can you buy for $x$ dollars?    **B**
(A) $\frac{x}{7}$   (B) $\frac{100x}{7}$   (C) $\frac{7}{x}$   (D) $\frac{7}{100x}$   (E) $\frac{x}{700}$

**16.** $4x$ is what percentage of $2x$?
(A) 2%    (B) 4%    (C) 200%   **C**
(D) 50%    (E) 400%

# 8 SIMPLIFYING RATIONAL EXPRESSIONS

## HISTORY

### Pythagoras
### 572 B.C.?–
### 495 B.C.?

*Pythagoras was a Greek mathematician of the fifth century B.C. He founded the famous Pythagorean school, which studied mathematics, philosophy, and natural sciences.*

The Pythagoreans took the first steps in the development of number theory and also laid the basis of future number mysticism. They discovered amicable or friendly numbers and one particular pair, 284 and 220, attained a mystical aura. Superstition maintained that two pendants bearing these numbers would seal a lifetime friendship between the wearers of the pendants. Other numbers with mystical connections and linked to Pythagoras are the perfect, abundant, and deficient numbers.

**Project**

Do a report on amicable, perfect, abundant, and deficient numbers. Include examples of each.

# RATIONAL EXPRESSIONS                                    8.1

*Objectives*    **To find the value(s), if any, for which a rational expression is undefined**
**To multiply rational expressions**

Recall that fractions, such as $\frac{2}{3}, \frac{0}{5}, \frac{-3}{4}$, and $\frac{7}{-5}$ are called *rational numbers*. You have seen that a rational number is a number of the form $\frac{a}{b}$ where $a$ and $b$ are integers and $b \neq 0$. Since the denominator of a fraction cannot be 0, a fraction like $\frac{16}{0}$ is *undefined*.

Consider the following fractions in which the numerator and denominator are polynomials. Such expressions are called *rational expressions*.

$$\frac{4}{x} \qquad \frac{a+7}{3} \qquad \frac{d-5}{d^2-7d+10}$$

| Definition: Rational expression | A **rational expression** is an expression of the form $\frac{a}{b}$ where $a$ and $b$ are polynomials and $b \neq 0$. |
|---|---|

*Example 1*    **Evaluate $\dfrac{x-4}{x-8}$ if $x = 8$.**

$\dfrac{x-4}{x-8} = \dfrac{8-4}{8-8}$   ◄ *Substitute 8 for x.*

$\qquad = \dfrac{4}{0}$   ◄ *The denominator of a fraction cannot be 0.*

**Thus, $\dfrac{x-4}{x-8}$ is undefined if $x = 8$ because the denominator cannot be 0.**

*Example 2*    **For what value of $x$ is $\dfrac{x+3}{2x+4}$ undefined?**

*Strategy:* A rational expression is undefined if the denominator is 0. Set the denominator, $2x + 4$, equal to 0 and solve the equation.
$$2x + 4 = 0$$
$$2x = -4$$
$$x = -2$$

**Thus, $\dfrac{x+3}{2x+4}$ is undefined if $x = -2$.**

In the rational expression $\dfrac{a + 7}{3}$, notice that the denominator, 3, is not 0.

Thus, $\dfrac{a + 7}{3}$ is defined for *all values* of $a$.

You have seen that a rational expression may be undefined for one value of the variable, or the rational expression may be defined for all values of the variable. A rational expression also may be undefined for more than one value of the variable.

*Example 3*  **For what values of $a$ is $\dfrac{5a + 3}{a^2 - 10a + 16}$ undefined?**

*Think:* The rational expression will be undefined if the denominator is 0.

$a^2 - 10a + 16 = 0$ ◀ *Set the denominator equal to 0.*

$(a - 8)(a - 2) = 0$ ◀ *Factor: $a^2 - 10a + 16 = (a - 8)(a - 2)$.*

$a - 8 = 0$ or $a - 2 = 0$ ◀ *Set each factor equal to 0.*

$a = 8$ or $a = 2$ ◀ *Solve each equation for $a$.*

Thus, $\dfrac{5a + 3}{a^2 - 10a + 16}$ is undefined if $a = 8$ or $a = 2$.

In arithmetic, the product of two fractions is a fraction of the form $\dfrac{product\ of\ the\ numerators}{product\ of\ the\ denominators}$. The product of two rational expressions has the same form.

Thus, $\dfrac{7}{x} \cdot \dfrac{4}{y} = \dfrac{7 \cdot 4}{x \cdot y} = \dfrac{28}{xy}$.

The definition is stated below.

---

**Product of two rational expressions**

If $\dfrac{a}{b}$ and $\dfrac{c}{d}$ are rational expressions, $b \neq 0$, $d \neq 0$, then

$$\frac{a}{b} \cdot \frac{c}{d} = \frac{a \cdot c}{b \cdot d}.$$

---

*Example 4*  **Multiply $\dfrac{2a^3}{5b^4} \cdot \dfrac{4a^6}{7b}$.**

$$\dfrac{2a^3}{5b^4} \cdot \dfrac{4a^6}{7b} = \dfrac{2a^3 \cdot 4a^6}{5b^4 \cdot 7b}$$ ◀ *Product of the numerators*
◀ *Product of the denominators*

$$= \dfrac{2 \cdot 4 \cdot a^3 \cdot a^6}{5 \cdot 7 \cdot b^4 \cdot b^1}$$ ◀ $b = b^1$

$$= \dfrac{8a^9}{35b^5}$$ ◀ $a^3 \cdot a^6 = a^{3+6}$, *or* $a^9$
◀ $b^4 \cdot b^1 = b^{4+1}$, *or* $b^5$

*Example 5*    **Multiply** $-\dfrac{x+1}{x+5} \cdot \dfrac{2x-3}{3x+1}$.

$$-\frac{x+1}{x+5} \cdot \frac{2x-3}{3x+1} = -\frac{(x+1)(2x-3)}{(x+5)(3x+1)}$$

$$= -\frac{2x^2 - x - 3}{3x^2 + 16x + 5} \qquad \blacktriangleleft \ (x+1)(2x-3) = 2x^2 - x - 3$$

You can rewrite 3 and $x-4$ in the form $\dfrac{a}{b}$. Thus, $3 = \dfrac{3}{1}$ and $x - 4 = \dfrac{x-4}{1}$.
You will find this procedure helpful in the next example.

*Example 6*    **Multiply.**

$$3 \cdot \frac{b-7}{b+5}$$

$$\frac{3}{1} \cdot \frac{b-7}{b+5}$$

$$\frac{3(b-7)}{1(b+5)}, \text{ or } \frac{3b-21}{b+5}$$

$$(x-4) \cdot \frac{2x+3}{x-6}$$

$$\frac{x-4}{1} \cdot \frac{2x+3}{x-6}$$

$$\frac{(x-4)(2x+3)}{1(x-6)}, \text{ or } \frac{2x^2 - 5x - 12}{x-6}$$

# Oral Exercises

**Which of the following are undefined?**

**1.** $\dfrac{3}{3}$    **2.** $\dfrac{7}{0}$    **3.** $\dfrac{0}{4}$    **4.** $\dfrac{8}{6-2}$    **5.** $\dfrac{7+3}{5-5}$    **6.** $\dfrac{4+3}{9-2}$    **7.** $\dfrac{6-1}{3-3}$

**Ex. 2, 5, and 7 are undefined fractions.**

**Multiply.**

**8.** $\dfrac{x^5}{y^2} \cdot \dfrac{x^4}{y^3} \cdot \dfrac{x^9}{y^5}$    **9.** $\dfrac{1}{m^5} \cdot \dfrac{1}{m^2} \cdot \dfrac{1}{m^7}$    **10.** $5 \cdot \dfrac{2a}{3b} \cdot \dfrac{10a}{3b}$    **11.** $x^3 \cdot \dfrac{x^5}{y^4} \cdot \dfrac{x^8}{y^4}$    **12.** $\dfrac{7}{a^3} \cdot \dfrac{6}{a^2}$

**13.** $\dfrac{3}{t^4} \cdot \dfrac{5}{t^6} \cdot \dfrac{15}{t^{10}}$    **14.** $\dfrac{b}{4} \cdot \dfrac{b^5}{5} \cdot \dfrac{b^6}{20}$    **15.** $\dfrac{m^3}{p^2} \cdot \dfrac{m}{p} \cdot \dfrac{m^4}{p^3}$    **16.** $y^3 \cdot \dfrac{y^2}{w^4} \cdot \dfrac{y^5}{w^4}$    **17.** $\dfrac{9}{a^3} \cdot \dfrac{6}{a}$

**12.** $\dfrac{42}{a^5}$    **17.** $\dfrac{54}{a^4}$

# Written Exercises

**For what value(s) of $x$ is the rational expression undefined?**

Ⓐ  **1.** $\dfrac{x+3}{2x-4}$  2    **2.** $\dfrac{5}{2x+10}$  −5  **3.** $\dfrac{7x+1}{3x-15}$  5    **4.** $\dfrac{11x}{2x+8}$  −4  **5.** $\dfrac{3x}{x-5}$  5

**6.** $\dfrac{2x+3}{x^2-7x+12}$  3, 4  **7.** $\dfrac{3x}{x^2-4}$  ±2  **8.** $\dfrac{2x+3}{x^2-5x+4}$  1, 4  **9.** $\dfrac{9x-5}{x^2-7x}$  0, 7 **10.** $\dfrac{12x+8}{9x^2-25}$

$$\pm\frac{5}{3}$$

Rational Expressions

**Multiply.** For Ex. 15–20 and 24–35, see TE Answer Section.

**11.** $\dfrac{2a^2}{3b^3} \cdot \dfrac{5a^3}{7b^2}$ **$\dfrac{10a^5}{21b^5}$**  
**12.** $\dfrac{7x^2}{3y^4} \cdot -\dfrac{4x^5}{5y^2}$ **$-\dfrac{28x^7}{15y^6}$**  
**13.** $-\dfrac{5x^3}{3y^7} \cdot \dfrac{4x^2}{7y^4}$ **$-\dfrac{20x^5}{21y^{11}}$**  
**14.** $\dfrac{6a^3}{5b^2} \cdot \dfrac{3a^5}{b^2}$ **$\dfrac{18a^8}{5b^4}$**

**15.** $-\dfrac{x+1}{x+2} \cdot \dfrac{2x-3}{2x+3}$  
**16.** $\dfrac{a+4}{a+1} \cdot \dfrac{3a-1}{3a+1}$  
**17.** $\dfrac{b+3}{b+5} \cdot -\dfrac{2b-5}{2b+7}$

**18.** $\dfrac{3p-1}{p+6} \cdot \dfrac{2p+3}{p-4}$  
**19.** $\dfrac{7d-2}{d-5} \cdot \dfrac{d+4}{d}$  
**20.** $\dfrac{r-4}{r} \cdot \dfrac{3r+5}{r+4}$

**21.** $4 \cdot \dfrac{x-7}{x+6}$ **$\dfrac{4x-28}{x+6}$**  
**22.** $3 \cdot \dfrac{2b-7}{3b-4}$ **$\dfrac{6b-21}{3b-4}$**  
**23.** $\dfrac{x+2}{2x+7} \cdot 8$ **$\dfrac{8x+16}{2x+7}$**

**24.** $(m-5) \cdot \dfrac{2m+1}{m-6}$  
**25.** $\dfrac{-a-2}{a+4} \cdot (2a-1)$  
**26.** $(2r-3) \cdot \dfrac{r+6}{3r-4}$

Ⓑ **27.** $\dfrac{a+b}{x-y} \cdot \dfrac{a-b}{x+3y}$  
**28.** $\dfrac{3m-n}{m+2n} \cdot \dfrac{2m+n}{3m+2n}$  
**29.** $\dfrac{5r-t}{r+3t} \cdot \dfrac{4r-t}{r+2t}$

**30.** $(x+y) \cdot \dfrac{x-y}{x+7y}$  
**31.** $\dfrac{a-3b}{x-2y} \cdot \dfrac{a+3b}{x+2y}$  
**32.** $\left(\dfrac{3x+2y}{m+n}\right)^2$

**33.** $\left(-\dfrac{2a+b}{3a+b}\right)^2$  
**34.** $\dfrac{a+5b}{3a-4b} \cdot (a-5b)$  
**35.** $-\dfrac{4a-b}{2a+b} \cdot \dfrac{a+2b}{a-3b}$

**36.** $\dfrac{x-4}{x+5} \cdot \dfrac{x^2-7x+2}{x^2+4x+2}$ **$\dfrac{x^3-11x^2+30x-8}{x^3+9x^2+22x+10}$**  
**37.** $\dfrac{2a-3}{a+4} \cdot \dfrac{a^2-2a+3}{a^2+5a-4}$ **$\dfrac{2a^3-7a^2+12a-9}{a^3+9a^2+16a-16}$**

**38.** $\dfrac{2p-1}{p-6} \cdot \dfrac{p^2-3p+1}{p^2+4p+2}$ **$\dfrac{2p^3-7p^2+5p-1}{p^3-2p^2-22p-12}$**  
**39.** $(3t-2) \cdot \dfrac{2t^2-5t+2}{3t-7}$ **$\dfrac{6t^3-19t^2+16t-4}{3t-7}$**

**For what value(s) of $x$ is each rational expression undefined?** (Ex. 40–43)

Ⓒ **40.** $\dfrac{x+2}{x^3-7x^2+12x}$  
**0, 3, 4**

**41.** $\dfrac{3x-2}{4x^4-109x^2+225}$  
**$\pm\frac{3}{2}, \pm 5$**

**42.** $\dfrac{3x-7}{x^5-13x^3+36x}$  
**0, ±2, ±3**

**43.** $\dfrac{2x+5}{x^5-18x^3+81x}$  
**0, ±3**

**44.** For what values of $x$, if any, will $\dfrac{x^{a^2}}{5} \cdot \dfrac{x^{5a}}{3} = \dfrac{x^{a^2+5a}}{15}$ be true? **all values**

# CUMULATIVE REVIEW

**Factor completely.**

**1.** $2a^3 + 9a^2 - 18a$  
$a(2a-3)(a+6)$

**2.** $4x^2 - 9y^2$  
$(2x+3y)(2x-3y)$

**3.** $ax - xm - 7a + 7m$  
$(a-m)(x-7)$

# CALCULATOR ACTIVITIES

Evaluate for the given value of $x$.

**1.** $\dfrac{x^2-7x+4}{2x+10}$ for $x = -35$  **$-24.56666$**

**2.** $\dfrac{3x^2+7x+2}{x^2+2x}$ for $x = 14$  **3.0714285**

**3.** $\dfrac{7x+4}{x^2-43x-90}$ for $x = 45$. What happened in Exercise 3? Why?

The calculator will indicate an error in some manner,
for example, E., EO., flashing 9's, or the like.
The denominator is zero.

# SIMPLEST FORM

*Objective*   **To simplify rational expressions**

You can rewrite $\frac{10}{15}$ as a product of two fractions. This is easy to see if you factor the numerator and the denominator into primes, as shown.

$$\frac{10}{15} = \frac{2 \cdot 5}{5 \cdot 3} \qquad \text{or} \qquad \frac{10}{15} = \frac{2 \cdot 5}{3 \cdot 5}$$

$$= \frac{2}{5} \cdot \frac{5}{3} \qquad\qquad\qquad = \frac{2}{3} \cdot \frac{5}{5}$$

You can now use the second product to rewrite $\frac{10}{15}$ in simplest form:

$$\frac{10}{15} = \frac{2}{3} \cdot \frac{5}{5} = \frac{2}{3} \cdot 1, \text{ or } \frac{2}{3}, \text{ because 1 is the } \textit{multiplicative identity}.$$

Since the greatest common factor of 2 and 3 is 1, 2 and 3 are *relatively prime*.

**Thus,** $\frac{2}{3}$ is in simplest form, or $\frac{2}{3}$ is reduced to lowest terms.

| Simplest form of a rational expression | If $c \neq 0$ and $b \neq 0$, then $\dfrac{a \cdot c}{b \cdot c} = \dfrac{a}{b} \cdot \dfrac{c}{c} = \dfrac{a}{b} \cdot 1 = \dfrac{a}{b}$. $\dfrac{a}{b}$ is in **simplest form** if $a$ and $b$ are relatively prime. |
|---|---|

The concept of simplest form of a rational expression can be extended to cases in which there are more than two factors in the numerator, or in the denominator, or in both the numerator and denominator, as shown in Example 1.

*Example 1*   **Simplify** $\dfrac{12m}{18m}$.

First, factor the numerator and denominator into primes.

$$\frac{12m}{18m} = \frac{2 \cdot 2 \cdot 3 \cdot m}{3 \cdot 3 \cdot 2 \cdot m}$$

Then, rearrange the factors so that like factors are over each other.

$$\frac{2 \cdot 2 \cdot 3 \cdot m}{3 \cdot 2 \cdot 3 \cdot m} = \frac{2}{3} \cdot \frac{2}{2} \cdot \frac{3}{3} \cdot \frac{m}{m}$$

$$= \frac{2}{3} \cdot 1 \cdot 1 \cdot 1$$

$$= \frac{2}{3} \cdot 1, \text{ or } \frac{2}{3}$$

Here is a more compact form.

$$\frac{12m}{18m} = \frac{2 \cdot 2 \cdot 3 \cdot m}{3 \cdot 3 \cdot 2 \cdot m}$$

$$= \frac{2 \cdot \overset{1}{2} \cdot \overset{1}{3} \cdot \overset{1}{m}}{3 \cdot \underset{1}{3} \cdot \underset{1}{2} \cdot \underset{1}{m}}$$

$$= \frac{2 \cdot 1 \cdot 1 \cdot 1}{1 \cdot 3 \cdot 1 \cdot 1}, \text{ or } \frac{2}{3}$$

You can simplify a more complicated rational expression such as $\dfrac{x^2 - 9x + 20}{2x^2 - 13x + 15}$ in basically the same way. You begin by factoring the numerator and the denominator, as shown in Example 2.

*Example 2*    **Simplify** $\dfrac{x^2 - 9x + 20}{2x^2 - 13x + 15}.$

$$\frac{x^2 - 9x + 20}{2x^2 - 13x + 15} = \frac{(x - 5)(x - 4)}{(2x - 3)(x - 5)} \quad \blacktriangleleft \text{ Factor } x^2 - 9x + 20 \text{ as } (x - 5)(x - 4).$$
$$\blacktriangleleft \text{ Factor } 2x^2 - 13x + 15 \text{ as } (2x - 3)(x - 5).$$

$$= \frac{\overset{1}{\cancel{(x - 5)}}(x - 4)}{(2x - 3)\underset{1}{\cancel{(x - 5)}}} = \frac{1(x - 4)}{(2x - 3)1}, \text{ or } \frac{x - 4}{2x - 3} \quad \blacktriangleleft \textit{ Divide out the common factors.}$$

Recall that factoring completely involves first looking for a GCF. This is illustrated in the next two examples.

*Example 3*    **Simplify** $\dfrac{a - 6}{3a - 18}.$

$$\frac{a - 6}{3a - 18} = \frac{a - 6}{3(a - 6)} \quad \blacktriangleleft \ a - 6 \textit{ is not factorable.}$$
$$\blacktriangleleft \ 3a - 18 = 3(a - 6); \ 3 \textit{ is the GCF.}$$

$$= \frac{\overset{1}{\cancel{a - 6}}}{3\underset{1}{\cancel{(a - 6)}}} = \frac{1}{3 \cdot 1}, \text{ or } \frac{1}{3}$$

*Example 4*    **Simplify** $-\dfrac{3x^3 - 48x}{2x^2 - 8x}.$

$$-\frac{3x^3 - 48x}{2x^2 - 8x} = -\frac{3 \cdot x(x^2 - 16)}{2 \cdot x(x - 4)} \quad \blacktriangleleft \ 3x^3 - 48x = 3(x^3 - 16x) = 3 \cdot x(x^2 - 16)$$
$$\blacktriangleleft \ 2x^2 - 8x = 2(x^2 - 4x) = 2 \cdot x(x - 4)$$

$$= -\frac{3 \cdot \overset{1}{\cancel{x}}\overset{1}{\cancel{(x - 4)}}(x + 4)}{2 \cdot \underset{1}{\cancel{x}}\underset{1}{\cancel{(x - 4)}}} \quad \blacktriangleleft \textit{ Factor } x^2 - 16, \textit{ the difference of two squares:}$$
$$x^2 - 16 = (x)^2 - (4)^2 = (x - 4)(x + 4).$$

$$= -\frac{3(x + 4)}{2}, \text{ or } -\frac{3x + 12}{2} \quad \blacktriangleleft \ 3(x + 4) = 3x + 12$$

Rational expressions such as $\dfrac{x + 3}{x + 5}$ cannot be simplified because there are no factors common to both the numerator and the denominator. Notice what happens as different values are substituted for $x$.

If $x = 0$, then $\dfrac{x + 3}{x + 5} = \dfrac{0 + 3}{0 + 5}$, or $\dfrac{3}{5}$. If $x = 2$, then $\dfrac{x + 3}{x + 5} = \dfrac{2 + 3}{2 + 5}$, or $\dfrac{5}{7}$.

Chapter Eight

## Oral Exercises

**Simplify, if possible.**

1. $\dfrac{5}{10}$   $\dfrac{1}{2}$

2. $\dfrac{3}{9}$   $\dfrac{1}{3}$

3. $\dfrac{2}{5}$   not possible

4. $\dfrac{5}{5(x+3)}$   $\dfrac{1}{x+3}$

5. $\dfrac{x+4}{(x+4)3}$   $\dfrac{1}{3}$

6. $\dfrac{(x+5)(x-4)}{(x-4)(x+2)}$   $\dfrac{x+5}{x+2}$

7. $\dfrac{2(m-4)}{m-4}$   2

8. $\dfrac{(5x-2)(x+4)}{(x+4)(5x-2)}$   1

9. $\dfrac{3(x+4)}{x+3}$   not possible

## Written Exercises

**Simplify, if possible.** For Ex. 5–8 and 13–42, see TE Answer Section.

(A)

1. $\dfrac{8m}{12m}$   $\dfrac{2}{3}$

2. $\dfrac{14k}{16}$   $\dfrac{7k}{8}$

3. $-\dfrac{27}{30b}$   $-\dfrac{9}{10b}$

4. $-\dfrac{15t}{24t}$   $-\dfrac{5}{8}$

5. $\dfrac{a^2-7a+12}{2a^2-9a+9}$

6. $\dfrac{m^2-7m-18}{2m^2+3m-2}$

7. $\dfrac{b^2-36}{b^2-8b+12}$

8. $-\dfrac{x^2-10x+21}{x^2-3x-28}$

9. $\dfrac{x-4}{2x-8}$   $\dfrac{1}{2}$

10. $-\dfrac{a+3}{a-5}$   $-\dfrac{a+3}{a-5}$

11. $-\dfrac{a-5}{a^2-5a}$   $-\dfrac{1}{a}$

12. $\dfrac{3b-12}{b^2-6b+8}$   $\dfrac{3}{b-2}$

13. $-\dfrac{x^2-4x+3}{2x-6}$

14. $\dfrac{5b-20}{b^2-16}$

15. $\dfrac{2a+3}{4a+3}$

16. $\dfrac{a^2-8a+12}{3a-18}$

17. $\dfrac{m^2-5m-24}{m^2+3m}$

18. $\dfrac{y^2-5y-14}{y^2-7y}$

19. $\dfrac{3x^2-12x}{2x^2-4x}$

20. $\dfrac{2x^3-50x}{3x^2-15x}$

21. $\dfrac{2n^2+3n-2}{n^2-4}$

22. $\dfrac{3a^2-a-14}{2a^2+3a-2}$

23. $\dfrac{2b^2-13b-7}{6b+3}$

24. $\dfrac{3k^2+10k-8}{k^2+4k}$

(B)

25. $\dfrac{4k^2-25}{6k^2+k-35}$

26. $\dfrac{10a^2+13a-3}{6a^2+7a-3}$

27. $\dfrac{2x^2-3x-20}{2x^2+3x-5}$

28. $\dfrac{9y^2+12y+4}{3y^2-16y-12}$

29. $\dfrac{6a^2-7a-20}{3a^2-14a-24}$

30. $\dfrac{2b^4-7b^3-30b^2}{6b^4+13b^3-5b^2}$

31. $\dfrac{16m^2-24m+9}{8m^2-2m-3}$

32. $\dfrac{3k^4+7k^3-20k^2}{6k^4-7k^3-5k^2}$

33. $\dfrac{6x^3-7x^2-3x}{2x^3+13x^2-24x}$

34. $\dfrac{x^2y+x^2-4y-4}{x^2-10x+16}$

35. $\dfrac{by+cy-br-cr}{b^2+6bc+5c^2}$

36. $\dfrac{a^2x^2-x^2-9a^2+9}{ax-a^2-x+a}$

(C)

37. $\dfrac{x^4-13x^2+36}{x^3+x^2-6x}$

38. $\dfrac{9p^6-145p^4+16p^2}{3p^4-11p^3-4p^2}$

39. $\dfrac{50y^6-58y^4+8y^2}{30y^2+18y-12}$

40. $\dfrac{x^4-34x^2y^2+225y^4}{7x^2+14xy-105y^2}$

41. $\dfrac{2b^4-6b^3-8b^2}{6b^6-102b^4+96b^2}$

42. $\dfrac{8a^2-48a+64}{4a^8-80a^6+256a^4}$

## CUMULATIVE REVIEW

**Solve.**

1. $7x-15=-2x^2$   $-5, \dfrac{3}{2}$

2. $7a=-a^2$   $0, -7$

3. $\dfrac{2}{3}x-5=\dfrac{1}{9}x$   9

# THE $-1$ TECHNIQUE

*Objective*     **To simplify rational expressions by using the $-1$ technique**

The quadratic equation $2x^2 + 3x + 1 = 0$ is in standard form. Notice that the polynomial, $2x^2 + 3x + 1$, is in descending order of exponents with the first coefficient, 2, positive. This polynomial is said to be in *convenient form*. It is easier to simplify a rational expression if every polynomial in the rational expression is in convenient form.

You can get the polynomial $5x - x^2 + 4$ in a more convenient form as follows: Write the polynomial in descending order of exponents: $-1x^2 + 5x + 4$. Then, since the first coefficient is negative, factor out $-1$: $-1(x^2 - 5x - 4)$. **Thus,** a more convenient form of $5x - x^2 + 4$ is $-1(x^2 - 5x - 4)$.

*Example 1*     **Simplify** $\dfrac{x-5}{25-x^2}$.

$\dfrac{x-5}{25-x^2} = \dfrac{x-5}{-1x^2+25}$  ◄ *Rewrite the rational expression with the denominator in convenient form. Rewrite $25 - 1x^2$ in descending order of exponents.*

$= \dfrac{x-5}{-1(x^2-25)}$  ◄ *Factor out $-1$ to make the first coefficient positive.*

$= \dfrac{x-5}{-1(x-5)(x+5)}$  ◄ *Factor.*

$= \dfrac{\overset{1}{\cancel{x-5}}}{-1\underset{1}{\cancel{(x-5)}}(x+5)}$, or $\dfrac{1}{-1(x+5)}$  ◄ *Divide out common factors.*

These illustrations should suggest two other ways to interpret a negative fraction.

$$-\frac{12}{3} \qquad\qquad \frac{-1(12)}{3} \qquad\qquad \frac{12}{-1(3)}$$
$$-(12 \div 3) = -4 \qquad = \frac{-12}{3} = -4 \qquad = \frac{12}{-3} = -4$$

This concept also applies to rational expressions.

| The negative sign in a rational expression | If $b \neq 0$, $-\dfrac{a}{b} = \dfrac{-1(a)}{b} = \dfrac{a}{-1(b)}$. |
|---|---|

The answer, $\dfrac{1}{-1(x+5)}$, in Example 1 can also be given as $-\dfrac{1}{x+5}$, or $\dfrac{-1}{x+5}$.

Although any one of the three forms, $-\dfrac{a}{b}$, $\dfrac{-1(a)}{b}$, or $\dfrac{a}{-1(b)}$, is acceptable, the form $-\dfrac{a}{b}$ will be used most of the time in this book.

**Example 2**  **Simplify** $\dfrac{b^2 + 2b - 24}{12 + b - b^2}$.

$$\dfrac{b^2 + 2b - 24}{12 + b - b^2} = \dfrac{b^2 + 2b - 24}{-1b^2 + b + 12}$$ ◀ *First, rewrite the denominator in convenient form.*

$$= \dfrac{b^2 + 2b - 24}{-1(b^2 - b - 12)}$$ ◀ *Factor out $-1$.*

$$= \dfrac{(b + 6)(b - 4)}{-1(b - 4)(b + 3)}$$ ◀ *Factor $b^2 + 2b - 24$.*
  ◀ *Factor $b^2 - b - 12$.*

$$= \dfrac{b + 6}{-1(b + 3)}, \text{ or } -\dfrac{b + 6}{b + 3}$$ ◀ $\dfrac{a}{-1(b)} = -\dfrac{a}{b}$

Sometimes, you may have to put both the numerator and the denominator in convenient form before you can simplify a rational expression. You will use this procedure in the next example.

**Example 3**  **Simplify** $\dfrac{-3x^2 + 7x + 6}{9 - 3x}$.

$$\dfrac{-3x^2 + 7x + 6}{9 - 3x} = \dfrac{-3x^2 + 7x + 6}{-3x + 9}$$

$$= \dfrac{-1(3x^2 - 7x - 6)}{-1(3x - 9)}$$ ◀ *Factor out $-1$ from the numerator and the denominator so that the first coefficient of each is positive.*

$$= \dfrac{-1(3x + 2)(x - 3)}{-1(3)(x - 3)}$$ ◀ *Factor.*
  ◀ *3 is the GCF of $3x - 9$.*

$$= \dfrac{-1(3x + 2)(x - 3)}{-1(3)(x - 3)} = \dfrac{3x + 2}{3}$$

# Reading in Algebra

1. Why is $6x - x^2 + 4$ not in convenient form? **It is not in descending order of exponents and the coefficient of $x^2$ is $-1$.**
2. What must you do to put $-x^2 - 5x + 4$ in convenient form? **Factor out $-1$.**
3. True or false: $-\dfrac{x + 5}{x^2 - 9}$ and $\dfrac{x + 5}{9 - x^2}$ are two other names for $\dfrac{-1(x + 5)}{x^2 - 9}$. **True**

## Oral Exercises

**Rename in convenient form.**

**1.** $-3x + 7$   $-1(3x - 7)$    **2.** $-x^2 - 4$   $-1(x^2 + 4)$    **3.** $-a^2 + 5a$      **4.** $-m^2 - 2m$

**5.** $-2x^2 + 5$      **6.** $-x^2 - x - 4$      **7.** $-a^2 - 3a + 2$      **8.** $-5p^2 - 4p - 2$

**9.** $-3b^2 - b + 1$    **10.** $-t^2 - 3t + 1$    **3.** $-1(a^2 - 5a)$   **4.** $-1(m^2 + 2m)$   **5.** $-1(2x^2 - 5)$

**6.** $-1(x^2 + x + 4)$   **7.** $-1(a^2 + 3a - 2)$   **8.** $-1(5p^2 + 4p + 2)$   **9.** $-1(3b^2 + b - 1)$   **10.** $-1(t^2 + 3t - 1)$

## Written Exercises

**5.** $-(m + 3)$   **9.** $-(x - 4)$   **10.** $-(p + 4)$

**Simplify.**

**11.** $-(x - 1)$   **13.** $-\dfrac{y + 6}{y + 2}$   **14.** $-\dfrac{a + 5}{a - 8}$   **15.** $-\dfrac{x + 2}{x - 2}$   **17.** $\dfrac{2x - 3}{x - 4}$

Ⓐ

**1.** $\dfrac{x - 2}{4 - x^2} \quad \dfrac{-1}{x + 2}$    **2.** $\dfrac{7 - b}{2b - 14} - \dfrac{1}{2}$    **3.** $\dfrac{3 - b}{4b - 12} - \dfrac{1}{4}$    **4.** $\dfrac{a + 4}{16 - a^2} - \dfrac{1}{a - 4}$

**5.** $\dfrac{m^2 - m - 12}{4 - m}$    **6.** $\dfrac{14 - 2x}{x^2 - 9x + 14} - \dfrac{2}{x - 2}$   **7.** $\dfrac{3 - c}{c^2 - 5c + 6} - \dfrac{1}{c - 2}$   **8.** $\dfrac{6 - 2b}{b^2 - 6b + 9} - \dfrac{2}{b - 3}$

**9.** $\dfrac{x^2 - 7x + 12}{3 - x}$    **10.** $\dfrac{p^2 - p - 20}{5 - p}$    **11.** $\dfrac{-x^2 + 5x - 4}{x - 4}$    **12.** $\dfrac{c^2 + 8c - 20}{16 - 8c} - \dfrac{c + 10}{8}$

**13.** $\dfrac{y^2 - y - 42}{14 + 5y - y^2}$   **14.** $\dfrac{a^2 + 10a + 25}{40 + 3a - a^2}$   **15.** $\dfrac{x^2 - 8x - 20}{-x^2 + 12x - 20}$   **16.** $\dfrac{36 - m^2}{-m^2 + 4m + 12} \quad \dfrac{m + 6}{m + 2}$

**17.** $\dfrac{-2x^2 - x + 6}{-x^2 + 2x + 8}$   **18.** $\dfrac{-m^2 + 16}{-12 + 7m - m^2}$   **19.** $\dfrac{-t^2 + 5t + 24}{21 - t^2 + 4t}$   **20.** $\dfrac{x^2 - 2x - 35}{21 + 11x - 2x^2} - \dfrac{x + 5}{2x + 3}$

Ⓑ **21.** $\dfrac{2a^2 + 8a - 64}{28 - 3a - a^2} - \dfrac{2(a + 8)}{a + 7}$   **22.** $\dfrac{15 + 2y - y^2}{2y^2 + 20y + 42} - \dfrac{y - 5}{2(y + 7)}$   **23.** $\dfrac{21 - 18x - 3x^2}{-15x^2 + 15} \quad \dfrac{x + 7}{5(x + 1)}$

**24.** $\dfrac{12 + n - n^2}{2n^2 - 18} - \dfrac{n - 4}{2(n - 3)}$   **25.** $\dfrac{2x^2 - 6x}{24 - 2x - 2x^2} - \dfrac{x}{x + 4}$   **26.** $\dfrac{-a^2 + b^2}{a^2 + 3ab + 2b^2} - \dfrac{a - b}{a + 2b}$

**27.** $\dfrac{2x^2 - xy - y^2}{-2x^2 + 2y^2} - \dfrac{2x + y}{2(x + y)}$   **28.** $\dfrac{4a^2 - 4b^2}{-4a^2 + 20ab - 16b^2}$   **29.** $\dfrac{-10 + 22a - 4a^2}{3a^2 - 18a + 15} - \dfrac{2(2a - 1)}{3(a - 1)}$

Ⓒ **30.** $\dfrac{x^4 - 10x^2 + 9}{27 - 3x^2} - \dfrac{x^2 - 1}{3}$   **31.** $\dfrac{-36 + 37x^2 - x^4}{-x^2 - 5x + 6}$   **32.** $\dfrac{81 - a^4}{36 + 5a^2 - a^4} \quad \dfrac{a^2 + 9}{a^2 + 4}$

**33.** $\dfrac{-a^2 - 4ab + 21b^2}{a^2 + 7ab - 3a - 21b}$   **34.** $\dfrac{-rx^2 + 4r - sx^2 + 4s}{rx - 2r + sx - 2s}$   **35.** $\dfrac{-t^3 + 3t^2 + 64t - 192}{24 + 5t - t^2}$

**18.** $\dfrac{m + 4}{m - 3}$   **19.** $\dfrac{t - 8}{t - 7}$   **28.** $-\dfrac{a + b}{a - 4b}$   **31.** $(x - 6)(x + 1)$   **33.** $-\dfrac{a - 3b}{a - 3}$   **34.** $-(x + 2)$   **35.** $\dfrac{(t - 3)(t + 8)}{t + 3}$

## CUMULATIVE REVIEW

**1.** Simplify $(-5ab^3)^2$.
     $25a^2b^6$

**2.** Evaluate $-ab^3$ for $a = -1$ and $b = -2$.   $-8$    **3.** Solve $x = -x^2$.
     $0, -1$

## NON-ROUTINE PROBLEMS

$x = 4$

$x^2 = 16$   ◀ *Square each side.*

$x^2 - x = 12$   ◀ *Subtract the first equation from the second equation and solve.*

$x^2 - x - 12 = 0$, so $x = 4$   or   $x = -3$

But, only one of the roots checks in the original equation. What's wrong?

**The first equation has one root, and the second equation has two different roots.**

# THE QUOTIENT OF POWERS PROPERTY    8.4

*Objective*    **To simplify rational expressions by using the quotient of powers property**

Using the meaning of exponents, you can simplify $\dfrac{x^5}{x^3}$ and $\dfrac{a^2}{a^6}$ as shown below.

$$\frac{x^5}{x^3} = \frac{x \cdot x \cdot x \cdot x \cdot x}{x \cdot x \cdot x}$$

$$= \frac{\overset{1}{\cancel{x}} \cdot \overset{1}{\cancel{x}} \cdot \overset{1}{\cancel{x}} \cdot x \cdot x}{\underset{1}{\cancel{x}} \cdot \underset{1}{\cancel{x}} \cdot \underset{1}{\cancel{x}}}$$

$$= \frac{x \cdot x}{1}, \text{ or } x^2$$

$$\frac{a^2}{a^6} = \frac{a \cdot a}{a \cdot a \cdot a \cdot a \cdot a \cdot a}$$

$$= \frac{\overset{1}{\cancel{a}} \cdot \overset{1}{\cancel{a}}}{\underset{1}{\cancel{a}} \cdot \underset{1}{\cancel{a}} \cdot a \cdot a \cdot a \cdot a}$$

$$= \frac{1}{a \cdot a \cdot a \cdot a}, \text{ or } \frac{1}{a^4}$$

Notice the patterns:

$$\frac{x^5}{x^3} = \frac{x^{5-3}}{1} = \frac{x^2}{1}, \text{ or } x^2 \quad \text{and} \quad \frac{a^2}{a^6} = \frac{1}{a^{6-2}} = \frac{1}{a^4}.$$

The patterns suggest the following property.

| Quotient of powers property | $\dfrac{x^m}{x^n} = \dfrac{x^{m-n}}{1}$, or $x^{m-n}$ if $m > n$ <br><br> $\dfrac{x^m}{x^n} = \dfrac{1}{x^{n-m}}$ if $m < n$. |
|---|---|

You can use this property to simplify rational expressions such as $\dfrac{x^7}{x^9}$ and $\dfrac{b^5}{b}$

without having to rewrite numerators and denominators in factored form. The process is illustrated in Example 1.

*Example 1*    **Simplify.**

$$\frac{x^7}{x^9}$$

$$\frac{1}{x^{9-7}}$$

$$\frac{1}{x^2}$$

$$\frac{b^5}{b}$$

$$\frac{b^5}{b^1} = \frac{b^{5-1}}{1}$$

$$= \frac{b^4}{1}, \text{ or } b^4$$

The rational expression $\dfrac{a^5 b^4}{a^2 b^6}$ can also be simplified by using the quotient of powers property. This process, as well as a more compact form, are illustrated in Example 2.

**Example 2**    **Simplify** $\dfrac{a^5 b^4}{a^2 b^6}$.

$$\frac{a^5 b^4}{a^2 b^6} = \frac{a^5}{a^2} \cdot \frac{b^4}{b^6}$$

$$= \frac{a^{5-2}}{1} \cdot \frac{1}{b^{6-4}}$$

$$= \frac{a^3}{1} \cdot \frac{1}{b^2}$$

$$= \frac{a^3}{b^2}$$

Here is a more compact form for arranging your work.

$$\frac{a^5 b^4}{a^2 b^6}$$

$$\frac{\overset{a^3}{\cancel{a^5}} \overset{1}{\cancel{b^4}}}{\underset{1}{\cancel{a^2}} \underset{b^2}{\cancel{b^6}}}$$

◄ *Think:* $\dfrac{a^5}{a^2} = \dfrac{a^{5-2}}{1} = \dfrac{a^3}{1}$ *and*

$$\frac{a^3 \cdot 1}{1 \cdot b^2}, \text{ or } \frac{a^3}{b^2} \qquad \frac{b^4}{b^6} = \frac{1}{b^{6-4}} = \frac{1}{b^2}.$$

**Example 3**    **Simplify** $\dfrac{18x^3 y^7}{20xy^{10}}$.

$$\frac{18x^3 y^7}{20xy^{10}} = \frac{3 \cdot 3 \cdot 2 \cdot x^3 y^7}{2 \cdot 2 \cdot 5 \cdot x^1 y^{10}}$$

$$= \frac{3 \cdot 3 \cdot \overset{1}{\cancel{2}} \cdot \overset{x^2}{\cancel{x^3}} \overset{1}{\cancel{y^7}}}{\underset{1}{\cancel{2}} \cdot 2 \cdot 5 \cdot \underset{1}{\cancel{x^1}} \cdot \underset{y^3}{\cancel{y^{10}}}}$$

◄ *Think:* $\dfrac{x^3}{x^1} = \dfrac{x^{3-1}}{1} = x^2$ *and* $\dfrac{y^7}{y^{10}} = \dfrac{1}{y^{10-7}} = \dfrac{1}{y^3}.$

$$= \frac{9x^2}{10y^3}$$

Sometimes you will have to combine the quotient of powers property with other methods you have used to simplify rational expressions. The process is illustrated in the next two examples.

**Example 4**    **Simplify** $\dfrac{x^6 y(2x^2 - 14x + 20)}{x^2 y^2(6x - 12)}$.

$$\frac{x^6 y(2x^2 - 14x + 20)}{x^2 y^2(6x - 12)} = \frac{x^6 \cdot y^1 \cdot 2(x^2 - 7x + 10)}{x^2 \cdot y^2 \cdot 6(x - 2)}$$

◄ *Look for common factors:*
$2x^2 - 14x + 20 = 2(x^2 - 7x + 10)$
*and* $6x - 12 = 6(x - 2)$.

$$= \frac{x^6 \cdot y^1 \cdot 2(x - 5)(x - 2)}{x^2 \cdot y^2 \cdot 2 \cdot 3(x - 2)}$$

◄ *Factor* $x^2 - 7x + 10$.

◄ $6 = 2 \cdot 3$

$$= \frac{\overset{x^4}{\cancel{x^6}} \cdot \overset{1}{\cancel{y^1}} \cdot \overset{1}{\cancel{2}}(x - 5)(\cancel{x - 2})}{\underset{1}{\cancel{x^2}} \cdot \underset{y^1}{\cancel{y^2}} \cdot \underset{1}{\cancel{2}} \cdot 3(\underset{1}{\cancel{x - 2}})}, \text{ or } \frac{x^4(x - 5)}{3y}$$

*Example 5*    **Simplify** $\dfrac{m^8 p^6 (y^2 + 9y + 20)}{m^{11} p^5 (-4 - y)}$.

$$\dfrac{m^8 p^6 (y^2 + 9y + 20)}{m^{11} p^5 (-4 - y)} = \dfrac{m^8 p^6 (y^2 + 9y + 20)}{m^{11} p^5 (-1y - 4)} \quad \blacktriangleleft \textit{Rewrite } -4 - y \textit{ in convenient form.}$$

$$= \dfrac{m^8 p^6 (y^2 + 9y + 20)}{m^{11} p^5 (-1)(y + 4)} \quad \blacktriangleleft \textit{Factor out } -1.$$

$$= \dfrac{\overset{1}{\cancel{m^8}} \cdot \overset{p^1}{\cancel{p^6}}(y + 5)\cancel{(y + 4)}}{\underset{m^3}{\cancel{m^{11}}} \cdot \underset{1}{\cancel{p^5}}(-1)\underset{1}{\cancel{(y + 4)}}}$$

$$= \dfrac{p^1 (y + 5)}{m^3 (-1)}$$

$$= -\dfrac{p(y + 5)}{m^3}$$

# Written Exercises

18. $\dfrac{y(m - 4)}{3}$   21. $\dfrac{a + 2}{2y}$

24. $-\dfrac{a(x - 1)}{2b}$   30. $-\dfrac{1}{m - 2}$   31. $-\dfrac{x - 2}{b^2(x + 4)}$   33. $-\dfrac{p - 2}{2p(p + 3)}$

**Simplify.**

Ⓐ 1. $\dfrac{a^4}{a^9} \quad \dfrac{1}{a^5}$   2. $\dfrac{x^8}{x^2} \quad x^6$   3. $\dfrac{b^{11}}{b^{13}} \quad \dfrac{1}{b^2}$   4. $\dfrac{a^7}{a^6} \quad a$   5. $\dfrac{r^5}{r^6} \quad \dfrac{1}{r}$   6. $\dfrac{a^3 b^4}{a^5 b^8} \quad \dfrac{1}{a^2 b^4}$   7. $\dfrac{x^4 y^4}{xy^3} \quad x^3 y$

8. $\dfrac{x^6 y^4}{x^8 y^3} \quad \dfrac{y}{x^2}$   9. $\dfrac{p^3 r^2}{pr^7} \quad \dfrac{p^2}{r^5}$   10. $\dfrac{a^5 b^9}{a^8 b^6} \quad \dfrac{b^3}{a^3}$   11. $\dfrac{m^6 t^3}{m^3 t} \quad m^3 t^2$

12. $\dfrac{6a^7 n^5}{8a^3 n^9} \quad \dfrac{3a^4}{4n^4}$   13. $\dfrac{14x^5 y^3}{21x^2 y^8} \quad \dfrac{2x^3}{3y^5}$   14. $\dfrac{8t^9 w^6}{10t^3 w^7} \quad \dfrac{4t^6}{5w}$   15. $\dfrac{15gh^3}{5g^2 h^4} \quad \dfrac{3}{gh}$

16. $\dfrac{x^5 (a^2 - 5a - 6)}{x^8 (2a - 12)} \quad \dfrac{a + 1}{2x^3}$   17. $\dfrac{a^7 (b^2 - 9)}{a^8 (b - 3)} \quad \dfrac{b + 3}{a}$   18. $\dfrac{y^3 (m^2 - 11m + 28)}{y^2 (3m - 21)}$

19. $\dfrac{m^4 (4x^2 - 64)}{m^2 (2x - 8)} \quad 2m^2 (x + 4)$   20. $\dfrac{c^3 (6x - 30)}{c^9 (2x^2 - 50)} \quad \dfrac{3}{c^6 (x + 5)}$   21. $\dfrac{y^3 (2a^2 + 22a + 36)}{y^4 (4a + 36)}$

22. $\dfrac{a^3 b^5 (2c^2 + 7c - 4)}{a^4 b^7 (1 - 2c)} \quad -\dfrac{c + 4}{ab^2}$   23. $\dfrac{x^3 y^3 (2x^2 + 5x - 3)}{x^2 y^7 (9 - x^2)} \quad -\dfrac{x(2x - 1)}{y^4 (x - 3)}$   24. $\dfrac{a^2 b (3x^2 - 12x + 9)}{ab^2 (18 - 6x)}$

Ⓑ 25. $\dfrac{x^2 y - x^2}{x^6} \quad \dfrac{y - 1}{x^4}$   26. $\dfrac{a^4}{a^3 b - a^3} \quad \dfrac{a}{b - 1}$   27. $\dfrac{x^7}{bx^2 + x^2 y} \quad \dfrac{x^5}{b + y}$

28. $\dfrac{a^2 (18 - 6x)}{3a^4 x^2 - 27a^4} \quad -\dfrac{2}{a^2 (x + 3)}$   29. $\dfrac{b^2 x^2 - b^2 x - 2b^2}{b^5 x^2 + 2b^5 x + b^5} \quad \dfrac{x - 2}{b^3 (x + 1)}$   30. $\dfrac{5a^4 - a^4 m}{a^4 m^2 - 7a^4 m + 10a^4}$

31. $\dfrac{b^3 x^2 - 7b^3 x + 10b^3}{20b^5 + b^5 x - b^5 x^2}$   32. $\dfrac{m^3 - 4m^2 + 3m}{m^7 - 6m^6 + 9m^5} \quad \dfrac{m - 1}{m^4 (m - 3)}$   33. $\dfrac{2p^3 - 14p^2 + 20p}{60p^2 + 8p^3 - 4p^4}$

Ⓒ 34. $\dfrac{12x^5 + 76x^4 - 160x^3}{240x^2 - 66x^3 - 12x^4}$   35. $\dfrac{81a^3 - a^3 x^4}{x^8 - 18x^6 + 81x^4}$   36. $\dfrac{-400xy + 41x^3 y - x^5 y}{x^3 y^2 - x^2 y^2 - 20xy^2}$

37. $\dfrac{x^2 y^2 + axy^2 - rx^2 y - arxy}{x^4 y^2 - a^2 x^2 y^2} \quad \dfrac{y - r}{xy(x - a)}$   38. $\dfrac{(ax + 3bx + 2ay + 6by)a^3 b^4}{a^3 b - 6a^2 b^2 - 27ab^3} \quad \dfrac{a^2 b^3 (x + 2y)}{a - 9b}$

34. $-\dfrac{2x(3x - 5)}{3(2x - 5)}$   35. $-\dfrac{a^3 (x^2 + 9)}{x^4 (x^2 - 9)}$   36. $-\dfrac{(x + 5)(x - 4)}{y}$

The Quotient of Powers Property                                                         **217**

# SIMPLEST FORM OF A PRODUCT                    8.5

*Objective*        **To multiply rational expressions and simplify the results**

To simplify a product like $\dfrac{y^2 - 4y + 3}{4y + 16} \cdot \dfrac{2y + 8}{y - 3}$, you should begin by

applying the definition of the product of two rational expressions, $\dfrac{a}{b} \cdot \dfrac{c}{d} = \dfrac{a \cdot c}{b \cdot d}$,

$b \neq 0$ and $d \neq 0$. Then, you should apply one or more of the techniques for simplifying rational expressions. The process is illustrated in the examples that follow.

*Example 1*      **Multiply. Simplify the result.**

$\dfrac{y^2 - 4y + 3}{4y + 16} \cdot \dfrac{2y + 8}{y - 3}$

$\dfrac{(y^2 - 4y + 3)(2y + 8)}{(4y + 16)(y - 3)}$ ◀ *Product of the numerators*
◀ *Product of the denominators*

$\dfrac{(y^2 - 4y + 3)(2)(y + 4)}{4(y + 4)(y - 3)}$ ◀ *Look for the greatest common monomial factors in the numerator and the denominator.*

$\dfrac{(y - 3)(y - 1)(2)(y + 4)}{2 \cdot 2(y + 4)(y - 3)}$ ◀ *Factor $y^2 - 4y + 3$ into two binomial factors.*
◀ *Factor 4 into primes.*

$\dfrac{\overset{1}{\cancel{(y - 3)}}(y - 1)\overset{1}{\cancel{(2)}}\overset{1}{\cancel{(y + 4)}}}{\underset{1}{\cancel{2}} \cdot 2\underset{1}{\cancel{(y + 4)}}\underset{1}{\cancel{(y - 3)}}}$, or $\dfrac{y - 1}{2}$ ◀ *Divide out the common factors.*

*Example 2*      **Simplify the product.**

$(x - 4) \cdot \dfrac{6x - 12}{4x^2 - 20x + 16}$

$\dfrac{x - 4}{1} \cdot \dfrac{6x - 12}{4x^2 - 20x + 16}$ ◀ *Rewrite $x - 4$ as $\dfrac{x - 4}{1}$.*

$\dfrac{(x - 4)(6x - 12)}{1(4x^2 - 20x + 16)}$ ◀ $\dfrac{a}{b} \cdot \dfrac{c}{d} = \dfrac{a \cdot c}{b \cdot d}$

$\dfrac{(x - 4)(6)(x - 2)}{1 \cdot 4(x^2 - 5x + 4)}$

$\dfrac{(x - 4) \cdot 3 \cdot 2(x - 2)}{1 \cdot 2 \cdot 2(x - 4)(x - 1)}$

$\dfrac{\overset{1}{\cancel{(x - 4)}} \cdot 3 \cdot \overset{1}{\cancel{2}}(x - 2)}{1 \cdot \underset{1}{\cancel{2}} \cdot 2\underset{1}{\cancel{(x - 4)}}(x - 1)}$, or $\dfrac{3(x - 2)}{2(x - 1)}$

In the next example, a combination of all of the techniques for simplifying rational expressions is used. The techniques that you will need are:
1. rewriting a polynomial in convenient form,
2. factoring out greatest common monomial factors in both numerator and denominator,
3. factoring a polynomial into two binomials,
4. factoring constants into primes,
5. using the quotient of powers property, and
6. dividing out the common factors in numerator and denominator.

*Example 3*   **Simplify the product.**

$$\frac{m^2 - 7m + 10}{5m^7p^3} \cdot \frac{20m^4p^2}{12 - 6m}$$

$$\frac{(m^2 - 7m + 10) \cdot 20 \cdot m^4 \cdot p^2}{5m^7p^3(12 - 6m)}$$

$$\frac{(m^2 - 7m + 10) \cdot 20 \cdot m^4 \cdot p^2}{5 \cdot m^7 \cdot p^3 \cdot -1(6m - 12)}$$ ◀ *Write $12 - 6m$ in convenient form.*

$$\frac{(m^2 - 7m + 10) \cdot 20 \cdot m^4 \cdot p^2}{5 \cdot m^7 \cdot p^3 \cdot -1 \cdot 6(m - 2)}$$ ◀ *6 is the GCF of $6m - 12$.*

$$\frac{(m - 5)(m - 2) \cdot 2 \cdot 2 \cdot 5 \cdot m^4 \cdot p^2}{5 \cdot m^7 \cdot p^3 \cdot -1 \cdot 3 \cdot 2(m - 2)}$$

$$\frac{(m - 5)(\cancel{m - 2}) \cdot \cancel{2} \cdot 2 \cdot \cancel{5} \cdot \cancel{m^4} \cdot \cancel{p^2}}{\cancel{5} \cdot \cancel{m^7} \cdot \cancel{p^3} \cdot -1 \cdot 3 \cdot \cancel{2}(\cancel{m - 2})}$$

$$\frac{(m - 5)2}{-1 \cdot 3m^3p^1}, \text{ or } -\frac{2(m - 5)}{3m^3p}$$

# Reading in Algebra

**Use the product** $\dfrac{a^2 - 25}{a^2 + 9a + 20} \cdot \dfrac{3a^2 + 12a}{5 - a}$ **to answer the following questions.**
1. Which polynomial must be written in convenient form?  **$5 - a$**
2. Which polynomial has a GCF? What is the GCF?  **$3a^2 + 12a$; $3a$**
3. Which polynomial is the difference of two squares?  **$a^2 - 25$**

# Written Exercises

**Simplify.**

Ⓐ **1.** $\dfrac{x^2 + 8x + 15}{x + 4} \cdot \dfrac{2x + 8}{x + 3}$   **$2(x + 5)$**

**2.** $\dfrac{a - 10}{a - 3} \cdot \dfrac{a^2 - 10a + 21}{3a - 30}$   **$\dfrac{a - 7}{3}$**

**3.** $\dfrac{5y + 20}{y^2 - 5y - 14} \cdot \dfrac{y - 7}{y + 4}$   **$\dfrac{5}{y + 2}$**

**4.** $(x - 2) \cdot \dfrac{5x - 15}{x^2 - 3x + 2}$   **$\dfrac{5(x - 3)}{x - 1}$**

**5.** $\dfrac{2b + 12}{4b^2 - 16b + 12} \cdot (b - 3)$   **$\dfrac{b + 6}{2(b - 1)}$**

**6.** $(m - 3) \cdot \dfrac{6m + 36}{3m^2 - 27}$   **$\dfrac{2(m + 6)}{m + 3}$**

Simplest Form of a Product

**219**

**7.** $\dfrac{7b^5x^6}{10} \cdot \dfrac{15}{21b^3x^7} \;\; \dfrac{b^2}{2x}$

**8.** $\dfrac{4}{9x^6y^8} \cdot \dfrac{15x^8y^7}{20} \;\; \dfrac{x^2}{3y}$

**9.** $\dfrac{5a^5b^2}{24} \cdot \dfrac{4}{25a^4b^4} \;\; \dfrac{a}{30b^2}$

**10.** $\dfrac{a^2 - 7a + 10}{7x^6y} \cdot \dfrac{15x^2y^5}{10 - 5a}$

**11.** $\dfrac{x^2 - 25}{5a^2b^4} \cdot \dfrac{9a^3b}{15 - 3x}$

**12.** $\dfrac{21 - 7x}{21x^4y^2} \cdot \dfrac{2x^3y^6}{x^2 + 5x - 24}$

**13.** $\dfrac{12x^3y^2}{4m - 16} \cdot \dfrac{16 - m^2}{9x^4y^7}$

**14.** $\dfrac{y^2 - y - 20}{4ab^2} \cdot \dfrac{6a^2b^8}{10 - 2y}$

**15.** $\dfrac{9a^3b}{7 - x} \cdot \dfrac{2x - 14}{3ab^7} \;\; \dfrac{6a^2}{b^6}$

**16.** $\dfrac{4c + 6}{27d^2e^9} \cdot \dfrac{9d^3e^4}{2c^2 + 5c + 3}$

**17.** $\dfrac{8a + 12}{a - 2} \cdot \dfrac{4 - a^2}{2a + 3}$

**18.** $\dfrac{b^2 + 6b - 16}{b^2 - 25} \cdot \dfrac{5 - b}{b + 8} \;\; \dfrac{b - 2}{b + 5}$

**10.** $-\dfrac{3y^4(a - 5)}{7x^4}$ **11.** $-\dfrac{3a(x + 5)}{5b^3}$ **12.** $-\dfrac{2y^4}{3x(x + 8)}$ **13.** $-\dfrac{m + 4}{3xy^5}$ **14.** $-\dfrac{3ab^6(y + 4)}{4}$ **16.** $\dfrac{2d}{3e^5(c + 1)}$

(B) **19.** $\dfrac{5 - x}{x^2 - x - 20} \cdot \dfrac{x^2 + x - 12}{x^2 - 2x - 3} \;\; -\dfrac{1}{x + 1}$

**20.** $\dfrac{2a^2 - 4a - 16}{a^2 - a - 12} \cdot \dfrac{a^2 + a - 6}{a^2 - 6a + 8} \;\; \dfrac{2(a + 2)}{a - 4}$

**21.** $\dfrac{y^2 - 5y + 6}{y^2 - 7y + 10} \cdot \dfrac{5 + 4y - y^2}{y^2 - 2y - 3} \;\; -1$

**22.** $\dfrac{35 - 2x - x^2}{x^3 - 9x} \cdot \dfrac{x^2 + 3x}{x + 7} \;\; -\dfrac{x - 5}{x - 3}$

**23.** $\dfrac{c^2 - 2c - 8}{c^2 - 6c - 16} \cdot \dfrac{c^2 - 3c - 40}{c^2 - c - 6} \;\; \dfrac{(c - 4)(c + 5)}{(c - 3)(c + 2)}$

**24.** $\dfrac{2b^2 + 11b - 21}{b^2 - b - 56} \cdot \dfrac{b^2 - 10b + 16}{2b^2 - 3b} \;\; \dfrac{b - 2}{b}$

**25.** $\dfrac{n^2 + 3n - 18}{n^2 - 36} \cdot \dfrac{1 - 4n^2}{2n^2 - 5n - 3} \;\; -\dfrac{2n - 1}{n - 6}$

**26.** $\dfrac{3a^2 + 14a - 5}{a^2 + 2a - 15} \cdot \dfrac{2a^2 - 15a - 8}{3a^2 + 11a - 4} \;\; \dfrac{(2a + 1)(a - 8)}{(a - 3)(a + 4)}$

**27.** $\dfrac{a^2 - 7ab - 18b^2}{-a + 9b} \cdot \dfrac{3a - 6b}{a^2 - 4b^2} \;\; -3$

**28.** $\dfrac{3x^2 - xy - 4y^2}{x^2 + 11xy + 18y^2} \cdot \dfrac{x^2 + 8xy - 9y^2}{y^2 - x^2} \;\; -\dfrac{3x - 4y}{x + 2y}$

**29.** $\dfrac{2x^2 - 13x + 20}{3x^2 - 10x - 8} \cdot \dfrac{-3 - x}{2x^2 + x - 15} \;\; -\dfrac{1}{3x + 2}$

**30.** $\dfrac{2z^2 - 12z - 14}{z^3 - 16z} \cdot \dfrac{-4 - 4z}{3z - 21} \;\; -\dfrac{8(z + 1)^2}{3z(z + 4)(z - 4)}$

**31.** $\dfrac{5b^2 + 41by + 8y^2}{2b^2 + 9by - 56y^2} \cdot \dfrac{2b^2 + 3by - 35y^2}{25b^2 - y^2}$

**32.** $\dfrac{2c^2 - 10cd + 12d^2}{3c^2 - 21cd + 30d^2} \cdot \dfrac{15c^2 + 12cd - 3d^2}{4c^2 - 8cd - 12d^2}$

For Ex. 17, 31–34, 36–39, see TE Answer Section.

(C) **33.** $\dfrac{4a^2 - 20a + 21}{21 - 4a - a^2} \cdot \dfrac{a^2 - 1}{2a^2 - 5a - 7} \cdot \dfrac{a^2 + 8a + 7}{4a^2 - 12a + 9}$

**34.** $\dfrac{b^2 - 100}{b^2 - b - 90} \cdot \dfrac{10b^3}{b^2 + 10b} \cdot \dfrac{4b + 6}{8b^3}$

**35.** $\dfrac{x^2 + xy + mx + my}{x^2 - 2mx - 3m^2} \cdot \dfrac{x^3}{x^2 + xy} \;\; \dfrac{x^2}{x - 3m}$

**36.** $\dfrac{ax - 3a + bx - 3b}{a^2 - b^2} \cdot \dfrac{a^2 - 7ab + 6b^2}{ax - 6bx - 3a + 18b}$

**37.** $\dfrac{8 + 2a - a^2}{a^2 - 2a - 3} \cdot \dfrac{a^2 - 3a}{4a - a^3} \cdot \dfrac{2a^2 + 9a + 7}{a^2 - a - 12}$

**38.** $\dfrac{ax + 7a + 3bx + 21b}{s^2 - r^2} \cdot \dfrac{rx + sx - 4r - 4s}{ax + 3bx - 4a - 12b}$

**39.** $\dfrac{3ax - 4y - 4x + 3ay}{-25 + 15a - 2a^2} \cdot \dfrac{4a^4 - 125a^2 + 625}{6a^2 + 7a - 20}$

**40.** $\dfrac{x^4 - 10x^2 + 9}{2x^2 - 38x + 96} \cdot \dfrac{64x - 4x^2}{x^2 + 2x - 3} \;\; -2x(x + 1)$

## CUMULATIVE REVIEW

**1.** Graph
$5 - (2 - 3x) > 7x + 11.$
$x < -2.$ Check students' graphs.

**2.** Solve
$0.002x - 1.15 = 0.3x - 0.554.$
$-2$

**3.** Solve
$-9x + 20 = -x^2.$ **4, 5**

## NON-ROUTINE PROBLEMS

A clock strikes the number of hours every hour. If it takes 7 seconds
to strike 8 o'clock, how many seconds will the clock be striking each day?   **132 s**

# MULTIPLYING AND DIVIDING 8.6

8.6

*Objectives* **To divide rational expressions**
**To simplify rational expressions that involve both multiplication and division**

Dividing two rational expressions is similar to dividing fractions.

| Dividing rational expressions | If $a$, $b$, $c$, and $d$ are rational expressions, $\dfrac{a}{b} \div \dfrac{c}{d} = \dfrac{a}{b} \cdot \dfrac{d}{c}$, if $b \neq 0$, $c = 0$, and $d = 0$. |
|---|---|

*Example 1* **Simplify** $\dfrac{x^2 - 10x + 16}{x^2 - 49} \div \dfrac{3x - 24}{x - 7}$.

$$\frac{x^2 - 10x + 16}{x^2 - 49} \div \frac{3x - 24}{x - 7} = \frac{x^2 - 10x + 16}{x^2 - 49} \cdot \frac{x - 7}{3x - 24} \quad \blacktriangleleft \frac{a}{b} \div \frac{c}{d} = \frac{a}{b} \cdot \frac{d}{c}$$

$$= \frac{(x^2 - 10x + 16)(x - 7)}{(x^2 - 49)(3x - 24)}$$

$$= \frac{\overset{1}{(x - 8)}(x - 2)\overset{1}{(x - 7)}}{\underset{1}{(x - 7)}(x + 7) \cdot 3\underset{1}{(x - 8)}}, \text{ or } \frac{x - 2}{3(x + 7)}$$

*Example 2* **Simplify** $\dfrac{4a^7 b^2}{10a - 6a^2} \div \dfrac{6a^5 b^7}{3a^2 - 17a + 20}$.

$$\frac{4a^7 b^2}{10a - 6a^2} \div \frac{6a^5 b^7}{3a^2 - 17a + 20} = \frac{4a^7 b^2}{10a - 6a^2} \cdot \frac{3a^2 - 17a + 20}{6a^5 b^7}$$

$$= \frac{4 \cdot a^7 \cdot b^2 (3a^2 - 17a + 20)}{-1(6a^2 - 10a) \cdot 6 \cdot a^5 \cdot b^7} \quad \blacktriangleleft \text{ Rewrite } 10a - 6a^2 \text{ in convenient form:}$$
$$10a - 6a^2 = -6a^2 + 10a = -1(6a^2 - 10a).$$

$$= \frac{2 \cdot 2 \cdot a^7 \cdot b^2 (3a - 5)(a - 4)}{-1 \cdot 2 \cdot a(3a - 5) \cdot 3 \cdot 2 \cdot a^5 \cdot b^7} \quad \blacktriangleleft \text{ Factor } 6a^2 - 10a = 2 \cdot a(3a - 5).$$

$$= \frac{\overset{1}{\cancel{2}} \cdot \overset{1}{\cancel{2}} \cdot \overset{a^2}{\cancel{a^7}} \cdot \overset{1}{\cancel{b^2}} (\overset{}{3a - 5})(a - 4)}{-1 \cdot \underset{1}{\cancel{2}} \cdot a(\underset{1}{3a - 5}) \cdot 3 \cdot \underset{1}{\cancel{2}} \cdot \underset{1}{\cancel{a^5}} \cdot \underset{b^5}{\cancel{b^7}}}$$

$$= \frac{\overset{a^1}{\cancel{a^2}}(a - 4)}{-1 \cdot \underset{1}{\cancel{a}} \cdot 3 \cdot b^5}$$

$$= \frac{a(a - 4)}{-1 \cdot 3 \cdot b^5}, \text{ or } -\frac{a(a - 4)}{3b^5}$$

The next two examples illustrate the simplifying process for rational expressions involving both multiplication and division. Remember that you should write the reciprocal of a rational expression only if it immediately follows a division symbol.

**Example 3**  **Simplify** $\dfrac{a^2 + 7a + 10}{8a^3} \cdot \dfrac{a + 3}{a^2 + 4a} \div \dfrac{a^2 + 5a + 6}{10a^2}$.

$\dfrac{a^2 + 7a + 10}{8a^3} \cdot \dfrac{a + 3}{a^2 + 4a} \div \dfrac{a^2 + 5a + 6}{10a^2}$ ◀ *This is the only rational expression that immediately follows a division symbol.*

$\dfrac{a^2 + 7a + 10}{8a^3} \cdot \dfrac{a + 3}{a^2 + 4a} \cdot \dfrac{10a^2}{a^2 + 5a + 6}$

$\dfrac{(a^2 + 7a + 10)(a + 3) \cdot 10 \cdot a^2}{8 \cdot a^3 \cdot (a^2 + 4a)(a^2 + 5a + 6)}$

$\dfrac{(a + 5)(a + 2)(a + 3) \cdot 5 \cdot 2 \cdot a^2}{2 \cdot 2 \cdot 2 \cdot a^3 \cdot a(a + 4)(a + 3)(a + 2)}$ ◀ *Factoring $a^2 + 4a$ involves factoring out the GCF, $a$.*

$\dfrac{(a + 5)\cancel{(a + 2)}\cancel{(a + 3)} \cdot 5 \cdot \cancel{2} \cdot \cancel{a^2}}{\cancel{2} \cdot 2 \cdot 2 \cdot \cancel{a^3} \cdot a(a + 4)\cancel{(a + 3)}\cancel{(a + 2)}}$

$\dfrac{(a + 5)5}{2 \cdot 2 \cdot a^1 \cdot a^1(a + 4)}$, or

$\dfrac{5(a + 5)}{4a^2(a + 4)}$

**Example 4**  **Simplify** $\dfrac{x}{x + 4} \div \dfrac{6x^2}{3x + 12} \cdot \dfrac{x^2 + 12x + 20}{x^2 - 4}$.

$\dfrac{x}{x + 4} \div \dfrac{6x^2}{3x + 12} \cdot \dfrac{x^2 + 12x + 20}{x^2 - 4} = \dfrac{x}{x + 4} \cdot \dfrac{3x + 12}{6x^2} \cdot \dfrac{x^2 + 12x + 20}{x^2 - 4}$

$= \dfrac{x(3x + 12)(x^2 + 12x + 20)}{(x + 4) \cdot 6 \cdot x^2(x^2 - 4)}$

$= \dfrac{x \cdot 3(x + 4)(x + 10)(x + 2)}{(x + 4) \cdot 3 \cdot 2 \cdot x^2(x - 2)(x + 2)}$

$= \dfrac{\cancel{x} \cdot \cancel{3}\cancel{(x + 4)}(x + 10)\cancel{(x + 2)}}{\cancel{(x + 4)} \cdot \cancel{3} \cdot 2 \cdot \cancel{x^2}(x - 2)\cancel{(x + 2)}}$, or $\dfrac{x + 10}{2x(x - 2)}$

## Oral Exercises

**Give the reciprocal.**

1. $\dfrac{3}{4}\quad\dfrac{4}{3}$

2. $7\quad\dfrac{1}{7}$

3. $\dfrac{1}{a - 2}\quad\dfrac{a - 2}{1}$

4. $x + 1\quad\dfrac{1}{x + 1}$

5. $\dfrac{b^2 - 7b + 10}{4b^2}\quad\dfrac{4b^2}{b^2 - 7b + 10}$

6. $\dfrac{8y^2z^3}{2y^2 - 18}\quad\dfrac{2y^2 - 18}{8y^2z^3}$

## Written Exercises

**Simplify.**
1. $\dfrac{(x-4)(x+2)}{3(x+7)(x-3)}$  5. $\dfrac{5}{3b^3(b+2)}$  6. $\dfrac{2(x+2)}{5x^2}$  9. $-\dfrac{9(m+6)}{n}$  10. $\dfrac{m+1}{2}$  11. $\dfrac{5(b-6)}{6(b-3)}$  12. $\dfrac{m}{m-3}$

**(A)**

1. $\dfrac{x^2-2x-8}{x^2-49} \div \dfrac{3x-9}{x-7}$

2. $\dfrac{5a-30}{9} \div \dfrac{a^2-36}{3a+18}$   $\dfrac{5}{3}$

3. $\dfrac{m^2-4}{a+1} \div \dfrac{2m+4}{7a+7}$   $\dfrac{7(m-2)}{2}$

4. $\dfrac{x^2-7x-18}{k^6} \div \dfrac{2x^2-18x}{k^4}$   $\dfrac{x+2}{2xk^2}$

5. $\dfrac{5b^4}{b^2-3b-10} \div \dfrac{9b^8}{3b^2-15b}$

6. $\dfrac{x^2-5x-14}{5x^7} \div \dfrac{3x^2-21x}{6x^6}$

7. $\dfrac{3a^3b^5}{15a-3a^2} \div \dfrac{a^4b}{5a-25}$   $-\dfrac{5b^4}{a^2}$

8. $\dfrac{16-4x}{xy^5} \div \dfrac{x^2-4x}{7x^5y^4}$   $-\dfrac{28x^3}{y}$

9. $\dfrac{36-m^2}{mn^6} \div \dfrac{3m^2-18m}{27m^2n^5}$

10. $\dfrac{m^2-4m+3}{m^2-1} \div \dfrac{2m-6}{m^2+2m+1}$

11. $\dfrac{5b^2-20}{b^2-5b+6} \div \dfrac{6b+12}{b-6}$

12. $\dfrac{m}{m^2-6m+9} \div \dfrac{1}{m-3}$

13. $\dfrac{4ab^3}{3a^2+5a} \div \dfrac{6a^5b^7}{3a^2+17a+20}$   $\dfrac{2(a+4)}{3a^5b^4}$

14. $\dfrac{8x^7y^5}{5x^2-10x} \div \dfrac{4x^8y^4}{5x^2+16x-52}$   $\dfrac{2y(5x+26)}{5x^2}$

15. $\dfrac{a^2-9}{b^2} \cdot \dfrac{b^3}{a+3} \div \dfrac{a-3}{5}$   $5b$

16. $\dfrac{k^2+7k+10}{9k} \cdot \dfrac{6k^2}{k+1} \div \dfrac{3k+6}{k^2+k}$   $\dfrac{2k^2(k+5)}{9}$

17. $\dfrac{x^2-4x+3}{x^4} \div \dfrac{x-3}{x^6} \cdot \dfrac{x^2-x}{2x-2}$   $\dfrac{x^3(x-1)}{2}$

18. $\dfrac{a+4}{2a^2} \div \dfrac{4}{a^2-5a} \cdot \dfrac{2a^3}{a^2-9a+20}$   $\dfrac{a^2(a+4)}{4(a-4)}$

19. $\dfrac{m^4}{m^2-7m} \div \dfrac{m^3}{m^2-8m+7} \cdot \dfrac{4m}{6m-6}$   $\dfrac{2m}{3}$

20. $\dfrac{b^2-5b}{b^4} \cdot \dfrac{b^3}{6x^2y} \div \dfrac{7b-35}{4xy^3}$   $\dfrac{2y^2}{21x}$

**For Ex. 26–36, see TE Answer Section.**

**(B)**

21. $\dfrac{a^2+3a-28}{a^2-10a+21} \div \dfrac{a^2+12a+35}{a^2+2a-15}$   $\dfrac{a-4}{a-7}$

22. $\dfrac{x^2+x-20}{x^2+5x-36} \div \dfrac{x^2-3x-40}{x^2+6x-27}$   $\dfrac{x-3}{x-8}$

23. $\dfrac{a^2+5ab}{2a^2+7ab-15b^2} \div \dfrac{a^7}{6a^2-7ab-3b^2}$   $\dfrac{3a+b}{a^6}$

24. $\dfrac{4x^2+11xy+6y^2}{x^2+2xy} \div \dfrac{16x^2-9y^2}{x^2}$   $\dfrac{x}{4x-3y}$

25. $\dfrac{y^3-4y}{3y^2-10y+8} \div \dfrac{10y+3y^2-y^3}{3y^2-19y+20}$   $-1$

26. $\dfrac{20+x-12x^2}{4x^3-16x^2-20x} \div \dfrac{6x^2-23x+20}{4x^4-25x^2}$

27. $\dfrac{x^2+3x-18}{x^2+x} \cdot \dfrac{x^2+x-20}{36-x^2} \div \dfrac{x^2+2x-15}{x^2-x-2}$

28. $\dfrac{y^2-dy-2d^2}{3y+9d} \div \dfrac{y^2-d^2}{y^2-4dy-5d^2} \cdot \dfrac{9y}{3y+15d}$

29. $\dfrac{x^2-5x-24}{x^2+7x} \div \dfrac{x^2-6x-16}{3x^2-9x} \cdot \dfrac{x^2+9x+14}{9-x^2}$

30. $\dfrac{x^2+4x-32}{x^2-12x+35} \div \dfrac{16x-4x^2}{x^2-4x-21} \cdot \dfrac{2x^2-10x}{x^2+11x+24}$

**(C)**

31. $\dfrac{b^2-9}{3b-b^2} \div \dfrac{2ab+6a+5b+15}{2a+5}$

32. $\dfrac{x^4-65x^2+64}{50x-40x^2-10x^3} \div \dfrac{x^2+9x+8}{5x^2+25x}$

33. $\dfrac{x^4-34x^2+225}{2x^3+2x^2-12x} \div \dfrac{x^2+2x-15}{12x^2-6x^3}$

34. $\dfrac{ab-7ac+3bd-21cd}{b^2-5bc-14c^2} \div \dfrac{a^2+2ad-3d^2}{4c^2-b^2}$

35. $\dfrac{ab+5ac-2bd-10cd}{b^2+7bc+10c^2} \div \dfrac{a^2+2ad-8d^2}{b^2-4c^2}$

36. $\dfrac{ax+3ay-3dx-9dy}{x^2+7xy+12y^2} \div \dfrac{a^2+ad-12d^2}{x^2-16y^2}$

## CUMULATIVE REVIEW

**Solve.**

1. A team played 20 games and won 15. The number of losses is what percent of the games played? **25%**

2. A number increased by $\dfrac{2}{3}$ of itself is 30. Find the number. **18**

Multiplying and Dividing    **223**

# GEOMETRY REVIEW

*Objective*    **To find the volume of a cylinder and a cone**

The cone and the cylinder in the diagram have the same height, $h$, and the same radius, $r$.

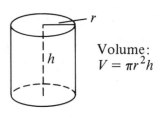

Volume:
$V = \pi r^2 h$

The cone can be filled with water, and the water can then be poured into the cylinder.

After filling the cone three times and then emptying it into the cylinder, the cylinder will be completely filled.

This shows that the volume of the cone is $\frac{1}{3}$ the volume of the cylinder.

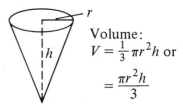

Volume:
$V = \frac{1}{3}\pi r^2 h$ or

$= \frac{\pi r^2 h}{3}$

*Example*    **Find the volume of a cone with height 16 cm and diameter 6 cm. Write the result in terms of $\pi$.**

$V = \dfrac{\pi r^2 h}{3}$

$V = \dfrac{\pi \cdot (3 \text{ cm})^2 \cdot 16 \text{ cm}}{3}$  ◀ *$d = 6$ cm; $r = \frac{1}{2}d$ or $\frac{1}{2} \cdot 6$ or 3 cm*

$V = \dfrac{\pi \cdot 144 \text{ cm}^3}{3}$  ◀ *$3 \cdot 3 \cdot 16 = 144$; cm $\cdot$ cm $\cdot$ cm $=$ cm$^3$*

$V = 48\pi \text{ cm}^3$

# Written Exercises

**Find the volume of each figure. Write the answer in terms of $\pi$.**

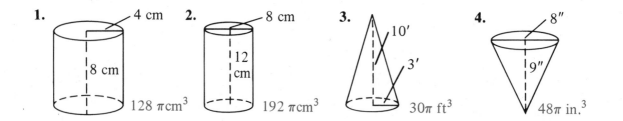

1. 4 cm, 8 cm  $128\,\pi\text{cm}^3$
2. 8 cm, 12 cm  $192\,\pi\text{cm}^3$
3. 10', 3'  $30\pi \text{ ft}^3$
4. 8", 9"  $48\pi \text{ in.}^3$

# RATIO AND PROPORTION

*Objectives*  **To identify the extremes and the means of a proportion**
**To solve a proportion**
**To solve problems using proportions**

You can use a fraction to compare a basketball team's 5 wins to its 2 losses, as shown below.

$$\text{wins} \to \frac{5}{2}$$
$$\text{losses} \to$$

You can say that the results are "in the ratio 5 to 2."
A ratio can be written in two ways: (1) as a fraction and (2) with a colon.

The ratio 5 to 2 can be written as $\frac{5}{2}$, or $5:2$.

| | |
|---|---|
| Definition: Ratio | A **ratio** is the comparison of two numbers $a$ and $b$, $b \neq 0$, by division. The ratio can be written in two ways: $$\frac{a}{b} \text{ or } a:b.$$ |

An equation such as $\frac{a}{2} = \frac{3}{5}$ states that two ratios are equal. Such an equation is called a *proportion*.

| | |
|---|---|
| Definition: Proportion | A **proportion** is an equation formed by two ratios. $$\frac{a}{b} = \frac{c}{d}$$ $a$ and $d$ are the **extremes.** $b$ and $c$ are the **means.** or $a:b = c:d$ with **extremes** $a$ and $d$, and **means** $b$ and $c$. |

*Example 1*  **Identify the extremes and the means of $\frac{a}{2} = \frac{3}{5}$.**

The extremes are $a$ and 5, the first and fourth terms of the proportion.
The means are 2 and 3, the second and third terms of the proportion.

You can solve the proportion $\frac{a}{2} = \frac{3}{5}$ by multiplying each side of the equation by the product of the denominators. This will suggest a pattern for solving proportions, as illustrated in the next example.

Ratio and Proportion

*Example 2*    **Solve the proportion.**

$$\frac{a}{2} = \frac{3}{5}$$

$$5 \cdot \overset{1}{\cancel{2}} \cdot \frac{a}{\underset{1}{\cancel{2}}} = \frac{3}{\underset{1}{\cancel{5}}} \cdot \overset{1}{\cancel{5}} \cdot 2 \quad \blacktriangleleft \textit{Multiply each side by } 5 \cdot 2.$$

$$5 \cdot a = 3 \cdot 2$$

$$5a = 6$$

$$a = \frac{6}{5}, \text{ or } 1\frac{1}{5}$$

**Thus,** the solution is $\frac{6}{5}$, or $1\frac{1}{5}$.

In solving the proportion $\frac{a}{2} = \frac{3}{5}$, notice that $5 \cdot a$, the product of the extremes, equals $2 \cdot 3$, the product of the means.

| Property of proportion | In a proportion $\frac{a}{b} = \frac{c}{d}$, $a \cdot d = b \cdot c$. <br> The product of the extremes equals the product of the means. |
|---|---|

*Example 3*    **Solve each proportion.**

$$\frac{x}{5} = \frac{3}{7}$$

$7 \cdot x = 5 \cdot 3 \quad \blacktriangleleft \textit{The product of the extremes equals}$
$7x = 15 \qquad\qquad \textit{the product of the means.}$

$$x = \frac{15}{7}, \text{ or } 2\frac{1}{7}$$

**Thus,** the solution is $\frac{15}{7}$, or $2\frac{1}{7}$.

$$\frac{x - 6}{3} = \frac{2}{5}$$

$$5(x - 6) = 3 \cdot 2$$

$$5x - 30 = 6$$

$$5x = 36$$

$$x = \frac{36}{5}, \text{ or } 7\frac{1}{5}$$

**Thus,** the solution is $\frac{36}{5}$, or $7\frac{1}{5}$.

To solve some proportions, you will have to use a quadratic equation.

*Example 4*    **Solve $\dfrac{b}{2} = \dfrac{2}{b + 3}$.**

$\dfrac{b}{2} = \dfrac{2}{b + 3} \quad \blacktriangleright$

$$b(b + 3) = 2 \cdot 2 \quad \blacktriangleleft \textit{The product of the extremes equals the}$$
$$b^2 + 3b = 4 \qquad\qquad \textit{product of the means.}$$
$$b^2 + 3b - 4 = 0 \quad \blacktriangleleft \textit{Add } -4 \textit{ to each side of the equation.}$$
$$(b + 4)(b - 1) = 0 \quad \blacktriangleleft \textit{Factor.}$$
$$b + 4 = 0 \quad \text{or} \quad b - 1 = 0 \quad \blacktriangleleft \textit{Set each factor equal to 0.}$$
$$b = -4 \quad \text{or} \qquad b = 1 \quad \blacktriangleleft \textit{Solve each equation for b.}$$

**Thus,** the solutions are $-4$ and $1$.

Frequently, comparisons are used in commercials seen on television to convince the public that many people use a certain product. This helps the manufacturer of the product to increase sales. The following example illustrates an application of proportion to such a real-life situation.

**Example 5**

**If 1 out of 6 people buy Elly Fant brand peanuts, how many people can be expected to buy this brand of peanuts in a city of 24,000 inhabitants?**

Let $p$ = the number of Elly Fant brand peanut buyers.

$$\frac{p}{24,000} = \frac{\text{Elly Fant buyers}}{\text{total population}}$$  ◀ *Write a proportion by setting the ratios equal.*

$$\frac{1}{6} = \frac{\text{Elly Fant buyers}}{\text{total population}}$$  ◀ *One out of 6 means $\frac{1}{6}$.*

So, $\dfrac{p}{24,000} = \dfrac{1}{6}$  ◀ *The two ratios are equal.*

$$6p = 24,000$$
$$p = 4,000$$

**So,** 4,000 people can be expected to buy Elly Fant brand peanuts.

## Written Exercises

For complete answers to Ex. 1–12, see TE Answer Section.

**Identify the means and the extremes of the proportion. Then solve the proportion.**

(A) 1. $\dfrac{7}{4} = \dfrac{a}{3}$ **21** $\dfrac{1}{4}$

2. $\dfrac{2}{5} = \dfrac{x}{4}$ **1$\frac{3}{5}$**

3. $\dfrac{3}{7} = \dfrac{m}{5}$ **2$\frac{1}{7}$**

4. $\dfrac{5}{b} = \dfrac{3}{11}$ **18$\frac{1}{3}$**

5. $\dfrac{x-5}{7} = \dfrac{3}{5}$ **9$\frac{1}{5}$**

6. $\dfrac{2}{x} = \dfrac{3}{x+6}$ **12**

7. $\dfrac{m-2}{4} = \dfrac{m+3}{5}$ **22**

8. $\dfrac{x-1}{4} = \dfrac{3+x}{2}$ **−7**

9. $\dfrac{a+3}{5} = \dfrac{14}{10}$ **4**

10. $\dfrac{x-1}{5} = \dfrac{2x+7}{3}$

11. $\dfrac{5}{2m+5} = \dfrac{2}{4m-1}$

12. $\dfrac{x-6}{2} = \dfrac{x+4}{3}$ **26**

10. **−5$\frac{3}{7}$**    11. **$\frac{15}{16}$**

**Solve each proportion.**

13. $\dfrac{12}{x} = \dfrac{x}{3}$ **±6**

14. $\dfrac{a}{16} = \dfrac{4}{a}$ **±8**

15. $\dfrac{x}{2} = \dfrac{x}{x+3}$ **0, −1**

16. $\dfrac{y}{3} = \dfrac{2}{y-5}$ **−1, 6**

(B) 17. $\dfrac{y-4}{1} = \dfrac{2}{y-3}$

18. $\dfrac{a-5}{9} = \dfrac{-3}{a+7}$

19. $\dfrac{3}{b+2} = \dfrac{b-8}{13}$

20. $\dfrac{-2}{p-2} = \dfrac{p-8}{4}$

**2, 5**    **−4, 2**    **−5, 11**    **4, 6**

**Solve each problem.**

21. In Stantonville, 2 out of 5 people belong to a union. How many union members are there if the population is 70,000? **28,000**

22. If 7 out of 8 people use Attack toothpaste, how many people use Attack in a city with a population of 40,000? **35,000**

23. Pam's batting average is 0.250 (250:1,000). During baseball season, how many hits should she get in 120 times at bat? **30**

24. Three out of 5 freshmen study algebra. How many study algebra in a freshman class of 500? **300**

Ratio and Proportion

# APPLICATIONS

| Read | → | Plan | → | Solve | → | Interpret |

The number of miles a car averages on a gallon of gas, mi/gal, is important in predicting the cost of the gas for an automobile trip that you might plan.

*Example*

**Anita's car averages 40 mi/gal. How many gallons of gas will she need for a 3,500-mi trip? Find the cost of the trip if gas sells at $1.25/gal.**

Let $g$ = the number of gallons of gas Anita needs.

$\dfrac{3,500}{g} = \dfrac{\text{number of miles}}{\text{number of gallons}}$ ◀ *Write the ratio, total miles to total gallons.*

$\dfrac{40}{1} = \dfrac{\text{number of miles}}{\text{number of gallons}}$ ◀ *Write 40 mi/gal as a ratio.*

$\dfrac{3,500}{g} = \dfrac{40}{1}$ ◀ *Set the two ratios equal to each other.*

$40g = 3,500$ ◀ *If $\dfrac{a}{b} = \dfrac{c}{d}$, then $a \cdot d = b \cdot c$.*

$g = 87\frac{1}{2}$

**So,** Anita needs approximately 88 gal of gas for the trip.
The cost of the gas is 88 ($1.25), or $110.

**Solve each problem.**

1. A sales representative's car averages 30 mi/gal. Find the cost of the gas for a 1,000-mi trip if gas sells at $1.15/gal. **$38.33**

2. Jan's car averages 30 mi/gal. Find the cost of the gas for a 2,500-mile cross-country trip if gas sells at $1.21/gal. **$100.83**

3. Bill budgets $50 for gas for a vacation. His car averages 40 mi/gal. If gas costs $1.24/gal, is a 1,500-mi trip within his gas budget? **Yes. Cost of gas is $46.50.**

4. Jane commutes 500 mi a week to work. Her car averages 45 mi/gal. If gas sells at $2.19/gal, how much does she spend on gas in 4 weeks? **$97.33**

# COMPUTER ACTIVITIES

## Evaluating Rational Algebraic Expressions

You can write a program to evaluate an algebraic expression for different values of its variable(s). However, you must be careful to avoid error messages. For example, if the computer were to attempt to evaluate

$$\frac{x^2 - 8x + 15}{2x - 4} \text{ for } x = 2,$$

it would lead to $\quad \dfrac{2^2 - 8 \cdot 2 + 15}{2 \cdot 2 - 4} = \dfrac{4 - 16 + 15}{4 - 4} = \dfrac{3}{0}.$

This is undefined since you cannot divide by 0. The computer would halt its operation, and print a message similar to the following: "ERROR – DIVISION BY 0." The program below avoids this type of error message by using the decision-making statement, IF . . . THEN.

```
10    FOR I = 1 TO 3
15    INPUT "TYPE IN A VALUE FOR X. ";X
20    LET N = X ^ 2 - 8 * X + 15
25    LET D = 2 * X - 4
30    IF D = 0 THEN 50
35    LET Q = N / D
40    PRINT "THE VALUE OF Q IS "Q"."
45    GOTO 55
50    PRINT "THERE IS NO DIVISION BY 0."
55    NEXT I
60    END
```

Line 10 sets up the program to perform 3 different evaluations.

Line 30 computes the expression's numerator and stores it in N.
Line 40 computes the denominator and stores it in D.
Line 50 checks the denominator for unacceptable values.

See the Computer Section beginning on page 488 for more information.

## Exercises      For Ex. 4, see TE Answer Section.

1. Evaluate the rational expression in the program above for the following values of $x$: 6, 4, and $-2$. Then enter and RUN the program for the same three values of $x$. Are the computer's decimal answers the same as the results you computed? $\quad \frac{3}{8}, -\frac{1}{4}, -4\frac{3}{8};$

**Consider the algebraic expression** $\dfrac{2x - 3}{x - 7} \cdot \dfrac{x + 5}{x + 4}.$

2. Find the value of the expression for $x = 5, 7, 2, -4, 6.$  $3\frac{8}{9}$, **undefined,** $-\frac{7}{30},$ **undefined,** $-9\frac{9}{10}$

3. Find the algebraic product of the expression and evaluate the resulting expression for the same values of $x$. $\frac{2x^2 + 7x - 15}{x^2 - 3x - 28},$ **same as Ex. 2**

4. Write a program to evaluate the expression first in its original form, and then in its algebraic product form.

5. Compare both sets of results computed by your program(s) with the values you found for both forms of the expression.

Activities

**229**

**Vocabulary**

convenient form [8.3]   rational expression [8.1]
extremes [8.7]   reciprocal [8.6]
means [8.7]   relatively prime [8.2]
proportion [8.7]   simplest form [8.2]
ratio [8.7]

For what value(s) of $x$ is the rational expression undefined? [8.1]

**1.** $\dfrac{6}{x-8}$  **8**

**2.** $\dfrac{3x+4}{2x+12}$  **−6**

**3.** $\dfrac{7x-5}{x^2-8x+12}$  **2, 6**

Multiply. [8.1]

**4.** $\dfrac{4a^3}{5b^4} \cdot \dfrac{3a^2}{7b}$  $\dfrac{12a^5}{35b^5}$

**5.** $(2x-3) \cdot \dfrac{x^2-7x+1}{3x-4}$  $\dfrac{2x^3-17x^2+23x-3}{3x-4}$

**6.** $\left(-\dfrac{3a+b}{a-5b}\right)^2$  $\dfrac{9a^2+6ab+b^2}{a^2-10ab+25b^2}$

**7.** $\dfrac{a-3b}{a+4b} \cdot \dfrac{a+3b}{2a-5b}$  $\dfrac{a^2-9b^2}{2a^2+3ab-20b^2}$

Simplify, if possible.

**8.** $\dfrac{5a-20}{3a-12}$  [8.2]  $\dfrac{5}{3}$

**9.** $\dfrac{a^2-5a-ab+5b}{2a^2-3a-35}$  $\dfrac{a-b}{2a+7}$

**10.** $\dfrac{a^2-b^2}{-2a^2-ab+b^2}$  [8.3]  $-\dfrac{a-b}{2a-b}$

**11.** $\dfrac{-10+22a-4a^2}{3a^2-16a+5}$  $-\dfrac{2(2a-1)}{3a-1}$

Simplify. [8.4]

**12.** $\dfrac{12a^3b^5}{18ab^7}$  $\dfrac{2a^2}{3b^2}$

**13.** $\dfrac{7c^3-c^3d}{c^3d^2-10c^3d+21c^3}$  $-\dfrac{1}{d-3}$

**14.** $\dfrac{x^2-3x+2}{x+5} \cdot \dfrac{2x+10}{x-2}$  [8.5]  $2(x-1)$

**15.** $\dfrac{45-4x-x^2}{x^3-4x} \cdot \dfrac{x^2+2x}{x+9}$  $-\dfrac{x-5}{x-2}$

**16.** $\dfrac{k^2-7k-18}{k^5} \div \dfrac{2k^2-18k}{6k}$  [8.6]  $\dfrac{3(k+2)}{k^5}$

**17.** $\dfrac{x^2-7x+12}{x^2-25} \div \dfrac{2x-8}{x-5}$  $\dfrac{x-3}{2(x+5)}$

**18.** $\dfrac{2x^2-x-15}{x^2-7x} \div \dfrac{x^2+5x+6}{49-x^2}$ $\cdot \dfrac{x^3+2x^2}{x^2+4x-21}$  $-\dfrac{x(2x+5)}{x+3}$

Identify the means and the extremes of the proportion. Then solve the proportion. [8.7]

**19.** $\dfrac{x}{4}=\dfrac{3}{2}$  **6**

**20.** $\dfrac{a+2}{4}=\dfrac{a-5}{3}$  **26**

**21.** $\dfrac{x+5}{4}=\dfrac{-1}{x}$

**22.** $\dfrac{a-7}{4}=\dfrac{-9}{a+6}$  **−2, 3**

21. **−4, −1**   For complete answers to Ex. 19–22,
Solve each problem.   see TE Answer Section.

**23.** Two out of 9 people use Gum toothbrushes. How many use Gum  **8,000** toothbrushes in a town of 36,000 people?

**24.** Frank's batting average is 0.125. In a season, how many hits should he get in 40 times at bat?  **5**

**25.** Four out of 5 freshmen study algebra. How many study algebra in a freshman class of 300?  **240**

Simplify.

**28.** $-\dfrac{x+4}{x+3}$

★ **26.** $\dfrac{y^4-10y^2+9}{y^3-2y^2-3y}$  $\dfrac{(y+3)(y-1)}{y}$
[8.2]

★ **27.** $\dfrac{r^2-r^2x^2-4+4x^2}{rx+2x-r-2}$  $-(x+1)(r-2)$
[8.3]

★ **28.** $\dfrac{x^4-20x^2+64}{20+3x-2x^2} \cdot \dfrac{2x^2-x-15}{x^4-13x^2+36}$
[8.5]

★ **29.** $\dfrac{m^4-82m^2+81}{2m^2-13m-45} \div \dfrac{m^2+5m-6}{2m^2+17m+30}$
[8.6] $\cdot \dfrac{3-4m}{4m^2+33m-27}$  $-(m+1)$

Identify the means and the extremes of the proportion. Then solve the proportion.

**1.** $\dfrac{t}{6} = \dfrac{5}{2}$ **15**

**2.** $\dfrac{x - 4}{3} = \dfrac{x + 6}{5}$ **19**

**3.** $\dfrac{x + 7}{6} = \dfrac{3}{x}$ **−9, 2**

**4.** $\dfrac{a + 8}{-8} = \dfrac{3}{a - 2}$ **−4, −2**

Multiply.

**5.** $\dfrac{3a^5}{7b^2} \cdot \dfrac{4a}{5b^4}$ **$\dfrac{12a^6}{35b^6}$**

**6.** $(3x - 2) \cdot \dfrac{x^2 - 7x + 3}{x + 4}$ **$\dfrac{3x^3 - 23x^2 + 23x - 6}{x + 4}$**

**7.** $\left(\dfrac{2a - b}{a + 4b}\right)^2$ **$\dfrac{4a^2 - 4ab + b^2}{a^2 + 8ab + 16b^2}$**

For what value(s) of $x$ is the rational expression undefined?

**8.** $\dfrac{2x + 7}{3x + 9}$ **−3**

**9.** $\dfrac{5x - 2}{x^2 - 8x + 12}$ **2, 6**

Solve each problem.

**10.** Seven out of 10 people use Smoothie Peanut Butter. How many use this brand in a city of 3,000? **2,100**

**11.** Janine's batting average is 0.250. In a season, how many hits can she expect to get in 60 times at bat? **15**

Simplify.

**12.** $\dfrac{5 - x}{x^2 - x - 20} \cdot \dfrac{x^2 + 8x + 16}{x^2 + 7x + 12}$ **$-\dfrac{1}{x + 3}$**

**13.** $\dfrac{b^2 - b - 20}{b^2 - 49} \cdot \dfrac{2b + 14}{2b + 8}$ **$\dfrac{b - 5}{b - 7}$**

**14.** $\dfrac{m^3 - m}{4m^2} \div \dfrac{2 - 2m}{m^8}$ **$-\dfrac{m^7(m + 1)}{8}$**

Simplify.

**15.** $\dfrac{9t}{2t - 5} \cdot \dfrac{8t - 20}{12} \div \dfrac{8t^4}{36}$ **$\dfrac{27}{2t^3}$**

**16.** $\dfrac{36 - m^2}{2m^2 - 8m - 24} - \dfrac{(6 + m)}{2(m + 2)}$

**17.** $\dfrac{x^6 y^4 (n^2 - 2n + 1)}{x^3 y^8 (1 - n)}$ **$-\dfrac{x^3(n - 1)}{y^4}$**

**18.** $\dfrac{x^2 - 7x + 12}{x^2 - 3x - 40} \div \dfrac{20 - x - x^2}{x^2 + 8x}$ $\cdot \dfrac{x^2 - 10x + 16}{9x^2 - x^4}$ **$\dfrac{(x + 8)(x - 2)}{x(x + 5)(x + 5)(x + 3)}$**

**19.** $\dfrac{2c^2 - 10cd + 12d^2}{3c^2 - 21cd + 30d^2} \cdot \dfrac{15c^2 + 12cd - 3d^2}{4c^2 - 8cd - 12d^2}$

**20.** $\dfrac{b^2 + 3b - 4}{b^2 + 6b + 8} \div \dfrac{b^2 + 4b + 3}{b^2 + 3b + 2}$ **$\dfrac{b - 1}{b + 3}$**

**21.** $\dfrac{20 - 4m}{am + bm - 5a - 5b} - \dfrac{4}{a + b}$

**22.** $\dfrac{a^2 - 7ab - 18b^2}{-a + 9b} \div \dfrac{3a + 9b}{a^2 - 4b^2}$

**22.** $-\dfrac{(a + 2b)(a + 2b)(a - 2b)}{3(a + 3b)}$ **19.** $\dfrac{5c - d}{2(c - 5d)}$

★ **23.** $\dfrac{x^4 - 17x^2 + 16}{4 + 3x - x^2} - (x + 4)(x - 1)$

★ **24.** $\dfrac{x^2 + 8x + 12}{49x^2 - x^4} \div \dfrac{2x^2 - 6x - 108}{x^2 + 7x}$ $\cdot \dfrac{x^2 - 13x + 36}{x^2 - 6x - 16} - \dfrac{x - 4}{2x(x - 7)(x - 8)}$

★ **25.** For what value(s) of $x$ is $\dfrac{3x + 4}{x^4 - 13x^2 + 36}$ undefined? **±2, ±3**

# CUMULATIVE REVIEW

## (Chapters 1–8)

Each exercise has five choices of answers. Choose the one best answer. (Ex. 1–9)

**1.** Compute $(-3)^3 \cdot 2^2$.  **B**
(A) $-36$  (B) $-108$  (C) $-23$  (D) $5$
(E) None of these

**2.** Factor $3x^2 + 9xy^3$  **C**
(A) $x(3x + 9y^3)$  (B) $3xy(x + 3y^2)$
(C) $3x(x + 3y^3)$  (D) $3x^2(1 + 9y^3)$
(E) None of these

**3.** The value of $x$ for which $\dfrac{3x - 9}{4x + 8}$  **C**

is undefined is
(A) $3$  (B) $2$  (C) $-2$  (D) $8$
(E) None of these

**4.** Evaluate $-2x + 4y$ if $x = -2$ and $y = -5$.  **C**
(A) $-80$  (B) $-24$  (C) $-16$  (D) $80$
(E) None of these

**5.** Solve $-7 \geq 8 + 3x$.  **C**
(A) $x < -5$  (B) $x \geq 5$  (C) $x \leq -5$
(D) $5 \geq x$  (E) None of these

**6.** Multiply $-7x^3y \cdot -4xy^5$.  **D**
(A) $28x^3y^5$  (B) $28x^4y^5$  (C) $-28x^4y^6$
(D) $28x^4y^6$  (E) None of these

**7.** Subtract $-y^2 - 4y$ from $y^3 + 3y^2 - y + 2$.  **B**
(A) $y^3 + 2y^2 + 3y + 2$
(B) $y^3 + 4y^2 + 3y + 2$
(C) $-y^3 - 4y^2 - 3y - 2$
(D) $y^3 + 4y^2 + 4y + 2$
(E) None of these

**8.** Simplify $(x - 6)^2$.  **A**
(A) $x^2 - 12x + 36$  (B) $x^2 - 36$
(C) $x^2 + 12x + 36$  (D) $2x - 12$
(E) $x^2 + 36$

**9.** Simplify $\dfrac{16 - x^2}{x^2 + 2x - 24}$.  **C**
(A) $\dfrac{x + 4}{x + 6}$  (B) $\dfrac{x - 4}{x + 6}$  (C) $-\dfrac{x + 4}{x + 6}$
(D) $-\dfrac{x - 4}{x + 6}$  (E) None of these

**10.** Find the reciprocal of $4y$.  $\frac{1}{4y}$ $(y \neq 0)$

Simplify.
**11.** $x(3x^3 - 7x + 2)$  $3x^4 - 7x^2 + 2x$
**12.** $a^3 - (-2a^2 - a + 1)$  $a^3 + 2a^2 + a - 1$
**13.** $(-4y^3)^3$  $-64y^9$
**14.** $\dfrac{6x^3}{18x^4}$  $\dfrac{1}{3x}$    **15.** $\dfrac{5m - 25}{m^2 - 25}$  $\dfrac{5}{m + 5}$

Solve.
**16.** $3x - 4 = 23$  $9$
**17.** $|2x - 4| = 16$  $10, -6$
**18.** $4a - (5 - a) = 20 - 10a$  $\frac{5}{3}$
**19.** $5m - 12 = -2m^2$  $\frac{3}{2}, -4$
**20.** $a^2 - 7a = 0$  $0, 7$
**21.** $2x^3 - 2x^2 - 40x = 0$  $0, 5, -4$
**22.** $(x - 5)^2 - 3(x - 5) + 2 = 0$  $6, 7$
                    **23.** $x > 4$ or $x < -4$
Graph the solution set on a number line.
**23.** $|x| > 4$       **24.** $|x| \leq 3$  $x \leq 3$ and $x \geq -3$
**25.** $|2a - 4| < 8$  $-2 < a < 6$
**26.** $2|x - 2| + 10 > 12$  $x > 3$ or $x < 1$
**27.** $y \neq -3$  {all real nos., but $-3$}
**28.** $3(a - 4) = a - (15 - 2a)$  $\emptyset$
**29.** $4(x - 2) = 2x - (-2x + 8)$
**30.** $x \not< |-3|$  $x \geq 3$       **29.** {all real nos.}
**31.** $-4a + 6 \leq -6a + 12$  $a \leq 3$

Graph the solution set of the compound inequality.
**32.** $x < -4$ or $x > 3$
**33.** $2a - 1 < 13$ and $3a + 5 \geq 35$
**34.** $2 < 2m - 8 < 6$
**35.** $6x - 2 < 8x + 14$ or $|x| > 3$

**33.** ——|—|—|—|—|—|—
       $-2$  $0$  $2$

**35.** ◄—|—|—|—|—|—|—|—►
       $-6\ -4\ -2\ 0\ 2\ 4\ 6$

Multiply.
**36.** $(2b - 5c)(4b - c)$  $8b^2 - 22bc + 5c^2$
**37.** $(a + 3)(a^2 - 4a + 2)$  $a^3 - a^2 - 10a + 6$
**38.** $3y^4(5y^3 - 7y^2 - y)$  $15y^7 - 21y^6 - 3y^5$

Tell whether the set is finite or infinite.
**39.** $\{-8, -10, -12, -14, \ldots\}$ **infinite**
**40.** $\{2, 4, 6, 11\}$ **finite**

**44.** $4x(x - 3)(x + 3)(x - 2)(x + 2)$
Factor completely.
**41.** $3a^2 + 12a - 15$  $3(a + 5)(a - 1)$
**42.** $2y^3 + 2y^2 - 12y$  $2y(y - 2)(y + 3)$
**43.** $4a^2 - 49b^2$  $(2a - 7b)(2a + 7b)$
**44.** $4x^5 - 52x^3 + 144x$
**45.** $a^2x^2 - 9a^2 - x^2 + 9$
$(a - 1)(a + 1)(x - 3)(x + 3)$

Classify each as a monomial, binomial, or trinomial.
**46.** $4a^2 - 7a + 5$  **47.** $3a^4$ **monomial**
**trinomial**
**48.** Find the missing factor $a^6 \cdot ? = a^{14}$.  $a^8$

**49.** Simplify. Then write in descending order of exponents.
$-1 - b^5 + 4b^4 - 2b^2 - 5b^4 - 3b^2 + 2b^3 - 7$
$-b^5 - b^4 + 2b^3 - 5b^2 - 8$
**50.** What is the degree of the polynomial $3x^4y^2 - 7x^3y^7$?  **10**

Match each item with the correct property below.
**51.** $7 \cdot a + 7 \cdot b = 7(a + b)$  **B**
**52.** $0 + a = a$  **A**
**53.** $(3a)(4b) = (4b)(3a)$  **E**
**54.** $a \cdot (b \cdot c) = (a \cdot b) \cdot c$  **C**
**55.** $-a = -1 \cdot a$  **F**
**56.** $(3a + 4)1 = 3a + 4$  **D**
**57.** $a + (-a) = 0$  **G**

**(A)** Add. identity  **(B)** Distributive
**(C)** Associative (mult.)  **(D)** Mult. identity
**(E)** Commutative (mult.)
**(F)** Mult. property of $-1$
**(G)** Add. inverse

Multiply.
**58.** $x + 3 \cdot \dfrac{4x + 36}{x^2 + 12x + 27}$  **4**
**59.** $\dfrac{y^2 - 1}{4y - 16} \cdot \dfrac{y^2 - 4}{y^2 + y - 2}$  $\dfrac{(y + 1)(y - 2)}{4(y - 4)}$
**60.** $\dfrac{5 - x}{x - 8} \cdot \dfrac{x^2 - 7x - 8}{x^2 - 25}$  $-\dfrac{x + 1}{x + 5}$

Divide.
**61.** $\dfrac{x^2 - 7x + 12}{3x - x^2} \div \dfrac{5x - 20}{x}$  $-\dfrac{1}{5}$
**62.** $\dfrac{a^2 - 7a}{5x^3y^2} \div \dfrac{a^2 - 49}{15x^2y^4}$  $\dfrac{3ay^2}{x(a + 7)}$

**63.** Simplify.
$\dfrac{m^2 - 5m}{m^2 - 6m - 5} \div \dfrac{m^2 + 5m}{m^2 - 1} \cdot \dfrac{2m + 10}{m + 1}$
$\dfrac{2m^2 - 12m + 10}{m^2 - 6m - 5}$

Solve each problem.
**64.** 15 kg less than twice Henry's mass is 125 kg. Find Henry's mass. **70 kg**

**65.** The square of a number is 16 less than 10 times the number. Find the number. (*Hint:* More than one number will work).
**8, 2**
**66.** Find two consecutive integers whose sum is 77. **38, 39**

**67.** The larger of two numbers is 6 less than twice the smaller. The sum of the numbers is 18. Find the two numbers. **8, 10**

**68.** The selling price of a stove is $500. The profit is 25% of the cost. Find the cost.
**400**
**69.** The length of a rectangle is 5 m more than 3 times the width. The perimeter is 66 m. Find the length and width. **7, 26**

**70.** If 4 more than twice a number is decreased by 7 less than 4 times the number, the result is the same as the number decreased by 13. Find the number.

8

# COLLEGE PREP TEST

Directions: Choose the letter of the correct answer for each problem.

**1.** On a road map, $1.5''$ represents 45 mi.
**C** How many miles does $1.7''$ represent?

(A) 62      (B) 60      (C) 51

(D) 30      (E) 28

**2.** If $x$ oranges sell for $d$ dollars, find the price in cents for $y$ oranges selling at the
**E** same rate.

(A) $\dfrac{100d}{y}$    (B) $\dfrac{100xy}{d}$    (C) $\dfrac{100y}{d}$

(D) $\dfrac{100xd}{y}$    (E) $\dfrac{100dy}{x}$

**3.** If 1 in.$^3$ of lead weighs $\frac{1}{2}$ oz, what part
**C** of 1 ft$^3$ of lead weighs 4 oz?

(A) $\dfrac{1}{36}$    (B) $\dfrac{1}{72}$    (C) $\dfrac{1}{216}$

(D) $\dfrac{1}{432}$    (E) $\dfrac{1}{864}$

**4.** How many minutes will it take a train traveling at the rate of 30 mph to cover
**D** a distance of $\frac{3}{5}$ mi?

(A) 50      (B) 18      (C) 12

(D) 1.2      (E) $\dfrac{5}{6}$

**5.** If the ratio $p$ to $q$ is $\frac{3}{5}$ and the ratio $q$ to $r$
**D** is $\frac{10}{13}$, then what is the ratio $p$ to $r$?

(A) $\dfrac{3}{13}$    (B) $\dfrac{3}{10}$    (C) $\dfrac{1}{2}$

(D) $\dfrac{6}{13}$    (E) $\dfrac{6}{10}$

**6.** What part of a quarter is the sum of
**E** three pennies, one nickel, and one dime?

(A) $\dfrac{1}{72}$    (B) $\dfrac{3}{25}$    (C) $\dfrac{1}{5}$

(D) $\dfrac{7}{25}$    (E) $\dfrac{18}{25}$

**7.** If $\dfrac{4}{5} \cdot \dfrac{5}{6} \cdot \dfrac{6}{7} \cdot \dfrac{7}{8} \cdot \dfrac{8}{9} \cdot \dfrac{9}{10} \cdot p = 1$,
**B** then what is the value of $p$?

(A) 10      (B) $\dfrac{10}{4}$      (C) 1

(D) $\dfrac{10}{11}$    (E) $\dfrac{4}{10}$

**8.** If $\dfrac{n}{5} > n$, then $n$ may take on which
**A** of the following values?

(A) $-20$      (B) 0      (C) 5

(D) 20      (E) none of these

**9.** If the operation $\star$ for positive numbers
**B** is defined as $a \star b = \dfrac{ab}{a + b}$, then

$4 \star (4 \star 4)$ is

(A) 2      (B) $\dfrac{4}{3}$      (C) $\dfrac{3}{4}$

(D) $\dfrac{1}{2}$      (E) $\dfrac{1}{4}$

**10.** After 15 L of gas were removed from a car's gas tank that was half full, the tank was $\frac{1}{8}$ full. What is the capacity
**C** of the tank in liters?

(A) $7\frac{1}{2}$      (B) 20      (C) 40

(D) 75      (E) 120

# 9 COMBINING RATIONAL EXPRESSIONS

## Problem Solving Formulation

Just as with large corporations, consumers have a growing need for financial planning. With all the options available for large purchases such as a car, people often need advice on how best to budget and manage their money.

**Project**

Find the total weekly cost of owning a car.

Use the following questions to help you formulate the problem.

How much will gas cost each week? (10,000 miles a year)
How much will insurance cost?
What are the regular maintenance charges?
What will the weekly car payment be?
What other expenses are involved in owning a car?

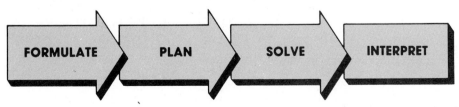

FORMULATE — PLAN — SOLVE — INTERPRET

# ADDITION: LIKE DENOMINATORS 9.1

*Objective*  **To add rational expressions with like denominators**

You know that $\frac{2}{9} + \frac{5}{9} = \frac{7}{9}$. Below is a demonstration of why this is true.

$$\frac{2}{9} + \frac{5}{9} = 2 \cdot \frac{1}{9} + 5 \cdot \frac{1}{9} \text{ by the property } \frac{x}{y} = x \cdot \frac{1}{y}$$

$$= (2 + 5)\frac{1}{9} \text{ by the distributive property in reverse}$$

$$= 7 \cdot \frac{1}{9}, \text{ or } \frac{7}{9} \text{ by the property } x \cdot \frac{1}{y} = \frac{x}{y}$$

This suggests a rule for adding two rational expressions with like denominators.

| Addition property for rational expressions | For all $a$ and $c$ and for each nonzero $b$, $\dfrac{a}{b} + \dfrac{c}{b} = \dfrac{a + c}{b}$. |
|---|---|

Assume that no denominator is 0 in all examples and exercises.

*Example 1*  **Add.**

$$\frac{4}{x} + \frac{5}{x}$$
$$\frac{4 + 5}{x} = \frac{9}{x}$$

$$\frac{2y}{y - 4} + \frac{3y}{y - 4}$$
$$\frac{2y + 3y}{y - 4} = \frac{5y}{y - 4}$$

Sometimes, you can rewrite the sum of two rational expressions in simpler form. You can do this by first factoring the numerator and the denominator and then dividing out the common factors. The process is illustrated below.

*Example 2*  **Add.**

$$\frac{3m}{5} + \frac{7m}{5}$$
$$\frac{3m + 7m}{5}$$
$$\frac{10m}{5}$$
$$\frac{\overset{1}{\cancel{5}} \cdot 2 \cdot m}{\underset{1}{\cancel{5}}} = \frac{2m}{1}, \text{ or } 2m$$

$$\frac{x^2}{2x - 6} + \frac{-9}{2x - 6}$$
$$\frac{x^2 - 9}{2x - 6}$$
$$\frac{(\overset{1}{\cancel{x - 3}})(x + 3)}{2(\underset{1}{\cancel{x - 3}})} \quad \blacktriangleleft \text{ Factor } x^2 - 9.$$
$$\qquad\qquad\quad \blacktriangleleft \text{ Factor } 2x - 6. \text{ The GCF is 2.}$$
$$\frac{x + 3}{2}$$

The addition property for rational expressions with like denominators can be extended to include more than two rational expressions. For example, $\frac{a}{b} + \frac{c}{b} + \frac{d}{b} = \frac{a + c + d}{b}$. This extension is illustrated in the next example.

## Example 3    Add.

$$\frac{4x + 11}{x^2 - 10x + 16} + \frac{-3x - 18}{x^2 - 10x + 16} + \frac{-1}{x^2 - 10x + 16}$$

$$\frac{4x + 11 - 3x - 18 - 1}{x^2 - 10x + 16}$$

$$\frac{x - 8}{x^2 - 10x + 16} \quad \blacktriangleleft \text{ Combine like terms in the numerator.}$$

$$\frac{\overset{1}{\cancel{x - 8}}}{(\cancel{x - 8})(x - 2)} = \frac{1}{x - 2} \quad \blacktriangleleft \text{ The sum in simplest form.}$$

# Written Exercises

**Add. Simplify, if possible.**

2. $\dfrac{9x}{x - 2}$   12. $\dfrac{1}{a - 2}$   4. $\dfrac{3a}{a + 5}$

**(A)**

1. $\dfrac{7}{m} + \dfrac{3}{m}$ **10** $\!\!/m$

2. $\dfrac{4x}{x - 2} + \dfrac{5x}{x - 2}$

3. $\dfrac{3x}{2} + \dfrac{4x}{2}$ **7x** $\!\!/$**2**

4. $\dfrac{2a}{a + 5} + \dfrac{a}{a + 5}$

5. $\dfrac{7m}{2} + \dfrac{m}{2}$ **4m**

6. $\dfrac{3}{10x} + \dfrac{5}{10x}$ **4** $\!\!/$**5x**

7. $\dfrac{5}{9a} + \dfrac{4}{9a}$ **1** $\!\!/a$

8. $\dfrac{7x}{15} + \dfrac{2x}{15}$ **3x** $\!\!/$**5**

9. $\dfrac{x^2}{2x - 12} + \dfrac{-36}{2x - 12}$ **x + 6** $\!\!/$**2**

10. $\dfrac{a}{5a - 20} + \dfrac{-4}{5a - 20}$ **1** $\!\!/$**5**

11. $\dfrac{x^2}{x - 5} + \dfrac{-25}{x - 5}$ **x + 5**

12. $\dfrac{a}{a^2 - 7a + 10} + \dfrac{-5}{a^2 - 7a + 10}$

13. $\dfrac{2x}{x^2 - 2x} + \dfrac{-4}{x^2 - 2x}$ **2** $\!\!/$**x**

14. $\dfrac{-10}{a^2 + 7a} + \dfrac{a + 17}{a^2 + 7a}$ **1** $\!\!/a$

**(B)**

15. $\dfrac{2x}{2x^2 + 13x + 15} + \dfrac{3}{2x^2 + 13x + 15}$ **1** $\!\!/$**x + 5**

16. $\dfrac{3a}{6a^2 + a - 12} + \dfrac{-4}{6a^2 + a - 12}$ **1** $\!\!/$**2a + 3**

17. $\dfrac{6y + 10}{3y^2 - 2y - 5} + \dfrac{-20}{3y^2 - 2y - 5}$ **2** $\!\!/$**y + 1**

18. $\dfrac{5p^2 - 10p}{2p^2 + 3p - 20} + \dfrac{-p^2}{2p^2 + 3p - 20}$ **2p** $\!\!/$**p + 4**

19. $\dfrac{2a^2}{6a^2 - 5a - 25} + \dfrac{a - 15}{6a^2 - 5a - 25}$ **a + 3** $\!\!/$**3a + 5**

20. $\dfrac{4b}{6b^2 + 11b - 35} + \dfrac{3b^2 - 15}{6b^2 + 11b - 35}$

21. $\dfrac{16a^2 - 7}{5a^2 + 26a - 24} + \dfrac{9a^2 - 9}{5a^2 + 26a - 24}$

22. $\dfrac{2x^2 - 5x}{3x^2 - x - 14} + \dfrac{x^2 - 2x + 5}{3x^2 - x - 14}$

23. $\dfrac{a^3}{a^2 + a} + \dfrac{4a^2}{a^2 + a} + \dfrac{3a}{a^2 + a}$ **a + 3**

24. $\dfrac{2x^2}{3x + 6y} + \dfrac{7xy}{3x + 6y} + \dfrac{6y^2}{3x + 6y}$ **2x + 3y** $\!\!/$**3**

**(C)**

25. $\dfrac{4x^2 - ax - 6a^2}{xy + bx + ay + ab} + \dfrac{-3x^2 - 3ax + a^2}{xy + bx + ay + ab}$

26. $\dfrac{5x^4 - 4x^3 - 20x^2}{3x^5 - 30x^3 + 27x} + \dfrac{x^4 - 8x^3 + 2x^2}{3x^5 - 30x^3 + 27x}$

27. $\dfrac{4ax + 2a - 4b}{ax^2 - bx^2 + 3ax - 3bx + 2a - 2b} + \dfrac{-3ax - bx - a + 3b}{ax^2 - bx^2 + 3ax - 3bx + 2a - 2b}$

20. $\dfrac{b + 3}{2b + 7}$   21. $\dfrac{5a + 4}{a + 6}$   22. $\dfrac{3x^2 - 7x + 5}{3x^2 - x - 14}$   25. $\dfrac{x - 5a}{y + b}$   26. $\dfrac{2x}{(x - 1)(x + 3)}$   27. $\dfrac{1}{x + 2}$

Addition: Like Denominators

# ADDITION: MONOMIAL DENOMINATORS    9.2

*Objective*    **To add rational expressions with unlike monomial denominators.**

The fractions $\frac{3}{5}$ and $\frac{6}{10}$ are equivalent because they have the same value. This can be shown by using the Multiplication Property of One as follows.

$$\frac{3}{5} = \frac{3}{5} \cdot 1$$

$$= \frac{3}{5} \cdot \frac{2}{2} = \frac{3 \cdot 2}{5 \cdot 2} = \frac{6}{10}$$

Multiplying the numerator and the denominator of $\frac{3}{5}$ by 2 does not change the value of the fraction $\frac{3}{5}$. This suggests the following property.

| | |
|---|---|
| Identity property for rational expressions | For all $a$ and for all nonzero $b$ and $c$, $\frac{a}{b} = \frac{a \cdot c}{b \cdot c}$.<br><br>Multiplying both the numerator and the denominator by the same nonzero number produces an equivalent rational expression. |

In the sum $\frac{1}{6} + \frac{2}{3}$, notice that the denominators are not the same. But, you can make them the same by using the property stated above. The procedure for doing this follows.

First, factor the 6 into primes.    $\frac{1}{3 \cdot 2} + \frac{2}{3}$

The denominators would be alike if the second denominator had the factor 2.

Then, multiply the numerator and the denominator of $\frac{2}{3}$ by 2.

$$\frac{1}{3 \cdot 2} + \frac{2}{3} = \frac{1}{3 \cdot 2} + \frac{2 \cdot 2}{3 \cdot 2} \text{ by the Identity Property}$$

$$= \frac{1}{6} + \frac{4}{6}$$

The denominators are now the same. **Thus,** $\frac{1}{6} + \frac{2}{3} = \frac{1}{6} + \frac{4}{6} = \frac{5}{6}$

For the sum $\frac{1}{6} + \frac{2}{3}$, $3 \cdot 2$ is the **least common multiple (LCM)** of the denominators. The procedure for adding rational expressions with unlike denominators is the same. You begin by finding the least common multiple of the denominators.

**Example 1**  Add $\dfrac{2b}{7} + \dfrac{3b}{14}$.

Find the LCM (least common multiple). Factor each denominator if possible.

$\dfrac{2b}{7} + \dfrac{3b}{7 \cdot 2}$  ◀ *7 is already prime; $14 = 7 \cdot 2$.*

The only factors present are 7 and 2. To have a common denominator, each denominator must have exactly the same factors, 7 and 2.

$$\dfrac{2b}{7} + \dfrac{3b}{7 \cdot 2}$$
needs 2     has 7 · 2

$\dfrac{2b \cdot 2}{7 \cdot 2} + \dfrac{3b}{7 \cdot 2}$  ◀ *Multiply the numerator and the denominator of $\dfrac{2b}{7}$ by the missing factor, 2.*

$\dfrac{4b}{7 \cdot 2} + \dfrac{3b}{7 \cdot 2}$  ◀ *The denominators are now the same.*

$\dfrac{7b}{7 \cdot 2}$

$\dfrac{\overset{1}{\cancel{7}} \cdot b}{\underset{1}{\cancel{7}} \cdot 2}$

$\dfrac{b}{2}$  ◀ *Write the answer in simplest form.*

**Example 2**  Add $\dfrac{x}{12} + \dfrac{2x}{3} + \dfrac{1}{6}$.

$$\dfrac{x}{2 \cdot 2 \cdot 3} + \dfrac{2x}{3} + \dfrac{1}{3 \cdot 2}$$
The LCM = 2 · 2 · 3.  ◀ *There are at most two 2's and one 3 in any denominator.*

$$\dfrac{x}{2 \cdot 2 \cdot 3} + \dfrac{2x}{3} + \dfrac{1}{3 \cdot 2}$$
has two 2's, one 3    needs two 2's   needs one 2

$\dfrac{x}{2 \cdot 2 \cdot 3} + \dfrac{2x \cdot 2 \cdot 2}{3 \cdot 2 \cdot 2} + \dfrac{1 \cdot 2}{3 \cdot 2 \cdot 2}$  ◀ *Multiply each numerator and denominator by the factors needed.*

$\dfrac{x}{12} + \dfrac{8x}{12} + \dfrac{2}{12}$

$\dfrac{1x + 8x + 2}{12}$  ◀ *Combine the numerators.*

$\dfrac{9x + 2}{12}$  ◀ *The sum is in simplest form.*

The next example illustrates the process for finding the least common multiple for rational expressions that contain variables in the denominators.

*Example 3*    **Add** $\dfrac{2y + 7}{4y} + \dfrac{2y - 1}{3y}$.

$$\frac{2y + 7}{2 \cdot 2 \cdot y} + \frac{2y - 1}{3 \cdot y}$$

The LCM = $2 \cdot 2 \cdot 3 \cdot y$. Each denominator needs $2 \cdot 2 \cdot 3 \cdot y$ to be the same.

$$\frac{(2y + 7)3}{2 \cdot 2 \cdot y \cdot 3} + \frac{(2y - 1)2 \cdot 2}{3 \cdot y \cdot 2 \cdot 2}$$

$$\frac{6y + 21}{12y} + \frac{8y - 4}{12y} \quad \blacktriangleleft \quad (2y - 1)2 \cdot 2 = (2y - 1)4 = 8y - 4$$

$$\frac{14y + 17}{12y} \quad \blacktriangleleft \quad \textit{Combine the numerators. The sum is in simplest form.}$$

## Reading in Algebra

**1. LCM represents the least common multiple.**
**2. To find the LCM of the denominators of two or more
fractions, factor each denominator into prime factors, and write down all factors as many times as they**

1. What does the LCM mean?    **occur at most in any denominator.**

2. Tell, in your own words, how to go about finding the LCM of the denominators for
two or more fractions. **3. No, the correct LCM is $2 \cdot 2 \cdot 3$. Two 2's are needed in the LCM since there
are two 2's in the denominator of the first fraction.**

3. Is $2 \cdot 3$ the LCM of the denominators in $\dfrac{1}{2 \cdot 2} + \dfrac{5}{2 \cdot 3}$? Why or why not?

4. A common multiple of the denominators in $\dfrac{1}{2} + \dfrac{1}{3 \cdot 2}$ is 12. However, 12 is not the

least common multiple. Why not? **LCM needs only one 2. There is only one 2 in any denominator.
LCM is $2 \cdot 3$ or 6.**

## Oral Exercises

**Find the LCM of the denominators.**   **5. $3 \cdot 3 \cdot 7$**    **6. $3 \cdot 5 \cdot 5 \cdot 7$**    **7. $2 \cdot 2 \cdot 5 \cdot 5 \cdot 5$**

1. $\dfrac{3}{5} + \dfrac{7}{2 \cdot 5}$ **$2 \cdot 5$**

2. $\dfrac{3}{7} + \dfrac{4}{5 \cdot 7}$ **$5 \cdot 7$**

3. $\dfrac{7a}{3 \cdot 5} + \dfrac{3a}{5}$ **$3 \cdot 5$**

4. $\dfrac{5b}{3} + \dfrac{11b}{3 \cdot 2} + \dfrac{b}{2 \cdot 3 \cdot 2}$ **$2 \cdot 3 \cdot 2$**

5. $\dfrac{5m}{3} + \dfrac{m}{7 \cdot 3} + \dfrac{4m}{3 \cdot 3 \cdot 7}$

6. $\dfrac{2a}{3 \cdot 5 \cdot 7} + \dfrac{a}{5 \cdot 5} + \dfrac{5a}{3 \cdot 7}$

7. $\dfrac{3p - 1}{2 \cdot 5 \cdot 5} + \dfrac{4p - 3}{5 \cdot 5 \cdot 5} + \dfrac{7p - 1}{2 \cdot 2}$

8. $\dfrac{5x - 1}{2 \cdot 7} + \dfrac{4x - 1}{7 \cdot 7 \cdot 2}$

9. $\dfrac{3b - 1}{5 \cdot 3} + \dfrac{2b}{3 \cdot 3} + \dfrac{4b - 1}{5 \cdot 2}$

10. $\dfrac{3a - 1}{4} + \dfrac{7a + 2}{2}$ **4**

11. $\dfrac{2x - 5}{3x} + \dfrac{2x - 5}{6x}$ **$6x$**

12. $\dfrac{2a - 1}{5} + \dfrac{4a + 3}{10} + \dfrac{6a - 7}{2}$

**8. $2 \cdot 7 \cdot 7$**    **9. $2 \cdot 3 \cdot 3 \cdot 5$**    **12. 10**

## Written Exercises

**Add. Simplify, if possible.**

Ⓐ 1. $\dfrac{t}{3} + \dfrac{t}{6}$ **$\dfrac{t}{2}$**

2. $\dfrac{b}{5} + \dfrac{3b}{10}$ **$\dfrac{b}{2}$**

3. $\dfrac{3d}{8} + \dfrac{d}{4}$ **$\dfrac{5d}{8}$**

4. $\dfrac{2b}{7} + \dfrac{b}{14}$ **$\dfrac{5b}{14}$**

5. $\dfrac{a}{7} + \dfrac{5a}{14}$ **$\dfrac{a}{2}$**

6. $\dfrac{2a}{3} + \dfrac{5a}{6}$ **$\dfrac{3a}{2}$**

7. $\dfrac{2x}{15} + \dfrac{3x}{5}$ **$\dfrac{11x}{15}$**

8. $\dfrac{5y}{6} + \dfrac{y}{3}$ **$\dfrac{7y}{6}$**

9. $\dfrac{y}{12} + \dfrac{2y}{3} + \dfrac{1}{6}$  $\dfrac{9y + 2}{12}$

10. $\dfrac{7m}{18} + \dfrac{7}{2} + \dfrac{5m}{6}$  $\dfrac{22m + 63}{18}$

11. $\dfrac{3m}{14} + \dfrac{3m}{7} + \dfrac{1}{4}$  $\dfrac{18m + 7}{28}$

12. $\dfrac{2x - 3}{6x} + \dfrac{5x - 2}{2x}$  $\dfrac{17x - 9}{6x}$

13. $\dfrac{3m - 2}{4m} + \dfrac{m + 6}{5m}$  $\dfrac{19m + 14}{20m}$

14. $\dfrac{m + 1}{3m} + \dfrac{m + 4}{4m}$  $\dfrac{7m + 16}{12m}$

15. $\dfrac{2b - 1}{6b} + \dfrac{4b + 1}{3b}$  $\dfrac{10b + 1}{6b}$

16. $\dfrac{3a}{4} + \dfrac{2a}{3} + \dfrac{5}{6}$  $\dfrac{17a + 10}{12}$

17. $\dfrac{b - 1}{9} + \dfrac{2b - 1}{4}$  $\dfrac{22b - 13}{36}$

18. $\dfrac{5y}{4} + \dfrac{y}{6} + \dfrac{7y}{12}$  $2y$

19. $\dfrac{2x - 3}{4} + \dfrac{4x + 5}{6}$  $\dfrac{14x + 1}{12}$

20. $\dfrac{6y}{5} + \dfrac{3y}{4} + \dfrac{y}{10}$  $\dfrac{41y}{20}$

Ⓑ 21. $\dfrac{3x - 1}{6} + \dfrac{1}{3} + \dfrac{5x + 2}{2}$  $\dfrac{18x + 7}{6}$

22. $\dfrac{2y - 3}{7} + \dfrac{3y + 1}{14} + \dfrac{4y - 3}{2}$  $\dfrac{35y - 26}{14}$

23. $\dfrac{5a - 4}{9a} + \dfrac{3a - 1}{3a} + \dfrac{7a + 4}{2a}$  $\dfrac{91a + 22}{18a}$

24. $\dfrac{5b - 1}{4b} + \dfrac{b + 2}{8b} + \dfrac{2b - 3}{2b}$  $\dfrac{19b - 12}{8b}$

25. $\dfrac{t - 4}{2} + \dfrac{3t - 5}{14} + \dfrac{4t - 3}{21}$  $\dfrac{38t - 105}{42}$

26. $\dfrac{a + 1}{3a} + \dfrac{5}{6a} + \dfrac{7a - 5}{4a}$  $\dfrac{25a - 1}{12a}$

27. $\dfrac{3k - 1}{15} + \dfrac{2k}{5} + \dfrac{4k - 1}{2}$  $\dfrac{78k - 17}{30}$

28. $\dfrac{5}{24b} + \dfrac{3b - 1}{6b} + \dfrac{b + 5}{4b}$  $\dfrac{18b + 31}{24b}$

29. $\dfrac{2a - 1}{9a} + \dfrac{3a + 5}{4a} + \dfrac{3a - 1}{3a}$  $\dfrac{71a + 29}{36a}$

30. $\dfrac{5y - 1}{5} + \dfrac{y}{6} + \dfrac{3y - 4}{10}$  $\dfrac{22y - 9}{15}$

Ⓒ 31. $\dfrac{3m^2 - 7m + 1}{2m} + \dfrac{4m - 1}{5m} + \dfrac{m^2}{6m}$

32. $\dfrac{7a + 4}{8} + \dfrac{3a^2 - a + 4}{5} + \dfrac{2a - 1}{6}$

33. $\dfrac{x + y}{6xy} + \dfrac{2x - y}{4y} + \dfrac{3x - 2y}{3x}$

34. $\dfrac{a - b}{3a} + \dfrac{a^2 - 2ab + b^2}{6ab} + \dfrac{2a + b}{5b}$

32. $\dfrac{72a^2 + 121a + 136}{120}$   34. $\dfrac{17a^2 + 6ab - 5b^2}{30ab}$

Simplify. [Hint: First, add within the parentheses.]

35. $\left( \dfrac{3a}{4} + \dfrac{1}{6} \right) \div \dfrac{81a^2 - 4}{36}$  $\dfrac{3}{9a - 2}$

36. $\left( \dfrac{a}{x} + \dfrac{b}{3x} \right) \dfrac{3x^2 + 3xy}{3ax + bx + 3ay + by}$  $1$

31. $\dfrac{50m^2 - 81m + 9}{30m}$   33. $\dfrac{6x^2 + 2x + 2y + 9xy - 8y^2}{12xy}$

# CUMULATIVE REVIEW

1. Simplify: $\dfrac{3a^3 b^5 c^6}{18ab^6 c^4} \cdot$  $\dfrac{a^2 c^2}{6b}$

2. Solve: $2x^2 - 20 = -3x.$  $\dfrac{5}{2}, -4$

3. Divide: $\dfrac{x^2 - 2x - 8}{x - 5} \div \dfrac{x^2 - 4x}{25 - x^2}.$  $-\dfrac{(x + 2)(x + 5)}{x}$

# CALCULATOR ACTIVITIES

Add. Simplify, if possible.

1. $\dfrac{385a}{3} + \dfrac{429a}{51}$  $\dfrac{6974a}{51}$

2. $\dfrac{215}{11m} + \dfrac{499}{5m} + \dfrac{176}{25m}$  $\dfrac{34756}{275m}$

3. $\dfrac{314b}{3} + \dfrac{218b + 19}{46} + \dfrac{2b - 1}{69}$  $\dfrac{15102b + 55}{138}$

4. $\dfrac{514y}{13} + \dfrac{318y}{7} + \dfrac{119y + 14}{35}$  $\dfrac{40207y + 182}{455}$

5. $\dfrac{305b}{3} + \dfrac{415b}{74} + \dfrac{519b}{2}$  $\dfrac{40712b}{111}$

6. $\dfrac{1,115}{2b} + \dfrac{118b - 2}{49b} + \dfrac{b + 118}{7b}$  $\dfrac{250b + 56283}{98b}$

Addition: Monomial Denominators

**241**

# ADDITION: POLYNOMIAL DENOMINATORS 9.3

*Objective*    **To add rational expressions with polynomial denominators**

In the sum $\dfrac{-3}{x^2 - 2x - 15} + \dfrac{4}{x + 3} + \dfrac{3}{x - 5}$, the denominators are polynomials. To find the LCM, you can use the same technique you used in the previous lesson. The technique is illustrated in Example 1.

*Example 1*    **Add** $\dfrac{-3}{x^2 - 2x - 15} + \dfrac{4}{x + 3} + \dfrac{3}{x - 5}$.

1. Find the LCM. Factor each denominator if possible.

$$\frac{-3}{(x - 5)(x + 3)} + \frac{4}{x + 3} + \frac{3}{x - 5} \quad \blacktriangleleft \quad \text{\textit{Factor } } x^2 - 2x - 15; \\ x + 3 \text{ \textit{and} } x - 5 \text{ \textit{cannot be factored.}}$$

The only factors present are $(x - 5)$ and $(x + 3)$. The LCM is $(x - 5)(x + 3)$. To have a common denominator, each must have exactly the same factors, $x - 5$ and $x + 3$.

$$\frac{-3}{(x - 5)(x + 3)} + \frac{4}{x + 3} + \frac{3}{x - 5}$$
$$\text{needs } (x - 5) \qquad \text{needs } (x + 3)$$

2. Multiply the numerator and the denominator of each by the missing factor.

$$\frac{-3}{(x - 5)(x + 3)} + \frac{4(x - 5)}{(x + 3)(x - 5)} + \frac{3(x + 3)}{(x - 5)(x + 3)} \quad \blacktriangleleft \quad \text{\textit{The denominators are now the same.}}$$
$$\frac{-3}{(x - 5)(x + 3)} + \frac{4x - 20}{(x - 5)(x + 3)} + \frac{3x + 9}{(x - 5)(x + 3)} \quad \blacktriangleleft \quad \begin{array}{l} 4(x - 5) = 4x - 20; \\ 3(x + 3) = 3x + 9 \end{array}$$

3. Combine the numerators.

$$\frac{-3 + 4x - 20 + 3x + 9}{(x - 5)(x + 3)} \quad \blacktriangleleft \quad \frac{a}{b} + \frac{c}{b} + \frac{d}{b} = \frac{a + c + d}{b}$$
$$\frac{7x - 14}{(x - 5)(x + 3)} \quad \blacktriangleleft \quad \text{\textit{Combine like terms in the numerator.}}$$

4. Now, factor the numerator to determine if the sum can be simplified.

$$\frac{7(x - 2)}{(x - 5)(x + 3)} \quad \blacktriangleleft \quad 7x - 14 = 7 \cdot x - 7 \cdot 2 = 7(x - 2)$$

Since there are no factors common to the numerator and the denominator, the sum

$$\frac{7x - 14}{(x - 5)(x + 3)}, \text{ or } \frac{7(x - 2)}{(x - 5)(x + 3)},$$

is in simplest form.

You should always check to see if a sum can be written in simpler form. The next example results in a sum that can be simplified.

**Example 2**    Add $\dfrac{9x + 14}{x^2 + 7x} + \dfrac{x}{x + 7}$.

When you factor, always look for a greatest common factor, GCF. Since the GCF of $x^2 + 7x$ is $x$, $x^2 + 7x = x(x + 7)$.

$\dfrac{9x + 14}{x(x + 7)} + \dfrac{x}{x + 7}$   ◀ *The LCM = $x(x + 7)$.*

$\dfrac{9x + 14}{x(x + 7)} + \dfrac{x}{x + 7}$

needs $x$

$\dfrac{9x + 14}{x(x + 7)} + \dfrac{x \cdot x}{(x + 7)x}$

$\dfrac{9x + 14}{x(x + 7)} + \dfrac{x^2}{x(x + 7)}$

$\dfrac{x^2 + 9x + 14}{x(x + 7)}$   ◀ *Combine like terms in the numerator. Write the terms in descending order of exponents.*

$\dfrac{(\cancel{x + 7})(x + 2)}{x(\cancel{x + 7})}$   ◀ *Factor the numerator and divide out common factors.*

$\dfrac{x + 2}{x}$   ◀ *The sum is now in simplest form.*

Sometimes, you will have to multiply two binomials, such as $(a + 6)(a - 4)$, to find a sum. This situation occurs in Example 3.

**Example 3**    Add $\dfrac{a}{a^2 - 36} + \dfrac{a - 4}{a^2 - 5a - 6}$

Notice that each denominator can be factored.

$a^2 - 36$ is the difference ▶ $\dfrac{a}{(a - 6)(a + 6)} + \dfrac{a - 4}{(a - 6)(a + 1)}$   ◀ *The LCM = $(a - 6)(a + 6)(a + 1)$.*
of two squares.

$\dfrac{a}{(a - 6)(a + 6)} + \dfrac{a - 4}{(a - 6)(a + 1)}$

needs $(a + 1)$    needs $(a + 6)$

$\dfrac{\cdot \; a(a + 1)}{(a - 6)(a + 6)(a + 1)} + \dfrac{(a - 4)(a + 6)}{(a - 6)(a + 1)(a + 6)}$

$a(a + 1)$ ▶ $\dfrac{a^2 + a}{(a - 6)(a + 6)(a + 1)} + \dfrac{a^2 + 2a - 24}{(a - 6)(a + 6)(a + 1)}$   ◀ $(a - 4)(a + 6) = a^2 + 2a - 24$

$\dfrac{a^2 + a^2 + 1a + 2a - 24}{(a - 6)(a + 6)(a + 1)}$

$2a^2 + 3a - 24$ cannot be factored, ▶ $\dfrac{2a^2 + 3a - 24}{(a - 6)(a + 6)(a + 1)}$
so the result is in simplest form.

Addition: Polynomial Denominators                                   **243**

## Reading in Algebra

**Determine whether each statement is always true, sometimes true, or never true.**

1. When two or more rational expressions are added, the factors of each denominator are also factors of the LCM of the denominators.  **A**
2. When two or more rational expressions are added, all the factors of the LCM of the denominators are also factors of each denominator.  **S**
3. When two or more rational expressions are added, the sum can be written in simpler form.  **S**
4. The LCM of the denominators of two rational expressions is the sum of the denominators.  **N**

## Oral Exercises

**Find the LCM of the denominators.**          1. $(x + 5)(x - 4)$     2. $a(a - 8)(a + 2)$

1. $\dfrac{7}{(x + 5)(x - 4)} + \dfrac{2}{x + 5} + \dfrac{3}{x - 4}$

2. $\dfrac{7}{a - 8} + \dfrac{3}{a(a - 8)} + \dfrac{2}{a + 2}$

3. $\dfrac{5}{(x - 5)(x + 2)} + \dfrac{4}{(x + 2)(x + 5)}$

4. $\dfrac{7}{a - 6} + \dfrac{4}{a + 6} + \dfrac{3}{(a + 6)(a - 6)}$

3. $(x - 5)(x + 2)(x + 5)$     4. $(a - 6)(a + 6)$

## Written Exercises

**Add. Simplify, if possible.**

Ⓐ

1. $\dfrac{3}{a - 3} + \dfrac{-18}{a^2 - 9} + \dfrac{5}{a + 3}$  $\dfrac{8}{a + 3}$

2. $\dfrac{-6}{b^2 - 7b + 10} + \dfrac{2}{b - 5} + \dfrac{3}{b - 2}$  $\dfrac{5}{b - 2}$

3. $\dfrac{8x + 15}{x^2 + 5x} + \dfrac{x}{x + 5}$  $\dfrac{x + 3}{x}$

4. $\dfrac{9y + 20}{y^2 + 4y} + \dfrac{y}{y + 4}$  $\dfrac{y + 5}{y}$

5. $\dfrac{3}{t + 4} + \dfrac{6}{t^2 - 16}$  $\dfrac{3(t - 2)}{(t - 4)(t + 4)}$

6. $\dfrac{7}{x^2 - x - 12} + \dfrac{3}{x - 4} + \dfrac{2}{x + 3}$  $\dfrac{5x + 8}{(x - 4)(x + 3)}$

7. $\dfrac{5}{m - 7} + \dfrac{3}{m^2 - 3m - 28}$  $\dfrac{5m + 23}{(m - 7)(m + 4)}$

8. $\dfrac{3}{k^2 - k - 2} + \dfrac{3}{k - 2} + \dfrac{7}{k + 1}$  $\dfrac{10k - 8}{(k - 2)(k + 1)}$

9. $\dfrac{a}{a - 5} + \dfrac{-3a - 10}{a^2 - 5a}$  $\dfrac{a + 2}{a}$

10. $\dfrac{4}{m - 4} + \dfrac{-16}{m^2 - 4m}$  $\dfrac{4}{m}$

$\dfrac{x^2 + 10x + 11}{(x + 6)(x - 8)}$        $\dfrac{2a^2 - 8a - 4}{(a - 5)(a + 3)(a + 1)}$        $\dfrac{3m^2 + 6m - 8}{m(3m + 5)}$

Ⓑ

11. $\dfrac{2x - 1}{x^2 - 2x - 48} + \dfrac{x + 2}{x - 8}$

12. $\dfrac{a - 4}{a^2 - 2a - 15} + \dfrac{a}{a^2 + 4a + 3}$

13. $\dfrac{4m - 3}{3m^2 + 5m} + \dfrac{m - 1}{m}$

14. $\dfrac{a}{a^2 - 9} + \dfrac{3a - 3}{a^2 - a - 6}$

15. $\dfrac{4x - 1}{x^2 - 7x} + \dfrac{2}{x^2 - 49}$

16. $\dfrac{2b}{b^2 - 7b + 10} + \dfrac{b - 1}{b^2 - 25}$

$\dfrac{4a^2 + 8a - 9}{(a - 3)(a + 3)(a + 2)}$        $\dfrac{4x^2 + 29x - 7}{x(x - 7)(x + 7)}$        $\dfrac{3b^2 + 7b + 2}{(b - 5)(b + 5)(b - 2)}$

Ⓒ

17. $\dfrac{x}{x^2 + 4x - 21} + \dfrac{4}{x^2 + 7x} + \dfrac{2x}{x^2 - 9}$  $\dfrac{3x^3 + 21x^2 - 36}{x(x + 7)(x - 3)(x + 3)}$

18. $\dfrac{3m}{2m^2 + 7m - 15} + \dfrac{6}{2m^2 - 3m} + \dfrac{2m + 3}{2m^2 + 17m + 35}$  $\dfrac{10m^3 + 33m^2 + 93m + 210}{m(2m - 3)(m + 5)(2m + 7)}$

# SOME SPECIAL CASES OF ADDITION　　9.4

*Objectives*　　**To add a whole number or a polynomial and a rational expression**

**To add rational expressions with nonfactorable denominators**

**To add rational expressions with monomial denominators containing different powers of a variable**

It is sometimes necessary to add a whole number, such as 7, and a rational expression, such as $\dfrac{2}{5m}$. You begin by writing 7 as $\dfrac{7}{1}$. The process is illustrated in Example 1.

*Example 1*　　**Add $7 + \dfrac{2}{5m}$.**

$$\frac{7}{1} + \frac{2}{5 \cdot m}$$

needs $5 \cdot m$　　◀ *The LCM = 5 · m.*

$$\frac{7 \cdot 5m}{1 \cdot 5m} + \frac{2}{5 \cdot m}$$

$$\frac{35m}{5m} + \frac{2}{5m}$$

$$\frac{35m + 2}{5m}$$　　◀ *The sum is in simplest form, for 35m + 2 cannot be factored.*

In the sum $\dfrac{5}{x + 2} + \dfrac{4}{x + 4}$, both denominators are polynomials. However, *neither* denominator can be factored. The LCM, therefore, must be $(x + 2)(x + 4)$, *the product of the two denominators.* You can now add the rational expressions, as shown in Example 2.

*Example 2*　　**Add $\dfrac{5}{x + 2} + \dfrac{4}{x + 4}$.**

$$\frac{5(x + 4)}{(x + 2)(x + 4)} + \frac{4(x + 2)}{(x + 4)(x + 2)}$$　　◀ *The LCM = (x + 2)(x + 4). In each, multiply the numerator and the denominator by the needed factor.*

$$\frac{5x + 20}{(x + 2)(x + 4)} + \frac{4x + 8}{(x + 2)(x + 4)}$$

$$\frac{9x + 28}{(x + 2)(x + 4)}$$　　◀ *There are no common factors, so the sum is in simplest form.*

In the sum $\dfrac{7}{2x^2} + \dfrac{-4}{5x} + \dfrac{3}{10x^3}$, notice that each denominator is the product of an integer and a power of $x$. You will find it easier to discover the LCM of the denominators if you rewrite them in factored form without exponents.

## Example 3  Add $\dfrac{7}{2x^2} + \dfrac{-4}{5x} + \dfrac{3}{10x^3}$.

$\dfrac{7}{2 \cdot x \cdot x} + \dfrac{-4}{5 \cdot x} + \dfrac{3}{5 \cdot 2 \cdot x \cdot x \cdot x}$ ◄ *At most, there are one 2, one 5, and 3 x's in*
$\quad\quad$ The LCM $= 2 \cdot 5 \cdot x \cdot x \cdot x.$ $\quad$ *any denominator.*

$\dfrac{7 \cdot 5 \cdot x}{2 \cdot x \cdot x \cdot 5 \cdot x} + \dfrac{-4 \cdot 2 \cdot x \cdot x}{5 \cdot x \cdot 2 \cdot x \cdot x} + \dfrac{3}{5 \cdot 2 \cdot x \cdot x \cdot x}$

$\dfrac{35x}{10x^3} + \dfrac{-8x^2}{10x^3} + \dfrac{3}{10x^3}$

$\quad\quad\quad \dfrac{-8x^2 + 35x + 3}{10x^3}$ ◄ *Combine the numerators and write the terms in descending order of exponents.*

## Example 4  Add $\dfrac{2a + 5}{4a} + \dfrac{a - 1}{6a^3} + \dfrac{3 - a}{3a^2}$.

$\dfrac{2a + 5}{2 \cdot 2 \cdot a} + \dfrac{a - 1}{2 \cdot 3 \cdot a \cdot a \cdot a} + \dfrac{3 - a}{3 \cdot a \cdot a}$ ◄ *At most, there are two 2's, one 3, and 3 a's*
$\quad\quad$ The LCM $= 2 \cdot 2 \cdot 3 \cdot a \cdot a \cdot a.$ $\quad$ *in any denominator.*

$\dfrac{(2a + 5)3 \cdot a \cdot a}{2 \cdot 2 \cdot a \cdot 3 \cdot a \cdot a} + \dfrac{(a - 1)2}{2 \cdot 3 \cdot a \cdot a \cdot a \cdot 2} + \dfrac{(3 - a)2 \cdot 2 \cdot a}{3 \cdot a \cdot a \cdot 2 \cdot 2 \cdot a}$

$\dfrac{(2a + 5)3a^2}{12a^3} + \dfrac{(a - 1)2}{12a^3} + \dfrac{(3 - a)4a}{12a^3}$ ◄ $3 \cdot a \cdot a = 3a^2 ; 2 \cdot 2 \cdot a = 4a$

$\dfrac{6a^3 + 15a^2}{12a^3} + \dfrac{2a - 2}{12a^3} + \dfrac{12a - 4a^2}{12a^3}$ ◄ *Use the distributive property.*

$\quad\quad\quad \dfrac{6a^3 + 11a^2 + 14a - 2}{12a^3}$ ◄ *Combine like terms in the numerator.*

## Reading in Algebra

**For each sum in the left column, find its correct match in the right column. An expression from the right column may be used more than once.**

**1.** $\dfrac{5}{x - 8} + \dfrac{7}{x + 3}$ **c**

**2.** $\dfrac{2}{x^2 - 4} + 3x$ **a**

**3.** $6 + \dfrac{5}{4x - 2}$ **a**

**4.** $\dfrac{7}{3x^2} + \dfrac{5}{4x^3} + \dfrac{7}{2x}$ **b**

a. The sum of a monomial and a rational expression

b. Each denominator is the product of an integer and a power of $x$.

c. No denominator can be factored.

# Written Exercises

16. $\dfrac{7m^2 + 3m + 4}{m^2}$

Add.

7. $\dfrac{22b - 28}{(3b - 5)(2b + 1)}$  10. $\dfrac{16x + 13}{(x - 2)(x + 3)}$  12. $\dfrac{11p^2 + 15p + 2}{6p^3}$  13. $\dfrac{-35y^2 + 18y + 2}{15y^3}$

(A)

1. $\dfrac{5}{m} + 3$  $\dfrac{5 + 3m}{m}$    2. $9 + \dfrac{4}{7b}$  $\dfrac{63b + 4}{7b}$    3. $7 + \dfrac{8}{m}$  $\dfrac{7m + 8}{m}$    4. $\dfrac{5}{x} + 1$  $\dfrac{5 + x}{x}$

5. $\dfrac{6}{a - 2} + \dfrac{5}{a - 4}$  $\dfrac{11a - 34}{(a - 2)(a - 4)}$    6. $\dfrac{7}{x + 2} + \dfrac{3}{x - 1}$  $\dfrac{10x - 1}{(x + 2)(x - 1)}$    7. $\dfrac{2}{3b - 5} + \dfrac{6}{2b + 1}$

8. $\dfrac{4}{t - 3} + \dfrac{7}{2t + 3}$  $\dfrac{15t - 9}{(t - 3)(2t + 3)}$    9. $\dfrac{2}{b - 9} + \dfrac{-3}{b + 3}$  $\dfrac{-b + 33}{(b - 9)(b + 3)}$    10. $\dfrac{9}{x - 2} + \dfrac{7}{x + 3}$

11. $\dfrac{7}{g^3} + \dfrac{3}{g} + \dfrac{5}{g^2}$  $\dfrac{3g^2 + 5g + 7}{g^3}$    12. $\dfrac{5}{2p^2} + \dfrac{11}{6p} + \dfrac{1}{3p^3}$    13. $\dfrac{6}{5y^2} + \dfrac{-7}{3y} + \dfrac{2}{15y^3}$

14. $\dfrac{2}{y} + \dfrac{3}{y^2} + 8$  $\dfrac{8y^2 + 2y + 3}{y^2}$    15. $5k + \dfrac{3}{2k}$  $\dfrac{10k^2 + 3}{2k}$    16. $\dfrac{4}{m^2} + \dfrac{3}{m} + 7$

17. $\dfrac{2}{3z} + \dfrac{7 - 4z}{6z^3} + \dfrac{3 + 2z}{4z^2}$  $\dfrac{14z^2 + z + 14}{12z^3}$    18. $\dfrac{1}{7b} + \dfrac{5 - 3b}{2b^2} + \dfrac{3b - 2}{14b^3}$  $\dfrac{-19b^2 + 38b - 2}{14b^3}$

19. $\dfrac{8 - x}{5x} + \dfrac{6 + 2x}{10x^3} + \dfrac{3 - 4x}{2x}$    20. $\dfrac{3w + 7}{3w^2} + \dfrac{w - 2}{4w^3} + \dfrac{5 - w}{6w}$

19. $\dfrac{-22x^3 + 31x^2 + 2x + 6}{10x^3}$    20. $\dfrac{-2w^3 + 22w^2 + 31w - 6}{12w^3}$

## Example    Add $x + 2 + \dfrac{5}{x - 4}$.

$$\dfrac{x + 2}{1} + \dfrac{5}{x - 4} = \dfrac{(x + 2)(x - 4)}{1(x - 4)} + \dfrac{5}{x - 4} \qquad \blacktriangleleft \ \textit{The LCM} = 1(x - 4), \textit{ or } x - 4.$$

$$(x + 2)(x - 4) \ \blacktriangleright \quad = \dfrac{x^2 - 2x - 8 + 5}{x - 4}, \textit{ or } \dfrac{x^2 - 2x - 3}{x - 4} \qquad \blacktriangleleft \ \textit{Combine numerators.}$$

(B)

21. $x + 3 + \dfrac{7}{x - 2}$  $\dfrac{x^2 + x + 1}{x - 2}$    22. $2m - 1 + \dfrac{4m}{m + 6}$  $\dfrac{2m^2 + 15m - 6}{m + 6}$    23. $\dfrac{7}{3a - 4} + a + 5$

24. $5x - 2 + \dfrac{x - 8}{2x - 3}$  $\dfrac{10x^2 - 18x - 2}{2x - 3}$    25. $\dfrac{a - 7}{2a + 5} + a - 4$  $\dfrac{2a^2 - 2a - 27}{2a + 5}$    26. $x + 6 + \dfrac{3x - 4}{x + 6}$

27. $3t - 7 + \dfrac{2t + 1}{t - 6}$  $\dfrac{3t^2 - 23t + 43}{t - 6}$    28. $\dfrac{5m + 1}{m - 4} + m + 4$  $\dfrac{m^2 + 5m - 15}{m - 4}$    29. $g + 6 + \dfrac{3g - 4}{g + 6}$

23. $\dfrac{3a^2 + 11a - 13}{3a - 4}$    26. $\dfrac{x^2 + 15x + 32}{x + 6}$    29. $\dfrac{g^2 + 15g + 32}{g + 6}$

(C)

30. $\dfrac{7}{a - 1} + \dfrac{3}{a + 4} + \dfrac{5}{a - 2}$  $\dfrac{15a^2 + 20a - 70}{(a - 1)(a + 4)(a - 2)}$    31. $\dfrac{6}{x - 5} + \dfrac{x + 4}{x - 6} + \dfrac{x}{x + 2}$  $\dfrac{2x^3 - 4x^2 - 16x - 112}{(x - 5)(x - 6)(x + 2)}$

32. $x^2 + 3x + 1 + \dfrac{4}{x - 5}$  $\dfrac{x^3 - 2x^2 - 14x - 1}{x - 5}$    33. $\dfrac{7}{b - 5} + b^2 + 2b - 3$  $\dfrac{b^3 - 3b^2 - 13b + 22}{b - 5}$

34. $x^2 + 16 + \dfrac{5}{x - 4} + \dfrac{4}{x + 4}$  $\dfrac{x^4 + 9x - 252}{(x - 4) \cdot (x + 4)}$    35. $\dfrac{7}{a - 2} + a^2 + \dfrac{4}{a + 2} + 4$  $\dfrac{a^4 + 11a - 10}{(a - 2)(a + 2)}$

# CUMULATIVE REVIEW

1. Solve: **104.6667** $0.15x - 7.2 = 8.5$.

2. Simplify: $\dfrac{4 - x}{x^2 - 5x + 4}$.    $-\dfrac{1}{x - 1}$

3. Solve:  **$\dfrac{3}{2}, -1$**  $-x - 3 = -2x^2$.

Some Special Cases of Addition

# SUBTRACTION

*Objective* **To subtract rational expressions and simplify the results**

It is possible to subtract two rational expressions by using (1) *the definition of subtraction* and (2) *the ways of interpreting the negative sign in a rational expression.* For example, the difference

$$\frac{3}{x} - \frac{7}{x} = \frac{3}{x} + \left(-\frac{7}{x}\right) \text{ by the definition of subtraction, } a - b = a + (-b);$$

$$= \frac{3}{x} + \frac{-7}{x} \text{ by } -\frac{a}{b} = \frac{-1 \cdot a}{b} = \frac{-a}{b}.$$

The subtraction exercise is now an addition exercise that you can complete easily. The process is now applied to more complicated rational expressions.

*Example 1* **Subtract** $\dfrac{x}{x^2 - 5x + 4} - \dfrac{2}{x - 4}$.

$$\frac{x}{x^2 - 5x + 4} + \frac{-1 \cdot 2}{x - 4} \qquad \blacktriangleleft \text{ Rewrite by using } a - b = a + (-b)$$
$$\text{and } -\frac{a}{b} = \frac{-1 \cdot a}{b}.$$

$$\frac{x}{(x - 4)(x - 1)} + \frac{-2}{x - 4} \qquad \blacktriangleleft \text{ Factor } x^2 - 5x + 4.$$

$$\frac{x}{(x - 4)(x - 1)} + \frac{-2(x - 1)}{(x - 4)(x - 1)} \qquad \blacktriangleleft \text{ The LCM} = (x - 4)(x - 1).$$

$$\frac{x}{(x - 4)(x - 1)} + \frac{-2x + 2}{(x - 4)(x - 1)} \qquad \blacktriangleleft -2(x - 1) = -2x + 2$$

$$\frac{1x - 2x + 2}{(x - 4)(x - 1)} = \frac{-x + 2}{(x - 4)(x - 1)} \qquad \blacktriangleleft \text{ Combine like terms in the numerator.}$$

*Example 2* **Subtract** $5 - \dfrac{2a - 1}{3a}$.

$$\frac{5}{1} + \frac{-1(2a - 1)}{3a} \qquad \blacktriangleleft \text{ Rewrite by using } a - b = a + (-b)$$
$$\text{and } -\frac{a}{b} = \frac{-1 \cdot a}{b}.$$

$$\frac{5}{1} + \frac{-2a + 1}{3 \cdot a}$$

$$\frac{5 \cdot 3 \cdot a}{1 \cdot 3 \cdot a} + \frac{-2a + 1}{3 \cdot a}$$

$$\frac{15a}{3a} + \frac{-2a + 1}{3a}$$

$$\frac{15a - 2a + 1}{3a} = \frac{13a + 1}{3a}$$

In subtraction, as in addition, it is sometimes possible to reduce the result to simpler terms. Such is the case in the next example.

*Example 3*    **Subtract $\dfrac{a^2 - 22}{a^2 - 9a + 20} - \dfrac{a - 2}{a - 5}$. Simplify the result.**

$\dfrac{a^2 - 22}{a^2 - 9a + 20} + \dfrac{-1(a - 2)}{a - 5}$   ◀ *Rewrite by using $a - b = a + (-b)$ and $-\dfrac{a}{b} = \dfrac{-1 \cdot a}{b}$.*

$\dfrac{a^2 - 22}{(a - 5)(a - 4)} + \dfrac{-1a + 2}{a - 5}$

$\dfrac{a^2 - 22}{(a - 5)(a - 4)} + \dfrac{(-1a + 2)(a - 4)}{(a - 5)(a - 4)}$

$\dfrac{a^2 - 22}{(a - 5)(a - 4)} + \dfrac{-1a^2 + 6a - 8}{(a - 5)(a - 4)}$   ◀ $(-1a + 2)(a - 4) = -1a^2 + 6a - 8$

$\dfrac{a^2 - 22 - 1a^2 + 6a - 8}{(a - 5)(a - 4)}$

$\dfrac{6a - 30}{(a - 5)(a - 4)}$

$\dfrac{\overset{1}{\cancel{6(a - 5)}}}{\underset{1}{\cancel{(a - 5)}}(a - 4)}$   ◀ *Factor $6a - 30$. Then divide out common factors.*

$\dfrac{6}{a - 4}$

## Oral Exercises

**Rename each subtraction as an addition.**

1. $\dfrac{4}{a^2 - 25} - \dfrac{2}{a - 5}$
   $\dfrac{4}{a^2 - 25} + \dfrac{-1(2)}{a - 5}$

2. $\dfrac{7}{x^2 - 5x + 4} - \dfrac{3}{x - 4}$
   $\dfrac{7}{x^2 - 5x + 4} + \dfrac{-1(3)}{x - 4}$

3. $\dfrac{7}{a^2 - 3a} - \dfrac{-2}{a - 3}$
   $\dfrac{7}{a^2 - 3a} + \dfrac{-1(-2)}{a - 3}$

4. $\dfrac{7}{b^2 - 4} - \dfrac{b + 3}{b - 2}$
   $\dfrac{7}{b^2 - 4} + \dfrac{-1(b + 3)}{b - 2}$

5. $\dfrac{6}{t^2 - 7t + 10} - \dfrac{t + 2}{t - 5}$
   $\dfrac{6}{t^2 - 7t + 10} + \dfrac{-1(t + 2)}{t - 5}$

6. $\dfrac{y}{y^2 - 8y + 7} - \dfrac{y - 2}{y - 1}$
   $\dfrac{y}{y^2 - 8y + 7} + \dfrac{-1(y - 2)}{y - 1}$

7. $\dfrac{2}{b^2 - 4b + 3} - \dfrac{6 + b}{b - 3}$
   $\dfrac{2}{b^2 - 4b + 3} + \dfrac{-1(6 + b)}{b - 3}$

8. $\dfrac{7}{x^2 - x} - \dfrac{-4}{x - 1}$
   $\dfrac{7}{x^2 - x} + \dfrac{-1(-4)}{x - 1}$

9. $\dfrac{g + 5}{g^2 - 4} - \dfrac{4g - 1}{g + 2}$
   $\dfrac{g + 5}{g^2 - 4} + \dfrac{-1(4g - 1)}{g + 2}$

10. $\dfrac{k}{k^2 - 9k + 14} - \dfrac{k + 3}{k - 2}$
    $\dfrac{k}{k^2 - 9k + 14} + \dfrac{-1(k + 3)}{k - 2}$

11. $\dfrac{2m}{m + 4} - \dfrac{m - 3}{m^2 - 16}$
    $\dfrac{2m}{m + 4} + \dfrac{-1(m - 3)}{m^2 - 16}$

12. $\dfrac{3a}{a^2 - 5a} - \dfrac{3 + a}{a}$
    $\dfrac{3a}{a^2 - 5a} + \dfrac{-1(3 + a)}{a}$

## Written Exercises

**Subtract. Simplify, if possible.**

(A) 1. $\dfrac{5}{x^2 - 25} - \dfrac{3}{x - 5}$

2. $\dfrac{7}{a^2 - 8a + 12} - \dfrac{4}{a - 6}$

3. $\dfrac{6}{b^2 - 3b + 2} - \dfrac{-8}{b - 2}$

4. $7 - \dfrac{2x - 8}{3x}$   $\dfrac{19x + 8}{3x}$

5. $3 - \dfrac{3b - 1}{2b}$   $\dfrac{3b + 1}{2b}$

6. $6 - \dfrac{2y + 5}{y - 2}$   $\dfrac{4y - 17}{y - 2}$

1. $\dfrac{-3x - 10}{(x - 5)(x + 5)}$

2. $\dfrac{15 - 4a}{(a - 6)(a - 2)}$

3. $\dfrac{8b - 2}{(b - 2)(b - 1)}$

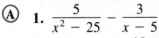

Subtraction

**249**

**7.** $\dfrac{4x + 22}{(x - 4)(x + 4)}$    **8.** $\dfrac{-w + 2}{(w - 3)(w - 1)}$    **9.** $\dfrac{-4b - 14}{(b - 6)(b + 4)}$    **10.** $\dfrac{2k + 19}{(2k - 1)(2k + 3)}$    **11.** $\dfrac{-2l - 22}{(3l + 5)(5l - 1)}$

**7.** $\dfrac{5}{x - 4} - \dfrac{x - 2}{x^2 - 16}$    **8.** $\dfrac{w}{w^2 - 4w + 3} - \dfrac{2}{w - 3}$    **9.** $\dfrac{2}{b^2 - 2b - 24} - \dfrac{4}{b - 6}$

**10.** $\dfrac{5}{2k - 1} - \dfrac{4}{2k + 3}$    **11.** $\dfrac{2}{3l + 5} - \dfrac{4}{5l - 1}$    **12.** $\dfrac{7}{m - 4} - \dfrac{3}{m + 5}\ \dfrac{4m + 47}{(m - 4)(m + 5)}$

**13.** $\dfrac{3}{8y^2} - \dfrac{5}{12y^3}\ \dfrac{9y - 10}{24y^3}$    **14.** $\dfrac{6}{5x^2} - \dfrac{-3}{10x}\ \dfrac{12 + 3x}{10x^2}$    **15.** $\dfrac{4}{9m^2} - \dfrac{5}{3m}\ \dfrac{4 - 15m}{9m^2}$

**16.** $\dfrac{6}{x - 4}$    **17.** $\dfrac{1}{m - 2}$    **19.** $\dfrac{t - 6}{(t - 3)(t - 2)}$    **22.** $\dfrac{-3a^2 - a + 17}{(a + 3)(a + 2)}$    **23.** $-\dfrac{y^2 + 2y + 4}{(y - 3)(y + 3)}$    **24.** $\dfrac{n^2 + 15n - 25}{(2n + 3)(n - 5)}$

**(B) 16.** $\dfrac{x^2 + 2}{x^2 - 5x + 4} - \dfrac{x - 2}{x - 1}$    **17.** $\dfrac{m^2 - 8}{m^2 - 8m + 12} - \dfrac{m + 1}{m - 6}$    **18.** $\dfrac{y^2 + 5}{y^2 - 5y} - \dfrac{y + 3}{y}\ \dfrac{2y + 20}{y(y - 5)}$

**19.** $\dfrac{t^2 - 8}{t^2 - 5t + 6} - \dfrac{t + 1}{t - 3}$    **20.** $\dfrac{a^2 + 9a - 49}{a^2 - 7a} - \dfrac{a + 2}{a - 7}\ \dfrac{7}{a}$    **21.** $\dfrac{x^2 + 3x - 10}{x^2 - 10x + 16} - \dfrac{x + 3}{x - 8}\ \dfrac{2}{x - 8}$

**22.** $\dfrac{7}{a + 3} - \dfrac{3a - 1}{a + 2}$    **23.** $\dfrac{2y - 1}{y^2 - 9} - \dfrac{y + 1}{y - 3}$    **24.** $\dfrac{3n^2}{2n^2 - 7n - 15} - \dfrac{2n - 5}{2n + 3}$

**25.** $\dfrac{y}{y^2 + y - 6} - \dfrac{y - 1}{y^2 - y - 12}$    **26.** $\dfrac{4}{b^2 - 9} - \dfrac{2b + 1}{b^2 - 3b}$    **27.** $\dfrac{x}{x^2 - 5x + 4} - \dfrac{x + 5}{x^2 - 2x - 8}$

**28.** $\dfrac{p + 2}{p + 1} - \dfrac{p - 1}{p^2 - 6p - 7}$    **29.** $\dfrac{3b + 1}{b} - \dfrac{2b + 1}{b^2 - 3b}$    **30.** $\dfrac{6}{m^2 - 8m + 15} - \dfrac{2m - 5}{m^2 - 3m}$

**25.** $\dfrac{-y - 2}{(y + 3)(y - 2)(y - 4)}$    **26.** $-\dfrac{2b^2 + 3b + 3}{b(b - 3)(b + 3)}$    **27.** $\dfrac{-2x + 5}{(x - 1)(x - 4)(x + 2)}$    **28.** $\dfrac{p^2 - 6p - 13}{(p - 7)(p + 1)}$

**31.** $\dfrac{x - 3}{x + 6} - \dfrac{x - 7}{2x - 5}\ \dfrac{x^2 - 10x + 57}{(x + 6)(2x - 5)}$    **32.** $\dfrac{x - 5}{2x - 5} - \dfrac{2x + 1}{x - 7}\ \dfrac{-3x^2 - 4x + 40}{(2x - 5)(x - 7)}$

**(C) 33.** $\dfrac{x^2}{x^4 - 17x^2 + 16} - \dfrac{6}{x^2 + 3x - 4}$    **34.** $\dfrac{2a^2}{a^4 - 13a^2 + 36} - \dfrac{3a^2}{a^4 - 10a^2 + 9}$

**35.** $\dfrac{5}{xy + ay - 2x - 2a} - \dfrac{a + 4}{y^2 - 3y + 2}$    **36.** $\dfrac{4}{xy + y^2 - bx - by} - \dfrac{7}{y^2 - b^2}$

**29.** $\dfrac{3b^2 - 10b - 4}{b(b - 3)}$    **30.** $-\dfrac{2m^2 - 21m + 25}{m(m - 5)(m - 3)}$    **33.** $\dfrac{-5x^2 + 18x + 24}{(x - 4)(x + 4)(x + 1)(x - 1)}$    **34.** $\dfrac{-a^4 + 10a^2}{(a^2 - 9)(a^2 - 4)(a^2 - 1)}$

## CUMULATIVE REVIEW

**35.** $\dfrac{5y - 5 - ax - a^2 - 4x - 4a}{(y - 2)(x + a)(y - 1)}$    **36.** $\dfrac{4b - 7x - 3y}{(y - b)(x + y)(y + b)}$

### Solve each problem.

**1.** The length of a rectangle is 4 m more than 3 times the width. The perimeter is 24 m. Find the length and the width of the rectangle. **10, 2**

**2.** The larger of two numbers is 6 more than twice the smaller number. The sum of the two numbers is 12. Find each number.

**2, 10**

---

## NON-ROUTINE PROBLEMS

When simplified, the product,

$$\left(1 - \frac{1}{3}\right)\left(1 - \frac{1}{4}\right)\left(1 - \frac{1}{5}\right)\left(1 - \frac{1}{6}\right) \cdots \left(1 - \frac{1}{n}\right),$$ becomes which of the following? **b**

   a. $\dfrac{1}{n}$    b. $\dfrac{2}{n}$    c. $\dfrac{2(n - 1)}{n}$    d. $\dfrac{3}{n(n + 1)}$

# SIMPLIFYING: THE −1 TECHNIQUE 9.6

*Objective*　**To simplify sums and differences of rational expressions by using the −1 technique**

It will be easier to simplify an expression such as $\dfrac{6x}{x^2-49}+\dfrac{3}{7-x}$

if you first rewrite the denominator, $7-x$, in *convenient form*. The procedure for doing this is reviewed in Example 1.

*Example 1*　**Simplify** $\dfrac{6x}{x^2-49}+\dfrac{3}{7-x}$.

1. Rewrite $7-x$ in convenient form.

$$\dfrac{6x}{x^2-49}+\dfrac{3}{-1(x-7)} \quad \blacktriangleleft 7-x=-x+7=-1(x-7)$$

2. Rewrite $\dfrac{3}{-1(x-7)}$ so that the factor $-1$ does not appear in the denominator.

$$\dfrac{6x}{x^2-49}+\dfrac{-1\cdot 3}{x-7} \quad \blacktriangleleft Use\ \dfrac{a}{-1b}=\dfrac{-1\cdot a}{b}.$$

3. Add the rational expressions.

$$\dfrac{6x}{(x-7)(x+7)}+\dfrac{-3}{x-7}$$

$$\dfrac{6x}{(x-7)(x+7)}+\dfrac{-3(x+7)}{(x-7)(x+7)} \quad \blacktriangleleft The\ LCM=(x-7)(x+7).$$

$$\dfrac{6x}{(x-7)(x+7)}+\dfrac{-3x-21}{(x-7)(x+7)}$$

$$\dfrac{6x-3x-21}{(x-7)(x+7)}$$

$$\dfrac{3x-21}{(x-7)(x+7)}$$

$$\dfrac{3\overset{1}{(\cancel{x-7})}}{\underset{1}{(\cancel{x-7})}(x+7)}$$

$$\dfrac{3}{x+7} \quad \blacktriangleleft The\ result\ is\ in\ simplest\ form.$$

You will find it easier to simplify an expression such as

$$\dfrac{-24}{a^2-7a+10}-\dfrac{a+3}{5-a}$$

if you first rewrite it as addition and then rewrite the denominator, $5-a$, in convenient form. This procedure is used in Example 2.

**Example 2**   **Simplify** $\dfrac{-24}{a^2 - 7a + 10} - \dfrac{a + 3}{5 - a}$.

$\dfrac{-24}{a^2 - 7a + 10} + \dfrac{-1(a + 3)}{5 - a}$  ◀ *Rewrite by using* $a - b = a + (-b)$ *and* $-\dfrac{a}{b} = \dfrac{-1 \cdot a}{b}$.

$\dfrac{-24}{a^2 - 7a + 10} + \dfrac{-1(a + 3)}{-1(a - 5)}$  ◀ *Rewrite* $5 - a$ *in convenient form:* $5 - a = -a + 5 = -1(a - 5)$.

$\dfrac{-24}{(a - 5)(a - 2)} + \dfrac{a + 3}{a - 5}$  ◀ $\dfrac{-1(a + 3)}{-1(a - 5)} = \dfrac{1(a + 3)}{1(a - 5)} = \dfrac{a + 3}{a - 5}$; *neg. ÷ neg. = pos.*

$\dfrac{-24}{(a - 5)(a - 2)} + \dfrac{(a + 3)(a - 2)}{(a - 5)(a - 2)}$  ◀ *The LCM* $= (a - 5)(a - 2)$.

$\dfrac{-24}{(a - 5)(a - 2)} + \dfrac{a^2 + a - 6}{(a - 5)(a - 2)}$  ◀ $(a + 3)(a - 2) = a^2 + a - 6$

$\dfrac{a^2 + a - 30}{(a - 5)(a - 2)} = \dfrac{(a + 6)\overset{1}{(a - 5)}}{\underset{1}{(a - 5)}(a - 2)}$, *or* $\dfrac{a + 6}{a - 2}$

*Tues 3/4*

# Written Exercises

Ⓐ  **Simplify.**

5. $\dfrac{-11}{(b - 6)(b + 5)}$  6. $\dfrac{-11y + 17}{5(y - 2)}$  11. $\dfrac{2m + 4}{(m - 5)(m - 3)}$  16. $\dfrac{8a + 10}{(a + 8)(a - 2)}$

19. $\dfrac{m^2 - 3m - 12}{(m + 6)(m - 4)(m - 6)}$  20. $\dfrac{-3b^2 + b + 24}{b(b - 5)(b + 4)}$  21. $\dfrac{6a^2 + 20a - 1}{(3a + 1)(a - 5)}$  23. $\dfrac{5a^2 + 25a - 1}{(a - 6)(a + 6)(a + 5)}$

1. $\dfrac{4x}{x^2 - 4} + \dfrac{2}{2 - x}$  $\dfrac{2}{x + 2}$

2. $\dfrac{4a}{a^2 - 25} + \dfrac{2}{5 - a}$  $\dfrac{2}{a + 5}$

3. $\dfrac{4b + 2}{b^2 - 2b - 8} + \dfrac{3}{4 - b}$  $\dfrac{1}{b + 2}$

4. $\dfrac{10a}{a^2 - 9} + \dfrac{5}{3 - a}$  $\dfrac{5}{a + 3}$

5. $\dfrac{2b - 1}{b^2 - b - 30} + \dfrac{2}{6 - b}$

6. $\dfrac{4y - 3}{5y - 10} + \dfrac{3y - 4}{2 - y}$

7. $\dfrac{2a}{a^2 - 6a + 8} - \dfrac{a}{2 - a}$  $\dfrac{a}{a - 4}$

8. $\dfrac{-9b - 27}{b^2 - 3b - 18} - \dfrac{b}{6 - b}$  $\dfrac{b - 9}{b - 6}$

9. $\dfrac{-2y - 3}{y^2 - 3y} - \dfrac{y}{3 - y}$  $\dfrac{y + 1}{y}$

10. $\dfrac{-7a}{a^2 - 3a - 10} - \dfrac{a}{5 - a}$  $\dfrac{a}{a + 2}$

11. $\dfrac{7}{m - 5} + \dfrac{5}{3 - m}$

12. $\dfrac{7}{m} - \dfrac{4 - m}{5m - m^2}$  $\dfrac{31 - 6m}{m(5 - m)}$

13. $\dfrac{3}{x - 1} + \dfrac{3x + 2}{x - x^2} + \dfrac{3}{x}$  $\dfrac{-5 + 3x}{x(x - 1)}$

14. $\dfrac{5}{a - 2} + \dfrac{3a - 2}{2a - a^2} + \dfrac{4}{a}$  $\dfrac{6a - 6}{a(a - 2)}$

15. $\dfrac{x^2 + 3x + 15}{x^2 + 5x - 24} - \dfrac{2}{3 - x} + \dfrac{5}{x + 8}$  $\dfrac{x + 2}{x - 3}$

16. $\dfrac{3}{a + 8} - \dfrac{2}{2 - a} + \dfrac{3a}{a^2 + 6a - 16}$

Ⓑ  17. $\dfrac{-2b - 13}{b^2 - 11b + 28} - \dfrac{b + 2}{7 - b}$  $\dfrac{b + 3}{b - 4}$

18. $\dfrac{4y - 22}{y^2 - 11y + 18} - \dfrac{y + 7}{9 - y}$  $\dfrac{y^2 + 9y - 36}{(y - 9)(y - 2)}$

19. $\dfrac{1}{m + 6} - \dfrac{3}{36 - m^2} + \dfrac{4}{m^2 + 2m - 24}$

20. $\dfrac{6}{b^2 - 5b} - \dfrac{2b + 1}{b^2 - b - 20} + \dfrac{1}{5 - b}$

21. $\dfrac{2a - 1}{3a^2 - 14a - 5} - \dfrac{2a + 5}{5 - a} + \dfrac{1}{3a + 1}$

22. $\dfrac{3a - 1}{a^2 - 49} - \dfrac{3a + 2}{14 + 5a - a^2}$  $\dfrac{6a^2 + 28a + 12}{(a - 7)(a + 7)(a + 2)}$

Ⓒ  23. $\dfrac{7}{a^2 - 36} - \dfrac{4a - 3}{30 + a - a^2} + \dfrac{a + 3}{a^2 + 11a + 30}$

24. $\dfrac{x^2 - 3x + 1}{x^2 - 4} - \dfrac{x^2 + 2x - 4}{2 - x} + \dfrac{3x - 4}{x + 2}$

25. $\dfrac{6}{50 + 23x^2 - x^4} - \dfrac{3}{x^3 - 5x^2 + 2x - 10}$

26. $\dfrac{2a - 3}{a^3 - 12a^2 + a - 12} + \dfrac{4a + 1}{144 + 143a^2 - a^4}$

24. $\dfrac{x^3 + 8x^2 - 13x + 1}{(x - 2)(x + 2)}$    25. $\dfrac{-3x - 21}{(x^2 + 2)(x - 5)(x + 5)}$

$\dfrac{2a^2 + 17a - 37}{(a^2 + 1)(a - 12)(a + 12)}$

**252**

# DIVIDING BY A MONOMIAL

**Objectives**    **To divide a monomial by a monomial**
                    **To divide a polynomial by a monomial**

Recall that $3\overline{)12}$ means $\dfrac{12}{3}$. Therefore, to find the quotient in $x^3\overline{)x^7}$, you can rewrite it as

$$\frac{x^7}{x^3}, \text{ which is equal to } x^{7-3}, \text{ or } x^4,$$

by the quotient of powers property, $\dfrac{a^m}{a^n} = a^{m-n}$, if $m > n$. However, you do not have to rewrite the division using the fraction bar. You can proceed as follows:    Think: $x^3\overline{)\overset{x^{7-3}}{x^7}}$ , or $x^3\overline{)\overset{x^4}{x^7}}$.

It is easy to check the result by multiplying: $x^3 \cdot x^4 = x^7$.

**Example 1**    **Divide.**

| | |
|---|---|
| $a^3\overline{)a^9}$ | $b\overline{)b^5}$ |
| $a^3\overline{)\overset{a^{9-3}}{a^9}}$ | $b^1\overline{)\overset{b^{5-1}}{b^5}}$ . |
| **So,** the quotient is $a^6$. | **So,** the quotient is $b^4$. |

The process can be extended to monomials with numerical coefficients other than 1. Since $3x^2\overline{)15x^6} = \dfrac{15x^6}{3x^2} = \dfrac{15 \cdot x^6}{3 \cdot x^2} = \dfrac{15}{3} \cdot \dfrac{x^6}{x^2} = 5 \cdot x^4$, you can think $3\overline{)\overset{5}{15}}$ and $x^2\overline{)\overset{x^4}{x^6}}$, but you would write $3x^2\overline{)\overset{5x^4}{15x^6}}$.

*Checking*, $3x^2 \cdot 5x^4 = 3 \cdot 5 \cdot x^2 \cdot x^4 = 15x^6$.

**Example 2**    **Divide $4a^2\overline{)-32a^5}$. Check the result.**

$4a^2\overline{)\overset{-8a^3}{-32a^5}}$  ◀  *Think:* $4\overline{)\overset{-8}{-32}}$ and $a^2\overline{)\overset{a^{5-2}}{a^5}}$ or $a^2\overline{)\overset{a^3}{a^5}}$.

*Check.* $4a^2(-8a^3) = 4(-8)a^2 \cdot a^3 = -32a^5$

**So,** the quotient is $-8a^3$.

In a division such as $3x^2\overline{)21x^2}$, the powers of $x$ are the same. You can think of simplifying the related rational expression, as shown in Example 3.

**Example 3**   **Divide $\dfrac{21x^2}{3x^2}$**

*Think:* $\dfrac{21x^2}{3x^2} = \dfrac{\overset{7}{\cancel{21}}}{\underset{1}{\cancel{3}}} \cdot \dfrac{\overset{1}{\cancel{x^2}}}{\underset{1}{\cancel{x^2}}} = \dfrac{7}{1}$, or 7.

**So,** $\dfrac{21x^2}{3x^2} = 7$.   ◀ *The quotient is 7.*

This suggests that for each nonzero number $a$, $\dfrac{a^m}{a^n} = 1$, if $m = n$.

The addition property for rational expressions allows you to write $\dfrac{15n^4 + 30n^3 - 10n^2}{5n^2}$ as $\dfrac{15n^4}{5n^2} + \dfrac{30n^3}{5n^2} + \dfrac{-10n^2}{5n^2}$.
Notice that you must divide each term of $15n^4 + 30n^3 - 10n^2$ by $5n^2$.

**Example 4**   **Divide $(15n^4 + 30n^3 - 10n^2) \div 5n^2$. Check.**

*Think:* $\dfrac{15n^4}{5n^2} + \dfrac{30n^3}{5n^2} - \dfrac{10n^2}{5n^2}$

$$5n^2\overline{)15n^4 + 30n^3 - 10n^2}\ ^{\displaystyle 3n^2 + 6n^1 - 2}$$   ◀ *Divide by $5n^2$.*

Check. $5n^2(3n^2 + 6n^1 - 2) = 15n^4 + 30n^3 - 10n^2$   ◀ *Distribute the $5n^2$.*
**So,** the quotient is $3n^2 + 6n^1 - 2$, or $3n^2 + 6n - 2$.

## Written Exercises

**Divide.**

(A)  **1.** $a^8 \div a^5$  **$a^3$**   **2.** $b^{13} \div b^4$  **$b^9$**   **3.** $m^7 \div m^6$  **$m$**   **4.** $a^9 \div a$  **$a^8$**

**5.** $24a^4 \div 6a^2$  **$4a^2$**   **6.** $21m^3 \div -7m$  **$-3m^2$**  **7.** $-8b^4 \div 2b$  **$-4b^3$**   **8.** $15x^4 \div 3x$  **$5x^3$**

**9.** $-24a^2 \div 4a^2$  **$-6$**   **10.** $35b^7 \div 5b^3$  **$7b^4$**   **11.** $-35x \div -7x$  **$5$**   **12.** $45b \div 5b$  **$9$**

**13.** $45a \div 3a$  **$15$**   **14.** $4x^5 \div x$  **$4x^4$**   **15.** $-48p^2 \div 6p^2$  **$-8$**  **16.** $-25b^5 \div -5b^4$  **$5b$**

**17.** $(9a^2 - 12a) \div 3a$  **$3a - 4$**   **18.** $(12x^2 - 16x) \div 4x$  **$3x - 4$**

**19.** $(4m^2 - 16m) \div 2m$  **$2m - 8$**   **20.** $(18p^3 - 36p^2) \div 3p^2$  **$6p - 12$**

**21.** $(20n^3 - 15n^2 + 15n) \div 5n$  **$4n^2 - 3n + 3$**  **22.** $(12a^3 - 24a^2 - 8a) \div 4a$  **$3a^2 - 6a - 2$**

**23.** $(18a^8 - 12a^6 + 30a^4) \div 6a^4$  **$3a^4 - 2a^2 + 5$**  **24.** $(9b^4 - 12b^3 - 3b^2) \div 3b^2$  **$3b^2 - 4b - 1$**

(B)  **25.** $(p^5 - p^4 + 2p^3 - 5p^2) \div p^2$  **$p^3 - p^2 + 2p - 5$**   **26.** $(4x^5 - 8x^4 - 12x^3 + 6x^2) \div -2x^2$  **$-2x^3 + 4x^2 + 6x - 3$**

**27.** $(p^5 - 4p^4 - 3p^2 + 5p) \div p$   **28.** $(48r^4 + 32r^3 - 64r^2 - 112r) \div -8r$

**29.** $(15g^7 - 45g^6 - 18g^5 - 105g^4) \div 3g^3$  **$5g^4 - 15g^3 - 6g^2 - 35g$**   **30.** $(32g^6 - 108g^5 - 48g^4 - 66g^3) \div 2g$  **$16g^5 - 54g^4 - 24g^3 - 33g^2$**

(C)  **31.** $(x^{3m} - x^{2m} + x^m) \div x^m$  **$x^{2m} - x^m + 1$**   **32.** $(x^3y^2 - x^2y^2 + xy^3) \div xy$  **$x^2y - xy + y^2$**

**27.** $p^4 - 4p^3 - 3p + 5$   **28.** $-6r^3 - 4r^2 + 8r + 14$

# DIVIDING BY A BINOMIAL 9.8

*Objective*  **To divide a polynomial by a binomial**

The procedure for dividing 785 by 25 is reviewed below. The comments in parentheses indicate the thinking process for each step.

$$
\begin{array}{r}
31 \\
25\overline{)785} \\
75\!\downarrow \\
\overline{\phantom{0}35} \\
25 \\
\overline{\phantom{0}10}
\end{array}
$$

(1. *Divide:* $2\overline{)7}$.)
(2. *Multiply:* $3 \cdot 25$.)
(3. *Subtract:* $78 - 75$.)

(4. *"Bring down"* the next digit, 5.)
(5. *Divide:* $2\overline{)3}$.)
(6. *Multiply:* $1 \cdot 25$.)
(7. *Subtract:* $35 - 25$. The *remainder* is 10.)

(8. *Check.*)   $31 \cdot 25 = 775$ and $775 + 10 = 785$

**So,** the quotient is 31 and the remainder is 10.

You follow the same procedure to divide a polynomial by a binomial. The process is illustrated, by steps, in Example 1.

*Example 1*  **Divide $(x^2 + 10x + 21) \div (x + 3)$. Check.**

*Step 1*

$$
\begin{array}{r}
x^1 \phantom{+ 10x + 21} \\
x + 3\overline{)x^2 + 10x + 21} \\
\underline{x^2 + \phantom{0}3x} \\
7x
\end{array}
$$

◀ *Divide:* $x\overline{)x^2}$; $x^2$ divided by x is $x^1$.
◀ *Multiply:* $x(x + 3) = x^2 + 3x$.
◀ *Subtract:* $x^2 + 10x - (x^2 + 3x) = 7x$.

*Step 2* "Bring down" the next term of the *dividend*, 21.

$$
\begin{array}{r}
x^1 + \phantom{0}7 \phantom{1} \\
x + 3\overline{)x^2 + 10x + 21} \\
\underline{x^2 + \phantom{0}3x} \\
7x + 21 \\
\underline{7x + 21} \\
0
\end{array}
$$

◀ *Divide:* $x\overline{)7x}$
◀ *Multiply:* $7(x + 3) = 7x + 21$.
◀ *Subtract:* $7x + 21 - (7x + 21) = 0$. The remainder is 0.

*Check.*  $(x + 7)(x + 3) + 0 = x^2 + 10x + 21$   ◀ $(x + 7)(x + 3)$

(quotient)(divisor) + remainder = dividend

**Thus,** the quotient is $x + 7$, and the remainder is 0.

In the next example, the remainder is a number other than 0. A more compact form for arranging your work is shown to the right of the completed division.

*Example 2*    **Divide $(6a^2 - 7a + 5) \div (2a - 5)$.**

*Step 1*
$$2a - 5 \overline{)6a^2 - 7a + 5} \quad \overset{3a}{}$$

$$\underline{6a^2 - 15a}$$

$$8a$$

◀ *Divide: $2a\overline{)6a^2}$.*
◀ *Multiply: $3a(2a - 5) = 6a^2 - 15a$.*
◀ *Subtract: $6a^2 - 7a - (6a^2 - 15a) = -7a - (-15a) = 8a$.*

*Step 2*
$$2a - 5 \overline{)6a^2 - 7a + 5} \quad \overset{3a + 4}{}$$

$$\underline{6a^2 - 15a} \qquad \downarrow$$

$$8a + 5$$

$$\underline{8a - 20}$$

$$25$$

◀ *Divide: $2a\overline{)8a}$.*
◀ *Multiply: $4(2a - 5) = 8a - 20$.*
◀ *Subtract: $8a + 5 - (8a - 20) = 25$.*

A more compact form:
$$2a - 5 \overline{)6a^2 - 7a + 5} \quad \overset{3a + 4}{}$$
$$\underline{6a^2 - 15a}$$
$$8a + 5$$
$$\underline{8a - 20}$$
$$25$$

*Check on your own.* ▶ **So,** the quotient is $3a + 4$, and the remainder is 25.

Sometimes, the polynomial in the dividend has a term missing. For example, the $x^2$ term is missing in the dividend $6x^3 - 15x + 18$. The next example shows you how to insert this missing term by using 0 as a coefficient.

*Example 3*    **Divide $(6x^3 - 15x + 18) \div (3x + 6)$.**

*First,* rewrite the dividend, $6x^3 - 15x + 18$, so that all terms appear. Write $0x^2$ between $6x^3$ and $-15x$. Then, divide.

$$3x + 6 \overline{)6x^3 + 0x^2 - 15x + 18} \quad \overset{2x^2}{}$$
$$\underline{6x^3 + 12x^2}$$
$$-12x^2$$

◀ *Divide: $3x\overline{)6x^3}$*
◀ *Multiply: $2x^2(3x + 6) = 6x^3 + 12x^2$.*
◀ *Subtract: $6x^3 + 0x^2 - (6x^3 + 12x^2)$*
  $= 0x^2 - (12x^2) = -12x^2$

$$3x + 6 \overline{)6x^3 + 0x^2 - 15x + 18} \quad \overset{2x^2 - 4x}{}$$
$$\underline{6x^3 + 12x^2} \qquad \downarrow$$
$$-12x^2 - 15x$$
$$\underline{-12x^2 - 24x}$$
$$9x$$

◀ *Divide: $3x\overline{)-12x^2}$.*
◀ *Multiply: $-4x(3x + 6) = -12x^2 - 24x$.*
◀ *Subtract: $-12x^2 - 15x - (-12x^2 - 24x)$*
  $= -15x - (-24x) = 9x$.

$$3x + 6 \overline{)6x^3 + 0x^2 - 15x + 18} \quad \overset{2x^2 - 4x + 3}{}$$
$$\underline{6x^3 + 12x^2}$$
$$-12x^2 - 15x$$
$$\underline{-12x^2 - 24x}$$
$$9x + 18$$
$$\underline{9x + 18}$$
$$0$$

◀ *Divide: $3x\overline{)9x}$.*

◀ *Subtract: $9x + 18$*
  $- (9x + 18) = 0$.

A more compact form:
$$3x + 6 \overline{)6x^3 + 0x^2 - 15x + 18} \quad \overset{2x^2 - 4x + 3}{}$$
$$\underline{6x^3 + 12x^2}$$
$$-12x^2 - 15x$$
$$\underline{-12x^2 - 24x}$$
$$9x + 18$$
$$\underline{9x + 18}$$
$$0$$

**So,** the quotient is $2x^2 - 4x + 3$.

## Reading in Algebra

**Refer to the worked-out example at the right for Exercises 1–6.**
1. How was the $b$ in the quotient obtained? **$2b^2$ is divided by $2b$**
2. How was the $14b - 15$ obtained?
**by subtracting $(2b^2 - 3b)$ from $2b^2 + 11b - 15$**
**Match each expression on the left with its correct name on the right.**
3. $b + 7$ **b**        5. $6$ **a**        a. remainder    c. divisor
4. $2b^2 + 11b - 15$ **d**  6. $2b - 3$ **c**   b. quotient    d. dividend

$$\begin{array}{r} b + 7 \\ 2b - 3\overline{)2b^2 + 11b - 15} \\ \underline{2b^2 - 3b} \\ 14b - 15 \\ \underline{14b - 21} \\ 6 \end{array}$$

## Oral Exercises

**What missing term should be inserted in each dividend?**    **4. 0x    5. 0b³**
1. $x + 5\overline{)x^3 + 3x - 40}$ **0x²**    2. $x - 6\overline{)x^2 - 12}$ **0x**    3. $2a - 1\overline{)4a^3 - 8a^2 + 10}$ **0a**
4. $3x - 2\overline{)15x^3 + 4x^2 - 5}$    5. $2b - 1\overline{)6b^4 - 5b^2 + 2b - 8}$    6. $3y - 1\overline{)12y^3 - 9y^2 + 7}$ **0y**
7. $(x^4 - 3x^2 + 7x - 2) \div (x - 6)$ **0x³**    8. $(m^4 - 7m^2 + 2m - 1) \div (m - 1)$ **0m³**
9. $(4k^3 - 8k + 6) \div (2k - 1)$ **0k²**

## Written Exercises

1. $x - 2$    2. $x - 9$    3. $x + 8$    *Thus*
**Divide. Check.** 4. $3x + 1$    5. $3x + 1$    6. $2b - 5$

(A)
1. $(x^2 - 6x + 8) \div (x - 4)$    2. $(x^2 - 6x - 27) \div (x + 3)$    3. $(x^2 + 3x - 40) \div (x - 5)$
4. $(3x^2 - 5x - 2) \div (x - 2)$    5. $(3x^2 - 8x - 3) \div (x - 3)$    6. $(2b^2 + 3b - 20) \div (b + 4)$
7. $(3y^2 + 7y - 4) \div (y + 3)$    8. $(6t^2 - 13t + 2) \div (3t - 2)$    9. $(8m^2 - 30m + 7) \div (4m - 1)$
    $3y - 2, r2$                        $2t - 3, r-4$                        $2m - 7$

**Divide.**
10. $(6a^2 - 9a - 12) \div (2a - 5)$ $3a + 3, r3$    11. $(20m^2 + 7m - 6) \div (5m - 2)$ $4m + 3$
12. $(8x^2 - 30x + 11) \div (4x - 1)$ $2x - 7, r4$    13. $(6x^2 + \quad - 24) \div (2x + 6)$ $3x - 4$
14. $(15k^2 - 2k - 8) \div (3k + 2)$ $5k - 4$    15. $(6b^2 + 5b - 25) \div (3b - 5)$ $2b + 5$
16. $(3x^2 - 5x - 30) \div (x - 4)$ $3x + 7, r-2$    17. $(8w^2 + 2w - 8) \div (2w + 5)$ $4w - 9, r37$
18. $(9y^2 + 3y - 4) \div (3y - 2)$ $3y + 3, r2$  19. $2x^2 - 4x + 6$    21. $5a^2 - a + 1$    22. $y^2 - 7y + 4, r2$
19. $(6x^3 - 2x^2 - 2x + 30) \div (3x + 5)$    20. $(3x^3 + 17x^2 - 8x - 12) \div (x + 6)$ $3x^2 - x - 2$
21. $(10a^3 - 22a^2 + 6a - 4) \div (2a - 4)$    22. $(3y^3 - 23y^2 + 26y - 6) \div (3y - 2)$
23. $(2x^3 - x^2 - 17x + 9) \div (x - 3)$    24. $(2p^3 - 9p^2 + 10p + 5) \div (2p - 1)$
        $2x^2 + 5x - 2, r3$                        $p^2 - 4p + 3, r8$

(B)
25. $(9x^3 - 4x + 2) \div (3x - 1)$ $3x^2 + x - 1, r1$    26. $(4a^3 - 7a - 2) \div (2a - 1)$ $2a^2 + a - 3, r-5$
27. $(6y^3 - 8y^2 - 16) \div (2y - 4)$ $3y^2 + 2y + 4$    28. $(3b^3 + 16b^2 + 9) \div (3b - 2)$ $b^2 + 6b + 4, r17$
29. $(9g^4 + 5g^2 - 12g - 6) \div (3g - 2)$    30. $(4x^3 - 52x - 48) \div (2x + 6)$ $2x^2 - 6x - 8$
31. $(2x^3 - 22x + 12) \div (x - 6)$    32. $(6n^3 - 22n + 1) \div (2n - 4)$ $3n^2 + 6n + 1, r5$
29. $3g^3 + 2g^2 + 3g - 2, r-10$    31. $2x^2 + 12x + 50, r312$    33. $3x^2 - 6x + 4, r-9$    34. $4x^2 + 3x + 21, r-3$

(C)
33. $(9x^3 - 12x^2 - 1) \div (3x + 2)$    34. $(4x^3 - 25x^2 - 150) \div (x - 7)$
35. $(8a^3 - 125) \div (2a - 5)$ $4a^2 + 10a + 25$    36. $(27y^3 - 64) \div (3y - 4)$ $9y^2 + 12y + 16$
37. $(6x^3 + 15x^2 - 4x - 5) \div (3x^2 - 2)$    38. $(9a^6 - 3a^3 + 3a^2 - 4) \div (3a^3 - 1)$
39. $(2x^3 - 12x^2 - 5x - 32) \div (2x^2 - 5)$    40. $(8a^5 + 20a^3 - 2a^2 - 5) \div (4a^3 - 1)$
    37. $2x + 5, r5$                    $x - 6, r-62$                    38. $3a^3, r3a^2 - 4$                    $2a^2 + 5$

Dividing by a Binomial                                                        **257**

# APPLICATIONS

Read → Plan → Solve → Interpret

**Solve.**

1. A basketball team won 15 games and lost 5 games. What percent of the games played did the team win? **75%**

2. A store manager lists the selling price of a stereo at $390. The profit is 30% of the cost. Find the cost of the stereo. **$300**

3. Juan budgeted $265 for renting a car. How far can he travel if the rental rate is $73 plus 12¢/km? **1,600 km**

4. This year, Duval's weekly take-home pay is $300. Last year, Duval's weekly pay was $250. The increase in pay is what percent of last year's pay? **20%**

5. In the morning, the temperature was 2°C. It rose 8° by 3:00 P.M., then dropped 15° by evening. What was the temperature in the evening? **−5°C**

6. Jane commutes 576 mi a week to work. Her car averages 18 mi/gal. If gas sells for $1.31/gal, how much does she spend on gas in 3 weeks? **$125.76**

7. In the freshman class, 3 out of 5 students study algebra. How many students study algebra in a class of 700 students? **420**

8. If 4 out of 5 people use Never Die Batteries, how many use this product in a city with a population of 50,000? **40,000**

9. Tina is paid $4.50 an hour for a 40-hour week plus a 6% commission on all sewing machine sales. What must her total sales be for her to earn $240 in a week? **$1,000**

10. Bolinda is paid a commission of 6% on all cameras she sells plus $4.50/h for 40 h a week. How much must Bolinda sell in cameras to earn a total of $210 a week? **$500**

## Finding Least Common Denominators

The least common denominator of two or more fractions is the smallest number that has each denominator as a factor. For example, the LCD of $\frac{1}{2}$, $\frac{1}{4}$, and $\frac{1}{3}$ is 12.

This program will find the least common denominator of two fractions. The method used is to form successive multiples of one denominator until one is found which is also a multiple of the other denominator. The program uses the INT function, which takes the integer part of what is in the parentheses. For example, INT(3.75) means 3.

```
100   PRINT "THIS PROGRAM FINDS THE LCD OF TWO"
110   PRINT "DENOMINATORS.  TYPE IN THE SMALLER"
120   PRINT "DENOMINATOR, A COMMA, THEN THE"
130   PRINT "LARGER DENOMINATOR."
140   INPUT A,B
150   REM  LINE 160 CHECKS IF A IS LESS THAN OR EQUAL
          TO B.
160   IF A > B THEN 100
170   REM   THE LOOP COMPUTES MULTIPLES OF A, BEGINNIN
          G
180   REM   WITH A * 1.   THEN IT CHECKS EACH MULTIPLE
190   REM   TO SEE IF B IS ALSO A DIVISOR.
200   FOR X = 1 TO B
210   LET M = X * A
220   IF  INT (M / B) = M / B THEN 240
230   NEXT X
240   PRINT M" IS THE LCD OF "A" AND "B"."
250   END
```

See the Computer Section beginning on page 488 for more information.

## Exercises      For Ex. 5–6, see TE Answer Section.

Solve each of the following. Check your answers on the computer.
1. INT(7.3896) **7**        2. INT(8.999) **8**          3. INT(−6.273) **−7**

4. Enter and RUN the program above. Use the denominators 3, 4 and then the denominators 7, 13. **12, 91**

5. Change this program to find the LCD of three numbers A, B, C. [*HINT*: Change lines 140, 200, and 220. Then add a line after 130 to check if C is a divisor of M.]

6. Write a program to find the greatest common factor of two numbers.

# CHAPTER NINE REVIEW

Ⓐ Exercises are 1–5, 8, 9, 11–13, 15, 19–22, 25, 26. Ⓒ Exercises are starred. The rest are Ⓑ Exercises.

**Vocabulary**

Addition Property for Rational Expressions [9.1]
dividend [9.8]     divisor [9.8]
Identity Property for Rational Expressions [9.2]
least common multiple [9.2]
quotient [9.7]     remainder [9.8]

Add. Simplify, if possible.

1. $\dfrac{a}{7a + 21} + \dfrac{3}{7a + 21}$ [9.1] $\dfrac{1}{7}$

2. $\dfrac{5y}{4} + \dfrac{3y}{8} + \dfrac{y}{6}$ [9.2] $\dfrac{43y}{24}$

3. $\dfrac{3}{x^2 - 8x + 15} + \dfrac{4}{x - 5} + \dfrac{2}{x - 3}$ [9.3]

4. $\dfrac{3}{4y^3} + \dfrac{5}{12y} + \dfrac{-7}{2y^2}$ $\dfrac{5y^2 - 42y + 9}{12y^3}$

5. $\dfrac{4}{a + 3} + \dfrac{5}{2a^2 + a - 15} + \dfrac{2}{2a - 5}$

6. $\dfrac{2x - 3}{x^2 - 2x - 24} + \dfrac{x + 5}{x + 4}$ $\dfrac{x^2 + x - 33}{(x + 4)(x - 6)}$

7. $\dfrac{-2x^2 - 5x}{x^2 + 7x} + \dfrac{x - 2}{x + 7} + \dfrac{2x - 3}{x}$ $\dfrac{x - 3}{x}$

3. $\dfrac{6x - 19}{(x - 3)(x - 5)}$     5. $\dfrac{10a - 9}{(a + 3)(2a - 5)}$

Add. [9.4]

8. $\dfrac{6}{a - 3} + \dfrac{7}{a + 2}$ $\dfrac{13a - 9}{(a - 3)(a + 2)}$

9. $\dfrac{7}{15y} + \dfrac{-3}{5y^3} + \dfrac{2}{3y^2}$ $\dfrac{7y^2 + 10y - 9}{15y^3}$

10. $\dfrac{a - 5}{2a + 3} + a - 2$ $\dfrac{2a^2 - 11}{2a + 3}$

Subtract. Simplify, if possible. [9.5]

11. $\dfrac{2}{a^2 + 7a + 10} - \dfrac{3}{a + 5}$ $\dfrac{-4 - 3a}{(a + 5)(a + 2)}$

12. $4 - \dfrac{2a + 3}{a - 2}$ $\dfrac{2a - 11}{a - 2}$

13. $\dfrac{7}{5x^2} - \dfrac{3}{2x}$ $\dfrac{14 - 15x}{10x^2}$

14. $\dfrac{5}{a + 3} - \dfrac{2a + 1}{a + 2}$ $\dfrac{-2a^2 + 2a - 7}{(a + 3)(a + 2)}$

Simplify. [9.6]

15. $\dfrac{7b - 9}{b^2 - 3b + 2} + \dfrac{5}{2 - b}$ $\dfrac{2}{b - 1}$

16. $\dfrac{2n - 1}{n + 3} - \dfrac{4n^2}{9 - n^2} + \dfrac{2}{n - 3}$ $\dfrac{6n^2 - 5n + 9}{n^2 - 9}$

17. $\dfrac{a^2 - 7}{a^2 - 4a + 3} - \dfrac{a + 2}{1 - a} + \dfrac{4}{a^2 - 3a}$

18. $\dfrac{3b + 1}{b^2 - 25} - \dfrac{2b + 1}{5 - b}$ $\dfrac{2b^2 + 14b + 6}{(b - 5)(b + 5)}$

17. $\dfrac{2a^3 - a^2 - 9a - 4}{a(a - 3)(a - 1)}$

Divide. [9.7]

19. $20m^4 \div 4m^3$ $5m$   20. $18b^7 \div 3b^2$ $6b^5$

21. $(6m^3 - 4m^2 + 8m) \div 2m$ $3m^2 - 2m + 4$

22. $(30m^4 - 25m^3 + 10m^2) \div 5m^2$

23. $\dfrac{8p^5 - 4p^4 + 6p^3 - 12p^2}{2p^2}$ $4p^3 - 2p^2 + 3p - 6$

24. $\dfrac{45q^5 - 15q^4 + 30q^3 + 12q^2 - 48q}{3q}$

24. $15q^4 - 5q^3 + 10q^2 + 4q - 16$

25. $\dfrac{3x^2 - 5x - 20}{x - 4}$ $3x + 7, r8$

26. $(6x^3 - 5x^2 - 13x + 20) \div (3x + 5)$ $2x^2 - 5x + 4$

27. $(a^3 - 12a - 12) \div (a + 3)$ $a^2 - 3a - 3, r-3$

Add. Simplify, if possible. [9.3]

★28. $\dfrac{x}{x^2 + 3x - 10} + \dfrac{3}{x^2 + 5x} + \dfrac{3x}{x^2 - 4}$ $\dfrac{4x^3 + 20x^2 - 12}{x(x + 5)(x - 2)(x + 2)}$

Subtract. Simplify, if possible. [9.5]

★29. $\dfrac{-8}{ab + b^2 - bc - ac} - \dfrac{9}{b^2 - c^2}$ $\dfrac{-9a - 17b - 8c}{(a + b)(b - c)(b + c)}$

Simplify. [9.6]

★30. $\dfrac{6}{-6 + 5m - m^2} - \dfrac{4}{m^4 - 13m^2 + 36}$

$-\dfrac{6m^2 + 30m + 40}{(m - 3)(m + 3)(m - 2)(m + 2)}$

# CHAPTER NINE TEST

Ⓐ Exercises are 1–3, 5, 6, 8, 9–12, 14. Ⓒ Exercises are starred. The rest are Ⓑ Exercises.

Divide.

**1.** $\dfrac{15a^6}{3a^2}$   **5a⁴**

**2.** $\dfrac{15b^3 - 12b^2 + 9b}{3b}$
  **5b² − 4b + 3**

**3.** $\dfrac{x^2 + 3x - 48}{x - 5}$
  **x + 8, r−8**

**4.** $\dfrac{4a^3 - 5a + 8}{2a - 1}$
  **2a² + a − 2, r6**

Add. Simplify, if possible.

**5.** $\dfrac{5}{3x^2} + \dfrac{-3}{7x^3} + \dfrac{2}{21x}$   $\dfrac{\textbf{2x}^2 + \textbf{35x} - \textbf{9}}{\textbf{21x}^3}$

**6.** $\dfrac{10p + 21}{p^2 + 3p} + \dfrac{p}{p + 3}$   $\dfrac{\textbf{p} + \textbf{7}}{\textbf{p}}$

**7.** $x + 6 + \dfrac{3}{x + 8}$   $\dfrac{\textbf{x}^2 + \textbf{14x} + \textbf{51}}{\textbf{x} + \textbf{8}}$

**8.** $\dfrac{7}{x^2 - 7x + 10} + \dfrac{4}{x^2 - 5x} + \dfrac{3}{x - 5}$
  $\dfrac{\textbf{3x}^2 + \textbf{5x} - \textbf{8}}{\textbf{x}(\textbf{x} - \textbf{5})(\textbf{x} - \textbf{2})}$

Subtract. Simplify, if possible.

**9.** $\dfrac{6}{a - 4} - \dfrac{3}{a + 2}$   $\dfrac{\textbf{3a} + \textbf{24}}{(\textbf{a} - \textbf{4})(\textbf{a} + \textbf{2})}$

**10.** $\dfrac{2x}{x - 4} - \dfrac{1}{2 - x}$   $\dfrac{\textbf{2x}^2 - \textbf{3x} - \textbf{4}}{(\textbf{x} - \textbf{4})(\textbf{x} - \textbf{2})}$

Simplify.

**11.** $\dfrac{3a - 4}{a^2 - 9a + 20} + \dfrac{4}{5 - a}$   $\dfrac{\textbf{a} - \textbf{12}}{(\textbf{a} - \textbf{5})(\textbf{a} - \textbf{4})}$

**12.** $\dfrac{x^2 - 8x - 28}{x^2 - 2x - 24} - \dfrac{4}{6 - x} + \dfrac{3}{x + 4}$   $\dfrac{\textbf{x} + \textbf{5}}{\textbf{x} + \textbf{4}}$

Add. Simplify, if possible.

**13.** $\dfrac{2b - 1}{b^2 + 3b - 10} + \dfrac{2}{b^2 - b - 30}$
  $\dfrac{\textbf{2b}^2 - \textbf{11b} + \textbf{2}}{(\textbf{b} + \textbf{5})(\textbf{b} - \textbf{2})(\textbf{b} - \textbf{6})}$

**14.** $\dfrac{5}{q^2 - 9q + 14} + \dfrac{7}{q - 7} + \dfrac{3}{q - 2}$
  $\dfrac{\textbf{10q} - \textbf{30}}{(\textbf{q} - \textbf{7})(\textbf{q} - \textbf{2})}$

**15.** $\dfrac{a + 5}{2a^2 - 5a - 12} + \dfrac{a - 1}{2a + 3}$   $\dfrac{\textbf{a}^2 - \textbf{4a} + \textbf{9}}{(\textbf{2a} + \textbf{3})(\textbf{a} - \textbf{4})}$

Simplify.

**16.** $\dfrac{3a - 2}{a + 4} - \dfrac{2a^2}{16 - a^2} + \dfrac{7}{a - 4}$
  $\dfrac{\textbf{5a}^2 - \textbf{7a} + \textbf{36}}{\textbf{a}^2 - \textbf{16}}$

Add. Simplify, if possible.

**17.** $\dfrac{3b - 4}{20} + \dfrac{7b}{5} + \dfrac{4b - 1}{2}$   $\dfrac{\textbf{71b} - \textbf{14}}{\textbf{20}}$

Simplify.

**★18.** $\dfrac{2}{12 + a - a^2} - \dfrac{5}{a^4 - 25a^2 + 144}$
  $\dfrac{\textbf{−2a}^2 - \textbf{2a} + \textbf{19}}{(\textbf{a} - \textbf{4})(\textbf{a} + \textbf{4})(\textbf{a} - \textbf{3})(\textbf{a} + \textbf{3})}$

**★19.** $\dfrac{7}{n - 2} - \dfrac{4}{6 - n} + \dfrac{3}{n + 5}$
  $\dfrac{\textbf{14n}^2 - \textbf{19n} - \textbf{214}}{(\textbf{n} - \textbf{2})(\textbf{n} - \textbf{6})(\textbf{n} + \textbf{5})}$

**★20.** $\dfrac{3a + 3b}{xa - 3ya + xb - 3yb} + \dfrac{5}{3y - x}$   $\dfrac{\textbf{−2}}{\textbf{x} - \textbf{3y}}$

# COLLEGE PREP TEST

Directions: In each item, you are to compare a quantity in Column 1 with a quantity in Column 2. Write the letter of the correct answer from these choices:
(A) The quantity in Column 1 is greater than the quantity in Column 2;
(B) The quantity in Column 2 is greater than the quantity in Column 1;
(C) The quantity in Column 1 is equal to the quantity in Column 2;
(D) The relationship cannot be determined from the given information.

Notes: Information centered over both columns refers to one or both of the quantities to be compared.
A symbol that appears in both columns has the same meaning in each column, and all variables represent real numbers.

Sample Question and Answer

| Column 1 | Column 2 |
|---|---|
| $x \neq 0$ and $y \neq 0$ | |
| $\dfrac{1}{x} - \dfrac{1}{y}$ | $\dfrac{y - x}{xy}$ |

Answer: C because

$$\frac{1}{x} - \frac{1}{y} = \frac{1}{x} + \frac{-1}{y}$$

$$= \frac{1 \cdot y}{x \cdot y} + \frac{-1 \cdot x}{y \cdot x}$$

$$= \frac{1y - 1x}{xy}, \text{ or } \frac{y - x}{xy}$$

| Column 1 | Column 2 |
|---|---|
| 1. $\dfrac{5}{9}$ | $\dfrac{2}{3}$ **B** |
| $x \neq 0$ and $y \neq 0$ | **C** |
| 2. $\dfrac{1}{x} + \dfrac{1}{y}$ | $\dfrac{x + y}{xy}$ |
| $x > 0, y \neq 0$, and $\dfrac{2}{x} = \dfrac{3}{y}$ | **B** |
| 3. $x$ | $y$ |
| $t > w > 0$ and $r > 0$ | **B** |
| 4. $\dfrac{r}{t}$ | $\dfrac{r}{w}$ |

| Column 1 | Column 2 |
|---|---|
| $x > y > z > 0$ | **A** |
| 5. $\dfrac{x}{y}$ | $\dfrac{z}{x}$ |
| $\dfrac{a}{d} < 1$ | **D** |
| 6. $d$ | $a$ |
| $x = -2$ | **A** |
| 7. $\dfrac{x^5}{x^3}$ | $\dfrac{x^7}{x^4}$ |
| $y \neq 0$ | **A** |
| 8. $\dfrac{x + y}{y}$ | $\dfrac{x}{y}$ |
| $-5 < x < -1$ | **B** |
| 9. $\dfrac{1}{x^3}$ | $\dfrac{1}{x^2}$ |
| $c = \dfrac{1}{a + b}$, $a > 0$, and $b > 0$ | **C** |
| 10. $\dfrac{2}{c}$ | $2a + 2b$ |

# 10 APPLICATIONS OF RATIONAL EXPRESSIONS

**Non-Routine Problem Solving**

**Problem 1**    A runner took three hours to run 30 mi. Another runner ran the same distance at the rate of 6 mi/h. What was the average rate for the two runners? **$7\frac{1}{2}$ mi/h**

**Problem 2**    Two runners who are 20 km apart start running toward each other at the same time. They travel on a long, straight road at the rate of 10 km/h. When they start, a fly begins flying back and forth from the forehead of one runner to the forehead of the other runner at the rate of 15 km/h. How far has the fly flown by the time the runners meet? **15 km**

# EQUATIONS WITH RATIONAL EXPRESSIONS

*Objective*    **To solve equations containing rational expressions**

An equation such as $\dfrac{-3a}{a^2 - 4a - 32} = \dfrac{2}{a - 8} + \dfrac{3}{a + 4}$ contains rational expressions. The easiest way to solve such an equation is to change it to an equivalent equation with no term in $\dfrac{a}{b}$ form. This can be done by multiplying each side of the equation by the least common multiple, LCM, of the denominators. A solution of the resulting equation will usually be a solution of the original equation.

*Example 1*    **Solve the equation.**

$$\frac{-3a}{a^2 - 4a - 32} = \frac{2}{a - 8} + \frac{3}{a + 4}$$

$$\frac{-3a}{(a - 8)(a + 4)} = \frac{2}{a - 8} + \frac{3}{a + 4} \quad \blacktriangleleft \text{ Factor: } a^2 - 4a - 32 = (a - 8)(a + 4).$$

$$(a - 8)(a + 4) \cdot \frac{-3a}{(a - 8)(a + 4)} = (a - 8)(a + 4)\left[\frac{2}{a - 8} + \frac{3}{a + 4}\right] \quad \blacktriangleleft \text{ LCM is } (a - 8)(a + 4)$$

$$\overset{1}{(a - 8)}\overset{1}{(a + 4)} \cdot \frac{-3a}{\underset{1}{(a - 8)}\underset{1}{(a + 4)}} = \overset{1}{(a - 8)}(a + 4)\frac{2}{\underset{1}{a - 8}} + (a - 8)\overset{1}{(a + 4)}\frac{3}{\underset{1}{a + 4}}$$

$$-3a = (a + 4)2 + (a - 8)3$$
$$-3a = 2a + 8 + 3a - 24$$
$$-3a = 5a - 16$$
$$-8a = -16$$
$$a = 2$$

Check to see whether 2 is a solution of the original equation.

*Check.*    
*Replace a with 2.* ▶

| $\dfrac{-3a}{a^2 - 4a - 32}$ | $\dfrac{2}{a - 8} + \dfrac{3}{a + 4}$ |
|---|---|
| $\dfrac{-3 \cdot 2}{2^2 - 4 \cdot 2 - 32}$ | $\dfrac{2}{2 - 8} + \dfrac{3}{2 + 4}$ |
| $\dfrac{-6}{4 - 8 - 32}$ | $\dfrac{2}{-6} + \dfrac{3}{6}$ |
| $\dfrac{-6}{-36}$ | $\dfrac{-2}{6} + \dfrac{3}{6} \quad \blacktriangleleft \dfrac{2}{-6} = \dfrac{-2}{6}$ |
| $\dfrac{1}{6}$ | $\dfrac{1}{6}$ |

$\dfrac{1}{6} = \dfrac{1}{6}$, true    **Thus, the solution is 2.**

Sometimes, when both sides of an equation containing rational expressions are multiplied by the LCM of the denominators, the resulting equation is a quadratic equation.

## Example 2

**Solve** $\dfrac{-20}{x^2 - 4x - 45} = \dfrac{x + 3}{x + 5} + \dfrac{-2}{9 - x}.$

$$\frac{-20}{x^2 - 4x - 45} = \frac{x + 3}{x + 5} + \frac{-2}{9 - x}$$

$$\frac{-20}{x^2 - 4x - 45} = \frac{x + 3}{x + 5} + \frac{-2}{-1(x - 9)}$$
◀ *Rewrite $9 - x$ in convenient form:*
$9 - x = -x + 9 = -1(x - 9)$.

$$\frac{-20}{x^2 - 4x - 45} = \frac{x + 3}{x + 5} + \frac{2}{x - 9}$$
◀ $\dfrac{-2}{-1(x - 9)} = \dfrac{-1(-2)}{x - 9} = \dfrac{2}{x - 9}$

$$\frac{-20}{(x - 9)(x + 5)} = \frac{x + 3}{x + 5} + \frac{2}{x - 9}$$
◀ *Factor:* $x^2 - 4x - 45 = (x - 9)(x + 5)$.

$$(x - 9)(x + 5) \cdot \frac{-20}{(x - 9)(x + 5)} = (x - 9)(x + 5)\left[\frac{x + 3}{x + 5} + \frac{2}{x - 9}\right]$$
◀ *LCM is* $(x - 9)(x + 5)$

$$(x - 9)(x + 5) \cdot \frac{-20}{(x - 9)(x + 5)} = (x - 9)(x + 5)\frac{x + 3}{x + 5} + (x - 9)(x + 5)\frac{2}{x - 9}$$

$$\overset{1}{\cancel{(x - 9)}}\,\overset{1}{\cancel{(x + 5)}} \cdot \frac{-20}{\underset{1}{\cancel{(x - 9)}}\,\underset{1}{\cancel{(x + 5)}}} = (x - 9)\overset{1}{\cancel{(x + 5)}}\frac{x + 3}{\underset{1}{\cancel{x + 5}}} + \overset{1}{\cancel{(x - 9)}}(x + 5)\frac{2}{\underset{1}{\cancel{x - 9}}}$$

$$-20 = (x - 9)(x + 3) + (x + 5)2$$
$$-20 = x^2 - 6x - 27 + 2x + 10 \quad ◀ \ (x - 9)(x + 3) = x^2 - 6x - 27$$
$$-20 = x^2 - 4x - 17$$

Put this quadratic equation in standard form. The coefficient of the $x^2$ term is positive. Now, get the polynomial equal to 0.

$$0 = x^2 - 4x + 3 \quad ◀ \ \textit{Add 20 to each side.}$$
$$0 = (x - 3)(x - 1) \quad ◀ \ \textit{Factor.}$$
$$x - 3 = 0 \quad \text{or} \quad x - 1 = 0$$
$$x = 3 \qquad\qquad x = 1$$

Check on your own in the original equation. Recall that you must replace $x$ with both 3 and 1.

**So,** the solutions are 3 and 1.

When both sides of an equation containing rational expressions are multiplied by the LCM of the denominators, the new equation may have a solution that *is not* a solution of the original equation. This situation is illustrated in Example 3 on the next page.

Equations with Rational Expressions

*Example 3*    **Solve** $\dfrac{x+5}{x-4} = \dfrac{3}{x} + \dfrac{36}{x^2-4x}$.

$$\frac{x+5}{x-4} = \frac{3}{x} + \frac{36}{x^2-4x}$$

$$\frac{x+5}{x-4} = \frac{3}{x} + \frac{36}{x(x-4)} \quad \blacktriangleleft \; \textit{Factor: } x^2 - 4x = x(x-4).$$

$$x(x-4)\frac{x+5}{x-4} = x(x-4)\frac{3}{x} + x(x-4)\frac{36}{x(x-4)} \quad \blacktriangleleft \; \textit{Multiply each side by the LCM of the}$$
*denominators. Use the distributive property*
*on the right side of the equation.*

$$x\overset{1}{(x-4)}\frac{x+5}{x-4} \underset{1}{} = \overset{1}{x}(x-4)\frac{3}{\underset{1}{x}} + \overset{1}{x}\overset{1}{(x-4)}\frac{36}{\underset{1}{x}\underset{1}{(x-4)}}$$

$$x(x+5) = (x-4)3 + 36$$
$$x^2 + 5x = 3x - 12 + 36$$
$$x^2 + 5x = 3x + 24 \quad \blacktriangleleft \; \textit{Put the quadratic equation in standard form.}$$
$$x^2 + 2x = 24 \quad \blacktriangleleft \; \textit{Subtract } 3x \textit{ from each side.}$$
$$x^2 + 2x - 24 = 0 \quad \blacktriangleleft \; \textit{Subtract } 24 \textit{ from each side.}$$
$$(x+6)(x-4) = 0$$
$$x + 6 = 0 \quad \text{or} \quad x - 4 = 0$$
$$x = -6 \qquad\qquad x = 4 \quad \blacktriangleleft \; \textit{Two solutions are possible. Check each in the original equation.}$$

*Check.*

*Replace x with −6.* $\blacktriangleright$

| $\dfrac{x+5}{x-4}$ | $\dfrac{3}{x} + \dfrac{36}{x^2-4x}$ |
|---|---|
| $\dfrac{-6+5}{-6-4}$ | $\dfrac{3}{-6} + \dfrac{36}{(-6)^2 - 4\cdot(-6)}$ |
| $\dfrac{-1}{-10}$ | $-\dfrac{3}{6} + \dfrac{36}{36+24}$ |
| $\dfrac{1}{10}$ | $-\dfrac{1}{2} + \dfrac{36}{60}$ |
| | $-\dfrac{1}{2} + \dfrac{3}{5}$ |
| | $-\dfrac{5}{10} + \dfrac{6}{10}$ |
| | $\dfrac{1}{10}$ |

$$\frac{1}{10} = \frac{1}{10}, \text{ true}$$

*Replace x with 4.* $\blacktriangleright$

| $\dfrac{x+5}{x-4}$ | $\dfrac{3}{x} + \dfrac{36}{x^2-4x}$ |
|---|---|
| $\dfrac{4+5}{4-4}$ | $\dfrac{3}{4} + \dfrac{36}{4^2 - 4\cdot 4}$ |
| $\dfrac{9}{0}$ | $\dfrac{3}{4} + \dfrac{36}{16-16}$ |
| | $\dfrac{3}{4} + \dfrac{36}{0}$ |

False. The fractions $\dfrac{9}{0}$ and $\dfrac{36}{0}$ are undefined because you cannot divide by 0. Therefore, 4 *is not* a solution of the original equation.

**Thus, −6 is the only solution.**

In Example 3, 4 was a solution of the *derived equation* but not of the original equation. The 4 is called an *extraneous solution*.

| Definition: Extraneous solution | An **extraneous solution** of an equation is a solution that does not check in the original equation. |
|---|---|

## Reading in Algebra

**1.** What two numbers would be extraneous solutions of $\dfrac{5}{x-4} = \dfrac{7}{x-3}$? **3, 4**

**2.** What is the greatest number of extraneous solutions that could occur for
an equation such as $\dfrac{5}{x+2} = \dfrac{7}{x-4} + \dfrac{3}{2x+10}$? **3**

**3.** Is a solution of a derived equation always a solution of the original
equation? Why, or why not? **No. Some solutions could be extraneous, i.e., they do
not check in the original equation.**

## Written Exercises

**Solve.**

Ⓐ **1.** $\dfrac{3x}{x^2 - 5x + 4} = \dfrac{2}{x-4} + \dfrac{3}{x-1}$ **x = 7**

**2.** $\dfrac{2}{x^2 - 9} + \dfrac{5}{x-3} = \dfrac{4}{x+3}$ **x = −29**

**3.** $\dfrac{3}{b-7} - \dfrac{2}{b+4} = \dfrac{6}{b^2 - 3b - 28}$ **b = −20**

**4.** $\dfrac{7}{m^2 - 5m} - \dfrac{3}{m-5} = \dfrac{4}{m}$ **m = $\dfrac{27}{7}$**

**5.** $\dfrac{4}{a^2 - 7a + 10} = \dfrac{4}{a-5} + \dfrac{3}{a-2}$ **a = $\dfrac{27}{7}$**

**6.** $\dfrac{5}{x^2 - 3x} - \dfrac{3}{x-3} = \dfrac{2}{x}$ **x = $\dfrac{11}{5}$**

**7.** $\dfrac{3}{y+4} + \dfrac{2}{6-y} = \dfrac{5}{y^2 - 2y - 24}$ **y = 31**

**8.** $\dfrac{14}{x^2 - 3x} - \dfrac{8}{x} = \dfrac{10}{3-x}$ **x = −19**

**9.** $\dfrac{2t-1}{t^2 - 9t + 20} = \dfrac{7}{t-5} + \dfrac{4}{4-t}$ **t = 7**

**10.** $\dfrac{4}{a-2} - \dfrac{10}{3-a} = \dfrac{4a+10}{a^2 - 5a + 6}$ **a = $\dfrac{21}{5}$**

**Solve. Check for extraneous solutions.**

Ⓑ **11.** $\dfrac{6}{a+2} + \dfrac{3}{a^2 - 4} = \dfrac{2a-7}{a-2}$ **a = −$\dfrac{1}{2}$, 5**

**12.** $\dfrac{-7}{b^2 - 9b + 20} = \dfrac{b}{b-4} + \dfrac{1}{b-5}$ **b = 3, 1**

**13.** $\dfrac{x+1}{x-3} = \dfrac{3}{x} + \dfrac{12}{x^2 - 3x}$ **x = −1**

**14.** $\dfrac{a}{a-4} + \dfrac{2}{a} = \dfrac{16}{a^2 - 4a}$ **a = −6**

**15.** $\dfrac{13}{y^2 - 4} + \dfrac{y}{2-y} = \dfrac{-2}{y+2}$ **y = ±3**

**16.** $x - \dfrac{4x}{x-6} = \dfrac{24}{6-x}$ **x = 4**

**17.** $\dfrac{a+1}{4} = \dfrac{4}{a+1}$ **a = −5, 3**

**18.** $\dfrac{x+3}{x+5} + \dfrac{2}{x-9} = \dfrac{-20}{x^2 - 4x - 45}$ **x = 3, 1**

**19.** $\dfrac{2b-3}{b-5} - \dfrac{b}{b+4} = \dfrac{20b-37}{b^2 - b - 20}$ **∅**

**20.** $\dfrac{2x-1}{x+4} - \dfrac{x}{x+3} = \dfrac{-1}{x^2 + 7x + 12}$ **x = −2, 1**

Ⓒ **21.** $\dfrac{7}{2x^2 + 3x - 20} - \dfrac{2}{x^2 - 2x - 24}$

$= \dfrac{4}{2x^2 - 17x + 30}$ **x = −48**

**22.** $\dfrac{6}{a^2 - 16} = \dfrac{7}{a^2 - 7a + 12} - \dfrac{2}{a^2 + a - 12}$

**a = 54**

**23.** $\dfrac{1}{y^2 + 5y + 6} - \dfrac{2}{y^2 - y - 6} = \dfrac{3}{9 - y^2}$ **y = $\dfrac{3}{2}$**

**24.** $\dfrac{5}{m^2 + 2m - 24} - \dfrac{2}{4m - m^2} = \dfrac{4}{m^2 + 6m}$

**m = $\dfrac{-28}{3}$**

## CUMULATIVE REVIEW

**1.** Find two consecutive even integers whose product is 48. **6, 8 or −8, −6**

**2.** Solve: $5x + 12 = 2x^2$. **x = −$\dfrac{3}{2}$ or x = 4**   **3.** Solve: $1.2x - 0.38 = 0.6x + 0.04$ **x = .7**

# APPLICATIONS

$$\boxed{\text{Read}} \longrightarrow \boxed{\text{Plan}} \longrightarrow \boxed{\text{Solve}} \longrightarrow \boxed{\text{Interpret}}$$

An **average**, or **arithmetic mean**, can be found by adding all the person's scores and then dividing by the number of scores. For example, suppose Joe's three bowling scores for the day are 140, 160, and 150. His average score is:

$$\frac{140 + 160 + 150}{3} = \frac{450}{3}, \text{ or } 150.$$

*Example*

**Marc's test scores in history are 65, 70, and 70. What score must he get on the fourth test to give him a passing average of 70?**

Let $x$ = the score he needs.

$$\frac{65 + 70 + 70 + x}{4} = 70 \qquad \blacktriangleleft \text{ } \textit{The average is the sum of the scores divided by the number of scores, 4.}$$

$$\frac{205 + x}{4} = \frac{70}{1}$$

$$\overset{1}{\cancel{4}} \cdot \frac{205 + x}{\underset{1}{\cancel{4}}} = 4 \cdot \frac{70}{1} \qquad \blacktriangleleft \text{ } \textit{Multiply each side by the LCM of the denominators, 4.}$$

$$205 + x = 280$$
$$x = 75 \qquad \blacktriangleleft \text{ } \textit{Add } -205 \text{ to each side.}$$

**So,** he needs a score of 75 to get a passing average of 70.

Solve each problem.

**1.** John's test grades are 65, 75, 85, and 70. What score must he get on the fifth test so that his average is 70? **55**

**2.** Mary needs a bowling average of 220 for the night. Her scores so far are 250 and 180. What must she bowl in the third game? **230**

**3.** A basketball team's average for 10 games last year was 66. So far this season, their scores are: 64, 80, 70, 60, 68, 48, 52, 80, and 82. What must the team score be in their last game this season to maintain last year's average? **56**

*Objective*     **To solve problems about work**

If Bill can mow a lawn in 4 h, then in 1 h he can mow $\frac{1}{4}$ of the lawn. His rate of work is $\frac{1}{4}$ of the job per hour. At this rate in 3 h, he should complete $3 \cdot \frac{1}{4}$, or $\frac{3}{4}$ of the job. Notice that this relationship can be

indicated as $\left( \begin{array}{c} \text{number of} \\ \text{hours worked} \end{array} \right) \times \left( \begin{array}{c} \text{rate} \\ \text{per hour} \end{array} \right) = \left( \begin{array}{c} \text{amount of} \\ \text{work completed} \end{array} \right).$

$$3 \quad \cdot \quad \frac{1}{4} \quad = \quad \frac{3}{4}$$

This suggests a formula that can be used to solve problems about work.

| Formula for work problems | $\underbrace{\text{Amount of work completed}}_{w} = \underbrace{(\text{number of units of time})}_{t} \cdot \underbrace{(\text{rate per unit of time})}_{r}.$ |
|---|---|

*Example 1*     **Pierre can paint a house in 7 days. What part of the house will he paint in 3 days? in 4 days? in *x* days?**

First, find his rate: 7 days to do the entire job means $\frac{1}{7}$ of the job completed per day.
Then, find the part of the job he would complete in:

3 days $\quad 3 \cdot \frac{1}{7} = \frac{3}{7}$ of the job, $\qquad$ ◀ *Use $w = t \cdot r$.*

4 days $\quad 4 \cdot \frac{1}{7} = \frac{4}{7}$ of the job, and $\qquad w = 3 \cdot \frac{1}{7}$

*x* days $\quad x \cdot \frac{1}{7} = \frac{x}{7}$ of the job.

In building a fence, suppose Julia can do $\frac{2}{5}$ of the job and Juanita can do $\frac{3}{5}$ of the job. If they work together, they will complete the job, for

$$\frac{2}{5} + \frac{3}{5} = 1.$$

If a job is completed, the sum of the fractional parts of the job done by each person working on the job must be 1. This is shown in Example 2.

*Example 2*    **It takes Regina 3 h to prepare surgical equipment. Another nurse, Carlo, can do it in 2 h. How long will it take to do the job if they work together?**

READ ▶    You are given: Regina takes 3 h to do the job alone.
                                    Carlo takes 2 h to do the job alone.
            You are asked to find how long it will take them to do the job together.

PLAN ▶    Represent the work done by each.

| | **Part of Job Done in 1 h** | **Number of Hours Working Together** | **Part of Job Completed** |
|---|---|---|---|
| Regina | $\dfrac{1}{3}$ | $x$ | $x \cdot \dfrac{1}{3} = \dfrac{x}{3}$ |
| Carlo | $\dfrac{1}{2}$ | $x$ | $x \cdot \dfrac{1}{2} = \dfrac{x}{2}$ |

◀ *Let x = the number of hours they worked together.*

Write an equation:
Part Regina did + part Carlo did = the whole job.

$$\frac{x}{3} + \frac{x}{2} = 1$$

◀ *The sum of the parts of a job is 1, the whole job.*

SOLVE ▶    Solve the equation.

$$\frac{x}{3} + \frac{x}{2} = 1$$

$$6\left(\frac{x}{3} + \frac{x}{2}\right) = 6 \cdot 1$$    ◀ *Multiply each side by the LCM of the denominators, 6.*

$$\overset{2}{\cancel{6}} \cdot \frac{x}{\cancel{3}}_1 + \overset{3}{\cancel{6}} \cdot \frac{x}{\cancel{2}}_1 = 6 \cdot 1$$    ◀ *Distribute the 6 and divide out the common factors.*

$$2x + 3x = 6$$    ◀ *Simplify each term.*
$$5x = 6$$
$$x = \frac{6}{5}$$

INTERPRET ▶    Part done by Regina + part done by Carlo = 1.

*Check.*

$$\begin{array}{c|c} x \cdot \dfrac{1}{3} + x \cdot \dfrac{1}{2} & 1 \\ \hline \dfrac{\overset{2}{\cancel{6}}}{5} \cdot \dfrac{1}{\cancel{3}}_1 + \dfrac{\overset{3}{\cancel{6}}}{5} \cdot \dfrac{1}{\cancel{2}}_1 & 1 \\ \dfrac{2}{5} + \dfrac{3}{5} & \\ 1 & \end{array}$$

*Replace x with* $\dfrac{6}{5}$. ▶

$$1 = 1, \text{ true}$$

**Thus,** the job will take $\dfrac{6}{5}$ h, or $1\dfrac{1}{5}$ h, if they work together.

If $x$ represents the number of hours to complete a job, then $\frac{1}{x}$ is the part of the job done in 1 h, and $\frac{n}{x}$ is the part of the job done in $n$ hours.

**Example 3**

**Working together, Pat and Pam can paint a house in 16 h. It would take Pam 40 h to do it alone. How long would it take Pat to do it alone?**

| | Part of Job Done in 1 h | Number of Hours Working Together | Part of Job Completed | |
|---|---|---|---|---|
| Pat | $\frac{1}{x}$ | 16 | $16 \cdot \frac{1}{x} = \frac{16}{x}$ | ◀ *Let $x$ = the number of hours it would take Pat to do the job alone.* |
| Pam | $\frac{1}{40}$ | 16 | $16 \cdot \frac{1}{40} = \frac{16}{40}$ | |

$$\frac{16}{x} + \frac{16}{40} = 1 \qquad ◀ \textit{The part Pat did + the part Pam did = 1.}$$

$$40 \cdot \overset{1}{\cancel{x}} \cdot \frac{16}{\cancel{x}} + \overset{1}{\cancel{40}} \cdot x \cdot \frac{16}{\cancel{40}} = 40 \cdot x \cdot 1 \qquad ◀ \textit{Multiply each side by the LCM of the denominators, } 40 \cdot x.$$

$$640 + 16x = 40x$$
$$640 = 24x$$
$$26\frac{2}{3} = x \qquad \textbf{So,} \text{ it would take Pat } 26\frac{2}{3} \text{ h, working alone.}$$

**Example 4**

**It takes Lois 3 times as long as Richie to mow a lawn. How long would it take each of them to do it alone, if together they can do it in 5 h?**

| | Part of Job Done in 1 h | Number of Hours Working Together | Part of Job Completed | |
|---|---|---|---|---|
| Richie | $\frac{1}{x}$ | 5 | $5 \cdot \frac{1}{x} = \frac{5}{x}$ | ◀ *Let $x$ = hours it would take Richie alone.* |
| Lois | $\frac{1}{3x}$ | 5 | $5 \cdot \frac{1}{3x} = \frac{5}{3x}$ | ◀ *Let $3x$ = hours it would take Lois alone.* |

$$\frac{5}{x} + \frac{5}{3x} = 1$$

$$3 \cdot \overset{1}{\cancel{x}} \cdot \frac{5}{\cancel{x}} + \overset{1}{\cancel{3}} \cdot \overset{1}{\cancel{x}} \cdot \frac{5}{\cancel{3x}} = 3 \cdot x \cdot 1$$

$$15 + 5 = 3x$$
$$20 = 3x$$
$$6\frac{2}{3} = x \qquad ◀ \textit{Richie's time, $x$, is } 6\frac{2}{3}. \textit{ Lois's time, $3x$, is } 3 \cdot 6\frac{2}{3} = 3 \cdot \frac{20}{3} = 20.$$

**So,** it would take Richie $6\frac{2}{3}$ h and Lois 20 h to do the job alone.

## Written Exercises

**Solve each problem.**

(A) 1. It takes Jack 5 h and Joan 10 h to paint a house. How long will it take them to do the job if they work together? $3\frac{1}{3}$ **h**

2. Bernie can prepare a dinner in 4 h. Belinda can do it in 5 h. How long will it take them to do the job if they work together? $2\frac{2}{9}$ **h**

3. Frank can build a fence in twice the time it would take Sandy. Working together, they can build it in 7 h. How long would it take each of them to do it alone? **Sandy: $10\frac{1}{2}$ h; Frank: 21 h**

4. Together, Fay and Rob can paint a house in 3 days. It would take Rob 3 times as long as it would take Fay to do it alone. How long would it take each of them, working alone? **Fay: 4 days, Rob: 12 days**

5. Working together, Salvatore and Marie can wallpaper an apartment in 14 h. It takes Marie 30 h to do it alone. How long would it take Salvatore to do it alone? $26\frac{1}{4}$ **h**

6. Together, Gail and Dick can address banquet invitations in 6 h. If it takes Gail 15 h to do it alone, how long would it take Dick to do the job alone? **10 h.**

7. It takes Florence 4 times as long as Ricardo to deliver newspapers each week. How long would it take each of them to do it alone if they can do the weekly job together in 5 h? **Ricardo, $6\frac{1}{4}$ h; Florence, 25 h**

8. Working together, Mr. Gassner and his son can build a house in 4 months. It takes Mr. Gassner twice as long as it takes his son to do it alone. How long would each take to do it alone? **Mr. Gassner, 12 months; Son, 6 months.**

(B) 9. To do a job alone, it would take Anita 3 h, Randy 2 h, and Yvonne 6 h. How long would it take to do the job if they all worked together? **1 h**

10. To do a job alone, it would take Shawn 3 days, Mona 4 days, and Harriet 7 days. How long would it take them if they work together to complete the job? $1\frac{23}{61}$ **days**

11. Eric can mow a lawn in 2 h, and Muna can do it in 4 h. How long would it take Barbara to do it if, working together, all three can do the job in 1 h? **4 h**

12. One pipe can fill a pool in 5 h. A second pipe can do it in 4 h, but a third pipe needs 7 h. How long would it take to fill the pool if all three pipes are used? $1\frac{57}{83}$ **h**

(C) 13. It takes Jake 14 h to repair a car's transmission. After he had worked for 7 h, Alice began to help him. Together, they finished the job in 3 more hours. How long would it take Alice to do the job alone? $10\frac{1}{2}$ **h**

14. Ray can assemble a television kit in 14 h. After he had worked for 3 h, Dodi joined him, and they finished the kit in 4 more hours. How long would it take Dodi to build the kit if she worked alone? **8 h**

15. A pipe can fill a swimming pool in 24 h, whereas an outlet pipe can empty the pool in 36 h. How long will it take to fill the now empty pool if both pipes are operating at the same time? **72 h**

16. Judy and Abe can do an inventory in $1\frac{1}{2}$ h if they work together. It takes Abe 3 h less than twice as long as it takes Judy to do it alone. How long would it take each of them to do it alone? **Abe: 3 h**

 **Judy: 3 h**

# COMPLEX RATIONAL EXPRESSIONS

*Objective*

## To simplify complex rational expressions in which the denominators of the individual rational expressions are monomials

The first two expressions are *complex fractions*; the third is a *complex rational expression*.

$$\dfrac{\dfrac{1}{2}}{8} \begin{matrix} \rbrace\ \text{numerator} \\[6pt] \text{denominator} \end{matrix} \qquad \dfrac{\dfrac{5}{2}}{3} \begin{matrix} \rbrace\ \text{numerator} \\[6pt] \rbrace\ \text{denominator} \end{matrix} \qquad \dfrac{2a + \dfrac{1}{6}}{\dfrac{a}{6} + \dfrac{2}{3}} \begin{matrix} \rbrace\ \text{numerator} \\[6pt] \rbrace\ \text{denominator} \end{matrix}$$

---

Definitions:
Complex
fraction,
Complex
rational
expression

A **complex fraction** is a fraction whose numerator, or denominator, or both numerator and denominator contain at least one fraction.

A **complex rational expression** is a rational expression whose numerator, or denominator, or both numerator and denominator contain at least one rational expression.

---

You can simplify a complex rational expression by using the property $\dfrac{a}{b} = \dfrac{a \cdot c}{b \cdot c}$, where $c$ is the LCM of all the denominators of the individual rational expressions contained in the complex rational expression.

*Example 1*   **Simplify** $\dfrac{\dfrac{2}{5}}{\dfrac{1}{3}}$.

$$\frac{\dfrac{2}{5}}{\dfrac{1}{3}} = \frac{3 \cdot 5 \cdot \dfrac{2}{5}}{3 \cdot 5 \cdot \dfrac{1}{3}}$$   ◀ *Multiply both the numerator and the denominator by 3 · 5, the LCM of all the denominators.*

$$= \frac{3 \cdot \cancel{5} \cdot \dfrac{2}{\cancel{5}}}{\cancel{3} \cdot 5 \cdot \dfrac{1}{\cancel{3}}}$$   ◀ *Divide out the common factors.*

$$= \frac{3 \cdot 1 \cdot 2}{1 \cdot 5 \cdot 1}$$

$$= \frac{6}{5}$$   ◀ $\dfrac{6}{5}$ *is equivalent to* $\dfrac{\dfrac{2}{5}}{\dfrac{1}{3}}$.

Sometimes, as in Example 2, it is helpful to rewrite a complex rational expression so that each term is in $\frac{a}{b}$ form.

## *Example 2*    **Simplify.**

$$\frac{2a + \dfrac{1}{6}}{\dfrac{a}{6} + \dfrac{2}{3}} = \frac{\dfrac{2a}{1} + \dfrac{1}{3 \cdot 2}}{\dfrac{a}{3 \cdot 2} + \dfrac{2}{3}}$$    ◄ *Rewrite 2a as $\dfrac{2a}{1}$, $\dfrac{a}{b}$ form. Factor all denominators.*

$$= \frac{3 \cdot 2\left(\dfrac{2a}{1} + \dfrac{1}{3 \cdot 2}\right)}{3 \cdot 2\left(\dfrac{a}{3 \cdot 2} + \dfrac{2}{3}\right)}$$    ◄ *Multiply both the numerator and the denominator by the LCM of all the denominators, 3 · 2.*

$$= \frac{3 \cdot 2 \cdot \dfrac{2a}{1} + \overset{1}{\cancel{3}} \cdot \overset{1}{\cancel{2}} \cdot \dfrac{1}{\underset{1 \ \ 1}{\cancel{3} \cdot \cancel{2}}}}{\underset{1 \ \ 1}{\cancel{3} \cdot \cancel{2}} \cdot \dfrac{a}{\underset{1 \ \ 1}{\cancel{3} \cdot \cancel{2}}} + \overset{1}{\cancel{3}} \cdot 2 \cdot \dfrac{2}{\underset{1}{\cancel{3}}}}$$    ◄ *Distribute 3 · 2 and divide out the common factors.*

$$= \frac{3 \cdot 2 \cdot 2a + 1 \cdot 1 \cdot 1}{1 \cdot 1 \cdot a + 1 \cdot 2 \cdot 2} = \frac{12a + 1}{a + 4}$$

In Example 3, the LCM of the denominators contains a power of a variable.

## *Example 3*    **Simplify.**

$$\frac{\dfrac{3}{5} - \dfrac{2}{a}}{\dfrac{1}{10a} + \dfrac{1}{2a^2}} = \frac{\dfrac{3}{5} + \dfrac{-2}{a}}{\dfrac{1}{5 \cdot 2 \cdot a} + \dfrac{1}{2 \cdot a \cdot a}}$$    ◄ $\dfrac{3}{5} - \dfrac{2}{a} = \dfrac{3}{5} + \dfrac{-1 \cdot 2}{a} = \dfrac{3}{5} + \dfrac{-2}{a}$

◄ *Factor the denominators.*

$$= \frac{5 \cdot 2 \cdot a \cdot a\left(\dfrac{3}{5} + \dfrac{-2}{a}\right)}{5 \cdot 2 \cdot a \cdot a\left(\dfrac{1}{5 \cdot 2 \cdot a} + \dfrac{1}{2 \cdot a \cdot a}\right)}$$    ◄ *The LCM of all the denominators is 5 · 2 · a · a. Multiply both the numerator and the denominator by the LCM.*

$$= \frac{\overset{1}{\cancel{5}} \cdot 2 \cdot a \cdot a \cdot \dfrac{3}{\underset{1}{\cancel{5}}} + 5 \cdot 2 \cdot a \cdot \overset{1}{\cancel{a}} \cdot \dfrac{-2}{\underset{1}{\cancel{a}}}}{\underset{1 \ \ 1 \ \ 1}{\cancel{5} \cdot \cancel{2} \cdot \cancel{a}} \cdot a \cdot \dfrac{1}{\underset{1 \ \ 1 \ \ 1}{\cancel{5} \cdot \cancel{2} \cdot \cancel{a}}} + 5 \cdot \underset{1 \ \ 1 \ \ 1}{\cancel{2} \cdot \cancel{a} \cdot \cancel{a}} \cdot \dfrac{1}{\underset{1 \ \ 1 \ \ 1}{\cancel{2} \cdot \cancel{a} \cdot \cancel{a}}}}$$

$$= \frac{2 \cdot a \cdot a \cdot 3 + 5 \cdot 2 \cdot a \cdot (-2)}{a \cdot 1 + 5 \cdot 1} = \frac{6a^2 - 20a}{a + 5}$$

Sometimes, the equivalent expression that results from simplifying a complex rational expression can be simplified further.

## Example 4 Simplify.

$$\frac{1 - \dfrac{7}{a} + \dfrac{10}{a^2}}{1 - \dfrac{25}{a^2}} = \frac{a^2\left(\dfrac{1}{1} + \dfrac{-7}{a} + \dfrac{10}{a^2}\right)}{a^2\left(\dfrac{1}{1} - \dfrac{25}{a^2}\right)}$$ ◀ *The LCM of all the denominators is $a \cdot a$, or $a^2$. Multiply both the numerator and the denominator by the LCM.*

$$= \frac{a^2 \cdot \dfrac{1}{1} + a^2 \cdot \dfrac{-7}{a} + a^2 \cdot \dfrac{10}{a^2}}{a^2 \cdot \dfrac{1}{1} + a^2 \cdot \dfrac{-25}{a^2}}$$

$$= \frac{a^2 - 7a + 10}{a^2 - 25} = \frac{(a - 5)(a - 2)}{(a + 5)(a - 5)}, \text{ or } \frac{a - 2}{a + 5}$$ ◀ *Factor numerator and denominator. Then, divide out common factors.*

## Written Exercises

**Simplify.**

(A) 1. $\dfrac{\dfrac{1}{5} - \dfrac{3}{10}}{\dfrac{2}{3}}$

2. $\dfrac{\dfrac{2}{3} + \dfrac{1}{6}}{\dfrac{1}{2} + \dfrac{1}{3}}$ $\quad 1$

3. $\dfrac{4a + \dfrac{1}{3}}{\dfrac{a}{5} + \dfrac{7}{15}}$ $\quad \dfrac{60a + 5}{3a + 7}$

4. $\dfrac{5a + \dfrac{1}{2}}{7a + \dfrac{1}{2}}$ $\quad \dfrac{10a + 1}{14a + 1}$

5. $\dfrac{\dfrac{2b}{7} + \dfrac{1}{14}}{\dfrac{b}{2} + \dfrac{3}{7}}$ $\quad \dfrac{4b + 1}{7b + 6}$

6. $\dfrac{\dfrac{6}{m} - \dfrac{2}{9}}{\dfrac{1}{3} + \dfrac{4}{m^2}}$ $\quad \dfrac{54m - 2m^2}{3m^2 + 36}$

7. $\dfrac{\dfrac{7}{2} - \dfrac{3}{b}}{\dfrac{5}{b^2} + \dfrac{11}{14}}$ $\quad \dfrac{49b^2 - 42b}{70 + 11b^2}$

8. $\dfrac{\dfrac{8}{x^2} - \dfrac{3}{x}}{\dfrac{11}{x} + \dfrac{1}{x^2}}$ $\quad \dfrac{8 - 3x}{11x + 1}$

(B) 9. $\dfrac{1 - \dfrac{5}{m} + \dfrac{4}{m^2}}{1 - \dfrac{16}{m^2}}$ $\quad \dfrac{m - 1}{m + 4}$

10. $\dfrac{1 + \dfrac{12}{x} + \dfrac{27}{x^2}}{\dfrac{1}{x} + \dfrac{9}{x^2}}$ $\quad x + 3$

11. $\dfrac{1 - \dfrac{9}{a} + \dfrac{14}{a^2}}{\dfrac{1}{a} - \dfrac{7}{a^2}}$ $\quad a - 2$

12. $\dfrac{b + \dfrac{1}{4}}{b^2 + \dfrac{3}{4}b + \dfrac{1}{8}}$ $\quad \dfrac{2}{2b + 1}$

13. $\dfrac{1 - \dfrac{4}{x} - \dfrac{45}{x^2}}{2 + \dfrac{7}{x} - \dfrac{15}{x^2}}$ $\quad \dfrac{x - 9}{2x - 3}$

14. $\dfrac{\dfrac{a^2}{6} + \dfrac{7a}{12} + \dfrac{1}{2}}{\dfrac{1}{2a} + \dfrac{1}{a^2}}$ $\quad \dfrac{2a^3 + 3a^2}{6}$

(C) 15. $\dfrac{\dfrac{a^2x}{y} + a^2 - \dfrac{9x}{y} - 9}{\dfrac{x^2}{y} + \dfrac{3x}{y} + x + 3}$ $\quad \dfrac{a^2 - 9}{x + 3}$

16. $\dfrac{1 - \dfrac{13}{x^2} + \dfrac{36}{x^4}}{\dfrac{1}{x^2} - \dfrac{1}{x^3} - \dfrac{6}{x^4}}$ $\quad (x + 3)(x - 2)$

17. $\dfrac{\dfrac{1}{x} - \dfrac{7}{x^2} + \dfrac{12}{x^3}}{\dfrac{1}{x} + \dfrac{4}{x^2} - \dfrac{32}{x^3}} - \dfrac{x^2 - 3x}{x + 8}$ $\quad \dfrac{-x^2 + 4x - 3}{x + 8}$

# COMPLEX RATIONAL EXPRESSIONS WITH POLYNOMIAL DENOMINATORS

*Objective*

**To simplify complex rational expressions in which the denominators of the individual rational expressions are polynomials**

Many complex rational expressions contain rational expressions with polynomial denominators. The procedure for simplifying such expressions is the same as that of the last lesson. Recall that first you factor each denominator, where possible, and then multiply the numerator and the denominator of the complex rational expression by the LCM of all the denominators.

*Example 1*     **Simplify** $\dfrac{\dfrac{4}{m-5}+\dfrac{6}{m+2}}{\dfrac{9}{m^2-3m-10}+\dfrac{1}{m-5}}$.

$$\dfrac{\dfrac{4}{m-5}+\dfrac{6}{m+2}}{\dfrac{9}{m^2-3m-10}+\dfrac{1}{m-5}} = \dfrac{\dfrac{4}{m-5}+\dfrac{6}{m+2}}{\dfrac{9}{(m-5)(m+2)}+\dfrac{1}{m-5}}$$  ◀ *Factor $m^2 - 3m - 10$.*

$$= \dfrac{(m-5)(m+2)\left(\dfrac{4}{m-5}+\dfrac{6}{m+2}\right)}{(m-5)(m+2)\left[\dfrac{9}{(m-5)(m+2)}+\dfrac{1}{m-5}\right]}$$  ◀ *Multiply both the numerator and the denominator by $(m-5)(m+2)$, the LCM of all the denominators.*

$$= \dfrac{(m-5)(m+2)\cdot\dfrac{4}{m-5} + (m-5)(m+2)\cdot\dfrac{6}{m+2}}{(m-5)(m+2)\cdot\dfrac{9}{(m-5)(m+2)} + (m-5)(m+2)\cdot\dfrac{1}{m-5}}$$  ◀ *Use the distributive property. Then, divide out common factors.*

$$= \dfrac{(m+2)4 + (m-5)6}{1\cdot 9 + (m+2)\cdot 1}$$

$$= \dfrac{4m+8+6m-30}{9+m+2}$$

$$= \dfrac{10m-22}{m+11}$$

When you simplify a complex rational expression, you should always check to see if the result can be simplified further. Recall that you can check by factoring the numerator and the denominator and then dividing out the common factors. This is illustrated in Example 2 on the next page.

## Example 2    Simplify.

$$\dfrac{\dfrac{2}{x} - \dfrac{10}{x^2 + 7x}}{\dfrac{5}{x+7} + \dfrac{2}{x}} = \dfrac{\dfrac{2}{x} + \dfrac{-10}{x(x+7)}}{\dfrac{5}{x+7} + \dfrac{2}{x}}$$ ◀ $-\dfrac{10}{x^2+7x} = \dfrac{-1(10)}{x^2+7x} = \dfrac{-10}{x^2+7x};\ x^2 + 7x = x(x+7)$

$$= \dfrac{x(x+7)\left[\dfrac{2}{x} + \dfrac{-10}{x(x+7)}\right]}{x(x+7)\left[\dfrac{5}{x+7} + \dfrac{2}{x}\right]}$$ ◀ *Multiply the numerator and the denominator by the LCM of all the denominators, $x(x+7)$.*

$$= \dfrac{\overset{1}{\cancel{x}}(x+7) \cdot \dfrac{2}{\underset{1}{\cancel{x}}} + \overset{1}{\cancel{x}}\overset{1}{(\cancel{x+7})} \cdot \dfrac{-10}{\underset{1}{\cancel{x(x+7)}}}}{x\overset{1}{\cancel{(x+7)}} \cdot \dfrac{5}{\underset{1}{\cancel{x+7}}} + \overset{1}{\cancel{x}}(x+7) \cdot \dfrac{2}{\underset{1}{\cancel{x}}}}$$ ◀ *Use the distributive property. Then, divide out common factors.*

$$= \dfrac{(x+7)2 + 1 \cdot (-10)}{x \cdot 5 + (x+7)2}$$

$$= \dfrac{2x + 14 - 10}{5x + 2x + 14}$$

$$= \dfrac{2x + 4}{7x + 14} = \dfrac{2\overset{1}{\cancel{(x+2)}}}{7\underset{1}{\cancel{(x+2)}}},\ \text{or}\ \dfrac{2}{7}$$ ◀ *Factor the numerator and the denominator. Divide out common factors.*

In the process of simplifying a complex rational expression, you may have to multiply two binomials. Recall that $(m+2)(m-5)$ *is not* $m^2 - 10$, *but* $m^2 - 3m - 10$.

A situation of this type occurs in Example 3.

## Example 3    Simplify.

$$\dfrac{m + 2 + \dfrac{3}{m-5}}{\dfrac{4}{m-5} + 1} = \dfrac{(m-5)\left[\dfrac{m+2}{1} + \dfrac{3}{m-5}\right]}{(m-5)\left[\dfrac{4}{m-5} + \dfrac{1}{1}\right]}$$ ◀ *Think: $m - 5$ cannot be factored, so multiply the numerator and the denominator by the LCM of all the denominators, $m - 5$.*

$$= \dfrac{(m-5) \cdot \dfrac{m+2}{1} + \overset{1}{\cancel{(m-5)}} \cdot \dfrac{3}{\underset{1}{\cancel{m-5}}}}{\overset{1}{\cancel{m-5}} \cdot \dfrac{4}{\underset{1}{\cancel{m-5}}} + (m-5) \cdot 1}$$

$$= \dfrac{(m-5)(m+2) + 3}{4 + m - 5} = \dfrac{m^2 - 3m - 10 + 3}{4 + m - 5},\ \text{or}\ \dfrac{m^2 - 3m - 7}{m - 1}$$ ◀ *Simplest form*

## Written Exercises

**Simplify.**

(A) 1. $\dfrac{\dfrac{4}{a^2 - 11a + 30} + \dfrac{3}{a - 6}}{\dfrac{2}{a - 6} + \dfrac{1}{a - 5}}$  $\dfrac{3a - 11}{3a - 16}$

2. $\dfrac{\dfrac{7}{x^2 - 4} + \dfrac{2}{x - 2}}{\dfrac{6}{x - 2} + \dfrac{5}{x + 2}}$  $\dfrac{2x + 11}{11x + 2}$

3. $\dfrac{\dfrac{2}{a} - \dfrac{16}{a^2 + 6a}}{\dfrac{4}{a + 6} - \dfrac{1}{a}}$  $\dfrac{2}{3}$

4. $\dfrac{\dfrac{3}{m} - \dfrac{12}{m^2 + 2m}}{\dfrac{2}{m + 2} - \dfrac{1}{m}}$  3

5. $\dfrac{b + 8 + \dfrac{5}{b - 3}}{1 - \dfrac{1}{b - 3}}$  $\dfrac{b^2 + 5b - 19}{b - 4}$

6. $\dfrac{a + 4 + \dfrac{2}{a - 3}}{\dfrac{5}{a - 3} + 1}$

7. $\dfrac{\dfrac{5}{m + 3} + \dfrac{-10}{m^2 + 3m}}{\dfrac{4}{m + 3} - \dfrac{1}{m}}$  $\dfrac{5m - 10}{3m - 3}$

8. $\dfrac{4 + \dfrac{3}{x - 8}}{2 + \dfrac{1}{x - 8}}$  $\dfrac{4x - 29}{2x - 15}$

9. $\dfrac{\dfrac{5b}{2b - 1} + \dfrac{1}{3}}{\dfrac{7}{3} + \dfrac{4}{2b - 1}}$  $\dfrac{17b - 1}{14b + 5}$

10. $\dfrac{\dfrac{6}{a - 3} + \dfrac{2}{2a + 1}}{\dfrac{7a}{2a^2 - 5a - 3}}$  2

11. $\dfrac{\dfrac{2}{x^2 - 4x - 21} - \dfrac{5}{x - 7}}{\dfrac{6}{x - 7} + \dfrac{3}{x + 3}}$  $\dfrac{-5x - 13}{9x - 3}$

12. $\dfrac{\dfrac{5y}{y^2 - 16}}{\dfrac{10}{y - 4} + \dfrac{10}{y + 4}}$  $\dfrac{1}{4}$

(B) 13. $\dfrac{\dfrac{x + 4}{x} - \dfrac{3}{x + 5}}{\dfrac{x - 1}{x^2 + 5x} + \dfrac{3}{x + 5}}$  $\dfrac{x^2 + 6x + 20}{4x - 1}$

14. $\dfrac{\dfrac{1}{a + 5} + \dfrac{1}{a - 3}}{\dfrac{2a^2 - 3a - 5}{a^2 + 2a - 15}}$  $\dfrac{2}{2a - 5}$

15. $\dfrac{\dfrac{x + 3}{x + 4} + \dfrac{x + 9}{x - 7}}{\dfrac{3}{x^2 - 3x - 28}}$

16. $\dfrac{\dfrac{x + 4}{x} - \dfrac{7}{x + 3}}{\dfrac{x + 4}{x^2 + 3x} + \dfrac{3}{x + 3}}$  $\dfrac{x^2 + 12}{4x + 4}$

17. $\dfrac{\dfrac{2a + 4}{a + 8} - \dfrac{a - 1}{a - 2}}{\dfrac{a^2 - 49}{a^2 + 6a - 16}}$  $\dfrac{a}{a + 7}$

18. $\dfrac{\dfrac{x + 4}{x - 6} + \dfrac{x + 1}{x + 2}}{\dfrac{4}{x^2 - 4x - 12}}$

(C) 19. $\dfrac{\dfrac{5}{a^2 - 5a + 6} + \dfrac{4}{a^2 + 5a + 6}}{\dfrac{7}{a^2 - a - 6} - \dfrac{2}{a^4 - 13a^2 + 36}}$

20. $\dfrac{\dfrac{1}{ax^2 - a + bx^2 - b} + \dfrac{2}{x - 1}}{\dfrac{4}{x - 1} - \dfrac{3}{ax + a + bx + b}}$

21. $\dfrac{\dfrac{3}{b^4 - 17b^2 + 16} + \dfrac{4}{b^2 - 3b - 4}}{\dfrac{7}{b^2 + 5b + 4} - \dfrac{2}{b^2 - 5b + 4}}$

22. $\dfrac{\dfrac{4}{ax + ay - bx - by} - \dfrac{3}{a - b}}{\dfrac{2}{a - b} + \dfrac{2}{x^2 + xy}}$

## NON-ROUTINE PROBLEMS

From where on the surface of the earth can you travel 100 km due south, then 100 km due west, and, finally, 100 km due north and arrive again at your starting point?

**The North Pole.**

19. $\dfrac{9a^2 + 5a + 54}{7a^2 + 7a - 44}$  20. $\dfrac{1 + 2ax + 2bx + 2a + 2b}{4ax + 4bx + 4a + 4b - 3x + 3}$  21. $\dfrac{4b^2 + 12b - 13}{5b^2 - 45b + 20}$  22. $\dfrac{-3x^2 + 4x - 3xy}{2x^2 + 2xy + 2a - 2b}$

# TRANSPORTATION

Unique conditions evolve unique design in the many vehicles used for transporting people, raw materials, and manufactured goods in our energy-conscious world.

Trucks at sunrise. Dispatchers for trucking firms are responsible for planning efficient routes and schedules for the drivers.

Subway control room.
Computerized control centers
monitor all train movements and
can predict possible accidents,
then prevent them automatically.

Pneumatic tube trains. An early
subway system used compressed
air and lever systems to start and
stop vehicles.

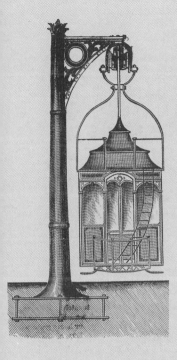

Stephenson's Suspension Street RR. In this illustration of a monorail, the carriage wheels ride in an overhead track supported by columns.

Memphis monorail. Modern monorails are electrically powered with gear-driven cables.

The power for this transport is provided by a horse, making it similar to horse-drawn canal barges.

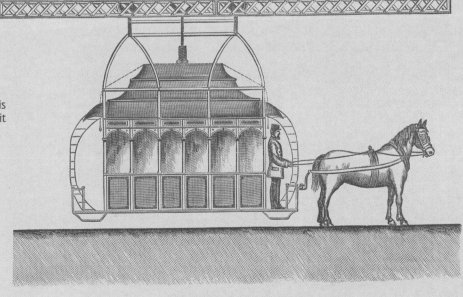

People biking alongside traffic. Bicycle transport is both flexible and efficient in traffic since variable and derailleur gear systems multiply human muscle power.

In transporting people into cities for work, subways average 16 mi/h; trolleys, 5–6 mi/h; and buses, up to 50 mi/h.

Carriages in New Orleans. Rutted, unpaved roads were very uncomfortable to ride on until heavier wagon springs were used to support large carriages.

One horse sashay. Road conditions and the kind of draft animals available determine some aspects of the design of carts and wagons.

Camel carts on way to Jaipur. People in Third World countries, which lack railroad systems and paved roads, must rely on traditional means of transportation.

Queen Elizabeth II. The screw motion of the large, nearly flat propellers allows a big passenger liner to maintain a constant speed on the ocean.

Steamboat Niagara. The transport of raw materials, manufactured goods, and passengers upstream became possible when the flat-bottom hull was combined with the paddle wheel in the steamboat.

Tankers today can be 1,150 ft long (350 m) and weigh 500,000 t, the largest of any humanmade structures other than the very largest skyscrapers.

Loading passengers on jet. Bernoulli's principle, the variation in pressure below and above an aerodynamically designed airplane wing, produces the lift for the heavier-than-air vehicles so vital to worldwide transport.

Orville Wright flying over Berlin, Germany. Early planes like Wright's were fragile and could carry only a pilot and one passenger, and had enough fuel for only one hour of flight.

Loading cargo on jet. The jet engine expels hot gases, making use of Newton's Third Law of Motion that for every action there is an equal but opposite reaction.

Freight yard. About 85% of a railroad's income comes from freight. An average train has over 70 cars and hauls so much tonnage that two locomotives may be required.

Train in mountains. It took five years to lay the track for the first transcontinental railroad, but by 1869 the versatile steam locomotive made possible safe passage and trade through the rugged mountains.

Truck on overpass. Modern trucks with diesel engines convert 35% of their fuel into useful energy, nearly twice the efficiency of a gasoline-fueled internal-combustion engine.

The capacity of early trucks was small but provided transportation to areas inaccessible by railroad.

# APPLICATIONS

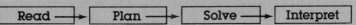

Read ➞ Plan ➞ Solve ➞ Interpret

A *parallel series circuit* is shown at the right. The battery is the source of power. When you turn on the power, voltage from the battery forces current to flow. The current flows to $A$, then splits. Part of the current flows through $R_1$, and part of it flows through $R_2$. The two branches join up again at $B$ and form a single current that flows back to the battery. Resistors control the flow of current such that the greater the resistance, the less the current.

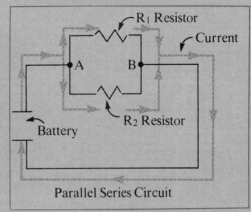

R₁ Resistor

Current

A   B

Battery

R₂ Resistor

Parallel Series Circuit

A volume control knob on a stereo set is a resistor. Reducing the sound resists the flow of current. The formula for the total resistance, $R$, for two resistors in a parallel series circuit is

$$R = \frac{1}{\dfrac{1}{R_1} + \dfrac{1}{R_2}}.$$

Resistance is measured in *ohms*. The symbol for ohms is $\Omega$.

*Example*

**Two resistors in a parallel series circuit have resistances of 5 $\Omega$ and 10 $\Omega$. Find the total resistance.**

Use $R = \dfrac{1}{\dfrac{1}{R_1} + \dfrac{1}{R_2}}$, where $R_1 = 5$ and $R_2 = 10$.

$R = \dfrac{1}{\dfrac{1}{5} + \dfrac{1}{10}}$ ◄ *Substitute 5 for $R_1$ and 10 for $R_2$.*

$= \dfrac{10 \cdot 1}{10\left(\dfrac{1}{5} + \dfrac{1}{10}\right)}$ ◄ *Multiply the numerator and the denominator by the LCM of all the denominators, 10.*

$= \dfrac{10 \cdot 1}{\overset{2}{\cancel{10}} \cdot \dfrac{1}{\underset{1}{5}} + \overset{1}{\cancel{10}} \cdot \dfrac{1}{\underset{1}{10}}} = \dfrac{10}{2 + 1} = \dfrac{10}{3}$

So, the total resistance is $3\frac{1}{3}$ $\Omega$, ($3\frac{1}{3}$ ohms).

Find the total resistance in a parallel series circuit with the given resistances.

1. 20 $\Omega$, 30 $\Omega$   2. 40 $\Omega$, 20 $\Omega$   3. 15 $\Omega$, 20 $\Omega$   4. 100 $\Omega$, 25 $\Omega$   5. 60 $\Omega$, 80 $\Omega$

1. 12 $\Omega$   2. 13.33 $\Omega$   3. 8.6 $\Omega$   4. 20 $\Omega$   5. 34.3 $\Omega$

# FORMULAS AND LITERAL EQUATIONS     10.5

*Objectives*     **To solve a formula, or literal equation, for one of its variables**
**To evaluate a formula for one of its variables, given the values of its other variables**

An equation such as $ax + b = c$ is called a **literal equation** because it contains only letters or variables. You can solve this equation for $x$ in the same way that you would solve the equation $3x + 2 = 7$.

*Example 1*     **Solve $ax + b = c$ for $x$.**

$$ax + b = c$$
$$ax + b - b = c - b$$
$$ax = c - b$$
$$x = \frac{c - b}{a}$$

*Think:*
$$3x + 2 = 7$$
$$3x + 2 - 2 = 7 - 2$$
$$3x = 5$$
$$x = \frac{5}{3}$$

Recall that a formula for the perimeter of a rectangle is $p = 2l + 2w$.
If you know the perimeter and the width of the rectangle and you want to find its length, you may find it desirable to solve the formula for $l$. This procedure is shown in Example 2.

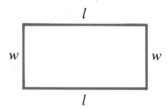

*Example 2*     **Solve $p = 2l + 2w$ for $l$. Find $l$ if $p = 48$ cm and $w = 6$ cm.**

$$p = 2l + 2w$$
$$p - 2w = 2l + 2w - 2w \quad \blacktriangleleft \text{ } Subtract\ 2w\ from\ each\ side\ to\ get\ 2l\ alone.$$
$$p - 2w = 2l$$
$$\frac{p - 2w}{2} = l \quad\quad\quad\quad \blacktriangleleft \text{ } Divide\ each\ side\ by\ 2.$$

You now have a formula for $l$ in terms of $p$ and $w$.

$$l = \frac{p - 2w}{2}$$
$$l = \frac{48 - 2 \cdot 6}{2} \quad \blacktriangleleft \text{ } Substitute\ 48\ for\ p\ and\ 6\ for\ w.$$
$$l = \frac{48 - 12}{2}$$
$$l = \frac{36}{2}, \text{ or } 18 \quad\quad \textbf{So,} \text{ the length of the rectangle is 18 cm.}$$

The formula $A = prt + p$ can be used to find: (1) the amount of money in an account after the interest has been credited and (2) the total amount due on a loan.

## Example 3

**Solve $A = prt + p$ for $r$.**

*Think:* Get the term containing $r$ alone on one side of the equation.

$$A = prt + p$$
$$A - p = prt + p - p \quad \blacktriangleleft \textit{ Subtract p from each side to get prt alone.}$$
$$A - p = p \cdot r \cdot t$$
$$\frac{A - p}{p \cdot t} = \frac{\cancel{p} \cdot r \cdot \cancel{t}}{\cancel{p} \cdot \cancel{t}} \quad \blacktriangleleft \textit{ Divide each side by p $\cdot$ t to get r alone.}$$
$$\frac{A - p}{pt} = r$$

If a formula, or literal equation, contains fractions, the easiest way to solve it for one of the variables is to *eliminate* the fractions first. You can eliminate the fractions by multiplying each side of the equation by the LCM of the denominators.

## Example 4

**Solve $k = \frac{7}{3}m + 20$ for $m$. Then, evaluate $m$ if $k = 62$.**

$$k = \frac{7}{3}m + 20$$
$$3 \cdot k = 3\left(\frac{7}{3}m + 20\right)$$
$$3k = \overset{1}{\cancel{3}} \cdot \frac{7}{\underset{1}{\cancel{3}}}m + 3 \cdot 20$$
$$3k = 7m + 60$$
$$3k - 60 = 7m \quad \blacktriangleleft \textit{ Get 7m alone by subtracting 60 from each side.}$$
$$\frac{3k - 60}{7} = m \quad \blacktriangleleft \textit{ Divide each side by 7.}$$

Now, substitute 62 for $k$ in $m = \dfrac{3 \cdot k - 60}{7}$

$$m = \frac{3 \cdot 62 - 60}{7}$$
$$m = \frac{186 - 60}{7}$$
$$m = \frac{126}{7}, \text{ or } 18$$

To solve an equation for a variable that appears in more than one term, get all of those terms on one side of the equation, then factor, as in example 5.

## Example 5　　Solve $ax = m - 4x$ for $x$.

$$ax = m - 4x$$
$$ax + 4x = m - 4x + 4x$$　◀ Add $4x$ to each side to get all $x$ terms alone on one side.
$$ax + 4x = m$$
$$x(a + 4) = m$$　◀ Factor: $ax + 4x = x(a + 4)$.

$$x = \frac{m}{a + 4}$$　◀ Divide each side by $a + 4$: $\dfrac{x(\overset{1}{\cancel{a + 4}})}{\cancel{a + 4}} = \dfrac{m}{a + 4}$; $x = \dfrac{m}{a + 4}$.

# Written Exercises

**Solve for $x$. (Ex. 1–20)**　　6. $x = m - 3c$　　10. $x = \frac{4m + 12}{b}$　　11. $x = \frac{4d - 5k}{7}$　　13. $x = \frac{5a + b}{2}$

Ⓐ　1. $bx = 5$　$x = \frac{5}{b}$　　　2. $rx = 3$　$x = \frac{3}{r}$　　　3. $4x = 4m$　$x = m$　　　4. $tw = 6x$　$x = \frac{tw}{6}$
　5. $x + 7b = 0$　$x = -7b$　6. $3c + x = m$　　　7. $x - 9t = 0$　$x = 9t$　　8. $3 = x - 9k$　$x = 3 + 9k$
　9. $5x - b = r$　$x = \frac{r + b}{5}$　10. $bx - 12 = 4m$　　11. $5k + 7x = 4d$　　12. $9x + 11a = 7t$　$x = \frac{7t - 11a}{9}$

13. $\dfrac{x}{5} = \dfrac{a}{2} + \dfrac{b}{10}$　　14. $\dfrac{n}{7} + \dfrac{t}{2} = \dfrac{x}{14}$　　15. $\dfrac{x}{3a} = \dfrac{t}{g}$　$x = \dfrac{3at}{g}$　　16. $\dfrac{x}{l} - k = b$　$x = l(b + k)$

17. $6 - 3x = kx$　　18. $bx = 21 - dx$　　19. $nx = r + px$　　20. $9ax = 2ax + g$　$x = \frac{g}{7a}$
14. $x = 2n + 7t$　　17. $x = \frac{6}{3 + k}$　　18. $x = \frac{21}{b + d}$　　19. $x = \frac{r}{n - p}$

21. Solve $t = 6s$ for $s$. Then, find $s$ if
　$t = 36$.　$s = \frac{t}{6}$; $s = 6$

22. Solve $I = prt$ for $r$. Then, find $r$ if
　$p = 60$, $I = 240$, and $t = 4$.　$r = \frac{I}{pt}$; $r = 1$

23. Solve $p = 3t + b$ for $t$. Then, find $t$ if
　$p = 48$ and $b = 12$.　$t = \frac{p - b}{3}$; $t = 12$

24. Solve $y = t + 14d$ for $d$. Then, find $d$ if
　$y = 62$ and $t = 6$.　$d = \frac{y - t}{14}$; $d = 4$

25. Solve $c = 2\pi r$ for $r$. Then, find $r$ if
　$c = 18.84$ and $\pi = 3.14$.　$r = \frac{c}{2\pi}$; $r = 3$

26. Solve $V = lwh$ for $w$. Then, find $w$ if
　$v = 270$, $l = 15$, and $h = 6$.

27. Solve $S = \frac{1}{5}tm^2$ for $t$. Then, find $t$ if
　$S = 45$ and $m = 3$.　$t = \frac{5s}{m^2}$; $t = 25$

28. Solve $V = \pi r^2 h$ for $h$. Then, find $h$ if
　$V = 25.12$, $\pi = 3.14$, and $r = 2$.
　26. $w = \frac{v}{lh}$; $w = 3$　　28. $h = \frac{v}{\pi r^2}$; 2

**Solve for $x$.**

Ⓑ　29. $7x - 5a = 4x + 16a$　　30. $11a + 4bx = 3c$　　31. $kx - b = 3d + 7b$　$x = \frac{3d + 8b}{k}$
32. $4bx - 2c = 9bx + 8c$　　33. $5 - 3bx = -2b + 2bx$　　34. $-5a + 4bx = 7a + 10bx$　$x = \frac{-2a}{b}$
35. $a^2 - ax + 10 = 5x - 7a$　　36. $5x - ax = a^2 - 25$　　37. $px + qx = p^2 - 6pq - 7q^2$

38. $\dfrac{1}{a} + \dfrac{1}{b} = \dfrac{1}{x}$　$x = \dfrac{ab}{b + a}$　　39. $a = \dfrac{x - v}{t}$　$x = at + v$　　40. $\dfrac{x}{a} + \dfrac{x}{b} = 1$　$x = \dfrac{ab}{b + a}$　$x = p - 7q$

　29. $x = 7a$　30. $x = \frac{3c - 11a}{4b}$　32. $x = \frac{-2c}{b}$　33. $x = \frac{5 + 2b}{5b}$　35. $x = a + 2$　36. $x = -a - 5$
**Solve each equation for the variable indicated.**

Ⓒ　41. $p = \dfrac{2}{3}(m - 16)$ for $m$　$m = \dfrac{3}{2}p + 16$　　42. $T = \pi(r + l)$ for $l$　$l = \dfrac{T}{\pi} - r$

43. $V = \dfrac{A}{x_1} - \dfrac{B}{x_2}$ for $x_2$　$x_2 = \dfrac{-Bx_1}{Vx_1 - A}$　　44. $w = -G\left(c - \dfrac{Q}{m}\right)$ for $Q$　$Q = m\left(\dfrac{w}{G} + c\right)$

45. $s = \dfrac{a - ar^2}{1 - r}$ for $a$　$a = \dfrac{s}{1 + r}$　　46. $a_1 = b\left(\dfrac{a_2}{b} + 4\right)$ for $a_2$　$a_2 = a_1 - 4b$

47. $\dfrac{rs}{a} + \dfrac{st}{b} = 1$ for $s$　$s = \dfrac{ab}{br + at}$　　48. $ry^2 - b^2r - ty^2 + b^2t = y^2 - b^2$ for $r$
　　　　　　　　　　　　　　　　　　　　　$r = 1 + t$

# DISTANCE PROBLEMS

*Objective*    **To solve problems about motion**

If a car travels 80 km/h, after 4 h it will have traveled 80 · 4, or 320 km. This suggests the following formula for finding the distance traveled at a given speed for a given length of time.

| Distance formula | Distance = (rate) (time). <br> $d = r \cdot t$ |
|---|---|

You can use this formula together with the four basic problem-solving steps to solve problems about motion.

*Example 1*    **Two cars traveled in opposite directions from the same starting point. The rate of one car was 20 km/h less than the rate of the other car. After 5 h, the cars were 700 km apart. Find the rate of each car.**

READ ▶    You are given the time and the total distance.
You are asked to find the rate of each car.

PLAN ▶    Make a sketch and represent each rate algebraically.

Let $x$ = the rate of the faster car.
Let $x - 20$ = the rate of the slower car.
Organize the data in a chart.

|  | **Rate** | **Time** | **Distance ($d = r \cdot t$)** |
|---|---|---|---|
| Faster car | $x$ | 5 | $5x$ |
| Slower car | $x - 20$ | 5 | $5(x - 20)$, or $5x - 100$ |

Write an equation: $\begin{pmatrix}\text{distance of} \\ \text{faster car}\end{pmatrix} + \begin{pmatrix}\text{distance of} \\ \text{slower car}\end{pmatrix} = \begin{pmatrix}\text{total distance} \\ \text{apart}\end{pmatrix}.$

$$5x \quad + \quad 5x - 100 = 700$$

SOLVE ▶
$$10x - 100 = 700$$
$$10x = 800$$
$$x = 80$$

The rate of the faster car, $x$, is 80 km/h.
The rate of the slower car, $x - 20$, is $80 - 20$, or 60 km/h.

INTERPRET ▶    *Check*.    Total distance apart is 700 km.

| $5 \cdot 80 + 5 \cdot 60$ | 700 |
|---|---|
| $400 + 300$ |  |
| $700$ | $700 = 700$, true |

Sometimes, problems about motion involve two people traveling the same distance. In such cases, you can write an equation by setting the algebraic representations for these distances equal to each other.

*Example 2*    **Irv started out in his car traveling at 60 km/h. Two hours later, Jerilyn left from the same point. She drove along the same road at 75 km/h. How many hours had she driven before she caught up with Irv?**

Make a sketch. Represent the time each drove algebraically.
    Let $x$ = the time Jerilyn drove.
Let $x + 2$ = the time Irv drove.  ◀ *He drove 2 h longer.*

*Represent the distance* ▶
*each traveled.*

| | Rate | Time | Distance ($d = r \cdot t$) |
|---|---|---|---|
| Irv | 60 | $x + 2$ | $60(x + 2)$, or $60x + 120$ |
| Jerilyn | 75 | $x$ | $75x$ |

(Irv's distance) = (Jerilyn's distance).  ◀ *Write an equation.*

$$60x + 120 = 75x$$
$$120 = 15x$$
$$8 = x$$  **So,** it took Jerilyn 8 h to catch up to Irv.  ◀ *Check on your own.*

*Example 3*    **The O'Learys drove to the beach at 75 km/h. They returned later, in heavy traffic, at 50 km/h. It took 1 h longer to return home than it did to get to the beach. How long did it take to get home?**

Make a sketch. Represent the time it took each way algebraically.
    Let $x$ = the time it took to go to the beach.
Let $x + 1$ = the time it took to return home.  ◀ *It took 1 h longer to get home.*

| | Rate | Time | Distance ($d = r \cdot t$) |
|---|---|---|---|
| To beach | 75 | $x$ | $75x$ |
| Return | 50 | $x + 1$ | $50(x + 1)$, or $50x + 50$ |

(Distance to beach) = (distance of return trip).  ◀ *Write an equation.*
$$75x = 50x + 50$$
$$25x = 50$$   ◀ *Solve.*
$$x = 2$$
The time it took to get to the beach, $x$, is 2 h.
The time it took to return home, $x + 1$, is $2 + 1$, or 3 h.
**So,** it took 3 h to return home.  ◀ *Check on your own.*

Chapter Ten

The formula $d = r \cdot t$ can be solved for $t$ by dividing each side by $r$, as shown.

$$\frac{d}{r} = \frac{\cancel{r} \cdot t}{\cancel{r}}, \text{ or } \frac{d}{r} = t$$

This formula is read $\dfrac{\text{distance}}{\text{rate}} = \text{time}$. You will find it more convenient to use this formula to solve motion problems in which each person traveled for the same amount of time.

*Example 4*   **It took Dave the same time to drive 210 km as it took Julep to drive 195 km. Julep's speed was 5 km/h less than Dave's speed. How fast did Dave drive?**

Let $x$ = Dave's rate.
Let $x - 5$ = Julep's rate.   ◀ *Represent the rate of each.*

|       | **Rate** | **Time** | **Distance** |
|-------|----------|----------|--------------|
| Dave  | $x$      | $\dfrac{210}{x}$ | 210 |
| Julep | $x - 5$  | $\dfrac{195}{x - 5}$ | 195 |

◀ $\dfrac{\text{distance}}{\text{rate}} = \text{time}.$

(Dave's time) = (Julep's time).   ◀ *The time is the same for each.*

$$\frac{210}{x} = \frac{195}{x - 5}$$

$$\overset{1}{\cancel{x}} \cdot (x - 5) \cdot \frac{210}{\underset{1}{\cancel{x}}} = x \cdot (\overset{1}{\cancel{x - 5}}) \cdot \frac{195}{\underset{1}{\cancel{x - 5}}}$$   ◀ *Multiply each side by the LCM of the denominators, $x(x - 5)$.*

$$(x - 5)\,210 = 195x$$
$$210x - 1{,}050 = 195x$$   ◀ *Use the distributive property.*
$$-1{,}050 = -15x$$   ◀ *Add $-210x$ to each side.*
$$70 = x$$   **So,** Dave drove at the rate of 70 km/h.

# Reading in Algebra

**John walked at the rate of 3 km/h for 2 h. He returned home over the same route, walking at 6 km/h for 1 h. Use the problem to choose the one correct ending for each statement below.**

**1.** The total time John walked was $\underline{\overset{?}{\phantom{xx}}}$
   a. 2 h          b. 1 h          c. 3 h   **c**

**2.** The total distance he walked was $\underline{\overset{?}{\phantom{xx}}}$
   a. 6 km          b. 3 km          c. 12 km   **c**

**3.** The average speed for the entire walk was $\underline{\overset{?}{\phantom{xx}}}$
   a. 4 km/h          b. 6 km/h          c. 9 km/h   **a**

## Oral Exercises

**Use the distance formula, $d = r \cdot t$, to find the indicated value.**

1. rate, 60 km/h
   time, 3 h
   distance? **180 km**

2. distance, 200 km
   rate, 40 km/h
   time? **5 h**

3. distance, 16 km
   time, 2 h
   rate? **8 km/h**

4. rate, 40 km/h
   time, $2\frac{1}{2}$ h
   distance? **100 km**

## Written Exercises

(A) **Solve each problem.** 3. 1st car: $62\frac{1}{2}$ km/h; 2nd car: $52\frac{1}{2}$ km/h.    4. Car 1: $67\frac{1}{2}$ km/h; Car 2: $52\frac{1}{2}$ km/h.

1. Two trains left the same station at the same time and traveled in opposite directions. The E train averaged 130 km/h. The A train's speed was 110 km/h. In how many hours were they 480 km apart? **2 hours**

2. Two trucks started toward each other at the same time from towns 270 km apart. One truck averaged 70 km/h, and the other averaged 65 km/h. After how many hours did they pass each other? **2 hours**

3. Two cars traveled in opposite directions from the same starting point. The rate of one car was 10 km/h faster than the rate of the other car. After 4 h the cars were 460 km apart. Find each car's rate.

4. Two cars traveled in opposite directions from the same starting point. The rate of one car was 15 km/h faster than the rate of the other car. After 2 h, the cars were 240 km apart. Find each car's rate.

5. Leroy started out in his car at the rate of 50 km/h. One hour later, Tina left from the same point driving along the same road at 75 km/h. How long did it take her to catch up with Leroy? **2 hours**

6. A train left a station traveling 100 km/h. A second train left 2 h later and headed in the same direction at 125 km/h. After how many hours did the second train overtake the first train? **8 hours**

7. The Brodskys drove to the mountains at 60 km/h. They returned at 40 km/h. The return trip took 2 h longer than the trip to the mountains. How long did it take them to get home? **6 hours**

8. The Joneses drove to their parents' house at 45 km/h. The return trip took 1 h less because they traveled at 60 km/h. How long did it take them to get home?

   **3 hours**

(B) 9. Jill hiked 20 km in the same time it took Marie to hike 28 km. Jill's rate was 2 km/h less than Marie's. How fast did each girl walk? **Jill: 5 km/h; Marie: 7 km/h.**

10. It took Bill the same time to drive 135 km as it took Harriet to drive $\frac{135}{x} = \frac{180}{x+15}$ 180 km. Harriet's speed was 15 km/h faster than Bill's. How fast did Bill drive? **45 km/h.**

11. Jake cycled 20 km to the beach. He returned cycling 1 km/h faster. The $x$ $\frac{20}{x}$ $20$ total time for the round trip was 9 h. Find his rate for each part of the $x+1$ $\frac{20}{x+1}$ $20$ trip. **Going rate: 4 km/h   Returning rate: 5 km/h**

    $$\frac{20}{x} + \frac{20}{x+1} = 9$$

(C) 12. Miguel walked 10 km into the country. He returned walking 3 km/h slower. The total time for the round trip was 7 h. How fast did he walk on each part of the trip? **Going rate: 5 km/h.**
    **Returning rate: 2 km/h.** $\frac{10}{x} + \frac{10}{x-3} = 7$

# PROBLEM SOLVING

1. Working together, it takes Bill and Gina 6 days to paint a house. It takes Bill twice as long as Gina to do it alone. How long would it take each to do the job alone?
**Gina: 9 days; Bill: 18 days**

2. A football team won 7 games and lost 2 games. What percentage of the games played did the team win? **$77\frac{7}{9}$%**

3. Two cars traveled in opposite directions from the same starting point. The rate of one car was 10 km/h less than the rate of the other car. After 10 h, the cars were 500 km apart. Find the rate of each car.
**Car 1: 20 km/h; Car 2: 30 km/h.**

4. Larry started out in his car traveling at 60 km/h. One hour later, Sheila left from the same point and drove along the same road at 80 km/h. How many hours did she drive before she caught up with Larry? **3 h**

5. To do a job alone, it would take Mike 4 days, Marsha 3 days, and Tim 9 days. How long would it take them if they all work together? **$1\frac{11}{25}$ days**

6. The length of a rectangle is 6 cm more than the width. The perimeter is 40 cm. Find the length and the width of the rectangle. **l: 13 cm, w: 7 cm.**

7. Brenda is paid $160 a week plus a commission of $30 on each television set she sells. How many must she sell to earn at least $310? **At least 5 televisions**

8. Six times the sum of a number and 8 is the same as 20 more than 10 times the number. Find the number. **7**

9. Find three consecutive even integers whose sum is 48. **14, 16, 18**

10. Find two consecutive odd integers whose product is 99. **9, 11 or −11, −9**

11. A 28-ft rope is cut into two pieces so that the longer piece is 3 times the smaller piece. Find the length of each piece. **7, 21**

12. In a high school with 1,200 students, 3 out of 5 students go out for sports. How many students are in the sports program? **720**

13. The smaller of two numbers is 4 less than the greater number. If the greater number is increased by 3 times the smaller, the result is 36. Find the two numbers. **12, 8**

14. One side of a triangle is twice as long as the second side. The remaining side is 4 m longer than the second side. The perimeter is 24 m. Find the length of each side of the triangle. **10 m, 5 m, 9 m**

**Vocabulary**

complex rational expression [10.3]
extraneous solution [10.1]
literal equation [10.5]

Solve. Check for extraneous solutions. [10.1]

**1.** $\dfrac{5a}{a^2 - 2a - 3} = \dfrac{4}{a - 3} + \dfrac{2}{a + 1}$  **a = 2**

**2.** $\dfrac{4}{y + 5} - \dfrac{2}{y - 8} = \dfrac{3}{y^2 - 3y - 40}$  **y = 22½**

**3.** $\dfrac{7}{m^2 - 3m} - \dfrac{4}{m} = \dfrac{5}{3 - m}$  **m = −19**

**4.** $\dfrac{3x}{x^2 - 5x + 4} = \dfrac{2}{x - 4} - \dfrac{3}{1 - x}$  **x = 7**

**5.** $\dfrac{x + 1}{x - 4} = \dfrac{4x + 4}{x^2 - 4x}$  **6.** $\dfrac{x^2 - 8x}{x - 5} = \dfrac{15}{5 - x}$

  **x = −1**    **x = 3**

Solve each problem. [10.2]

**7.** Working together, two carpenters can build a house in 7 months. It takes one of them twice as long as it takes the other to do it alone. How long would it take each of them to do it alone? **10½ mo; 21 mo.**

**8.** To do a job alone, it would take Rose 4 h, Bob 3 h, and Shelley 5 h. How long would it take to do the job if they all worked together? **1 13/47 h.**

Simplify. [10.3]  **11.** x − 2  **12.** $\dfrac{2x}{8x - 29}$

**9.** $\dfrac{\frac{2}{3} + \frac{1}{2}}{\frac{1}{6} + \frac{1}{10}}$  **35/8**  **10.** $\dfrac{\frac{4}{b^2} + \frac{3}{b}}{\frac{2}{b} + \frac{5}{3b^2}}$  **$\frac{12 + 9b}{6b + 5}$**

**11.** $\dfrac{1 - \frac{9}{x} + \frac{14}{x^2}}{\frac{1}{x} - \frac{7}{x^2}}$  **12.** $\dfrac{\frac{6}{x^2 - 7x + 12} + \frac{2}{x - 4}}{\frac{3}{x - 4} + \frac{5}{x - 3}}$

**13.** $\dfrac{x - 5 + \frac{3}{x - 2}}{\frac{3}{x - 2} + 2}$  **14.** $\dfrac{\frac{1}{x + 3} + \frac{1}{x - 3}}{\frac{2x^2 + x - 10}{x^2 - 4x - 21}}$

**13.** $\dfrac{x - 7x + 13}{2x - 1}$  **14.** $\dfrac{2x(x - 7)}{(2x + 5)(x - 2)(x - 3)}$

Solve for x. [10.5]  **15.** $x = \dfrac{4y + 5}{a}$

**15.** $ax - 5 = 4y$   **16.** $7 - 5x = tx$

**17.** $\dfrac{x}{5} = \dfrac{b}{10} + \dfrac{c}{2}$   **18.** $4x - ax = 16 - a^2$

**19.** $ax + bx = a^2 - b^2$  **x = a − b**

**16.** $x = \dfrac{7}{t + 5}$   **17.** $x = \dfrac{b + 5c}{2}$   **18.** $x = 4 + a$

**20.** Solve $p = 2s + b$ for $s$. Then, find $s$ if $p = 52$ and $b = 4$.  **$s = \frac{p - b}{2}$, s = 24**

**21.** Solve $k = \frac{1}{4}fd^2$ for $f$. Then, find $f$ if $k = 7$ and $d = 2$.  **$f = \frac{4k}{d^2}$, f = 7**

Solve each problem. [10.6]  **80 km/h, 100 km/h**

**22.** Two trains traveled in opposite directions from the same starting point. The rate of one train was 20 km/h faster than the rate of the other. After 2 h, they were 360 km apart. Find the rate of each train.

**23.** Two cars started toward each other at the same time from towns 170 km apart. One car's rate was 45 km/h, and the other car's rate was 40 km/h. After how many hours did they meet? **2 h**

★ **24.** Solve. Check for extraneous solutions. [10.1]  **x = 1**

$\dfrac{1}{3x^2 + 5x - 2} = \dfrac{4}{3x^2 + 8x - 3} - \dfrac{4}{x^2 + 5x + 6}$

★ **25.** Simplify: [10.4]  **$\dfrac{b^2 + b - 3}{-b^2 + 4b + 12}$**

$\dfrac{\dfrac{6}{b^4 - 13b^2 + 36} + \dfrac{2}{b^2 - b - 6}}{\dfrac{1}{b^2 - 5b + 6} - \dfrac{3}{b^2 + b - 6}}.$

★ **26.** Solve for $x$: $\dfrac{xy}{a} - \dfrac{tx}{b} = \dfrac{4}{ab}$. [10.5]  **$x = \dfrac{4}{by - at}$**

★ **27.** Solve the problem: [10.2]
A pipe can fill a tank in 8 h, whereas an outlet pipe can empty the tank in 12 h. How long will it take to fill the now empty tank if both pipes are operating at the same time? **24 h.**

# CHAPTER TEN TEST

(A) Exercises are 1, 2, 3, 6–8, 10, 11, 13–15, 17. (C) Exercises are starred. The rest are (B) Exercises.

Simplify.

**1.** $\dfrac{\frac{2}{3} + \frac{3}{5}}{\frac{1}{5} + \frac{7}{15}}$   $\dfrac{19}{10}$

**2.** $\dfrac{\frac{4}{a^2} + \frac{3}{a}}{\frac{7}{a} + \frac{5}{3a^2}}$   $\dfrac{9a + 12}{21a + 5}$

**3.** $\dfrac{1 - \frac{2}{x} - \frac{15}{x^2}}{\frac{1}{x} - \frac{5}{x^2}}$   $x + 3$

**4.** $\dfrac{\frac{7y}{y^2 - 25}}{\frac{3}{y - 5} + \frac{2}{y + 5}}$   $\dfrac{7y}{5y + 5}$

**5.** $\dfrac{\frac{1}{x + 6} + \frac{1}{x + 2}}{\frac{x^2 + 11x + 28}{x^2 + 8x + 12}}$   $\dfrac{2}{x + 7}$

Solve. Check for extraneous solutions.

**6.** $\dfrac{3}{x - 6} - \dfrac{4}{x + 2} = \dfrac{5}{x^2 - 4x - 12}$   $x = 25$

**7.** $\dfrac{6}{a + 3} - \dfrac{7}{8 - a} = \dfrac{1}{a^2 - 5a - 24}$   $a = \dfrac{28}{13}$

**8.** $\dfrac{4}{a^2 - 16} + \dfrac{2}{a - 4} = \dfrac{5}{a + 4}$   $a = \dfrac{32}{3}$

**9.** $x - \dfrac{5x}{x - 4} = \dfrac{20}{4 - x}$   $x = 5$

Solve for $x$.

**10.** $7 - 2x = mx$   $x = \frac{7}{m + 2}$

**11.** $\dfrac{x}{5a} = \dfrac{p}{r}$   $x = \dfrac{5ap}{r}$

**12.** $3bx - 5c = 7bx + 4c$   $x = \frac{-9c}{4b}$

**13.** Solve $a = mpq$ for $p$. Then, find $p$ if $a = 32$, $m = 2$, and $q = 7$.   $p = \dfrac{a}{mq}$;   $p = \dfrac{16}{7}$

**14.** Solve $l = \dfrac{1}{5}mp^2$ for $m$. Then, find $m$ if $p = 2$ and $l = 20$.   $m = \dfrac{5l}{p^2}$, $m = 25$

Solve each problem.

**15.** Working together, Maurice and Marni can mow a lawn in 7 h. It would take Maurice 10 h to do it alone. How long would it take Marni to do it alone?   $23\frac{1}{3}$ h

**16.** To do a job alone, it would take Hank 3 days, Alicia 4 days, and Jaime 6 days. How long should it take them to do the job if they work together?   $1\frac{1}{3}$ **days.**

**17.** The Jacksons drove to a resort at 80 km/h. They returned that night at 60 km/h. It took 1 h less to get to the resort than it did to return. How long did the trip home take? **4 h.**

**18.** It took Tony the same time to drive 275 km as it took Tricia to drive 325 km. Tony's rate was 10 km/h less than Tricia's. How fast was Tricia driving?   **65 km/h**

★ **19.** Solve. Check for extraneous solutions.
$$\dfrac{4}{3x^2 - 11x - 4} - \dfrac{4}{2x^2 - 7x - 4} = \dfrac{-2}{6x^2 + 5x + 1}$$
$x = -4$

★ **20.** Solve for $x$.
$xy^2 - b^2x = cy - bm - bc + my$   $x = \dfrac{c + m}{y + b}$

# COMPUTER ACTIVITIES

## Solving a Formula for Any Given Variable

You can use a computer to evaluate any variable in an equation, given values for the other variables. The formula for the perimeter of a rectangle is $p = 2l + 2w$. This formula is solved below for each of the three variables in BASIC notation.

$$P = 2 * L + 2 * W, \quad L = (P - 2 * W)/2, \quad W = (P - 2 * L)/2$$

```
100    PRINT "THIS PROGRAM SOLVES PERIMETER PROBLEMS."
110    PRINT "DO YOU WANT TO SOLVE FOR P?"
120    PRINT "YES = Y, NO = N"
130    INPUT A$
140    IF A$ = "Y" THEN 240
150    PRINT "DO YOU WANT TO SOLVE FOR L?"
160    INPUT A$
170    IF A$ = "Y" THEN 290
180    PRINT "TO SOLVE FOR W, TYPE IN VALUES FOR P, L.
       "
190    INPUT P,L
200    LET W = (P - 2 * L) / 2
210    PRINT "W = "W".";
220    PRINT " THE FORMULA IS W = (P - 2 * L)/2."
230    GOTO 330
240    PRINT "TYPE IN VALUES FOR L, W."
250    INPUT L,W
260    LET P = 2 * L + 2 * W
270    PRINT "P = "P".   THE FORMULA IS P =2 * L + 2 *
       W."
280    GOTO 330
290    PRINT "TYPE IN VALUES FOR P, W."
300    INPUT P,W
310    LET L = (P - 2 * W) / 2
320    PRINT "L = "L".   THE FORMULA IS L = (P - 2 * W)
       /2."
330    END
```

See the Computer Section beginning on page 488 for more information.

## Exercises   For Ex. 4, see TE Answer Section.

**Type in the program above and RUN it three times for the values:**

1. L = 3, W = 2 **10**          2. P = 20, L = 5 **5**

3. P = 18, W = 6 **3**

4. Write a program that will solve for each of the variables in the formula $k = \frac{1}{2}fd^2$ or in BASIC notation K = .5 * F * (D $\wedge$ 2).

Directions: Choose the one best answer to each question.

**1.** By how much does $a - b$ exceed $b - a$?

*B*  (A) $2(b - a)$  (B) $2(a - b)$  (C) $-1$
   (D) $\dfrac{a - b}{b - a}$   (E) $a - b$

**2.** If $x$ pencils cost $y$ cents, how many pencils can be bought for $d$ dollars?

*A*  (A) $\dfrac{100dx}{y}$  (B) $\dfrac{100xy}{d}$  (C) $\dfrac{100x}{dy}$
   (D) $\dfrac{dx}{y}$  (E) $\dfrac{100y}{dx}$

**3.** If $m = \dfrac{ab}{a + c}$, then $a = \underline{\ ?\ }$

*C*  (A) $\dfrac{b}{c}$  (B) $-\dfrac{c}{b}$  (C) $\dfrac{cm}{b - m}$
   (D) $\dfrac{cm}{m - b}$  (E) $\dfrac{ab - cm}{m}$

**4.** If $x = 5$, then $x^2 + 5 = \underline{\ ?\ }$

*B*  (A) 125  (B) 30  (C) 15
   (D) 5  (E) 0

**5.** What is the value of $x$ if $\dfrac{t + w}{t - w} = \dfrac{x}{w - t}$?

*E*  (A) $t - w$  (B) $t + w$  (C) $t^2 - w^2$
   (D) $-1$  (E) $-t - w$

**6.** If Marc makes a profit of 20% on the selling price of a hamburger, what percentage profit does he make on the cost?

*B*  (A) 20%  (B) 25%  (C) 40%
   (D) 80%  (E) none of these

**7.** If $\dfrac{1}{x} = \dfrac{1}{a} + \dfrac{1}{b}$, then the value of $x$ is $\underline{\ ?\ }$

*D*  (A) $a + b$  (B) $\dfrac{2ab}{a + b}$  (C) $\dfrac{a + b}{ab}$
   (D) $\dfrac{ab}{a + b}$  (E) $ab$

**8.** The length of a rectangle is increased by 50%. By what percentage would the width have to be decreased to keep the same area?

*A*  (A) $33\frac{1}{3}\%$  (B) 50%  (C) $66\frac{2}{3}\%$
   (D) 150%  (E) 200%

**9.** If $x = 27$ and $\dfrac{x}{a} = \dfrac{3}{b}$, then $\dfrac{b}{a} = \underline{\ ?\ }$

*A*  (A) $\frac{1}{9}$  (B) 9  (C) 24
   (D) 30  (E) 81

**10.** What is the value of $x$ in terms of $y$ and $z$ if $\dfrac{5}{x} = \dfrac{\dfrac{5}{y}}{z}$?

*E*  (A) $y - z$  (B) $\dfrac{1}{yz}$  (C) $\dfrac{z}{y}$
   (D) $\dfrac{y}{z}$  (E) $yz$

**11.** A formula for the circumference of a circle is $C = 2\pi r$, where $r$ is the radius. What is the maximum number of glass tumblers, each with a circumference of $4\pi''$, that can be placed on a table $24''$ by $28''$?

*B*  (A) 36  (B) 42  (C) 56
   (D) 96  (E) 192

# 11 GRAPHING IN A COORDINATE PLANE

## HISTORY

### René Descartes 1596–1650

*René Descartes was born in France in the sixteenth century. As a philosopher, his basic premise was "I think, therefore I am."*

As a mathematician, Descartes' contributions to algebra include improved notation and the rule of signs. Perhaps his greatest contribution to mathematics was his application of algebra to geometry. This marked the beginning of modern analytic geometry.

Descartes developed the early ideas about coordinates. The one-to-one correspondence between the set of all ordered pairs of real numbers and the set of all points in a plane is called a Cartesian coordinate system in honor of Descartes. The plane is called a Cartesian plane.

*Project*    Prepare a report on the life and accomplishments of René Descartes.

# COORDINATES OF POINTS IN A PLANE    11.1

*Objectives*    **To give the coordinates of a point in a plane**
**To identify the quadrant in which a given point is located**

You have used a number to name the coordinate of a point on a number line. For example, on the number line below,

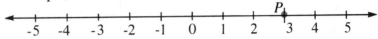

3 is the coordinate of point $P$.

Suppose you wanted to find a seat in a theater. If the usher told you that your seat was number three, you would probably ask for the row number. You need two numbers to locate the seat. You can use an *ordered pair of numbers* to represent the 3rd seat in the 20th row, as shown below.

$$(3, \quad 20)$$

3rd seat————↑   ↑————20th row

The order of the numbers in the pair is important. Because the first number refers to the seat and the second number refers to the row, the ordered pair (20, 3) refers to the 20th seat in the 3rd row. You will use ordered pairs of numbers to name the coordinates of points in a plane. To do so, you will have to use two number lines, as described below.

Draw a horizontal number line. At the point with coordinate 0, draw a vertical number line. The two number lines are *perpendicular* to each other. (They form right angles.) Each line has the same zero point, called the *origin*.

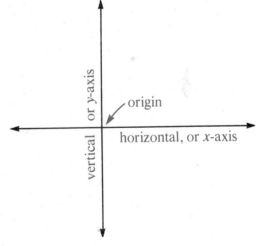

The horizontal number line is called the *x-axis*. The *positive* direction is to the *right* on the x-axis.

The vertical number line is called the *y-axis*. The *positive* direction is *upward* on the y-axis.

At the right, notice that two coordinates, 4 on the x-axis and 2 on the y-axis, correspond to $P$. This suggests that you must use an ordered pair of numbers to describe the location of a point in a plane. The technique is illustrated in Example 1 on the next page.

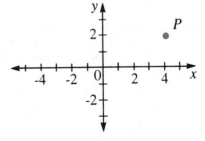

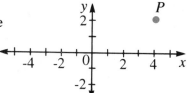

## Example 1

**What ordered pair of numbers describes the location of point *P*?**

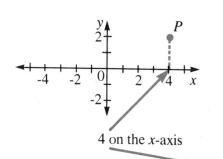

4 on the *x*-axis

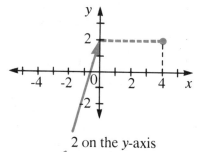

2 on the *y*-axis

(4,2)

The location of point *P* is described by the ordered pair:

(*x*-coordinate, *y*-coordinate).

(4, 2)

**Definitions:**
**Abscissa**
**ordinate**

The **abscissa** is the *x*-coordinate of an ordered pair and is always given first. The **ordinate** is the *y*-coordinate of an ordered pair and is always given second. For the ordered pair (4, 2), 4 is the abscissa and 2 is the ordinate.

The axes divide a coordinate plane into four *quadrants,* as shown at the right. Each point, except the origin, is located in one of the quadrants or on one of the axes. The origin, (0, 0), is on both axes. *P*(5, 3) means point *P* with *x*-coordinate 5 and *y*-coordinate 3.

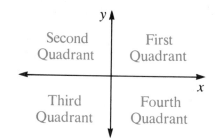

## Example 2

**Give the ordered pair and the quadrant for each of these points: *P*, *Q*, *R*, and *S*.**

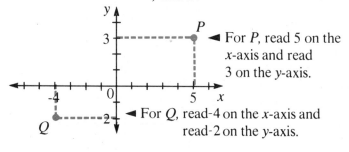

◄ For *P*, read 5 on the *x*-axis and read 3 on the *y*-axis.

◄ For *Q*, read-4 on the *x*-axis and read-2 on the *y*-axis.

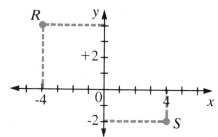

*P*(5, 3), first quadrant; *Q*(−4, −2), third quadrant;
*R*(−4, 4), second quadrant; *S*(4, −2), fourth quadrant

# Reading in Algebra

**Determine whether each statement is always true, sometimes true, or never true. Give a reason for your answer.**

1. The ordinate of an ordered pair is the $x$-coordinate. **N, Ordinate is $y$-coordinate.**
2. The origin is in the first quadrant. **N, Origin is at intersection of axes.**
3. The $x$-axis is the horizontal axis. **A**
4. A point is located in one of four quadrants. **S, A point at the origin is not in a quadrant.**
5. The abscissa of an ordered pair is positive. **S, Abscissa is positive in first and fourth quadrants.**

# Oral Exercises

1. abscissa, 3; ordinate 5  2. abscissa −4; ordinate, 2
3. abscissa, 6; ordinate, −5  4. abscissa, −8; ordinate, −1  5. abscissa, 2; ordinate, −3

**For each ordered pair, tell which number is the abscissa and which is the ordinate.**

1. $(3, 5)$   2. $(-4, 2)$   3. $(6, -5)$   4. $(-8, -1)$   5. $(2, -3)$   6. $(5, -1)$
7. $(-8, -2)$   8. $(0, 0)$   9. $(3, 6)$   10. $(5, 5)$   11. $(-3, 0)$   12. $(-4, -4)$

6. abscissa, 5; ordinate, −1  7. abscissa, −8; ordinate −2  8. abscissa, 0; ordinate, 0
9. abscissa, 3; ordinate, 6  10. abscissa, 5; ordinate, 5  11. abscissa, −3; ordinate, 0  12. abscissa, −4; ordinate, −4

# Written Exercises

**What ordered pair of numbers describes the location of the point?**

(A)
1. $A$ $(1, 1)$
2. $B$ $(-3, 3)$
3. $C$ $(-2, -1)$
4. $D$ $(3, -2)$
5. $E$ $(-5, -3)$
6. $F$ $(1, -2)$
7. $G$ $(-2, 1)$
8. $H$ $(5, 3)$

9. $I$ $(2, 1)$
10. $J$ $(-3, 2)$
11. $K$ $(-5, -1)$
12. $L$ $(-4, -3)$
13. $M$ $(5, -2)$
14. $N$ $(4, -3)$
15. $O$ $(4, 3)$
16. $P$ $(-1, 1)$

**Which points in the diagrams above are in the given quadrant?**

17. Quadrant 1   18. Quadrant 2   19. Quadrant 3   20. Quadrant 4
   *A, H; I, O*       *B, G; J, P*       *C, E; K, L*       *F, D; N, M*

**Using the graph at the right, find the point(s) whose coordinates satisfy the given condition.**

(B)
21. The abscissa is 3. **B**
22. The ordinate is −2. **F**
23. The abscissa is negative. **E, F**
24. The sum of the abscissa and the ordinate is 4. **B**
25. The ordinate is 2. **A, C, D, E**

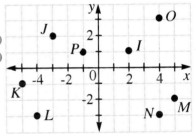

**Form an ordered pair of numbers such that the abscissa is the solution of the first equation and the ordinate is the solution of the second equation.**

(C)
26. $-3x + 5 = 23$; $-16 = 4y + 12$ $(-6, -7)$   27. $3x + 5 = x - (4 - x)$; $7y + 12 = -(6 - y)$
    $(-9, -3)$
28. $\dfrac{2x - 5}{3} = \dfrac{x - 1}{2}$; $\dfrac{5}{y + 2} = \dfrac{10}{y^2 + 2y}$   29. $2|x - 6| - 4 = 8$; $\dfrac{8}{y - 6} = 2$   $(12, 10)$; $(0, 10)$
    $(7, 2)$

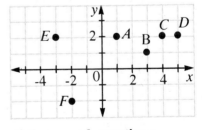

Coordinates of Points in a Plane

# POINT PLOTTING

*Objectives*     **To plot a point given its coordinates**
**To solve problems involving coordinates of points**

You will find it helpful to use graph paper to *plot*, or *graph*, points.
However, you will have to draw and label your own axes on the graph
paper. To plot a point such as $B(5, 2)$, you will have to take two steps, as
shown below.

*First,* start at the origin and move 5
units to the right.

*Second,* move 2 units up from
there and place a dot at the point
at which you stop. Label this
point.

$B$ is the graph of the ordered
pair $(5, 2)$.

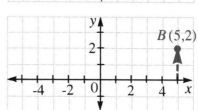

*Example 1*     **Plot $M(-4, 3)$.**

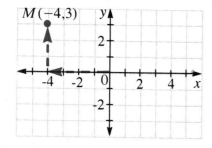

Start at the origin.
Move to the left 4 units.     ◀     *−4 is to the left*
Then move up 3 units.              *of the origin.*
Label the point.

From the illustrations above, you have probably noticed that the signs of the
coordinates tell you in which directions to move horizontally and vertically
from the origin. Study the pattern below.

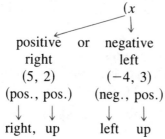

*Example 2*     **Plot the points: $A(-3, -2)$, $B(-1, 3)$, and $C(4, -1)$.**

$B(-1, 3)$ ▶

↙      ↘

*left 1      up 3*

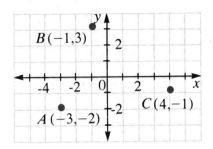

$A(-3, -2)$ ▶

↙      ↘

*left 3    down 2*

◀ $C(4, -1)$

↙      ↘

*right 4    down 1*

Sometimes, one of the coordinates of a point is 0. An example of this is $(-3, 0)$. The $-3$ tells you to move to the left 3 units. The 0 indicates that you are to move neither up nor down. Example 3 illustrates how to plot a point with a zero coordinate.

*Example 3*     **Plot $A(-3, 0)$.**                    **Plot $B(0, -4)$.**

Begin at the origin: $(-3, 0)$                Begin at the origin: $(0, -4)$

Move left 3. Move neither up nor down.        Move neither left nor right. Move down 4.

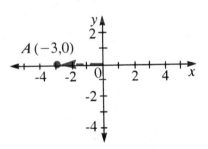

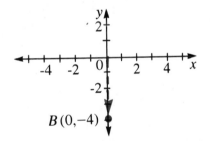

Example 3 and the two illustrations below suggest a pattern about the coordinates of points on the axes.

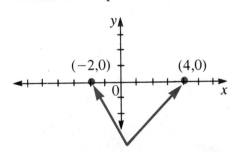

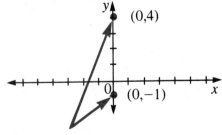

Any point on the $x$-axis has 0 as its $y$-coordinate (the second coordinate).

Any point on the $y$-axis has 0 as its $x$-coordinate (the first coordinate).

1. From origin, move 4 units to the right and 5 units up. 2. From origin, move 3 units to the right and 4
units down. 3. From origin, move 4 units to the left and 6 units up.
4. From origin, move 1 unit to the left and 6 units down.

## Oral Exercises

**For each coordinate tell in which direction and how far you would move
to plot the given point.**

| | | | | |
|---|---|---|---|---|
| **1.** $A(4, 5)$ | **2.** $B(3, -4)$ | **3.** $C(-4, 6)$ | **4.** $D(-1, -6)$ | **5.** $E(-8, 4)$ |
| **6.** $F(0, 1)$ | **7.** $P(9, 0)$ | **8.** $R(0, -10)$ | **9.** $S(-2, 0)$ | **10.** $T(-5, 5)$ |

5. From origin, move 8 units to the left and 4 units up. 6. From origin, 1 unit up.
7. From origin, 9 units to the right. 8. From origin, 10 units down. 9. From origin, 2 units to the left.
10. From origin, 5 units to the left and 5 units up.

## Written Exercises

Ⓐ **Plot. See TE Answer Section.**

| | | | | |
|---|---|---|---|---|
| **1.** $A(2, 2)$ | **2.** $B(-5, 1)$ | **3.** $C(-4, -6)$ | **4.** $D(5, -3)$ | **5.** $E(8, 0)$ |
| **6.** $F(0, 0)$ | **7.** $G(-6, 0)$ | **8.** $H(0, -1)$ | **9.** $K(0, -5)$ | **10.** $M(-7, 2)$ |
| **11.** $P(5, -8)$ | **12.** $Q(0, 4)$ | **13.** $R(7, 0)$ | **14.** $S(-9, 0)$ | **15.** $T(-5, -5)$ |

Ⓑ **16.** $W\left(4\frac{1}{2}, 3\frac{1}{2}\right)$    **17.** $Z\left(-3\frac{1}{2}, 2\frac{1}{2}\right)$    **18.** $N(3.5, -4.5)$    **19.** $U\left(-1\frac{1}{2}, 0\right)$

**20.** $L(-5.5, 0)$    **21.** $I\left(-2\frac{1}{2}, -2\frac{1}{2}\right)$    **22.** $J\left(1\frac{1}{2}, -4\frac{1}{2}\right)$    **23.** $V(2.5, -1.5)$

*Example*    $A(3, 1)$, $B(8, 1)$, and $C(8, 3)$ are
three vertices of a rectangle. Plot
the points. Then, find the coordinates
of the fourth vertex, $D$.

*D has the same x-coordinate
as A and the same y-coordinate* ▶
*as C.*

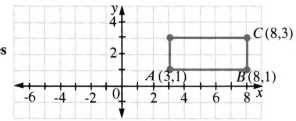

**So,** $D(3, 3)$ is the fourth vertex of the rectangle.

**$A$, $B$, and $C$ are three vertices of a rectangle. Plot the points. Find the
coordinates of the fourth vertex to complete the rectangle.**

**24.** $A(2, 4)$, $B(9, 4)$, and $C(9, 7)$   **D(2, 7)**     **25.** $A(-4, 2)$, $B(5, 2)$, and $C(5, -1)$   **D(-4, -1)**

**26.** $A(5, -1)$, $B(8, -1)$ and $C(8, -6)$   **D(5, -6)**   **27.** $A(-8, 6)$, $B(-5, 6)$, and $C(-5, -4)$

                                                             **D(-8, -4)**

## CUMULATIVE REVIEW

**1.** Write the ratio $3:5$ as a fraction. $\frac{3}{5}$   **2.** Evaluate $-3a^5$ if $a = -2$.   $^{96}$   **3.** Solve $|4x - 2| = 12$.   $\frac{7}{2}, -\frac{5}{2}$

## NON-ROUTINE PROBLEMS

What is the size of the angle formed by the hands of a clock when the time is 2:15? $22\frac{1}{2}°$

# SLOPE

*Objectives*

**To find the slope of a nonvertical line segment**
**To express slopes in terms of variables**

If the *x*-axis represents level ground and the segments $\overline{AB}$ and $\overline{AC}$ represent paths up a hill, $\overline{AC}$ would be more difficult to climb than $\overline{AB}$. The reason for this is that $\overline{AC}$ has a steeper, or greater, *slope* than $\overline{AB}$.

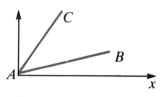

A mathematical description for the slope of a line segment is suggested by a carpenter's description for the slope of a roof of a house.

If the roof rises 3 units for every 4 units of horizontal run, a carpenter expresses the slope of the roof as the ratio *rise* to *run*.

At the right, the slope is $\dfrac{3}{4}$. $\begin{array}{l}\leftarrow \text{rise} \\ \leftarrow \text{run}\end{array}$

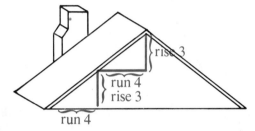

If you know the coordinates of two points on a line segment, you can describe the slope of the line segment in terms of rise over run, as shown in Example 1.

*Example 1*

**Find the slope of the line segment joining $A(3, 2)$ and $B(8, 6)$.**

*First,* plot the two points. *Then,* connect them with a line segment.

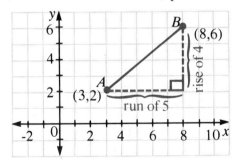

◀ *Sketching the right triangle will help you visualize the rise and the run.*

**So,** the slope of $\overline{AB} = \dfrac{4}{5}$. $\begin{array}{l}\blacktriangleleft\ rise \\ \blacktriangleleft\ run\end{array}$

Note that the slope of $\overline{AB}$ can be expressed as $\dfrac{\text{the difference of the } y\text{-coordinates}}{\text{the difference of the } x\text{-coordinates}}$

because $\begin{array}{l}y\text{-coord. of } B \rightarrow \\ x\text{-coord. of } B \rightarrow\end{array} \dfrac{6 - 2}{8 - 3} \begin{array}{l}\leftarrow y\text{-coord. of } A \\ \leftarrow x\text{-coord. of } A\end{array} = \dfrac{4}{5}$.

This suggests that the slope of a nonvertical segment can be found directly from the coordinates of the endpoints of the segment.

| Definition: Slope | $\text{slope} = \dfrac{\text{the difference of the } y\text{-coordinates}}{\text{the difference of the } x\text{-coordinates}}.$ |

*Example 2*   **Find the slope of $\overline{AB}$ for $A(6, 3)$ and $B(-2, -5)$.**

You can find the ratio $\dfrac{\text{the difference of the } y\text{-coordinates}}{\text{the difference of the } x\text{-coordinates}}$ in two ways.

First way

$$\dfrac{\text{slope}}{\text{of } \overline{AB}} = \dfrac{y\text{-coord. of } B - y\text{-coord. of } A}{x\text{-coord. of } B - x\text{-coord. of } A}$$

$$= \dfrac{-5 - 3}{-2 - 6}$$

$$= \dfrac{-8}{-8}$$

$$= 1$$

Second way

$$\dfrac{\text{slope}}{\text{of } \overline{AB}} = \dfrac{y\text{-coord. of } A - y\text{-coord. of } B}{x\text{-coord. of } A - x\text{-coord. of } B}$$

$$= \dfrac{3 - (-5)}{6 - (-2)}$$

$$= \dfrac{3 + 5}{6 + 2}$$

$$= \dfrac{8}{8}$$

$$= 1$$

Both ways give the same slope.

**Thus,** the slope of $\overline{AB}$ is 1.

The slope of a line segment can be a negative number, as in Example 3.

*Example 3*   **Find the slope of $\overline{CD}$ for $C(7, 1)$ and $D(-3, 5)$.**

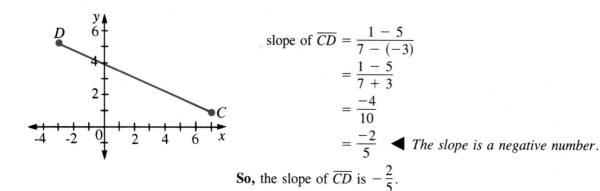

$$\text{slope of } \overline{CD} = \dfrac{1 - 5}{7 - (-3)}$$

$$= \dfrac{1 - 5}{7 + 3}$$

$$= \dfrac{-4}{10}$$

$$= \dfrac{-2}{5} \quad \blacktriangleleft \; \textit{The slope is a negative number.}$$

**So,** the slope of $\overline{CD}$ is $-\dfrac{2}{5}$.

You have seen that the slope of a nonvertical line segment can be either positive or negative. The slope of a nonvertical line segment can also be zero, as you will see in Example 4 on the next page. Vertical line segments will be discussed in the next lesson.

*Example 4*  **Find the slope of the horizontal line segment $\overline{CD}$ for $C(-2, 3)$ and $D(4, 3)$.**

*Plot C and D. Then draw $\overline{CD}$.* ▶

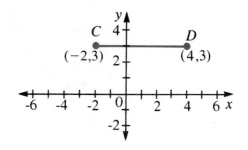

slope of $\overline{CD} = \dfrac{3 - 3}{4 - (-2)}$

$= \dfrac{0}{6}$, or 0

**So,** the slope of $\overline{CD}$ is 0.

## Written Exercises

**Find the slope of $\overline{AB}$ for the given points.**

(A)  **1.** $A(1, 1)$, $B(5, 4)$ $\frac{3}{4}$  **2.** $A(5, 7)$, $B(9, 8)$ $\frac{1}{4}$  **3.** $A(3, 6)$, $B(6, 8)$ $\frac{2}{3}$
  **4.** $A(-2, -3)$, $B(9, 3)$ $\frac{6}{11}$  **5.** $A(-4, -2)$, $B(5, 1)$ $\frac{1}{3}$  **6.** $A(4, 3)$, $B(1, 9)$ $-2$
  **7.** $A(5, 2)$, $B(4, 7)$ $-5$  **8.** $A(-5, -2)$, $B(4, -1)$ $\frac{1}{9}$  **9.** $A(7, -1)$, $B(-5, -4)$ $\frac{1}{4}$
  **10.** $A(-5, 3)$, $B(7, 3)$ **0**  **11.** $A(0, -4)$, $B(5, -1)$ $\frac{3}{5}$  **12.** $A(-1, -2)$, $B(6, -1)$ $\frac{1}{7}$
  **13.** $A(-8, 3)$, $B(7, -1)$ $-\frac{4}{15}$  **14.** $A(-3, -6)$, $B(2, -6)$ **0**  **15.** $A(0, 4)$, $B(7, 4)$ **0**

**Express the slope of $\overline{AB}$ in terms of variables (Ex. 16–23).**

(B)  **16.** $A(3b, 7k)$, $B(4b, 9k)$ $\frac{2k}{b}$  **17.** $A(4p, 3t)$, $B(3p, -2t)$ $\frac{5t}{p}$
  **18.** $A(2c, -5d)$, $B(-4c, 3d)$ $-\frac{4d}{3c}$  **19.** $A(-7b, 6k)$, $B(b, -k)$ $-\frac{7k}{8b}$
  **20.** $A(3a - 1, 4b - 2)$, $B(-2a + b, b - 1)$  **21.** $A(-2b + 5, 3c)$, $B(-b - 7, 3c - 4)$ $\frac{4}{-b + 12}$
  **22.** $A(-k - 1, m - 4)$, $B(k - 4, -m + 4)$  **23.** $A(4l - 2n, -6l - 3n)$, $B(5l + n, -l - n)$
  **20.** $\frac{3b - 1}{5a - b - 1}$  **22.** $\frac{2m - 8}{-2k + 3}$  **23.** $\frac{5l + 2n}{l + 3n}$
  **24.** Plot the points $A(1, 1)$, $B(4, 7)$, and $C(4, 10)$. Connect $A$ and $B$. Connect $A$ and $C$.
    Which segment appears to have the steeper slope, $\overline{AB}$ or $\overline{AC}$? **$\overline{AC}$**
    Justify your conclusion by comparing the slope of $\overline{AB}$ with the slope of $\overline{AC}$.
    **slope of $\overline{AB}$: 2  slope of $\overline{AC}$: 3; the slope of $\overline{AB}$ is less than the slope of $\overline{AC}$.**

**Determine the value(s) of $a$ so that $\overline{AB}$ has the given slope.**

(C)  **25.** $A(3, 4)$, $B(a, 7)$; slope $= \dfrac{3}{5}$ **8**  **26.** $A(0, -3)$, $B(4, a)$; slope $= \dfrac{5}{4}$ **2**

  **27.** $A(2, 3a)$, $B(3, a^2)$; slope $= 0$ **0, 3**  **28.** $A(3, -10)$, $B(6, a^2 - 7a)$; slope $= 0$ **5, 2**

## CUMULATIVE REVIEW

**1.** Solve: $|2a - 4| = 6$.  **2.** Solve: $\dfrac{x - 4}{3} = \dfrac{x + 1}{2}$.  **3.** Simplify: $\dfrac{18m^4 n^3 (b^2 - b - 12)}{12mn^4 (4 - b)}$.
  **5, -1**  **-11**
  $-\dfrac{3m^3 (b + 3)}{2n}$

# SLOPE OF A LINE

*Objectives*  **To find the slope of a line, using the slope of one of its segments**
**To determine the slant of a line, using its slope**
**To determine if two lines are parallel, using their slopes**
**To determine if three points are on the same line**

If you walk up a flight of stairs, you will notice that the slope, or steepness, of the staircase is the same no matter which section of steps you are climbing. $\frac{\text{Rise}}{\text{run}}$ is always $\frac{1}{3}$ in the flight of stairs pictured at the right. The same is true for a line, as you will see in Example 1.

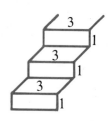

*Example 1*  **A, B, C, and D are points on the same line. Find the slopes of $\overline{AB}$, $\overline{CD}$, and $\overline{AD}$.**

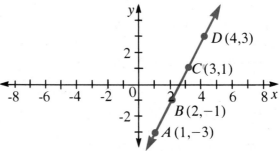

slope of $\overline{AB} = \dfrac{-1 - (-3)}{2 - 1}$

$= \dfrac{-1 + 3}{1}$

$= \dfrac{2}{1}$

slope of $\overline{CD} = \dfrac{3 - 1}{4 - 3}$

$= \dfrac{2}{1}$

slope of $\overline{AD} = \dfrac{3 - (-3)}{4 - 1}$

$= \dfrac{6}{3}$

$= \dfrac{2}{1}$

The slope of each of the three segments is the same.  **Thus,** the slope of the line, $\overleftrightarrow{AB}$, is $\frac{2}{1}$, or 2.

Example 1 suggests that you can use any two points of a line to find the slope of that line. It also suggests that all segments of a line have the same slope.

*Example 2*  **Find the slope of $\overleftrightarrow{PQ}$ for $P(5, 2)$ and $Q(2, 4)$.**

slope of $\overleftrightarrow{PQ} = \dfrac{4 - 2}{2 - 5}$

$= \dfrac{2}{-3}$

$= -\dfrac{2}{3}$

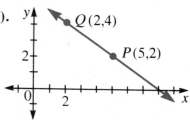

Notice that a line with positive slope slants up to the right (Example 1) and that a line with negative slope slants down to the right (Example 2). The next two examples involve the slope of a horizontal line and the slope of a vertical line, respectively.

*Example 3*     **Find the slope of $\overleftrightarrow{AB}$.**

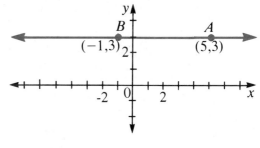

slope of $\overleftrightarrow{AB} = \dfrac{3 - 3}{5 - (-1)}$

$= \dfrac{0}{6}$

$= 0$

Notice that the line slants neither up nor down.

**Thus,** the slope of the horizontal line $\overleftrightarrow{AB}$ is 0.

*Example 4*     **Find the slope of $\overleftrightarrow{CD}$.**

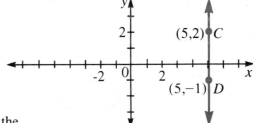

slope of $\overleftrightarrow{CD} = \dfrac{2 - (-1)}{5 - 5}$

$= \dfrac{3}{0}$

The quotient is undefined because the denominator is 0.

**Thus,** the slope of the vertical line $\overleftrightarrow{CD}$ is undefined.

The results of Examples 3 and 4 are generalized below.

**Slopes of horizontal and vertical lines**     The **slope of a horizontal line** is zero.
The **slope of a vertical line** is undefined.

Examples 1–4 have shown that the slope of a line determines its slant.

*Summary*

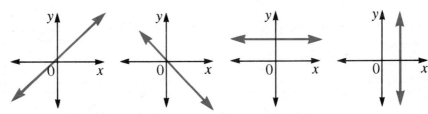

Positive slope: slants up to the right.

Negative slope: slants down to the right.

Zero slope: horizontal line.

Undefined slope: vertical line.

*Example 5*   **Find the slope of $\overleftrightarrow{CD}$ for $C(-1, -3)$ and $D(6, 9)$. Describe the slant of $\overleftrightarrow{CD}$.**

slope of $\overleftrightarrow{CD} = \dfrac{9 - (-3)}{6 - (-1)} = \dfrac{12}{7}$   ◀ *Slope is positive.*

**So,** $\overleftrightarrow{CD}$ slants up to the right.

If two lines are *parallel,* they have no points in common. Parallel lines slant in the same direction. Moreover, parallel lines have the same slope, as shown in Example 6.

*Example 6*   **$\overleftrightarrow{PQ}$ is parallel to $\overleftrightarrow{AB}$. Find the slope of each line.**

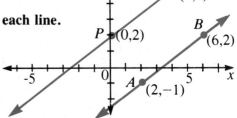

slope of $\overleftrightarrow{AB}$
$\dfrac{2 - (-1)}{6 - 2} = \dfrac{3}{4}$

slope of $\overleftrightarrow{PQ}$
$\dfrac{5 - 2}{4 - 0} = \dfrac{3}{4}$

**Thus,** the slope of each line is $\dfrac{3}{4}$.

| Slopes of parallel lines | Lines that are **parallel** have the same slope. |
| --- | --- |

# Written Exercises

1. undefined, vertical     2. and 4., down to the right     7. undefined, vertical

**Find the slope of $\overleftrightarrow{PQ}$ for the given points. Describe the slant of $\overleftrightarrow{PQ}$.**

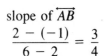

**1.** $P(4, 5), Q(4, 8)$       **2.** $P(4, 1), Q(-3, 5)$ $-\frac{4}{7}$       **3.** $P(4, 6), Q(7, 6)$ **0**
**4.** $P(-4, -2), Q(8, -3)$ $-\frac{1}{12}$   **5.** $P(2, 2), Q(-6, -3)$ $\frac{5}{8}$   **6.** $P(-9, 1), Q(7, 1)$ **0**
**7.** $P(-6, -1), Q(-6, 5)$     **8.** $P(-9, -2), Q(4, 2)$ $\frac{4}{13}$   **9.** $P(3, -4), Q(-6, -4)$ **0**
      5. and 8., up to the right     3., 6., 9., horizontal

**From the given points, determine whether $\overleftrightarrow{PQ}$ is parallel to $\overleftrightarrow{RS}$.**

**10.** $P(3, 6), Q(4, 8)$, PQ ∦ RS   **11.** $P(5, 9), Q(8, 11)$,  PQ ∥ RS   **12.** $P(4, 6), Q(5, 8)$, PQ ∦ RS
$R(3, 7), S(1, 4)$               $R(5, 3), S(11, 7)$                   $R(3, 4), S(5, 14)$

**Determine whether $A$, $B$, and $C$ lie on the same line. [Hint: Does slope of $\overline{AB}$ = slope of $\overline{BC}$?]**

**13.** $A(2, 2), B(-2, -6), C(6, 10)$ collinear   **14.** $A(2, 5), B(0, 7), C(3, 2)$ not on the same line
**15.** $A(4, -1), B(0, -5), C(2, -1)$           **16.** $A(4, 3), B(-6, -2), C(8, 5)$ collinear
**17.** $A(2, 1), B(-4, -8), C(6, 7)$ collinear   **18.** $A(1, 1), B(2, -2), C(-1, 5)$ not on the same line
      15. not on the same line

**Find the value(s) of $a$ so that $\overleftrightarrow{PQ}$ will be parallel to $\overleftrightarrow{RS}$.**

**19.** $P(2, 4), Q(3, 6), R(8, 1), S(10, a)$ **5**   **20.** $P(3, -1), Q(7, 11), R(-1, -1), S(1, a)$ **5**
**21.** $P(5, -1), Q(6, 11), R(6, -4a), S(7, a^2)$   **22.** $P(5, -2), Q(3, 6), R(-4, a^2), S(-3, 3a)$
                              $-6, 2$                                             $4, -1$

# EQUATIONS OF LINES

*Objectives*
**To write an equation of a line given two of its points**
**To verify that a point lies on a line given an equation of the line**
**To write an equation of a line given one of its points and its slope**

The line through the two points $R(-2, -5)$ and $S(1, 4)$ is shown at the right. $G(x, y)$ represents any other point on the same line. You can write an equation that is satisfied by all points $(x, y)$ on a line by using the slope of the line. Since the slope of a line is the same as the slope of any segment, you can form an equation of $\overrightarrow{RG}$ as follows: (1) Find the slope of $\overline{GS}$. (2) Find the slope of $\overline{RS}$. (3) Set the two slopes equal to each other.

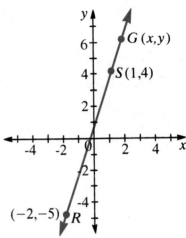

*Example 1*    **Write an equation for $\overrightarrow{RS}$ (shown above).**

(1) Find the slope of $\overline{GS}$.

$$\frac{y - 4}{x - 1}$$

(2) Find the slope of $\overline{RS}$.

$$\frac{4 - (-5)}{1 - (-2)} = \frac{9}{3}, \text{ or } \frac{3}{1}$$

(3) Set the two slopes equal to each other.

$$\frac{y - 4}{x - 1} = \frac{3}{1}$$

The equation $\dfrac{y - 4}{x - 1} = \dfrac{3}{1}$, a proportion, is an equation of the line. However, the equation can be written in a more convenient form by solving it for $y$.

$1(y - 4) = 3(x - 1)$  ◀ *Product of extremes equals product of means.*
$y - 4 = 3x - 3$  ◀ *Distribute 3: $3(x - 1) = 3x - 3$.*
$y = 3x + 1$  ◀ *Add 4 to each side to get y alone.*

You can show that $y = 3x + 1$ is an equation of $\overrightarrow{RS}$ by showing that the coordinates of the points $R(-2, -5)$ and $S(1, 4)$ satisfy the equation. This is done by substituting in the equation, as shown below.

For $R(-2, -5)$,
$x = -2$ and $y = -5$.

| $y$ | $3x + 1$ |
|---|---|
| $-5$ | $3 \cdot -2 + 1$ |
| | $-6 + 1$ |
| | $-5$ |

$-5 = -5$, true

For $S(1, 4)$,
$x = 1$ and $y = 4$.

| $y$ | $3x + 1$ |
|---|---|
| $4$ | $3 \cdot 1 + 1$ |
| | $3 + 1$ |
| | $4$ |

$4 = 4$, true

**Thus, $y = 3x + 1$ is an equation for $\overrightarrow{RS}$ with $R(-2, -5)$ and $S(1, 4)$.**

Equations of Lines

Since the slope of a line is the same for any two segments on the line, you can find an equation for the line in more than one way. This is illustrated in Example 2.

*Example 2*   **Write an equation for $\overleftrightarrow{PQ}$, given $P(1, -3)$ and $Q(-1, 1)$.**

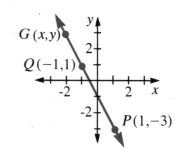

Plot $P(1, -3)$ and $Q(-1, 1)$ and draw a line through the points.

Let $G(x, y)$ represent any other point on the line. Set the slope of $\overline{GQ}$ equal to the slope of $\overline{PQ}$, or set the slope of $\overline{GP}$ equal to the slope of $\overline{PQ}$.

slope of $\overline{GQ}$ = slope of $\overline{PQ}$

$$\frac{y - 1}{x - (-1)} = \frac{1 - (-3)}{-1 - 1}$$

$$\frac{y - 1}{x + 1} = \frac{4}{-2} \quad \blacktriangleleft \text{ Solve the proportion.}$$

$$-2(y - 1) = 4(x + 1)$$
$$-2y + 2 = 4x + 4$$
$$-2y = 4x + 2 \quad \blacktriangleleft \text{ Subtract 2 from each side.}$$
$$y = \frac{4x + 2}{-2} \quad \blacktriangleleft \text{ Divide each side by } -2.$$
$$y = -2x - 1$$

**Thus,** $y = -2x - 1$ is an equation of $\overline{PQ}$.

slope of $\overline{GP}$ = slope of $\overline{PQ}$

$$\frac{y - (-3)}{x - 1} = \frac{1 - (-3)}{-1 - 1}$$

$$\frac{y + 3}{x - 1} = \frac{4}{-2}$$

$$-2(y + 3) = 4(x - 1)$$
$$-2y - 6 = 4x - 4$$
$$-2y = 4x + 2$$
$$y = \frac{4x + 2}{-2}$$
$$y = -2x - 1$$

The next example illustrates how you can use an equation of a line to show that a given point lies on the line.

*Example 3*   **Show that the point $R(2, -5)$ is on $\overleftrightarrow{PQ}$ of Example 2.**

*Strategy:* See if the coordinates of $R$ satisfy the equation of $\overleftrightarrow{PQ}$.

$$
\begin{array}{c|l}
y & -2x - 1 \\
\hline
-5 & -2 \cdot 2 - 1 \quad \blacktriangleleft \text{ Substitute 2 for } x \text{ and } -5 \text{ for } y. \\
& -4 - 1 \\
& -5
\end{array}
$$

$-5 = -5$, true

You can check by drawing the line passing through $P(1, -3)$ and $Q(-1, 1)$ and then plotting $(2, -5)$, as shown at the right. Notice that the line also passes through the point $R(2, -5)$.

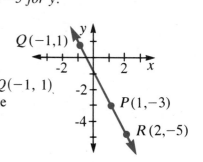

**Thus,** $R(2, -5)$ is on $\overleftrightarrow{PQ}$.

Moreover, any point whose coordinates satisfy the equation of a line will lie on that line.

*Example 4*   **Find an equation of the line passing through the point $Q(3, 4)$ and having slope $\dfrac{2}{3}$.**

Plot the point $Q(3, 4)$. Use the slope $\dfrac{2}{3}$ to plot

another point: move 2 up and 3 to the right.
Draw a line through the two points. Let $G(x, y)$
represent any other point on the line.

Set the slope of $\overleftrightarrow{GQ}$ equal to $\dfrac{2}{3}$, the given slope.

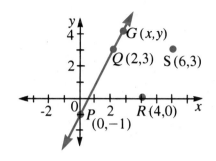

$$\frac{y - 4}{x - 3} = \frac{2}{3}$$

$3(y - 4) = 2(x - 3)$   ◀ *Solve the proportion.*
$3y - 12 = 2x - 6$
$3y = 2x + 6$   ◀ *Add 12 to each side.*
$y = \dfrac{2x + 6}{3}$   ◀ *Divide each side by 3.*

$y = \dfrac{2}{3}x + 2$

# Written Exercises

7. $y = 4x - 5$   8. $y = -3x + 12$   9. $y = -2x - 8$   10. $y = -3x + 3$   13. $y = -3x + 4$   14. $y = -5x + 3$

**Write an equation for $\overleftrightarrow{PQ}$ given the coordinates of its two points.**

(A)   1. $P(3, 3)$, $Q(4, 5)$  $y = 2x - 3$   2. $P(4, 8)$, $Q(6, 14)$  $y = 3x - 4$   3. $P(2, 0)$, $Q(3, 2)$  $y = 2x - 4$
   4. $P(1, -2)$, $Q(4, 1)$  $y = x - 3$   5. $P(-2, 2)$, $Q(2, 6)$  $y = x + 4$   6. $P(-3, -8)$, $Q(4, 6)$  $y = 2x - 2$
   7. $P(-1, -9)$, $Q(2, 3)$   8. $P(-1, 15)$, $Q(2, 6)$   9. $P(-3, -2)$, $Q(1, -10)$
   10. $P(-2, 9)$, $Q(3, -6)$   11. $P(3, 9)$, $Q(0, 0)$  $y = 3x$   12. $P(-2, -12)$, $Q(5, 2)$  $y = 2x - 8$
   13. $P(3, -5)$, $Q(0, 4)$   14. $P(0, 3)$, $Q(2, -7)$   15. $P(3, -13)$, $Q(4, 1)$  $y = 14x - 55$

16. $-4 = 3(-2) + 2$; $-4 = -4$, true   17. $0 = 2(-3) + 6$; $0 = 0$, true   18. $-1 = -(-4) - 5$; $-1 = -1$, true

**Show that $P$ lies on the line with the given equation.**

   16. $P(-2, -4)$; $y = 3x + 2$   17. $P(-3, 0)$; $y = 2x + 6$   18. $P(-4, -1)$; $y = -x - 5$

19. $y = 3x - 5$; $R$ lies on $\overleftrightarrow{PQ}$   20. $y = -2x + 3$; $R$ lies on $\overleftrightarrow{PQ}$   21. $y = 2x + 1$; $R$ lies on $\overleftrightarrow{PQ}$

**Write an equation for $\overleftrightarrow{PQ}$. Then, show whether $R$ lies on $\overleftrightarrow{PQ}$.**   22. $y = 3x - 7$; $R$ lies on $\overleftrightarrow{PQ}$

(B)   19. $P(1, -2)$, $Q(0, -5)$, $R(4, 7)$   20. $P(-3, 9)$, $Q(-2, 7)$, $R(4, -5)$
   21. $P(0, 1)$, $Q(2, 5)$, $R(5, 11)$   22. $P(0, -7)$, $Q(2, -1)$, $R(4, 5)$
   23. $P(-2, -9)$, $Q(0, -1)$, $R(2, 7)$   24. $P(-2, -9)$, $Q(1, -6)$, $R(2, -11)$
   25. $P(-2, -11)$, $Q(-1, -7)$, $R(2, 5)$   26. $P(-3, 8)$, $Q(0, -1)$, $R(1, -4)$

23. $y = 4x - 1$; $R$ lies on $\overleftrightarrow{PQ}$   24. $y = x - 7$; $R$ is not on $\overleftrightarrow{PQ}$

**Write an equation of the line passing through the given point and having the given slope.**   25. $y = 4x - 3$; $R$ lies on $\overleftrightarrow{PQ}$   26. $y = -3x - 1$; $R$ lies on $\overleftrightarrow{PQ}$
   27. $P(-4, 1)$; slope, $3$  $y = 3x + 13$   28. $P(-2, 5)$; slope, $-2$  $y = -2x + 1$
   29. $P(0, 0)$; slope, $-5$  $y = -5x$   30. $P(1, -4)$; slope, $0$  $y = -4$
(C)   31. $P(0, b)$; slope, $4$  $y = 4x + b$   32. $P(\frac{2}{3}, 0)$; slope, $6k$  $y = 6kx - 4k$

Equations of Lines

# USING EQUATIONS OF LINES                     11.6

*Objectives*    **To write an equation for a line whose slope is not an integer**
**To find the y-coordinate of a point on a line, given the x-coordinate of that point and the equation of the line**
**To write a table of values, given an equation of a line**

For the lines in the previous lesson, you wrote equations such as $y = 3x - 2$ in which the x-coefficient, 3, and the constant, $-2$, are integers. However, Example 1 shows that the x-coefficient and the constant may not be integers.

*Example 1*    **Write an equation for $\overleftrightarrow{AB}$, given $A(-1, -1)$ and $B(9, 3)$.**

*First,* plot the points $A$ and $B$ and draw the line passing through them. *Then,* indicate $G(x, y)$, a general point, on $\overleftrightarrow{AB}$.

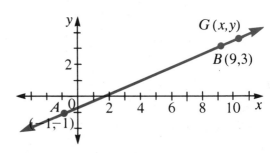

slope of $\overline{BG}$ = slope of $\overline{AB}$

$$\frac{y - 3}{x - 9} = \frac{3 - (-1)}{9 - (-1)}$$

$$\frac{y - 3}{x - 9} = \frac{4}{10}$$

$10(y - 3) = 4(x - 9)$   ◀ *The product of the extremes = the product of the means.*
$10y - 30 = 4x - 36$   ◀ *Use the distributive property.*
$10y = 4x - 6$   ◀ *Add 30 to each side.*

$$y = \frac{4x - 6}{10}$$   ◀ *Divide each side by 10.*

$$y = \frac{4}{10}x - \frac{6}{10}$$

$$y = \frac{2}{5}x - \frac{3}{5}$$   ◀ *Rewrite each fraction in simplest form.*

You can use an equation of a line to determine the y-coordinate of a point if you are given the x-coordinate of the point, as shown in Example 2.

*Example 2*    **$C(14, y)$ lies on $\overleftrightarrow{AB}$ in Example 1. Find y.**

$$y = \frac{2}{5}x - \frac{3}{5}$$   ◀ *Equation of $\overleftrightarrow{AB}$*

$$y = \frac{2}{5} \cdot 14 - \frac{3}{5}$$   ◀ *Let x = 14 in the equation.*

$$y = \frac{28}{5} - \frac{3}{5} = \frac{25}{5}, \text{ or } 5$$

**So, $C(14, 5)$ lies on $\overleftrightarrow{AB}$, and y is 5.**

You can use the equation of a line to find the coordinates of any number of points that lie on the line. Suppose the equation of $\overleftrightarrow{MN}$ is $y = 2x - 3$. You can choose some values for $x$ and then substitute them in the equation to find the corresponding $y$-values. These $x$- and $y$-values are the $x$- and $y$-coordinates. Display your work in a table, as shown in Example 3 below.

**Example 3** **An equation of $\overleftrightarrow{MN}$ is $y = 2x - 3$. Write a table of values and give the coordinates of three points on the line.**

Choose any three numbers for $x$, say 1, 0, and $-1$.

| $x$ | $2x - 3$ | $y$ |
|---|---|---|
| 1 | $2 \cdot 1 - 3$ | $-1$ |
| 0 | $2 \cdot 0 - 3$ | $-3$ |
| $-1$ | $2 \cdot (-1) - 3$ | $-5$ |

◀ *Substitute 1 for x in $y = 2x - 3$.*

$y = 2 \cdot 1 - 3 = 2 - 3$, or $-1$

**So, $(1, -1)$, $(0, -3)$, and $(-1, -5)$ are three points on $\overleftrightarrow{MN}$.**

# Written Exercises

1. $y = \frac{3}{2}x + \frac{1}{2}$   2. $y = \frac{1}{3}x + \frac{2}{3}$   3. $y = \frac{1}{4}x + \frac{5}{2}$

**Write an equation for $\overrightarrow{PQ}$ for the given points.**

Ⓐ 1. $P(5, 8)$, $Q(7, 11)$       2. $P(1, 1)$, $Q(4, 2)$       3. $P(2, 3)$, $Q(6, 4)$
4. $P(0, 7)$, $Q(4, 8)$       5. $P(4, 4)$, $Q(-1, 6)$       6. $P(-1, -2)$, $Q(-6, -4)$
     $y = \frac{1}{4}x + 7$              $y = -\frac{2}{5}x + \frac{28}{5}$              $y = \frac{2}{5}x - \frac{8}{5}$

**Write an equation for $\overleftrightarrow{AB}$. Then, use the equation to find the $y$-coordinate of point $C$ on $\overleftrightarrow{AB}$.**

7. $A(3, 3)$, $B(0, 5)$, $C(6, y)$  $y = -\frac{2}{3}x + 5$; 1       8. $A(-2, -4)$, $B(4, 0)$, $C(1, y)$  $y = \frac{2}{3}x - \frac{8}{3}$; $-2$
9. $A(0, 1)$, $B(-5, -1)$, $C(-5, y)$  $y = \frac{2}{5}x + 1$; $-1$ 10. $A(3, 6)$, $B(7, 12)$, $C(-6, y)$  $y = \frac{3}{2}x + \frac{3}{2}$; $-\frac{15}{2}$

**Write a table of values and give the coordinates of three points on the line whose equation is given.**       12. $(-1, -5)$; $(0, -2)$; $(1, 1)$ **Answers may vary.**

Ⓑ 11. $y = x + 5$       12. $y = 3x - 2$       13. $y = -3x + 4$       14. $y = \frac{1}{2}x - 6$
11. $(-1, 4)$; $(0, 5)$; $(1, 6)$       13. $(-1, 7)$; $(0, 4)$; $(1, 1)$       14. $(-2, -7)$; $(0, -6)$; $(2, -5)$

**Write an equation for $\overrightarrow{PQ}$. Then, use the equation to set up a table of values and give the coordinates of three other points on $\overrightarrow{PQ}$.**

15. $P(0, -1)$, $Q(4, 3)$ $y = x - 1$ 16. $P(7, 9)$, $Q(13, 12)$       17. $P(-7, -9)$, $Q(17, 7)$ $y = \frac{2}{3}x - \frac{13}{3}$
     $(-1, -2)$; $(1, 0)$; $(2, 1)$       $y = \frac{1}{2}x + \frac{11}{2}$  $(-1, 5)$; $(0, \frac{11}{2})$; $(1, 6)$       $(-1, -5)$; $(0, -\frac{13}{3})$; $(1, -\frac{11}{3})$

**Find the value of $b$ so that the given point lies on the line described by the equation.**

Ⓒ 18. $y = bx - 4$; $(4, 8)$ **3**       19. $y = bx + 1$; $(5, -3)$ $-\frac{4}{5}$   20. $y = bx - \frac{5}{2}$; $\left(2\frac{1}{2}, 5\right)$ **3**

# CALCULATOR ACTIVITIES

**Write an equation for $\overrightarrow{PQ}$ for the given coordinates.**   2. $y = -5x - 3$
1. $P(3, 176)$, $Q(5, 528)$ 1. $y = 176x - 352$     2. $P(3.04, -18.2)$, $Q(-0.05, -2.75)$
3. $P(3.2, 6.02)$, $Q(-2.4, -0.14)$   3. $y = 1.1x + 2.5$

# SLOPE-INTERCEPT FORM OF AN EQUATION OF A LINE

*Objectives*

**To find the slope and *y*-intercept of a line, given its equation**
**To write an equation of a line given its slope and *y*-intercept**
**To graph a line, given its equation in slope-intercept form**

The line passing through $P(0, 2)$ and $Q(3, 4)$ has a slope of $\dfrac{4-2}{3-0}$, or $\dfrac{2}{3}$.

There is a relationship between the slope and the equation of the line. This relationship is illustrated below.

Let $G(x, y)$ be any point on $\overleftrightarrow{PQ}$.
Then the slope of $\overline{GQ}$ equals the slope of $\overline{PQ}$.

$$\frac{y-4}{x-3} = \frac{4-2}{3-0}$$

$$\frac{y-4}{x-3} = \frac{2}{3}$$

$$3y - 12 = 2x - 6$$

$$3y = 2x + 6$$

$$y = \frac{2}{3}x + 2$$

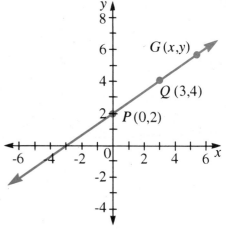

Notice that the slope of $\overleftrightarrow{PQ}$ is $\dfrac{2}{3}$. The slope is the coefficient of $x$.

The graph of $\overleftrightarrow{PQ}$, shown above, reveals another relationship. The line intersects the *y*-axis at $(0, 2)$. The constant, 2, in the equation

$$y = \frac{2}{3}x + \overset{\downarrow}{2}$$

is the *y*-coordinate of the point of intersection of $\overleftrightarrow{PQ}$ with the *y*-axis.

| Definition: y-intercept | The **y-intercept** is the *y*-coordinate of the point of intersection of a line with the *y*-axis. |
|---|---|

This equation for $\overleftrightarrow{PQ}$ is said to be written in *slope-intercept* form:

$$y = \frac{2}{3}x + 2$$

The slope of $\overleftrightarrow{PQ}$ is $\dfrac{2}{3}$ and the *y*-intercept is 2.

The generalization below follows from page 310.

---

**Equation of a line: slope-intercept form**

$y = mx + b$ is an equation of a line with

slope $m$ and $y$-intercept $b$.

---

You can use this generalization to find the slope and the $y$-intercept of a line from an equation of the line.

*Example 1*   **Find the slope and the $y$-intercept of the line with the given equation.**

$$y = \frac{3}{7}x - 10$$

The slope is $\frac{3}{7}$, and the $y$-intercept is $-10$.

$$y = -4x$$

The slope is $-4$, and the $y$-intercept is 0.  ◀ *Think:* $y = -4x + 0$.

The technique of Example 1 also works in reverse. If you are given the slope and the $y$-intercept of a line, you can write an equation of the line. This procedure is illustrated below.

Since $b$ is the $y$-intercept, the point $P(0, b)$ is on the line. Let $G(x, y)$ represent any other point on the line.

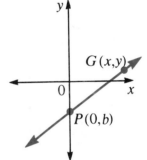

$$\text{slope of } \overline{PG} = m$$
$$\frac{y - b}{x - 0} = \frac{m}{1}$$
$$y - b = mx$$
$$y = mx + b \quad ◀ \text{ Add } b \text{ to each side.}$$

This suggests the following generalization.

---

**Equation of a line with slope $m$ and $y$-intercept $b$**

If the slope of a line is $m$ and the $y$-intercept is $b$, then the equation of the line is $y = mx + b$.

---

*Example 2*   **Write an equation of a line with slope $-6$ and $y$-intercept $-4$.**

The slope is $-6$.         The $y$-intercept is $-4$.

$$y = mx + b$$

The equation is $y = -6x - 4$.

The *y*-intercept and the slope of a line can be used to graph the line. Therefore, if you are given an equation in slope-intercept form, you can graph the line described by that equation. For example, the equation $y = \frac{2}{5}x + 1$ tells you that the line crosses the *y*-axis at the point (0, 1). This gives you one point of the line. You can use the slope, $\frac{2}{5}$, to locate another point. The line can now be drawn through the two points. This technique is illustrated in Example 3.

## *Example 3*    **Graph: $y = \frac{2}{5}x + 1$.**

(1)  The *y*-intercept is 1. Plot (0, 1).

(2)  Use the slope, $\frac{2}{5}$, to plot another point.

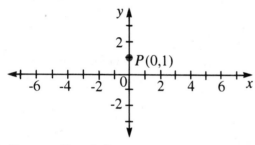

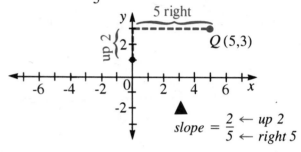

(3)  Connect *P* and *Q*.

(4)  The line through the points $P$ (0, 1) and $Q$ (5, 3) is the graph of $y = \frac{2}{5}x + 1$.

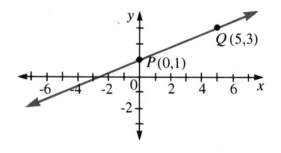

## *Example 4*    **Graph $y = -\frac{2}{3}x + 4$.**

Rewrite the equation as $y = \frac{-2}{3}x + 4$.
slope————↗        ↑——*y*-intercept

Plot (0, 4). Then, use the slope to plot a second point.

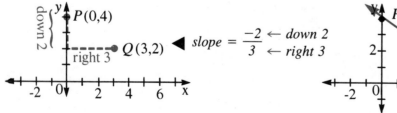

slope $= \dfrac{-2}{3} \begin{array}{l} \leftarrow \text{down 2} \\ \leftarrow \text{right 3} \end{array}$

◀ Connect the two points with a line.

## Oral Exercises

1. From origin, move up 2 units. Mark a point. Move up 1 more unit and to the right 3 units for second pt.
**Tell how to graph the line with the given equation.**                    Connect points.

**1.** $y = \frac{1}{3}x + 2$      **2.** $y = \frac{2}{5}x + 4$      **3.** $y = -\frac{4}{5}x$      **4.** $y = -\frac{1}{2}x - 5$

2. Move up 4 units. Mark a point. Move up 2 units and to the right 5 units. Connect points.   3. From origin,

## Written Exercises
move down 4 units and to the right 5 units. Connect origin to point.

4. Move down 5 units. Mark a point. Move down 1 unit and 2 units to the right. Connect points.

**Give the slope and the y-intercept of the line with the given equation.**

Ⓐ   **1.** $y = \frac{1}{3}x + 6$   $\frac{1}{3}$; 6      **2.** $y = \frac{3}{4}x + 5$   $\frac{3}{4}$; 5      **3.** $y = -\frac{4}{5}x + 9$   $-\frac{4}{5}$; 9

    **4.** $y = 3x$   **3; 0**      **5.** $y = -\frac{2}{3}x - 7$   $-\frac{2}{3}$; $-7$      **6.** $y = x + 5$   **1; 5**

**Write an equation for the line with the given slope, m, and y-intercept, b.**

**7.** $m = 3, b = 5$   $y = 3x + 5$   **8.** $m = -4, b = -6$   $y = -4x - 6$   **9.** $m = \frac{2}{3}, b = -1$   $y = \frac{2}{3}x - 1$

**Graph the line described by the given equation. (Ex. 10–22)** See TE Answer Section.

**10.** $y = \frac{2}{3}x + 5$      **11.** $y = -\frac{4}{5}x + 3$      **12.** $y = -\frac{3}{4}x - 4$      **13.** $y = 2x + 4$

**14.** $y = 4x - 2$      **15.** $y = -3x + 2$      **16.** $y = -4x - 1$      **17.** $y = -2x + 6$

**18.** $y = \frac{-2}{3}x - 5$      **19.** $y = \frac{5}{2}x - 1$      **20.** $y = -\frac{2}{3}x$

Ⓑ   **21.** $y = x$ [Hint: Rewrite in $y = mx + b$ form.]      **22.** $y = 5x$

**23.** Write an equation of the line with slope $\frac{3}{4}$ and y-intercept the same as that

    of the line described by the equation $y = -\frac{2}{5}x - 6$.   $y = \frac{3}{4}x - 6$

**24.** Graph the line with slope $-3$ and y-intercept the same as that of the line
    described by the equation $y = 4x - 8$.

**Graph the line as described below. For Ex. 24–28 see TE Answer Section.**

Ⓒ   **25.** y-intercept $-2$ and parallel to the line with     **26.** y-intercept $-4$ and parallel to the line with

    equation $y = \frac{2}{3}x + 1$            equation $y = -\frac{2}{5}x + 6$

**27.** passing through the point $(1, -3)$ and     **28.** y-intercept the same as that of the line with
    parallel to the line with equation $y = x$          equation $y = 3x$ and parallel to the line for
                                           $y = -2x - 4$

## CUMULATIVE REVIEW

**1.** The length of a rectangle is 2 cm more than     **2.** Two trucks started toward each other at the
    3 times the width. The perimeter is 36 cm.        same time from towns 500 km apart. One
    Find the length and the width of the          truck traveled at 65 km/h, the other at 60
    rectangle. **14, 4**                      4 km/h. After how many hours did they meet?

# GRAPHING NONVERTICAL LINES    11.8

*Objective*    **To graph a line, given its equation**

The equation $y = \frac{2}{3}x + 5$ is the equation of a line. Its slope is $\frac{2}{3}$ and its y-intercept is 5. This equation is of the form $y = mx + b$. In the previous lesson, you saw that it is easy to graph a line when its equation is given in slope-intercept form. When an equation of a line, such as $5x - 3y = 9$, is not in slope-intercept form, you can write it in this form by solving for $y$.

$$5x - 3y = 9$$
$$-5x + 5x - 3y = -5x + 9$$
$$-3y = -5x + 9$$
$$y = \frac{-5x + 9}{-3}$$
$$y = \frac{5}{3}x - 3$$

The equation is now in slope-intercept form. The slope is $\frac{5}{3}$, and the y-intercept is $-3$.

To graph an equation such as $y = -1x + 3$, it is convenient to rewrite the x-term, as $\frac{-1}{1}x$, so that the coefficient is in the form $\frac{a}{b}$. This procedure is used in Example 1.

*Example 1*    **Graph: $-y - x = -3$.**

*Think:* Put the equation in the form $y = mx + b$.

$-1y - 1x = -3$    ◀ $-y = -1y$ and $-x = -1x$.
$-1y = 1x - 3$    ◀ *Add 1x to each side.*
$y = -1x + 3$    ◀ *Divide each side by $-1$.*
$y = \frac{-1}{1}x + 3$    ◀ *Rewrite the coefficient of x as a fraction: $-1 = \frac{-1}{1}$.*

The slope is $\frac{-1}{1}$, and the y-intercept is 3.

Plot (0, 3). Then, use the slope to plot another point.

Draw $\overleftrightarrow{PQ}$.

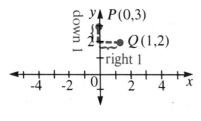

◀ $slope = \dfrac{-1}{1} \begin{array}{l}\leftarrow down\ 1 \\ \leftarrow right\ 1\end{array}$

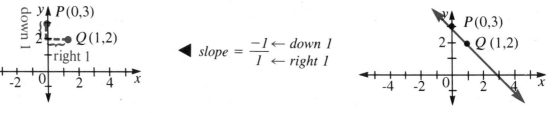

The *y*-intercept of a line can be 0. This situation occurs in the next example.

*Example 2*    **Graph: $x + 4y = 0$.**

$$1x + 4y = 0$$
$$4y = -1x \quad \blacktriangleleft \textit{ Subtract 1x from each side.}$$
$$y = \frac{-1}{4}x \quad \blacktriangleleft \textit{ Divide each side by 4.}$$
$$\text{or } y = \frac{-1}{4}x + 0$$

The slope is $-\dfrac{1}{4}$, and the *y*-intercept is 0.

Since the *y*-intercept is 0, the line passes through the origin (0, 0).
Plot (0, 0). Then, use the slope to plot another point.

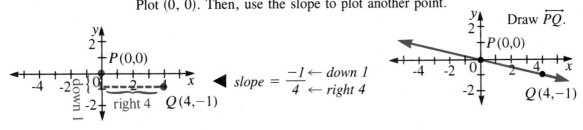

$$\blacktriangleleft \quad slope = \frac{-1}{4} \begin{array}{l} \leftarrow down\ 1 \\ \leftarrow right\ 4 \end{array}$$

The next example illustrates the case when the slope is 0.

*Example 3*    **Graph: $y = -2$.**

Rewrite $y = -2$ so that the equation will be in slope-intercept form.
$$y = 0x - 2 \quad \blacktriangleleft \textit{ Since there is no x-term, use 0x.}$$
$$y = \frac{0}{1}x - 2$$

The slope is $\dfrac{0}{1}$, and the *y*-intercept is $-2$.

Plot (0, $-2$). Then, use the slope to plot another point.    Draw $\overleftrightarrow{PQ}$.

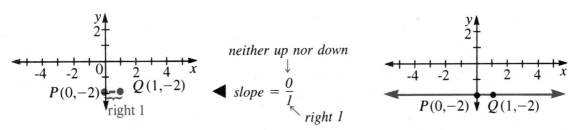

$$\blacktriangleleft \quad slope = \frac{0}{1} \begin{array}{l} \textit{neither up nor down} \\ \downarrow \\ \\ \textit{right 1} \end{array}$$

$\overleftrightarrow{PQ}$ is a horizontal line. A horizontal line has zero slope.

Graphing Nonvertical Lines

*Example 4*    **Graph: $6y - 18 = 0$.**

*Solve for y.* ▶    $6y - 18 = 0$
$6y = 18$    ◀ *Add 18 to each side.*
$y = 3$    ◀ *Divide each side by 6.*
or $y = 0x + 3$
The graph will be a horizontal line.

*Plot (0, 3) because the* ▶    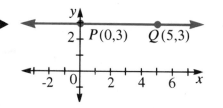    ◀ *Then plot any other point*
*y-intercept is 3.*                                         *with y-coordinate 3,*
*such as Q(5, 3).*
*Draw a horizontal line*
*through these points.*

# Reading in Algebra

**For each statement or phrase, select the expression(s) that do not apply.**

**1.** The equation is in $y = mx + b$ form. **a, c**
    a. $5x - 2y = 10$    b. $y = 3x + 4$    c. $3y = 9x + 15$    d. $y = 4x$
**2.** For the equation $y = 3x - 4$, **a, b**
    a. slope $= 4$.    b. y-intercept $= 3$.    c. slope $= \frac{3}{1}$.    d. y-intercept $= -4$.
**3.** The graph is a horizontal line. **c**
    a. $y = 3$    b. $3y - 0x = 4$    c. $y = 2x - 1$    d. $4y - 6 = -14$

# Written Exercises

**For Ex. 1–23 see TE Answer Section.**

Graph.

Ⓐ **1.** $3x - 4y = 8$    **2.** $4y + 3x = 20$    **3.** $7y + 3x = 21$
**4.** $y - 6 = x$    **5.** $y + x = 4$    **6.** $3y - x = 9$
**7.** $y = -2$    **8.** $y - 4 = 6$    **9.** $2y - 6 = 10$
**10.** $3y - x = 12$    **11.** $y = 5$    **12.** $6x + 2y = 10$

Ⓑ **13.** $4x - 2y = 1$    **14.** $2x + 6y = -3$    **15.** $4x - 2y = 9$
**16.** $3x - (4 - y) = 8$    **17.** $6x - (4 - x) = 2y - 8$    **18.** $7 - (3 - 3y) = -5 + 2x$

**Write a table of values and give the coordinates of three points on the line whose equation is given. Graph the line. Answers may vary.**

**19.** $6x - 3y = 9$    **20.** $2x + 5y = 10$    **21.** $3x - 2y = 8$
$(-1, -5); (0, -3); (1, -1)$    $(-5, 4); (0, 2); (5, 0)$    $(0, -4); (-2, -7); (2, -1)$
Ⓒ **22.** Graph $ky - 3x = 4$ if the line is parallel    **23.** Graph $y - 6 = kx$ if the line is parallel to
    to the line with equation $6x + 8y = 24$.    the line with equation $y - (3 - x) = 4$.

# GRAPHING VERTICAL LINES

*Objective*     **To graph a vertical line, given an equation of the line**

The graph of $\overleftrightarrow{PQ}$ at the right is a vertical
line. Observe two patterns:

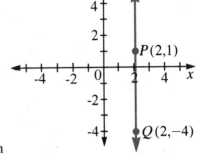

   (1) The slope of $\overleftrightarrow{PQ}$ is $\dfrac{1 - (-4)}{2 - 2}$, or $\dfrac{5}{0}$. Therefore the slope
      is undefined, because the denominator is 0.

   (2) $\overleftrightarrow{PQ}$ has no $y$-intercept because it does not cross the
      $y$-axis. It is parallel to the $y$-axis.

A vertical line such as $\overleftrightarrow{PQ}$ does not have an equation of the form

$$y = mx + b$$

because the slope is undefined and there is no $y$-intercept for a vertical line.

However, if you label several more points on the line, you will notice a
pattern that can be described by an equation, as shown in Example 1 below.

*Example 1*     **Copy the graph of $\overleftrightarrow{PQ}$ above. Label several more points. Write an
equation for $\overleftrightarrow{PQ}$.**

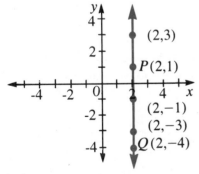

Observe that $\overleftrightarrow{PQ}$ is parallel to the $y$-axis and
is 2 units to the right of the $y$-axis.

Observe, also, that every point on $\overleftrightarrow{PQ}$ has the same
$x$-coordinate, 2.

$\overleftrightarrow{PQ}$ contains all points such that $x = 2$.

**So,** an equation of $\overleftrightarrow{PQ}$ is $x = 2$.

Notice that $y$ does not appear in the equation, $x = 2$, for $\overleftrightarrow{PQ}$ in Example 1.
$y$ can be any number, but $x$ is always 2. Example 1 suggests that a vertical
line can be described by an equation of the form $x = c$, where $c$ is a constant.

## Example 2    Graph: $x = -4$.

Use any two points with $x$-coordinate
$-4$ to graph the line. For example, use
$P(-4, -1)$ and $Q(-4, 2)$.

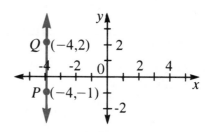

Draw a line through the two points,
$P$ and $Q$.

The graph of the equation $10 - 2x = 0$ is a vertical line because only the
variable $x$ appears in the equation. In situations of this type, you first will
have to solve the equation for $x$. The procedure is illustrated in the next
example.

## Example 3    Graph: $10 - 2x = 0$.

*First*, solve the equation for $x$.
$$10 - 2x = 0$$
$$-2x = -10 \quad \blacktriangleleft \; Subtract\ 10\ from\ each\ side.$$
$$x = 5 \quad \blacktriangleleft \; Divide\ each\ side\ by\ -2.$$

*Then*, graph the vertical line that contains all points with $x$-coordinate 5.

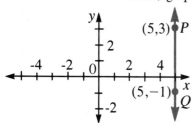

$\blacktriangleleft$ *Think: y can be any number, but x is
always 5. Use any two points with
x-coordinate 5, such as (5, 3) and
(5, −1), to draw the line.*

# Written Exercises

**Graph. For Ex. 1–21 see TE Answer Section.**

Ⓐ  **1.** $x = -3$  **2.** $x = -1$  **3.** $x = 3$  **4.** $x = -6$
 **5.** $x = -5$  **6.** $x = 6$  **7.** $x = 0$  **8.** $x = -7$
 **9.** $3x = -6$  **10.** $x = 9$  **11.** $-4x = -16$  **12.** $8 = -2x$

Ⓑ  **13.** $3x - 9 = 0$  **14.** $12 + 2x = 0$  **15.** $6x - 18 = 0$
 **16.** $-5 - x = 4$  **17.** $-x + 9 = 8$  **18.** $6 - x = 4$

Ⓒ  **19.** $|x| = 4$  **20.** $|4 - 2x| = 6$  **21.** $2|6 - x| - 8 = 12$

# GRAPHING INEQUALITIES

*Objective*

**To graph an inequality in two variables**

The graph of $y = \frac{1}{3}x + 1$ *separates* the coordinate plane into two regions—one above the line, the other below the line.

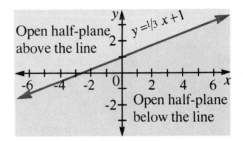

Each region is referred to as an *open half-plane*. The line is called the *boundary* of the two half-planes.

Notice that the point $P(3, 2)$ is on the line. Points $R(3, 3)$ and $S(3, 5)$ are above the line. $R$ and $S$ have a $y$-coordinate greater than the $y$-coordinate 2 of $P(3, 2)$.

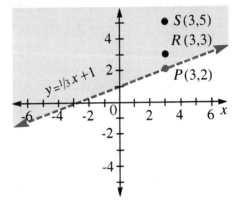

**Thus,** $y > \frac{1}{3}x + 1$ has as its graph the upper open half-plane (shaded in color). The boundary is dashed to indicate that points on the line are not part of the graph of $y > \frac{1}{3}x + 1$.

The graph of $y > \frac{1}{3}x + 1$ is the upper open half-plane.

An inequality may contain any one of the four symbols: $<$, $>$, $\geq$, or $\leq$. The graphs below illustrate the four possible inequality situations.

| $y < mx+b$ | $y > mx+b$ | $y \geq mx+b$ | $y \leq mx+b$ |
| $y < mx+b$ | $y > mx+b$ | $y \geq mx+b$ | $y \leq mx+b$ |

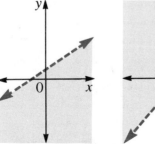
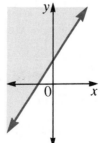
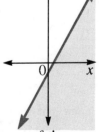

The boundary line is not part of the graph. It is drawn as a dashed line.

The boundary line is part of the graph. It is drawn as a solid line.

*Example 1*     **Graph: $y > 2x - 1$.**

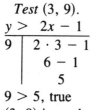

           *slope*        *y-intercept*

◀ *First, graph $y = \dfrac{2}{1}x - 1$ with a dashed line.*

◀ *Then, shade the region above the line.*

**So,** the graph is the half-plane above the line.

You can use the equation for the graph of an inequality to determine if a given point is included on the graph.

*Example 2*     **Determine whether the points, (3, 9), (3, 5), and (3, 2), are on the graph of $y > 2x - 1$.**

*Test (3, 9).*

$$
\begin{array}{c|l}
y > & 2x - 1 \\
\hline
9 & 2 \cdot 3 - 1 \\
& 6 - 1 \\
& 5
\end{array}
$$

$9 > 5$, true

**Thus,** (3, 9) is on the graph of $y > 2x - 1$.

*Test (3, 5).*

$$
\begin{array}{c|l}
y > & 2x - 1 \\
\hline
5 & 2 \cdot 3 - 1 \\
& 6 - 1 \\
& 5
\end{array}
$$

$5 > 5$, false

**Thus,** (3, 5) is not on the graph of $y > 2x - 1$.

*Test (3, 2).*

$$
\begin{array}{c|l}
y > & 2x - 1 \\
\hline
2 & 2 \cdot 3 - 1 \\
& 6 - 1 \\
& 5
\end{array}
$$

$2 > 5$, false

**Thus,** (3, 2) is not on the graph of $y > 2x - 1$.

The inequality $-x - 2y \geq -8$ is not in convenient form for graphing. You first should solve the inequality for $y$. Recall that dividing each side of an inequality by a negative number *reverses* the order of the inequality.

*Example 3*     **Graph: $-x - 2y \geq -8$.**

*First* solve for $y$.

$\quad -1x - 2y \geq -8$      ◀ $-x = -1x$

$\quad\quad -2y \geq 1x - 8$      ◀ *Add $1x$ to each side.*

$\quad\quad\quad y \leq -\dfrac{1}{2}x + 4$      ◀ *Divide each side by $-2$ and reverse the order of the inequality.*

*Second,* graph the line $y = -\dfrac{1}{2}x + 4$.

      *slope*        *y-intercept*

*Then,* shade the region below the line.

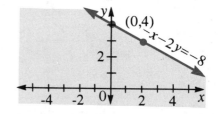

**So,** the graph of $-x - 2y \geq -8$ is the line and the shaded region below the line.

The graph of an inequality might not always be the region above or below a line. For example, a vertical line separates the plane into two half-planes one to the left and one to the right of the line. This is illustrated in Example 4.

*Example 4*     **Graph: $x < 4$.**

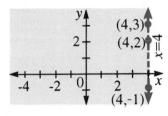

*First,* graph $x = 4$ with a dashed line:
The graph of $x = 4$ is a vertical line.
The $x$-coordinate of every point on the line is 4.

Then shade the region ◄ *The x-coordinate of every point*
to the left of the line.     *to the left of the line is less than 4.*

**So,** the graph of $x < 4$ is the shaded region to the left of the line.

If the inequality in Example 4 were $x > 4$, the graph would be the shaded half-plane to the right of the line described by $x = 4$. This suggests the following generalization.

**Graphs:** $x < a$, $x = a$, $x > a$

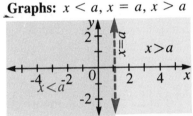

The graph of $x = a$ is a vertical line.

The graph of $x > a$ is the half-plane to the right of the line.

The graph of $x < a$ is the half-plane to the left of the line.

1. $y = 5x + 1$; below; solid 2. $y = x + 4$; above; dashed 3. $y = x + 2$; below; dashed 4. $y = 2x - 6$; above; solid 5. $x = 4$; left; dashed 6. $x = 6$; right; solid 7. $x = 0$; left; dashed

## Oral Exercises

**Give the equation that should be graphed first in order to graph the inequality. Tell whether the shaded half-plane will be above, below, to the left, or to the right of the line. Tell whether the line will be solid or dashed in the graph.**

**1.** $y \le 5x + 1$     **2.** $y > x + 4$     **3.** $y < x + 2$     **4.** $y \ge 2x - 6$
**5.** $x < 4$     **6.** $x \ge 6$     **7.** $x < 0$     **8.** $y \ge -9$
                                                                          $y = -9$; above; solid
**9.** $y < \frac{2}{3}x$     **10.** $y > x + 1$     **11.** $x \le -2$     **12.** $x > 6$

9. $y = \frac{2}{3}x$; below; dashed 10. $y = x + 1$; above; dashed
11. $x = -2$; left; solid     12. $x = 6$; right; dashed

## Written Exercises

**Graph. For Ex. 1–24 see TE Answer Section.**

Ⓐ  **1.** $y < 2x - 1$     **2.** $y \ge x - 4$     **3.** $y < -x + 1$     **4.** $y \le -x + 4$

**5.** $y \ge 3x$     **6.** $y \le \frac{2}{3}x - 5$     **7.** $y \ge \frac{1}{2}x + 1$     **8.** $y \le -x - 9$

**9.** $y > \frac{2}{3}x + 1$     **10.** $y \le 3x - 2$     **11.** $y \ge -2x - 2$     **12.** $y < \frac{1}{3}x - 6$

Graphing Inequalities     **321**

**13.** $y < -\dfrac{1}{2}x - 2$    **14.** $y \geq -\dfrac{1}{4}x - 7$    **15.** $y \leq -\dfrac{3}{4}x - 10$    **16.** $y > -\dfrac{1}{6}x - 5$

**17.** $y > x - 9$     **18.** $y \leq -2$       **19.** $y \geq -5$       **20.** $y < -6$

**21.** $x < 5$       **22.** $x \geq -2$       **23.** $x \leq 0$        **24.** $x \geq 5$

**Graph. Then, determine which of the given points belong to the graph.** See TE Answer Section.

**25.** $y > x - 4$; (3, 5), (3, 6) **(3, 5)** & **(3, 6)**    **26.** $y \leq x - 4$; (6, 2), (10, 5) **(6, 2)** & **(10, 5)**

**27.** $y \leq 2x + 1$; (4, 10), (4, 0) **(4, 0)**       **28.** $y < 2x - 3$; (3, 3), (3, 2) **(3, 2)**

**Solve for $y$.** 29. $y < -x + 4$   30. $y < x - 2$   31. $y < x + 1$   32. $y \geq 3x - 5$   33. $y > -\frac{2}{3}x$   34. $y < -5x + 4$

Ⓑ **29.** $x + y < 4$       **30.** $x - y > 2$       **31.** $x + 1 > y$        **32.** $3x - y \leq 5$

**33.** $2x + 3y > 0$      **34.** $5x + y < 4$      **35.** $-2x + y \geq 9$     **36.** $4y < x$   $y < \frac{1}{4}x$

**37.** $3x - 2y \geq 10$     **38.** $5x - y < 4$      **39.** $6x - 3y > 18$     **40.** $-x - y < 1$

**41.** $5 - y > x$       **42.** $-x - 6 \geq -2y$    **43.** $2x + 5y < 10$     **44.** $4 - y \geq 8$   $y \leq -4$

35. $y \geq 2x + 9$ 37. $y \leq \frac{3}{2}x - 5$ 38. $y > 5x - 4$ 39. $y < 2x - 6$ 40. $y > -x - 1$ 41. $y < -x + 5$

**Graph.** See TE Answer Section.

42. $y \geq \frac{1}{2}x + 3$ 43. $y < -\frac{2}{5}x + 2$

**45.** $3x + 2y \geq 2$      **46.** $2x - 3y \leq 9$      **47.** $4x - 5y > 20$      **48.** $5x - 3y \leq -9$

**49.** $-7y \geq 14 - 3x$    **50.** $3y - 5x < 15$     **51.** $2x - 9y > 18$     **52.** $6x - 9y \leq -9$

**53.** $7 - (2 - y) < -3x$      **54.** $6 - (4 - 2x) \geq 2 - 4y$      **55.** $3x - (6 - 2y) > x - (4 - y)$

Ⓒ **56.** $7 - [5 - (4 - y)] > 2 - x$        **57.** $6 - (4 - y) \leq y - [5 - (4 - 2y)]$

**58.** $7 < y \leq 8$                   **59.** $9 < x \leq 14$

**60.** $|x| > 3$                         **61.** $|y| \leq 4$

# CUMULATIVE REVIEW

**1.** Evaluate $-x^3 - x$ if $x = -2$.    **2.** Solve: $-5x + 4 = -x^2$.    **3.** Add: $\dfrac{7x - 38}{x^2 - 8x + 12} + \dfrac{6}{2 - x}$.

1. 10                                 2. 4, 1

3. $\frac{1}{x-6}$

## NON-ROUTINE PROBLEMS

1. If you omit the digit common to both numerator and denominator of $\dfrac{26}{65}$, you don't change its value because $\dfrac{26}{65} = \dfrac{2}{5}$. What other fractions with two digits in numerator and denominator can be simplified in the same manner? $\dfrac{16}{64}, \dfrac{49}{98}$

2. At the time of the Alaskan Gold Rush, three prospectors struck it rich and had to cross the Yukon River. However, they could find only one small rowboat with room for either two of the men or one man and a huge bag of gold nuggets. Each man owned a bag of gold, but the contents were not of the same value. Jake's gold was worth $8,000; Bill's, $5,000; and Hank's, $3,000. They did not trust each other. Finally, they agreed that no one should be left alone or in the boat with more gold than he owned. How did they get themselves and the gold across the river in no more than 13 crossings if returns across the river count as crossings? **See TE Answer Section.**

# APPLICATIONS

$$\boxed{\text{Read}} \longrightarrow \boxed{\text{Plan}} \longrightarrow \boxed{\text{Solve}} \longrightarrow \boxed{\text{Interpret}}$$

Scientists record data from experiments with the hope of eventually finding patterns that will enable them to predict future results. For example, in testing the effectiveness of a new antibacterial spray, a chemist might record the number of bacteria present in a tissue culture after the spray had been used for different periods of time. He might record the data in this table.

| Number of Hours | 4 | 6 | 8 | 10 | 12 | 14 | 16 |
|---|---|---|---|---|---|---|---|
| Number of Bacteria | 80 | 70 | 58 | 52 | 36 | 33 | 20 |

The data can be presented as a set of ordered pairs and could then be plotted.
(4, 80), (6, 70), (8, 58), (10, 52), (12, 36), (14, 33), (16, 20).

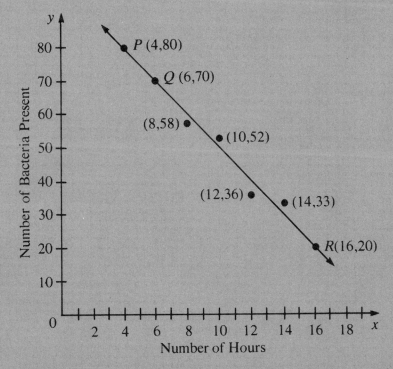

Notice that a line can be drawn through three of the given points, $P(4, 80)$, $Q(6, 70)$, and $R(16, 20)$. Since the other points are very close to the line, it can be used as an approximation of the set of all data for the experiment. The line is called a **linear model of the experiment.** Using any two of the three points that are actually on the line, you can write an equation for the linear model. You can then use the equation to predict the approximate number of bacteria present after a given number of hours. The procedure is illustrated in the example that follows on the next page.

*Example*    **Use the points $P(4, 80)$ and $Q(6, 70)$ to write an equation for the linear model on page 323. Then use the equation to predict the number of bacteria present after 8 h.**

$\dfrac{y - 80}{x - 4} = \dfrac{70 - 80}{6 - 4}$ ◀ *Slope of $\overline{PW}$ = slope of $\overline{PQ}$*

$\dfrac{y - 80}{x - 4} = \dfrac{-5}{1}$ ◀ $\dfrac{70 - 80}{6 - 4} = \dfrac{-10}{2} = \dfrac{-5}{1}$

$(y - 80)1 = -5(x - 4)$

$y - 80 = -5x + 20$

$y = -5x + 100$

Find $y$ if $x = 8$.

$y = -5x + 100$

$y = -5 \cdot 8 + 100$ ◀ *Substitute 8 for $x$.*

$y = 60$

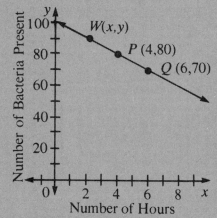

Thus, $y = -5x + 100$ is the equation for the model and about 60 bacteria are present after 8 h.

The population of ants in an anthill is tallied at the end of each month. The table below sums up the data.

| Number of Months | 2 | 3 | 4 | 5 | 6 | 7 |
|---|---|---|---|---|---|---|
| Population at the End of the Period | 8,000 | 9,500 | 10,000 | 11,000 | 13,000 | 13,500 |

1. Write the data as a set of ordered pairs and graph the corresponding points. Draw a line to contain as many points as possible.

2. Write an equation for the linear model of the data. **$y = 1,000x + 6,000$**

3. Use the equation to predict the approximate number of ants at the end of 8 mo.

4. Do the coordinates of the point $(0, 6,000)$ satisfy the equation of the linear model? How many ants were there at the beginning of the experiment?
   **3. 14,000   4. yes; 6,000**

The monthly cost of owning a car is dependent upon the number of miles the car is driven, as shown in this table.

| Number of Miles per Month | 400 | 500 | 800 | 1,000 | 1,200 | 1,500 |
|---|---|---|---|---|---|---|
| Operating Cost per Month | $50 | $60 | $65 | $70 | $75 | $80 |

5. Construct a linear model for the data.

6. Write an equation for the linear model. **$y = \dfrac{1}{40}x + 45$**

7. Use the equation to approximate the cost for driving a car 3,000 mi/mo.

8. Approximately how many miles per month can one drive at a cost of $100?
   **7. $120   8. 2,200 mi**

# GEOMETRY REVIEW

*Objectives*    **To find the measure of alternate interior angles**
**To find the measure of corresponding angles**

When a transversal intersects parallel lines, special pairs of angles have equal measures. Two such pairs are *alternate interior* and *corresponding*.

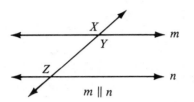

$m \parallel n$

The mathematical symbol for parallel is $\parallel$.

$\angle Y$ and $\angle Z$ are alternate interior angles.
$$m\angle Y = m\angle Z$$
$\angle X$ and $\angle Z$ are corresponding angles.
$$m\angle X = m\angle Z$$

*Example*    **In the diagram, the lines are parallel.**
**Find $t$ and the measure of each angle.**

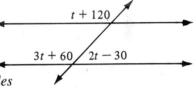

$3t + 60 = t + 120$  ◀  *Corresponding angles*
$2t + 60 = 120$           *have equal measure.*
$\quad\quad 2t = 60$
$\quad\quad\ t = 30$

| $3t + 60$ | $t + 120$ | $2t - 30$ |
|---|---|---|
| $3 \cdot 30 + 60$ | $30 + 120$ | $2 \cdot 30 - 30$ |
| $90 + 60$ | $150$ | $60 - 30$ |
| $150$ | | $30$ |

# Written Exercises

**In each of the exercises below, the lines are parallel. Find $x$ and the measure of any angle that contains $x$.**

1.
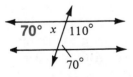

$70°$  $x$  $110°$
$70°$

2.
$100°$
$80°$  $x$  $100°$

3.    $120°$
$60°$  $4x + 20$  $60°$
$x = 10$

4.
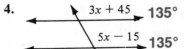

$3x + 45$  $135°$
$5x - 15$  $135°$

4. $x = 30$

5.
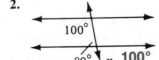

$20°$  $7x - 50$  $10x + 60$  $160°$
$5x - 30$  $20°$

5. $x = 10$

6.
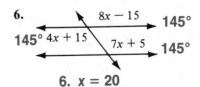

$8x - 15$  $145°$
$145°$  $4x + 15$  $7x + 5$  $145°$

6. $x = 20$

# CHAPTER ELEVEN REVIEW

Ⓐ Exercises are 1–6, 10–12, 16–21, 24–27, 30–36, 38–42. Ⓒ Exercises are starred. The rest are Ⓑ Exercises.

## Vocabulary

abscissa [11.1]  graph [11.2]
horizontal line [11.1]  intercept [11.7]
open half-plane [11.10]  ordered pair [11.1]
ordinate [11.1]  origin [11.1]
parallel [11.4]  quadrant [11.1]
slope [11.3]  vertical line [11.1]

What ordered pair of numbers describes the location of the point? [11.1]

1. $A$ **(2, 0)**
2. $B$ **(1, 2)**
3. $C$ **(−1, 0)**
4. $D$ **(0, −1)**
5. $E$ **(5, −1)**
6. $F$ **(−3, 3)**

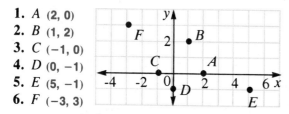

Using the graph above, find the point whose coordinates satisfy the given condition. [11.1]

7. The abscissa is −2. **None**

8. The ordinate is 2. **B**

9. The point is in quadrant 2. **F**

Plot. [11.2] **See TE Answer Section**
10. $A(3, 3)$            11. $B(−4, 2)$
12. $C(0, −4)$           13. $D\left(1\frac{1}{2}, -4\frac{1}{2}\right)$

Determine the slope of $\overline{AB}$.
14. $A(2k, 5k)$, $B(−5k, 3k)$ $\frac{2}{7}$
15. $A(2a − 1, 5b − 2)$, $B(−3a + 6, −b + 1)$ $\frac{6b-3}{5a-7}$

16. $\frac{4}{5}$, **up to the right** 17. **−1, down to the right**
Find the slope of $\overrightarrow{AB}$ for the given points.
Describe the slant of $\overrightarrow{AB}$. [11.4]
16. $A(4, 3)$, $B(9, 7)$   17. $A(4, −2)$, $B(1, 1)$
18. $A(3, −4)$, $B(3, 5)$  19. $A(−5, 3)$, $B(−1, 3)$
18. **undefined, vertical** 19. **0, horizontal**
From the given points, determine whether $\overrightarrow{PQ}$
is parallel to $\overrightarrow{RS}$. [11.4]   **20. $\overrightarrow{PQ} \parallel \overrightarrow{RS}$**
20. $P(1, −5)$, $Q(4, 1)$ $R(6, 13)$, $S(−4, −7)$
21. $P(5, 2)$, $Q(−5, −4)$   21. $\overrightarrow{PQ} \parallel \overrightarrow{RS}$
    $R(10, 7)$, $S(−15, −8)$

Determine whether $A$, $B$, and $C$ lie on the same line. [11.4] **22. A, B, C collinear**
**23. A, B, C not on the same line**
22. $A(0, 5)$, $B(−2, 11)$, $C(2, −1)$
23. $A(−1, −6)$, $B(0, −4)$, $C(4, −1)$

Write an equation for $\overrightarrow{PQ}$ with the given coordinates.
24. $P(5, 9)$, $Q(7, 15)$ [11.5] **$y = 3x − 6$**
25. $P(−2, −12)$, $Q(5, 2)$ **$y = 2x − 8$**
26. $P(−2, 3)$, $Q(1, 4)$ [11.6] **$y = \frac{1}{3}x + \frac{11}{3}$**
27. Write an equation for the line passing through the point $Q(2, 5)$ and with slope $\frac{3}{4}$. [11.5] **$y = \frac{3}{4}x + \frac{7}{2}$**

28. Write an equation for $\overrightarrow{PQ}$, given $P(3, 5)$ and $Q(4, 7)$. Then, show that $R(5, 9)$ lies on $\overrightarrow{PQ}$. [11.5] **$y = 2x − 1$; R lies on $\overrightarrow{PQ}$**
29. Write an equation for $\overrightarrow{AB}$ for $A(9, 11)$ and $B(15, 17)$. Then, use the equation to set up a table of values and give the coordinates of three other points on $\overrightarrow{AB}$. [11.6] **See TE Answer Section.**
**$y = x + 2$; (−1, 1), (0, 2), (1, 3)**

Give the slope and the $y$-intercept of the line with the given equation. Graph the line.
30. $y = −\frac{2}{3}x + 6$ [11.7] $−\frac{2}{3}$, 6  31. $y = 4x$ 4, 0
32. $x − 4y = 20$ [11.8]   33. $3x − 4y = 12$
34. $3x − 2(5 − y) = 12$   35. $x = −4$ [11.9]
36. $14 + 2x = 6$ **undefined, none** 34. **$−\frac{3}{2}$, 11**
32. $\frac{1}{4}$, −5  33. $\frac{3}{4}$, −3  35. **undefined, none**
37. Write an equation for the line with slope $\frac{2}{3}$ and $y$-intercept the same as that of the line with equation $5x − 2y = 12$. [11.7]
37. $y = \frac{2}{3}x − 6$
Graph. [11.10] **See TE Answer Section.**
38. $y < 3x + 1$    39. $x ≤ −1$    40. $y > 2$
41. $3x − 4y > −8$  42. $5x − 3y ≤ −12$
★ 43. Determine the value(s) of $a$ so that the slope of $\overline{AB}$ is 0, given $A(6, a)$ and $B(7, a^2)$. [11.3] **0, 1**
★ 44. Find the value(s) of $a$ so that $\overrightarrow{PQ}$ is parallel to $\overrightarrow{RS}$, given $P(6, 6)$, $Q(7, 5)$, $R(7, −4a − 1)$, and $S(8, a^2 + 1)$. [11.4]

**44. −3, −1**

Ⓐ **Exercises are 1, 2, 7–13, 18–26, 30–33.** Ⓒ **Exercises are starred. The rest are** Ⓑ **Exercises.**

Write an equation for $\overleftrightarrow{PQ}$ with the given coordinates.

**1.** $P(5, 10)$, $Q(7, 14)$   **2.** $P(-5, 4)$, $Q(0, 2)$

**1.** $y = 2x$      **2.** $y = -\frac{2}{5}x + 2$

Write an equation for $\overleftrightarrow{PQ}$. Then, show whether $R$ lies on $\overleftrightarrow{PQ}$.

**3.** $P(4, 5)$, $Q(6, 6)$, $R(8, -1)$ $y = \frac{1}{2}x + 3$
 **$R$ is not on $\overleftrightarrow{PQ}$.**

Plot. **See TE Answer Section**

**4.** $A(3, 5)$   **5.** $B(-4, 0)$   **6.** $C\left(3\frac{1}{2}, -1\frac{1}{2}\right)$

From the given points, determine whether $\overrightarrow{PQ}$ is parallel to $\overrightarrow{RS}$.

**7.** $P(6, 3)$, $Q(-3, -3)$, $R(0, 4)$, $S(3, 6)$
$\overrightarrow{PQ} \parallel \overrightarrow{RS}$

What ordered pair of numbers describes the location of the point?

**8.** $A$ **(3, 2)**
**9.** $B$ **(0, -2)**
**10.** $C$ **(-1, 2)**
**11.** $D$ **(3, -1)**
**12.** $E$ **(5, 1)**
**13.** $F$ **(-2, -1)**

Using the graph above, find the point(s) whose coordinates satisfy the given condition.

**14.** The abscissa is 3. **A, D**

**15.** The ordinate is $-2$. **B**

**16.** The sum of the abscissa and the ordinate is 2. **D**

**17.** The point is in quadrant 3. **F**

Find the slope of $\overleftrightarrow{AB}$ for the given points. Describe the slant of $\overleftrightarrow{AB}$.

**18.** $A(4, 5)$, $B(9, 11)$ $\frac{6}{5}$, **up and to the right**

**19.** $A(-5, -3)$, $B(2, -3)$ **0, horizontal**

**20.** $A(-2, 5)$, $B(-2, -1)$ **undefined, vert.**

**21.** $A(-4, -2)$, $B(-6, 5)$ $-\frac{7}{2}$, **down and to the right**

Give the slope and the $y$-intercept of the line with the given equation. Graph the line. **See TE Answer Section**

**22.** $y = \frac{3}{4}x - 1$ $\frac{3}{4}, -1$ **23.** $4x - 3y = 9$ $\frac{4}{3}, -3$

**24.** $6 + 2y = 8$ **0, 1**  **25.** $-3x - 7 = 2$

**26.** $2x - 2(4 - y) = 14$  **undefined, none**
 $-1, 11$

**27.** Write an equation for the line with slope $-\frac{1}{3}$ and $y$-intercept the same as that of the line with equation $4x - 8y = 20$. $y = -\frac{1}{3}x - \frac{5}{2}$

**28.** Write an equation for $\overleftrightarrow{AB}$ for $A(0, -1)$ and $B(-2, 5)$. Then, use the equation to set up a table of values and give the coordinates of three other points on $\overleftrightarrow{AB}$.
 $y = -3x - 1;\ (-1, 2),\ (-2, 5),\ (1, -4)$

Express the slope of $\overleftrightarrow{AB}$ in terms of variables.

**29.** $A(-3a - 1, b + 4)$, $B(-5a + 4, -b + 2)$
 $\frac{2b + 2}{2a - 5}$

Graph. **See TE Answer Section.**

**30.** $y < 2x + 1$          **31.** $x \geq -1$

**32.** $5x - 3y \leq 6$

**33.** $3x - (4 + y) < x + 1$

**34.** Determine whether $A(0, 6)$, $B(-1, -3)$, and $C(-2, 0)$ lie on the same line.
 **A, B, and C do not lie on the same line**

★ **35.** Graph the line with $y$-intercept $-2$ and parallel to the line with equation $4x - 3y = 15$. **See TE Answer Section.**

★ **36.** Find the value of $b$ so that $(-3, 1)$ lies on the line with equation $y = bx - 4$. $-\frac{5}{3}$

★ **37.** Find the value(s) of $a$ so that $\overrightarrow{PQ}$ is parallel to $\overrightarrow{RS}$, given $P(5, -4)$, $Q(6, 6)$, $R(6, -2a - 1)$, and $S(7, a^2 + 1)$. $-4, 2$

# COLLEGE PREP TEST

Directions: Choose the best answer for each question.

**1.** If $2x + 7 = 23$, then $x^2 + 8 = $ __?__ . **B**

 (A) 16  (B) 72  (C) 24
 (D) 18  (E) none of these

**2.** If $x = 4$, which of the following is not an even integer? **D**

 (A) $x + 4$  (B) $x - 4$  (C) $x$
 (D) $x + 1$  (E) $x + 6$

**3.** The nine regions in the diagram are to be colored so that no two regions bordering each other have the same color. What is the least number of colors you will have to use? **B**

 (A) 2
 (B) 3
 (C) 4
 (D) 5
 (E) 6

**4.** In a group of 87 people, 58 have brown hair; 63 have brown eyes; and 49 have both brown hair and brown eyes. How many have neither brown hair nor brown eyes? **E**

 (A) 34  (B) 2  (C) 14  (D) 9  (E) 15

**5.** When is $(a + b)^2 = a^2 + b^2$ a true statement? **A**

 (A) sometimes  (B) always  (C) never
 (D) only with fractions
 (E) cannot be determined

**6.** Which of the following is the greatest? **A**

 (A) $\frac{2}{3}$  (B) $\frac{4}{9}$  (C) $\frac{2}{5}$  (D) $\frac{2}{4}$  (E) $\frac{8}{15}$

**7.** Eight telephone poles are spaced 15 ft apart in a row. What is the distance from the first to the last pole? **D**

 (A) 30 ft  (B) 60 ft  (C) 85 ft
 (D) 105 ft  (E) 120 ft

**8.** A student's test scores are 70, 90, 65, 85, and 75. What must the student score on the next test for an average of 80? **E**

 (A) 73  (B) 81  (C) 90  (D) 92  (E) 95

**9.** A cashier gave a customer change for a dollar in half dollars and dimes. How many coins did the customer receive? **C**

 (A) 4  (B) 7  (C) 6  (D) 9  (E) 10

**10.** $ABCD$ is a square with $AE = 2$ and $GC = 8$. If the shaded area is 44, then the area of $FBEJ$ is __?__ . **B**

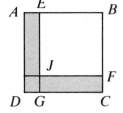

 (A) 36  (B) 56  (C) 64  (D) 68  (E) 80

**11.** How many thirds are there in $\frac{5}{8}$? **B**

 (A) $\frac{5}{24}$  (B) $\frac{15}{8}$  (C) $\frac{8}{15}$
 (D) $4\frac{1}{3}$  (E) 15

**12.** If $\frac{1}{2} + \frac{2}{3} + \frac{3}{y} = \frac{23}{12}$, then $y = $ __?__ . **C**

 (A) 2  (B) 3  (C) 4  (D) 9  (E) 12

# 12 SYSTEMS OF EQUATIONS

An important aspect of hotel or motel management is maintenance. The manager must plan and budget for the cost of a painting project and decide whether the hotel's maintenance staff or an outside group should do the work.

*Project*

A wing of the Siesta motel that contains 27 units needs to be repainted. Decide on the dimensions and specifications of the rooms to be repainted. (Each unit has the same size rooms.) Determine the cost of the project.

Formulate the questions you need to answer before you solve the problem.

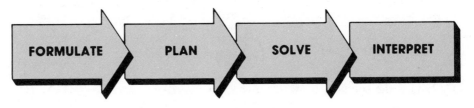

FORMULATE → PLAN → SOLVE → INTERPRET

# GRAPHING SYSTEMS OF EQUATIONS  12.1

*Objectives*  **To solve a system of equations in two variables by graphing**
**To determine whether a system is independent and consistent,**
**inconsistent, or dependent and consistent**

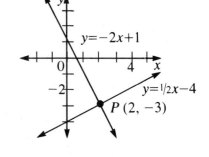

The graphs of $y = -2x + 1$ and $y = \frac{1}{2}x - 4$ are shown at the
right. Notice that one point, $P(2, -3)$, lies on both lines.
$P(2, -3)$ is the **point of intersection** of the lines. Therefore, the
coordinates of $P$ must satisfy both equations. You can verify this
by showing that $(2, -3)$ is a solution of each equation, as follows.
Substitute 2 for $x$ and $-3$ for $y$ in each equation and determine
if the result is true.

| $y$ | $-2x + 1$ |
|-----|-----------|
| $-3$ | $-2 \cdot 2 + 1$ |
|  | $-4 + 1$ |
|  | $-3$ |
| $-3 = -3$, true | |

| $y$ | $\frac{1}{2}x - 4$ |
|-----|-----------|
| $-3$ | $\frac{1}{2} \cdot 2 - 4$ |
|  | $1 - 4$ |
|  | $-3$ |
| $-3 = -3$, true | |

**Thus,** $(2, -3)$ is the solution of the system of equations $y = \frac{1}{2}x - 4$ and
$y = -2x + 1$. This suggests the procedure for solving a system of equations
by graphing.

| Solving a system of equations by graphing | To solve a system of equations by graphing: (1) Graph each equation on the same set of axes. (2) Find the coordinates of the point of intersection of the lines. |
|---|---|

*Example 1*   **Solve the system by graphing.**
$$y = \phantom{-}2x - 3$$
$$y = -\frac{1}{2}x + 2$$

Graph each equation on the same set of axes.

*Graph* $y = 2x - 3$
*or* $y = \dfrac{2}{1}x - 3$.
        *slope*   *y-intercept*

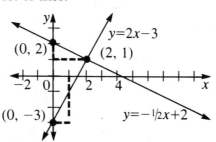

*Graph* $y = -\dfrac{1}{2}x + 2$
        *slope*   *y-intercept*

**So,** $(2, 1)$ is the solution.  ◀ *The two lines intersect at (2, 1).*

*Example 2*      **Solve the system $3x - 2y = 6$ by graphing. Check the solution.**
$$x + y = 2$$

*First* write each equation in slope-intercept form: $y = mx + b$.

| $3x - 2y = 6$ | $x + y = 2$ |
|---|---|
| $-2y = -3x + 6$ | $y = -x + 2$ |
| $y = \frac{3}{2}x - 3$ | or $y = -\frac{1}{1}x + 2$ |

*Notice* that $m$ is different in each equation and $b$ is also different.

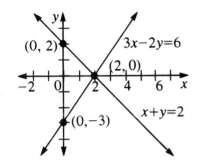

◀ *Graph each equation on the same set of axes.*
◀ *The point of intersection is (2, 0).*

*Check.* Substitute 2 for $x$ and 0 for $y$ in each equation.

| $3x - 2y$ | $6$ |
|---|---|
| $3 \cdot 2 - 2 \cdot 0$ | $6$ |
| $6 - 0$ | |
| $6$ | |

$6 = 6$, true

| $x + y$ | $2$ |
|---|---|
| $2 + 0$ | $2$ |
| $2$ | |

$2 = 2$, true

**Thus,** $(2, 0)$ satisfies both equations and is the solution.

Example 3 illustrates that there is no solution for a system of equations whose graph is a pair of parallel lines.

*Example 3*      **Solve the system $y = 2x + 1$ by graphing.**
$$y = 2x - 3$$

*Notice* that $m$ is the same in each equation but $b$ is different.

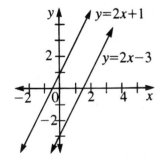

◀ *Graph each line on the same set of axes.*
  *The lines are parallel.*

◀ *Since the lines have no point in common,*
  *there is* no solution *of this system.*

A system of equations with no solution is called an **inconsistent system.**

**So,** $y = 2x + 1$ and $y = 2x - 3$ is an inconsistent system of two equations.

Example 4 illustrates that there is an infinite number of solutions for a system of equations whose graphs coincide.

*Example 4*    **Solve the system $2x - y = 1$ by graphing.**
                      $$4x - 2y = 2$$

*First* write each equation in slope-intercept form: $y = mx + b$.

$$\begin{array}{c|c}
2x - y = 1 & 4x - 2y = 2 \\
2x - 1y = 1 & -2y = -4x + 2 \\
-1y = -2x + 1 & y = 2x - 1 \\
y = 2x - 1 &
\end{array}$$

*Notice* that the resulting equations are the same, $y = 2x - 1$.

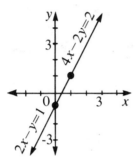

So each equation has the same values for $m$ and $b$.

Since the equations are the same, their graphs *coincide*. Any point on the line will have coordinates that satisfy both equations.

A system of equations that has an infinite number of solutions is called a **consistent and dependent system**.

When solving a system of equations graphically, write each equation in $y = mx + b$ form and graph. Comparing $m$ and $b$ will help you determine how many solutions there are.

| $m$ | $b$ | Type of System | Graph | Number of Solutions |
|---|---|---|---|---|
| different | same or different | consistent independent | intersecting lines | 1 |
| same | different | inconsistent | parallel lines | 0 |
| same | same | consistent dependent | coinciding lines | infinite |

# Written Exercises

**For Ex. 1–12, see TE Answer Section.**

**Solve by graphing. Check the solution.**

Ⓐ **1.** $y = -4x + 5$   **2.** $y = x + 6$   **3.** $y = -3x + 7$   **4.** $y = 4x - 3$
      $y = 3x - 9$ **(2, −3)**   $y = -2x$ **(−2, 4)**   $y = 2x - 3$ **(2, 1)**   $y = -2x + 9$ **(2, 5)**

**5.** $3x - 2y = 4$   **6.** $3x - 2y = 6$   **7.** $x + 3y = 6$   **8.**   $x + y = 4$
      $y = -2x + 5$ **(2, 1)**   $x - y = 2$ **(2, 0)**   $x - 3y = 6$ **(6, 0)**   $2x - 2y = 10$
                                                                                 $\left(4\tfrac{1}{2}, -\tfrac{1}{2}\right)$

**Graph. Indicate whether the system is *independent and consistent*, *inconsistent*, or *dependent and consistent*.**

Ⓑ **9.** $3x + 2y = 8$   **10.**   $x + y = 6$   **11.**   $-x + y = 4$   **12.** $x - 4y = 16 - 2y$
      $x - 4y = -9$         $-3x - 3y = -18$         $-2x + 2y = 10$          $x + y = 4$
      **independent,**      **dependent,**            **inconsistent**         **independent, consistent**
      **consistent**        **consistent**

# GRAPHING SYSTEMS OF INEQUALITIES      12.2

*Objectives*      **To solve a system of inequalities in two variables by graphing**
**To verify the solution of a system of inequalities**

In the last lesson, you solved a system of equations by graphing. The following is an example of a system of inequalities.

$$y \le 2$$
$$x > 3$$

The solution of the system is the set of all ordered pairs whose coordinates satisfy both inequalities. To find the solution of the system, graph each inequality on the same set of axes. The solution will be the region where the two graphs overlap, as shown in Example 1 below.

*Example 1*      **Solve the system** $y \le 2$ **by graphing.**
$$x > 3$$

*First,* graph $y = 2$ to be a line.
*Think:* The $y$-coordinate is always 2; so the line is horizontal. Shade the region below the line.

*Now,* graph $x > 3$ on the same set of axes. Graph $x = 3$ with a dashed line. *Think:* The $x$-coordinate is always 3; so the line is vertical. Shade the region to the right of the line.

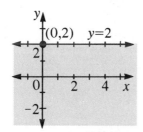

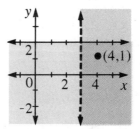

The *double-shaded region* contains all the points that are solutions of both inequalities. One point in this region, (4, 1), is indicated on the graph.
**So,** the solution is the double-shaded region.

To verify that the double-shaded region is the solution of the system, pick any point in the double-shaded region and show that its coordinates satisfy both inequalities.

*Example 2*      **Verify that (4, 1) satisfies** $y \le 2$ **in Example 1.**
$$x > 3$$

*Test* (4, 1).

| $y$ | 2 |
|---|---|
| 1 | 2 |

| $x$ | 3 |
|---|---|
| 4 | 3 |

$1 \le 2$, true      $4 > 3$, true      **Thus,** (4, 1) satisfies both inequalities.

*Example 3*　　　**Solve the system $3x - 5y < 10$ by graphing.**
$$y < -\tfrac{2}{3}x + 1$$

$3x - 5y < 10$　　　◀ *Solve $3x - 5y < 10$ for y.*
$\quad -5y < -3x + 10$
$\qquad y > \tfrac{3}{5}x - 2$　　◀ *Dividing each side by $-5$ reverses the order from $<$ to $>$.*

Use the slope, $m$, and $y$-intercept, $b$, of $y = mx + b$ to graph.

Graph $y = \tfrac{3}{5}x - 2$ with a dashed line.
Shade the region above the line.

*Now,* graph $y = -\tfrac{2}{3}x + 1$ on the same set of axes.
Graph $y = -\tfrac{2}{3}x + 1$ with a dashed line.
Shade the region below the line.

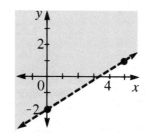
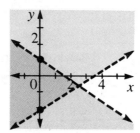

The double-shaded region contains all the points that are solutions of both inequalities.

**So,** the solution of the system is the double-shaded region.

## Written Exercises

**For Ex. 1–15, see TE Answer Section.**

**Solve by graphing. Then, pick a point in each double-shaded region and show that it satisfies both inequalities.**

(A)　**1.** $x < 4$
$\quad\ y \le 3$ **(2, 2)**

**2.** $x \ge 2$
$\quad y < -4$ **(3, −5)**

**3.** $y < x + 1$
$\quad y > -x + 5$ **(4, 3)**

**4.** $y \le 2x - 4$
$\quad y > \tfrac{1}{2}x - 5$ **(4, 1)**

**Solve by graphing.**

**5.** $x < -4$
$\quad\ y \ge 3$

**6.** $x < 6$
$\quad y \ge -5$

**7.** $y > x + 4$
$\quad x \le 1$

**8.** $3x - y \le 6$
$\quad\ y < 4$

(B)　**9.** $3x - 4y \le 12$
$\quad\ x - y > 1$

**10.** $3x - 2y < 2$
$\quad 2y > 3x - 2$

**11.** $y - 2x > -3$
$\quad x - 3y < 9$

**12.** $y - x < 0$
$\quad 3y \ge 2x - 15$

**Graph the three inequalities on the same set of axes. Which region of the graph is the solution of the system?**

(C)　**13.** $x + y \le 2$
$\qquad x \ge 0$
$\qquad y \ge 4$

**14.** $2x - y \le 15$
$\qquad x \ge 1$
$\qquad y > -2$

**15.**　　$x + 4y < 8$
$\quad 3x - 2y + 4 > 0$
$\qquad\qquad y \ge 0$

For advertising information call: 1-800-8 COUPON

ANAHEIM HILLS
CAR WASH
$1⁰⁰ Off
ANY CAR WASH
LA PALMA AT IMPERIAL
777-6605
We Recycle and Reclaim Our Water
One wash per coupon
H8

For advertising information call: 1-800-8 COUPON

$3 OFF
Any Dry Cleaning
Order Of $8 Or More

$7 OFF
Any Dry Cleaning
Order Of $21 Or More

PRESENT COUPON WITH INCOMING ORDER

EL RANCHO

128 South Fairmont At Santa Ana Canyon Road
(Across from Hughes)
921-9958
H8

For advertising information call: 1-800-8 COUPON

50¢ OFF
Movie Rental
(Excludes 3 for 2 offer)
Limit 1 coupon per customer per day

VCR - Stereo - TV Repair
"Our Specialty" $19.95
VCR Hand (With Coupon)
Cleaning (Reg. $39.95)

VIDEO SUPER SHOP

### system of equations by substitution

graphing, you can use algebraic methods to solve systems of
ually, the algebraic methods take less time, and they yield
results, particularly if the solutions are not integers. One of
methods involves *substitution*, which is the method you will
son.

tem $y = 3x$     **algebraically. Check the solution.**
$x + y = 12$

uation, $y$ is equal to $3x$. Since $y$ must be the same in the second
can substitute $3x$ for $y$ in $x + y = 12$.

$$x + 3x = 12$$
$$4x = 12$$
$$x = 3$$

◀ *Substitute $3x$ for $y$ to get an equation with only one variable. Then, solve the resulting equation for $x$.*

To find $y$, replace $x$ with 3 in either equation. Then, solve the resulting
equation for $y$.

$y = 3x$     or     $x + y = 12$
$y = 3 \cdot 3$            $3 + y = 12$
$y = 9$                 $y = 9$

So, $x = 3$ and $y = 9$.

*Check.*

| $y$ | $3x$ | | $x + y$ | $12$ |
|-----|------|---|---------|------|
| $9$ | $3 \cdot 3$ | | $3 + 9$ | $12$ |
| | $9$ | | $12$ | |
| | $9 = 9$, true | | $12 = 12$, true | |

**Thus,** $(3, 9)$ is a solution of the system.

Now, consider the system $2x + 3y = 7$.
$3x - y = 5$

In order to solve this system by substitution, you first must solve one of
the equations for one of its variables. It is easier to choose a variable whose
coefficient is 1 or $-1$. Therefore, you should begin by solving the second
equation, $3x - y = 5$, for $y$. The complete solution is shown in Example 2
on the next page.

*Example 2*

**Solve** $2x + 3y = 7$. **Check the solution.**
$$3x - y = 5$$

*First,* solve $3x - y = 5$ for $y$.
$$3x - 1y = 5 \qquad \blacktriangleleft \; -y = -1y$$
$$-1y = -3x + 5 \qquad \blacktriangleleft \; \textit{Add } -3x \textit{ to each side.}$$
$$y = 3x - 5 \qquad \blacktriangleleft \; \textit{Divide each side by } -1.$$

*Now,* replace $y$ with $3x - 5$ in the other equation, $2x + 3y = 7$.
$$2x + 3y = 7$$
$$\downarrow$$
$$2x + 3(3x - 5) = 7$$
$$2x + 9x - 15 = 7 \qquad \blacktriangleleft \; 3(3x - 5) = 9x - 15$$
$$11x - 15 = 7$$
$$11x = 22$$
$$x = 2$$

To find $y$, replace $x$ with 2 in any one of the equations. It is easier to use $y = 3x - 5$ because it is already solved for $y$.
$$y = 3x - 5$$
$$y = 3 \cdot 2 - 5 \qquad \blacktriangleleft \; \textit{Substitute 2 for } x.$$
$$y = 6 - 5$$
$$y = 1$$

So, $x = 2$ and $y = 1$

*Check.*

| $2x + 3y$ | 7 |
|---|---|
| $2 \cdot 2 + 3 \cdot 1$ | 7 |
| $4 + 3$ | |
| 7 | |
| $7 = 7$, true | |

| $3x - y$ | 5 |
|---|---|
| $3 \cdot 2 - 1$ | 5 |
| $6 - 1$ | |
| 5 | |
| $5 = 5$, true | |

$\blacktriangleleft$ *Replace x with 2 and y with 1 in each equation.*

**Thus,** $(2, 1)$ is a solution.

*Summary*

To solve a system of equations by substitution:

(1) Solve one of the equations for one of its variables (if neither equation is already solved for one of its variables).

(2) Substitute for that variable in the other equation.

(3) Solve the resulting equation.

(4) Find the value of the second variable by substituting the result from (3) in any one of the equations.

(5) Check the solution.

## Oral Exercises

**In each system, tell for which variable it is easier to solve.**

**1.** $3x - y = 7$
$4x + 2y = 6$ **y**

**2.** $5x - 3y = 2$
$x + 4y = 5$ **x**

**3.** $x + 4y = 9$
$3x - 5y = -7$ **x**

**4.** $7x - 3y = 1$
$3x - y = 4$ **y**

## Written Exercises

**Solve by substitution. Check.**

Ⓐ **1.** $y = 2x$
$x + y = 12$ **(4, 8)**

**2.** $x + y = 8$
$y = x - 2$ **(5, 3)**

**3.** $y = 2x - 6$
$x - y = 4$ **(2, −2)**

**4.** $3x - 4y = 4$
$x + 2y = 8$ **(4, 2)**

**5.** $3x - 7y = 5$
$2x - y = 7$ **(4, 1)**

**6.** $x + 3y = 4$
$2x - 5y = -3$ **(1, 1)**

**Solve by substitution.**

**7.** $x + y = 15$
$y = x + 3$ **(6, 9)**

**8.** $x - y = -1$
$-2x + 3y = 5$ **(2, 3)**

**9.** $3x + 2y = 14$
$4 - 2x = y$ **(−6, 16)**

**10.** $x - y = 1$
$3x + 4y = 10$ **(2, 1)**

**11.** $x - y = 16$
$x + y = 74$ **(45, 29)**

**12.** $5x - 7y = -3$
$3x + y = 19$ **(5, 4)**

Ⓑ **13.** $3x + y = 2$ $(\frac{1}{2}, \frac{1}{2})$
$x - y = 0$

**14.** $3x + y = 1$
$5x - y = 1$ $(\frac{1}{4}, \frac{1}{4})$

**15.** $2x + y = 1$
$10x - 4y = 2$ $(\frac{1}{3}, \frac{1}{3})$

**16.** $3x - 3y = 18$
$x + y = 3$ $\left(4\frac{1}{2}, -1\frac{1}{2}\right)$

**17.** $x + y = 8$
$2x - y = -6$ $(\frac{2}{3}, 7\frac{1}{3})$

**18.** $x + 2y = 2$
$4x + 3y = 5$ $(\frac{4}{5}, \frac{3}{5})$

**19.** $2x + 2y = 4$
$3x - 3y = 18$ **(4, −2)**

**20.** $6x - 3y = 8$
$6x - 2y = 12$ $(3\frac{1}{3}, 4)$

**21.** $3y = 5 - 2x$
$4x + 3y = 1$ **(−2, 3)**

Ⓒ **22.** $\dfrac{x}{4} = \dfrac{y}{2}$ $(-3\frac{3}{7}, -1\frac{5}{7})$
$x + 5y = -12$

**23.** $\dfrac{y}{2} = \dfrac{12 - 3x}{3}$ **(3, 2)**
$y = 3x - 7$

**24.** $x + y = 5$
$\dfrac{x}{3} + \dfrac{y}{2} = 2$ **(3, 2)**

**25.** $3x - 2y = 8$ **(4, 2)**
$2x + 3y = 14$

**26.** $2x - 3y = -12$
$\dfrac{x + y}{4} = 6$ **(12, 12)**

**27.** $\dfrac{x}{3} + \dfrac{y}{5} = \dfrac{8}{15}$ **(1, 1)**
$3x + y = 4$

## CUMULATIVE REVIEW

**1.** Find the slope of $\overrightarrow{PQ}$ for $P(2, 5)$ and $Q(-4, -1)$. **1**

**2.** Graph $3x - 4y = 8$.
**See TE Answer Section.**

**3.** Evaluate $-ab^3$ for $a = 2$ and $b = -3$. **54**

## CALCULATOR ACTIVITIES

Solve by substitution.

**1.** $y = 3.14x - 7.2$ **(2.298, 0.01572)**
$5.2x + 3.04y = 12.004$

**2.** $x - 2.06y = 9.602$ **(7.138, −1.196)**
$4.3x - 0.056y = 30.762$

# PROBLEM SOLVING

*Objective*   **To solve problems involving systems of equations by the substitution method**

You have already solved several types of problems by using an equation in one variable. Frequently, problems involving two quantities can be solved by using a system of equations in two variables. As shown in Example 1, the basic steps for solving any problem still apply.

*Example 1*   **Mary's age is 5 times Jose's age. The sum of their ages is 18. How old is each?**

READ ▶   You are asked to find two ages.

PLAN ▶   Use a different variable to represent each age.
Let $x$ = Mary's age.
Let $y$ = Jose's age.
Write an equation for each of the first two sentences.

Mary's age is 5 times Jose's age.   |   The sum of their ages is 18.

$$x = 5y \qquad\qquad x + y = 18$$

You now have a system of equations.

SOLVE ▶   Solve the system by substitution.

$$x = 5y \qquad (1)$$
$$x + y = 18 \qquad (2)$$

$5y + y = 18$   ◀ *Replace x with 5y in Equation (2).*
$6y = 18$
$y = 3$   ◀ *Jose's age*

To find $x$, replace $y$ with 3 in Equation (1).
$$x = 5y \qquad (1)$$
$$x = 5(3)$$
$$x = 15 \quad ◀ Mary's age.$$

INTERPRET ▶   *Check.*   Mary's age is 5 times Jose's age.   |   The sum of their ages is 18.

| 15 | 5 · 3 | | 15 + 3 | 18 |
| | 15 | | 18 | |

15 = 15, true   |   18 = 18, true

**Thus,** Jose's age is 3, and Mary's age is 15.

If a problem involves a geometric concept, it is helpful to draw a sketch, as illustrated in Example 2 on the next page.

*Example 2*     **The length of a rectangle is 4 m less than twice the width. The perimeter is 28 m. Find the length and the width.**

Let $l$ = the length of the rectangle.
Let $w$ = the width of the rectangle.

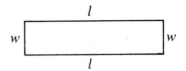

The perimeter is 28 m.
$$l + l + w + w = 28$$
or     $$2l + 2w = 28$$

The length is 4 m less than twice the width.
$$l = 2w - 4$$

Solve $l = 2w - 4$     (1)
$2l + 2w = 28$     (2)
$2(2w - 4) + 2w = 28$     ◄ *Replace $l$ with $2w - 4$ in Equation (2).*
$4w - 8 + 2w = 28$
$6w = 36$
$w = 6$
$l = 2w - 4$     (2)     ◄ *Find $l$ by replacing $w$ with 6 in Equation (2).*
$l = 2 \cdot 6 - 4$
$l = 12 - 4$, or 8     **So,** the length is 8 m, and the width is 6 m.   Check on your own.

Before you can use the substitution method in the next example, you will have to solve one of the equations for one of its variables.

*Example 3*     **A class has a total of 25 students. Twice the number of girls is equal to 3 times the number of boys. How many boys and how many girls are there in the class?**

Let $x$ = the number of girls.
Let $y$ = the number of boys.
$x + y = 25$     (1)     ◄ *The total is 25.*
$2x = 3y$     (2)     ◄ *Twice the number of girls = 3 times the number of boys.*
*First,* solve $x + y = 25$ for $x$.
$x = 25 - y$     (3)
$2x = 3y$     (2)
$2(25 - y) = 3y$     ◄ *Replace $x$ with $25 - y$ in Equation (2).*
$50 - 2y = 3y$
$50 = 5y$
$10 = y$     ◄ *Number of boys*

$x = 25 - y$     (3)     ◄ *Replace $y$ with 10 in Equation (3).*
$x = 25 - 10$     ◄ *Number of girls*
$x = 15$     **So,** there are 15 girls and 10 boys in the class.   Check on your own.

# Written Exercises

**Solve each problem.**

**(A)** 1. Mona's age is 4 times Billy's age. The sum of their ages is 15. Find the age of each. **3, 12**

2. Jane drove twice as far as Merv did. The difference between their distances was 75 km. How far did each drive? **75, 150**

3. Wanda is 5 years older than Marian. The sum of their ages is 19. Find the age of each. **7, 12**

4. Beth scored 3 times as many points as Nick did in basketball. The difference between their scores was 10. How many points did each score? **5, 15**

5. Gary Chin has a total of 27 tapes and records. The number of tapes is 3 less than twice the number of records. How many of each does he have? **17 tapes, 10 records.**

6. The length of a rectangle is 2 m less than 3 times the width. The perimeter is 68 m. Find the length and the width of the rectangle. **25, 9**

7. The length of a rectangle is 5 m more than twice the width. The perimeter is 52 m. Find the length and the width of the rectangle. **19, 7**

8. The length of a rectangle is 8 km more than 4 times the width. The perimeter is 46 km. Find the length and the width of the rectangle. **20, 3**

9. In October, Yvonne sold 5 times as many cars as Pablo did. Together, they sold 24 cars. How many cars did each sell? **Pablo 4, Yvonne 20**

10. Rico scored 6 more points than twice as many as Running Brook did. Their combined score was 27 points. How many points did each score? **7, 20**

**(B)** 11. The sum of two numbers is 35. Twice the first number is equal to five times the second number. Find the two numbers. **25, 10**

12. The difference between two numbers is 4. Three times the first number, increased by the second number, is 20. Find the two numbers. **4, 8 or 6, 2**

13. The sum of the ages of Ron and his sister is 20. Ron's age, decreased by twice his sister's age, is 2. Find the age of each. **14, 6**

14. The larger of two numbers is 4 less than 3 times the smaller number. If 3 more than twice the larger number is decreased by twice the smaller number, the result is 11. Find the numbers. **8, 4**

**(C)** 15. Half the sum of the distances traveled by Clarissa and Desmond is 640 km. If $\frac{1}{4}$ of the distance Clarissa traveled is 5 km more than $\frac{1}{6}$ of the distance Desmond traveled, how far did each travel? **Clarissa, 524; Desmond, 756**

16. During a holiday weekend, the number of campers admitted to a park was 5 less than $\frac{1}{2}$ the number of trailers admitted. If 4 more trailers had been admitted, there would have been a total of 89 trailers and campers. How many were there of each kind? **trailers, 60; campers, 25**

17. Gary is five times as old as Frank. One third of the sum of their ages is equal to 32. Find their ages. **Frank: 16 y; Gary: 80 y**

# THE ADDITION METHOD 12.5

Objectives | **To solve a system of equations by addition**
**To solve problems**

In this lesson, you will learn another algebraic method for solving a system of equations. This method makes use of the Addition Property for Equations.

Consider the system $5x - 2y = 30$.
$$x + 2y = 6$$
To find the solution for this system, you first must find a third equation that contains only one variable, just as you did in using the substitution method. Notice that adding the two equations
$$5x - 2y = 30$$
$$x + 2y = \phantom{0}6$$
results in the equation $6x + \phantom{0}0 = 36$, or $6x = 36$ which has only one variable, $x$.
The complete solution is illustrated in Example 1 below.

Example 1 | **Solve the system $5x - 2y = 30$ by addition. Check.**
$$x + 2y = 6$$

First, add the two equations.
$$5x - 2y = 30 \qquad (1)$$
$$\underline{x + 2y = \phantom{0}6} \qquad (2)$$
$$6x + 0 \phantom{0}= 36 \quad \blacktriangleleft \;\; \textit{−2y and +2y are opposites, or additive inverses, of}$$
$$6x = 36 \qquad \textit{each other.}$$
$$x = 6 \quad \blacktriangleleft \;\; \textit{Solve for x.}$$

To find $y$, replace $x$ with 6 in either Equation (1) or (2).
$$5x - 2y = 30 \qquad (1) \qquad\qquad \text{or} \quad x + 2y = 6 \qquad (2)$$
$$5 \cdot 6 - 2y = 30 \quad \blacktriangleleft \; \textit{Replace x with 6.} \qquad 6 + 2y = 6$$
$$30 - 2y = 30 \quad \blacktriangleleft \; \textit{Solve for y.} \qquad\qquad 2y = 0$$
$$-2y = 0 \qquad\qquad\qquad\qquad\qquad\qquad y = 0$$
$$y = 0$$
So, $x = 6$ and $y = 0$.

*Check.*
*Replace x with 6,* ▶
*and y with 0*
*in each equation.*

| $x + 2y$ | 6 |
|---|---|
| $6 + 2(0)$ | 6 |
| $6 + 0$ | |
| 6 | |

$6 = 6$, true

| $5x - 2y$ | 30 |
|---|---|
| $5(6) - 2(0)$ | 30 |
| $30 - 0$ | |
| 30 | |

$30 = 30$, true

**Thus,** (6, 0) is a solution of the system.

In the system of equations $7x - y = 3$, the $x$ terms are opposites, or
$$-7x - 3y = 1$$
additive inverses, of each other. Adding the two equations results in an equation with only one variable term, a $y$ term, as shown in Example 2.

*Example 2*    **Solve $7x - y = 3$ by addition. Check.**
$$-7x - 3y = 1$$

Opposites $\begin{cases} 7x - y = 3 \quad (1) \\ -7x - 3y = 1 \quad (2) \end{cases}$

$\phantom{Opposites}\overline{\quad 0 - 4y = 4\quad}$ ◀ *Add the two equations.*

$\phantom{Opposites\quad 0} -4y = 4$ ◀ *The resulting equation contains only one variable.*

$\phantom{Opposites\quad 0 -4} y = -1$

To find $x$, replace $y$ with $-1$ in Equation (1) or (2).
$$7x - y = 3 \qquad (1)$$
$$7x - (-1) = 3 \quad ◀ \text{ } Replace\ y\ with\ -1\ in\ Equation\ (1).$$
$$7x + 1 = 3$$
$$7x = 2$$
$$x = \frac{2}{7}$$

So, $x = \frac{2}{7}$ and $y = -1$.

Check by substituting $\frac{2}{7}$ for $x$ and $-1$ for $y$ in each equation.

| $7x - y$ | $3$ |
|---|---|
| $\overset{1}{7} \cdot \dfrac{2}{\underset{1}{7}} - (-1)$ | $3$ |
| $2 + 1$ | |
| $3$ | |

$3 = 3$, true

| $-7x - 3y$ | $1$ |
|---|---|
| $\overset{-1}{-7} \cdot \dfrac{2}{\underset{1}{7}} - 3(-1)$ | $1$ |
| $-2 + 3$ | |
| $1$ | |

$1 = 1$, true

**Thus,** $\left(\frac{2}{7}, -1\right)$ is a solution of the system.

The *addition method* works best if both variables are on one side in each equation. In the system below, this is not the case in Equation (1).
$$2x = -9y + 24 \qquad (1)$$
$$-2x + 5y = 4 \qquad (2)$$

You should rewrite the first equation so that both variables will be on the same side.

This procedure is illustrated in Example 3 on the next page.

Chapter Twelve

## Example 3

**Solve** $2x = -9y + 24$.
$-2x + 5y = 4$

*First*, get both variables on one side in Equation (1).

$2x = -9y + 24$    (1)

$2x + 9y = 24$         (3)   ◀ *Add 9y to each side.*

*Now*, solve the system consisting of Equations (3) and (2).

$$\begin{array}{rl} 2x + 9y = 24 & (3) \\ \underline{-2x + 5y = \phantom{0}4} & (2) \\ 14y = 28 & ◀ \textit{Add the two equations.} \\ y = 2 \end{array}$$

To find $x$, replace $y$ with 2 in Equation (1) or (2) or (3).

$2x = -9y + 24$    ◀ *Substitute in (1).*

$2x = -9 \cdot 2 + 24$

$2x = 6$

$x = 3$

So, $x = 3$ and $y = 2$.

$(3, 2)$ is a solution of the system.   ◀ *Check (3, 2) in each of the original equations.*

# Written Exercises

**Solve by addition. Check.**

                                                          **4.** $(0, 1)$

Ⓐ   **1.** $x + y = 6$ **(9, −3)**    **2.** $-3x + y = -3$ **(2, 3)**   **3.** $3x - 2y = 2$   **(2, 2)**   **4.** $-7x + y = 1$

          $x - y = 12$                  $3x + y = 9$             $7x + 2y = 18$           $7x + 3y = 3$

**Solve by addition.**

                                                        **(2, −2)**

     **5.** $x + 3y = -7$ **(2, −3)**   **6.** $5x - 4y = 1$    **(1, 1)**   **7.** $-5x + 2y = 1$ $\left(\frac{7}{5}, 4\right)$   **8.** $5x + 3y = 4$

         $-x - 7y = 19$          $7x + 4y = 11$            $5x - 3y = -5$          $4x - 3y = 14$

     **9.** $5x - 3y = 14$ $\left(2, -\frac{4}{3}\right)$   **10.**    $2x + 7y = 16$ **(1, 2)** **11.** $3x - 7y = 24$ $\left(5, -\frac{9}{7}\right)$   **12.** $-9x - 5y = 17$

         $8x + 3y = 12$            $-2x + 8y = 14$           $3x + 7y = 6$           $9x + 2y = 4$

              **13. (5, 2)**                 **14. (1, −1)**                                  **12. (2, −7) 16. (2, 2)**

Ⓑ   **13.** $4x - 3y = 14$      **14.** $x - 4y = 5$       **15.** $3x + 2y = 13$ **(3, 2)** **16.** $2y = -x + 6$

         $5x = -3y + 31$          $2x = -4y - 2$           $5x = 2y + 11$          $5x - 2y = 6$

**Solve each problem.**

**17.** The sum of the ages of two sisters is 32. The difference of their ages is 4. Find the age of each. **14 and 18**

**18.** The perimeter of a rectangle is 48 m. If the length is decreased by twice the width, the result is 12 m. Find the length and the width of the rectangle. **20 m, 4 m**

**Solve by addition.**

                                        **20. (6, 30)**

Ⓒ   **19.** $3x - (6 - y) = 1$        **20.** $0.05x - 0.13y = -3.6$    **21.**    $\frac{3}{5}x - \frac{1}{3}y = 1$ **(5, 6)**

              $4x - y = 7$ **(2, 1)**           $-0.05x + 0.25y = 7.2$             $-\frac{3}{5}x + \frac{4}{3}y = 5$

# THE MULTIPLICATION–ADDITION METHOD

*Objectives*    **To solve systems of equations by the addition method**
**To solve problems**

The addition method for solving systems of equations requires that one of the variables be eliminated when the equations are added.
If the equations of the system $4x - y = 7$ are added,
$$5x + 3y = 13$$
the resulting equation, $\overline{9x + 2y = 20}$, still contains both variables.
However, if you multiply each side of Equation (1) by 3 and then add the resulting equation to Equation (2), the sum will be an equation with only one variable. A detailed solution follows in Example 1 below.

*Example 1*    **Solve the system $4x - y = 7$ by addition.**
$$5x + 3y = 13$$

$$
\begin{array}{ll}
4x - y = 7 & (1) \\
3(4x - y) = 3 \cdot 7 & \blacktriangleleft \text{ Multiply each side of Equation (1) by 3.} \\
12x - 3y = 21 & (3) \quad \blacktriangleleft \text{ Now, the } y \text{ terms are opposites of each other.} \\
\underline{5x + 3y = 13} & (2) \\
17x \quad\quad = 34 & \blacktriangleleft \text{ Add the two equations.} \\
\quad\quad x = 2 &
\end{array}
$$

To find $y$, replace $x$ with 2 in either Equation (1) or (2).
$$
\begin{array}{ll}
4x - y = 7 & (1) \\
4 \cdot 2 - y = 7 & \blacktriangleleft \text{ Replace } x \text{ with 2 in Equation (1).} \\
8 - y = 7 & \\
-1y = -1 & \\
y = 1 & \quad\quad\quad \textit{Check.}
\end{array}
$$
So, $x = 2$ and $y = 1$.

| $4x - y$ | 7 |
|---|---|
| $4 \cdot 2 - 1$ | 7 |
| $8 - 1$ | |
| 7 | |

$7 = 7$, true

| $5x + 3y$ | 13 |
|---|---|
| $5 \cdot 2 + 3 \cdot 1$ | 13 |
| $10 + 3$ | |
| 13 | |

$13 = 13$, true

**Thus,** $(2, 1)$ is a solution of the system.

You can solve any system of equations in one of two ways: (1) by eliminating the $y$ terms or (2) by eliminating the $x$ terms. In using the Addition Method, you first may have to use two multipliers, a different one for each equation, in order to make the $y$ terms or the $x$ terms opposites of each other. This technique and the two ways for solving the system $5x - 2y = 8$ and $2x + 7y = 11$ are illustrated in Examples 2 and 3.

*Example 2*  **Solve the system $5x - 2y = 8$   by eliminating the $y$ terms.**
**$2x + 7y = 11$**

$5x - 2y = 8$ (1)  ◀ *Think: To make the y terms opposites of each other,*
$2x + 7y = 11$ (2)     *multiply each side of Equation (1) by 7 and each*
                       *side of Equation (2) by 2.*

$7(5x - 2y) = 7 \cdot 8$  ◀ *Multiply each side of Equation (1) by 7.*

$2(2x + 7y) = 2 \cdot 11$  ◀ *Multiply each side of Equation (2) by 2.*

$35x - 14y = 56$  ◀ *Now the y terms are opposites of each other.*

$\underline{4x + 14y = 22}$

$39x \qquad = 78$  ◀ *Add the two equations.*

$\qquad x = 2$

To find $y$, replace $x$ with 2 in either Equation (1) or (2).

$5x - 2y = 8$   (1)

$5 \cdot 2 - 2y = 8$   ◀ *Replace x with 2 in Equation (1).*

$10 - 2y = 8$

$-2y = -2$

$y = 1$

**So,** $x = 2$ and $y = 1$.


*Example 3*  **Solve the system $5x - 2y = 8$ by eliminating the $x$ terms.**
**$2x + 7y = 11$**

$5x - 2y = 8$ (1)  ◀ *Think: To make the x terms opposites of each other, multiply*
$2x + 7y = 11$ (2)     *each side of Equation (1) by 2 and each side of Equation (2)*
                       *by $-5$ so the coefficient of the x term will be negative.*

$2(5x - 2y) = 2 \cdot 8$  ◀ *Multiply each side of Equation (1) by 2.*

$-5(2x + 7y) = -5 \cdot 11$  ◀ *Multiply each side of Equation (2) by $-5$.*

$10x - 4y = 16$  ◀ *Now the x terms are opposites of each other.*

$\underline{-10x - 35y = -55}$

$-39y = -39$  ◀ *Add the two equations.*

$y = 1$

To find $x$, replace $y$ with 1 in either Equation (1) or (2).

$2x + 7y = 11$   (2)

$2x + 7 \cdot 1 = 11$   ◀ *Replace y with 1 in Equation (2).*

$2x + 7 = 11$

$2x = 4$

$x = 2$

**So,** $x = 2$ and $y = 1$.


The solution is the *same* whether you first make the $y$ terms opposites, as in
Example 2, or first make the $x$ terms opposites, as in Example 3.


The Multiplication–Addition Method                                    **345**

# Oral Exercises

**Select a multiplier for each equation so that the result of adding the equations is an equation in only one variable.**

1. $3x - 5y = 1$
   $x + y = 7$ **5**

2. $4x - 5y = 13$
   $3x + y = 6$ **5**

3. $2x - 3y = 1$
   $x + 5y = 7$ **−2**

4. $3x + y = 4$ **−2**
   $x + 2y = 3$

5. $2x + 3y = 2$ **−3**
   $3x + 5y = 2$ **2**

6. $5x - 3y = 11$ **2**
   $7x + 2y = 34$ **3**

7. $3x + 2y = 8$ **3**
   $5x - 3y = 4$ **2**

8. $3x - 5y = -1$ **−2**
   $2x + 4y = 14$ **3**

# Written Exercises

**Solve by addition.**

Ⓐ 
1. $4x - 3y = 1$ **x = 1**
   $2x + y = 3$ **y = 1**

2. $6x - 2y = 6$ **x = 2**
   $x + 4y = 14$ **y = 3**

3. $x + y = 4$ **x = 3**
   $2x + 3y = 9$ **y = 1**

4. $5x + 2y = 7$ **x = 1**
   $x + 4y = 5$ **y = 1**

5. $5x - 2y = 3$ **x = 1**
   $2x + 7y = 9$ **y = 1**

6. $3x + 2y = 12$ **x = 4**
   $2x + 5y = 8$ **y = 0**

7. $3x + 5y = 11$ **x = 2**
   $2x - 3y = 1$ **y = 1**

8. $7x + 3y = 13$ **x = 1**
   $3x - 2y = -1$ **y = 2**

9. $5x - 4y = 14$ **x = 2**
   $3x + 3y = 3$ **y = −1**

10. $3x + 9y = 0$ **x = 3**
    $11x - 2y = 35$ **y = −1**

11. $5x + 2y = 7$ **x = 1**
    $3x + 7y = 10$ **y = 1**

12. $2x - 7y = 3$ **x = −2**
    $5x - 4y = -6$ **y = −1**

Ⓑ 
13. $3x = -2y + 13$ **x = 3**
    $2x + 5y = 16$ **y = 2**

14. $-2y = -3x + 6$ **x = 4**
    $5x + 7y = 41$ **y = 3**

15. $3x = -3y - 6$ **x = 1**
    $x - 2y = 7$ **y = −3**

16. $5x - 2y = 8$ **x = 2**
    $3x = 5y + 1$ **y = 1**

**Solve each problem.**

17. The difference of Randi's age and Saul's age is 6 years. The sum of 3 times Randi's age and 4 times Saul's age is 102. How old is each? **Randi: 18   Saul: 12**

18. The sum of two numbers is 14. Twice the larger number, increased by three times the smaller number, is 34. Find the two numbers. **6 and 8**

**Solve by addition.**

Ⓒ 
19. $\dfrac{x - 2y}{8} = \dfrac{1}{2}$ **x = 2**
    $3x + 2y = 4$ **y = −1**

20. $\dfrac{6}{7}x - y = \dfrac{3}{7}$ **x = $\frac{1}{3}$**
    $\dfrac{9}{7}x + 2y = \dfrac{1}{7}$ **y = $-\dfrac{1}{7}$**

21. **x = 2; y = 1**
    21. $6x - 3(y - 2x) = 3x + 15$
    $y - (4y - x) = 7x - 15$

# CUMULATIVE REVIEW

1. Solve $11x - 4 = -3x^2$.
   $\frac{1}{3}$; −4

2. Graph $3x - 2y = 12$.
   **For Ex. 2, see TE Answer Section.**

3. Solve $3x - 4 < 5x + 10$.
   $x > -7$

## NON-ROUTINE PROBLEMS

When $\frac{1}{4}$ of the adults left a beach party, the ratio of adults to children was $1:2$. When 30 children then left, the ratio of children to adults was then $4:3$. How many people remained at the beach? **105 (45 adults, 60 children)**

# COIN AND MIXTURE PROBLEMS 12.7

*Objectives*  **To solve problems about coins**
**To solve problems about mixtures and combinations**

When you solve problems involving different types of coins, you will find it helpful to represent the total value in the same unit of money, usually cents. For example, the total value of 3 nickels and 4 dimes is *not* 7 because nickels are worth 5 cents each and dimes are worth 10 cents each. The value in cents of 3 nickels and 4 dimes is

$$3 \cdot 5 + 4 \cdot 10 = 15 + 40, \text{ or } 55 \text{ cents.}$$

Similarly, the value in cents of $h$ half-dollars and $q$ quarters is

$$h \cdot 50 + q \cdot 25, \text{ or } 50h + 25q.$$

Notice how this information is used with the basic problem-solving strategy to solve the problem about coins in Example 1.

*Example 1*  **The number of quarters that Eleanor has is 3 times the number of nickels. She has \$1.60 in all. How many coins of each type does she have?**

READ ▶  You are asked to find the number of quarters and the number of nickels that Eleanor has.

PLAN ▶  Represent the data. Let $q$ = the number of quarters.
Let $n$ = the number of nickels.
Write an equation for each of the first two sentences.
$q = 3n$  (1)  ◀ *No. of quarters is 3 times no. of nickels.*
$q \cdot 25 + n \cdot 5 = 160$, or $25q + 5n = 160$  (2)  ◀ *Value of quarters + value of nickels in cents*

SOLVE ▶
$$q = 3n \quad (1)$$
$$25q + 5n = 160 \quad (2)$$
$$25 \cdot 3n + 5n = 160 \qquad ◀ \textit{Substitute } 3n \textit{ for } q \textit{ in Equation (2).}$$
$$75n + 5n = 160$$
$$80n = 160$$
$$n = 2$$

To find $q$, replace $n$ with 2 in Equation (1).
$$q = 3n$$
$$q = 3 \cdot 2$$
$$q = 6$$

INTERPRET ▶

| The no. of quarters is 3 times the no. of nickels. | | The total value is 160 cents. | |
|---|---|---|---|
| 6 | 3 · 2 | $6 \cdot 25 + 2 \cdot 5$ | 160 |
| | 6 | 160 | |
| 6 = 6, true | | 160 = 160, true | |

**Thus,** Eleanor has 2 nickels and 6 quarters.

Sometimes a problem about coins leads to a system of equations that is easier to solve by the addition method. This situation occurs in Example 2.

**Example 2**     **Paul has 30 coins in dimes and quarters. Their total value is $4.50. How many coins of each type does he have?**

Let $d$ = the number of dimes.     ◄ *Represent the data.*
Let $q$ = the number of quarters.

Write an equation for the first two sentences.
$$d + q = 30 \quad (1)$$ ◄ *He has 30 coins, or no. of dimes + no. of quarters is 30.*
$$d \cdot 10 + q \cdot 25 = 450, \text{ or } 10d + 25q = 450 \quad (2)$$ ◄ *Their total value is $4.50, or 450¢.*
                                                   *Value of d dimes + value of q quarters is 450.*

Solve the system of equations.
$$d + q = 30 \quad (1)$$ ◄ *The multiplication–addition method is easier because*
$$10d + 25q = 450 \quad (2)$$    *neither equation is solved for d or q.*

$$-10(d + q) = -10(30)$$ ◄ *Make the d terms opposites by multiplying each side of (1)*
                                   *by −10.*

$$\begin{array}{rl} -10d - 10q = -300 & (3) \\ 10d + 25q = \phantom{-}450 & (2) \\ \hline 15q = \phantom{-}150 & \\ q = 10 & \end{array}$$

◄ *Solve the system of equations (3) and (2) by addition.*

To find $d$, replace $q$ with 10 in Equation (1).
$$\begin{array}{r} d + q = 30 \\ d + 10 = 30 \\ d = 20 \end{array}$$

*Check.*

| The number of coins is 30. | |
|---|---|
| 20 + 10 | 30 |
| 30 | 30 |
| | 30 = 30, true |

| The total value of the coins is 450 cents. | |
|---|---|
| 20 · 10 + 10 · 25 | 450 |
| 200 + 250 | |
| 450 | |
| | 450 = 450, true |

**Thus,** Paul has 20 dimes and 10 quarters.

You can easily adapt the methods for solving problems involving mixtures, or combinations, of coins to solving problems involving any other mixtures or combinations.

Example 3 involves combinations of tickets to a school play.

*Example 3*    **The cost of an adult ticket for a school play was $2.00. The cost of a student ticket was $1.50. The total income from the sale of tickets was $550. The number of $2.00 tickets sold was 100 less than 3 times the number of $1.50 tickets. How many tickets of each type were sold?**

Let $x$ = the number of $2.00 tickets.    ◀ *Represent the data.*
Let $y$ = the number of $1.50 tickets.

Write two equations. One equation will involve the total income. The second equation will relate the number of each type of ticket sold.

$$\left(\begin{array}{c}\text{the no. of}\\ \text{\$2.00 tickets}\end{array}\right)\left(\begin{array}{c}\text{the cost}\\ \text{in ¢}\end{array}\right) + \left(\begin{array}{c}\text{the no. of}\\ \text{\$1.50 tickets}\end{array}\right)\left(\begin{array}{c}\text{the cost}\\ \text{in ¢}\end{array}\right) = \text{the total income in ¢}$$

$$x \cdot 200 \quad + \quad y \cdot 150 \quad = 55,000 \quad ◀ \; \$550 = 55,000 ¢$$

or $200x + 150y = 55,000$   (1)

The no. of $2.00 tickets is 100 less than 3 times the no. of $1.50 tickets.

$$x \quad = \quad 3y - 100 \qquad (2)$$

Solve the system $200x + 150y = 55,000$    (1)   by substitution because Equation (2) is
$\qquad\qquad\qquad\qquad x = 3y - 100$    (2)
already solved for $x$.
$200(3y - 100) + 150y = 55,000$   ◀ *Substitute $3y - 100$ for x in Equation (1).*
$600y - 20,000 + 150y = 55,000$
$\qquad 750y - 20,000 = 55,000$
$\qquad\qquad\quad 750y = 75,000$
$\qquad\qquad\qquad y = 100$

To find $x$, replace $y$ with 100 in Equation (2).
$x = 3y - 100$    (2)
$x = 3 \cdot 100 - 100$
$x = 300 - 100$, or 200

**So,** 200 tickets at $2.00 each and 100 tickets at $1.50 each were sold.
Check on your own.

# Reading in Algebra

**Choose the one correct ending for each statement.**
 1. A collection of 6 dimes and 3 nickels has a total value in cents of ?
    (a) $6 + 3$    (b) $6 \cdot 5 + 3 \cdot 10$    (c) $6 \cdot 10 + 3 \cdot 5$    (d) $200(6 + 3)$ **c**
 2. The total income, in cents, from the sale of $x$ tickets at $3.00 each and $y$ tickets at $1.25 each is?
    (a) 425    (b) $3x + 125y$    (c) $100(x + y)$    (d) $300x + 125y$ **d**

# Written Exercises

**Solve each problem.**

(A) 1. The number of quarters that John has is 4 times the number of nickels. He has $2.10 in all. How many coins of each type does he have? **8 quarters   2 nickels**

2. The number of nickels that Christine has is 5 times the number of dimes. Their value is $1.05. How many coins of each type does she have? **15 nickels   3 dimes**

3. Bob has 6 more dimes than quarters. He has $1.65 in all. How many coins of each type does he have? **9 dimes   3 quarters**

4. Jill has $2.30 in dimes and quarters. The number of dimes is 5 less than the number of quarters. How many coins of each type does she have? **3 dimes   8 quarters**

5. Marie has 24 coins in half-dollars and dimes. Their total value is $3.60. How many coins of each type are there? **3 half-dollars   21 dimes**

6. A cash register contains 15 coins in dimes and nickels. The total value is $1.25. How many coins of each type are there? **10 dimes   5 nickels**

7. The cost of an adult ticket to a football game was $1.75. The cost of a student ticket was $1.25. The number of student tickets sold was twice the number of adult tickets. The total income from the sale of tickets was $850. How many tickets of each type were sold? **Adult: 200   Student: 400**

8. A $31.50 box of nuts contains almonds mixed with pecans. Almonds cost $7.50/kg and pecans cost $9.00/kg. The number of kilograms of almonds is 3 times the number of kilograms of pecans. Find the number of kilograms of each kind of nut in the mixture. **Almonds: 3 kg   Pecans: 1 kg**

9. Mr. Lead wants to mix pencils costing $0.25 each with pencils costing $0.20 each. The mixture will cost $3.60. If the number of $0.20 pencils is 9 more than the number of $0.25 pencils, how many of each type should be in the package? **13 and 4**

10. Cashews cost $11.25/kg. Pecans cost $13.00/kg. The number of kilograms of cashews is 3 less than the number of kilograms of pecans. How many kilograms of each should be mixed to make a box that will sell for $136.00? **Cashews: 4   Pecans: 7**

(B) 11. A parking meter contains $6.25 in dimes and quarters. If the number of dimes is 2 more than 3 times the number of quarters, how many of each coin are in the parking meter? **11 quarters   35 dimes**

12. Angelo has $1.90 in dimes and nickels. If he has 4 less nickels than 5 times the number of dimes, how many dimes does he have? **6**

13. Paper clips costing $0.30 each are to be boxed with paper clips costing $0.25 each. The box of clips will cost $5.70. If four times the number of $0.30 clips is the same as three times the number of $0.25 clips, find the number of each type in the box. **9 @ $0.30   12 @ $0.25**

14. A $7.35 assortment of pads contains $0.45 pads and $0.50 pads. If the number of $0.45 pads is 3 less than $\frac{1}{2}$ the number of $0.50 pads, find the number of each type of pad in the package. **3 @ $0.45   12 @ $0.50**

(C) 15. Mr. Weiss has $0.95 in dimes and nickels. The total number of coins is one more than twice the number of dimes. How many coins of each type does he have? **7 nickels   6 dimes**

16. Ms. Garcia wants to sell a box of fruit that contains cherries and plums for $19.20. The cherries sell at $2.40/kg; the plums at $3.60/kg. The total number of kilograms of fruit is 3 kg more than twice the number of kilograms of plums. How many kilograms of each type of fruit are in the box? **5 kg cherries   2 kg plums**

# DIGIT PROBLEMS 12.8

*Objective*  **To solve problems about two-digit numbers**

The number 35 is a two-digit number that can be written as

$$10 \cdot 3 + 5.$$

The tens digit is 3.    The units digit is 5.

In general, any two-digit number can be written as

$$10t + u,$$

where $t$ is the tens digit and $u$ is the units digit.
The tens digit can be any integer from 1 through 9.
The units digit can be any integer from 0 through 9.

Examples 1 and 2 will help you represent the data in problems involving two-digit numbers.

*Example 1*    **Find the sum of the digits of each two-digit number.**

65
$65 = 10 \cdot 6 + 5$
tens digit —↑    ↑— units digit
**So,** the sum of the digits is $6 + 5$, or 11.

$10t + u$
tens digit —↑    ↑— units digit
**So,** the sum of the digits is $t + u$.

*Example 2*    **Find the number formed by reversing the digits of each two-digit number.**

*Reverse the tens digit* ▶
*and the units digit.*

$75 = 10 \cdot 7 + 5$
$57 = 10 \cdot 5 + 7$
**So,** $10 \cdot 5 + 7$, or 57, is the number formed by reversing the digits.

$10 \cdot t + u$
$10 \cdot u + t$
**So,** $10 \cdot u + t$, or $10u + t$, is the number formed by reversing the digits.

*Summary*    For any two-digit number,
if $t$ represents the tens digit of the number and
if $u$ represents the units digit of the number,
then $10t + u$ represents the number

and $10u + t$ represents the number with the digits *reversed*.

You can now use this summary and the basic problem-solving strategy to solve all problems involving two-digit numbers, as illustrated on the next two pages.

*Example 3*    **The sum of the digits of a two-digit number is 15. If the digits are reversed, the new number is 27 less than the original number. Find the original number.**

You are given: The sum of the digits of a two-digit number is 15.
The number with its digits reversed is 27 less than the original number.

READ ▶    You are asked to find a two-digit number.

PLAN ▶    Write an equation for each of the first two sentences.
Represent the digits, the original number, and the new number.
Let $t$ = the tens digit of the original number.
Let $u$ = the units digit of the original number.
$10t + u$ = the original number.

$10u + t$ = the new number.    ◀ *the original number with its digits reversed*
The sum of the digits is 15.    |    The new number is 27 less than the original number.

$$t + u = 15$$    |    $$10u + t = (10t + u) - 27$$
Rewrite this equation with all the variables on one side.
$$10u + t = 10t + u - 27$$
$$-9t + 10u = u - 27$$    ◀ *Add $-10t$ to each side.*
$$-9t + 9u = -27$$    ◀ *Add $-u$ to each side.*
$$-t + u = -3$$    ◀ *Divide each side by 9.*

SOLVE ▶
$$t + u = 15 \quad (1)$$    ◀ *Solve the system by addition because $t$ and $-t$ are*
$$\underline{-t + u = -3} \quad (2)$$    *opposites.*
$$2u = 12$$    ◀ *Add the two equations.*
$$u = 6$$

To find $t$, replace $u$ with 6 in Equation (1).
$$t + u = 15 \quad (1)$$
$$t + 6 = 15$$
$$t = 9$$    So, the original number is $10 \cdot 9 + 6$, or 96.

INTERPRET ▶

| The sum of the digits is 15. | | The new number is 27 less than the original. | |
|---|---|---|---|
| 9 + 6 | 15 | 69 | 27 less than 96 |
| 15 | | | 96 − 27 |
| | 15 = 15, true | | 69 |
| | | | 69 = 69, true |

**Thus,** the original two-digit number is 96.
The multiplication–addition method is applied in the next example.

*Example 4*    **Four times the tens digit of a two-digit number, increased by the units digit, is 18. If the digits are reversed, the new number is 9 less than twice the original number. Find the original number.**

Let $t$ = the tens digit of the original number.
Let $u$ = the units digit of the original number.
$10t + u$ = the original number.
$10u + t$ = the new number.  ◀ *the original number with the digits reversed*

Write an equation for each of the first two sentences.

4 times the tens digit increased by the units digit is 18.
$$4t \qquad + \qquad u \qquad = 18 \quad (1)$$

The new number is 9 less than twice the original number.

$$10u + t \quad = \quad 2(10t + u) - 9$$
$$10u + t = 20t + 2u - 9$$
$$8u - 19t = -9 \quad ◀ \textit{Get all variables on one side. Add } -20t \textit{ and } -2u \textit{ to each side.}$$
$$\text{or } -19t + 8u = -9 \quad (2)$$

Solve the system    $4t + u = 18$    (1)   by the addition method
$\qquad\qquad\qquad -19t + 8u = -9$    (2)

$$-8(4t + u) = -8 \cdot 18 \quad ◀ \textit{Multiply each side of Equation (1) by } -8.$$
$$-32t - 8u = -144$$

$$
\begin{array}{ll}
-32t - 8u = -144 & (3) \\
\underline{-19t + 8u = \quad -9} & (2) \\
-51t \qquad\quad = -153 & ◀ \textit{Add the two equations.} \\
\quad t = 3 &
\end{array}
$$

To find $u$, replace $t$ with 3 in Equation (1).
$$4t + u = 18$$
$$4 \cdot 3 + u = 18$$
$$12 + u = 18$$
$$u = 6$$

The tens digit, $t$, is 3, and the units digit, $u$, is 6.

**So,** the two-digit number is 36.  ◀ *Check in each of the original equations.*

## Reading in Algebra

**Indicate whether each statement is true or false. If false, tell why it is false.**
1. In the number 36, the digits are 30 and 6.  **False; digits are 3 and 6; 30 is not a digit.**
2. In the two-digit number $10t + u$, the sum of the digits is $t + u$.  **True**
3. If the digits of 48 are reversed, the new units digit is 8.  **False; new units digit is 4.**
4. Three times the two-digit number $10t + u$ is $30t + u$.  **False; correct answer is $30t + 3u$.**

Digit Problems

# Written Exercises

**Solve each problem.**

(A) **1.** The sum of the digits of a two-digit number is 10. If the digits are reversed, the new number is 18 less than the original number. Find the original number. **64**

**2.** The sum of the digits of a two-digit number is 7. If the digits are reversed, the new number is 27 more than the original number. Find the original number. **25**

**3.** The tens digit of a two-digit number is 3 times the units digit. If the digits are reversed, the new number is 36 less than the original number. Find the original number. **62**

**4.** The units digit of a two-digit number is 4 times the tens digit. If the digits are reversed, the new number is 54 more than the original number. Find the original number. **28**

**5.** The tens digit of a two-digit number is 3 more than twice the units digit. If the digits are reversed, the new number is 54 less than the original number. Find the original number. **93**

**6.** Three times the tens digit of a two-digit number, increased by the units digit, is 21. If the digits are reversed, the new number is 9 more than the original number. Find the original number. **56**

**7.** Twice the tens digit of a two-digit number, increased by the units digit, is 19. If the digits are reversed, the new number is 45 less than the original number. Find the original number. **83**

**8.** Four times the units digit of a two-digit number is 1 less than the tens digit. If the digits are reversed, the new number is 63 less than the original number. Find the original number. **92**

(B) **9.** The units digit of a two-digit number is 1 more than 4 times the tens digit. If the digits are reversed, the new number is 5 more than 3 times the original number. Find the original number. **29**

**10.** The sum of the digits of a two-digit number is 13. If the digits are reversed, the new number is 4 less than twice the original number. Find the original number. **49**

**11.** The tens digit of a two-digit number is 3 more than 5 times the units digit. If the digits are reversed, the new number is 2 more than twice the tens digit of the original number. Find the original number. **81**

**12.** The units digit of a two-digit number is 12 less than twice the tens digit. If the digits are reversed, the new number is 3 less than 8 times the tens digit of the original number. Find the original number. **96**

(C) **13.** The denominator of a fraction is a two-digit number whose digits add up to ten. If the digits are reversed, the new number is the numerator. If 41 is subtracted from the numerator, the value of the resulting fraction is $\frac{1}{2}$. Find the original fraction. $\frac{64}{46}$

# CUMULATIVE REVIEW

**1.** Graph $3x - 4y = 8$.     **2.** Solve $|2x - 4| = 24$.     **3.** $(x^3 - 9x^2 + 26x - 24) \div (x - 2)$.

**For Ex. 1, see TE Answer Section.**          $x = 14, x = -10$                    $x^2 - 7x + 12$

# MIXED PROBLEM SOLVING

**Solve each problem.**

1. The sum of the digits of a two-digit number is 14. If the digits are reversed, the new number is 18 more than the original number. Find the original number. **68**

2. The perimeter of a rectangle is 22 m. The length of the rectangle is 1 m less than 3 times the width. Find the length and width of the rectangle. **8, 3**

3. The difference between Cassandra's age and Borg's age is 3 years. The sum of 8 times Cassandra's age and twice Borg's age is 94 years. How old is each? **B, 7; C, 10**

4. In June Mr. Jobey sold 6 times as many cars as Ms. Sánchez did. Together they sold 21 cars. How many cars did each salesperson sell? **Ms. S, 3; Mr. J, 18**

5. Adina has a total of $1.65 in nickels and dimes. The number of dimes is 6 more than the number of nickels. How many coins are there of each type? **nickels, 7 dimes, 13**

6. The sum of the digits of a two-digit number is 12. The units digit is twice the tens digit. Find the two-digit number. **48**

7. Mr. Harvey wants to mix screws costing $0.25 apiece with bolts costing $0.40 each. The mixture will cost $3.10. The number of $0.25 pieces is 2 more than the number of $0.40 pieces. How many of each type should be included in the package? **4 bolts 6 screws**

8. The larger of two numbers is 3 more than twice the smaller. If twice the larger is decreased by 5 times the smaller, the result is zero. Find each of the two numbers. **6, 15**

9. The units digit of a two-digit number is twice the tens digit. If the digits are reversed, the new number is 18 more than the original number. Find the original number. **24**

10. Jacques wants to make up packs of $0.45 comic books mixed with $0.75 ones to sell at $9.00 per pack. The number of $0.45 comic books is 4 more than the number of $0.75 comic books. Find the number of each type in the pack. **6 @ 75¢, 10 @ 45¢**

11. The perimeter of a rectangle is 40 m. If the width is increased by twice the length, the result is 32 m. Find the length and width of the rectangle. **12, 8**

12. The sum of the ages of Flavia and Ingrid is 36 years. The difference between 3 times Flavia's age and twice Ingrid's age is 28 years. How old is each? **Flavia, 20; Ingrid, 16**

13. Jane scored 4 times as many points as Lenny did in basketball. The sum of their scores was 35. How many points did each score? **Lenny 7, Jane 28**

14. The difference between two numbers is 6. Twice the first number, decreased by the second number, is 30. Find the two numbers. **36, 42 & 24, 18**

15. The larger of two numbers is 7 less than twice the smaller. If two more than 3 times the smaller is decreased by the larger, the result is 23. Find the numbers. **14, 21**

16. Jamie has $1.55 in dimes and nickels. The total number of coins is 7 more than twice the number of dimes. How many coins of each type does Jamie have? **8 dimes, 15 nickels**

# APPLICATIONS

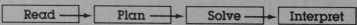

Read → | Plan → | Solve → | Interpret

Suppose that the Radex Radio Company wishes to produce a new small radio. The company will have to sell a certain number of radios just to pay for the initial expenses of producing the radios. The point at which this happens is called the **break-even point.** Then, from that point on, the income from the sales (revenue) will be greater than the production cost. The company will then begin to make a profit. To find the break-even point, Radex proceeds as follows. Radex uses this formula to determine the cost of producing $x$ radios: $y = 40x + 1000$. Since Radex plans to sell the radios for \$50 each the company uses this formula to determine the revenue from the sale of $x$ radios: $y = 50x$. Radex then graphs the system of equations on the same set of axes.

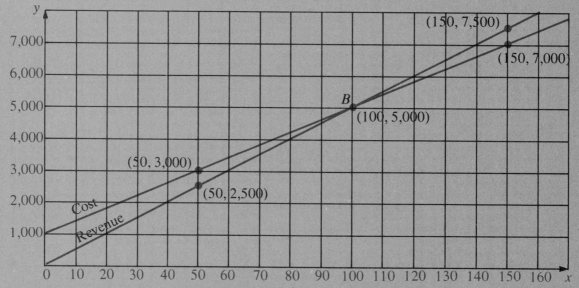

When $x < 100$, the points on the cost line are above the points on the revenue line. This means that the cost of production is greater than the income from the sales. The company is losing money.
But, when $x > 100$, the points on the revenue line are above the points on the cost line. This means that the income from the sales is greater than the cost of production. Thus, the company is now making a profit.
Point $B$ on the graph is the *break-even point.* Notice that when $x = 100$, or when 100 radios are sold, the cost of production and the income are the same, \$5,000. At this point, Radex has recovered its initial expenses and has broken even. You can find the break-even point by setting up and solving a system of equations. One of the equations involves the cost of production, the other, revenue. You can solve the system graphically as shown above or algebraically as shown in the example on the next page.

*Example*    The formula for finding the cost of producing $x$ radios is $y = 40x + 1,000$. The formula for finding the revenue from selling $x$ radios at $50 each is $y = 50x$. Find, algebraically, the number of radios that must be sold to break even. Then, find the revenue from these sales. Will the company make a profit if 300 radios are sold? Why or why not?

Solve the system $y = 40x + 1,000$ (1) by substitution
$y = 50x$ (2)

$y = 40x + 1,000$ (1)

$50x = 40x + 1,000$ ◄ *Substitute $50x$ for $y$ in Equation (1).*
$10x = 1,000$ ◄ *Add $-40x$ to each side.*
$x = 100$

To find $y$, replace $x$ with 100 in Equation (1) or (2).
$y = 50x$ (2) ◄ *Use Equation (2).*
$y = 50 \cdot 100$
$y = 5,000$ ◄ *Revenue from selling 100 radios*

The cost of producing 100 radios is $40 \cdot 100 + 1,000$, or $5,000. So, the company must sell 100 radios to break even, since the revenue will be $5,000, the same as the cost of production. The company will make a profit on the sale of 300 radios, since 300 is more than 100, the break-even number.

# Written Exercises

**Use the following information for Exercises 1–6.**
Jacob makes pine end tables. He uses $y = 80x + 100$ as the cost formula for producing $x$ tables. Each table sells for $100, so he uses $y = 100x$ as the revenue formula.

1. How many tables must he sell to break even? **5**

2. How much revenue does he receive if he breaks even? **$500**

3. Will he make a profit if he makes and sells 3 tables? Why or why not? **no; loss of $40**

4. Will he make a profit if he makes and sells 10 tables? Why or why not? **yes; profit of $100**

5. Find his profit if he sells 15 tables. **$200**

6. Will he make a profit if he makes and sells 5 tables? Why or why not? **no; 5 is break-even point**

**Use the following information for Exercises 7–11.**
Kate produces chairs. For figuring the total cost of producing $x$ chairs, she uses the formula $y = 40x + 200$. Each chair sells for $48. The revenue formula she uses is $y = 48x$.

7. How many chairs must she sell to break even? **25**

8. How much revenue does she receive if she breaks even? **$1200**

9. Will she make a profit if she sells 30 chairs? **yes, $40**

10. Find her profit if she sells 50 chairs. **$200**

11. Does it pay for Kate to produce the product if it is known beforehand that there will be a demand for 24 units? Why or why not? **no, it will lose money**

Applications

**Vocabulary**

consistent [12.1]    independent [12.1]
dependent [12.1]    parallel [12.1]
digit [12.8]    substitution [12.3]
inconsistent [12.1]

Solve by graphing. [12.1]

**1.** $y = 2x$ **(4, 8)**     **2.** $y = 2x + 1$ **(0, 1)**
$x + y = 12$           $x + y = 1$

**3.** $3x - 4y = -4$   For Ex. 1–9, see
$y = x$ **(4, 4)**     **TE Answer Section.**

Graph. Indicate whether each system is independent and consistent, inconsistent, or dependent and consistent.

**4.** $3x - 2y = 4$   **5.** $2x - y = 5$
$6x - 4y = 1$       $-6x + 3y = -15$

**6.** $x - y = 4$
$x + y = 6$

Solve by graphing. [12.2]

**7.** $x < -8$        **8.** $y > x + 1$
$y \geq 2$           $y < -x - 7$

**9.** $2x - 5y \leq 10$
$x - y > 2$

Solve by substitution. [12.3]

**10.** $y = 3x$ **x = -4**  **11.** $5x - 3y = 1$ **x = $\frac{1}{2}$**
$x - y = 8$           $x + y = 1$   **y = $\frac{1}{2}$**
    **y = -12**      **12. length: 18; width: 4**

Solve each problem. [12.4]

**12.** The length of a rectangle is 6 m more than 3 times the width. The perimeter is 44 m. Find the length and the width.

**13.** Barbara's age is 5 years less than her brother's age. If three times her brother's age is increased by Barbara's age, the result is 51. Find each of their ages.
    **Barbara: 9   brother: 14**

Solve by addition.  **14. x = 7   y = -3**

**14.** $x - y = 10$          **15.** $x - 3y = 1$
$x + y = 4$ [12.5] [12.6] $2x = -6y + 14$

**16.** $2x = -3y + 3$ **x = 3** **17.** $5x - 3y = 8$
$3x + 4y = 5$ **y = -1**      $2x = 7y + 9$

**15. x = 4; y = 1   17. x = 1; y = -1**

Solve each problem.

**18.** Mark has 6 more quarters than dimes. He
[12.7] has \$3.25 in all. How many coins of each type does he have? **11 quarters   5 dimes**

**19.** The cost of an adult ticket to a school play was \$2.00. A student ticket cost \$1.50. The number of student tickets sold was 50 more than twice the number of adult tickets. The total income from the tickets was \$800. How many tickets of each type were sold?  **Student: 340; Adult: 145**

**20.** The sum of the digits of a two-digit
[12.8] number is 8. If the digits are reversed, the new number is 36 less than the original number. Find the original number. **62**

**21.** Three times the tens digit of a two-digit number, increased by the units digit, is 17. If the digits are reversed, the new number is 27 less than the original number. Find the original number. **52**

★ **22.** Solve the system $y - x \leq 4$
    For Ex. ★ 22, see TE        $x < -2$
    Answer Section.            $y \geq -3$
    by graphing.

★ **23.** Solve the system $\frac{x}{2} = \frac{y}{3}$
                          $x - 4y = -10$ **x = 2**
                                        **y = 3**
    by substitution.

★ **24.** Solve the system $x - 3(y - x) = 2x + 3$
                          $y - (-3y - x) = -2x + 13$
    by addition. **x = 3   y = 1**

Ⓐ Exercises are 1–4, 7, 9, 12, 14. Ⓒ Exercises are starred. The rest are Ⓑ Exercises.

Solve by addition.

**1.** $x + y = 4$  $x = 2$
   $3x - 2y = 2$  $y = 2$

**2.** $5x = 2y + 8$
   $3x + 4y = 10$
   $x = 2$  $y = 1$

Solve by graphing.

**3.** $y = -2x + 3$
   $3x + y = -2$
3. $x = -5; y = 13$

**4.** $3x - 2y = 6$
   $x - y = 2$
4. $x = 2; y = 0$

Graph. Indicate whether each system is independent and consistent, inconsistent, or dependent and consistent.

**5.** $5x - 2y = 7$
   $-5x + 2y = 8$

**6.** $3x - y = 4$
   $-9x + 3y = -12$

**For Ex. 5 and 6, see TE Answer Section.**

Solve by substitution.

**7.** $y = 5x$
   $x + 3y = 12$  $x = \frac{3}{4}$
   $y = \frac{15}{4}$

**8.** $3x - 4y = 19$
   $x + y = 4$
   $x = 5; y = -1$

Solve by graphing.

**9.** $x \le 5$
   $y > -1$

**10.** $y > x - 6$
   $y \le -x + 4$

**For Ex. 9 and 10, see TE Answer Section.**

Solve each problem.

**11.** The sum of the digits of a two-digit number is 6. The number with its digits reversed is 18 less than the original number. Find the original number. **42**

**12.** The length of a rectangle is 5 m less than 4 times the width. The perimeter is 50 m. Find the length and the width of the rectangle. **19 m, 6 m**

Solve each problem.

**13.** Two varieties of gift-wrap paper are to be mixed. The assortment will cost $2.00. One sells at $0.40 per sheet, the other at $0.30 per sheet. The number of sheets at $0.30 per sheet is to be twice the number of sheets at $0.40 per sheet. How many of each type should be included? **4 @ $0.30**  **2 @ $0.40**

**14.** A parking meter contains 3 times as many nickels as quarters. The meter contains $3.20 in all. How many coins of each type are there? **8 quarters, 24 nickels**

**15.** An assortment of nuts costs $36. A special brand of pecans selling at $7.50/kg is mixed with almonds selling at $6/kg. How many kilograms of each type are in a 5 kg box? **Pecans: 4 kg  Almonds: 1 kg**

**16.** The larger of two numbers is 4 more than twice the smaller. Three times the smaller, increased by the larger, is 14. Find the two numbers. **8 and 2**

★ **17.** Solve the system $\frac{x}{4} = \frac{y}{2}$ by substitution.
   $x + 3y = -7$
   $x = \frac{-14}{5}$
   $y = \frac{-7}{5}$

★ **18.** For which value of $m$ will the system
   $12x - 3y = 9$
   $m^2 x - y = 4$
   be inconsistent? $m = \pm 2$

# COMPUTER ACTIVITIES

## Graphing Equations of Lines

Using the equation for a line in BASIC notation $Y = M * X + B$, you can quickly compute a table of X, Y values on the computer.

This program will print a list of $(x, y)$ values for graphing a line of the form $y = mx + b$.

```
10   PRINT "THIS PROGRAM FINDS A LIST OF VALUES FOR G
       RAPHING A LINE."
15   PRINT "TYPE IN THE SLOPE OF THE LINE."
20   INPUT M
25   PRINT "TYPE IN THE Y-INTERCEPT OF THE LINE."
30   INPUT B
35   PRINT "X","Y"
40   FOR X = 0 TO 5 STEP .5
45   LET Y = M * X + B
50   PRINT X,Y
55   NEXT X
60   END
```

See the Computer Section beginning on page 488 for more information.

## Exercises  For Ex. 1–3 see TE Answer Section.

1. Type in the program above and RUN it for the lines $y = 2x - 1$ and $y = 6x - 1$. Use each table of values generated by the program to graph the lines $y = 2x - 1$ and $y = 6x - 1$.

2. Alter the program above to find a table of $x$, $y$ values for $x = -5$ to 0.

   **Change line 40 to: FOR X = −5 TO Ø STEP .5**

★3. Type in the program below to see how the line $x = y$ looks on the computer. The $xy$-plane is rotated 90° clockwise, causing the negative $y$-axis to become the positive $x$-axis.

   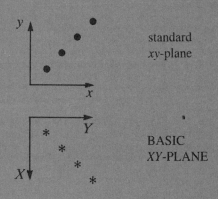

   ```
    5   REM PROGRAM PLOTS A LINE
   10   FOR X = 1 TO 10
   20   PRINT TAB (X); "*"
   30   NEXT X
   40   END
   ```

   TAB causes the printer to move X spaces, then print an *.

# CUMULATIVE REVIEW

(Chapters 1–12)

Each exercise has five choices of answers.
Choose the one best answer.

1. Evaluate $-5(x + 2) - 3x$ if $x = -2$.
   (A) 26   (B) 6   (C) $-6$   (D) 14
   (E) None of these   **B**

2. Solve $-3x + 2 \leq -13$.
   (A) $x \leq 5$   (B) $x \geq -5$   (C) $x \geq 5$
   (D) $x \geq \frac{11}{3}$   (E) None of these   **C**

3. Subtract $a^2 - a + 2$ from $3a^2 - a - 5$.
   (A) $2a^2 - 7$   (B) $-2a^2 + 7$
   (C) $4a^2 - 2a - 3$   (D) $2a^2 - 2a - 7$
   (E) None of these   **A**

4. Simplify $(2x + 5)^2$.
   (A) $4x^2 + 25$   (B) $4x^2 + 10x + 25$
   (C) $4x^2 + 20x + 25$   (D) $100x^2$
   (E) None of these   **C**

5. Factor $4x^3 - 12x^2 + 8x$ completely.
   (A) $4(x^3 - 3x^2 + 2x)$   (B) $x(4x^2 - 12x + 8)$
   (C) $4x(x^2 - 3x + 2)$   (D) $4x(x - 1)(x - 2)$
   (E) None of these   **D**

6. Find the value(s) of $x$ for which $\dfrac{x^2 - 4}{x^2 - 25}$
   is undefined.
   (A) $2, -2$   (B) $5, -5$   (C) $4$   (D) $25$
   (E) None of these   **B**

7. Solve $|x - 3| = 9$.
   (A) $3, 9$   (B) $12, -6$   (C) $12, 6$
   (D) $-12, -6$   (E) None of these   **B**

8. Find the reciprocal of $7d$.
   (A) $\dfrac{7}{d}$   (B) $\dfrac{d}{7}$   (C) $\dfrac{7d}{1}$   (D) $-\dfrac{1}{7d}$
   (E) None of these   **E**

9. Find the slope of the line $y = \frac{2}{3}x - 5$.
   (A) $\frac{2}{3}$   (B) $-5$   (C) $\frac{3}{2}$   (D) $2$
   (E) None of these   **A**

10. Add $\dfrac{4x - 3}{2} + \dfrac{2x - 1}{3}$.
    (A) $\dfrac{6x - 4}{5}$   (B) $\dfrac{6x - 4}{6}$   (C) $\dfrac{16x - 4}{6}$
    (D) $16x - 11$   (E) None of these   **E**

Simplify.

11. $\dfrac{9 - y^2}{y^2 - 5y - 24}$   $\dfrac{-(y - 3)}{y - 8}$

12. $\dfrac{3n - 1}{n + 4} - \dfrac{2n^2}{16 - n^2}$   $\dfrac{5n^2 - 13n + 4}{(n + 4)(n - 4)}$

13. $\dfrac{x - 7}{x} \div \dfrac{x^2 - 3x - 28}{x^2 + 4x}$   $1$

14. $3x - 3 + \dfrac{2}{x + 1}$   $\dfrac{3x^2 - 1}{x + 1}$

15. $\dfrac{1 - \dfrac{4}{m} - \dfrac{12}{m^2}}{\dfrac{1}{m} - \dfrac{6}{m^2}}$   $m + 2$

16. $\dfrac{\dfrac{4}{x^2 - x - 6} - \dfrac{2}{x - 3}}{\dfrac{5}{x + 2}}$   $\dfrac{-2x}{5(x - 3)}$

Solve.

17. $5x - 28 = -3x^2$   17. $x = \frac{7}{3}$

18. $x - y = 4$   $x + 2y = 10$   $x = 6$   $y = 2$

19. $4|x - 2| + 6 = 14$   19. $x = 0$   $x = 4$   $x = -4$

20. $\dfrac{x}{2} - \dfrac{x}{3} = 4$   $x = 24$

21. $0.02(2 - 0.5x) = 0.32 - 0.03x$   $x = 14$

22. $\dfrac{3}{5}(x - 3) + \dfrac{x + 7}{5} = \dfrac{x + 17}{10}$   $x = 3$

23. $\dfrac{6}{a + 2} + \dfrac{3}{a^2 - 4} = \dfrac{2a - 7}{a - 2}$   $a = -\frac{1}{2}$   $a = 5$

24. $\dfrac{x + 1}{x - 3} = \dfrac{3}{x} + \dfrac{12}{x^2 - 3x}$   $x = -1$

25. $-6x + 8 > 23 - 3x$   $x < -5$

26. $8c - 12 = -4c + 12$   $c = 2$

27. $4(5 - 2x) - 9 = 7 - 4(5 + 4x)$   $x = -3$

# CUMULATIVE REVIEW
## (Continued)

Solve for $x$.   28. $\dfrac{5a + 7}{p}$   29. $x = \dfrac{2b}{a - p}$   30. $-2$

**28.** $px - 7 = 5a$      **29.** $ax + b = px + 3b$

**30.** Find the slope of $\overline{CD}$. $C(4, -3); D(2, 1)$

**31.** Write an equation for $\overrightarrow{PQ}$. $P(2, 3); Q(-1, 4)$

$$y = -\tfrac{1}{3}x + \tfrac{11}{3}$$

Graph the following.

**32.** $3x - 2y = 8$      **33.** $2x - 4 = 8$

**34.** $y = -2$      **35.** $y < -2x + 1$

For Ex. 32–35, see TE Answer Section.

Divide.

**36.** $2b \overline{) 8b^3 - 14b^2 + 10b}$   $4b^2 - 7b + 5$

**37.** $2m - 1 \overline{) 8m^3 - 4m + 12}$

$4m^2 + 2m - 1$  remainder 11

**38.** $x - 5 \overline{) x^2 + 2x - 16}$  $x + 7$  remainder 19

Use the diagram to answer questions 39–42.

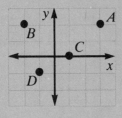

$(-2, 2)$

**39.** What are the coordinates of point $B$?

**40.** Which point has an abscissa of 1? **C**

**41.** For which point is the sum of the abscissa and ordinate 0? **B**

**42.** Which point is in quadrant III? **D**

**43.** Write an equation for $\overrightarrow{PQ}$ for $P(0, -1)$, $Q(2, 3)$. Then use the equation to set up a table of values for three more points.

$$y + 1 = 2x \ (1, 1), (3, 5), (4, 7)$$

Solve by graphing. For Ex. 44–45, see TE Answer Section

**44.** $x - y = 4$      **45.** $y > x - 2$

$3x + 2y = 11$         $y \le -x + 4$

Determine whether each system is independent and consistent, inconsistent, or dependent and consistent. **47. dependent and consistent**

**46.** $x - 2y = 8$      **47.** $2x - y = 7$

$-3x + 2y = 5$         $-10x + 5y = -35$

independent and consistent

**48.** Write an equation for the line with slope $-\tfrac{2}{3}$ and $y$-intercept the same as that of $4x - 2y = 8$. **$3y + 2x + 12 = 0$**

Multiply.

**49.** $(3b - 4c)(3b + 4c)$  $9b^2 - 16c^2$

**50.** $(x + 5)(x^2 - 2x + 1)$  $x^3 + 3x^2 - 9x + 5$

Which property is illustrated?

**51.** $a \cdot (b + c) = ab + ac$ **distributive law**

**52.** $(3x)(5y) = (5y)(3x)$

**53.** $a + 0 = a$ **additive property of zero**

**54.** $x - x = 0$ **additive inverse**

**52. commutative property of multiplication**

Graph the solution on a number line.

**55.** $|x| > 3$  For Ex. 55–57, see TE Answer Section

**56.** $x < 3$ or $x > 5$      **57.** $4 < 2x - 6 < 8$

Solve.

**58.** Moira wants to sell a box of two varieties of birthday cards for $11.60. The number of $.30 cards is 6 more than the number of $.40 cards. Find the number of cards of each type. **20 @ $0.30   14 @ $0.40**

**59.** The perimeter of a rectangle is 56 m. The width is 2 m less than the length. Find the length and width.

**length: 15 m   width: 13 m**

**60.** One number is 8 more than 3 times the other. If the larger is decreased by twice the smaller, the result is 10. Find the numbers. **14 and 2**

# COLLEGE PREP TEST

Directions: In each item, you are to compare a quantity in Column 1 with a quantity in Column 2. Write the letter of the correct answer from these choices:

(A) The quantity in Column 1 is greater than the quantity in Column 2;
(B) The quantity in Column 2 is greater than the quantity in Column 1;
(C) The quantity in Column 1 is equal to the quantity in Column 2;
(D) The relationship cannot be determined from the given information.

*Notes:* Information centered over both columns refers to one or both of the quantities to be compared. A symbol that appears in both columns has the same meaning in each column, and all variables represent real numbers.

| | Column 1 | Column 2 | |
|---|---|---|---|
| **1.** | | $x + y = 4$ $x - y = 2$ | **A** |
| | $x$ | $2y$ | |
| **2.** | $(y + 10)$ $- (y - 2x - 30)$ | $(x + 160)$ $- (120 - x)$ | **C** |
| **3.** | | $b \neq 0$ | |
| | $\dfrac{a + b}{b}$ | $\dfrac{a}{b} + 1$ | **C** |
| **4.** | $a : b = 1, a \neq 0,$ and $b \neq 0$ | | |
| | $a$ | $b$ | **C** |
| **5.** | | $0 < x < 10$ $0 < y < 12$ | |
| | $x$ | $y$ | **D** |
| **6.** | | $3x - y = 4$ $2x + y = 6$ | |
| | $y^x$ | $x^y$ | **C** |

| | Column 1 | Column 2 | |
|---|---|---|---|
| **7.** | $x = 3, y = -3,$ and $t \neq 0$ | | |
| | $\dfrac{r(x + y)}{t}$ | $\dfrac{5r(x + y)}{t}$ | **C** |
| **8.** | $\left(\dfrac{1}{0.03}\right)^2$ | $\dfrac{1}{3}$ | **A** |
| **9.** | | $x = 3y - 1$ $x + y = 7$ | |
| | $x - y$ | $xy$ | **B** |
| **10.** | $a = 3$ and $b = 5$ | | |
| | $\dfrac{ab}{\dfrac{1}{a} + \dfrac{1}{b}}$ | $\dfrac{1}{8}$ | **A** |
| **11.** | | $x < 0$ | |
| | $x^5 - 4$ | $2$ | **B** |
| **12.** | $B = 0, A > 2,$ and $C > 2$ | | |
| | $5B(C + A)$ | $A(C + B)$ | **B** |

# 13 FUNCTIONS AND RELATIONS

**Non-Routine Problem Solving**

**Problem 1**

To determine the weight of a large fish, two fishers set up a plank as a seesaw. They kept moving the plank until it was perfectly balanced when each of them stood on one end. Then they changed places and the lighter one put the fish on her side of the plank to maintain a perfect balance. The weights of the fishers were 120 lb and 150 lb. What was the weight of the fish? **$67\frac{1}{2}$ lb**

**Problem 2**

A man spent $\frac{1}{6}$ of his lifetime as a child, $\frac{1}{12}$ as a youth, and $\frac{1}{7}$ as a bachelor. Five years after he was married, his daughter was born. When his daughter died, she was $\frac{1}{2}$ the age her father was when he died. The father died four years after his daughter. How old was he when he died? **84**

# RELATIONS AND FUNCTIONS

*Objectives*

**To determine the domain and the range of a relation**
**To graph a relation on a coordinate plane**
**To determine whether or not a relation is a function**

The graph at the right consists of four points. It can be described by a *set of ordered pairs, C* = {(−2, 4), (−2, 1), (0, −3), (3, −1)}, which gives the coordinates of the points of the graph. Set *C* is called a **relation.**

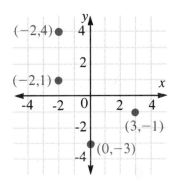

For *C*, the *set of all first coordinates*, {−2, 0, 3}, is called the **domain** of the relation. Notice that −2 is listed only once. The *set of all second coordinates* of *C*, {4, 1, −3, −1}, is called the **range** of the relation.

---

**Definitions:**
**Relation,**
**Domain, and**
**Range**

A **relation** is a set of ordered pairs.

The **domain**, *D*, is the set of all first coordinates of the ordered pairs in the relation.

The **range**, *R*, is the set of all second coordinates of the ordered pairs in the relation.

---

*Example 1*

**Graph *A* = {(6, 0), (−2, 3), (4, 3), (−4, 2), (−2, −1)}. Then, give the domain and the range of *A*.**

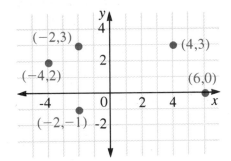

◀ *Graph of the relation, A*

The domain of *A* is {−4, −2, 4, 6}.    ◀ *In the domain, list −2 only once.*
The range of *A* is {−1, 2, 3, 0}.    ◀ *In the range, list 3 only once.*

Shown below is a special kind of relation, called a **function**.
$$G = \{(3, 7), (1, -4), (-5, 6), (8, 6)\}$$
Notice that no two first coordinates are the same. In a function, the second coordinates can be the same, but the first coordinates *cannot* be the same.

| Definition: Function | A **function** is a relation such that no two ordered pairs have the same first coordinate. |
|---|---|

*Example 2* **Graph $B = \{(0, -2), (2, 3), (-5, 2), (-4, -1), (5, 4)\}$. Then, give the domain and the range of relation $B$. Is $B$ a function?**

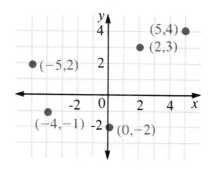

◀ *No two x-coordinates, first coordinates, are the same.*

The domain of $B$ is $\{0, 2, -5, -4, 5\}$. The range of $B$ is $\{-2, 3, 2, -1, 4\}$.
$B$ is a function, since no two first coordinates, $x$-coordinates, are the same.

*Example 3* **List the ordered pairs in relation $C$ graphed below. Give the domain and the range of $C$. Is $C$ a function?**

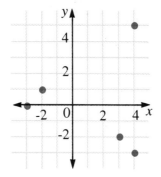

$C = \{(-3, 0), (-2, 1), (3, -2), (4, -3), (4, 5)\}$
The domain of $C$ is $\{-3, -2, 3, 4\}$.
The range of $C$ is $\{0, 1, -2, -3, 5\}$.

◀ *(4, −3), (4, 5)*
*same x-coordinate*

Since 4 is the first coordinate, $x$-coordinate, of two different ordered pairs in $C$, the relation $C$ is not a function.

It is not necessary to graph a relation in order to determine whether or not it is a function. You can use the definition of function, as shown below.

## Example 4    Determine whether or not the relation is a function.

$A = \{(2, 4), (3, 5), (6, 5), (-1, 3)\}$
$A$ is a function because no two ◀ *Use function* ▶ $B$ is not a function because $(1, 9)$ and $(1, 6)$
first coordinates are the same.    *definition.*    have the same first coordinate.

$B = \{(1, 9), (-2, 5), (1, 6), (-4, -3), (2, 8)\}$

## Written Exercises

**List the ordered pairs and give the domain and the range of the relation.**
**Is the relation a function? For Ex. 1–3, see TE Answer Section.**

Ⓐ  **1.**

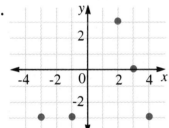

function

**2.**

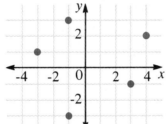

not a function

**3.**

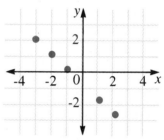

function

**Graph. Give the domain and the range of the relation. Is it a function?**
  **4.** $\{(0, 2), (-3, 1), (4, -5), (-5, 1)\}$ **function**  **5.** $\{(4, 0), (-4, 0), (0, -4), (0, 4)\}$
  **6.** $\{(4, 3), (5, 3), (0, 3), (1, 3)\}$ **function**  **7.** $\{(3, -1), (3, -2), (3, 0), (3, 1)\}$

**Determine whether or not the relation is a function.**  5. not a function  7. not a function
  **8.** $\{(-2, -2), (-3, -3), (-1, -1), (0, 0), (1, 1)\}$ **9.** $\{(5, -1), (1, 1), (4, 3), (5, -5), (2, -3)\}$
  **10.** $\{(-4, -2), (-4, -1), (-4, 0), (-4, 1), (-4, 2)\}$ **11.** $\{(-3, 2), (-2, 2), (-1, 2), (0, 2), (1, 2)\}$
  **8.** function  **9.** not a function  **10.** not a function  **11.** function

## Example    For what values of *k* will $R = \{(3k + 2, 8), (k + 6, 7)\}$ *not* be a function?

*Think:* $R$ will *not* be a function if the first coordinates of the two ordered pairs are the same. So, find the value of $k$ for which $3k + 2 = k + 6$.

*Solve the equation.* ▶    $3k + 2 = k + 6$
                     $2k = 4$   ◀ *Add $-k$ to each side. Then, add $-2$ to each side.*
                     $k = 2$
So, $R$ will *not* be a function if $k$ is 2.  ◀ *Substitute 2 for k: $3 \cdot 2 + 2 = 8$ and $2 + 6 = 8$.*

**For what value(s) of *k* will the relation *not* be a function?** 14. −11  17. 6, 2
Ⓑ  **12.** $\{(2k + 1, 2), (k + 3, 6)\}$ **2**  **13.** $\{(3k + 1, 2), (k + 3, 7)\}$ **1**  **14.** $\{(2k - 6, 4), (3k + 5, 6)\}$
  **15.** $\{(4k - 4, 5), (6k - 5, 6)\}$ $\frac{1}{2}$  **16.** $\{(3k - 6, 7), (-2k, 4)\}$ $1\frac{1}{5}$  **17.** $\{(-12, 2), (k^2 - 8k, 5)\}$

Ⓒ  **18.** $\{(|k + 1| + 2, 4), (8, 7)\}$ **5, −7**        **19.** $\{(k^3 - 5k^2 + 3k, -5), (-k, 4)\}$ **0, 1, 4**

# IDENTIFYING FUNCTIONS    13.2

*Objectives*    **To determine if graphs are functions by using the vertical-line test**
**To identify linear functions and constant linear functions**

In the graph at the right, two points (2, 3) and (2, −3) are labeled. However, the graph consists of an infinite number of points. One could never label all of them. To determine whether a graph is the graph of a function, you can draw vertical lines through the graph. Notice that the vertical line drawn through the graph of the relation at the right crosses it in the two points (2, 3) and (2, −3). Since two different ordered pairs have the same *x*-coordinate, 2, the relation is not a function. This method is known as the **vertical-line test.**

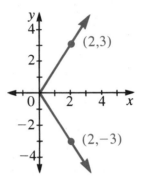 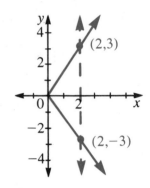

| Vertical-line test | If *no vertical line* crosses the graph of a relation in more than one point, the relation is a function. |
| --- | --- |

*Example 1*    **Determine which of the following is the graph of a function.**

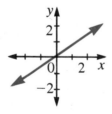    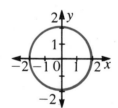

*Think:* Use the vertical-line test to decide.

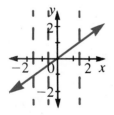    The graph is a function because no vertical line will ever cross the graph in more than one point.

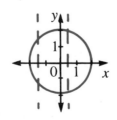    The graph is not a function because there is at least one vertical line that crosses the graph in more than one point.

Sometimes, an equation, or a formula, is used to describe the graph of a relation. Such is the case in Example 2.

*Example 2*    **Graph $y = \frac{2}{3}x + 1$. Is it a function?**

*First,* use the slope and the y-intercept to graph the equation.

$$y = \frac{2}{3}x + 1$$

slope    y-intercept

Each point on the line corresponds to an ordered pair. A set of ordered pairs is a relation.

*Second,* use the vertical-line test to determine whether the relation is a function.

No vertical line crosses the graph in more than one point.   **So,** $y = \frac{2}{3}x + 1$ is a function.

Example 2 shows a function whose graph is a line. Therefore, it is called a **linear function.**

| Definition: Linear function | A **linear function** is one whose graph is a *nonvertical* line. |
| --- | --- |

*Example 3*    **Is $y = -2$ a function? If so, is it a linear function?**

*Think:* Graph $y = -2$. The $x$ term does not appear in the equation. The slope is 0, and the line is *horizontal*. The $y$-coordinate is always $-2$. Draw the horizontal line $y = -2$.

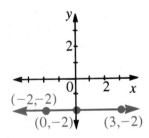

**So,** $y = -2$ is a function because no two ordered pairs have the same $x$-coordinate. It is a linear function because its graph is a nonvertical line.

In Example 3, notice that the $y$-coordinate of every point on the graph of $y = -2$ is the same, or constant. **So,** $y = -2$ is called a **constant linear function.**

Identifying Functions

| Definition: Constant linear function | A **constant linear function** is one whose graph is a *horizontal* line. |

*Example 4*

**Is $x = 4$ a function? If so, is it a linear function or a constant linear function?**

*Think:* Graph $x = 4$. In the equation $x = 4$, $y$ does not appear. The graph is a *vertical* line. The $x$-coordinate is always 4. Draw the vertical line $x = 4$.

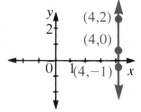

At least two ordered pairs have the same $x$-coordinate. **So, $x = 4$ *is not* a function.**

Vertical lines are graphs of relations which are not functions.

## Written Exercises

**Which relations are functions? Which functions are linear functions?**

(A)

**1.**

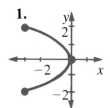

not a function

**2.**

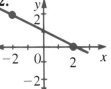

linear function

**3.**

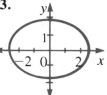

not a function

**4.**

not a function

**Graph. Which relations are functions? Which functions are linear (LF) functions? Which functions are constant linear functions? (CLF)**

For complete answers, see TE Answer Section.

**5.** $y = 2x$ **LF**     **6.** $y = x$ **LF**     **7.** $y = x - 4$ **LF**     **8.** $y = \frac{2}{3}x - 5$ **LF**

**9.** $y = 4$ **CLF**     **10.** $x = -4$     **11.** $y = -1$ **CLF**     **12.** $y = 1$ **CLF**

**13.** $y = -\frac{2}{3}x - 4$ **LF**    **14.** $x - 4 = 2$     **15.** $y + 3 = -1$ **CLF**    **16.** $3x - 4 = 17$

**17.** $y = -x + 4$ **LF**     **18.** $4 - y = 0$ **CLF**     **19.** $x - 1 = -3$     **20.** $3y - 4 = 8$ **CLF**

10. not a function   14. not a function      not a function     16. not a function

**Graph. Which relations are functions?**   21. function   22. function   23. function   24. not a function

(B) **21.** $3x - 2y = 4$     **22.** $4y - 3x = 8$     **23.** $4x + 4y = -16$     **24.** $-7x - 4 = 10$

**25.** $4 - (6 - y) = 8$    **26.** $8x - (5 - x) = 14$   **27.** $3x - (2x - y) = 4$   **28.** $\frac{x}{4} = y$

**29.** $\frac{y - 4}{2} = x - 2$    **30.** $\frac{x - 4}{2} = \frac{3}{2}$    25. function   26. not a function

27. function   28. function   29. function   30. not a function

# STANDARD FUNCTION NOTATION          13.3

*Objectives*   **To use the standard notation, *f(x)*, for a function**
**To determine the range of a function for a given domain**

A lowercase letter is usually used to name a function.
For example, $f = \{(0, 2), (-5, 6), (3, -4)\}$.
The domain of $f$ is $\{0, -5, 3\}$, and the range of $f$ is $\{2, 6, -4\}$. Each
element of the range is called a *value* of the function. For $(0, 2)$, the value of
the function $f$ at 0 is 2. This can be abbreviated $f(0) = 2$.

The value of $f$ at 0 is 2.

Similarly, for $(-5, 6)$, $f(-5) = 6$. This is read as "the value of $f$ at $-5$
is 6."

In general, the notation $f(x)$ means the *value* of $f$ at $x$.

| Definition: Value of a function | For any ordered pair $(x, y)$ of a function $f$, **the value of $f$ at $x$ is $y$, which** is denoted by $f(x) = y$. |
|---|---|

*Example 1*   **For $g = \{(0, -1), (-4, 3), (2, -3)\}$, find $g(0)$, $g(-4)$, and $g(2)$.**

For $(0, -1)$,       For $(-4, 3)$,       For $(2, -3)$,
$g(0) = -1$.         $g(-4) = 3$.         $g(2) = -3$.

Sometimes, a function is described by an equation, or a formula. This is
shown in the next example.

*Example 2*   **For $f(x) = 3x - 4$, find $f(2)$, $f(-1)$, and $f(6)$.**

Find $f(2)$.
$f(x) = 3x - 4$
$f(2) = 3 \cdot 2 - 4$ ◀ *Substitute 2*
$\quad = 6 - 4$            *for x.*
$\quad = 2$

Find $f(-1)$.
$f(x) = 3x - 4$
$f(-1) = 3 \cdot -1 - 4$
$\quad\quad = -3 - 4$
$\quad\quad = -7$

Find $f(6)$.
$f(x) = 3x - 4$
$f(6) = 3 \cdot 6 - 4$
$\quad = 18 - 4$
$\quad = 14$

If you are given an equation, or a formula, and the domain of a function,
you can use the definition, $f(x) = y$, to find the range of the function.

*Example 3*     **For $h(x) = -x^2 - 3$ and domain $D = \{-2, -1, 4\}$, find the range of $h(x)$.**

| Find $h(-2)$. | Find $h(-1)$. | Find $h(4)$. |
|---|---|---|
| $h(x) = -1x^2 - 3$ | $h(x) = -1x^2 - 3$ | $h(x) = -1x^2 - 3$ |
| $h(-2) = -1(-2)^2 - 3$ | $h(-1) = -1(-1)^2 - 3$ | $h(4) = -1(4)^2 - 3$ |
| $= -4 - 3 = -7$ | $= -1 - 3 = -4$ | $= -16 - 3 = -19$ |

**So,** the range of $h(x)$ is $\{-7, -4, -19\}$. ◀ *The range is the set of all values for a given domain of the function.*

# Written Exercises

**For $g = \{(-5, 2), (-4, -1), (3, -8), (-2, 0), (4, -9)\}$, find the indicated value.**

Ⓐ  **1.** $g(-4)$ **−1**     **2.** $g(4)$ **−9**     **3.** $g(3)$ **−8**     **4.** $g(-5)$ **2**     **5.** $g(-2)$ **0**

**For $g(x) = 4x - 3$, find the indicated value.**

**6.** $g(-2)$ **−11**     **7.** $g(0)$ **−3**     **8.** $g(-3)$ **−15**     **9.** $g(4)$ **13**     **10.** $g(-20)$ **−83**

**For $h(x) = x^2 - 4$, find the indicated value.**

**11.** $h(0)$ **−4**     **12.** $h(-2)$ **0**     **13.** $h(4)$ **12**     **14.** $h(-1)$ **−3**     **15.** $h(6)$   **32**
**17.** $\{5, 3, -11\}$ **18.** $\{-3, -4, -9\}$ **19.** $\{2, 7, 32\}$ **20.** $\{-10, -15, -22\}$ **21.** $\{12, 26, 44\}$ **22.** $\{7, 1\}$

**Find the range of each function for the given domain.**

**16.** $f(x) = 6x - 1$     $D = \{0, 2, 4\}$ $\{-1, 11, 23\}$ **17.** $h(x) = -2x - 1$     $D = \{-3, -2, 5\}$
**18.** $k(x) = -x - 5$     $D = \{-2, -1, 4\}$     **19.** $r(x) = -5x + 2$     $D = \{0, -1, -6\}$
**20.** $f(x) = -x^2 - 6$     $D = \{-3, 2, 4\}$     **21.** $g(x) = 2x^2 - 6$     $D = \{-3, -4, 5\}$

Ⓑ  **22.** $f(x) = x^2 + x + 1$     $D = \{-3, -1, 0\}$     **23.** $r(x) = -x^2 - 4x - 6$     $D = \{-2, 5, 3\}$
**24.** $g(x) = \dfrac{x^2 - 4}{x - 2}$     $D = \{-1, 0, 4\}$ $\{1, 2, 6\}$ **25.** $h(x) = \dfrac{x - 7}{x^2 - 2x - 35}$     $D = \{-1, 0, 3\}$
**26.** $f(x) = |x|$     $D = \{-3, 0, 3\}$ $\{0, 3\}$     **27.** $f(x) = (3x - 2)^2$     $D = \{-3, -2, -1\}$
**28.** $f(x) = (x - 1)^2$     $D = \{-4, 0, 4\}$ $\{25, 9, 1\}$ **29.** $f(x) = -3x - 9$     $D = \{0.3, 0.1, 0.4\}$
**23.** $\{-2, -27, -51\}$ **27.** $\{121, 64, 25\}$ **29.** $\{-9.3, -9.9, -10.2\}$ **25.** $\left\{\frac{1}{4}, \frac{1}{5}, \frac{1}{8}\right\}$   **32.** $x^4 + 6x^3 + 9x^2$

**For $f(x) = 3x - 1$ and $g(x) = x^2 + 1$, find the indicated value.**

Ⓒ  **30.** $f(2) + g(2)$ **10**     **31.** $f(-3) + g(-3)$ **0**     **32.** $[f(x) + g(x)]^2$     **33.** $\dfrac{f(x)}{g(x)}$ $\dfrac{3x - 1}{x^2 + 1}$

**34.** $f[g(1)]$ **5**     **35.** $f[g(-2)]$ **14**     **36.** $f[g(-4)]$ **50**     **68 37.** $f[g(-5) - 3]$

# CALCULATOR ACTIVITIES

**Find the range of each function for the given domain.** **1.** $\{0.0004, 225.150025, 48149.5249\}$
**1.** $f(x) = x^2$     $D = \{0.02, 219.43, -15.005\}$     **2.** $\{-16.2436, -4.539916, 5.075924\}$
**2.** $g(x) = x^2 - 7x - 4$     $D = \{3.42, -1.118, 0.078\}$
**3.** $h(x) = 3.14x^2 - 7.8x$     $D = \{12.43, -1.18, 5.04\}$
**4.** $r(x) = \frac{1}{3}\pi x$, where $\pi = 3.14$     $D = \{6.1, 8.3, 14.3\}$

**3.** $\{388.19138, 13.576136, 40.449024\}$     **4.** $\{25.538666, 34.749333, 59.869333\}$

# DIRECT VARIATION

*Objectives*
**To determine if a relation is a direct variation**
**To determine the constant of variation in a direct variation**
**To solve problems involving direct variation**

Frequently in an equation one variable is directly related to another so that the ratio of the variables is always the same, or constant. Such a relation is illustrated in the table below. The table shows the number of pretzels, $y$, produced in $x$ minutes. Notice that the ratio $y:x$ is always the same.

| Pretzel Production | |
|---|---|
| **Number of pretzels $y$** | **Time in minutes $x$** |
| 30 | 3 |
| 40 | 4 |
| 100 | 10 |
| 200 | 20 |

$\dfrac{y}{x} = \dfrac{30}{3} = \dfrac{10}{1} = 10; \quad \dfrac{y}{x} = \dfrac{40}{4} = \dfrac{10}{1} = 10$

$\dfrac{y}{x} = \dfrac{100}{10} = \dfrac{10}{1} = 10; \dfrac{y}{x} = \dfrac{200}{20} = \dfrac{10}{1} = 10$

The proportion $\dfrac{y}{x} = \dfrac{10}{1}$ can also be written as $y \cdot 1 = x \cdot 10$, or $y = 10x$, by using the product of the means equals the product of the extremes. The function $\dfrac{y}{x} = 10$, or $y = 10x$, is called a **direct variation.** A direct variation is a function in which the ratio of two variables is a *constant*. The 10 is called the **constant of variation.**

**Definition: Direct variation**

A **direct variation** is a *function* defined by an equation of the form $\dfrac{y}{x} = k$, or $y = kx$, where $k$ is a *nonzero* constant. $\dfrac{y}{x} = k$ or $y = kx$ indicates that $y$ *varies directly* as $x$, or $y$ *is directly proportional* to $x$.

*Example 1*

**From the table, determine if $y$ varies directly as $x$. If so, find the constant of variation.**

| $x$ | $y$ |
|---|---|
| 3 | $-9$ |
| 5 | $-15$ |
| $-1$ | 3 |

See if $\dfrac{y}{x} = k$ is a constant.

$\dfrac{-9}{3} = -3 \quad | \quad \dfrac{-15}{5} = -3 \quad | \quad \dfrac{3}{-1} = -3 \quad | \quad \dfrac{y}{x} = -3$ for all pairs $(x, y)$.

**So,** $y$ varies directly as $x$, and the constant of variation is $-3$.

Suppose two ordered pairs of a direct variation are $(x_1, y_1)$ and $(x_2, y_2)$.
Read: ($x$ sub 1, $y$ sub 1) and ($x$ sub 2, $y$ sub 2).

If $x_1 \neq 0$ and $x_2 \neq 0$, then $\dfrac{y_1}{x_1} = k$ and $\dfrac{y_2}{x_2} = k$

and $\dfrac{y_1}{x_1} = \dfrac{y_2}{x_2}$ by substitution.

This equation is a proportion, and $y$ is said to be directly proportional to $x$. This proportion can be used to solve problems involving direct variation, as illustrated in Example 2.

*Example 2*    **$y$ varies directly as $x$. $y$ is 24 when $x$ is 3. Find $y$ when $x$ is 4.**

Use    $\dfrac{y_1}{x_1} = \dfrac{y_2}{x_2}$

$\dfrac{24}{3} = \dfrac{y}{4}$    ◀ *Let $(x_1, y_1) = (3, 24)$ and $(x_2, y_2) = (4, y)$.*

$3 \cdot y = 24 \cdot 4$    ◀ *Solve the proportion.*

$3y = 96$

$y = 32$

**So,** $y$ is 32 when $x$ is 4.

Applications of direct variation are numerous in science and other fields. The proportion definition of direct variation

$$\frac{y_1}{x_1} = \frac{y_2}{x_2}$$

will be very useful in solving problems involving practical applications of direct variation.

*Example 3*    **At a given time and place, the height of an object varies directly as the length of its shadow. If a flagpole 6 m high casts a shadow 10 m long, find the height of a building that casts a shadow 45 m long.**

Let $x$ = the height of the building.
Since the height of the building varies directly as the length of its shadow,

use the proportion:    $\dfrac{h_1}{s_1} = \dfrac{h_2}{s_2}$.

Then    $\dfrac{6}{10} = \dfrac{x}{45}$

$6 \cdot 45 = 10 \cdot x$

$270 = 10x$

$27 = x$

**So,** the height of the building is 27 m.

Finding distances on a road map is an application of direct variation. The actual distance between two cities varies directly as their distance on a map.

*Example 4*

**On a map, a distance of 135 km is represented by 3 cm. How many kilometers are represented by 2 cm?**

Let $x$ = the number of kilometers represented by 2 cm.
Since the distance in kilometers varies directly as the distance in centimeters, use the proportion:

$$\frac{k_1}{c_1} = \frac{k_2}{c_2}$$

$$\frac{135}{3} = \frac{x}{2}$$

$270 = 3x$ ◀ *135 · 2 = 3 · x, or 270 = 3x*

$90 = x$ **So,** 2 cm represents 90 km.

## Reading in Algebra

**For the equation $\frac{y}{x} = \frac{2}{3}$, select the statements that do *not* apply. Explain why they do not apply.**

**1.** $y$ varies directly as $x$.

**2.** The constant of variation is 2.

**3.** $y$ is directly proportional to $x$.

**4.** $y = \frac{2}{3}x$

**5.** $y$ is directly proportional to $\frac{2}{3}$.

**2, 5; $\frac{2}{3}$ is constant of var.; $y$ is directly proportional to a variable, never a number.**

## Oral Exercises

**Name the constant of variation for each direct variation.**

**1.** $\frac{y}{x} = 2$  **2**    **2.** $y = \frac{3}{4}x$  **$\frac{3}{4}$**    **3.** $\frac{3}{5} = \frac{y}{x}$  **$\frac{3}{5}$**    **4.** $y = -2x$  **−2**    **5.** $\frac{5}{6} = \frac{y}{x}$  **$\frac{5}{6}$**    **6.** $4x = y$  **4**

## Written Exercises

**1. Yes, 5    2. No    3. Yes, −3    4. No    5. Yes, −4**

**Determine if $y$ varies directly as $x$. If so, find the constant of variation.**

**1.**

| $x$ | $y$ |
|---|---|
| 3 | 15 |
| 2 | 10 |
| 4 | 20 |

**2.**

| $x$ | $y$ |
|---|---|
| 3 | 6 |
| 4 | 8 |
| 5 | 15 |

**3.**

| $x$ | $y$ |
|---|---|
| −3 | 9 |
| 6 | −18 |
| 5 | −15 |

**4.**

| $x$ | $y$ |
|---|---|
| 1 | 7 |
| −2 | 14 |
| −5 | 35 |

**5.**

| $x$ | $y$ |
|---|---|
| 4 | −16 |
| 3 | −12 |
| −5 | 20 |

Direct Variation

**In each of the following, *y* varies directly as *x*.**

**6.** *y* is 32 when *x* is 8. Find *y* when *x* is 6. **24**

**7.** *y* is −48 when *x* is −8. Find *y* when *x* is 7. **42**

**8.** *y* is 3 when *x* is 21. Find *y* when *x* is 35. **5**

**9.** *y* is −4 when *x* is 32. Find *y* when *x* is 3. $-\dfrac{3}{8}$

**Solve each problem.**

Ⓑ **10.** The number of pencils sold varies directly as the cost. If 5 pencils cost $0.45, find the cost of 7 pencils. **$0.63**

**11.** The cost of gold varies directly as its mass. If 3 g of gold cost $225, find the cost of 7 g of gold. **$525**

**12.** On a scale drawing, 2 cm represents 30 m. How many meters are represented by 3 cm? **45 m**

**13.** On a map, 180 km are represented by 4 cm. How many kilometers are represented by 6 cm? **270 km**

**14.** At a given time and place, the height of an object varies directly as the length of its shadow. An 8-m flagpole casts a 10-m shadow. Find the height of a building that casts a 200-m shadow. **160 m**

**15.** On a blueprint, the length of a hallway that is 15 m long is represented by 2 cm. Find the dimensions of a room represented by a rectangle 3 cm by 5 cm.

**22.5 m by 37.5 m**

**16.** The bending of a beam varies directly as its mass. A beam is bent 20 mm by a mass of 40 kg. How much will the beam bend with a mass of 100 kg? **50 mm**

**17.** The weight of an object on the moon varies directly as its weight on earth. On earth, an object's weight is 90 kg, but on the moon, its weight is 14.4 kg. What would the weight of a 120-kg object be on the moon?

**19.2 kg**

**18.** Gas consumption of a car is directly proportional to the distance traveled. A car uses 20 L of gas to travel 200 km. How much gas will the car use on a 500-km trip? **50 L**

**19.** The number of kilograms of water in a person's body varies directly as the person's mass. A person with a mass of 93 kg contains 62 kg of water. How many kilograms of water are in a person with a mass of 105 kg? **70 kg**

Ⓒ **20.** *y* varies directly as the square of *x*. If *y* is 25 when *x* is 3, find *y* when *x* is 2. **$11\frac{1}{9}$**

**21.** *y* varies directly as the square of *x*. If *y* is 36 when *x* is 5, find *y* when *x* is 4. **23.04**

**22.** The distance needed to stop a car varies directly as the square of its speed. It requires 120 m to stop a car at 70 km/h. What distance is required to stop a car at 80 km/h? **156.73 m**

**23.** The distance that a free falling body falls varies directly as the square of the time it falls. A stone falls 44 m in 4 s. How far will it fall in 20 s? **1100 m**

## *CUMULATIVE REVIEW*

**1.** Simplify $\dfrac{3x^7 y^{10} z}{6x^6 y^{12} z^4}$.

$\dfrac{x}{2y^2 z^3}$

**2.** Solve $x^3 - 16x = 0$.

**{0, 4, −4}**

**3.** Multiply $\dfrac{a^2 - 25}{a^2 + a - 12} \cdot \dfrac{a + 4}{3a - 15}$.

$\dfrac{a + 5}{3(a - 3)}$

# INVERSE VARIATION                                    13.5

*Objectives*   **To determine if a relation is an inverse variation**
**To determine the constant of variation in an inverse variation**
**To solve problems involving inverse variation**

Recall that the area of a rectangle is found by multiplying the length by the width, that is, $A = l \cdot w$.

There are many rectangles, such as the rectangles below, that have the same area but have different lengths and widths.

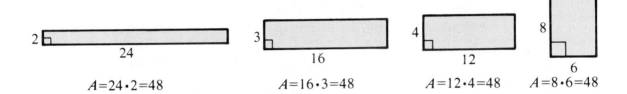

$A=24\cdot2=48$          $A=16\cdot3=48$      $A=12\cdot4=48$   $A=8\cdot6=48$

This relation is a variation in which the product of two factors is a constant.

In the table at the right, notice that as the
width increases, the length decreases.
In each case, the product $l \cdot w = 48$.
The table describes an *inverse variation*
in which the length varies inversely as the
width. The *constant of variation* is 48.

| $l$ | $w$ |
|----|----|
| 24 | 2 |
| 16 | 3 |
| 12 | 4 |
| 8 | 6 |

---

**Definition:**
**Inverse**
**variation**

An **inverse variation** is a relation in which the *product* of two factors is always the same, or constant.

$xy = k$, or $y = \dfrac{k}{x}$, defines an inverse variation, where $k$ is the nonzero *constant* of *variation*.

$xy = k$, or $y = \dfrac{k}{x}$, is read: *y varies inversely as x, or y is inversely proportional to x.*

---

You can use a table to determine whether $y$ varies inversely as $x$ by checking to see if the product of each pair of factors, $x$ and $y$, is the same constant. This is illustrated in Example 1 on the next page.

*Example 1*

**From the table, determine if *y* varies inversely as *x*. If so, find the constant of variation.**

| *x* | *y* |
|---|---|
| 3 | 8 |
| 6 | 4 |
| −12 | −2 |
| −24 | −1 |

See if $x \cdot y = k$, a constant.

$3 \cdot 8 = 24$

$6 \cdot 4 = 24$

$-12 \cdot -2 = 24$

$-24 \cdot -1 = 24$

$x \cdot y = 24$ for all pairs $(x, y)$.

**So,** *y* varies inversely as *x*. The constant of variation is 24.

Suppose two ordered pairs of an inverse variation are $(x_1, y_1)$ and $(x_2, y_2)$, then $x_1 \cdot y_1 = k$ and $x_2 \cdot y_2 = k$ by the definition of inverse variation, and $x_1 \cdot y_1 = x_2 \cdot y_2$ by substitution.

You can use this equation to solve problems that involve inverse variation.

*Example 2*

**y varies inversely as *x*, and *y* is 4 when *x* is 22. Find *x* when *y* is −11.**

Let $(x_1, y_1) = (22, 4)$ and $(x_2, y_2) = (x, -11)$.

Then

$22 \cdot 4 = x \cdot -11$  ◀ *Use $x_1 \cdot y_1 = x_2 \cdot y_2$.*

$88 = -11x$

$-8 = x$

**So,** *x* is −8 when *y* is −11.

Many problems in science and business are applications of inverse variation. One such application is illustrated in the next example.

*Example 3*

**The current in an electric circuit varies inversely as the resistance. When the current is 25 amps, the resistance is 16 ohms. Find the current when the resistance is 10 ohms.**

Let $c$ = the current when the resistance is 10 ohms.

$c_1 \cdot r_1 = c_2 \cdot r_2$  ◀ *The current varies inversely as the resistance.*

$25 \cdot 16 = c \cdot 10$

$400 = 10c$

$40 = c$

**So,** the current is 40 amps when the resistance is 10 ohms.

## Oral Exercises

**Which formulas describe inverse variations? For each inverse variation,
give the constant of variation.** 1. inverse, 45  2. inverse, −24  3. not inverse  4. inverse, 20

**1.** $r \cdot t = 45$      **2.** $x \cdot y = -24$      **3.** $\dfrac{C}{d} = 3.14$      **4.** $\dfrac{20}{y} = x$      **5.** $\dfrac{x}{5} = y$

5. not inverse

## Written Exercises

**Determine if $y$ varies inversely as $x$. If so, find the constant of variation.**

Ⓐ

**1.**

| $x$ | $y$ |
|---|---|
| 8 | 5 |
| −10 | 4 |
| 2 | 20 |
| −40 | −1 |

**2.**

| $x$ | $y$ |
|---|---|
| 2 | 2 |
| 4 | 1 |
| 0 | 4 |
| −2 | −2 |

**3.**

| $x$ | $y$ |
|---|---|
| 4 | −25 |
| −10 | 10 |
| 20 | 5 |
| −2 | −50 |

**4.**

| $x$ | $y$ |
|---|---|
| −7 | 8 |
| 28 | −2 |
| 14 | −4 |
| −1 | 56 |

**5.**

| $x$ | $y$ |
|---|---|
| 8 | −9 |
| 6 | −12 |
| −2 | 36 |
| 24 | −3 |

1. not inverse     2. not inverse     3. not inverse     4. inverse, −56     5. inverse, −72

**In each of the following, $y$ varies inversely as $x$.**

**6.** $y$ is 23 when $x$ is 8. **46**
Find $y$ when $x$ is 4.

**7.** $y$ is 40 when $x$ is −3. **15**
Find $y$ when $x$ is −8.

**8.** $y$ is −9 when $x$ is 16. **48**
Find $y$ when $x$ is −3.

**9.** $y$ is 12 when $x$ is −8. **4**
Find $y$ when $x$ is −24.

**10.** $y$ is 14 when $x$ is 8. **16**
Find $y$ when $x$ is 7.

**11.** $y$ is 30 when $x$ is $\frac{2}{3}$. **5**
Find $y$ when $x$ is 4.

**Solve each problem.**

Ⓑ
**12.** The length of a rectangle with a constant area varies inversely as the width. When the length is 18, the width is 6. Find the length when the width is 4. **27**

**13.** The volume of a gas varies inversely as the pressure. If the volume is 80 m³ under 4 kg of pressure, find the volume under 10 kg of pressure. **32 m³**

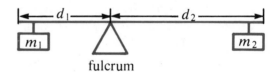

fulcrum

**14.** John's mass is 40 kg, and he is sitting 2 m from the fulcrum of a seesaw. Jane's mass is 20 kg. How far from the fulcrum must she sit to balance the seesaw? [Hint: The distance from the fulcrum varies inversely as the mass.] **4 m**

**15.** Laura has a mass of 60 kg and is sitting 265 cm from the fulcrum of a seesaw. Bill has a mass of 50 kg. How far from the fulcrum must he be to balance the seesaw? **318 cm**

**16.** José is sitting 125 cm from the fulcrum of a seesaw. Mona has a mass of 75 kg and is sitting 200 cm from the fulcrum. What is José's mass if the seesaw is balanced? **120 kg**

**17.** Time varies inversely as speed if the distance is constant. A trip takes 4 h at 80 km/h. How long does it take at 64 km/h? **5 h**

**18.** In an electric circuit, the current varies inversely as the resistance. The current is 40 amps when the resistance is 12 ohms. Find the current when the resistance is 20 ohms. **24 amps**

19. The length of the base of a triangle with constant area varies inversely as the height. When the base is 18 cm long, the height is 7 cm. Find the length of the base when the height is 6 cm. **21 cm**

20. Marsha has enough money to buy 3 m of fabric priced at $7.20/m. How many meters of fabric priced at $5.40/m can she buy with the same amount of money? **4 m**

21. The number of vibrations a string makes under constant tension is inversely proportional to its length. If a string 30 cm long vibrates 510 times per second, what length should the string be to vibrate 720 times per second? **21.25 cm**

22. When two meshed gears revolve, their speed is inversely proportional to the number of teeth they have. If a gear with 70 teeth revolves at a speed of 2,900 rev/min, at what speed should a gear with 112 teeth revolve? **1812.5 rev/min**

23. The number of hours required to do a job varies inversely as the number of people working. It takes 8 hours for 4 people to paint the inside of a house. How long would it take 5 people to do the job? **6.4 h**

24. The mass that a horizontal beam made of a certain kind of material can support varies inversely as the length of the beam. If a 10-m beam can support 1,400 kg, how many kilograms can an 8-m beam support? **1750 kg**

25. The height of a cylinder of constant volume varies inversely as the square of the length of the radius of the base. The height of a cylinder is 8 m, and the radius of the base is 6 m. Find the height of a cylinder of the same volume with a base radius of 4 m. **18 m**

26. The brightness of the illumination of an object varies inversely as the square of the distance of the object from the source of illumination. If a light meter reads 36 luxes at a distance of 4 m from a light source, find the reading at a distance of 3 m from the light source. **64 luxes**

27. Tina and Wilt are sitting 4 m apart on a seesaw. Tina has a mass of 48 kg and Wilt has a mass of 80 kg. How far from the fulcrum must Tina be sitting if the seesaw is balanced? **2.5 m**

28. The weight of a body at, or above, the Earth's surface varies inversely as the square of the body's distance from the Earth's center. An object has a weight of 220 kg when it is at the Earth's surface. What is the weight of the object when it is 370 km above the Earth's surface? (Use 6,500 km as the Earth's radius.) **196.9 kg to the nearest tenth of a kilogram**

## NON-ROUTINE PROBLEMS

The owner of a bicycle store was a bit eccentric when he took inventory. Instead of counting the number of bicycles and tricycles he had in the store, he chose to count the number of pedals and the number of wheels. One time, he came up with 186 wheels and 144 pedals. How many bicycles and how many tricycles did he have? **30 bicycles, 42 tricycles**

# APPLICATIONS

Read ──▶ Plan ──▶ Solve ──▶ Interpret

Solve each problem.

1. To approximate the number of deer in a forest, a conservationist caught 650 deer, tagged them, and then let them loose. Later, she caught 216 deer and found that 54 of them were tagged. How many deer were in the forest? (Assume that the number of tagged deer caught varies directly as the number of deer later caught.) **2,600 deer**

2. To determine the number of fish in a lake, a game warden caught 125 fish, tagged them, and threw them back into the lake. Later, he caught 65 fish and found that 13 of them were tagged. How many fish were in the lake? (Assume that the number of tagged fish caught varies directly as the number of fish later caught.) **625 fish**

3. The amount of money a family spends on food varies directly as their income. A family making $14,800 a year will spend $3,700 on food. How much will a family making $11,600 spend on food? **$2900.**

4. The amount of money that a family spends on car expenses varies directly as their income. A family making $16,000 a year will spend $2,400 a year for car expenses. How much will a family making $20,000 spend on car expenses? **$3,000**

5. A person's earnings vary directly as the number of hours worked. For working 45 hours, the earnings were $168.75. Find the earnings for 35 hours of work. **$131.25**

6. The profit for manufacturing a certain item is given by $P(x) = -1,000 + 0.55x$, where $x$ is the number of items manufactured. Find $P(100)$, $P(1,000)$, $P(5,000)$, and $P(30,000)$. When is $P(x) = 0$? How do you interpret your answer to $P(100)$? **−945, −450, 1750, 15,500, 1818.18, not enough items made to result in profit**

7. The cost of operating a television set varies directly as the number of hours it is in operation. It costs $8.00 to operate a certain set continuously for 40 days. At this rate, find cost of operating this set for one day. **$0.20**

# AGE PROBLEMS

*Objective*  **To solve problems about age**

Frequently, problems about ages involve relationships between the ages of people not only at the present time but also in the past or in the future. In Example 1, when you use the basic strategy for solving problems, you will have to represent the ages of the two people now and 6 years from now.

*Example 1*  **Meg is 4 times as old as Carlo. In 6 years, she will be twice as old as he will be then. How old is each now?**

READ ▶  You are asked to find each age now. You are given information about their ages *now* and 6 years *from now*.

|  | Now | 6 years from now |
|---|---|---|
| Meg's age | $x$ | $x + 6$ |
| Carlo's age | $y$ | $y + 6$ |

PLAN ▶  *Use two variables to represent their* ▶ *ages now and 6 years from now.*

Write an equation for each of the first two sentences.

| **Now** | **In 6 years** |
|---|---|
| Meg is 4 times Carlo. | Meg will be twice Carlo. |
| $x = 4y$   (1) | $x + 6 = 2(y + 6)$ |
|  | or $x + 6 = 2y + 12$   (2) |

SOLVE ▶  Solve the system of equations by substitution.

$$x = 4y \quad (1)$$
$$x + 6 = 2y + 12 \quad (2)$$
$$\downarrow$$

$4y + 6 = 2y + 12$  ◀ *Replace x with 4y in Equation (2).*
$2y + 6 = 12$  ◀ *Subtract 2y from each side.*
$\quad\;\, 2y = 6$  ◀ *Subtract 6 from each side.*
$\quad\;\;\, y = 3$  ◀ *Carlo's age now*

To find $x$, let $y = 3$ in either Equation (1) or (2).

$\quad x = 4y \quad (1)$  ◀ *Use Equation (1).*
$\quad x = 4 \cdot 3$
$\quad x = 12$  ◀ *Meg's age now*

INTERPRET ▶

| Meg's age now is 4 times Carlo's age now. | Her age 6 years from now will be twice his age. |
|---|---|
| 12 $\quad$ 4 · 3 | 12 + 6 $\quad$ 2(6 + 3) |
| 12 | 18 $\quad$ 2(9) |
| 12 = 12, true | 18 |
|  | 18 = 18, true |

**Thus,** Meg is 12, and Carlo is 3.

In the next example, you will have to represent the two people's ages in the past as well as the present time.

## Example 2

**Sylvia is 16 years younger than Martin. Twenty years ago, he was 5 times as old as she was then. How old is each now?**

|  | **Now** | **20 years ago** |
|---|---|---|
| Sylvia's age | $x$ | $x - 20$ |
| Martin's age | $y$ | $y - 20$ |

◀ *Represent the age of each now and 20 years ago.*

**Now**

Sylvia is 16 years younger than Martin.

$$x = y - 16 \quad (1)$$

**20 years ago**

Martin was 5 times Sylvia.

$$y - 20 = 5(x - 20)$$
$$\text{or } y - 20 = 5x - 100 \quad (2)$$

◀ *Write a system of equations.*

$$y - 20 = 5(y - 16) - 100$$
$$y - 20 = 5y - 80 - 100$$
$$y - 20 = 5y - 180$$
$$-4y - 20 = -180$$
$$-4y = -160$$
$$y = 40$$

◀ *Solve the system by substitution. Replace x with y − 16 in Equation (2).*

$$x = y - 16 \quad (1)$$
$$x = 40 - 16, \text{ or } 24$$

◀ *To find x, let y = 40 in Equation (1).*

**So,** Sylvia is 24, and Martin is 40. ◀ *Check on your own.*

## Written Exercises

**For each problem, find the age of each person now.**

1. Ralph's age is 3 times Peg's age. In 6 years, he will be twice as old as she will be then. **6, 18**

2. Irving is 8 years younger than Rhoda. Two years ago, she was 3 times as old as he was then. **14, 6**

3. Pedro is twice as old as Mona. Four years ago, he was 4 times as old as Mona was then. **6, 12**

4. Mr. Whitecloud is 24 years older than his son. Five years from now, he will be 3 times as old as his son will be then. **7, 31**

5. Andy is 40 years younger than Selma. In 10 years, Selma will be 3 times as old as Andy will be then. **50, 10**

6. Lee is 7 years older than Phil. Eleven years ago, Lee was twice as old as Phil was then. **18, 25**

7. Rudy is twice as old as Verna. In 4 years, he will be 8 years older than she will be then. **8, 16**

8. Lola is 6 times as old as Paul. Last year, she was 10 years older than he was then. **2, 12**

9. Mr. O'Leary is 3 times as old as his son. Fifteen years ago, he was 6 times as old as his son was then. **25, 75**

10. Mrs. Roth is 7 times as old as her daughter. In 5 years, she will be only 4 times as old as her daughter will be then. **5, 35**

11. Steve is 16 years older than Denise. In 3 years, the sum of their ages will be 32. **5, 21**

12. In 4 years, Nan will be 3 times as old as Dan will be then. The sum of their ages now is 56. **12, 44**

Age Problems

**Vocabulary**
constant linear function [13.2]
constant of variation [13.4]
direct variation [13.4]   domain [13.1]
function [13.1]   inverse variation [13.5]
linear function [13.2]   range [13.1]
relation [13.1]   vertical-line test [13.2]

List the ordered pairs and give the domain and the range of the relation. Is it a function? [13.1]
**For Ex. 1–3, see TE Answer Section.**

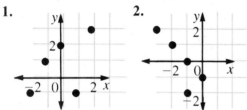

**1.**

function

**2.**

not a function

Graph. Give the domain and the range of the relation. Is it a function?

**3.** $\{(-3, -3), (-2, -2), (0, 0), (-2, 3), (-4, 4)\}$
**3.** $D = \{-3, -2, 0, -4\}$, $R = \{-3, -2, 0, 3, 4\}$, **no**

For what value(s) of $k$ will the relation *not* be a function?

**4.** $\{(2k - 9, -3), (-4k + 3, 6)\}$ **2**

**5.** $\{(10, 2), (k^2 - 3k, 4)\}$ **−2, 5**

**Complete answers for Ex. 6–11 are in TE Answer**
Graph. Which relations are functions? **Section.**
Which functions are linear functions? Which
functions are constant linear functions? [13.2]

**6.** $y = 3x$ **LF**        **7.** $x = 7$ **not a function**

**8.** $y - 3 = 4$ **CLF**    **9.** $y - (4 - x) = 2$

**10.** $8y - (6 - y) = 21$ **11.** $4(x - 4) = 2y$
        **9. LF   10. CLF   11. LF**
For each function, find the indicated value. [13.3]
**12.** $g = \{(4, 1), (-6, -2)\}$, $g(-6)$ **−2**
**13.** $f(x) = x^2 + 5$, $f(-5)$ **30**

For $f(x) = 2x - 1$ and $g(x) = x^2 + 2$, find
the indicated value.
★**14.** $f(4) + g(-1)$ **10** ★ **15.** $f[g(-3)]$ **21**

Find the range of each function for the given
domain. **16.** $\{-5, -13, -17\}$ **18.** $\left\{\frac{4}{7}, \frac{7}{8}, 20\right\}$
**16.** $f(x) = 4x - 5$    $D = \{0, -2, -3\}$
**17.** $f(x) = |x|$    $D = \{-5, -1, 6\}$ **{5, 1, 6}**
**18.** $f(x) = \dfrac{3x - 2}{x + 5}$    $D = \{2, 3, -6\}$

**19. directly, 6   20. neither   21. inversely, 40**
Determine if $y$ varies directly as $x$ or if $y$
varies inversely as $x$. If so, find the constant of
variation. [13.4, 13.5]

**19.**

| $x$ | $y$ |
|---|---|
| 1 | 6 |
| −2 | −12 |
| 3 | 18 |

**20.**

| $x$ | $y$ |
|---|---|
| −1 | 2 |
| 2 | −4 |
| −3 | 9 |

**21.**

| $x$ | $y$ |
|---|---|
| 4 | 10 |
| −8 | −5 |
| 20 | 2 |

Solve each problem.
**22.** $y$ varies directly as $x$. $y$ is $-8$ when $x$ is
4. Find $y$ when $x$ is $-6$. [13.4] **12**

**23.** The cost of a certain metal varies directly
as its mass. If 5 kg cost $15, find the cost
of 8 kg. **$24**

**24.** On a map, 150 km are represented by
6 cm. How many kilometers are
represented by 9 cm on the map? **225 km**

**25.** $y$ varies inversely as $x$. $y$ is 20 when $x$ is
$-4$. Find $y$ when $x$ is $-16$. [13.5] **5**

**26.** The length of a rectangle of constant area
varies inversely as the width. If the length
is 32 when the width is 2, find the width
when the length is 4. **16**

**27.** The time required to do a job varies
inversely as the number of people
working. If it takes 5 painters 9 days to
paint a house, how long will it take
3 painters to do the job? **15 days**

★ **28.** Two cylinders have the same volume.
Their heights vary inversely as the squares
of the length of the radii of the bases. One
cylinder's height is 4 m and base radius is
3 m. Find the height of the other cylinder
if its base radius is 2 m. **9 m**

# CHAPTER THIRTEEN TEST

(A) Exercises are 1–3, 6–11, 14, 16, 19, 20.   (C) Exercises are starred.   The rest are (B) Exercises.

Graph. Which relations are functions? Which functions are linear functions? Which functions are constant linear functions?  **2. not a function**

1. $y = 5$ **CLF**
2. $x - 7 = 15$
3. $x + y = 12$ **LF**
4. $7x - (4 - x) = 12$
5. $2(y - 6) = 10$ **CLF**   **4. not a function**

**For Ex. 1–5, see TE Answer Section.**

Determine if $y$ varies directly as $x$ or if $y$ varies inversely as $x$. If so, find the constant of variation.

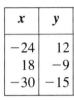

6.
| $x$ | $y$ |
|---|---|
| 9 | 12 |
| -36 | -3 |
| 54 | 2 |

7.
| $x$ | $y$ |
|---|---|
| -24 | 12 |
| 18 | -9 |
| -30 | -15 |

8.
| $x$ | $y$ |
|---|---|
| 64 | -32 |
| -80 | 40 |
| 48 | -24 |

inversely, 108        neither        directly, $-\frac{1}{2}$

List the ordered pairs and give the domain and the range of the relation. Is it a function?
**For Ex. 9–11, see TE Answer Section.**

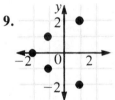

9.

10.

not a function        function

Graph. Give the domain and the range of the relation. Is it a function?
11. $\{(-6, -6), (-5, -5), (1, 1),$
$(8, 8), (4, 1)\}$
$D = \{-6, -5, 1, 4, 8\}, R = \{-6, -5, 1, 8\}$, **yes**

For what value(s) of $k$ will the relation *not* be a function?
12. $\{(3k - 7, 6), (-5k + 9, 2)\}$ **2**
13. $\{(2, 1), (k^2 - k, 0)\}$ **2, −1**

For each function, find the indicated value.
14. $g = \{(-1, 3), (2, 5), (-8, 4)\}, g(-8)$ **4**
15. $f(x) = x^2 - 6, f(-2)$ **−2**

Find the range of each function for the given domain.
16. $f(x) = -x^2 + 5$    $D = \{-2, 3, 2\}$
17. $f(x) = (x - 4)^2$    $D = \{-1, 0, 2\}$
18. $f(x) = \dfrac{x - 4}{x + 2}$    $D = \{2, 4, 5\}$

16. $\{1, -4\}$   17. $\{25, 16, 4\}$   18. $\left\{-\frac{1}{2}, 0, \frac{1}{7}\right\}$

Solve each problem. (Ex. 19–23)

19. $y$ varies directly as $x$. $y$ is 40 when $x$ is $-8$. Find $y$ when $x$ is $-3$. **15**

20. $y$ varies inversely as $x$. $y$ is 22 when $x$ is $-4$. Find $x$ when $y$ is $-11$. **8**

21. The cost of a certain metal varies directly as its mass. If 6 g cost \$9, find the cost of 15 g **\$22.50**

22. Jane has a mass of 45 kg and is sitting 2 m from the fulcrum of a seesaw. Juan has a mass of 50 kg. How far from the fulcrum must Juan sit to balance the seesaw? **1.8 m**

23. The current through a circuit varies inversely as the resistance. When the current is 25 amps, the resistance is 4 ohms. Find the current when the resistance is 5 ohm. **20 amps**

★ 24. If $f(x) = 2x - 5$ and $g(x) = x^2$, find $f[g(2)]$. **3**

Solve each problem.
★ 25. $y$ varies directly as the square of $x$. $y$ is 15 when $x$ is 4. Find $y$ when $x$ is 6. **33.75**

★ 26. The distance that a free-falling body falls varies directly as the square of the time it falls. A stone falls 66 m in 5 s. How far will it fall in 10 s? **264 m**

## Finding Domain and Range

For a function $f = \{(0, 1), (-4, 3), (2, -3)\}$ the domain of $f = \{0, -4, 2\}$ and the range of $f = \{1, 3, -3\}$.

For the function $f(x) = ax^2 + bx + c$, the computer can find values for the range. The domain and the coefficients $a$, $b$, and $c$ can be any real numbers.

This program gives values for the domain and range of a quadratic or linear function.

```
10   PRINT "THIS PROGRAM GIVES VALUES FOR THE DOMAIN
     AND RANGE."
15   REM  VALUES FOR A, B, AND C ARE IN THE DATA STAT
     EMENT.
20   FOR X = 1 TO 3
25   READ A,B,C
30   LET F = (A * X ^ 2) + (B * X) + C
35   PRINT X,F
40   NEXT X
45   PRINT "FIRST VALUES ARE VALUES FOR THE DOMAIN."
50   PRINT "SECOND VALUES ARE VALUES FOR THE RANGE."
55   DATA  0, 4, -3, 2, -3, -2, 3, -9, 0
60   END
```

See the Computer Section beginning on page 488 for more information.

## Exercises

Type in the program above and RUN it for the following equations:

1. $y = 4x - 3$
   domain: 1, range: 1
2. $y = 2x^2 - 3x - 2$
   domain: 2, range: 0
3. $y = 3x^2 - 9x$
   domain: 3, range: 0

Enter a new DATA statement in the program above using the following:
DATA 4, −3, 0, −4, 1, −1, 3. RUN the program with the new DATA line
to find the domain and range for the following equations: 55 DATA 4, −3, 0, −4, 1, −1, 3

4. $y = 3x^2 - 1$
   D = {1}; R = {2}
5. $y = x^2 + 4x$
   D = {2}; R = {12}
6. $y = 4x^2 - 3x + 1$
   D = {3}; R = {28}

Alter the program to have the values A, B, and C entered with an INPUT
statement instead of the READ . . . DATA pair. RUN the new program for
the following equations: For Ex. 7–10, see TE Answer Section.

7. $y = x^2 - x - 1$
   D = {1}; R = {−1}
8. $y = x^2 + 2x + 1$
   D = {2}; R = {9}
9. $y = 4x$
   D = {3}; R = {12}

10. Write a program to determine if a function varies directly. Use for the
    input the domain and range of each function.

# COLLEGE PREP TEST

Directions: Choose the one best answer for each question.

**1.** For the system $4y - x - 10 = 0$, $xy = \dfrac{?}{\quad}$
$3x = 2y$ **E**

(A) $\dfrac{2}{3}$     (B) 1     (C) $\dfrac{4}{3}$

(D) 5     (E) 6

**2.** A family owned $\dfrac{5}{8}$ of an interest in a house. They sold $\dfrac{1}{5}$ of their interest, at cost, for $10,000. What is the total value of the house? **D**

(A) $30,000    (B) $50,000    (C) $60,000

(D) $80,000    (E) $90,000

**3.** If a carload contains from 12 to 18 crates, what is the least number of crates in 4 carloads? **C**

(A) 24     (B) 36     (C) 48

(D) 60     (E) 72

**4.** There were 396 people in a theater. If the ratio of women to men was $2:3$ and the ratio of men to children was $1:2$, how many men were in the theater? **C**

(A) 36     (B) 72     (C) 108

(D) 132     (E) 198

**5.** If $4a - 3b = 6$ and $5a + 2b = 4$, then $18a - 2b = \dfrac{?}{\quad}$ **C**

(A) 2     (B) 10     (C) 20

(D) 30     (E) none of these

**6.** What is the next number in the sequence 5, 11, 23, 41, 65, . . .? **C**

(A) 71     (B) 89     (C) 95

(D) 106     (E) 131

**7.** A dinner check was to be divided equally among 8 people. When one refused to pay, each of the others had to pay 20¢ more. What was the total amount of the check? **B**

(A) $9.80    (B) $11.20    (C) $12.80

(D) $14.40    (E) $16.20

**8.** If $x$ is between 0 and 1, which of the following increases as $x$ increases? **C**

(A) $\dfrac{1}{x}$     (B) $1 - x^2$     (C) $x - 1$

(D) $1 - x$     (E) $\dfrac{1}{x^2}$

**9.** If the average of four numbers is 8, and three of the four numbers are 4, 7, and 9, what is the fourth number? **A**

(A) 12     (B) 7     (C) $6\dfrac{2}{3}$

(D) 5     (E) 2

**10.** Which fraction is closest to $\dfrac{1}{3}$? **B**

(A) $\dfrac{1}{2}$     (B) $\dfrac{3}{10}$     (C) $\dfrac{1}{4}$

(D) $\dfrac{4}{15}$     (E) $\dfrac{1}{30}$

**11.** The indicator of an oil tank showed that the tank was $\dfrac{1}{7}$ full. After a truck delivered 240 gal of oil, the indicator showed that the tank was $\dfrac{4}{7}$ full. What is the capacity of the tank in gallons? **A**

(A) 560     (B) 420     (C) 240

(D) 60     (E) none of these

# 14 RADICALS

Leonhard Euler's name is associated with every branch of mathematics. Several of his works are devoted to mathematical recreations. He also published extensively in areas of applied mathematics including hydraulics, shipbuilding, artillery, and theory of music.

In 1736, Euler resolved the famous Königsberg bridge problem. His method of solution led to the development of a new branch of mathematics, topology.

Many of the great mathematicians who came after Euler were deeply indebted to his intuitive and inventive genius.

**Project**

Research and report on the Königsberg bridge problem and Euler's solution of it. You may wish to extend your project by exploring topology.

# REPEATING DECIMALS AS FRACTIONS     14.1

*Objective*     **To write a repeating decimal as a fraction**

Recall that a rational number can be expressed in the form $\frac{a}{b}$ where $a$ and $b$ are integers and $b \neq 0$. Any rational number can be expressed as a terminating or repeating decimal. For example, in $\frac{3}{4} = 0.75$, the decimal terminates; in $\frac{3}{7} = 0.428571428571 \ldots$, or $0.\overline{428571}$, the decimal repeats. The examples below illustrate the technique for changing a repeating decimal to a fraction.

*Example 1*     **Find the fraction for the repeating decimal $0.\overline{6}$.**

Let $n = 0.666\ldots$ (1)     ◀ *$0.\overline{6} = 0.666\ldots$ because the bar indicates that the 6 repeats forever.*
  $10n = 10(0.666\ldots)$     ◀ *One digit repeats, so multiply each side by 10.*
  $10n = 6.666\ldots$ (2)     ◀ *Multiplying by 10 moves the decimal point one place to the right.*

  $\underline{1n = 0.666\ldots \quad (1)}$
  $9n = 6.000\ldots$     ◀ *Subtract Equation (1) from Equation (2).*
or $9n = 6$
  $n = \frac{6}{9}$, or $\frac{2}{3}$     ◀ *Check.* $3)\overline{2.000}$ $\;\; 0.666\ldots$

**So, $\frac{2}{3}$ is the fraction for $0.\overline{6}$.**

*Example 2*     **Find the fraction for the repeating decimal $1.8\overline{2}$.**

Let $n = 1.8222\ldots$ (1)     ◀ *The bar indicates that the 2 repeats forever.*
  $10n = 10(1.8222\ldots)$
  $10n = 18.222\ldots$ (2)
  $\underline{1n = 1.822\ldots \quad (1)}$
  $9n = 16.400$     ◀ *Subtract Equation (1) from Equation (2).*
or $9n = 16.4$
  $n = \frac{16.4}{9}$     ◀ *Multiply by $\frac{10}{10}$ to eliminate the decimal.*
  $n = \frac{164}{90}$     ◀ $\frac{16.4}{9} = \frac{16.4(10)}{9(10)} = \frac{164}{90}$
  $n = \frac{82}{45}$     ◀ *Write the fraction in simplest form.*

**So, $\frac{82}{45}$ is the fraction for $1.8\overline{2}$.**

When a repeating decimal contains a block of two or more repeating digits, the technique for finding the corresponding fraction is similar. However, a different multiplier will be used. Notice that the number of zeroes in the multiplier is the same as the number of repeating digits.

*Example 3*  **Find the fraction for $1.\overline{64}$.**

Let $n = 1.646464\ldots$ (1) ◀ *The digits 64 repeat.*
$100n = 100(1.646464\ldots)$ ◀ *Two digits repeat, so multiply each side by 100.*
$100n = 164.6464\ldots$ (2) ◀ *This moves the decimal point two places to the right.*
$\underline{1n = \quad 1.6464\ldots}$ (1)
$99n = 163.0000\ldots$ ◀ *Subtract Equation (1) from Equation (2).*
$99n = 163$
$n = \dfrac{163}{99}$

**So,** $\dfrac{163}{99}$ is the fraction for $1.\overline{64}$.

You have seen in this lesson that a repeating decimal can be written as a rational number or fraction. Recall, however, that a decimal that does *not* repeat or terminate is *not* rational. A decimal such as $3.454554555\ldots$ is **irrational**. In the next lesson you will learn about square roots; certain square roots result in irrational numbers.

## Written Exercises

**Find the fraction for each repeating decimal. (Ex. 1–23)**

(A) **1.** $0.\overline{3}$ $\frac{1}{3}$     **2.** $0.9\overline{4}$ $\frac{17}{18}$     **3.** $2.\overline{6}$ $\frac{8}{3}$     **4.** $0.7\overline{5}$ $\frac{34}{45}$     **5.** $0.\overline{9}$ $\frac{9}{9}$ **or 1**   **6.** $2.4\overline{3}$ $\frac{73}{30}$

**7.** $1.\overline{78}$ $\frac{59}{33}$     **8.** $0.\overline{62}$ $\frac{62}{99}$     **9.** $0.\overline{14}$ $\frac{14}{99}$     **10.** $2.\overline{75}$ $\frac{91}{33}$     **11.** $3.\overline{81}$ $\frac{42}{11}$     **12.** $12.\overline{56}$ $\frac{1,244}{99}$

(B) **13.** $0.\overline{143}$ $\frac{143}{999}$     **14.** $56.\overline{185}$ $\frac{56,129}{999}$     **15.** $2.8\overline{542}$ $\frac{28,540}{9,999}$     **16.** $13.\overline{242}$ $\frac{13,229}{999}$

**17.** $12.\overline{356}$ $\frac{12,344}{999}$     **18.** $49.\overline{3824}$ $\frac{493,775}{9,999}$     **19.** $15.\overline{3534}$ $\frac{51,173}{3,333}$     **20.** $204.\overline{449}$ $\frac{204,245}{999}$

(C) **21.** $2.3\overline{456}$ $\frac{23,222}{9,900}$     **22.** $12.7\overline{145}$ $\frac{6,993}{550}$     **23.** $109.35\overline{426}$ $\frac{3,641,497}{33,300}$

**24.** Give an illustration to show that the sum of two repeating decimals is the same as the sum of their fractions. [*Hint:* Use $0.\overline{3}$ and $0.\overline{6}$.]
**25.** What relationship exists among the repeating decimals for $\frac{1}{7}$, $\frac{2}{7}$, $\frac{3}{7}$, $\frac{4}{7}$, $\frac{5}{7}$, and $\frac{6}{7}$?

**24.** $0.\overline{3} = \frac{1}{3}$; $0.\overline{6} = \frac{2}{3}$; $0.\overline{3} + 0.\overline{6} = 0.\overline{9} = 1$; $\frac{1}{3} + \frac{2}{3} = 1$ **25.** **Each uses the digits 1, 4, 2, 8, 5, 7.**

## CUMULATIVE REVIEW

**1.** The larger of two numbers is twice the smaller. Find the numbers if their sum is 24. **16, 8**

**2.** The length of a rectangle is 4 m longer than the width. The perimeter is 56 m. Find the dimensions of the rectangle. **16, 12**

# FAMOUS MATHEMATICIANS

A Mathematician can be artist, scientist, engineer, inventor, or more simply, independent thinker—and is frequently more than one of these at once.

Sylvia Earle with her invention, "JIM," a deep-sea diving suit. The design of a sophisticated apparatus, such as "JIM," requires the engineering of many systems to ensure a safe environment under high pressure.

Cubi XIX. On a majestic scale, David Smith's "Cubi" is a series of sculptured geometric solids of stainless steel, each one so counterbalanced as to seem buoyant.

Louise Nevelson. Noted for her constructions of boxes and forms, Louise Nevelson's work is highly geometric.

Vasili Kandinsky, Russian postimpressionist painter. Kandinsky, one of the first abstract painters, used color and shape as symbols, just as mathematicians use operation signs and numerals for systems of numbers.

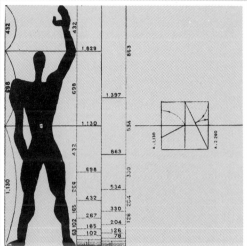

Le Corbusier's modular unit of
measurement based on the
human figure and mathematics.
Le Corbusier used his concept of
modular human beings to design
buildings based on anatomical
geometric proportions.

John Augustus Roebling.
A pioneer in the design of
suspension bridges, John Roebling
designed the Brooklyn Bridge in
New York City.

Leonardo da Vinci,
self-portrait in red chalk.
Da Vinci filled notebooks
with sketches showing
precise anatomical detail
and designs for inventions,
such as the helicopter,
that were inconceivable
for the fifteenth century.

Dr. Robert Goddard. Dr. Goddard's pioneering work with rockets—he first achieved an altitude of 41 ft and a range of 184 ft in 2½ sec—is directly responsible for the NASA space program.

John Wheeler, famous cosmologist. As an astrophysicist whose specialty is black hole physics, Wheeler's calculations are based on the minute variations of large-scale numbers obtained from astronomical observations.

Galileo. In the late sixteenth century, Galileo's invention of the first astronomical telescope provided proof of Copernicus' theory that the earth revolves about the sun.

Nan Davis, a paraplegic, stands with inventor Jerrold Petrofsky. Petrofsky was head of the research effort that enabled Ms. Davis to stand and walk with the aid of an experimental stimulation-feedback system.

Marie Curie. To identify radium, Marie Curie had to purify tons of pitchblende to extract one tenth of 1 g of the new element. She subsequently derived many of the properties of radioactivity from this effort.

Euclid teaching some of his students. Euclid's *Elements* are the oldest existing mathematical works. Thirteen volumes, written 2300 years ago, established arithmetic and geometric principles, and the procedure for formally stating a problem.

Nicolo Tartaglia. Tartaglia was the first to apply mathematics to the solution of artillery problems, and to describe the path of a projectile as curved rather than rectilinear.

Archimedes. Archimedes' principle states that a body immersed in a fluid is "buoyed up" by a force equal to the weight of the displaced fluid in a direction through the body's own center of gravity.

Ptolemy. A mathematical system of astronomy and geography where the sun, planets, and stars revolved around the earth was developed by Ptolemy. It was in use from the second until the sixteenth century, when Galileo's proof of the Copernican theory became accepted.

René Descartes. Referred to as the founder of analytical geometry, Descartes devised the Cartesian coordinates, and other important algebraic concepts.

The Mathematicians painting by Giorgio de Chirico. The symbolic mathematical and drafting tools, and mechanical objects that Chirico combined in his paintings, form a surreal collage against a background of the city of Turin, Italy.

Karl Friedrich Gauss, Mathematician. Gauss developed the modern theory of numbers, including quadratic forms, and was the first to represent the complex numbers in the form $a \pm bi$.

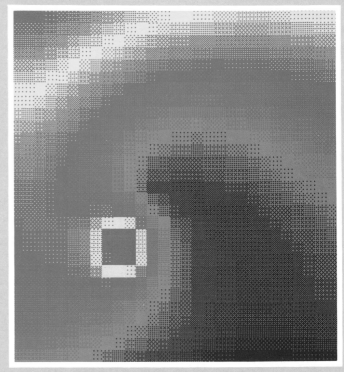

*Coordinated Colors—Cool to Calorific* by Miles Color Art. This colorful and unusual printout is an example of color graphics that can be generated on a computer by an algebraic function.

Sir Isaac Newton. The founder of differential and integral calculus, Isaac Newton also formulated the theory of gravitation.

*Electric Prism* by Sonia Terk-Delaunnay. The disk is a major recurrent form, symbolizing modernity in many of Sonia Terk-Delaunnay's works.

# THE SQUARE ROOT OF A NUMBER          14.2

*Objectives*  **To find the principal square root of a perfect square**
**To find approximations of square roots in a square root table**
**To solve area problems that lead to the use of square roots**

Recall that the inverse of an operation "undoes" the operation. For example, the inverse of addition is subtraction. The opposite, or inverse, of squaring a number is finding the *square root* of the number.

The square of 5 is $5^2$, or 25, and the square of $-5$ is $(-5)^2$, or 25. Therefore, the square roots of 25 are 5 and $-5$. Every positive real number has *two* square roots, one positive and one negative.

To find the two square roots of 49, you must find the numbers which when squared give 49. Since $7 \cdot 7$ is 49 and $-7 \cdot -7$ is 49, the square roots of 49 are 7 and $-7$.

| Definition: Principal square root | The **principal square root** of a positive real number is the positive square root of the number. The principal square root is indicated by the symbol $\sqrt{\phantom{x}}$, which is called a **radical sign.** |
|---|---|

The expressions $\sqrt{9} = 3$ and $-\sqrt{9} = -3$ indicate the principal square root of 9 and the negative square root of 9, respectively. The **radicand** is the expression within the radical sign. The radicand is 9 in the expression $\sqrt{9}$.

Notice that since $\sqrt{\phantom{x}}$ indicates principal square root, which is positive, then
$$\sqrt{9} = \sqrt{3^2} = 3 \text{ and } \sqrt{9} = \sqrt{(-3)^2} = 3 \text{ or } |-3|.$$

In general, $\sqrt{a^2} = |a|$. In the remainder of this text it will be assumed that $a \geq 0$ for an expression like $\sqrt{a^2}$. This eliminates the need to write the absolute value each time. Instead of $\sqrt{a^2} = |a|$ you can write $\sqrt{a^2} = a$.

*Example 1*  **Simplify.**

$$\sqrt{0.16} \qquad\qquad \sqrt{\frac{4}{25}}$$

$\sqrt{0.16} = 0.4$ because $(0.4)^2 = 0.16$. $\qquad$ $\sqrt{\frac{4}{25}} = \frac{2}{5}$ because $\left(\frac{2}{5}\right)^2 = \frac{4}{25}$.

The Square Root of a Number                                    **391**

Numbers such as 16 and 0.49 are *perfect squares*. This is true because $\sqrt{16} = 4$ and $\sqrt{0.49} = 0.7$ and both 4 and 0.7 are rational numbers.

| Definition:<br>Perfect<br>square | A **perfect square** is a rational number whose principal square root is a rational number. |
|---|---|

The number 8 is *not* a perfect square because there is no rational number whose square is exactly 8. There are some rational numbers whose squares are very close to 8, as shown below.

$$(2.8)^2 = 7.84$$
$$(2.82)^2 = 7.9524$$
$$(2.828)^2 = 7.997584$$

Since there is no rational number whose square is exactly 8, 8 is an *irrational number*. This statement can be proved by using advanced mathematics techniques.

The table on page 532 contains the square roots of the numbers from 1 to 100. If a number is irrational, the table gives the approximate square root correct to three decimal places.

*Example 2*  **Find an approximation of $\sqrt{8}$ by using the square root table. A portion of the table is shown below.**

*Read: number, square, square root* ▶

*Read down the column.* ▶

*Read across the column.* ▶

| Table of Roots and Powers | | |
|---|---|---|
| **No.** | **Sq.** | **Sq. Root** |
| 1 | 1 | 1.000 |
| 2 | 4 | 1.414 |
| . | . | . |
| . | . | . |
| 7 | 49 | 2.646 |
| 8 | 64 | 2.828 |

**So, $\sqrt{8} \doteq 2.828$.**  ◀  $\doteq$ *means "is approximately equal to."*

Many problems involving geometric situations, such as area, include applications of square roots. This is illustrated in Example 3 on the next page.

## Example 3

The area of a rectangle is 24 m². The length is three times the width. Find the length and the width of the rectangle, correct to the nearest tenth of a meter.

*Represent the data.* ▶  Let $x$ = the width of the rectangle in meters.
Let $3x$ = the length of the rectangle in meters.

*Formula for the area of a rectangle* ▶

$$l \cdot w = A$$
$$3x \cdot x = 24$$
$$3x^2 = 24$$
$$x^2 = 8$$
$$x = \sqrt{8}$$
$$x \doteq 2.828 \quad ◀ \text{ From the table, } \sqrt{8} \doteq 2.828.$$

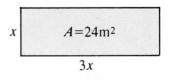

The width, $x$, $\doteq 2.828$; the length, $3x$, $\doteq 3(2.828) \doteq 8.484$.
**So,** the length is 8.5 m, and the width is      ◀ *Round 8.4$\boxed{8}$4 to 8.5, and*
2.8 m, correct to the nearest tenth of a meter.      *2.8$\boxed{2}$8 to 2.8.*

## Oral Exercises

**Is the number rational or irrational? Give a reason for your answer.**

**1.** $\sqrt{21}$ I   **2.** $\sqrt{64}$ R   **3.** $\sqrt{0.81}$ R   **4.** $\sqrt{1.44}$ R   **5.** $\sqrt{17}$ I   **6.** $\sqrt{12.1}$ I

1., 5., 6. **No rational square root.**   2. $\sqrt{64} = 8$   3. $\sqrt{0.81} = 0.9$   4. $\sqrt{1.44} = 1.2$

## Written Exercises

**Find the indicated principal square root.**

Ⓐ **1.** $\sqrt{25}$ 5   **2.** $\sqrt{81}$ 9   **3.** $\sqrt{\dfrac{100}{9}}$ $\dfrac{10}{3}$   **4.** $\sqrt{\dfrac{25}{16}}$ $\dfrac{5}{4}$   **5.** $\sqrt{\dfrac{1}{36}}$ $\dfrac{1}{6}$   **6.** $\sqrt{\dfrac{4}{81}}$ $\dfrac{2}{9}$

**Find an approximation of the indicated principal square root. Use the table on page 532.**

**7.** $\sqrt{95}$ 9.747   **8.** $\sqrt{33}$ 5.745   **9.** $\sqrt{26}$ 5.099   **10.** $\sqrt{57}$ 7.550   **11.** $\sqrt{66}$ 8.124   **12.** $\sqrt{72}$ 8.485

**For each square with the given area, find the length of a side of the square, correct to the nearest tenth of a unit.**

**13.** $A = 89$ cm²   **14.** $A = 62$ cm²   **15.** $A = 99$ cm²   **16.** $A = 17$ cm²   **17.** $A = 14$ cm²
      9.4 cm          7.9 cm          9.9 cm          4.1 cm          3.7 cm

**Solve each problem. Round answers to the nearest tenth of a unit.**

Ⓑ **18.** The length of a rectangle is three times the width. The area is 330 m². Find the length and the width of the rectangle. **31.5 m, 10.5 m**

**19.** The area of a rectangle is 156 m². The length is twice the width. Find the length and the width of the rectangle. **17.7 m, 8.8 m**

**20.** Find $r$ if $A = 25.12$ m²
[*Hint:* $A = \pi r^2$ and $\pi = 3.14$.] **2.8 m**

**21.** Find the height, $h$, of the triangle.
[*Hint:* $A = \frac{1}{2}bh$.]
      **17.0 m**

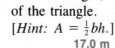

The Square Root of a Number

# APPROXIMATING SQUARE ROOTS          14.3

*Objective*          **To approximate the square root of a number by using the divide-and-average method**

The divide and average technique is one method that can be used to approximate the square root of a whole number that is not a perfect square.

*Example*          **Approximate $\sqrt{58}$ to the nearest tenth.**

*Step 1*          Estimate the square root.
Since $7 \cdot 7 = 49$ and $8 \cdot 8 = 64$, $\sqrt{58}$ is between 7 and 8. Estimate 7.5.

*Step 2*          Divide the estimated number, 7.5, into 58. Carry the division
to hundredths, but do not round.

$$
\begin{array}{r}
7.73 \\
7.5\overline{)58.0.00} \\
\underline{52\ 5} \\
5\ 5\ 0 \\
\underline{5\ 2\ 5} \\
2\ 50 \\
\underline{2\ 25} \\
25
\end{array}
$$

◄ *$\sqrt{58}$ is between 7.5 and 7.73, which is written*
*$7.5 < \sqrt{58} < 7.73$.*

*Step 3*          Average the divisor and quotient. Round to the nearest tenth.

$$
\begin{array}{r}
7.5 \\
+\ 7.73 \\
\hline
15.23
\end{array}
\qquad
\begin{array}{r}
7.61 = 7.6. \\
2\overline{)15.23}
\end{array}
$$
◄ *The average is 7.6.*

Check to see if the average is the same as the previous divisor. No, $7.6 \neq 7.5$.

*Step 4*          Divide the average, 7.6, into the number. Carry the division
to hundredths, but do not round.

$$
\begin{array}{r}
7.63 \\
7.6\overline{)58.0.00} \\
\underline{53\ 2} \\
4\ 8\ 0 \\
\underline{4\ 5\ 6} \\
2\ 40 \\
\underline{2\ 28} \\
12
\end{array}
$$

This time, the *divisor* and the *quotient* are the same in the tenths place:
$7.6 = 7.6$.     **So, $\sqrt{58} \doteq 7.6$ to the nearest tenth.**

If the divisor and the quotient had not been the same in the tenths place in
Step 4, you would have had to repeat Steps 3 and 4.

*Summary*

**The Divide-and-Average Method for Approximating Square Roots**
The basic steps for approximating a square root to the nearest tenth are:
Step 1    Estimate the square root.
Step 2    Divide the number by this estimate. Carry the division to hundredths, but do not round the result.
Step 3    Find the average of the divisor and the quotient. Round the result to the nearest tenth.
Step 4    Divide the number by this average. Carry the division to hundredths, but do not round the result.

Repeat Steps 3 and 4 until either:
(1) the average and the divisor are the same in Step 3, or
(2) the divisor and the quotient are the same in the tenths place in Step 4.

## Oral Exercises

**Between which two consecutive whole numbers is the indicated square root?**

1. $\sqrt{85}$   2. $\sqrt{27}$   3. $\sqrt{51}$   4. $\sqrt{34}$   5. $\sqrt{86}$   6. $\sqrt{48}$
   9 and 10      5 and 6        7 and 8        5 and 6        9 and 10       6 and 7

## Written Exercises

**Approximate to the nearest tenth. Use the divide-and-average method. (Ex. 1–31)**

(A)  1. $\sqrt{34}$ 5.8   2. $\sqrt{86}$ 9.3   3. $\sqrt{63}$ 7.9   4. $\sqrt{19}$ 4.4   5. $\sqrt{61}$ 7.8   6. $\sqrt{27}$ 5.2   7. $\sqrt{88}$ 9.4
8. $\sqrt{53}$ 7.3   9. $\sqrt{6}$ 2.4   10. $\sqrt{38}$ 6.2   11. $\sqrt{77}$ 8.8   12. $\sqrt{32}$ 5.7   13. $\sqrt{83}$ 9.1   14. $\sqrt{65}$ 8.1
15. $\sqrt{98}$ 9.9   16. $\sqrt{47}$ 6.9   17. $\sqrt{12}$ 3.5   18. $\sqrt{28}$ 5.3   19. $\sqrt{80}$ 8.9   20. $\sqrt{55}$ 7.4   21. $\sqrt{44}$ 6.6

(B)  22. $\sqrt{165}$ 12.8   23. $\sqrt{112}$ 10.6   24. $\sqrt{300}$ 17.3   25. $\sqrt{128}$ 11.3   26. $\sqrt{150}$ 12.2
27. $\sqrt{365}$ 19.1   28. $\sqrt{142}$ 11.9   29. $\sqrt{449}$ 21.2   30. $\sqrt{579}$ 24.1   31. $\sqrt{333}$ 18.2

(C)  32. What changes must be made in the four-step process described above so that the square root of a whole number can be approximated to the nearest hundredth? **In step 2 carry out division to 1,000ths; in step 3, round to 100ths; in step 4 carry out division to 1,000ths; stop when ave. = divisor.**

**Approximate to the nearest hundredth. Use the divide-and-average method.**
33. $\sqrt{79}$ 8.89   34. $\sqrt{275}$ 16.58   35. $\sqrt{473}$ 21.75   36. $\sqrt{1,152}$ 33.94

# CALCULATOR ACTIVITIES

**Estimate to the nearest whole number. Check your answers with a calculator.**
1. $\sqrt{219}$ 15   2. $\sqrt{758}$ 28   3. $\sqrt{2,341}$ 48   4. $\sqrt{2,000}$ 45   5. $\sqrt{3,096}$ 56

# SIMPLIFYING SQUARE ROOTS 14.4

*Objectives*      **To simplify an expression involving square roots**
**To simplify an expression involving square roots and then
to approximate the result to the nearest tenth**

By studying the illustration below, you should recognize an important property that is necessary for simplifying square roots.

| $\sqrt{25 \cdot 4}$ | $\sqrt{25} \cdot \sqrt{4}$ |
|---|---|
| $\sqrt{100}$ | $5 \cdot 2$ |
| 10 | 10 |

Notice that the results are the same.
The illustration suggests that the square root of the product of two numbers is the product of their square roots.

| **Product property for square roots** | For all numbers $a \geq 0$ and $b \geq 0$, $\sqrt{a \cdot b} = \sqrt{a} \cdot \sqrt{b}$ and $\sqrt{a} \cdot \sqrt{b} = \sqrt{a \cdot b}$. |
|---|---|

*Example 1*      **Simplify $\sqrt{3} \cdot \sqrt{7}$.**

$$\sqrt{3} \cdot \sqrt{7} = \sqrt{21} \quad \blacktriangleleft \quad \sqrt{a} \cdot \sqrt{b} = \sqrt{a \cdot b}$$

Another important property concerns the product of two square roots in which the radicands are the same.

*Example 2*      **Simplify $\sqrt{6} \cdot \sqrt{6}$ and $\sqrt{9} \cdot \sqrt{9}$.**

$$\sqrt{6} \cdot \sqrt{6} = \sqrt{6 \cdot 6} \qquad \sqrt{9} \cdot \sqrt{9} = \sqrt{9 \cdot 9}$$
$$= \sqrt{36} \qquad\qquad = \sqrt{81}$$
$$= 6 \qquad\qquad\quad = 9$$

**So,** $\sqrt{6} \cdot \sqrt{6} = 6$ and $\sqrt{9} \cdot \sqrt{9} = 9$.

Example 2 suggests that the product of two square roots with the same radicand is that radicand without a radical sign.

| **Product property for square roots of the same number** | For each number $a \geq 0$, $\sqrt{a^2} = \sqrt{a} \cdot \sqrt{a} = a$. |
|---|---|

You can use $\sqrt{a \cdot b} = \sqrt{a} \cdot \sqrt{b}$ and $\sqrt{a} \cdot \sqrt{a} = a$ to simplify square roots.

## Example 3

**Simplify $\sqrt{49 \cdot 4}$.**

$$\sqrt{49 \cdot 4} = \sqrt{49} \cdot \sqrt{4} = 7 \cdot 2 = 14$$

A square root is simplified when the radicand does not contain a factor that is a perfect square. Two methods for simplifying a square root in which the radicand is not a perfect square are shown in Example 4.

## Example 4

**Simplify $\sqrt{48}$.**

**One way: Factor 48 into primes.**

$$\sqrt{48} = \sqrt{4 \cdot 12}$$
$$= \sqrt{2 \cdot 2 \cdot 2 \cdot 2 \cdot 3}$$
$$= \underbrace{\sqrt{2} \cdot \sqrt{2}} \cdot \underbrace{\sqrt{2} \cdot \sqrt{2}} \cdot \sqrt{3} \quad \blacktriangleleft \text{ } Group \text{ } pairs \text{ } of$$
$$= \quad 2 \quad \cdot \quad 2 \quad \cdot \sqrt{3} \quad \quad \quad the \text{ } same \text{ } factors.$$
$$= 4\sqrt{3} \quad \blacktriangleleft \text{ } The \text{ } result \text{ } is \text{ } in \text{ } simplest \text{ } form,$$
$$for \text{ } the \text{ } radicand \text{ } 3 \text{ } does \text{ } not$$
$$have \text{ } a \text{ } perfect \text{ } square \text{ } factor.$$

**Another way: Find the greatest perfect square factor.**

$$\sqrt{48} = \sqrt{16 \cdot 3}$$
$$= \sqrt{16} \cdot \sqrt{3}$$
$$= 4\sqrt{3} \quad \blacktriangleleft \text{ } Simplest \text{ } radical$$
$$form$$

In applications involving square roots, it is often desirable to give the results in decimal form. To do this, you would first simplify the radical and then approximate the square root.

## Example 5

**Simplify $3\sqrt{60}$. Then, approximate the result to the nearest tenth.**

**One way**

$$3\sqrt{60} = 3\sqrt{10 \cdot 6}$$
$$3\sqrt{5 \cdot 2 \cdot 3 \cdot 2} \quad \blacktriangleleft \text{ } Factor \text{ } into \text{ } primes.$$
$$= 3\sqrt{5} \cdot \sqrt{2} \cdot \sqrt{3} \cdot \sqrt{2}$$
$$= 3\underbrace{\sqrt{2} \cdot \sqrt{2}} \cdot \sqrt{5} \cdot \sqrt{3} \quad \blacktriangleleft \text{ } Group$$
$$like \text{ } factors.$$
$$= 3 \cdot \underbrace{2} \quad \cdot \sqrt{5} \cdot \sqrt{3}$$
$$= \quad 6 \quad \sqrt{15} \quad \blacktriangleleft \text{ } Simplest \text{ } radical \text{ } form$$

**Another way**    *Find the greatest*

$$3\sqrt{60} = 3\sqrt{4 \cdot 15} \quad \blacktriangleleft \text{ } perfect \text{ } square$$
$$factor.$$
$$= 3\sqrt{4} \cdot \sqrt{15}$$
$$= 3 \cdot 2 \cdot \sqrt{15}$$
$$= 6\sqrt{15}$$

Now, use the table to approximate $6\sqrt{15}$ to the nearest tenth.
$$6\sqrt{15} \doteq 6(3.873)$$
$$\doteq 23.238, \text{ or } 23.2 \quad \blacktriangleleft \text{ } Round \text{ } to \text{ } the \text{ } nearest \text{ } tenth.$$
**So, $3\sqrt{60} = 6\sqrt{15} \doteq 23.2$.**

# Reading in Algebra

**For each expression, which of the results below it have the same value as the expression? If the value is not the same, tell why it is not.**

**1.** $\sqrt{18}$ **a, b**

  **a.** $\sqrt{10 + 8}$
  **b.** $\sqrt{3} \cdot \sqrt{3} \cdot \sqrt{2}$   **c.** $9 \cdot 2$
  **d.** $3\sqrt{6}$   **e.** $3\sqrt{3} \cdot \sqrt{2}$

**2.** $3\sqrt{8}$ **d**

  **a.** $\sqrt{24}$   **b.** $3 \cdot 2 \cdot 2 \cdot 2$
  **c.** $3 \cdot 4$   **d.** $3\sqrt{2} \cdot \sqrt{2} \cdot \sqrt{2}$
  **e.** $5\sqrt{2}$

**3.** $\sqrt{x \cdot y}$ **a, c**

  **a.** $\sqrt{y} \cdot \sqrt{x}$   **b.** $x \cdot y$
  **c.** $\sqrt{x} \cdot \sqrt{y}$   **d.** $x\sqrt{y}$
  **e.** $\sqrt{x} \cdot \sqrt{x} \cdot \sqrt{y} \cdot \sqrt{y}$

# Written Exercises

**Simplify.**

(A)
**1.** $\sqrt{3} \cdot \sqrt{5}$ $\sqrt{15}$   **2.** $\sqrt{5} \cdot \sqrt{2}$ $\sqrt{10}$   **3.** $\sqrt{7} \cdot \sqrt{11}$ $\sqrt{77}$   **4.** $\sqrt{13} \cdot \sqrt{3}$ $\sqrt{39}$   **5.** $\sqrt{11} \cdot \sqrt{11}$ **11**

**6.** $\sqrt{9 \cdot 4}$ **6**   **7.** $\sqrt{25 \cdot 4}$ **10**   **8.** $\sqrt{36 \cdot 49}$ **42**   **9.** $\sqrt{9 \cdot 25}$ **15**   **10.** $\sqrt{121 \cdot 81}$ **99**

**11.** $\sqrt{8}$ $2\sqrt{2}$   **12.** $\sqrt{20}$ $2\sqrt{5}$   **13.** $\sqrt{12}$ $2\sqrt{3}$   **14.** $\sqrt{28}$ $2\sqrt{7}$   **15.** $\sqrt{50}$ $5\sqrt{2}$

**16.** $\sqrt{40}$ $2\sqrt{10}$   **17.** $\sqrt{44}$ $2\sqrt{11}$   **18.** $\sqrt{27}$ $3\sqrt{3}$   **19.** $\sqrt{45}$ $3\sqrt{5}$   **20.** $\sqrt{32}$ $4\sqrt{2}$

**21.** $2\sqrt{24}$ $4\sqrt{6}$   **22.** $5\sqrt{90}$ $15\sqrt{10}$   **23.** $3\sqrt{80}$ $12\sqrt{5}$   **24.** $7\sqrt{54}$ $21\sqrt{6}$   **25.** $5\sqrt{72}$ $30\sqrt{2}$

**26.** $2\sqrt{17}$; 8.2   **27.** $8\sqrt{30}$; 43.8   **28.** $18\sqrt{3}$; 31.2   **29.** $50\sqrt{2}$; 70.7   **30.** $12\sqrt{33}$; 68.9

**Simplify. Then, approximate the result to the nearest tenth.**

(B)
**26.** $\sqrt{68}$   **27.** $4\sqrt{120}$   **28.** $3\sqrt{108}$   **29.** $5\sqrt{200}$   **30.** $6\sqrt{132}$

**31.** $-3\sqrt{180}$   **32.** $2\sqrt{160}$   **33.** $5\sqrt{240}$   **34.** $4\sqrt{400}$   **35.** $-2\sqrt{250}$

**31.** $-18\sqrt{5}$; $-40.2$   **32.** $8\sqrt{10}$; 25.3   **33.** $20\sqrt{15}$; 77.5   **34.** 80   **35.** $-10\sqrt{10}$; $-31.6$

**Simplify.**

(C)
**36.** $\sqrt{675}$   **37.** $\sqrt{686}$   **38.** $7\sqrt{2,016}$   **39.** $2\sqrt{3,179}$   **40.** $-6\sqrt{3,825}$

**36.** $15\sqrt{3}$   **37.** $7\sqrt{14}$   **38.** $84\sqrt{14}$   **39.** $34\sqrt{11}$   **40.** $-90\sqrt{17}$

**41.** Simplify $\sqrt{16 + 9}$ and $\sqrt{16} + \sqrt{9}$. What conclusion can you draw about the relationship between $\sqrt{a + b}$ and $\sqrt{a} + \sqrt{b}$. **5, 7; they are not equal**

# CUMULATIVE REVIEW

**1.** A cash box contained $3.75 in dimes and nickels. The number of dimes was twice the number of nickels. How many coins of each type were in the box?

**30 dimes, 15 nickels**

**2.** The perimeter of a rectangle is 88 cm. The length of the rectangle is 4 cm more than the width of the rectangle. Find the length and the width of the rectangle. **24 cm, 20 cm**

## NON-ROUTINE PROBLEMS

The Henrys started a trip between 8:00 A.M. and 9:00 A.M. when the hands of a clock were together. They arrived at their destination between 2:00 P.M. and 3:00 P.M. when the hands of the clock were exactly 180° apart. How long did the trip take? **6 h**

# RADICALS WITH VARIABLES 14.5

*Objectives*  **To evaluate and simplify square roots with variables in the radicand**
**To find the values of a variable for which a radical expression is defined**
**To simplify square roots in which the radicands contain variables with even exponents**

Recall that $(-4)^2 = 16$ and $4^2 = 16$. In fact, the square of any real number, except 0, is always positive. This suggests that the *radicand* can never be a negative number if its square root is to be a real number. For example, $\sqrt{-25}$ is neither 5 nor $-5$ because $5^2 = 25$, *not* $-25$, and $(-5)^2 = -5 \cdot -5 = 25$, *not* $-25$. Therefore, $\sqrt{-25}$ is not a real number.

| | |
|---|---|
| The values of $a$ for which $\sqrt{a}$ is a real number | $\sqrt{a}$ is a real number *if and only if* $a \geq 0$. <br><br> The square root of a negative number is not a real number. |

*Example 1*  **Evaluate $\sqrt{3a - 15}$ for $a = 2$ and $a = 9$. Simplify, if possible.**

$\sqrt{3a - 15}$
$\sqrt{3 \cdot 2 - 15}$  ◀ *Substitute 2 for a.*
$\sqrt{6 - 15}$
$\sqrt{-9}$
$\sqrt{-9}$ is not a real number.

$\sqrt{3a - 15}$
$\sqrt{3 \cdot 9 - 15}$
$\sqrt{27 - 15}$
$\sqrt{12}$
$\underbrace{\sqrt{2} \cdot \sqrt{2}} \cdot \sqrt{3}$
$2\sqrt{3}$
$2\sqrt{3}$ is a real number.

You can find the values of $x$ that will make an expression such as $\sqrt{4x - 8}$ a real number by finding the values of $x$ that will make the radicand, $4x - 8$, positive or zero.

*Example 2*  **For what values of $x$ will $\sqrt{4x - 8}$ be a real number?**

Solve $4x - 8 \geq 0$.  ◀ *The radicand **cannot be** negative.*
$\qquad 4x \geq 8$  $\qquad$ *It **must be** positive or zero.*
$\qquad\quad x \geq 2$

**So,** $\sqrt{4x - 8}$ will be a real number if $x \geq 2$.

You can simplify radicals containing variables by using the technique of factoring and the product property of powers: $a^m \cdot a^n = a^{m+n}$. Using this property, $a^7 \cdot a^3 = a^{7+3}$ or $a^{10}$. In reverse, an expression such as $a^{10}$ can be written in a variety of ways: $a^{10} = a^7 \cdot a^3$, or $a^8 \cdot a^2$, or $a^6 \cdot a^4$, or $a^5 \cdot a^5$, and so on. An expression such as $\sqrt{x^{10}}$ can be simplified by factoring $x^{10}$ into equal powers of $x$: $\sqrt{x^{10}} = \sqrt{x^5 \cdot x^5} = \sqrt{x^5} \cdot \sqrt{x^5}$, or $x^5$.

*Example 3*    **Simplify $\sqrt{a^8 b^{14}}$.**

$$
\begin{aligned}
\sqrt{a^8 b^{14}} &= \sqrt{a^8} \cdot \sqrt{b^{14}} \\
&= \sqrt{a^4 \cdot a^4} \cdot \sqrt{b^7 \cdot b^7} \\
&= \underbrace{\sqrt{a^4} \cdot \sqrt{a^4}} \cdot \underbrace{\sqrt{b^7} \cdot \sqrt{b^7}} \\
&= \qquad a^4 \qquad \cdot \qquad b^7 \qquad \text{, or } a^4 b^7
\end{aligned}
$$

From Example 3, you can see that a short-cut method exists for simplifying a square root when its variables have even exponents. The pattern is further illustrated by:

$$\sqrt{a^{20}} = a^{10}, \quad \sqrt{y^{16}} = y^8, \quad \text{and} \quad \sqrt{x^4 y^{12}} = \sqrt{x^4} \cdot \sqrt{y^{12}} = x^2 y^6.$$

| Square root property of even exponents | If $x$ is an even number, $\sqrt{a^x} = a^{\frac{x}{2}}$. |
| --- | --- |

*Example 4*    **Simplify $\sqrt{a^8 b^4 c^6}$ and $-\sqrt{49 x^{10} y^2}$.**

$$
\begin{aligned}
\sqrt{a^8 b^4 c^6} &= \sqrt{a^8} \cdot \sqrt{b^4} \cdot \sqrt{c^6} \\
&= a^{\frac{8}{2}} \cdot b^{\frac{4}{2}} \cdot c^{\frac{6}{2}} \\
&= a^4 b^2 c^3
\end{aligned}
\qquad
\begin{aligned}
-\sqrt{49 x^{10} y^2} &= -\sqrt{49} \cdot \sqrt{x^{10}} \cdot \sqrt{y^2} \\
&= -7 \cdot x^{\frac{10}{2}} \cdot y^{\frac{2}{2}} \\
&= -7 x^5 y
\end{aligned}
$$

If the coefficient of the variable in the radicand is not a perfect square, then you can use one of the methods presented in the last lesson.

*Example 5*    **Simplify $\sqrt{18 x^{16} y^{10}}$.**

**One way**

$$
\begin{aligned}
\sqrt{18 x^{16} y^{10}} &= \sqrt{18} \cdot \sqrt{x^{16}} \cdot \sqrt{y^{10}} \\
&= \sqrt{3 \cdot 3 \cdot 2} \cdot \sqrt{x^{16}} \cdot \sqrt{y^{10}} \\
&= \sqrt{3} \cdot \sqrt{3} \cdot \sqrt{2} \cdot \sqrt{x^{16}} \cdot \sqrt{y^{10}} \\
&= 3 \cdot \sqrt{2} \cdot x^8 \cdot y^5 \\
&= 3 x^8 y^5 \sqrt{2}
\end{aligned}
$$

**Another way**

$$
\begin{aligned}
\sqrt{18 x^{16} y^{10}} &= \sqrt{18} \cdot \sqrt{x^{16}} \cdot \sqrt{y^{10}} \\
&= \sqrt{9 \cdot 2} \cdot \sqrt{x^{16}} \cdot \sqrt{y^{10}} \\
&= \sqrt{9} \cdot \sqrt{2} \cdot x^8 \cdot y^5 \\
&= 3 \cdot \sqrt{2} \cdot x^8 \cdot y^5 \\
&= 3 x^8 y^5 \sqrt{2}
\end{aligned}
$$

## Oral Exercises

**Simplify, if possible.**

**1.** $\sqrt{9}$ **3**　　**2.** $\sqrt{-9}$　　**3.** $-\sqrt{9}$ **−3**　**4.** $\sqrt{49}$ **7**　　**5.** $\sqrt{-49}$　　**6.** $-\sqrt{36}$ **−6**

not a real number　　　　　　　　　　not a real number

## Written Exercises

**Evaluate for the given value of the variable. Simplify, if possible.**

Ⓐ　**1.** $\sqrt{x-5}$, $x = 14$　　**2.** $\sqrt{x+16}$, $x = 9$　　**3.** $\sqrt{2x-15}$, $x = 3$　　**4.** $\sqrt{5x-9}$, $x = -8$
　　　　**3**　　　　　　　　　　**5**　　　　　　$\sqrt{-9}$ not a real number　　$\sqrt{-49}$ not a real number

**For what values of $x$ will the radical be a real number?**

**5.** $\sqrt{2x-10}$ $x \geq 5$　　**6.** $\sqrt{6x+24}$ $x \geq -4$　**7.** $\sqrt{3x-7}$ $x \geq \frac{7}{3}$　　**8.** $\sqrt{-3x-15}$

　　　　　　　　　　　　　　　　　　　　　　　　　　**8.** $x \leq -5$

**Simplify.**

**9.** $\sqrt{a^4}$ $a^2$　　　　　　　**10.** $\sqrt{x^8}$ $x^4$　　　　　　**11.** $\sqrt{a^{24}}$ $a^{12}$

**12.** $-\sqrt{4a^6}$ $-2a^3$　　　**13.** $\sqrt{49y^{12}}$ $7y^6$　　　**14.** $\sqrt{16b^{10}}$ $4b^5$

**15.** $\sqrt{9b^6c^2}$ $3b^3c$　　　**16.** $\sqrt{25x^4y^{10}}$ $5x^2y^5$　　**17.** $-\sqrt{4c^{12}d^6}$ $-2c^6d^3$

**18.** $-\sqrt{16x^2y^{12}}$ $-4xy^6$　**19.** $\sqrt{49x^{14}y^2}$ $7x^7y$　　**20.** $-\sqrt{36c^{10}d^{18}}$ $-6c^5d^9$

**21.** $\sqrt{a^6b^4c^{10}}$ $a^3b^2c^5$　**22.** $-\sqrt{x^8y^2z^{20}}$ $-x^4yz^{10}$　**23.** $\sqrt{c^4m^{12}p^{14}}$ $c^2m^6p^7$

**24.** $-\sqrt{100a^2b^8c^{10}}$ $-10ab^4c^5$ **25.** $\sqrt{81x^{12}y^2z^{30}}$ $9x^6yz^{15}$　**26.** $\sqrt{100x^4y^4z^2}$ $10x^2y^2z$

**27.** $\sqrt{144a^4b^{12}c^{24}}$ $12a^2b^6c^{12}$　**28.** $-\sqrt{169a^{18}b^{16}c^2}$ $-13a^9b^8c$ **29.** $\sqrt{121x^{30}y^{24}z^{26}}$ $11x^{15}y^{12}z^{13}$

Ⓑ　**30.** $\sqrt{18x^6y^{16}}$ $3x^3y^8\sqrt{2}$　　**31.** $\sqrt{12x^4y^{10}}$ $2x^2y^5\sqrt{3}$　　**32.** $\sqrt{45a^4b^{12}}$ $3a^2b^6\sqrt{5}$

**33.** $\sqrt{8a^8b^2}$ $2a^4b\sqrt{2}$　　　**34.** $\sqrt{20n^2m^6}$ $2nm^3\sqrt{5}$　　**35.** $\sqrt{27a^{10}b^{12}}$ $3a^5b^6\sqrt{3}$

**36.** $\sqrt{24a^4b^4c^{12}}$ $2a^2b^2c^6\sqrt{6}$　**37.** $\sqrt{48a^{12}b^{14}c^2}$ $4a^6b^7c\sqrt{3}$ **38.** $\sqrt{32x^{16}y^{22}z^4}$ $4x^8y^{11}z^2\sqrt{2}$

**39.** $\sqrt{28x^2y^4z^8}$ $2xy^2z^4\sqrt{7}$　**40.** $\sqrt{98a^4b^8c^{16}}$ $7a^2b^4c^8\sqrt{2}$ **41.** $\sqrt{243x^{10}y^2z^{28}}$ $9x^5yz^{14}\sqrt{3}$

**42.** $\sqrt{56x^{40}y^{22}}$ $2x^{20}y^{11}\sqrt{14}$　**43.** $\sqrt{128x^2y^4z^8}$ $8xy^2z^4\sqrt{2}$　**44.** $\sqrt{125m^{40}n^{20}p^2q^{12}}$

　　　　　　　　　　　　　　　　　　　　　　　　　　$5m^{20}n^{10}pq^6\sqrt{5}$

**For what values of $x$ will the radical be a real number?**

Ⓒ　**45.** $\sqrt{x^2}$ **all $x$**　　　　**46.** $\sqrt{x^2-4}$ $x \geq 2$, $x \leq -2$　**47.** $\sqrt{x^2-7x+12}$ $x \geq 4$, $x \leq 3$

**48.** $\sqrt{x^3}$ $x \geq 0$　　　　**49.** $\sqrt{x^4-16x^2+64}$ **all $x$**　**50.** $\sqrt{x^4-81}$ $x \geq 3$, $x \leq -3$

**Simplify.**

**51.** $\sqrt{0.09x^4y^8}$ $0.3x^2y^4$　　**52.** $\sqrt{\dfrac{49}{36}x^{6m+2}}$ $\frac{7}{6}x^{3m+1}$　　**53.** $\sqrt{\sqrt{64}}$ $2\sqrt{2}$

## CUMULATIVE REVIEW

**1.** Simplify $\dfrac{\dfrac{5}{m-3}+\dfrac{7}{m+2}}{\dfrac{7}{m^2-m-6}+\dfrac{1}{m-3}}$.　**1.** $\dfrac{12m-11}{m+9}$　　**2.** Solve for $x$　　**3.** Solve

　　　　　　　　　　　　　　　　　　　　　　　　　$ax - c = yx + b$.　　　$5 - 0.03x = 0.7x - 0.11$.

　　　　　　　　　　　　　　　　　　　　　　　　$x = \dfrac{b+c}{a-y}$　　　　　　$x = 7$

Radicals with Variables

# ODD POWERS OF VARIABLES 14.6

*Objective*  **To simplify square roots in which the radicands contain variables with odd exponents**

You can simplify a radical such as $\sqrt{x^7}$ by first rewriting the radicand as a product of two factors, one of which is $x^1$. The other factor will then be an even power of $x$, and you can use the property $\sqrt{x^n} = x^{\frac{n}{2}}$ if $n$ is even to simplify the square root. The process is illustrated in Examples 1 and 2.

*Example 1*  **Simplify $\sqrt{x^7}$.**

$$
\begin{aligned}
\textit{Rewrite.} \blacktriangleright \quad \sqrt{x^7} &= \sqrt{x^6 \cdot x^1} \\
&= \sqrt{x^6} \cdot \sqrt{x^1} \\
&= x^3 \cdot \sqrt{x^1} \quad \blacktriangleleft \sqrt{x^6} = x^{\frac{6}{2}} = x^3 \\
&= x^3 \sqrt{x}
\end{aligned}
$$

*Example 2*  **Simplify $\sqrt{a^5 b^9}$.**

$$
\begin{aligned}
\sqrt{a^5 b^9} &= \sqrt{a^4 \cdot a^1 \cdot b^8 \cdot b^1} \\
&= \sqrt{a^4} \cdot \sqrt{b^8} \cdot \sqrt{a^1} \cdot \sqrt{b^1} \quad \blacktriangleleft \textit{Group perfect squares.} \\
&= a^2 \cdot b^4 \cdot \sqrt{a^1 b^1} \quad \blacktriangleleft \sqrt{a^1} \text{ and } \sqrt{b^1} \textit{ cannot be simplified, so} \\
&= a^2 b^4 \sqrt{ab} \qquad\qquad \textit{rewrite } \sqrt{a^1} \cdot \sqrt{b^1} \textit{ as } \sqrt{a^1 b^1}, \textit{ or } \sqrt{ab}.
\end{aligned}
$$

If the radicand includes a numerical coefficient, you will also have to simplify it by using the procedures that you have already learned.

*Example 3*  **Simplify $\sqrt{28 x y^{11}}$.**

$$
\begin{aligned}
\sqrt{28 x y^{11}} &= \sqrt{28 \cdot x^1 \cdot y^{10} \cdot y^1} \\
&= \sqrt{4 \cdot 7 \cdot x^1 \cdot y^{10} \cdot y^1} \quad \blacktriangleleft \textit{The greatest perfect square factor of 28 is 4.} \\
&= \sqrt{4} \cdot \sqrt{y^{10}} \cdot \sqrt{7} \cdot \sqrt{x^1} \cdot \sqrt{y^1} \quad \blacktriangleleft \textit{Group perfect squares.} \\
&= 2y^5 \cdot \sqrt{7} \cdot \sqrt{x^1} \cdot \sqrt{y^1} \quad \blacktriangleleft \sqrt{7}, \sqrt{x^1}, \textit{ and } \sqrt{y^1} \textit{ cannot be simplified.} \\
&= 2y^5 \sqrt{7xy} \quad \blacktriangleleft \textit{Multiply: } \sqrt{7} \cdot \sqrt{x^1} \cdot \sqrt{y^1} = \sqrt{7 \cdot x^1 \cdot y^1} = \sqrt{7xy}.
\end{aligned}
$$

The process for simplifying an expression such as $-5x^3 y \sqrt{18 x^5 y^6}$ can be completed in three steps: (1) simplify the radical; (2) multiply all factors that do not contain radicals; and (3) multiply all radicals that cannot be simplified further. These steps are shown in detail in Example 4.

## Example 4 Simplify $-5x^3y\sqrt{18x^5y^6}$.

$-5x^3y\sqrt{18x^5y^6} = -5x^3y^1\sqrt{9\cdot 2\cdot x^4\cdot x^1\cdot y^6}$ ◄ *Simplify the radicand.*
$= -5x^3y^1\cdot\sqrt{9}\cdot\sqrt{x^4}\cdot\sqrt{y^6}\cdot\sqrt{2}\cdot\sqrt{x^1}$
$= -5x^3y^1\cdot 3\cdot x^2\cdot y^3\cdot\sqrt{2}\cdot\sqrt{x^1}$
$= -5\cdot 3\cdot x^3\cdot x^2\cdot y^1\cdot y^3\cdot\sqrt{2}\cdot\sqrt{x^1}$ ◄ *Multiply the factors that*
$= -15x^5y^4\cdot\sqrt{2}\cdot\sqrt{x^1}$     *do not contain radicals.*
$= -15x^5y^4\sqrt{2x}$ ◄ *Multiply the radicals that cannot be simplified.*

# Reading in Algebra

**Tell whether the statement is true or false. If false, tell why it is false.**

**1.** $\sqrt{x^9} = x^3$ **F**     **2.** $\sqrt{x^5} = x^4\sqrt{x^1}$ **F**     **3.** $\sqrt{a^{11}}$ can be simplified. **T**
**4.** The numerical coefficient of the radicand of $\sqrt{48a^3}$ is a perfect square.
**5.** $\sqrt{a^1a^1}$ cannot be simplified because each power of the variable is an odd number.
4. F, 48 is not a perfect square     5. F, $\sqrt{a^1\cdot a^1} = a$

# Oral Exercises

**Read each radicand as the product of two factors such that one factor is
the first power of the variable.**

**1.** $\sqrt{x^{13}}$     **2.** $\sqrt{a^7}$     **3.** $\sqrt{b^{21}}$     **4.** $\sqrt{m^5}$     **5.** $\sqrt{h^3}$     **6.** $\sqrt{q^{15}}$
$x^{12}\cdot x^1$     $a^6\cdot a^1$     $b^{20}\cdot b^1$     $m^4\cdot m^1$     $h^2\cdot h^1$     $q^{14}\cdot q^1$

# Written Exercises

13. $2x^2y\sqrt{5xy}$   14. $3x^3y^4\sqrt{3xy}$   16. $-3my^3\sqrt{5my}$   18. $-2x^2y^3\sqrt{2x}$   19. $-5x^4y\sqrt{2y}$

**Simplify.**

Ⓐ **1.** $\sqrt{x^5}$ $x^2\sqrt{x}$     **2.** $\sqrt{y^3}$ $y\sqrt{y}$     **3.** $\sqrt{4b^7}$ $2b^3\sqrt{b}$     **4.** $\sqrt{36c}$ $6\sqrt{c}$
**5.** $\sqrt{12m^3}$ $2m\sqrt{3m}$     **6.** $\sqrt{24a^{11}}$ $2a^5\sqrt{6a}$     **7.** $\sqrt{5y^3}$ $y\sqrt{5y}$     **8.** $-\sqrt{3x^5}$ $-x^2\sqrt{3x}$
**9.** $\sqrt{xy^9}$ $y^4\sqrt{xy}$     **10.** $\sqrt{a^4b^7}$ $a^2b^3\sqrt{b}$     **11.** $-\sqrt{c^3d^5}$ $-cd^2\sqrt{cd}$     **12.** $\sqrt{m^{12}n}$ $m^6\sqrt{n}$
**13.** $\sqrt{20x^5y^3}$     **14.** $\sqrt{27x^7y^9}$     **15.** $\sqrt{18x^{11}y}$ $3x^5\sqrt{2xy}$     **16.** $-\sqrt{45m^3y^7}$
**17.** $\sqrt{40x^4y}$ $2x^2\sqrt{10y}$     **18.** $-\sqrt{8x^5y^6}$     **19.** $-\sqrt{50x^8y^3}$     **20.** $\sqrt{32x^9y^4}$
20. $4x^4y^2\sqrt{2x}$     23. $-8b^2c^5\sqrt{6b}$     25. $-21x^2y^7\sqrt{7y}$     26. $42e^3f^7\sqrt{2e}$
Ⓑ **21.** $2ab^2\sqrt{12a^3b^8}$ $4a^2b^6\sqrt{3a}$     **22.** $3xy\sqrt{18x^3y^4}$ $9x^2y^3\sqrt{2x}$     **23.** $-4bc^2\sqrt{24b^3c^6}$
**24.** $6a^2b\sqrt{75a^8b^3}$ $30a^6b^2\sqrt{3b}$     **25.** $-7xy^4\sqrt{63x^2y^7}$     **26.** $6e^2f^2\sqrt{98e^3f^{10}}$
**27.** $3\sqrt{44xy^2z}$ $6y\sqrt{11xz}$     **28.** $-5xyz^2\sqrt{90x^5y^4z^3}$     **29.** $4abc^3\sqrt{99a^6bc^{11}}$
     $-15x^3y^3z^3\sqrt{10xz}$     $12a^4bc^8\sqrt{11bc}$
Ⓒ **30.** $\sqrt{x^{2m+1}}$ $x^m\sqrt{x}$     **31.** $\sqrt{a^{4m}b^{6m+1}}$ $a^{2m}b^{3m}\sqrt{b}$     **32.** $\sqrt{(x+y)^3}$ $(x+y)\sqrt{x+y}$

**For what value of $a$ will the equation be true?**
**33.** $\sqrt{x^{3a+1}} = x^6\sqrt{x}$ **4**     **34.** $\sqrt{x^{2a+3}y^5} = x^5y^2\sqrt{xy}$ **4**

# ADDING AND SUBTRACTING RADICALS   14.7

**To simplify expressions containing radicals by combining like terms**

You know that the terms $5xy^2$ and $-2xy^2$ are like terms because their variable factors are the same. Similarly, $5\sqrt{3}$ and $-2\sqrt{3}$ are like terms because their radical factors are the same.

You can use the distributive property to combine like radicals. For example, the expression $5\sqrt{3} + 7\sqrt{3}$ has $\sqrt{3}$ as a common factor. So,
$$5\sqrt{3} + 7\sqrt{3} = (5 + 7)\sqrt{3} = 12\sqrt{3}.$$
However, you can combine the like terms immediately as you would with $5a + 7a$ to get $12a$. So, $5\sqrt{3} + 7\sqrt{3} = 12\sqrt{3}$.

*Example 1*   **Simplify $2\sqrt{3} - 9\sqrt{3} + 4\sqrt{3}$.**

$2\sqrt{3} - 9\sqrt{3} + 4\sqrt{3}$
$(2\sqrt{3} + 4\sqrt{3}) - 9\sqrt{3}$   ◀ *Group the positive terms.*
$\quad 6\sqrt{3} - 9\sqrt{3} = -3\sqrt{3}$

*Example 2*   **Simplify $6\sqrt{5} - 8\sqrt{11} + 4\sqrt{5}$.**

$6\sqrt{5} - 8\sqrt{11} + 4\sqrt{5} = (6\sqrt{5} + 4\sqrt{5}) - 8\sqrt{11}$   ◀ *Group like terms.*
$\qquad\qquad\qquad\quad = 10\sqrt{5} - 8\sqrt{11}$   ◀ *$10\sqrt{5}$ and $8\sqrt{11}$ are unlike terms.*

*Example 3*   **Simplify $2\sqrt{5} - 3\sqrt{20}$.**

**One Way**
$2\sqrt{5} - 3\sqrt{20}$
$2\sqrt{5} - 3\underbrace{\sqrt{2} \cdot \sqrt{2}} \cdot \sqrt{5}$   ◀ *Factor $\sqrt{20}$ into primes.*
$\quad 2\sqrt{5} - 3 \cdot 2 \cdot \sqrt{5}$
$\quad 2\sqrt{5} - 6\sqrt{5} = -4\sqrt{5}$

**Another Way**
$2\sqrt{5} - 3\sqrt{20}$
$2\sqrt{5} - 3\sqrt{4} \cdot \sqrt{5}$   ◀ *Look for perfect square factors.*
$\qquad\qquad \downarrow$
$2\sqrt{5} - 3 \cdot 2 \cdot \sqrt{5}$
$2\sqrt{5} - 6\sqrt{5} = -4\sqrt{5}$

Sometimes, as in Examples 4 and 5 on the next page, you will have to simplify each radical expression before you will find any like terms to combine.

## Example 4

**Simplify $4\sqrt{27} - 5\sqrt{12} + 3\sqrt{75}$.**

First, simplify each radical.

$$4\sqrt{27} - 5\sqrt{12} + 3\sqrt{75}$$

$$4 \cdot \underbrace{\sqrt{3} \cdot \sqrt{3}} \cdot \sqrt{3} - 5 \cdot \underbrace{\sqrt{2} \cdot \sqrt{2}} \cdot \sqrt{3} + 3 \cdot \underbrace{\sqrt{5} \cdot \sqrt{5}} \cdot \sqrt{3}$$

$$4 \cdot 3 \cdot \sqrt{3} \qquad - 5 \cdot 2 \cdot \sqrt{3} \qquad + 3 \cdot 5 \cdot \sqrt{3}$$

$$12\sqrt{3} - 10\sqrt{3} + 15\sqrt{3} = 17\sqrt{3}$$

## Example 5

**Simplify $\sqrt{25cd^3} + 4d\sqrt{9cd} - \sqrt{49cd^3}$.**

$$\sqrt{25cd^3} + 4d\sqrt{9cd} - \sqrt{49cd^3}$$

$$\sqrt{25} \cdot \sqrt{c} \cdot \sqrt{d^2} \cdot \sqrt{d^1} + 4 \cdot d \cdot \sqrt{9} \cdot \sqrt{c} \cdot \sqrt{d} - \sqrt{49} \cdot \sqrt{c} \cdot \sqrt{d^2} \cdot \sqrt{d^1}$$

$$5 \cdot \sqrt{c} \cdot d \cdot \sqrt{d} + 4 \cdot d \cdot 3 \cdot \sqrt{c} \cdot \sqrt{d} - 7 \cdot \sqrt{c} \cdot d \cdot \sqrt{d}$$

$$5d\sqrt{cd} + 12d\sqrt{cd} - 7d\sqrt{cd} = 10d\sqrt{cd} \qquad \blacktriangleleft \quad 5d + 12d - 7d = 10d$$

## Written Exercises

**Simplify.**

Ⓐ 1. $3\sqrt{7} + 5\sqrt{7}$   **$8\sqrt{7}$**

2. $5\sqrt{2} - 3\sqrt{2}$   **$2\sqrt{2}$**

3. $7\sqrt{11} + 6\sqrt{11}$   **$13\sqrt{11}$**

4. $8\sqrt{3} - 7\sqrt{3}$   **$\sqrt{3}$**

5. $6\sqrt{10} + 7\sqrt{10} - 4\sqrt{10}$   **$9\sqrt{10}$**

6. $8\sqrt{7} - 11\sqrt{7} + 4\sqrt{7}$   **$\sqrt{7}$**

7. $4\sqrt{5} - 2\sqrt{5} + 7\sqrt{11}$   **$2\sqrt{5} + 7\sqrt{11}$**

8. $6\sqrt{2} - 9\sqrt{5} - 4\sqrt{5}$   **$6\sqrt{2} - 13\sqrt{5}$**

9. $9\sqrt{2} - 3\sqrt{8}$   **$3\sqrt{2}$**

10. $5\sqrt{28} + 3\sqrt{7}$   **$13\sqrt{7}$**

11. $3\sqrt{2} + \sqrt{50}$   **$8\sqrt{2}$**

12. $4\sqrt{12} + 6\sqrt{3}$   **$14\sqrt{3}$**

13. $5\sqrt{2} + \sqrt{18} + \sqrt{32}$   **$12\sqrt{2}$**

14. $2\sqrt{27} - 3\sqrt{3} + \sqrt{48}$   **$7\sqrt{3}$**

15. $2\sqrt{44} + 6\sqrt{11} + \sqrt{99}$   **$13\sqrt{11}$**

16. $-\sqrt{32} + 6\sqrt{18} - 7\sqrt{98}$   **$-35\sqrt{2}$**

17. $\sqrt{20} - 2\sqrt{45} - 9\sqrt{5}$   **$-13\sqrt{5}$**

18. $-2\sqrt{63} - \sqrt{28} + 4\sqrt{7}$   **$-4\sqrt{7}$**

19. $\sqrt{b} + 9\sqrt{b}$   **$10\sqrt{b}$**

20. $7\sqrt{x} + 4\sqrt{x} - \sqrt{x}$   **$10\sqrt{x}$**    22. **$30\sqrt{ab}$**

21. $6\sqrt{xy} - 16\sqrt{xy} - 2\sqrt{9xy}$   **$-16\sqrt{xy}$**

22. $4\sqrt{25ab} + 6\sqrt{49ab} - 3\sqrt{100ab} - \sqrt{4ab}$

23. $4x\sqrt{25x^2} + 6x\sqrt{36x^2}$   **$56x^2$**

24. $3\sqrt{cd} - 7\sqrt{9cd} - \sqrt{25cd}$   **$-23\sqrt{cd}$**

25. $3\sqrt{12ab} - 5\sqrt{3ab} + 4\sqrt{27ab}$   **$13\sqrt{3ab}$**

26. $5\sqrt{32tw} - 4\sqrt{2tw} + 2\sqrt{50tw}$   **$26\sqrt{2tw}$**

29. **$-23ab\sqrt{ab}$**

Ⓑ 27. $\sqrt{a^3b} + a\sqrt{36ab}$   **$7a\sqrt{ab}$**

28. $\sqrt{16x^3y^2} + xy\sqrt{25x}$   **$9xy\sqrt{x}$**

29. $3\sqrt{a^3b^3} + 2b\sqrt{a^3b} - 7a\sqrt{16ab^3}$

30. $\sqrt{cd^3} + \sqrt{9cd^3} - 5d\sqrt{16cd}$   **$-16d\sqrt{cd}$**

31. $\sqrt{3x^3y} + x\sqrt{48xy} - \sqrt{75x^3y}$   **$0$**

32. $\sqrt{8x^4y} + x^2\sqrt{2y} - \sqrt{50x^4y}$   **$-2x^2\sqrt{2y}$**

33. $3b\sqrt{ab} + \sqrt{4ab^3} + \sqrt{ab^3}$   **$6b\sqrt{ab}$**

34. $7\sqrt{m^2n^3} + 2mn\sqrt{49n}$   **$21mn\sqrt{n}$**

35. $y\sqrt{27x^3y} - x\sqrt{6xy^3} + \sqrt{54x^3y^3}$

     **$3xy\sqrt{3xy} + 2xy\sqrt{6xy}$**

36. $7\sqrt{80x^3y^3} - 6y\sqrt{125x^2y}$   **$28xy\sqrt{5xy} - 30xy\sqrt{5y}$**

Ⓒ 37. $\sqrt{0.09xy^4} + y^2\sqrt{0.49x}$   **$y^2\sqrt{x}$**

38. $5\sqrt{0.16a^5b^3} + a^2b\sqrt{0.25ab}$   **$2.5a^2b\sqrt{ab}$**

# MULTIPLYING RADICALS

**Objective**     **To simplify products of radical expressions**

To simplify expressions containing square roots, you have used two important properties: $\sqrt{a} \cdot \sqrt{b} = \sqrt{ab}$ and $\sqrt{a} \cdot \sqrt{a} = a(a \geq 0, b \geq 0)$.
Thus, $\sqrt{7} \cdot \sqrt{3} = \sqrt{21}$ and $\sqrt{5} \cdot \sqrt{5} = 5$.
You can now apply these two properties to simplify more complex products.

**Example 1**     **Simplify $5\sqrt{7} \cdot 6\sqrt{2}$.**

$$
\begin{aligned}
5\sqrt{7} \cdot 6\sqrt{2} &= (5 \cdot 6)(\sqrt{7} \cdot \sqrt{2}) \quad \blacktriangleleft \; \text{Group whole-number factors and} \\
&= 30\sqrt{7 \cdot 2} \qquad\qquad\qquad \text{group radical factors.} \\
&= 30\sqrt{14}
\end{aligned}
$$

Since $\sqrt{a} \cdot \sqrt{a} = a$, you will find it helpful to look for factors that are the same when you are required to simplify a product. For example in $-8\sqrt{10x} \cdot 3\sqrt{2x}$, notice that $\sqrt{10x}$ can be factored into two radicals, one of which is $\sqrt{2x}$.

**Example 2**     **Simplify $-8\sqrt{10x} \cdot 3\sqrt{2x}$.**

$$
\begin{aligned}
-8\sqrt{10x} \cdot 3\sqrt{2x} &= -8 \cdot 3\sqrt{10x} \cdot \sqrt{2x} \\
&= -8 \cdot 3\sqrt{5 \cdot 2x} \cdot \sqrt{2x} \\
&= -8 \cdot 3 \cdot \sqrt{5} \cdot \underbrace{\sqrt{2x} \cdot \sqrt{2x}} \\
&= \quad -8 \cdot 3 \cdot \sqrt{5} \cdot 2x \quad \blacktriangleleft \; \sqrt{a} \cdot \sqrt{a} = a \\
&= \quad -8 \cdot 3 \cdot 2x \cdot \sqrt{5} \\
&= \quad -48x\sqrt{5}
\end{aligned}
$$

You can use the distributive property to simplify a product of a monomial and a polynomial such as $\sqrt{7}(\sqrt{5} + 4\sqrt{7})$.

**Example 3**     **Simplify $\sqrt{7}(\sqrt{5} + 4\sqrt{7})$.**

$$
\begin{aligned}
\sqrt{7}(\sqrt{5} + 4\sqrt{7}) &= \sqrt{7} \cdot \sqrt{5} + \sqrt{7} \cdot 4\sqrt{7} \quad \blacktriangleleft \; \textit{Distribute the } \sqrt{7}. \\
&= \sqrt{7} \cdot \sqrt{5} + 4 \cdot \sqrt{7} \cdot \sqrt{7} \\
&= \sqrt{35} + 4 \cdot 7 \quad \blacktriangleleft \; \sqrt{a} \cdot \sqrt{a} = a \\
&= \sqrt{35} + 28
\end{aligned}
$$

*Example 4*     **Simplify $-5\sqrt{6}\,(2\sqrt{2} - 4\sqrt{3})$.**

$$-5\sqrt{6}\,(2\sqrt{2} - 4\sqrt{3}) = -5\sqrt{6}\cdot 2\sqrt{2} + (-5\sqrt{6})(-4\sqrt{3})$$
$$= -10\sqrt{12} + 20\sqrt{18}$$
$$= -10\sqrt{2}\cdot\sqrt{2}\cdot\sqrt{3} + 20\sqrt{3}\cdot\sqrt{3}\cdot\sqrt{2}$$
$$= -10\cdot 2\cdot\sqrt{3} + 20\cdot 3\cdot\sqrt{2}$$
$$= -20\sqrt{3} + 60\sqrt{2}$$

You can simplify a product of two binomials that contain radicals in the same way as you would simplify the product of any two binomials. Examples 5 and 6 illustrate multiplying binomials using the FOIL method.

*Example 5*     **Simplify $(6\sqrt{3} - 2\sqrt{2})(4\sqrt{3} + 5\sqrt{2})$.**

$$(6\sqrt{3} - 2\sqrt{2})(4\sqrt{3} + 5\sqrt{2})$$
$$6\sqrt{3}\cdot 4\sqrt{3} + 6\sqrt{3}\cdot 5\sqrt{2} - 2\sqrt{2}\cdot 4\sqrt{3} - 2\sqrt{2}\cdot 5\sqrt{2}$$
$$6\sqrt{3}\cdot 4\sqrt{3} + 30\sqrt{6} - 8\sqrt{6} - 2\sqrt{2}\cdot 5\sqrt{2}$$
$$6\sqrt{3}\cdot 4\sqrt{3} \quad +22\sqrt{6} \quad\quad -2\sqrt{2}\cdot 5\sqrt{2}$$
$$6\cdot 4\cdot\sqrt{3}\cdot\sqrt{3} +22\sqrt{6} \quad\quad - 2\cdot 5\cdot\sqrt{2}\cdot\sqrt{2}$$
$$24\cdot 3 \quad\quad +22\sqrt{6} \quad\quad - 10\cdot 2$$
$$72 \quad\quad +22\sqrt{6} \quad\quad - 20$$
$$52 + 22\sqrt{6}$$

*Example 6*     **Simplify $(2\sqrt{5} - 3\sqrt{6})(2\sqrt{5} + 3\sqrt{6})$.**

$$(2\sqrt{5} - 3\sqrt{6})(2\sqrt{5} + 3\sqrt{6})$$
$$2\sqrt{5}\cdot 2\sqrt{5} + 2\sqrt{5}\cdot 3\sqrt{6} - 3\sqrt{6}\cdot 2\sqrt{5} - 3\sqrt{6}\cdot 3\sqrt{6}$$
$$2\sqrt{5}\cdot 2\sqrt{5} + 6\sqrt{30} - 6\sqrt{30} - 3\sqrt{6}\cdot 3\sqrt{6}$$
$$2\sqrt{5}\cdot 2\sqrt{5} \quad +0\sqrt{30} \quad\quad - 3\sqrt{6}\cdot 3\sqrt{6}$$
$$2\cdot 2\cdot\sqrt{5}\cdot\sqrt{5} +0 \quad\quad - 3\cdot 3\sqrt{6}\cdot\sqrt{6}$$
$$4\cdot 5 \quad\quad\quad - 9\cdot 6$$
$$20 \quad\quad\quad\quad - 54$$
$$-34$$

## Oral Exercises

**Simplify.**

**1.** $\sqrt{3}\cdot\sqrt{3}$    **2.** $\sqrt{7}\cdot\sqrt{5}$    **3.** $\sqrt{6}\cdot\sqrt{6}$    **4.** $\sqrt{8}\cdot\sqrt{8}$    **5.** $\sqrt{5}\cdot\sqrt{5}$    **6.** $\sqrt{11}\cdot\sqrt{2}$

1. 3    2. $\sqrt{35}$    3. 6    4. 8    5. 5    6. $\sqrt{22}$

**7.** $3\cdot 4\sqrt{6}$    **8.** $\sqrt{7}\cdot 2\sqrt{3}$    **9.** $7\cdot 4\sqrt{7}$    **10.** $\sqrt{3}\cdot 4\sqrt{3}$    **11.** $\sqrt{8}\cdot 3\sqrt{8}$    **12.** $\sqrt{2}\cdot 5\sqrt{2}$

7. $12\sqrt{6}$    8. $2\sqrt{21}$    9. $28\sqrt{7}$    10. 12    11. 24    12. 10

Multiplying Radicals

# Written Exercises

**Simplify.**

(A)

1. $3\sqrt{5} \cdot 7\sqrt{2}$ **$21\sqrt{10}$**
2. $4\sqrt{3} \cdot 2\sqrt{5}$ **$8\sqrt{15}$**
3. $9\sqrt{7} \cdot 3\sqrt{5}$ **$27\sqrt{35}$**
4. $3\sqrt{7} \cdot 5\sqrt{7}$ **105**
5. $2\sqrt{4} \cdot 4\sqrt{5}$ **$16\sqrt{5}$**
6. $3\sqrt{2} \cdot 8\sqrt{2}$ **48**
7. $8\sqrt{y} \cdot \sqrt{y}$ **$8y$**
8. $3\sqrt{d} \cdot 4\sqrt{d}$ **$12d$**
9. $6\sqrt{5x} \cdot \sqrt{5x}$ **$30x$**
10. $4\sqrt{5} \cdot 2\sqrt{10}$ **$40\sqrt{2}$**
11. $2\sqrt{7} \cdot 3\sqrt{14}$ **$42\sqrt{2}$**
12. $6\sqrt{6} \cdot 3\sqrt{3}$ **$54\sqrt{2}$**
13. $-5\sqrt{12a} \cdot 2\sqrt{2a}$ **$-20a\sqrt{6}$**
14. $8\sqrt{2x} \cdot 5\sqrt{6x}$ **$80x\sqrt{3}$**
15. $6\sqrt{8y} \cdot \sqrt{2y}$ **$24y$**
16. $\sqrt{5}(\sqrt{2} + \sqrt{3})$ **$\sqrt{10} + \sqrt{15}$**
17. $3\sqrt{5}(2\sqrt{7} - 4\sqrt{2})$ **$6\sqrt{35} - 12\sqrt{10}$**
18. $-4\sqrt{6}(3\sqrt{2} - 5\sqrt{3})$ **$-24\sqrt{3} + 60\sqrt{2}$**
19. $-6\sqrt{2}(\sqrt{10} - 7\sqrt{8})$ **$-12\sqrt{5} + 168$**
20. $5\sqrt{2}(4\sqrt{2} - \sqrt{6})$ **$40 - 10\sqrt{3}$**
21. $7\sqrt{8}(3\sqrt{2} + 4\sqrt{3})$ **$84 + 56\sqrt{6}$**
22. $(2\sqrt{6} + \sqrt{5})(8\sqrt{6} - \sqrt{5})$ **$91 + 6\sqrt{30}$**
23. $(\sqrt{5} - \sqrt{2})(3\sqrt{5} + 2\sqrt{2})$ **$11 - \sqrt{10}$**
24. $(\sqrt{7} + \sqrt{3})(\sqrt{7} - \sqrt{3})$ **4**
25. $(\sqrt{6} - \sqrt{5})(\sqrt{6} + \sqrt{5})$ **1**
26. $(3\sqrt{2} + 4\sqrt{3})(3\sqrt{2} - 4\sqrt{3})$ **$-30$**
27. $(4\sqrt{7} - 3\sqrt{3})(4\sqrt{7} + 3\sqrt{3})$ **85**
28. $(6\sqrt{2} + 2\sqrt{5})(6\sqrt{2} - 2\sqrt{5})$ **52**
29. $(2\sqrt{11} - 3\sqrt{2})(2\sqrt{11} + 3\sqrt{2})$ **26**

(B)

30. $(7\sqrt{2} - 4\sqrt{12})(3\sqrt{2} - \sqrt{12})$ **$90 - 38\sqrt{6}$**
31. $(5\sqrt{24} - 2\sqrt{2})(4\sqrt{24} - \sqrt{2})$ **$484 - 52\sqrt{3}$**
32. $(8\sqrt{20} - 3\sqrt{2})(\sqrt{20} + 5\sqrt{2})$ **$130 + 74\sqrt{10}$**
33. $(7\sqrt{6} + \sqrt{12})(4\sqrt{6} - 3\sqrt{12})$ **$132 - 102\sqrt{2}$**

## Example    Simplify $(2\sqrt{3} + \sqrt{8})^2$.

$$(2\sqrt{3} + \sqrt{8})^2 = (2\sqrt{3} + \sqrt{8})(2\sqrt{3} + \sqrt{8})$$

$$\underbrace{2\sqrt{3} \cdot 2\sqrt{3}}_{4\sqrt{3} \cdot \sqrt{3}} + \underbrace{2\sqrt{24} + 2\sqrt{24}}_{+4\sqrt{24}} + \underbrace{\sqrt{8} \cdot \sqrt{8}}_{+\sqrt{8} \cdot \sqrt{8}}$$

| $4 \cdot 3$ | $+ 4 \cdot 2\sqrt{6}$ | $+ 8$ | ◀ $\sqrt{24} = \sqrt{2} \cdot \sqrt{2} \cdot \sqrt{3} \cdot \sqrt{2} = 2\sqrt{6}$ |
| 12 | $+ 8\sqrt{6}$ | $+ 8$ | $= 20 + 8\sqrt{6}$ |

**Simplify.**

34. $(4\sqrt{6} + \sqrt{3})^2$
35. $(7\sqrt{2} - 4\sqrt{10})^2$
36. $(6\sqrt{5} - 3\sqrt{8})^2$
37. $(9\sqrt{8} - 3\sqrt{6})^2$

34. $99 + 24\sqrt{2}$    35. $258 - 112\sqrt{5}$    36. $252 - 72\sqrt{10}$    37. $702 - 216\sqrt{3}$

(C)

38. $(\sqrt{3} + \sqrt{2} - \sqrt{5})^2$
39. $(\sqrt{3} + \sqrt{2})(\sqrt{3} + \sqrt{6} + \sqrt{2})$ **$2\sqrt{6} + 2\sqrt{3} + 3\sqrt{2} + 5$**
40. $(\sqrt{5} + \sqrt{8})^3$ **$46\sqrt{2} + 29\sqrt{5}$**
41. $(\sqrt{7} - \sqrt{5})(4\sqrt{6} + \sqrt{2})(\sqrt{7} + \sqrt{5})$ **$8\sqrt{6} + 2\sqrt{2}$**
42. $(\sqrt{3} - \sqrt{2})^2(\sqrt{3} + \sqrt{2})^2$ **1**
43. $(\sqrt{12} - \sqrt{18} + \sqrt{20})(\sqrt{3} + \sqrt{2} - \sqrt{5})$

38. $10 + 2(\sqrt{6} - \sqrt{15} - \sqrt{10})$    43. $-10 + 5\sqrt{10} - \sqrt{6}$

# CUMULATIVE REVIEW

**Solve by graphing.**

1. $5x + 2y = 7$
   $2x + 8y = 10$
   **(1, 1)**
2. $x = -3$
   $y = -2$
   **(−3, −2)**
3. $0.5y = x + 2$
   $0.5x = 1 - 0.25y$
   **(0, 4)**
4. $x = 4y$
   $24 - 3x = 12y$
   **(4, 1)**

# DIVIDING RADICALS 14.9

Objective · **To simplify quotients that involve radicals**

The product property for square roots, $\sqrt{a} \cdot \sqrt{b} = \sqrt{ab}$, can be extended to quotients of square roots. These illustrations suggest a generalization.

$$\frac{\sqrt{9}}{\sqrt{64}} = \frac{3}{8} \qquad \sqrt{\frac{9}{64}} = \sqrt{\frac{3}{8} \cdot \frac{3}{8}} = \sqrt{\frac{3}{8}} \cdot \sqrt{\frac{3}{8}} = \frac{3}{8}$$

So, $\dfrac{\sqrt{9}}{\sqrt{64}} = \sqrt{\dfrac{9}{64}}$.

---

**Quotient property for square roots**

For all positive numbers $a$ and $b$, $\dfrac{\sqrt{a}}{\sqrt{b}} = \sqrt{\dfrac{a}{b}}$.

---

Example 1 · **Simplify $\dfrac{\sqrt{20}}{\sqrt{2}}$ and $\dfrac{\sqrt{50x^5}}{\sqrt{2x^3}}$**

$$\frac{\sqrt{20}}{\sqrt{2}} = \sqrt{\frac{20}{2}} \qquad\qquad \frac{\sqrt{50x^5}}{\sqrt{2x^3}} = \sqrt{\frac{50x^5}{2x^3}}$$
$$= \sqrt{10} \qquad\qquad\qquad\quad = \sqrt{25x^2}$$
$$= \sqrt{25} \cdot \sqrt{x^2}$$
$$= 5x$$

◀ $\dfrac{x^5}{x^3} = \dfrac{\overset{x^2}{\cancel{x^5}}}{\underset{1}{\cancel{x^3}}} = x^2$

You can use the property $\dfrac{a}{b} = \dfrac{a \cdot c}{b \cdot c}$, where $b \neq 0$ and $c \neq 0$, to rewrite an expression such as $\dfrac{3}{\sqrt{5}}$ so that there will be no radical in the denominator. Since $\sqrt{5} \cdot \sqrt{5} = 5$, you can multiply both the numerator and the denominator of $\dfrac{3}{\sqrt{5}}$ by $\sqrt{5}$. Thus, $\dfrac{3}{\sqrt{5}} = \dfrac{3 \cdot \sqrt{5}}{\sqrt{5} \cdot \sqrt{5}} = \dfrac{3\sqrt{5}}{5}$. There is no radical in the denominator, and $\dfrac{3\sqrt{5}}{5}$ is in simplest form. This process, illustrated in Example 2, is called *rationalizing the denominator*.

---

**Definition: Rationalize**

To **rationalize** the denominator of an expression means to rewrite the expression with no radicals in the denominator.

---

*Example 2*    **Simplify by rationalizing the denominator:** $\dfrac{3}{\sqrt{18}}$.

**One Way**

Multiply the numerator and the denominator by the square root in the denominator.

$$\frac{3}{\sqrt{18}} = \frac{3 \cdot \sqrt{18}}{\sqrt{18} \cdot \sqrt{18}}$$

$$= \frac{3\sqrt{18}}{18}$$

$$= \frac{3 \cdot \sqrt{3} \cdot \sqrt{3} \cdot \sqrt{2}}{18}$$

$$= \frac{\overset{1}{\cancel{3}} \cdot \overset{1}{\cancel{3}} \cdot \sqrt{2}}{\underset{1}{\cancel{3}} \cdot \underset{1}{\cancel{3}} \cdot 2}$$

$$= \frac{\sqrt{2}}{2}$$

Both ways give the same result.

**Another Way**

Multiply the numerator and the denominator by the least square root needed to make the denominator a perfect square.

$$\frac{3}{\sqrt{18}} = \frac{3 \cdot \sqrt{2}}{\sqrt{18} \cdot \sqrt{2}}$$

$$= \frac{3\sqrt{2}}{\sqrt{36}}$$

$$= \frac{3\sqrt{2}}{6}$$

$$= \frac{\overset{1}{\cancel{3}}\sqrt{2}}{\underset{2}{\cancel{6}}}$$

$$= \frac{\sqrt{2}}{2}$$

The technique for rationalizing denominators that contain variables within a radical sign is the same. For example, you can simplify $\sqrt{\dfrac{5}{b^3}}$ by first rewriting it as $\dfrac{\sqrt{5}}{\sqrt{b^3}}$. Since $b^3$ is not a perfect square, you can then multiply the numerator and the denominator by $\sqrt{b^1}$ so that the radicand in the denominator is a perfect square. Thus, $\dfrac{\sqrt{5}}{\sqrt{b^3}} = \dfrac{\sqrt{5} \cdot \sqrt{b^1}}{\sqrt{b^3} \cdot \sqrt{b^1}} = \dfrac{\sqrt{5b}}{\sqrt{b^4}}$, or $\dfrac{\sqrt{5b}}{b^2}$.

*Example 3*    **Simplify** $\sqrt{\dfrac{15}{24y^5}}$.

$$\sqrt{\frac{15}{24y^5}} = \sqrt{\frac{5}{8y^5}} = \frac{\sqrt{5}}{\sqrt{8y^5}} \qquad \blacktriangleleft \quad \sqrt{\frac{15}{24y^5}} = \sqrt{\frac{3 \cdot 5}{3 \cdot 8 \cdot y^5}} = \sqrt{\frac{5}{8y^5}}; \text{ 3 is the GCF of 15 and 24.}$$

$$= \frac{\sqrt{5} \cdot \sqrt{2} \cdot \sqrt{y^1}}{\sqrt{8y^5} \cdot \sqrt{2} \cdot \sqrt{y^1}} \qquad \blacktriangleleft \quad \sqrt{8} \cdot \sqrt{2} = \sqrt{16}; \text{ 16 is a perfect square.}$$

$$= \frac{\sqrt{5} \cdot \sqrt{2} \cdot \sqrt{y^1}}{\sqrt{8} \cdot \sqrt{2} \cdot \sqrt{y^5} \cdot \sqrt{y^1}} \qquad \sqrt{y^5} \cdot \sqrt{y^1} = \sqrt{y^6}; \text{ } y^6 \text{ is a perfect square.}$$

$$= \frac{\sqrt{10y}}{\sqrt{16}\sqrt{y^6}}$$

$$= \frac{\sqrt{10y}}{4y^3}$$

## Example 4  Simplify $\dfrac{12xy^3}{\sqrt{18xy^4}}$.

$$\dfrac{12xy^3}{\sqrt{18xy^4}} = \dfrac{12xy^3}{\sqrt{18}\cdot\sqrt{x^1}\cdot\sqrt{y^4}}$$ ◄ $y^4$ is already a perfect square.

$$= \dfrac{12x^1y^3\cdot\sqrt{2}\cdot\sqrt{x^1}}{\sqrt{18}\cdot\sqrt{x^1}\cdot\sqrt{y^4}\cdot\sqrt{2}\cdot\sqrt{x^1}}$$ ◄ $18\cdot 2$ is a perfect square; $\sqrt{x^1}\cdot\sqrt{x^1}$ is $x$.

$$= \dfrac{12x^1y^3\sqrt{2x^1}}{\sqrt{36}\cdot\sqrt{x^1}\cdot\sqrt{x^1}\cdot\sqrt{y^4}}$$

$$= \dfrac{12x^1y^3\sqrt{2x}}{6\cdot x\cdot y^2} = \dfrac{\overset{2}{\cancel{12}}\cdot\cancel{x^1}\cdot\overset{y^1}{\cancel{y^3}}\sqrt{2x}}{\underset{1}{\cancel{6}}\cdot\underset{1}{\cancel{x}}\cdot\underset{1}{\cancel{y^2}}},\ \text{or } 2y\sqrt{2x}$$

## Written Exercises

**Simplify.**

(A)

1. $\dfrac{\sqrt{15}}{\sqrt{3}}$  $\sqrt{5}$   
2. $\dfrac{\sqrt{32x^3}}{\sqrt{8x}}$  $2x$   
3. $\dfrac{\sqrt{20}}{\sqrt{2}}$  $\sqrt{10}$   
4. $\dfrac{\sqrt{45x^5}}{\sqrt{5x^3}}$  $3x$   
5. $\dfrac{\sqrt{28a^3}}{\sqrt{7a}}$  $2a$

6. $\dfrac{5}{\sqrt{7}}$  $\dfrac{5\sqrt{7}}{7}$   
7. $\dfrac{1}{\sqrt{3}}$  $\dfrac{\sqrt{3}}{3}$   
8. $\dfrac{3}{\sqrt{3}}$  $\sqrt{3}$   
9. $\dfrac{14}{\sqrt{7}}$  $2\sqrt{7}$   
10. $\dfrac{8}{\sqrt{2}}$  $4\sqrt{2}$

11. $\dfrac{3}{\sqrt{12}}$  $\dfrac{\sqrt{3}}{2}$   
12. $\dfrac{6}{\sqrt{20}}$  $\dfrac{3\sqrt{5}}{5}$   
13. $\dfrac{2}{\sqrt{10}}$  $\dfrac{\sqrt{10}}{5}$   
14. $\dfrac{6}{\sqrt{8}}$  $\dfrac{3\sqrt{2}}{2}$   
15. $\dfrac{15}{\sqrt{18}}$  $\dfrac{5\sqrt{2}}{2}$

16. $\sqrt{\dfrac{14x}{24x^2}}$  $\dfrac{\sqrt{21x}}{6x}$   
17. $\sqrt{\dfrac{5x^2}{15x^7}}$  $\dfrac{\sqrt{3x}}{3x^3}$   
18. $\sqrt{\dfrac{22y}{14y^3}}$  $\dfrac{\sqrt{77}}{7y}$   
19. $\sqrt{\dfrac{12a^5}{20a^9}}$  $\dfrac{\sqrt{15}}{5a^2}$   
20. $\sqrt{\dfrac{14b^6}{20b^7}}$  $\dfrac{\sqrt{70b}}{10b}$

21. $\dfrac{3}{\sqrt{x^7}}$  $\dfrac{3\sqrt{x}}{x^4}$   
22. $\dfrac{8}{\sqrt{y^3}}$  $\dfrac{8\sqrt{y}}{y^2}$   
23. $\dfrac{4}{\sqrt{20a^3}}$  $\dfrac{2\sqrt{5a}}{5a^2}$   
24. $\dfrac{6}{\sqrt{8a^5}}$  $\dfrac{3\sqrt{2a}}{2a^3}$   
25. $\dfrac{3x}{\sqrt{12x^3}}$  $\dfrac{\sqrt{3x}}{2x}$

(B)

26. $\dfrac{18ab}{\sqrt{a^6b}}$  $\dfrac{18\sqrt{b}}{a^2}$   
27. $\dfrac{8y^2}{\sqrt{12y}}$  $\dfrac{4y\sqrt{3y}}{3}$   
28. $\dfrac{5x}{\sqrt{x^4y}}$  $\dfrac{5\sqrt{y}}{xy}$   
29. $\dfrac{4c^4d}{\sqrt{cd}}$  $4c^3\sqrt{cd}$   
30. $\dfrac{20x^5y^4}{\sqrt{xy}}$

31. $\dfrac{8x^2y}{\sqrt{8xy^2}}$  $2x\sqrt{2x}$   
32. $\dfrac{9y^3}{\sqrt{18y}}$  $\dfrac{3y^2\sqrt{2y}}{2}$   
33. $\dfrac{9x^5y^2}{\sqrt{6x^3y}}$   
34. $\dfrac{12m^3n^5}{\sqrt{20m^7n^2}}$ ✶   
35. $\dfrac{3x^5y^4}{\sqrt{6x^3y^2}}$

## Example  Simplify $\dfrac{3}{4-\sqrt{2}}$.

30. $20x^4y^3\sqrt{xy}$  33. $\dfrac{3x^3y\sqrt{6xy}}{2}$  34. $\dfrac{6n^4\sqrt{5m}}{5m}$  35. $\dfrac{x^3y^3\sqrt{6x}}{2}$

$\dfrac{3}{4-\sqrt{2}}\cdot\dfrac{4+\sqrt{2}}{4+\sqrt{2}}$ ◄ Multiply the numerator and the denominator by $4+\sqrt{2}$.
$4+\sqrt{2}$ is called the **conjugate** of $4-\sqrt{2}$.

$\dfrac{3(4+\sqrt{2})}{(4-\sqrt{2})(4+\sqrt{2})} = \dfrac{12+3\sqrt{2}}{14}$ ◄ $(4-\sqrt{2})(4+\sqrt{2}) = 4^2-(\sqrt{2})^2 = 16-2 = 14$

**Simplify.**

36. $\dfrac{35+7\sqrt{3}}{22}$  37. $\dfrac{10+5\sqrt{2}}{2}$  38. $\dfrac{-21+7\sqrt{3}}{6}$  39. $\dfrac{-2\sqrt{3}-3\sqrt{2}}{3}$  40. $\dfrac{15\sqrt{5}+5}{44}$

(C)

36. $\dfrac{7}{5-\sqrt{3}}$   
37. $\dfrac{5}{2-\sqrt{2}}$   
38. $\dfrac{-7}{3+\sqrt{3}}$   
39. $\dfrac{\sqrt{2}}{\sqrt{6}-3}$   
40. $\dfrac{5}{3\sqrt{5}-1}$

Dividing Radicals

**411**

# THE PYTHAGOREAN PROPERTY 14.10

*Objectives*
**To find the length of a side of a right triangle by using the Pythagorean property**
**To determine whether a triangle is a right triangle given the lengths of the three sides of the triangle**

The triangle at the right is a right triangle. The longest side of the triangle, $c$, is called the **hypotenuse.** The other two sides of the triangle, $a$ and $b$, are called **legs.** About 500 B.C., Pythagoras, a Greek philosopher and mathematician, discovered the following relationship among the sides of a right triangle: $a^2 + b^2 = c^2$.

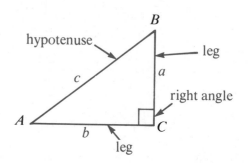

| Pythagorean property | In any right triangle, the sum of the squares of the lengths of the legs equals the square of the length of the hypotenuse. This property can also be stated as: If a triangle is a *right triangle*, then $a^2 + b^2 = c^2$, where $a$ and $b$ are the lengths of the legs and $c$ is the length of the hypotenuse. |
|---|---|

You can use this property to find the length of a side of a right triangle if you are given the lengths of the other two sides.

*Example 1*

**Find the length of the hypotenuse if the lengths of the two legs of a right triangle are 6 m and 8 m.**

Use the Pythagorean property: $a^2 + b^2 = c^2$.
$$8^2 + 6^2 = c^2$$
$$100 = c^2$$
$$\sqrt{100} = c$$
$$10 = c$$
**So,** the length of the *hypotenuse* is 10 m.

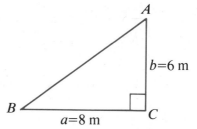

For the remainder of this lesson, you should note that $a$ and $b$ refer to the lengths of the legs of a right triangle and $c$ refers to the length of the hypotenuse. You will find it helpful to sketch a right triangle, as shown at the right, for each problem.

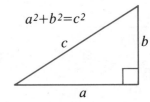

*Example 2*   **For each right triangle, find the missing length. Give the answer in simplest radical form: $a = 6$ and $b = 3$; $a = 6$ and $c = 8$.**

Sketch the triangle. Label the sides you know. Then, use the Pythagorean property to find the length of the remaining side.

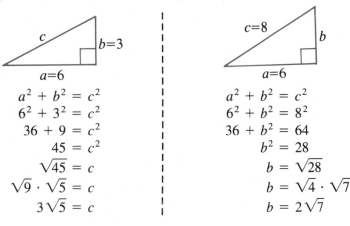

$$a^2 + b^2 = c^2$$
$$6^2 + 3^2 = c^2$$
$$36 + 9 = c^2$$
$$45 = c^2$$
$$\sqrt{45} = c$$
$$\sqrt{9} \cdot \sqrt{5} = c$$
$$3\sqrt{5} = c$$

$$a^2 + b^2 = c^2$$
$$6^2 + b^2 = 8^2$$
$$36 + b^2 = 64$$
$$b^2 = 28$$
$$b = \sqrt{28}$$
$$b = \sqrt{4} \cdot \sqrt{7}$$
$$b = 2\sqrt{7}$$

If you are given the lengths of the three sides of a triangle, you can determine whether or not the triangle is a right triangle by using the *converse* of the Pythagorean property.

To form the *converse* of a statement, you *interchange* the *if* and the *then* parts of the statement as follows:

*Statement:* If a triangle is a right triangle, then $a^2 + b^2 = c^2$.

*Converse:* If $a^2 + b^2 = c^2$, then the triangle is a right triangle.

Both the statement and its converse can be proved by using advanced mathematical techniques. You should keep in mind, however, that the converse of a true statement is not necessarily true.

Using the converse, you can show that a triangle with sides having lengths of 5, 13, and 12 is a right triangle as follows:

| $a^2 + b^2$ | $c^2$ |
|---|---|
| $5^2 + 12^2$ | $13^2$    (The longest side is $c$, the hypotenuse.) |
| $25 + 144$ | $169$ |
| $169$ | |

$$169 = 169, \text{ true}$$

| Converse of the Pythagorean property | If $a^2 + b^2 = c^2$, then the triangle is a right triangle. If the sum of the squares of the lengths of two sides of a triangle equals the square of the length of the third side, then the triangle is a right triangle. |
|---|---|

*Example 3*    **Is the triangle with sides of lengths 5, 9, and 6 a right triangle?**

$$\frac{a^2 + b^2}{5^2 + 6^2} \bigg| \frac{c^2}{9^2}$$

| $a^2 + b^2$ | $c^2$ |
|---|---|
| $5^2 + 6^2$ | $9^2$ |
| $25 + 36$ | $81$ |
| $61$ | |

◀ *The length of the longest side is 9.*

$$61 \neq 81$$

**Thus,** the triangle is *not* a right triangle.

The Pythagorean property can be applied to problems arising in many diverse areas, such as construction and navigation. One such practical application involving carpentry is illustrated in Example 4.

*Example 4*    **A carpenter wants to build a brace for a gate that is 2 m wide and 3 m high. Find the length of the brace to the nearest tenth of a meter.**

Sketch the gate. The brace is the diagonal of a rectangle. It is also the hypotenuse of a right triangle.
Let $c$ represent the length of the hypotenuse, the brace.

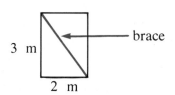

$$c^2 = 3^2 + 2^2$$
$$c^2 = 9 + 4$$
$$c^2 = 13$$
$$c = \sqrt{13}$$
$$c \doteq 3.606 \quad ◀ \textit{Use the square root table on page 532.}$$
$$c \doteq 3.6 \quad ◀ \textit{Round to the nearest tenth.}$$

**So,** the brace is 3.6 m long, to the nearest tenth of a meter.

# Oral Exercises

**The lengths of the three sides of a triangle are given below. For each triangle, tell if it is a right triangle. If it is a right triangle, tell which is the length of the hypotenuse.**

1. 6, 8, 10      2. 15, 9, 12      3. 5, 7, 4      4. 10, 20, 30      5. 12, 13, 5      6. 40, 30, 50
   **yes; 10**      **yes; 15**      **no**      **no**      **yes; 13**      **yes; 50**

# Written Exercises

**Solve each problem.**

(A) 1. Find the length of the hypotenuse if the lengths of the legs of a right triangle are 3 m and 4 m. **5 m**

2. Find the length of the hypotenuse if the lengths of the legs of a right triangle are 5 cm and 12 cm. **13 cm**

**3.** The length of the hypotenuse of a right triangle is 4 m. The length of one leg is 2 m. Find the length of the other leg. Give the answer in simplest radical form. **$2\sqrt{3}$ m**

**4.** The length of the hypotenuse of a right triangle is 6 cm. The length of one leg is 4 cm. Find the length of the other leg. Give the answer in simplest radical form. **$2\sqrt{5}$ cm**

**For each right triangle, find the missing length. Give the answer in simplest radical form.**

**5.** $a = 12, b = 16$ **$c = 20$**     **6.** $a = 9, c = 15$ **$b = 12$**     **7.** $a = 15, c = 17$ **$b = 8$**
**8.** $a = 3, b = 3$ **$c = 3\sqrt{2}$**     **9.** $a = 6, b = 6$ **$c = 6\sqrt{2}$**     **10.** $b = 2, c = 6$ **$a = 4\sqrt{2}$**
**11.** $a = 7, c = 25$ **$b = 24$**     **12.** $a = 8, b = 2$ **$c = 2\sqrt{17}$**     **13.** $c = 4, b = 2$ **$a = 2\sqrt{3}$**
**14.** $a = 4, b = 6$ **$c = 2\sqrt{13}$**     **15.** $a = 2, b = 2$ **$c = 2\sqrt{2}$**     **16.** $b = 2, c = 8$ **$a = 2\sqrt{15}$**

**Is the triangle with the given lengths for its three sides a right triangle?**

**17.** $3, 4, 2$ **no**          **18.** $10, 20, 16$ **no**          **19.** $6, 7, 8$ **no**          **20.** $41, 9, 40$ **yes**

**For each right triangle, find the missing length. If the answer is not a rational number, give it in simplest radical form.**

Ⓑ **21.** $b = 4, c = 4\sqrt{5}$ **$a = 8$**     **22.** $a = 2\sqrt{2}, b = 2\sqrt{3}$     **23.** $a = \sqrt{3}, c = \sqrt{12}$ **$b = 3$**
**24.** $a = 2\sqrt{3}, b = 2\sqrt{3}$     **25.** $b = 4\sqrt{2}, c = 8\sqrt{2}$     **26.** $b = 2\sqrt{5}, c = 8$ **$a = 2\sqrt{11}$**
**27.** $c = 8\sqrt{2}, a = 8$ **$b = 8$**     **28.** $c = 8\sqrt{3}, b = 4\sqrt{3}$     **29.** $a = 2\sqrt{5}, c = 2\sqrt{6}$ **$b = 2$**

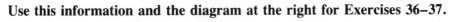
**22.** $c = 2\sqrt{5}$   **24.** $c = 2\sqrt{6}$   **25.** $a = 4\sqrt{6}$   **28.** $a = 12$

**Solve each problem. Round each answer to the nearest tenth of the unit.**

**30.** The size of the screen of a television set is usually given in terms of its diagonal. Find the diagonal length of a screen 15 cm wide and 20 cm long. **25 cm**

**31.** A rectangular field is 50 m wide by 100 m long. How long is the diagonal path connecting two opposite corners of the field? **111.8 m**

**32.** A mover must try to fit a thin circular mirror, 2 m in diameter, through a doorway measuring 1 m by 1.8 m. Will the mirror fit through the doorway? **yes**

**33.** A 4-m ladder is 1 m from the base of a building. How high up the building will the ladder reach? **3.9 m**

**34.** A 12-m loading ramp covers 9 m of ground. How high does it rise? **7.9 m**

**35.** A surveyor walked 8 km north, then 3 km west. How far was the surveyor from the starting point? **8.5 km**

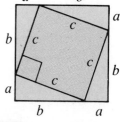

**Use this information and the diagram at the right for Exercises 36–37.**

The diagram at the right shows a square with each side of length $(a + b)$. The four-sided figure inside the square is a square with each side of length $c$.

Ⓒ **36.** Use the concept of area to tell why $(a + b)^2 = 4(\frac{1}{2}ab) + c^2$ must be true. **area of large square = 4(area of triangle) + area of small square**

**37.** Use the equation in Exercise 36 to verify the Pythagorean property.

**For 37 see TE answer section.**

# ZERO AND NEGATIVE INTEGRAL EXPONENTS

14.11

*Objectives*
**To simplify expressions containing zero and negative exponents**
**To write numbers in scientific notation**

Recall that to divide powers of the same base, you subtract the powers of the base. Notice what happens when the powers are the same.

$$\frac{4^7}{4^7} = 4^{7-7} = 4^0 \qquad \text{But } \frac{4^7}{4^7} = 1$$

because any number divided by itself is 1. Using the Substitution Property, you can conclude that $4^0 = 1$.

| | |
|---|---|
| Definition: Zero exponent | For each nonzero number $x$, $x^0 = 1$. |

You can also extend the Quotient of Powers Property to include negative exponents.

$$\frac{4^3}{4^5} = 4^{3-5} = 4^{-2} \quad \text{But } \frac{4^3}{4^5} = \frac{\overset{1}{\cancel{4}} \cdot \overset{1}{\cancel{4}} \cdot \overset{1}{\cancel{4}}}{\underset{1}{\cancel{4}} \cdot \underset{1}{\cancel{4}} \cdot \underset{1}{\cancel{4}} \cdot 4 \cdot 4} = \frac{1}{4^2}$$

This implies that $4^{-2} = \dfrac{1}{4^2}$.

Consider what happens when a negative exponent appears in the denominator.

$$\frac{1}{5^{-2}} = 1 \div 5^{-2} = 1 \cdot \frac{1}{5^{-2}} = 1 \cdot 5^2 = 5^2$$

**Thus,** $\dfrac{1}{5^{-2}} = 5^2$.

| | |
|---|---|
| Definition: Negative exponents | For each nonzero number $x$ and each positive integer $n$, $$x^{-n} = \frac{1}{x^n} \text{ and } \frac{1}{x^{-n}} = x^n.$$ |

*Example 1*
**Simplify. Write the result with positive exponents only.**

$$
\begin{array}{l}
8 \cdot 3^0 \\
8 \cdot 1 \\
\quad 8
\end{array}
\qquad
\begin{array}{l}
5b^{-7} \\
5 \cdot b^{-7} \\
5 \cdot \dfrac{1}{b^7} = \dfrac{5}{b^7}
\end{array}
$$

$$
\frac{-4x^7 y^{-2}}{8x^{-3} y^4} = \frac{-4 \cdot x^7 \cdot y^{-2}}{8 \cdot x^{-3} \cdot y^4}
$$

$$
= \frac{1 \cdot x^7 \cdot x^3}{2 \cdot y^2 \cdot y^4} = -\frac{1x^{10}}{2y^6}
$$

Powers of 10 are used to write very large and very small numbers in a more convenient form called *scientific notation*.

---

| Definition:<br>Scientific<br>notation | A number is in **scientific notation** when it is in the form $n \times 10^m$ where $1 \le n < 10$ and $m$ is an integer. |
| --- | --- |

---

| Number | Decimal Moved | | Scientific Notation |
| --- | --- | --- | --- |
| 93,000,000 | 9 3,000,000. | 7 places | $9.3 \times 10^7$ |
| 0.0000064 | 0.000006 4 | 6 places | $6.4 \times 10^{-6}$ |
| 4.890007 | 4.890007 | 0 places | $4.890007 \times 10^0$ |
| 3402.8000 | 3 402.8 | 3 places | $3.4028 \times 10^3$ |
| 0.0000289 | 0.00002 89 | 5 places | $2.89 \times 10^{-5}$ |

When a number $n$ is written in scientific notation, the power of 10 is negative if $n < 1$, zero if $1 \le n < 10$, and positive if $n \ge 10$.

*Example 2*   **Simplify** $\dfrac{(2,700,000)(0.12)}{0.00000045}$. **Write the result in scientific notation.**

$$\frac{(2,700,000)(0.12)}{0.00000045} = \frac{(2.7 \times 10^6)(1.2 \times 10^{-1})}{4.5 \times 10^{-7}}$$

◀ *Write each number in scientific notation.*

$$= \frac{2.7 \times 1.2 \times 10^6 \times 10^{-1} \times 10^7}{4.5}$$

◀ *Get all powers of 10 in the numerator.*

$$= \frac{3.24 \times 10^{12}}{4.5}$$

◀ $2.7 \times 1.2 = 3.24$

$$= 0.72 \times 10^{12}$$

◀ $3.24 \div 4.5 = 0.72$

$$= 7.2 \times 10^{-1} \times 10^{12}$$

◀ *Write 0.72 in scientific notation.*

$$= 7.2 \times 10^{11}$$

# Oral Exercises

**Simplify.**

1. $7^0$ **1**     2. $19 \cdot 2^0$ **19**     3. $5^{-2}$ $\dfrac{1}{25}$     4. $9^2 \cdot 9^0$ **81**     5. $4^0 \cdot 8^{-2}$ $\dfrac{1}{64}$

**Express each with positive exponents only. No variable is zero.**

6. $x^{-5} \cdot x^8$ $\boldsymbol{x^3}$     7. $m^{-6} \cdot n^{-4} \cdot m^7 \cdot n^4$ $\boldsymbol{m}$     8. $\dfrac{b^4}{b^{-2}}$ $\boldsymbol{b^6}$     9. $\dfrac{a^{-6}}{a^3}$ $\dfrac{1}{\boldsymbol{a^9}}$     10. $\dfrac{x^{-3}}{x^{-7}}$ $\boldsymbol{x^4}$

**Express each in scientific notation.**

11. 165,000       12. 0.0000067       13. 76,879,000,000       14. 0.00000000567
    $\boldsymbol{1.65 \times 10^5}$       $\boldsymbol{6.7 \times 10^{-6}}$       $\boldsymbol{7.6879 \times 10^{10}}$       $\boldsymbol{5.67 \times 10^{-9}}$

*Zero and Negative Integral Exponents*       **417**

# Written Exercises

3. $\dfrac{2d^{18}}{3c^3}$  4. $\dfrac{4}{3}m^3$  5. $\dfrac{3}{4}x^3y^3$  7. $\dfrac{-x^{10}y^2}{2z^8}$  8. $\dfrac{2r^6t^6}{s^{10}}$

**Simplify. Write each with positive exponents only. No variable is zero.**

1. $\dfrac{a^5b^{-6}}{c^3d^{-7}}$  $\dfrac{a^5d^7}{c^3b^6}$

2. $\dfrac{a^6b^{-9}}{a^{11}b^{-12}}$  $\dfrac{b^3}{a^5}$

3. $\dfrac{8c^{-7}d^9}{12c^{-4}d^{-9}}$

4. $\dfrac{20m^{-11}n^{-10}}{15m^{-14}n^{-10}}$

5. $\dfrac{6x^{-5}y^{-1}}{8x^{-8}y^{-4}}$

6. $(4a^4)(-7b^5)(a^{-4})(b^{-7})$  $\dfrac{-28}{b^2}$

7. $\dfrac{-4x^7y^{-2}z^{-3}}{8x^{-3}y^{-4}z^5}$

8. $\dfrac{12r^4s^{-7}t^0}{6r^{-2}s^3t^{-6}}$

9. $\dfrac{(5abc)^0}{5^{-2}ab^3c}$  $\dfrac{25}{ab^3c}$

**Write each number in scientific notation.**

10. $300{,}000$  $3 \times 10^5$

11. $\dfrac{4}{100{,}000}$  $4 \times 10^{-5}$

12. $27$ billion  $2.7 \times 10^{10}$

13. $9$ thousandths  $9 \times 10^{-3}$

**Simplify. Write each result in scientific notation.**

14. $\dfrac{360{,}000}{1.5(0.024)}$  $10^7$

15. $\dfrac{72{,}000(0.002)}{0.18}$  $8 \times 10^2$

16. $\dfrac{288{,}000(0.0003)}{0.000000036}$  $2.4 \times 10^9$

17. $\dfrac{0.000024}{0.000024(0.00125)}$  $8 \times 10^2$

18. $\dfrac{1{,}575{,}000(0.000071)}{0.000035}$  $3.195 \times 10^6$

19. $\dfrac{0.00231(0.72)}{0.055(3{,}600)}$  $8.4 \times 10^{-6}$

# GEOMETRY REVIEW

*Objective*  **To find the measure of angles of a triangle**

An exterior angle of a triangle is formed by extending one of its sides.

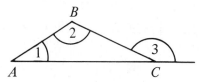

Extend side $\overline{AC}$ of triangle $ABC$.
$\angle 3$ is an exterior angle of $\triangle ABC$.
$$m\angle 1 + m\angle 2 = m\angle 3$$

*Example*  **Find the measure of $\angle x$.**

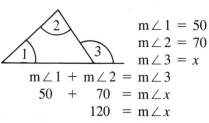

$m\angle 1 = 50$
$m\angle 2 = 70$
$m\angle 3 = x$

$m\angle 1 + m\angle 2 = m\angle 3$
$50 \ + \ 70 \ = m\angle x$
$120 \ = m\angle x$

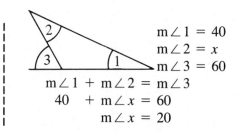

$m\angle 1 = 40$
$m\angle 2 = x$
$m\angle 3 = 60$

$m\angle 1 + m\angle 2 = m\angle 3$
$40 \ + m\angle x = 60$
$m\angle x = 20$

# Written Exercises

**Find the measure of $\angle x$.**

1.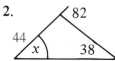

2.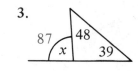

3.

4.

# APPLICATIONS

Read ⟶ Plan ⟶ Solve ⟶ Interpret

Two students, Tanya and Jacob, compared their grades for the first marking period in algebra. Each received the same mark for the marking period, 80. Tanya argued that her scores were more consistent than Jacob's since they were all close to 80. Even though Jacob had two A's, he also scored 65 on two tests.

| Tanya's Scores | Jacob's Scores |
|---|---|
| 80, 82, 78, 80 | 100, 65, 90, 65 |
| 80 | 100 |
| 82 | 65 |
| 78 | 90 |
| 80 | 65 |
| 320 | 320 |

Average: $\dfrac{320}{4} = 80$     Average: $\dfrac{320}{4} = 80$

There is a valuable tool for measuring the consistency of scores in finding an average, or mean. This commonly used statistic is called the **standard deviation.** It takes into consideration the amount by which each score *deviates, or differs, from the mean.* Below is an outline of how to find the standard deviation of Tanya's and Jacob's scores.

|  |  | Tanya | Jacob |
|---|---|---|---|
| Step 1 | Find the mean. | $\dfrac{80 + 82 + 78 + 80}{4} = 80$ | $\dfrac{100 + 65 + 90 + 65}{4} = \dfrac{320}{4} = 80$ |
| Step 2 | Find the deviation of each score from the mean. | $80 - 80 = 0$<br>$82 - 80 = 2$<br>$78 - 80 = -2$<br>$80 - 80 = 0$ | $100 - 80 = 20$<br>$65 - 80 = -15$<br>$90 - 80 = 10$<br>$65 - 80 = -15$ |
| Step 3 | Square the deviations. | $0^2 = 0, 2^2 = 4, (-2)^2 = 4$ | $20^2 = 400, (-15)^2 = 225, 10^2 = 100,$<br>$(-15)^2 = 225$ |
| Step 4 | Find the mean of the squares. | $\dfrac{0 + 4 + 4 + 0}{4} = \dfrac{8}{4} = 2$ | $\dfrac{400 + 225 + 100 + 225}{4} = 237.5$ |
| Step 5 | Find the square root of the answer. Approximate to three decimal places. | $\sqrt{2} \doteq 1.414$ | $\sqrt{237.5}$ rounds to $\sqrt{238}$.<br>$\sqrt{238} \doteq 15.427$ |

The standard deviation for Tanya's scores is 1.414 and for Jacob's scores, 15.427. Since the standard deviation for Tanya's scores is less than the standard deviation for Jacob's scores, her scores are more consistent than his.

The five-step process for computing standard deviation can be summarized by the formula on the next page.

| Formula for standard deviation | Let $m$ be the mean of the numbers $S_1, S_2, S_3, \ldots, S_n$, then the formula for standard deviation (s.d.) is |
|---|---|

$$\text{s. d.} = \sqrt{\frac{(S_1 - m)^2 + (S_2 - m)^2 + \cdots + (S_n - m)^2}{n}}.$$

## Example

**Marsha is on a basketball team. The number of baskets she scored in the last 5 games were: 10, 12, 8, 6, and 14. Compute the standard deviation. Another player, Joe, had a standard deviation of 3.642 for 5 games. Which player was more consistent?**

First, find the mean for Marsha's scores:

$$m = \frac{10 + 12 + 8 + 6 + 14}{5} = \frac{50}{5}, \text{ or } 10$$

Use the formula for standard deviation.

$$\text{s.d.} = \sqrt{\frac{(S_1 - m)^2 + (S_2 - m)^2 + (S_3 - m)^2 + (S_4 - m)^2 + (S_5 - m)^2}{5}} \quad \blacktriangleleft \text{\textit{There are 5 scores.}}$$

$$= \sqrt{\frac{(10 - 10)^2 + (12 - 10)^2 + (8 - 10)^2 + (6 - 10)^2 + (14 - 10)^2}{5}} \quad \blacktriangleleft \text{\textit{Substitute } } S_1 = 10, S_2 = 12, \text{ etc.}$$

$$= \sqrt{\frac{0^2 + 2^2 + (-2)^2 + (-4)^2 + 4^2}{5}}$$

$$= \sqrt{\frac{0 + 4 + 4 + 16 + 16}{5}} = \sqrt{\frac{40}{5}} = \sqrt{8}$$

Marsha's standard deviation is $\sqrt{8} \doteq 2.828$.

Marsha was more consistent since her standard deviation was less than Joe's. $\quad \blacktriangleleft \quad \underset{\text{\textit{Marsha}}}{2.828} < \underset{\text{\textit{Joe}}}{3.642}$

## Written Exercises

**Find the standard deviation for each set of scores.**

1. 90, 85, 95, 90, 100 **5.1**
2. 150, 145, 155, 170 **9.4**
3. 10, 8, 12, 6, 8, 16 **3.3**
4. 5, 8, 6, 7, 4 **1.4**

**Solve each problem.**

5. Washington's scores on his Spanish tests were 80, 85, 80, 85, and 75. Debby's scores were 80, 82, 83, 79, and 81. Who had the higher average? Who was more consistent? **both averaged 81; Debby**

**both averaged 90; Jane**

6. Jane's algebra scores were 90, 86, 88, 88, and 98. Wayne's scores were 87, 93, 80, 100, and 90. Who had the higher average? Who was more consistent?

7. On a math test, Mr. Koliba's class had a mean of 88 with a standard deviation of 12.3. Ms. Beekman's class had an average of 86 with a standard deviation of 3.4. Which class was more consistent?

**Ms. Beekman's**

8. Last year's average monthly temperature for two cities, $A$ and $B$, follow.

| | J | F | M | A | M | J | J | A | S | O | N | D |
|---|---|---|---|---|---|---|---|---|---|---|---|---|
| A | 23 | 25 | 34 | 47 | 59 | 68 | 73 | 71 | 60 | 53 | 35 | 30 |
| B | 24 | 30 | 35 | 50 | 57 | 73 | 74 | 76 | 64 | 57 | 44 | 28 |

Which had more consistent temperatures?

**A was slightly more consistent**

# COMPUTER ACTIVITIES

## Using the Pythagorean Theorem

The Pythagorean theorem $a^2 + b^2 = c^2$ in BASIC notation is
$A \wedge 2 + B \wedge 2 = C \wedge 2$.

This program finds the hypotenuse of a right triangle. In BASIC, SQR(V)
is the function to find the square root of a given variable V.

```
10    PRINT "THIS PROGRAM FINDS THE HYPOTENUSE OF A RI
         GHT TRIANGLE."
20    INPUT "ENTER THE LENGTH OF THE FIRST LEG. ";A
30    INPUT "ENTER THE LENGTH OF THE SECOND LEG. ";B
40    REM   LINE 50 COMPUTES S = C ^ 2 AND LINE 60 FIND
         S C.
50    LET S = A ^ 2 + B ^ 2
60    LET C =  SQR (S)
70    PRINT "THE LENGTH OF THE HYPOTENUSE IS "C"."
80    END
```

See the Computer Section beginning on page 488 for more information.

## Exercises   For Ex. 7–8, see TE Answer Section.

**Enter and RUN the program above for the following values:**

1. $a = 3, b = 4$ **5**  2. $a = 6, b = 8$ **10**  3. $a = 5, b = 12$ **13**
4. $a = 8, b = 15$ **17**  5. $a = 1, b = 1$  6. $a = 10, b = 20$ **22.3606798**

   **1.41421356**

7. Write a program to find the length of a side of a right triangle given the
   other two sides. Use the SQR function in BASIC and an INPUT statement
   to assign values to the variables C and B. To stop the program when it
   is completed, use a question.

8. Write a program to determine if a triangle is a right triangle given the
   length of the three sides.

9. Complete and run the program below for finding the square root of 5, 7,
   and 11.

```
10    INPUT ___X___
20    LET N = X/2
30    LET R = ___X/N___
40    LET S = R + ___N___
50    LET G = S/2
60    IF ABS (G − N) < .001 THEN 90
70    LET N = G
80    GO TO ___30___
90    PRINT " SQUARE ROOT OF " X " IS " G
100   ___END___            roots are 2.236068, 2.645751, 3.316625
```

Ⓐ Exercises are 1, 2, 5, 6, 9, 10, 13–15, 18, 20, 21, 23, 24, 26–28, 31–33, 36, 37.

# CHAPTER FOURTEEN REVIEW

Ⓒ Exercises are starred.          The rest are Ⓑ Exercises.

**Vocabulary**

hypotenuse [14.10]          rationalize [14.2]
principal square root [14.2]          square
Pythagorean property [14.10]          root [14.2]
radical [14.2]
radicand [14.2]

Find the fraction for the repeating decimal. [14.1]
**1.** $0.\overline{4}$ $\frac{4}{9}$  **2.** $0.6\overline{3}$ $\frac{19}{30}$  **3.** $0.\overline{162}$ $\frac{6}{37}$ **4.** $3.\overline{21}$ $3\frac{7}{33}$

Find an approximation of the indicated principal square root. Use the table on page 532. [14.2]
**5.** $\sqrt{77}$ **8.775**          **6.** $\sqrt{68}$ **8.246**

**7.** Find the height, $h$, to the nearest tenth of a meter. [14.2] **9.8 m**

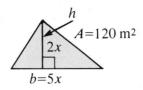

**8.** The length of a rectangle is 4 times the width. The area is 240 m². Find the length to the nearest tenth of a meter. [14.2] **31.0 m**

Approximate to the nearest tenth. Use the divide-and-average method. [14.3]
**9.** $\sqrt{35}$  **10.** $\sqrt{28}$  **11.** $\sqrt{113}$  **12.** $\sqrt{201}$
  **5.9**      **5.3**      **10.6**      **14.2**

Simplify. [14.4]
**13.** $\sqrt{7} \cdot \sqrt{3}$   **14.** $\sqrt{36} \cdot \sqrt{49}$   **15.** $\sqrt{24}$
  $\sqrt{21}$        **42**        $2\sqrt{6}$

Simplify. Then approximate the result to the nearest tenth. [14.4]
**16.** $\sqrt{84}$ $2\sqrt{21}$, **9.2**      **17.** $\sqrt{140}$ $2\sqrt{35}$, **11.8**

Simplify.  **19.** $2a^2b^4\sqrt{7}$        **21.** $2x^5y^7\sqrt{10x}$
**18.** $\sqrt{4x^6y^{12}}$ $2x^3y^6$       **19.** $\sqrt{28a^4b^8}$ [14.5]
**20.** $\sqrt{a^9}$ $a^4\sqrt{a}$          **21.** $\sqrt{40x^{11}y^{14}}$
**22.** $-5xy^3\sqrt{63x^4y^3}$ [14.6] $-15x^3y^4\sqrt{7y}$
**23.** $3\sqrt{5} - 4\sqrt{5} + 9\sqrt{5}$ [14.7] $8\sqrt{5}$
**24.** $2\sqrt{50ab} - 4\sqrt{8ab} + 5\sqrt{18ab}$ $17\sqrt{2ab}$
**25.** $7a\sqrt{27a^3b^3} + b\sqrt{48a^5b}$ $25a^2b\sqrt{3ab}$
**26.** $\sqrt{3}(\sqrt{7} + \sqrt{3})$ $3 + \sqrt{21}$
**27.** $-4\sqrt{2}(\sqrt{12} + 3\sqrt{8})$ [14.8] $-8\sqrt{6} - 48$
**28.** $(4\sqrt{3} - 2\sqrt{6})(2\sqrt{3} + 5\sqrt{6})$ $-36 + 48\sqrt{2}$
**29.** $(3\sqrt{8} + 2\sqrt{3})^2$    **30.** $(7\sqrt{2} - 3\sqrt{6})^2$
**29.** $84 + 24\sqrt{6}$   **30.** $152 - 84\sqrt{3}$   **33.** $\frac{1}{3x}$
**31.** $\frac{1}{\sqrt{5}}$ $\frac{\sqrt{5}}{5}$ **32.** $\frac{15}{\sqrt{20}}$ $\frac{3\sqrt{5}}{2}$**33.** $\sqrt{\frac{2x}{18x^3}}$ [14.9]
**34.** $\frac{20ab}{\sqrt{8a^6b}}$ $\frac{5\sqrt{2b}}{a^2}$  **35.** $\frac{20m^4n^3}{\sqrt{45m^7n^2}}$ $\frac{4n^2\sqrt{5m}}{3}$

**36.** Is the triangle with sides of lengths 5, 12, and 13 a right triangle? [14.10] **yes**

For each right triangle, find the missing length in radical form. [14.10]
**37.** $a = 4$, $c = 6$ $2\sqrt{5}$ **38.** $b = 6$, $c = 3\sqrt{5}$ **3**

**39.** An 8-m ladder is 2 m from the base of a building. How high does it rise? Give the answer to the nearest tenth of a meter. [14.10] **7.7 m**

★ **40.** Simplify $3\sqrt{2,028}$. Then approximate the result to the nearest tenth. [14.4]
  $78\sqrt{3}$, **135.1**

For what values of $x$ will the radical be a real number? [14.5]
★ **41.** $\sqrt{x^2 - 11x + 24}$ $x \geq 8$, $x \leq 3$

(A) Exercises are 1, 3, 4, 6, 8–13, 15, 16, 18, 19, 22, 23, 28, 29.

# CHAPTER FOURTEEN TEST

(C) Exercises are starred.          The rest are (B) Exercises.

Find the fraction for the repeating decimal.

**1.** $0.\overline{9}$   $\frac{1}{1}$       **2.** $0.\overline{132}$   $\frac{44}{333}$

Find an approximation of the indicated principal square root. Use the table on page 532.

**3.** $\sqrt{66}$   **8.124**       **4.** $\sqrt{43}$   **6.557**

**5.** Find the height, $h$, to the nearest tenth of a meter. **12.0 m**

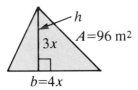

Approximate to the nearest tenth. Use the divide-and-average method.

**6.** $\sqrt{62}$   **7.9**       **7.** $\sqrt{302}$   **17.4**

Simplify.

**8.** $\sqrt{5} \cdot \sqrt{11}$   $\sqrt{55}$   **9.** $\sqrt{81} \cdot \sqrt{4}$   **18**

**10.** $\sqrt{60}$   $2\sqrt{15}$       **11.** $\sqrt{44}$   $2\sqrt{11}$

**12.** $\sqrt{9x^{10}y^4}$   $3x^5y^2$   **13.** $\sqrt{24x^7y^8}$   $2x^3y^4\sqrt{6x}$

**14.** $-4ab^2\sqrt{72a^5b^{10}}$   $-24a^3b^7\sqrt{2a}$

**15.** $7\sqrt{3} - 4\sqrt{3} + 2\sqrt{3}$   $5\sqrt{3}$

**16.** $3\sqrt{12ab} - 7\sqrt{3ab} + 4\sqrt{27ab}$   $11\sqrt{3ab}$

**17.** $3m^2\sqrt{20m^5n^5} + m^4\sqrt{45mn^5}$   $9m^4n^2\sqrt{5mn}$

**18.** $\sqrt{5}(\sqrt{6} + 2\sqrt{5})$   $\sqrt{30} + 10$

**19.** $-2\sqrt{3}(\sqrt{6} + 4\sqrt{12})$   $-6\sqrt{2} - 48$

**20.** $(2\sqrt{3} - \sqrt{8})(\sqrt{3} + 2\sqrt{8})$   $-10 + 6\sqrt{6}$

**21.** $(\sqrt{3} - 2\sqrt{6})^2$   $27 - 12\sqrt{2}$

Simplify.

**22.** $\dfrac{1}{\sqrt{7}}$   $\dfrac{\sqrt{7}}{7}$       **23.** $\dfrac{10}{\sqrt{12}}$   $\dfrac{5\sqrt{3}}{3}$

**24.** $\dfrac{12bc}{\sqrt{b^4c}}$   $\dfrac{12\sqrt{c}}{b}$    **25.** $\dfrac{12x^5y^3}{\sqrt{27x^5y^4}}$   $\dfrac{4x^2y\sqrt{3x}}{3}$

Simplify. Then approximate the result to the nearest tenth.

**26.** $\sqrt{72}$   $6\sqrt{2}$, **8.5**       **27.** $\sqrt{120}$   $2\sqrt{30}$, **11.0**

**28.** Is the triangle with sides of lengths 4, 5, and 6 a right triangle? **no**

For each right triangle, find the missing length in radical form.

**29.** $a = 1, c = 8$     **30.** $b = 2, c = 4\sqrt{3}$
      $3\sqrt{7}$            $2\sqrt{11}$

Solve each problem.

**31.** The length of a rectangle is 5 times the width. The area is 140 m². Find the length to the nearest tenth of a meter. **26.5 m**

**32.** A screen on a television set is 12 cm by 10 cm. What is the diagonal length of the screen to the nearest tenth of a centimeter? **15.6 cm**

★**33.** Simplify $4\sqrt{1,080}$. Then approximate the result to the nearest tenth. $24\sqrt{30}$, **131.5**

★**34.** For what values of $x$ will $\sqrt{x^2 - 5x + 4}$ be a real number? $x \geq 4, x \leq 1$

Simplify.   $-4 + \sqrt{15} + \sqrt{35}$

★**35.** $(\sqrt{3} + \sqrt{7})(\sqrt{3} + \sqrt{5} - \sqrt{7})$

★**36.** $\dfrac{-26}{2\sqrt{3} - 5}$   $4\sqrt{3} + 10$

# COLLEGE PREP TEST

Directions: In each item, you are to compare a quantity in Column 1 with a quantity in Column 2. Write the letter of the correct answer from these choices:

    A. The quantity in Column 1 is greater than the quantity in Column 2;

    B. The quantity in Column 2 is greater than the quantity in Column 1;

    C. The quantity in Column 1 is equal to the quantity in Column 2;

    D. The relationship cannot be determined from the given information.

*Notes:* Information centered over both columns refers to one or both of the quantities to be compared. A symbol that appears in both columns has the same meaning in each column, and all variables represent real numbers.

| | Column 1 | Column 2 | |
|---|---|---|---|
| 1. | $\sqrt{49}$ | $\sqrt{7}$ | A |
| 2. | $\sqrt{16 + 9}$ | $\sqrt{16} + \sqrt{9}$ | B |
| 3. | $\sqrt{\dfrac{1}{0.25}}$ | 4 | B |
| 4. | 0.425 | $3\sqrt{0.0196}$ | A |
| 5. | $\sqrt{9} + 1^4$ | $2\sqrt{4}$ | C |
| 6. | $\sqrt{\dfrac{1}{9} + \dfrac{1}{16}}$ | $\sqrt{\dfrac{1}{16}} + \sqrt{\dfrac{1}{9}}$ | B |
| 7. | $x\sqrt{0.09} = 3.0$ | | |
| | $x$ | 10 | C |

| | Column 1 | Column 2 | |
|---|---|---|---|
| 8. | $\dfrac{1}{x} = \sqrt{0.25}$ | | C |
| | $x$ | $\sqrt{\dfrac{64}{16}}$ | |
| 9. | $\sqrt{3}$ | $\dfrac{3}{\sqrt{3}}$ | C |
| 10. | $(0.4)^2$ | $\sqrt{0.16}$ | B |
| 11. | $2^x = 8$ | | |
| | $x$ | 1 | A |
| 12. | $x > 13$ | | |
| | $\sqrt{x + 3}$ | 4 | A |

# 15 QUADRATIC FUNCTIONS

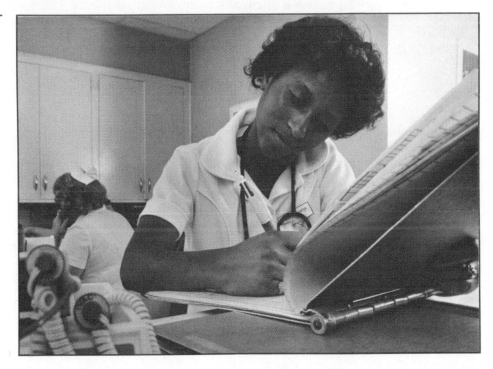

In many hospitals, one nurse at a floor station is responsible for administering medications. Sometimes the exact dosage prescribed by a doctor for a patient is not readily available. A nurse must use mathematics to calculate the correct dosage for each patient.

**Problem 1**

A pediatrician has ordered a certain medication 80 mg q. 4 h p.r.n. (every 4 h as needed). The nurse has the medication available but in a solution labeled 60 mg = 0.6 mL. How many mL of this solution should the nurse give the patient? **0.8 mL**

**Problem 2**

A doctor has ordered 3,000 mL of 5% G/W (glucose and water) to run 24 h for a patient. The IV (intravenous) set the nurse has available has a "drop factor" of 10 drops = 1 mL. How many drops per minute should the patient receive? **$20\frac{5}{6}$ drops**

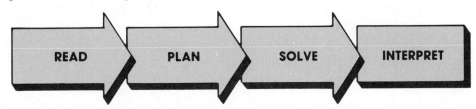

READ    PLAN    SOLVE    INTERPRET

# EQUATIONS CONTAINING RADICALS     15.1

*Objectives*     **To solve equations containing radicals**
**To solve word problems that lead to equations containing radicals**

Notice that the equation $x = 4$ has only one solution, 4.
If you square each side of the equation, you get:
$$x^2 = 4^2, \text{ or}$$
$$x^2 = 16$$
This equation has two solutions, 4 and $-4$. One of them, 4, is also a solution of the original equation, but $-4$ is not a solution of $x = 4$. Thus, the squaring of each side of an equation can result in an equation whose roots are not solutions of the original equation.

You can solve equations that contain radicals. For example, the equation $\sqrt{x} = 5$ can be changed to an equation that has no radicals by:
1. squaring each side, then
2. solving the resulting equation.

Be sure to check whether the solution is indeed a solution of the original equation. Equations containing radicals will be called *radical equations*.

*Example 1*     **Solve $\sqrt{x} = 5$.**

$$\sqrt{x} = 5$$
$$(\sqrt{x})^2 = 5^2$$
$$x = 25 \quad \blacktriangleleft \ (\sqrt{x})^2 = \sqrt{x} \cdot \sqrt{x} = x$$

**Thus,** the solution is 25.

*Check.*

| $\sqrt{x}$ | 5 |
|---|---|
| $\sqrt{25}$ | 5 |
| 5 | |
| | $5 = 5$, true |

To solve an equation such as $\sqrt{4x + 12} + 7 = 1$, first get the radical on one side. Then square each side. There might be no solution.

*Example 2*     **Solve $\sqrt{4x + 12} + 7 = 1$.**

$$\sqrt{4x + 12} + 7 = 1 \quad \blacktriangleleft \ Add\ -7\ to\ each\ side.$$
$$\sqrt{4x + 12} = -6 \quad \blacktriangleleft \ The\ radical\ is\ now\ by\ itself.$$
$$(\sqrt{4x + 12})^2 = (-6)^2 \quad \blacktriangleleft \ Square\ each\ side.$$
$$4x + 12 = 36 \quad \blacktriangleleft \ \sqrt{4x + 12} \cdot \sqrt{4x + 12} = 4x + 12$$
$$4x = 24$$
$$x = 6$$

*Check.*

| $\sqrt{4x + 12} + 7$ | 1 |
|---|---|
| $\sqrt{4 \cdot 6 + 12} + 7$ | 1 |
| $\sqrt{24 + 12} + 7$ | |
| $\sqrt{36} + 7$ | |
| $6 + 7$ | |
| 13 | |

But $13 \neq 1$
So, 6 is not a solution.

**Thus,** there is no solution of $\sqrt{4x + 12} + 7 = 1$.

## Example 3    Solve $\sqrt{6a} = 2\sqrt{3}$.

$$\sqrt{6a} = 2\sqrt{3}$$
$$(\sqrt{6a})^2 = (2\sqrt{3})^2$$
$$6a = 12 \quad \blacktriangleleft \; (2\sqrt{3})^2 = 2 \cdot 2 \cdot \sqrt{3} \cdot \sqrt{3} = 4 \cdot 3 = 12$$
$$a = 2$$

**Thus,** the solution is 2.

*Check.*

| $\sqrt{6a}$ | $2\sqrt{3}$ |
|---|---|
| $\sqrt{6 \cdot 2}$ | $2\sqrt{3}$ |
| $\sqrt{12}$ | |
| $2\sqrt{3}$ | |

$2\sqrt{3} = 2\sqrt{3}$, true

Sometimes squaring each side of a radical equation results in a quadratic equation. The two roots of the quadratic equation might not both be solutions of the original equation. This is illustrated in Example 4.

## Example 4    Solve $x + 1 = \sqrt{17 - 4x}$.

$$x + 1 = \sqrt{17 - 4x}$$
$$(x + 1)^2 = (\sqrt{17 - 4x})^2$$
$$x^2 + 2x + 1 = 17 - 4x \quad \blacktriangleleft \; (x + 1)^2 = (x + 1)(x + 1) = x^2 + 2x + 1$$
$$x^2 + 6x + 1 = 17 \quad \blacktriangleleft \; \text{Add } 4x \text{ to each side.}$$
$$x^2 + 6x - 16 = 0 \quad \blacktriangleleft \; \text{Add } -17 \text{ to each side.}$$
$$(x + 8)(x - 2) = 0 \quad \blacktriangleleft \; \text{Factor the left side of the equation.}$$
$$x + 8 = 0 \quad \text{or} \quad x - 2 = 0$$
$$x = -8 \qquad\qquad x = 2$$

Check $-8$.

| $x + 1$ | $\sqrt{17 - 4x}$ |
|---|---|
| $-8 + 1$ | $\sqrt{17 - 4 \cdot -8}$ |
| $-7$ | $\sqrt{17 + 32}$ |
| | $\sqrt{49}$ |
| | $7$ |

$-7 \neq 7$
So, $-8$ is not a solution.

Check 2.

| $x + 1$ | $\sqrt{17 - 4x}$ |
|---|---|
| $2 + 1$ | $\sqrt{17 - 4 \cdot 2}$ |
| $3$ | $\sqrt{17 - 8}$ |
| | $\sqrt{9}$ |
| | $3$ |

$3 = 3$, true
So, 2 is a solution.

**Thus,** the only solution is 2.

For the example above, $-8$ was a solution of the quadratic equation but not a solution of the original equation. Recall that an extraneous solution is a solution which does not check in the original equation. If squaring both sides of a radical equation produces a solution, such as $-8$, that does not check in the original equation, the solution is called an extraneous solution.

Equations Containing Radicals

## Oral Exercises

**What equation do you get when you square each side?**

1. $\sqrt{y} = 3$ **$y = 9$**   2. $6 = 3\sqrt{m}$ **$36 = 9m$**   3. $3 = \sqrt{3x}$ **$9 = 3x$**   4. $\sqrt{2x} = 5$ **$2x = 25$**
5. $7 = \sqrt{y}$ **$49 = y$**   6. $\sqrt{5x} = \sqrt{2}$ **$5x = 2$**   7. $4 = \sqrt{2x + 5}$   8. $\sqrt{x - 1} = 5$
9. $\sqrt{2x - 3} = \sqrt{x}$   10. $\sqrt{x - 1} = \sqrt{3x + 7}$   **$16 = 2x + 5$**   **$x - 1 = 25$**
    **$2x - 3 = x$**       **$x - 1 = 3x + 7$**

## Written Exercises

**Solve.**

**Ⓐ**
1. $\sqrt{y} = 6$ **36**   2. $10 = \sqrt{x}$ **100**   3. $\sqrt{3x} = 6$ **12**   4. $48 = \sqrt{3m}$ **768**
5. $\sqrt{x} + 1 = 7$ **36**   6. $\sqrt{z} + 3 = 4$ **1**   7. $\sqrt{2x - 1} = 3$ **5**   8. $4 = \sqrt{2x - 2}$ **9**
9. $\sqrt{2x - 5} + 8 = 7$ **∅**   10. $\sqrt{6x + 5} = 2$ **∅**   11. $\sqrt{3x + 2} = 8$ **12**
12. $5 - \sqrt{x} = 9$ **∅**   13. $6 = 10 + \sqrt{y + 2}$ **∅**   14. $\sqrt{3x + 1} - 2 = 3$ **8**
15. $\sqrt{3x - 4} = \sqrt{x + 8}$ **6**   16. $\sqrt{2x + 7} = \sqrt{4x + 5}$ **1**   17. $\sqrt{x - 5} - 8 = 0$ **69**
18. $\sqrt{2y} = 4\sqrt{5}$ **40**   19. $\sqrt{x} = 4\sqrt{3}$ **48**   20. $\sqrt{28} = 2\sqrt{x}$ **7**

**Ⓑ**
21. $x - 2 = \sqrt{19 - 6x}$ **3**   22. $\sqrt{x^2 - 34} = -3$ **∅**   23. $x + 1 = \sqrt{17 - 4x}$ **2**
24. $\sqrt{2p - 20} = p - 10$ **10, 12**   25. $\sqrt{y^2 + 1} = y - 1$ **∅**   26. $3 + \sqrt{3t + 1} = t$ **8**

27. Three times the square root of a number is 6. Find the number. **4**
28. Three more than the square root of a number is 7. Find the number. **16**
29. The square root of 4 more than 3 times a number is 5. Find the number. **7**
30. The square root of a number, decreased by 2, is 3. Find the number. **25**
31. The formula for finding the altitude $h$ of an equilateral triangle with side $s$ is $h = \frac{\sqrt{3}}{2} s$. Find $s$ if $h = 6$. **$4\sqrt{3}$**
32. A number is increased by 6. If the square root of this sum is multiplied by 3, the result is 12. Find the number. **10**

## Example   Solve $\sqrt{x} + 3 = \sqrt{x + 27}$.

$$\sqrt{x} + 3 = \sqrt{x + 27}$$
$$(\sqrt{x} + 3)^2 = (\sqrt{x + 27})^2$$
$$x + 6\sqrt{x} + 9 = x + 27 \quad \blacktriangleleft (\sqrt{x} + 3)^2 = x + 6\sqrt{x} + 9$$
$$6\sqrt{x} = 18 \quad \blacktriangleleft \text{Add } -x \text{ and } -9 \text{ to each side.}$$
$$(6\sqrt{x})^2 = 18^2$$
$$36x = 324 \quad \blacktriangleleft (6\sqrt{x})^2 = 6 \cdot 6 \cdot \sqrt{x} \cdot \sqrt{x} = 36x$$
$$x = 9$$

*Check.*

| $\sqrt{x} + 3$ | $\sqrt{x + 27}$ |
|---|---|
| $\sqrt{9} + 3$ | $\sqrt{9 + 27}$ |
| $3 + 3$ | $\sqrt{36}$ |
| $6$ | $6$ |

$6 = 6$, true
So, 9 is a solution.

**Ⓒ**
33. $\sqrt{x} + 2 = \sqrt{x + 20}$ **16**   34. $\sqrt{3x} - 3 = \sqrt{x} + 1$ **4**   35. $\sqrt{b} - 2 = \sqrt{2b - 28}$ **16**
36. $-\sqrt{t} + 15 = \sqrt{t} - 5$ **1**   37. $\sqrt{x} + 3 = \sqrt{x} - 9$ **∅**   38. $\sqrt{4y - 3} = 2 + \sqrt{2y - 5}$

**7, 3**

637-7879 Mackey
Debbie
4-6

rm $(kx + b)^2 = c$ for $x$

16 can be solved two ways.
adratic equation. For example,

$= 0$
$= -4$

way, by using the square root properties.

 suggests the following general solution for an equation like $x^2 = a$.

| Solution of $x^2 = a$ | If $x^2 = a$, then $x = \sqrt{a}$ or $x = -\sqrt{a}$ for each $a \geq 0$. |
|---|---|

*Example 1*     **Solve $y^2 = 28$.**

$$y^2 = 28 \quad \blacktriangleleft \text{ Take the square root of each side.}$$

$y = \sqrt{28}$, or $y = -\sqrt{28}$
$y = 2\sqrt{7}$, or $y = -2\sqrt{7}$  $\quad \blacktriangleleft \sqrt{28} = \sqrt{4} \cdot \sqrt{7} = 2\sqrt{7}$
**Thus,** the solutions are $2\sqrt{7}$ and $-2\sqrt{7}$.

The same technique can be used to solve a complex equation, such as $(2y + 6)^2 = 16$. Think of $2y + 6$ as a single quantity. This is shown in Example 2.

*Example 2*     **Solve $(2y + 6)^2 = 16$.**

$$(2y + 6)^2 = 16 \quad \blacktriangleleft \text{ Take the square root of each side.}$$

*Check.*

$2y + 6 = \sqrt{16}$, or $2y + 6 = -\sqrt{16}$
$2y + 6 = 4$,     or $2y + 6 = -4$
$\quad 2y = -2$,   or     $2y = -10$
$\quad\quad y = -1$,   or     $y = -5$
The possible solutions are $-1$ and $-5$.
**Thus,** the solutions are $-1$ and $-5$.

| $(2y + 6)^2$ | 16 |
|---|---|
| $(2 \cdot -1 + 6)^2$ | 16 |
| $(-2 + 6)^2$ | |
| $4^2$ | |
| 16 | |

| $(2y + 6)^2$ | 16 |
|---|---|
| $(2 \cdot -5 + 6)^2$ | 16 |
| $(-10 + 6)^2$ | |
| $(-4)^2$ | |
| 16 | |

Both solutions check.

The Solution of $x^2 = a$

You might get solutions that have to be rationalized, as shown in Example 3.

## Example 3

**Solve $3x^2 = 5$.**

$$3x^2 = 5$$

$$x^2 = \frac{5}{3} \qquad \blacktriangleleft \text{ Divide each side by 3.}$$

$$x = \sqrt{\frac{5}{3}}, \text{ or } x = -\sqrt{\frac{5}{3}}$$

$$x = \frac{\sqrt{15}}{3}, \text{ or } x = -\frac{\sqrt{15}}{3} \qquad \blacktriangleleft \quad \sqrt{\frac{5}{3}} = \frac{\sqrt{5}}{\sqrt{3}} = \frac{\sqrt{5} \cdot \sqrt{3}}{\sqrt{3} \cdot \sqrt{3}} = \frac{\sqrt{15}}{3}$$

**Thus,** the solutions are $\dfrac{\sqrt{15}}{3}$ and $-\dfrac{\sqrt{15}}{3}$.

Sometimes the solutions involve radicals that can be simplified as in the case of an equation such as $(x - 2)^2 = 18$. This is illustrated in Example 4.

## Example 4

**Solve $(x - 2)^2 = 18$.**

$$(x - 2)^2 = 18 \qquad \blacktriangleleft \text{ Take the square root of each side.}$$

$$x - 2 = \sqrt{18}, \qquad \text{or } x - 2 = -\sqrt{18}$$

$$x = 2 + \sqrt{18}, \text{ or } \qquad x = 2 - \sqrt{18} \qquad \blacktriangleleft \text{ Add 2 to each side.}$$

$$x = 2 + 3\sqrt{2}, \text{ or } \qquad x = 2 - 3\sqrt{2} \qquad \blacktriangleleft \quad \sqrt{18} = \sqrt{9} \cdot \sqrt{2} = 3\sqrt{2}$$

**Thus,** the solutions are $2 + 3\sqrt{2}$ and $2 - 3\sqrt{2}$.
The solution is sometimes written as $2 \pm 3\sqrt{2}$.

# Reading in Algebra

**Answer each of the following questions with true or false. If false, tell why it is false.**

 1. If $x^2 = 25$, the solution is $x = 5$. **False, Also x = −5**
 2. The solution $x = 4 + \sqrt{5}$ or $x = 4 - \sqrt{5}$ can be written as $4 \pm \sqrt{5}$. **True**
 3. The expression $3 + 5\sqrt{7}$ can be simplified to $8\sqrt{7}$. **False, terms are not alike.**

# Oral Exercises

**Give the solution of each.**

 1. $x^2 = 36$ **6, −6**     2. $x^2 = 49$ **7, −7**     3. $x^2 = 100$ **10, −10**     4. $x^2 = 1$ **1, −1**
 5. $x^2 = 64$ **8, −8**     6. $x^2 = 25$ **5, −5**     7. $x^2 = \frac{4}{9}$ **$\frac{2}{3}, -\frac{2}{3}$**     8. $x^2 = \frac{16}{81}$ **$\frac{4}{9}, -\frac{4}{9}$**

# Written Exercises

**Solve.**

**(A)**

**1.** $x^2 = 14$  $\pm\sqrt{14}$
**2.** $m^2 = 13$  $\pm\sqrt{13}$
**3.** $a^2 = 3$  $\pm\sqrt{3}$

**4.** $x^2 = 12$  $\pm 2\sqrt{3}$
**5.** $p^2 = 24$  $\pm 2\sqrt{6}$
**6.** $g^2 = 40$  $\pm 2\sqrt{10}$

**7.** $y^2 = 32$  $\pm 4\sqrt{2}$
**8.** $a^2 = 44$  $\pm 2\sqrt{11}$
**9.** $d^2 = 50$  $\pm 5\sqrt{2}$

**10.** $(y + 4)^2 = 49$ **3, −11**
**11.** $(x - 3)^2 = 9$ **6, 0**
**12.** $(x - 8)^2 = 25$ **13, 3**

**13.** $(2a + 4)^2 = 16$ **0, −4**
**14.** $(3y - 6)^2 = 81$ **5, −1**
**15.** $(4b + 8)^2 = 64$ **0, −4**

**16.** $3x^2 = 7$  $\pm\frac{\sqrt{21}}{3}$
**17.** $5a^2 = 2$  $\pm\frac{\sqrt{10}}{5}$
**18.** $7m^2 = 4$  $\pm\frac{2\sqrt{7}}{7}$

**19.** $5k^2 = 3$  $\pm\frac{\sqrt{15}}{5}$
**20.** $2a^2 = 3$  $\pm\frac{\sqrt{6}}{2}$
**21.** $3b^2 = 4$  $\pm\frac{2\sqrt{3}}{3}$

**22.** $7k^2 = 63$  $\pm 3$
**23.** $3y^2 = 36$  $\pm 2\sqrt{3}$
**24.** $4b^2 = 112$  $\pm 2\sqrt{7}$

**(B)**

**25.** $(x - 4)^2 = 8$ **4 ± 2√2**
**26.** $(d + 6)^2 = 12$ **−6 ± 2√3**
**27.** $(t - 8)^2 = 32$ **8 ± 4√2**

**28.** $(m - 7)^2 = 28$ **7 ± 2√7**
**29.** $(x + 7)^2 = 44$ **−7 ± 2√11**
**30.** $(e - 5)^2 = 50$ **5 ± 5√2**

**31.** $(x + 5)^2 = 40$ **−5 ± 2√10**
**32.** $(b - 8)^2 = 48$ **8 ± 4√3**
**33.** $(k + 7)^2 = 72$ **−7 ± 6√2**

**34.** $(y - 1)^2 = 60$ **1 ± 2√15**
**35.** $(b - 3)^2 = 68$ **3 ± 2√17**
**36.** $(l + 9)^2 = 80$ **−9 ± 4√5**

**Solve each problem.**

**37.** If 3 is added to a number, the square of the result is 49. Find the number. **4, −10**

**38.** The square of 2 less than a number is 64. Find the number. **10, −6**

**(C)** **Solve.**

**39.** $\dfrac{-3 \pm 2\sqrt{3}}{2}$
**40.** $\dfrac{3 \pm 2\sqrt{3}}{2}$
**41.** $\dfrac{1 \pm \sqrt{5}}{2}$
**42.** $\dfrac{1 \pm 2\sqrt{5}}{2}$

**39.** $3(2x + 3)^2 - 7 = 29$
**40.** $3(2x - 3)^2 + 8 = 44$
**41.** $5(4y - 2)^2 - 6 = 94$

**42.** $5(2x - 1)^2 - 60 = 40$
**43.** $4(2p + 1)^2 - 70 = 42$
**44.** $3(2b - 5)^2 - 7 = 5$ **$\frac{7}{2}, \frac{3}{2}$**

**43.** $\dfrac{-1 \pm 2\sqrt{7}}{2}$

**For the equation $a(x - b)^2 = c$, answer each of the following questions.**

**45.** If $a > 0$, what should be the value of $c$ to result in only one solution? **0**
**46.** If $a < 0$, what value should $c$ have to result in no solution? **c > 0**

# CUMULATIVE REVIEW

**1.** Factor $x^2 + 18x + 81$.
 **$(x + 9)^2$**

**2.** Simplify $\dfrac{x + 2}{x^2 - 9} - \dfrac{2}{3 - x}$.
 **$\dfrac{3x + 8}{(x + 3)(x - 3)}$**

**3.** Simplify $\sqrt{8} + 3\sqrt{2}$.
 **$5\sqrt{2}$**

## NON-ROUTINE PROBLEMS

Cass' son was pleased with the three bass his father had caught. He wanted to know how much they weighed. Cass said the largest weighed as much as the other two, but the smallest weighed 3 lb less than one-half the other two together, and the whole bunch weighed 18 lb. What were the weights of the three bass? **4, 5, and 9 lbs.**

# COMPLETING THE SQUARE 15.3

*Objective*    **To solve quadratic equations by completing the square**

In Lesson 15.2 you learned how to solve quadratic equations in which one side of the equation was a perfect square. Notice that the left side of the equation, $(x + 5)^2 = 9$, involves squaring a binomial.

Recall the formulas for squaring a binomial.

$$(a + b)^2 = a^2 + 2ab + b^2$$
$$(a - b)^2 = a^2 - 2ab + b^2$$

The product is a perfect square trinomial. There is a pattern common to all perfect square trinomials that allows you to predict what number must be added to an expression such as $a^2 + 14a$ to produce a trinomial that is a perfect square. In this lesson you will learn to complete the square.

Recall the pattern common to any perfect square trinomial, $x^2 + 2bx + b^2$. The coefficient of the second term is $2b$, and the third, or constant, term is the square of $b$.

$$x^2 + 10x + 25 \qquad a^2 - 8a + 16 \qquad m^2 + 12m + 36$$

$$2(5) \quad 5^2 \qquad\qquad 2(-4) \quad -4^2 \qquad\qquad 2(6) \quad 6^2$$

Example 1 shows how to use this pattern to complete the square.

*Example 1*    **What number must you add to $y^2 - 6y$ to make a perfect square trinomial?**

$$y^2 - 6y + \underset{?}{\underline{\phantom{?}}}$$

$2(-3)$  ◄ $-6 = 2(-3)$. *So, $b = -3$.*

$(-3)^2 = 9$

So, $b^2 = 9$.

Therefore, 9 must be added to $y^2 - 6y$ to make a perfect square trinomial, $y^2 - 6y + 9$.

In Example 1, you were shown a method called *completing the square*. The expression $y^2 - 6y$ was completed to a perfect square trinomial $y^2 - 6y + 9$, or $(y - 3)^2$, by adding 9 to $y^2 - 6y$.

You can use the method of completing the square to solve an equation such as $x^2 + 10x = -16$. The equation can be rewritten so that the left side is a perfect square trinomial. Then the left side can be expressed as the square of a binomial. The complete solution of $x^2 + 10x = -16$ by completing the square is shown in Example 2.

*Example 2*    **Solve $x^2 + 10x = -16$ by completing the square.**

What number must be added to $x^2 + 10x$ to complete the square?

$x^2 + 10x + \underline{?}$        ◀ *$10 = 2(5)$, so add $(5)^2$, or 25.*

$x^2 + 10x + 25 = -16 + 25$  ◀ *Add 25 to each side.*

$\qquad (x + 5)^2 = 9$       ◀ *$x^2 + 10x + 25 = (x + 5)(x + 5) = (x + 5)^2$*

$x + 5 = \sqrt{9}$, or $x + 5 = -\sqrt{9}$   *Check.*

| $x^2 + 10x$ | $-16$ | $x^2 + 10x$ | $-16$ |
|---|---|---|---|
| $(-2)^2 + 10 \cdot -2$ | $-16$ | $(-8)^2 + 10 \cdot -8$ | $-16$ |
| $4 - 20$ | | $64 - 80$ | |
| $-16$ | | $-16$ | |

$x + 5 = 3$, or $\quad x + 5 = -3$

$\qquad x = -2$, or $\qquad x = -8$

So, the solutions are $-2$ and $-8$.

You can use the technique of completing the square to solve a quadratic equation such as $x^2 - 8x + 3 = 0$. First rewrite the equation so that $x^2 - 8x$ is on one side and the constant term is on the other side. Then proceed as in Example 2.

*Example 3*    **Solve $x^2 - 8x + 3 = 0$ by completing the square.**

$x^2 - 8x + 3 = 0$

$\qquad x^2 - 8x = -3$     ◀ *Add $-3$ to each side.*

$x^2 - 8x + 16 = -3 + 16$  ◀ *$-8 = 2(-4)$, so add $(-4)^2$, or 16, to each side.*

$x^2 - 8x + 16 = 13$

$\qquad (x - 4)^2 = 13$     ◀ *$x^2 - 8x + 16 = (x - 4)^2$*

$x - 4 = \sqrt{13}$, or $x - 4 = -\sqrt{13}$

$x = 4 + \sqrt{13}$, or $x = 4 - \sqrt{13}$

The solutions can be written as $4 \pm \sqrt{13}$.

An equation such as $2x^2 + 5x + 3 = 0$ can be solved by completing the square. However, first each side must be divided by 2 so that the resulting equation will have an $x^2$ coefficient of 1. The solution will involve fractions, as shown in Example 4 on the next page.

## Example 4    Solve $2y^2 + 5y + 3 = 0$.

$2y^2 + 5y + 3 = 0$

$y^2 + \dfrac{5}{2}y + \dfrac{3}{2} = 0$ ◀ *Divide each side by 2.*

$y^2 + \dfrac{5}{2}y + \underline{?} = -\dfrac{3}{2} + \underline{?}$

$y^2 + \dfrac{5}{2}y + \dfrac{25}{16} = -\dfrac{3}{2} + \dfrac{25}{16}$ ◀ $y^2 + \dfrac{5}{2}y = y^2 + 2 \cdot \dfrac{5}{4}y.$ Add $\left(\dfrac{5}{4}\right)^2$, or $\dfrac{25}{16}$, to each side.

$\left(y + \dfrac{5}{4}\right)^2 = \dfrac{1}{16}$ ◀ $-\dfrac{3}{2} + \dfrac{25}{16} = -\dfrac{3 \cdot 8}{2 \cdot 8} + \dfrac{25}{16} = -\dfrac{24}{16} + \dfrac{25}{16} = \dfrac{1}{16}$

$y + \dfrac{5}{4} = \sqrt{\dfrac{1}{16}}$, or $y + \dfrac{5}{4} = -\sqrt{\dfrac{1}{16}}$

$y + \dfrac{5}{4} = \dfrac{1}{4}$    or $y + \dfrac{5}{4} = -\dfrac{1}{4}$ ◀ $\sqrt{\dfrac{1}{16}} = \dfrac{1}{4}$

$y = -1,$    or $y = -\dfrac{3}{2}$

**Thus,** the solutions are $-1$ and $-\dfrac{3}{2}$.

## Written Exercises

**Solve by completing the square.**

(A)  1. $x^2 - 6x = 27$  9, −3    2. $a^2 + 4a = -3$  −1, −3    3. $b^2 - 2b = 24$  6, −4
 4. $y^2 + 2y = 3$  1, −3    5. $x^2 - 2x = 15$  5, −3    6. $b^2 - 10b = -21$  7, 3
 7. $g^2 + 12g = -32$  −4, −8    8. $p^2 - 6p = -5$  5, 1    9. $y^2 + 16y + 55 = 0$  −5, −11
10. $x^2 - 4x - 21 = 0$  7, −3    11. $a^2 - 4a + 4 = 0$  2    12. $d^2 + 6d - 7 = 0$  1, −7
13. $l^2 - 18l + 72 = 0$  12, 6    14. $t^2 + 20t + 51 = 0$  −3, −17    15. $x^2 - 18x + 77 = 0$  11, 7
16. $r^2 - 24r + 80 = 0$  20, 4    17. $x^2 + 20x - 4 = 0$    18. $x^2 - 8x + 5 = 0$  $4 \pm \sqrt{11}$
19. $w^2 + 6w - 10 = 0$    20. $x^2 + 4x + 2 = 0$  $-2 \pm \sqrt{2}$  21. $g^2 + 2g - 6 = 0$  $-1 \pm \sqrt{7}$
22. $a^2 - 6a - 4 = 0$  $3 \pm \sqrt{13}$  23. $x^2 + 2x - 5 = 0$  $-1 \pm \sqrt{6}$  24. $b^2 + 14b + 6 = 0$  $-7 \pm \sqrt{43}$
        19. $-3 \pm \sqrt{19}$        17. $-10 \pm 2\sqrt{26}$

(B) 25. $x^2 + 3x - 10 = 0$  2, −5    26. $a^2 + 3a - 40 = 0$  5, −8    27. $y^2 - 3y + 2 = 0$  2, 1
28. $p^2 - 7p - 8 = 0$  8, −1    29. $t^2 - 15t + 56 = 0$  8, 7    30. $x^2 + x - 30 = 0$  5, −6
31. $g^2 - g = 6$  3, −2    32. $b^2 + 3b + 2 = 0$  −1, −2    33. $2a^2 + 7a - 4 = 0$  $\frac{1}{2}$, −4
34. $3a^2 + 2a = 1$  $\frac{1}{3}$, −1    35. $2x^2 + 7x + 5 = 0$  $-1, -\frac{5}{2}$  36. $6m^2 - m = 1$  $\frac{1}{2}, -\frac{1}{3}$

**Solve each problem.**

(C) 37. The square of a number is 2 less than 6 times the number. Find the number.
        $3 \pm \sqrt{7}$

38. If 5 times a number is added to the square of the number, the result is 1. Find the number. $\dfrac{-5 \pm \sqrt{29}}{2}$

**Solve for $x$ in terms of $b$ and $c$.**

39. $x^2 - 4x + c = 0$
    $2 \pm \sqrt{-c + 4}$

40. $x^2 + 2bx + c = 0$
    $-b \pm \sqrt{-c + b^2}$

41. $x^2 - bx + 3 = 0$
    $\dfrac{b \pm \sqrt{b^2 - 12}}{2}$

# THE QUADRATIC FORMULA 15.4

*Objective* **To solve a quadratic equation using the quadratic formula**

In the last lesson you learned to solve a quadratic equation by completing the square. If you use this method to solve the general quadratic equation $ax^2 + bx + c = 0$, you will get a formula that can be used to solve any quadratic equation. The solution of both $3x^2 + x - 1 = 0$ and the general equation $ax^2 + bx + c = 0$ are shown below. Following the steps in solving $3x^2 + x - 1 = 0$ will assist you in following the parallel steps for solving the general quadratic equation $ax^2 + bx + c = 0$.

$$3x^2 + 1x - 1 = 0 \qquad\qquad ax^2 + bx + c = 0$$

$$x^2 + \frac{1}{3}x - \frac{1}{3} = 0 \qquad\qquad x^2 + \frac{b}{a}x + \frac{c}{a} = 0$$

$$x^2 + \frac{1}{3}x = \frac{1}{3} \qquad\qquad x^2 + \frac{b}{a}x = -\frac{c}{a}$$

$$x^2 + \frac{1}{3}x + \frac{1}{36} = \frac{1}{3} + \frac{1}{36} \qquad\qquad x^2 + \frac{b}{a}x + \frac{b^2}{4a^2} = -\frac{c}{a} + \frac{b^2}{4a^2}$$

$$x^2 + \frac{1}{3}x + \frac{1}{36} = \frac{12}{36} + \frac{1}{36} \qquad\qquad x^2 + \frac{b}{a}x + \frac{b^2}{4a^2} = -\frac{4ac}{4a^2} + \frac{b^2}{4a^2}$$

$$x^2 + \frac{1}{3}x + \frac{1}{36} = \frac{13}{36} \qquad\qquad x^2 + \frac{b}{a}x + \frac{b^2}{4a^2} = \frac{b^2 - 4ac}{4a^2}$$

$$\left(x + \frac{1}{6}\right)^2 = \frac{13}{36} \qquad\qquad \left(x + \frac{b}{2a}\right)^2 = \frac{b^2 - 4ac}{4a^2}$$

$$x + \frac{1}{6} = \sqrt{\frac{13}{36}}, \text{ or } x + \frac{1}{6} = -\sqrt{\frac{13}{36}} \qquad x + \frac{b}{2a} = \sqrt{\frac{b^2 - 4ac}{4a^2}}, \text{ or } x + \frac{b}{2a} = -\sqrt{\frac{b^2 - 4ac}{4a^2}}$$

$$x = -\frac{1}{6} + \sqrt{\frac{13}{36}}, \text{ or } x = -\frac{1}{6} - \sqrt{\frac{13}{36}} \qquad x = -\frac{b}{2a} + \sqrt{\frac{b^2 - 4ac}{4a^2}}, \text{ or } x = -\frac{b}{2a} - \sqrt{\frac{b^2 - 4ac}{4a^2}}$$

$$x = -\frac{1}{6} + \frac{\sqrt{13}}{6}, \text{ or } x = -\frac{1}{6} - \frac{\sqrt{13}}{6} \qquad x = -\frac{b}{2a} + \frac{\sqrt{b^2 - 4ac}}{2a}, \text{ or } x = -\frac{b}{2a} - \frac{\sqrt{b^2 - 4ac}}{2a}$$

$$x = \frac{-1 + \sqrt{13}}{6}, \text{ or } x = \frac{-1 - \sqrt{13}}{6} \qquad x = \frac{-b + \sqrt{b^2 - 4ac}}{2a}, \text{ or } x = \frac{-b - \sqrt{b^2 - 4ac}}{2a}$$

**Thus,** $x = \dfrac{-1 \pm \sqrt{13}}{6}.$ $\qquad\qquad$ **Thus,** $x = \dfrac{-b \pm \sqrt{b^2 - 4ac}}{2a}.$

---

| The quadratic formula | The solutions of a quadratic equation of the form $ax^2 + bx + c = 0$ can be found by the formula $$x = \frac{-b \pm \sqrt{b^2 - 4ac}}{2a} \text{ for } a \neq 0,\ b^2 - 4ac > 0.$$ |
|---|---|

You can now use the quadratic formula to solve any quadratic equation.

*Example 1*    **Solve $x^2 - 7x + 10 = 0$ by the quadratic formula.**

$1x^2 - 7x + 10 = 0$    ◀ $x^2 = 1x^2$

$a = 1 \quad b = -7 \quad c = 10$

$x = \dfrac{-b \pm \sqrt{b^2 - 4ac}}{2a}$    ◀ *Use the quadratic formula.*

$x = \dfrac{-(-7) \pm \sqrt{(-7)^2 - 4 \cdot 1 \cdot 10}}{2 \cdot 1}$    ◀ *Substitute 1 for a, $-7$ for b, and 10 for c.*

$x = \dfrac{7 \pm \sqrt{9}}{2}$    ◀ $-(-7) = 7$

$x = \dfrac{7 \pm 3}{2}$    ◀ $\sqrt{9} = 3$

$x = \dfrac{7 + 3}{2}$, or $x = \dfrac{7 - 3}{2}$

$x = 5, \quad \text{or} \quad x = 2$    ◀ $\dfrac{7+3}{2} = \dfrac{10}{2} = 5; \dfrac{7-3}{2} = \dfrac{4}{2} = 2$

So, the solutions are 5 and 2.

Notice that $x^2 - 7x + 10 = 0$ could have been solved by factoring. Obviously, the quadratic formula is really necessary for quadratic equations that cannot be solved by factoring as in Example 2. To solve equations such as $x^2 - 1 = x$ and $4x = -2x^2 - 3$, first rewrite the equations in the form $ax^2 + bx + c = 0$. Then use the quadratic formula.

*Example 2*    **Solve by the quadratic formula.**

$x^2 - 1 = x$

$1x^2 - 1x - 1 = 0$    ◀ $a = 1, b = -1, c = -1$

$x = \dfrac{-b \pm \sqrt{b^2 - 4ac}}{2a}$

$x = \dfrac{-(-1) \pm \sqrt{(-1)^2 - 4 \cdot 1 \cdot -1}}{2 \cdot 1}$

$x = \dfrac{1 \pm \sqrt{1 + 4}}{2}$

$x = \dfrac{1 \pm \sqrt{5}}{2}$

---

$4x = -2x^2 - 3$

$2x^2 + 4x + 3 = 0$    ◀ $a = 2, b = 4, c = 3$

$x = \dfrac{-b \pm \sqrt{b^2 - 4ac}}{2a}$

$x = \dfrac{-4 \pm \sqrt{4^2 - 4 \cdot 2 \cdot 3}}{2 \cdot 2}$

$x = \dfrac{-4 \pm \sqrt{-8}}{4}$

There is no real solution because $\sqrt{-8}$ is not a real number.

It is easier to solve a quadratic equation such as $\dfrac{2}{3}x = -\dfrac{1}{6}x^2 - \dfrac{1}{3}$ by first multiplying each side by the LCM, as shown in Example 3.

*Example 3*    **Solve $\frac{2}{3}x = -\frac{1}{6}x^2 - \frac{1}{3}$.**

$$6 \cdot \frac{2}{3}x = 6\left(-\frac{1}{6}x^2 - \frac{1}{3}\right)$$

◀ *Multiply each side by the LCM, 6.*

$$4x = -1x^2 - 2$$

◀ $\overset{2}{\cancel{6}} \cdot \frac{2}{\underset{1}{\cancel{3}}} = 4; \overset{1}{\cancel{6}} \cdot -\frac{1}{\underset{1}{\cancel{6}}} = -1; \overset{2}{\cancel{6}} \cdot -\frac{1}{\underset{1}{\cancel{3}}} = -2$

$$1x^2 + 4x + 2 = 0$$

◀ *Put into standard form.*

$$a = 1 \quad b = 4 \quad c = 2$$

$$x = \frac{-4 \pm \sqrt{4^2 - 4 \cdot 1 \cdot 2}}{2 \cdot 1}$$

◀ *Use the quadratic formula*
$x = \dfrac{-b \pm \sqrt{b^2 - 4ac}}{2a}$

$$x = \frac{-4 \pm \sqrt{8}}{2}$$

$$x = \frac{-4 \pm 2\sqrt{2}}{2}$$

◀ $\sqrt{8} = \sqrt{4} \cdot \sqrt{2} = 2\sqrt{2}$

$$x = \frac{\overset{1}{\cancel{2}}(-2 \pm 1\sqrt{2})}{\underset{1}{\cancel{2}}}$$

◀ *Factor out common factor 2.*
*Divide out the common factors.*

$$x = -2 \pm \sqrt{2}$$

◀ *The roots are irrational.*

# Written Exercises

**Solve by using the quadratic formula. Leave irrational roots in simplest radical form.**

Ⓐ
**1.** $x^2 + 4x + 3 = 0$ **−1, −3**
**2.** $x^2 - 5x + 6 = 0$ **3, 2**
**3.** $x^2 - 6x + 8 = 0$ **4, 2**
**4.** $x^2 - 9x + 14 = 0$ **7, 2**
**5.** $x^2 - 9x + 8 = 0$ **8, 1**
**6.** $x^2 + 2x - 3 = 0$ **−3, 1**
**7.** $x^2 - 5x + 2 = 0$ $\frac{5 \pm \sqrt{17}}{2}$
**8.** $x^2 - 7x + 3 = 0$ $\frac{7 \pm \sqrt{37}}{2}$
**9.** $x^2 - 7x - 1 = 0$ $\frac{7 \pm \sqrt{53}}{2}$
**10.** $2x^2 - 3x + 1 = 0$ **1, $\frac{1}{2}$**
**11.** $x^2 + x - 1 = 0$ $\frac{-1 \pm \sqrt{5}}{2}$
**12.** $x^2 + 5x - 3 = 0$ $\frac{-5 \pm \sqrt{37}}{2}$
**13.** $x^2 + 8 = 3x$ **∅**
**14.** $x^2 = 7x - 11$ $\frac{7 \pm \sqrt{5}}{2}$
**15.** $2x + 7 = -x^2$ **∅**

Ⓑ
**16.** $-2x - 1 = -x^2$ **1 ± $\sqrt{2}$**
**17.** $x^2 = -4x + 6$ **−2 ± $\sqrt{10}$**
**18.** $2x - 2 = -x^2$ **−1 ± $\sqrt{3}$**
**19.** $x^2 + 2 = -4x$ **−2 ± $\sqrt{2}$**
**20.** $-x^2 + 6x = -2$ **3 ± $\sqrt{11}$**
**21.** $-x^2 = 4 + 2x$ **∅**
**22.** $x^2 + \frac{3}{2}x - \frac{5}{2} = 0$ **1, $-\frac{5}{2}$**
**23.** $\frac{1}{15}x^2 + \frac{5}{3} = \frac{2}{3}x$ **5**
**24.** $\frac{1}{2}x^2 + \frac{1}{2}x - 2 = 0$
$\frac{-1 \pm \sqrt{17}}{2}$

**Solve each problem.**

**25.** The square of John's age is 200 less than 30 times his age. How old is he?
**20 or 10 yrs.**

**26.** Ten times a number, increased by the square of the number, is 2. Find the number in simplest radical form. **−5 ± 3$\sqrt{3}$**

**Solve.**

Ⓒ
**27.** $\frac{4}{x} - 1 = \frac{4}{3x + 4}$ $\frac{2 \pm 2\sqrt{13}}{3}$

**28.** $\frac{3}{x - 1} = 1 + \frac{x - 4}{x - 3}$ **4, 2**

**29.** $\sqrt{6}x^2 - 4x - 2\sqrt{6} = 0$ $\sqrt{6}, \frac{-\sqrt{6}}{3}$

**30.** $x^2 - 6 = 2\sqrt{2}x$ **3$\sqrt{2}$, −$\sqrt{2}$**

**31.** For what value of $a$ will there be no solution for $x^2 + 2x + a = 0$? **a > 1**

# PROBLEMS ABOUT AREA

*Objective*   **To solve problems about areas that lead to quadratic equations**

The formula for the area of a rectangle is $A = l \cdot w$.

$$A = l \cdot w \qquad w$$
$$l$$

A word problem concerning the area of a rectangle might lead to a quadratic equation. However, if one of the two solutions of the quadratic equation is negative, it is rejected, because the length or width of a rectangle cannot be a negative number. This is illustrated in Example 1.

*Example 1*   **The length of a rectangle is 4 cm more than the width. The area is 45 cm². Find the length and width.**

READ ▶   You are given the area. You are asked to find the length and width.

PLAN ▶   Represent the length and width.           Draw a sketch
  Let $x$ = width in cm
  $x + 4$ = length in cm
  Use $A = l \cdot w$ to write an equation.

$$A = l \cdot w \qquad x$$
$$x + 4$$

  $45 = (x + 4)x$

SOLVE ▶
$$45 = x^2 + 4x \qquad \blacktriangleleft (x + 4)\,x = x \cdot x + 4 \cdot x = x^2 + 4x$$
$$0 = x^2 + 4x - 45 \qquad \blacktriangleleft \text{Put in standard form.}$$
$$0 = (x + 9)(x - 5) \qquad \blacktriangleleft \text{Factor.}$$
$$x + 9 = 0 \text{ or } x - 5 = 0$$
$$x = {}^-9 \text{ or } x = 5$$

Reject $^-9$ because a rectangle cannot have a negative number as its width. Therefore, the width, $x$, is 5, and the length is $x + 4 = 5 + 4$, or 9.

INTERPRET ▶

| $A$ | $l \cdot w$ |
|-----|-------------|
| 45  | $9 \cdot 5$ |
|     | 45          |

$45 = 45$
True

| Length | is 4 more than width. |
|--------|-----------------------|
| 9      | 4 more than 5 |
|        | 9 |

$9 = 9$
True

**Thus,** the length is 9 cm, and the width is 5 cm.

The formula for the area of a triangle is
$A = \frac{1}{2} \cdot b \cdot h$, where $b$ is the base and $h$ is the
height.

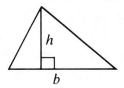

Sometimes area problems lead to quadratic equations that cannot be solved by factoring. Then the quadratic formula must be used, as illustrated in Example 2.

**Example 2**   **The base of a triangle is 6 m less than the height. The area is 12 m².**
**Find the base and height to the nearest tenth.**

Represent the base and height algebraically. Let $x$ = height in meters, $x - 6$ = base in meters.
Use $A = \frac{1}{2} \cdot b \cdot h$ to write an equation.

$$12 = \frac{1}{2}(x - 6)x$$

$$12 = \frac{1}{2}(x^2 - 6x)$$

$$2 \cdot 12 = \not{2} \cdot \frac{1}{\underset{1}{\not{2}}}(x^2 - 6x) \quad \blacktriangleleft \textit{Multiply each side by the LCM, 2.}$$

$$24 = x^2 - 6x$$

$$0 = 1x^2 - 6x - 24 \quad \blacktriangleleft \textit{Put in standard form.}$$

Solve by the quadratic formula because $x^2 - 6x - 24$ is not factorable.

$$x = \frac{-b \pm \sqrt{b^2 - 4ac}}{2a}$$

$$x = \frac{-(-6) \pm \sqrt{(-6)^2 - 4 \cdot 1 \cdot -24}}{2 \cdot 1}$$

$$x = \frac{6 \pm \sqrt{132}}{2}$$

$$x = \frac{6 \pm 2\sqrt{33}}{2} \quad \blacktriangleleft \sqrt{132} = \sqrt{2} \cdot \sqrt{66} = \sqrt{2} \cdot \sqrt{2} \cdot \sqrt{33} = 2\sqrt{33}$$

$$x = \frac{\overset{1}{\not{2}}(3 \pm 1\sqrt{33})}{\underset{1}{\not{2}}} \quad \blacktriangleleft \textit{Factor out 2. Divide out common factors.}$$

$$x = 3 \pm \sqrt{33}$$

$x = 3 + \sqrt{33} \quad$ or $\quad x = 3 - \sqrt{33}$

$x = 3 + 5.745 \quad$ or $\quad x = 3 - 5.745 \quad \blacktriangleleft \textit{From the table, } \sqrt{33} \doteq 5.745.$

$x = 8.745 \quad$ or $\quad x = -2.745 \quad \blacktriangleleft -2.745 \textit{ is not a height.}$

Therefore, the height, $x$, is 8.745 and the base is $x - 6 = 8.745 - 6$, or 2.745.

**Thus,** the height is 8.7 m and the base is 2.7 m, each rounded to the nearest tenth.
Check on your own, rounding results to the nearest whole number.

Problems About Area

## Oral Exercises

**Give the area of each rectangle in terms of $x$.**

**1.** $l = 5x$, $w = 4x$  $20x^2$      **2.** $l = x + 5$, $w = x$  $x^2 + 5x$      **3.** $l = 4x$, $w = x + 7$

**3.** $4x^2 + 28x$

**4.** $l = x + 7$, $w = x - 7$      **5.** $l = 3x + 4$, $w = 3x - 4$      **6.** $l = x$, $w = \dfrac{2}{3}x + 5$

**4.** $x^2 - 49$                    **5.** $9x^2 - 16$

**6.** $\dfrac{2}{3}x^2 + 5x$

## Written Exercises

**Solve.**

Ⓐ
**1.** The length of a rectangle is 2 m more than the width. The area is 15 m². Find the length and width. **3 m, 5 m**

**2.** The length of a rectangle is 7 cm less than twice the width. The area is 72 cm². Find the length and width. **9 cm, 8 cm**

**3.** The length of a rectangle is 2 cm more than 3 times the width. The area is 16 cm². Find the length and width. **8 cm, 2 cm**

**4.** The length of a rectangle is twice the width. The area is 32 km². Find the length and width. **8 km, 4 km**

**5.** The length of a rectangle is 3 m more than twice the width. The area is 27 m². Find the length and width. **9 m, 3 m**

**6.** The length of a rectangle is 1 cm less than 4 times the width. The area is 95 cm². Find the length and width. **19 cm, 5 cm**

**7.** The base of a triangle is 2 cm more than the height. The area is 24 cm². Find the base and height. **8 cm, 6 cm**

**8.** The height of a triangle is 3 km less than the base. The area is 27 km². Find the base and height. **9 km, 6 km**

**Solve. Use a square root table. Round each answer to the nearest tenth.**

Ⓑ
**9.** The length of a rectangle is 2 m more than the width. The area is 6 m². Find the length and width. **3.6 m, 1.6 m**

**10.** The base of a triangle is 4 cm more than the height. The area is 4 cm². Find the base and height. **5.5 cm, 1.5 cm**

**11.** The length of a rectangle is 8 cm more than the width. The area is 50 cm². Find the length and width. **12.1 cm, 4.1 cm**

**12.** The base of a triangle is 1 km more than 3 times the height. The area is 4 km². Find the base and height. **5.5 km, 1.5 km**

**13.** The height of a triangle is 3 m more than twice the base. The area is 3 m². Find the base and height. **1.1 m, 5.2 m**

**14.** The width of a rectangle is 5 cm less than twice the length. The area is 8 cm². Find the length and width. **3.6 cm, 2.2 cm**

**Solve.**

Ⓒ
**15.** The perimeter of a rectangle is 36 cm. The area is 80 cm². Find the length and width. **10 cm, 8 cm**

**16.** The length of a rectangle is 5 cm less than twice the width. The area is 28 cm. Find the width and the length. **5.2 cm, 5.4 cm**

## CALCULATOR ACTIVITIES

**Solve each equation. Round to the nearest thousandth.**

$0.140, -1.609$

**1.** $125x^2 - 225x - 207 = 0$ **2.470, −0.670**      **2.** $49x^2 + 72x - 11 = 0$

**3.** $5x^2 - 23x = 112$ **7.562, −2.962**      **4.** $-10.7x + 1 = -1.8x^2$

**5.849, 0.0950**

# PROBLEMS ABOUT WET MIXTURES 15.6

**Objective**

**To solve problems about wet mixtures**

Frequently, a liquid solution is a combination of two elements. For example, a solution of 48 L contains water and alcohol. If you know the part or percentage of the solution that is water, you can then find the number of liters of alcohol in the solution. This is illustrated in the following example.

**Example 1**

**A solution of 48 L contains water and alcohol. 25% of the solution is water. Find the number of liters of alcohol.**

Think: Total % $-$ % of water = % of alcohol
$$100\% \quad - \quad 25\% \quad = \quad 75\%$$
So, the amount of alcohol is 75% of the total of 48 L.
$$\text{liters of alcohol} = 0.75(48)$$
$$= 36$$

**Thus,** there are 36 L of alcohol in the 48-L solution. There are 12 L of water in the solution.

Notice that 75% of the 48 L above is alcohol. Adding 24 L of water gives a solution of 72 L. But 36 L of this new solution will still be alcohol. Below is a comparison of the original solution with the new solution.

| *Original Solution* | | *New Solution* | |
|---|---|---|---|
| alcohol + water = total | | alcohol + water + added water = total | |
| 36 L + 12 L = 48 L | | 36 L + 12 L + 24 L = 72 L | |

The amount of alcohol is the same in both solutions, 36 L. But the percentage of alcohol has changed.

The original solution contains 75% alcohol.

In the new solution, 36 is $\frac{1}{2}$ of, or 50% of, 72.

Thus, even though both solutions contain 36 L of alcohol, the percentage of alcohol has been reduced from 75% to 50%.

These concepts, together with the four basic steps of solving word problems, will be used in the examples that follow to solve word problems about liquid mixtures.

*Example 2*　**How many liters of water must be added to 32 L of a 25% sulfuric acid solution (sulfuric acid and water) to make a 10% sulfuric acid solution?**

READ ▶　The amount of sulfuric acid does not change. You are asked to find the amount of water added to reduce the percentage of sulfuric acid.

PLAN ▶　Let　$x$ = number of liters of water to be added
$32 + x$ = total number of liters in the new solution

Write an equation. The amount of sulfuric acid in the new and original solutions is the same.

$$\left(\begin{array}{c}\text{liters of sulfuric acid}\\\text{in original solution}\end{array}\right) = \left(\begin{array}{c}\text{liters of sulfuric acid}\\\text{in new solution}\end{array}\right)$$

$$0.25(32) \quad = \quad 0.10(32 + x)$$

◀ *25% = 0.25*
*10% = 0.10*

SOLVE ▶

$$8 = 3.2 + 0.10x$$
$$100 \cdot 8 = 100(3.2 + 0.10x)$$
$$800 = 100(3.2) + 100(0.10x)$$
$$800 = 320 + 10x$$
$$480 = 10x$$
$$48 = x$$

◀ *Multiply each side by 100.*

INTERPRET ▶

| liters of sulfuric acid in original solution | liters of sulfuric acid in new solution |
|---|---|
| 25% of 32 | 10% of (32 + 48) |
| 0.25(32) | 0.10(80) |
| 8 | 8 |

$$8 = 8$$
True

**Thus,** 48 L of water must be added.

In the next example you will be given information about the percentage of antifreeze in a solution of water and antifreeze. When more antifreeze is added, the amount of water will not change in the new solution. However, to write an equation you will first have to find the percentage of water in both the original and in the new solution. Since the total percentage of water and antifreeze is 100%, you can easily find the percentage of water if you are given the percentage of antifreeze. This is illustrated in Example 3 on the next page.

*Example 3*  **How many liters of antifreeze must be added to 30 L of a 20% antifreeze solution (antifreeze and water) to make a 30% antifreeze solution?**

*Think:* The amount of water does not change.
Original % of water: $100\% - 20\%$ antifreeze $= 80\%$ water
   New % of water: $100\% - 30\%$ antifreeze $= 70\%$ water
  Let $x =$ number of liters of antifreeze to be added
$30 + x =$ total number of liters in the new solution

Write an equation.
$$\binom{\text{liters of water in}}{\text{original solution}} = \binom{\text{liters of water in}}{\text{new solution}}$$
$$0.80\,(30) = 0.70\,(30 + x)$$
$$24 = 21 + 0.70x$$
$$2{,}400 = 2{,}100 + 70x \quad \blacktriangleleft \text{ Multiply each side by 100.}$$
$$300 = 70x$$
$$4\tfrac{2}{7} = x$$

Thus, $4\tfrac{2}{7}$ L of antifreeze must be added.

# Written Exercises

(A) **1.** How many liters of water must be added to 42 L of a 25% sulfuric acid solution to reduce it to a 20% solution? **10.5 L**

**2.** How many liters of water must a chemist add to 12 L of a 40% alcohol solution to make it a 30% alcohol solution? **4 L**

**3.** A pharmacist has 200 dL of a 20% peroxide in water solution. How much peroxide must be added to obtain a 40% peroxide solution? **$66\tfrac{2}{3}$ dL**

**4.** Dr. Sickler has 7 L of a 90% salt solution. How many liters of water must be added to reduce it to a 30% salt solution? **14 L**

**5.** How many liters of water must be added to 40 L of a 20% antifreeze solution to dilute it to a 16% antifreeze solution? **10 L**

**6.** A pharmacist has 5 dL of cough medicine that is 20% water. How much water must be added to make it 30% water so that it may be given to children? **$\tfrac{5}{7}$ dL**

**7.** How many cubic centimeters of alcohol must be added to 80 cm³ of a 25% alcohol solution to make it a 45% alcohol solution? **$29\tfrac{1}{11}$ cc**

**8.** A solution of 60 mL is 50% acid. How many milliliters of water must be added to dilute it to a 30% acid solution? **40 mL**

(B) **9.** Dr. Blue has one solution that is 40% salt and another that is 60% salt. How much of the 60% solution must be added to 15 dL of the 40% solution to obtain a solution that is 50% salt? **15 dL**

**10.** A chemist wants to add some 20% acid solution to 60 dL of an 80% acid solution to reduce it to a 60% acid solution. How much of the 20% solution must be added? **30 dL**

# PROBLEM SOLVING

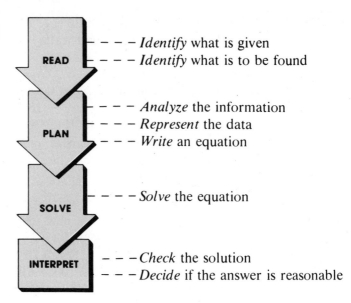

READ − − − *Identify* what is given
− − − *Identify* what is to be found

PLAN − − − *Analyze* the information
− − − *Represent* the data
− − − *Write* an equation

SOLVE − − − *Solve* the equation

INTERPRET − − − *Check* the solution
− − − *Decide* if the answer is reasonable

**Solve.**

1. The length of a rectangle is 3 m more than twice the width. The perimeter is 54 m. Find the length and width. **19 m, 8 m**

2. Twelve years more than twice Chuck's age is at most 40. How old is he?
   **at most 14 yrs.**

3. Martin has 5 more nickels than dimes. Their total value is $2.05. How many coins of each type does he have? **12 dimes, 17 nickels**

4. The selling price of a television is $240. The profit is 20% of the cost. Find the cost. **$200**

5. The tens digit of a two-digit number is three times the units digit. If the digits are reversed, the new number is 36 less than the original number. Find the original number. **62**

6. Two cars started toward each other at the same time from towns 460 km apart. One car traveled at a rate of 60 km/h and the other at 55 km/h. After how many hours did they meet? **4 hr.**

7. Working together, Mike and Tina can clean a house in 8 hours. It takes Mike 3 times longer than Tina to do it alone. How long would it take each alone?
   **32 hr for Mike; $10\frac{2}{3}$ hr for Tina.**

8. A 9-kg bag of coffee beans contains two varieties. One sells at $1.40/kg and the other at $1.50/kg. The mixture costs $13.00. How many kilograms of each are in the bag? **4 @ $1.50/kg, 5 @ $1.40/kg**

9. The cost of a certain metal varies directly as its mass. If 7 kg cost $21, find the cost of 9 kg. **$27**

10. Four out of seven people use a certain product. How many people use this product in a town of 35,000? **20,000**

11. Find three consecutive odd integers such that twice the second, added to the last, is 29. **7, 9, 11**

12. Find three consecutive integers such that the square of the first is 10 more than the last. **4, 5, 6 or −3, −2, −1**

13. The volume of a gas varies inversely as the pressure. If the volume is 60 m³ under 6 kg of pressure, find the volume under 9 kg of pressure. **40 m³**

14. The length of a rectangle is 3 m less than twice the width. The area is 20 m². Find the length and width. **5 m, 4 m**

# THE PARABOLA                                    15.7

*Objectives*   **To draw the graph of a quadratic equation (parabola)**
**To determine the coordinates of the minimum or maximum point of a parabola**

An equation such as $y = x^2 - 6x + 8$ describes a quadratic function. You can sketch a graph of this function by first setting up a table of values to find the coordinates of several points and then drawing a smooth curve through them, as illustrated in Example 1.

*Example 1*   **Graph the function $y = x^2 - 6x + 8$. Find the coordinates of the lowest point on the graph.**

Table of Values

| $x$ | $x^2 - 6x + 8$ | $y$ | Points on the graph |
|---|---|---|---|
| 0 | $0^2 - 6 \cdot 0 + 8$ | 8 | (0, 8) |
| 1 | $1^2 - 6 \cdot 1 + 8$ | 3 | (1, 3) |
| 2 | $2^2 - 6 \cdot 2 + 8$ | 0 | (2, 0) |
| 3 | $3^2 - 6 \cdot 3 + 8$ | −1 | (3, −1) |
| 4 | $4^2 - 6 \cdot 4 + 8$ | 0 | (4, 0) |
| 5 | $5^2 - 6 \cdot 5 + 8$ | 3 | (5, 3) |
| 6 | $6^2 - 6 \cdot 6 + 8$ | 8 | (6, 8) |

The graph is called a **parabola**. The lowest point on the graph is (3, −1).
This point is called the **turning point** of the parabola.

In Example 1 the dotted vertical line through the turning point (3, −1) is called the **axis of symmetry.** The equation of the axis of symmetry is $x = 3$. If the parabola were folded along this line $x = 3$, the two halves would coincide. The axis of symmetry is halfway between any two points with the same $y$-coordinate. For example, (2, 0) and (4, 0) have the same $y$-coordinate, 0. The line $x = 3$ is halfway between these two points (2, 0) and (4, 0).

Notice that the number halfway between 2 and 4 is $\dfrac{2 + 4}{2} = \dfrac{6}{2} = 3$, the $x$-coordinate of any point on the axis of symmetry. Also, 3 is the $x$-coordinate of the turning point of the parabola. This suggests a way for predicting the coordinates of the turning point of the parabola algebraically without first having to sketch the parabola, shown in Example 2.

The Parabola                                      **445**

*Example 2*     **Find the x-coordinate of the turning point of the parabola**
**$y = x^2 - 6x + 8$ algebraically.**

First notice that the turning point is on
the axis of symmetry.

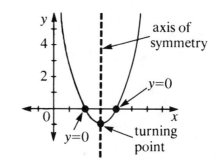

1. Let $y = 0$. $y = x^2 - 6x + 8 = 0$
2. Solve $1x^2 - 6x + 8 = 0$ for $x$.

$$x = \frac{-(-6) \pm \sqrt{(-6)^2 - 4 \cdot 1 \cdot 8}}{2 \cdot 1}$$

$$x = \frac{6 \pm \sqrt{4}}{2}$$

$$x = \frac{6 + 2}{2} \text{ or } x = \frac{6 - 2}{2}$$

$$x = 4 \qquad \text{or } x = 2$$

3. The axis of symmetry is halfway
   between 2 and 4.

$$x = \frac{1}{2}(2 + 4) = \frac{1}{2}(6) = 3$$

**Thus,** the x-coordinate of the turning point of $y = x^2 - 6x + 8$ is 3.

Notice that the x-coordinate of the turning point of the parabola
$y = x^2 - 6x + 8$ is halfway between the two points whose x coordinates are
2 and 4. In general, the graph of $y = ax^2 + bx + c$ is a parabola. The
x-coordinate of the turning point is halfway between the x-intercepts. This
can be determined algebraically, as suggested in the C exercises of this
lesson. The formula for the x-coordinate of the turning point is given below.

| Formula for x-coordinate of turning point of a parabola | For the parabola $y = ax^2 + bx + c$, the x-coordinate of the turning point is $$x = \frac{-b}{2a}.$$ |
|---|---|

*Example 3*     **Find the coordinates of the turning point of the parabola**
**$y = -2x^2 + 8x - 3$.**

$y = -2x^2 + 8x - 3$ ◀ $a = -2, b = 8$

$x = \dfrac{-b}{2a} = \dfrac{-8}{2 \cdot -2} = \dfrac{-8}{-4} = 2$ ◀ *Use* $x = \dfrac{-b}{2a}$ *to find the x-coordinate.*

To find the y-coordinate of the turning point, let $x = 2$ in the equation of the
parabola: $y = -2x^2 + 8x - 3$

$$y = -2(2)^2 + 8 \cdot 2 - 3$$
$$y = -8 + 16 - 3 = 5 \quad ◀ \textit{The y-coordinate is 5.}$$

**Thus,** the turning point is (2, 5).

It is easy to sketch a parabola once you know its turning point. You can use two or three values of $x$ on each side of the $x$-coordinate of the turning point to find several points through which the parabola can be drawn.

*Example 4*    **Draw the graph for $y = -2x^2 + 8x - 3$.**

From Example 3, you have already found that the turning point is (2, 5). Now set up a table of values using three values of $x$ on each side of 2. Then graph the parabola.

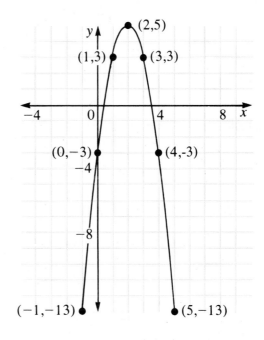

| $x$ | $-2x^2 + 8x - 3$ | $y$ | Points on the graph |
|---|---|---|---|
| $-1$ | $-2(-1)^2 + 8 \cdot -1 - 3$ | $-13$ | $(-1, -13)$ |
| $0$ | $-2(0)^2 + 8 \cdot 0 - 3$ | $-3$ | $(0, -3)$ |
| $1$ | $-2(1)^2 + 8 \cdot 1 - 3$ | $3$ | $(1, 3)$ |
| $2$ | $-2(2)^2 + 8 \cdot 2 - 3$ | $5$ | $(2, 5)$ ← Turning point |
| $3$ | $-2(3)^2 + 8 \cdot 3 - 3$ | $3$ | $(3, 3)$ |
| $4$ | $-2(4)^2 + 8 \cdot 4 - 3$ | $-3$ | $(4, -3)$ |
| $5$ | $-2(5)^2 + 8 \cdot 5 - 3$ | $-13$ | $(5, -13)$ |

Plot the points. Draw a smooth curve through the seven points.

The sketches of the two parabolas in Examples 1 and 4 suggest some interesting patterns.

$y = 1x^2 - 6x + 8$
↑
$x^2$ coefficient is positive.
The graph opens upward.

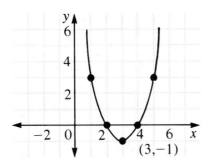

The turning point is a minimum. The minimum value of the function is $-1$.

$y = -2x^2 + 8x - 3$
↑
$x^2$ coefficient is negative.
The graph opens downward.

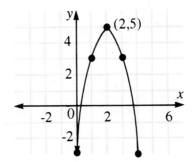

The turning point is a maximum. The maximum value of the function is 5.

The Parabola

| *Summary* | The graph $y = ax^2 + bx + c$ $(a \neq 0)$ is a parabola. |
|---|---|

If $a > 0$, the parabola opens upward. The turning point is a minimum.

If $a < 0$, the parabola opens downward. The turning point is a maximum.

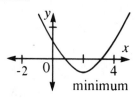

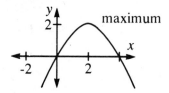

## Oral Exercises

**Will the function have a maximum or minimum value?**

1. $y = -x^2 + 4x + 2$ **max**
2. $y = x^2 + 6x + 1$ **min**
3. $y = -x^2 - 5x + 8$ **max**
4. $y = -4x^2$ **max**
5. $y = x^2$ **min**
6. $y = -2x^2 - 10x + 1$ **max**

## Written Exercises

Ⓐ **Determine the coordinates of the turning point of each parabola.**

1. $y = x^2 + 6x + 8$ **−3, −1**
2. $y = -x^2 - 2x - 4$ **−1, −3**
3. $y = -x^2 + 4x - 3$ **2, 1**
4. $y = -2x^2 + 8x - 5$ **2, 3**
5. $y = 3x^2 + 6x - 21$ **−1, −24**
6. $y = 2x^2 + 4x + 7$ **−1, 5**

**Draw the graph for each of the following. Label the turning point.** **See TE Answer Section.**

7. $y = x^2 - 4x + 6$ **2, 2**
8. $y = -2x^2 + 8x + 1$ **2, 9**
9. $y = 4x^2 - 16x + 6$ **2, −10**
10. $y = -3x^2 + 6x + 5$ **1, 8**
11. $y = 2x^2 - 8x - 6$ **2, −14**
12. $y = -x^2 - 2x - 8$ **−1, −7**

*Example*   **Find the minimum value of $y$ for $y = x^2 - 4x + 3$.**

The minimum value occurs at the turning point.

$$x = \frac{-b}{2a} = \frac{-(-4)}{2 \cdot 1} = \frac{4}{2} = 2$$   ◄ *Use $x = \dfrac{-b}{2a}$ to find the x-coordinate*

$$y = x^2 - 4x + 3$$   *of the turning point.*

$$y = (2)^2 - 4 \cdot 2 + 3 = -1$$   ◄ *Substitute 2 for x in $y = x^2 - 4x + 3$.*

**Thus,** the minimum value is $-1$.

**Find the minimum or maximum value of $y$.**

Ⓑ 13. Find the minimum value of $y$ if $y = x^2 + 8x + 1$. **−15**
14. Find the maximum value of $y$ if $y = -2x^2 - 8x + 6$. **14**

Ⓒ 15. Verify algebraically that $x = \dfrac{-b}{2a}$ is the x-coordinate of the turning point of the parabola $y = ax^2 + bx + c$. (HINT: The value is halfway between the two roots of the equation $ax^2 + bx + c = 0$.)

16. Find a formula for the y-coordinate of the turning point. $y = \dfrac{-b^2 + 4ac}{4a}$

# APPLICATIONS

$$\boxed{\text{Read}} \rightarrow \boxed{\text{Plan}} \rightarrow \boxed{\text{Solve}} \rightarrow \boxed{\text{Interpret}}$$

It is necessary for a business person to be able to predict the number of items that must be sold to keep costs down to a minimum, or profits to a maximum. Such an application is illustrated in the example below.

**Example**

**Ms. Jones runs a sandwich shop. She uses the following formula to approximate her profits: $p = -x^2 + 50x - 350$, where $p$ is the profit from selling $x$ units of sandwiches. How many units must she sell to make a maximum profit? What is the maximum profit?**

The graph of $p = -x^2 + 50x - 350$ will have a maximum since the coefficient of $x^2$ is negative. Use the formula $x = \dfrac{-b}{2a}$ to find the value of $x$ that makes $p$ a maximum.

$x = \dfrac{-b}{2a} = \dfrac{-50}{2 \cdot -1}$  ◀ *For $p = -x^2 + 50x - 350$, $a = -1$ and $b = 50$.*

$\phantom{x} = \dfrac{-50}{-2} = 25$  ◀ *Number of units to give maximum profit*

Now find the maximum profit by letting $x = 25$ in $p = -x^2 + 50x - 350$.
$p = -1x^2 + 50x - 350$
$p = -1(25)^2 + 50 \cdot 25 - 350$
$p = -1(625) + 1{,}250 - 350 = 275$
Thus, she must sell 25 units to make a maximum profit of $275.

## Written Exercises

**6; $24**

1. The formula for the cost of running a taco stand is $c = x^2 - 12x + 60$. How many units $x$ of tacos must be sold to keep costs at a minimum? Find the minimum cost.

3. The formula for the height ($s$) reached by a rocket fired straight up from the ground with an initial velocity of 96 ft/sec is $s = -16t^2 + 96t$. Find the time ($t$) for the rocket to reach a maximum height. Find this height. **3 seconds; 144 ft**

5. A hairdryer manufacturer determines that the total profit $p$ (in dollars) of manufacturing $x$ hairdryers is $p = -0.18x^2 + 36x + 4000$. How many must be sold to realize a maximum profit? What is the maximum profit? **100; $5800**

**4. 400; $160,200**

2. The cost $c$ of producing $x$ units of radios is $c = x^2 - 12x + 72$. How many radios should be made to produce the minimum cost? What is this minimum cost? **6; $36**

4. A new company finds their profit ($p$) is related to the number of items sold ($x$) by $p = 200 + 800x - x^2$. How many items must be sold to produce the maximum profit? Find the maximum profit.

6. A biologist's formula to predict the number of impulses fired after stimulation of a nerve is $i = -x^2 + 30x - 50$ where $i$ is the number of impulses per millisecond and $x$ is the number of milliseconds after stimulation. Find the time for the maximum number of impulses. **15 milliseconds**

**Vocabulary**

axis of symmetry [15.7]
completing the square [15.3]
extraneous [15.1]
maximum [15.7]
minimum [15.7]
parabola [15.7]
quadratic formula [15.4]
turning point [15.7]

Solve. [15.1]

1. $\sqrt{5x} = 2$  $\frac{4}{5}$
2. $\sqrt{m} + 2 = 5$  9
3. $\sqrt{3x - 4} + 6 = 5$  Ø
4. $\sqrt{2a + 1} - 1 = 2$  4
5. $x - 2 = \sqrt{19 - 6x}$  3
6. $x - 1 = \sqrt{9 - 4x}$  2

Solve.

7. If 6 is added to twice a number, the square root of the result is 4. Find the number. 5

Solve. [15.2]

8. $x^2 = 49$  ±7
9. $5y^2 = 100$  ±2√5
10. $a^2 = 30$  ±√30
11. $(2x + 4)^2 = 25$  $\frac{1}{2}, -\frac{9}{2}$
12. $(x + 4)^2 = 32$  −4 ± 4√2
13. $(x - 6)^2 = 40$  6 ± 2√10

Solve.

14. The square of 5 more than a number is 49. Find the number. 2, −12

Solve by completing the square. [15.3]

15. $x^2 + 10x = -24$  −4, −6
16. $x^2 - 8x + 16 = 0$  4
17. $x^2 + 2x - 24 = 0$  4, −6
18. $x^2 - 6x - 2 = 0$  3 ± √11
19. $x^2 + 5x = 6$  1, −6
20. $x^2 + 7x + 10 = 0$  −2, −5

Solve by using the quadratic formula. Leave irrational roots in simplest radical form. [15.4]

21. $x^2 + 8x + 12 = 0$
22. $x^2 - 9x = -20$  5, 4
23. $x^2 = 5x - 1$
24. $4x + 3 = -2x^2$  Ø
25. $-6x = -x^2 - 4$
26. $\frac{1}{2}x - 1 = -\frac{1}{4}x^2$

21. −2, −6
Solve. [15.5]  23. $\frac{5 \pm \sqrt{21}}{2}$  25. 3 ± √5   26. −1 ± √5

27. The length of a rectangle is 3 m more than the width. The area is 88 m². Find the length and width. **11 m, 8 m**
28. The height of a triangle is 3 cm less than twice the base. The area is 10 cm². Find the base and height. **4 cm, 5 cm**

Solve. Use a square root table. Round off each answer to the nearest tenth.

29. The width of a rectangle is 2 m less than the length. The area is 4 m². Find the length and the width. **3.2 m, 1.2 m**
30. The height of a triangle is 3 km more than twice the base. The area is 3 km². Find the base and height. **1.1 km, 5.2 km**

Determine the coordinates of the turning point of each parabola. [15.6]

31. $y = x^2 - 10x + 2$  (5, −23)
32. $y = -2x^2 - 8x + 4$  (−2, 12)

Graph each parabola. **See TE Answer Section.**

33. $y = -x^2 + 6x + 1$   34. $y = 3x^2 - 12x$

35. Find the minimum value of $y$ if $y = 2x^2 - 16x + 8$. **−24**

36. Find the maximum value of $y$ if $y = -x^2 - 12x + 8$. **44**

★ 37. Solve $\sqrt{2a + 1} = \sqrt{a} + 1$. [15.1] **4, 0**

★ 38. Solve $3(2x - 8)^2 - 60 = 36$. [15.2]
**4 ± 2√2**

★ 39. Solve for $x$ in terms of $b$ and $c$.
$x^2 + 6bx + c = 0$ [15.3]  **−3b ± √9b² − c**

★ 40. Solve $3x^2 - 3\sqrt{2}x + 1 = 0$. [15.4]

$$\frac{3\sqrt{2} \pm \sqrt{6}}{6}$$

Solve.

**1.** $\sqrt{3t} = 4$   $\frac{16}{3}$

**2.** $\sqrt{2p - 4} + 5 = 7$   **4**

**3.** $x - 2 = \sqrt{7 - 2x}$   **3**

**4.** $x^2 = 100$   **±10**

**5.** $5a^2 = 40$   **±2√2**

**6.** $(x + 2)^2 = 38$   $-2 \pm \sqrt{38}$

Solve by completing the square.

**7.** $p^2 + 14p = -48$   **−6, −8**

**8.** $y^2 + 8y = 2$   $-4 \pm 3\sqrt{2}$

**9.** $m^2 + 6m = -1$   $-3 \pm 2\sqrt{2}$

**10.** $2x^2 - 3x + 1 = 0$   $1, \frac{1}{2}$

Solve by using the quadratic formula. Leave irrational roots in simplest radical form.

**11.** $x^2 + 11x + 28 = 0$   **−4, −7**

**12.** $y^2 - 3y = -2$   **2, 1**

**13.** $-12a - 4 = -2a^2$   $3 \pm \sqrt{11}$

**14.** $\frac{1}{3}x - 1 = -\frac{1}{6}x^2$   $-1 \pm \sqrt{7}$

Solve.

**15.** If 4 is added to a number, the square of the result is 36. Find the number. **2, −10**

**16.** The length of a rectangle is 4 m more than twice the width. The area is 30 m². Find the length and width. **10 m, 3 m**

**17.** The base of a triangle is 6 km less than the height. The area is 20 km². Find the base and height. **4 km, 10 km**

**18.** The length of a rectangle is 10 cm more than the width. The area is 50 cm². Find the length and width to the nearest tenth. **13.7 cm, 3.7 cm**

**19.** Determine the coordinates of the turning point of the parabola $y = x^2 - 14x + 2$. **(7, −47)**

**20.** Graph the parabola $y = -x^2 - 6x + 1$. **See TE Answer Section.**

**21.** Find the maximum value of $y$ if $y = -x^2 - 10x + 4$. **29**

**22.** Find the minimum value of $y$ if $y = 2x^2 + 12x - 5$. **−23**

★ **23.** For what value of $c$ will the equation $x^2 - 4x + c = 0$ have no real solution? **c > 4**

Solve.

★ **24.** $5(2x - 4)^2 - 12 = 88$   $2 \pm \sqrt{5}$

★ **25.** $\sqrt{3x + 4} = \sqrt{x} + 2$   **0, 4**

## Solving Quadratic Equations

A quadratic equation in standard form $ax^2 + bx + c = 0$ becomes
$A * X \wedge 2 + B * X + C = 0$ in BASIC notation.

The solutions of a quadratic equation in standard notation are:

$$\frac{-b + \sqrt{b^2 - 4ac}}{2a} \quad \text{and} \quad \frac{-b - \sqrt{b^2 - 4ac}}{2a}$$

The solutions of a quadratic equation in BASIC notation are:
X1 = (−B + SQR(B $\wedge$ 2 − 4 * A * C))/(2 * A) and,
X2 = (−B − SQR(B $\wedge$ 2 − 4 * A * C))/(2 * A).

Complete the program below to find the roots of a quadratic equation given the coefficients A, B, and C.

```
. . . . . . . .
. . . . . . . .  (HINT: Print instructions.)
20   INPUT A,B,C
40   LET S1 = B ^ 2 - (4 * A * C)
50   LET S2 =  SQR (S1)
60   LET X1 = ( - B + S2) / (2 * A)
70   LET X2 =
. . . . . . . .
. . . . . . . .
. . . . . . . .  (HINT: Print solutions X1, X2, and equation in standard form.)
. . . . . . . .  (HINT: Inquire when to stop.)
999  END
```

See the Computer Section beginning on page 488 for more information.

## Exercises    For Ex. 1, 4, and 5, see TE Answer Section.

1. Complete the program above and RUN it for A = 1, B = −9, and C = 14.

**roots are 2, 7**

**RUN the completed program above to solve the following quadratic equations:**

2. $x^2 + 6x + 5 = 0$    **roots are −1, −5**

3. $3x^2 + 1 = -4x$    **roots are −1, −.333333 $(-\frac{1}{3})$**

4. Write a program to determine if an equation will have two, one, or no real solutions. Use the quadratic formula and print out the number of solutions and their values.

5. Write a program to print quadratic equations in standard form given the coefficients A, B, and C.

# COLLEGE PREP TEST

Directions: Choose the best answer for each problem.

**1.** For what value of $x$ is the statement
$\dfrac{7^x}{7^5} > 1$ true? **C**

    (A) $x > 1$      (B) $x < 5$      (C) $x > 5$
    (D) $x = 5$      (E) none of these

**2.** If $\neq$ means *is not equal to*, which value
of $x$ makes $8x - 2 \neq 7x$ true? **B**

    (A) $x = 2$      (B) $x \neq 2$      (C) $x > 0$
    (D) $x < 3$      (E) $x = -2$

**3.** Solve for $x$. $4 = \dfrac{2 - \dfrac{x}{y}}{6}$   **E**

    (A) $22y$      (B) $24y - 2$      (C) $\dfrac{2y - x}{6}$
    (D) $2 - 24y$      (E) $-22y$

**4.** Bill drove 200 mi at 40 mph. If he had
driven 10 mph faster, how many hours
would he have saved? **A**

    (A) 1      (B) 4      (C) 5
    (D) 9      (E) 20

**5.** If the operation $\star$ is defined to be
$x \star y = \dfrac{x + y}{xy}$, then $4 \star (4 \star 4)$ is **D**

    (A) 64      (B) $4\frac{1}{2}$      (C) $2\frac{1}{2}$
    (D) $\dfrac{9}{4}$      (E) $\dfrac{1}{2}$

**6.** If $\frac{1}{7}$ of a number is 5, what is $\frac{1}{5}$ of the
number? **B**

    (A) 35      (B) 7      (C) 5      (D) 1
    (E) It cannot be determined from the
       given information.

**7.** $\left(\dfrac{2}{5} \div \dfrac{5}{3}\right) - \left(\dfrac{1}{25} \div \dfrac{1}{6}\right) =$ **A**

    (A) 0      (B) $\dfrac{1}{36}$      (C) $\dfrac{99}{150}$
    (D) $\dfrac{6}{5}$      (E) $\dfrac{63}{50}$

**8.** If $a$ and $b$ are positive integers and
$a - b = 7$, what is the least possible
value of $a + b$? **A**

    (A) 9   (B) 8   (C) 7   (D) 6   (E) 5

**9.** If $\frac{4}{5}m = \frac{5}{4}p$ and $p \neq 0$, then $\dfrac{m}{p} =$ **E**

    (A) $\dfrac{16}{25}$      (B) $\dfrac{4}{5}$      (C) 1
    (D) $\dfrac{5}{4}$      (E) $\dfrac{25}{16}$

**10.** If $5x + 5x + 5x = 45$, then
$(5x - 12)^2 =$ **A**

    (A) 9   (B) 6   (C) 5   (D) 3   (E) $\dfrac{3}{5}$

**11.** If $x - y = 8$ and $x + y = 4$, the value
of $y^3$ is **A**

    (A) $-8$      (B) $-6$      (C) $-2$
    (D) 18      (E) 216

**12.** For what value of $k$ will the two solutions
of $x^2 - 6x + k = 0$ be the same? **A**

    (A) 9      (B) 6      (C) 3
    (D) 2      (E) none of these

**13.** If two halves of $3\frac{1}{2}$ is added to $3\frac{1}{2}$, the
result is                       **D**

    (A) 1      (B) $3\frac{1}{2}$      (C) $5\frac{1}{4}$
    (D) 7      (E) $10\frac{1}{2}$

# 16 PLANE TRIGONOMETRY

**CAREER**

*Landscape Architect*

In many places where homes, stores, or offices are being built, landscaping is a part of the overall plan. Landscape architects often utilize mathematical concepts in their work. They must deal with the area of the land, the height of the buildings, the slope of the terrain, and other considerations for which mathematics is needed.

*Project*

Pretend that you are a landscape architect hired to landscape the grounds leading up to and surrounding a new office complex. Make a scale drawing of the proposed layout. Determine the cost of the finished project.

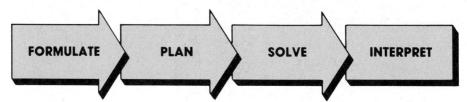

FORMULATE · PLAN · SOLVE · INTERPRET

# ANGLE PROPERTIES                                    16.1

*Objectives*    **To solve problems about complementary angles**
**To solve problems about supplementary angles**
**To solve problems about the angles of a triangle**
**To solve problems about the acute angles of a right triangle**

The diagrams below show pairs of angles, the sum of whose measures is 90°.

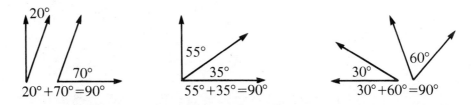

| | | |
|---|---|---|
| $20° + 70° = 90°$ | $55° + 35° = 90°$ | $30° + 60° = 90°$ |

| **Definition:** **Complementary** **angles** | Two angles the sum of whose measures is 90° are called **complementary angles.** Each angle is the *complement* of the other. |
|---|---|

You can use the definition of complementary angles to solve word problems, as shown in Example 1.

*Example 1*    **The measure of an angle is 10° more than 3 times the measure of its complement. Find the measure of the angle and its complement.**

Let $x$ = the degree measure of the complement.
Let $3x + 10$ = the degree measure of the angle.    ◀ *10 more than 3x*
Use the fact that the angles are complementary to write an equation.
The sum of the measures of complementary angles is 90°.

$$x + 3x + 10 = 90$$
$$4x + 10 = 90$$
$$4x = 80$$
$$x = 20 \quad ◀ \text{ Measure of the complement.}$$

The degree measure of the complement, $x$, is 20°.
The degree measure of the angle, $(3x + 10)$, is $3 \cdot 20 + 10$, or 70°.

| The angles are complementary. | | The angle's measure is 10 more than 3 times its complement's. |
|---|---|---|
| 70 + 20   90 | | 70      10 more than 3 times 20 |
| 90 | | $3 \cdot 20 + 10$ |
| 90 = 90, true | | 70 |
| | 70 = 70, true | |

**Thus,** the angle measures 70° and its complement measures 20°.

Sometimes, the relationship between two complementary angles is defined in terms of a ratio, as shown in Example 2.

**Example 2**    **The measures of two complementary angles are in the ratio 5:4. Find the measure of each angle.**

Let $5x$ = the degree measure of one angle.
Let $4x$ = the degree measure of the other angle.
Write an equation; $5x + 4x = 90$.    ◀ *The sum of the measures of*
$$9x = 90$$    *complementary angles is 90°.*
$$x = 10$$
The degree measure of one angle, $5x$, is $5 \cdot 10°$, or $50°$.
The degree measure of the other angle, $4x$ is $4 \cdot 10°$, or $40°$.

**Thus,** the angle measures are 50° and 40°.   ◀ *Check. 50° + 40° = 90°*
*and 50°:40° = 5:4.*

The diagrams below show pairs of angles, the sum of whose measures is 180°.

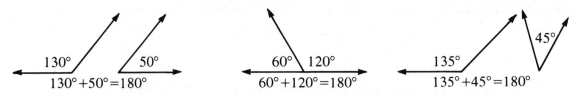

130°    50°         60° 120°        135°    45°
130°+50°=180°      60°+120°=180°      135°+45°=180°

---

**Definition:**
**Supplementary**
**angles**

Two angles the sum of whose measures is 180° are called **supplementary angles.** Each angle is the *supplement* of the other.

---

**Example 3**    **The measure of an angle is twice the measure of its supplement. Find the measure of the angle and its supplement.**

Let $x$ = the degree measure of the supplement.
Let $2x$ = the degree measure of the angle.
Use the fact that the angles are supplementary to write an equation.
The sum of the measures of supplementary angles is 180°.

$$x + 2x = 180$$
$$3x = 180$$
$$x = 60$$

The degree measure of the supplement, $x$, is 60°.
The degree measure of the angle, $2x$, is $2 \cdot 60°$, or 120°.
**So,** the angle measures 120° and its supplement measures 60°.

The diagrams below show an important property of the three angles of any triangle.

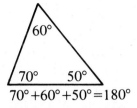

$70° + 60° + 50° = 180°$

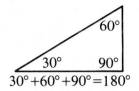

$30° + 60° + 90° = 180°$

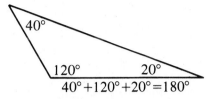

$40° + 120° + 20° = 180°$

| Sum of the measures of the angles of a triangle | For any triangle, the sum of the measures of the three angles is 180°. |
| --- | --- |

*Example 4*

**The second angle of a triangle measures three times the first angle. The third angle measures 80° more than the first angle. Find the measure of each angle.**

Let $x$ = the degree measure of the first angle.
Let $3x$ = the degree measure of the second angle.
Let $x + 80$ = the degree measure of the third angle.
Write an equation. Use the fact that the sum of the measures of the three angles is 180°.

$$x + 3x + x + 80 = 180$$
$$5x + 80 = 180$$
$$x = 20$$

The measure of the first angle, $x$, is 20°.
The measure of the second angle, $3x$, is $3 \cdot 20°$, or 60°.
The measure of the third angle, $x + 80°$, is $20° + 80°$, or 100°.
**So,** the angles measure 20°, 60°, and 100°. ◀ *This checks because 20° + 60° + 100° = 180°.*

The triangle shown is a right triangle because it contains a right angle, a 90° angle. Since the sum of the measures of the three angles of a triangle is 180°, you can write the equation $x + y + 90° = 180°$. If you subtract 90° from each side, then $x + y = 90°$, the sum of the measures of angles $A$ and $B$. Since the sum of the measures of angles $A$ and $B$ is 90°, the angles are complementary. Angles $A$ and $B$ are *acute angles* because each has a measure less than 90°.

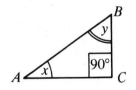

| Property of acute angles of a right triangle | The acute angles of a right triangle are complementary. The sum of their measures is 90°. |
| --- | --- |

*Example 5*  **One of the acute angles of a right triangle measures twice the other acute angle. Find the measure of each acute angle.**

Let $x$ = the measure of one acute angle.
Let $2x$ = the measure of the other acute angle.
$x + 2x = 90$ ◀ *The sum of the measures of the acute angles*
$\quad\quad 3x = 90$ $\quad$ *in a right triangle is 90°.*
$\quad\quad\ x = 30$

**So,** the measure of one acute angle, $x$, is 30°; the measure of the other acute angle, $2x$, is 2 · 30°, or 60°.

## Reading in Algebra

**Indicate whether each of the following statements is sometimes (S), always (A), or never (N) true.**
  1. The sum of the measures of the angles of a right triangle is 180°. **A**
  2. The complement of an angle is 120°. **N**
  3. Two angles of a triangle are complementary. **S**
  4. A triangle can have two of its angles each with a measure of 90°. **N**

## Written Exercises

**Solve each problem.**

(A) 1. An angle measures 70° less than its complement. Find the measure of the angle and its complement. **10°, 80°**

  2. An angle measures twice its complement. Find the measure of the angle and its complement. **30°, 60°**

  3. The measures of two complementary angles are in the ratio 2:1. Find the measure of each angle. **30°, 60°**

  4. An angle measures 60° more than twice its supplement. Find the measure of the angle and its supplement. **140°, 40°**

  5. The measures of two supplementary angles are in the ratio 7:5. Find the measure of each angle. **105°, 75°**

  6. An angle measures 30° less than twice its supplement. Find the measure of each angle. **110°, 70°**

  7. The second angle of a triangle measures twice the first. The third angle measures 30° more than the first. Find the measure of each angle. **37.5°, 75°, 67.5°**

  8. The measures of the angles of a triangle are consecutive even integers. Find the measure of each angle. **58°, 60°, 62°**

(B) 9. The first angle of a triangle measures 25° less than the second angle. The third angle measures 70° less than the sum of the measures of the first two angles. Find the measure of each angle. **50°, 75°, 55°**

  10. The supplement of an angle measures 10° more than twice the complement of that angle. Find the measure of the angle, its supplement, and its complement. **10°, 170°, 80°**

# PROPERTIES OF SIMILAR TRIANGLES 16.2

*Objectives*   **To find lengths of sides of similar triangles**
**To solve word problems related to similar triangles**

In the triangle at the right, the
degree measure of angle $A$ is 70°.

This is written as m $\angle A$ = 70.
Side $a$ is opposite $\angle A$.
It is customary to use a capital letter to represent
an angle and a small letter to represent the side
opposite that angle.

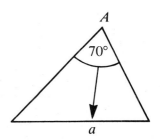

A pair of similar triangles is shown below.

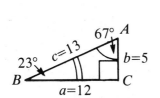
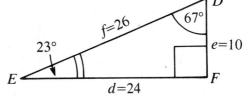

Notice the patterns:

$$m \angle A = m \angle D$$

$$m \angle B = m \angle E$$

$$m \angle C = m \angle F$$

Side opp. $\angle A$ ▶   $\dfrac{a}{d} = \dfrac{12}{24} = \dfrac{1}{2}$
Side opp. $\angle D$ ▶

$\dfrac{b}{e} = \dfrac{5}{10} = \dfrac{1}{2}$

$\dfrac{c}{f} = \dfrac{13}{26} = \dfrac{1}{2}$

Three pairs of corresponding angles
have equal degree measure.

Lengths of corresponding sides (opposite
corresponding angles of equal degree
measure) have the same ratio, $\frac{1}{2}$.

$$\triangle ABC \text{ is similar to } \triangle DEF,$$

$$\text{or } \triangle ABC \sim \triangle DEF.$$

This suggests two general conditions for similar triangles.

| Properties of similar triangles | Two triangles are similar if:<br>1. the corresponding angles have equal degree measure, and<br>2. the lengths of the corresponding sides have the same ratio. |
|---|---|

You can now use proportions to find the lengths of sides in similar triangles. This is illustrated in Example 1.

**Example 1**    $\triangle ABC \sim \triangle PQR$, with $m \angle A = m \angle P$, $m \angle B = m \angle Q$, $m \angle C = m \angle R$, $a = 6$, $p = 8$, and $c = 9$. Find $r$.

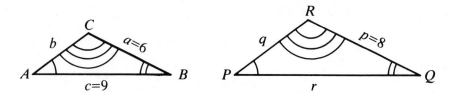

Use the fact that the lengths of the corresponding sides of similar triangles have the same ratio to write a proportion.

$$\frac{6}{8} = \frac{9}{r} \qquad \blacktriangleleft \quad \frac{a}{p} = \frac{c}{r}$$
$$6 \cdot r = 8 \cdot 9$$
$$6r = 72$$
$$r = 12$$

**So,** the length of side $\overline{PQ}$, $r$, is 12.

Similar triangles can be used to find "inaccessible" heights (heights that are too difficult to actually measure physically). For example, you can find the height of a flagpole (you might not want to climb it to measure its height) by using similar triangles. The similar triangles will involve the height of the pole and the length of its shadow, and a known height, such as your height, and the length of your shadow. The procedure is illustrated in the next example.

**Example 2**    **A person 2 m tall casts a shadow 6 m long. How high is a nearby flagpole if its shadow is 54 m?**

Let $x =$ the height of the flagpole.
Write a proportion for the two similar triangles.

$$\frac{x}{2} = \frac{54}{6}$$
$$6x = 108$$
$$x = 18$$

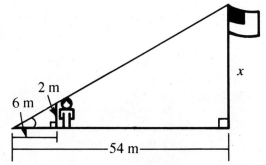

**So,** the height of the flagpole is 18 m.

# Written Exercises

**In Exercises 1–10, $\triangle ABC \sim \triangle DEF$. Find the indicated measures.**

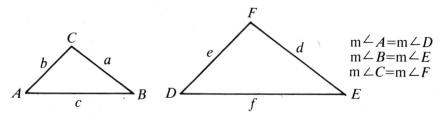

$m\angle A = m\angle D$
$m\angle B = m\angle E$
$m\angle C = m\angle F$

5. 3, 56   7. $13\frac{1}{2}$, $10\frac{1}{2}$   9. $4\frac{4}{5}$, 15   6. 18, $2\frac{2}{3}$   8. $6\frac{2}{3}$, $7\frac{1}{2}$   10. 24, $16\frac{1}{2}$

Ⓐ 1. $a = 6$, $d = 12$, and $c = 4$. Find $f$.  **8**
3. $b = 9$, $c = 14$, and $e = 14$. Find $f$.  **$21\frac{7}{9}$**
5. $c = 14$, $a = 5$, $d = 20$, $e = 12$. Find $b$, $f$.
7. $b = 9$, $c = 7$, $a = 10$, $d = 15$. Find $e$, $f$.
9. $b = 4$, $a = 6$, $e = 10$, $f = 12$. Find $c$, $d$.

2. $b = 9$, $c = 5$, and $e = 3$. Find $f$.  **$1\frac{2}{3}$**
4. $d = 4$, $a = 8$, and $c = 9$. Find $f$.  **$4\frac{1}{2}$**
6. $b = 8$, $c = 6$, $d = 6$, $f = 2$. Find $a$, $e$.
8. $c = 6$, $b = 5$, $f = 9$, $d = 10$. Find $a$, $e$.
10. $f = 14$, $d = 16$, $e = 11$, $c = 21$. Find $a$, $b$.

**Solve each problem.**

11. A tree 3 m tall casts a shadow 5 m long while a tower casts a shadow 120 m long. How tall is the tower?  **72 m**

12. A vertical meter stick casts a 6-m shadow while a telephone pole casts a 36-m shadow. How tall is the telephone pole?  **6 m**

Ⓑ 13. Bill walked 12 m up a ramp and was 5 m above the ground. If he walked 18 m farther up the ramp, how far above the ground would he be?  **$12\frac{1}{2}$ m**

14. Maria was lying 14 m from the base of a tree 11 m tall. She could see the top of a 600-m building just beyond the top of the tree. How far was she from the base of the building?  **763 m**

15. $\triangle ABC \sim \triangle EDC$ with $m\angle B = m\angle D$ and $m\angle A = m\angle E$. Find the width of the river, $\overline{BA}$.  **96 m**

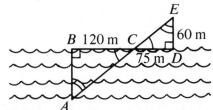

16. A 20-m ladder touches the side of a building at a height of 16 m. At what height would an 8-m ladder touch the building if it makes the same angle with the ground?  **$6\frac{2}{5}$ m**

---

## NON-ROUTINE PROBLEMS

A perfect number is one such that the sum of its factors (except itself) is the number itself. For example, the factors of 6 other than 6 itself are 1, 2, and 3. Since $1 + 2 + 3 = 6$, 6 is a perfect number. The next perfect number is less than 40. Find that number.  **28**

# INTRODUCTION TO TRIGONOMETRIC RATIOS

**Objectives**
**To identify the three sides of a right triangle in terms of one of the acute angles of the right triangle**
**To compute the sine, cosine, and tangent of an acute angle of a right triangle**

*Trigonometry* (from the Greek language) means "triangle measurement." Recall that in a right triangle, the side opposite the right angle is called the *hypotenuse*. The sides opposite the acute angles are called *legs* and are referred to as *opposite* or *adjacent* to an acute angle. This is illustrated below.

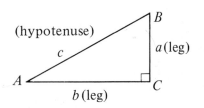

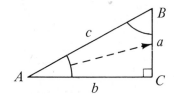

Leg *a* is opposite $< A$
and adjacent to $< B$.

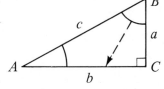

Leg *b* is opposite $< B$
and adjacent to $< A$.

Certain ratios are the same in all right triangles regardless of the lengths of the sides. Consider these similar right triangles.

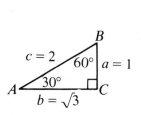

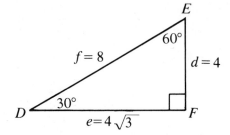

By the similar triangle property,

$$\frac{a}{d} = \frac{c}{f} \qquad\qquad \frac{b}{e} = \frac{c}{f} \qquad\qquad \frac{a}{d} = \frac{b}{e}$$

Using the property of proportions that states if $\frac{a}{b} = \frac{c}{d}$, then $\frac{a}{c} = \frac{b}{d}$, the following ratios are also true proportions.

$$\frac{a}{c} = \frac{d}{f} \qquad\qquad \frac{b}{c} = \frac{e}{f} \qquad\qquad \frac{a}{b} = \frac{d}{e}$$

$$\frac{1}{2} = \frac{4}{8} \qquad\qquad \frac{\sqrt{3}}{2} = \frac{4\sqrt{3}}{8} \qquad\qquad \frac{1}{\sqrt{3}} = \frac{4}{4\sqrt{3}}$$

These ratios are called *sine*, *cosine*, and *tangent*.

These common trigonometric ratios are defined below.

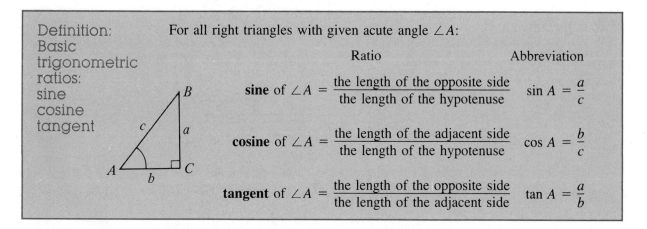

| | | Ratio | Abbreviation |
|---|---|---|---|

**Definition:**
**Basic trigonometric ratios:**
sine
cosine
tangent

For all right triangles with given acute angle $\angle A$:

**sine** of $\angle A = \dfrac{\text{the length of the opposite side}}{\text{the length of the hypotenuse}}$    $\sin A = \dfrac{a}{c}$

**cosine** of $\angle A = \dfrac{\text{the length of the adjacent side}}{\text{the length of the hypotenuse}}$    $\cos A = \dfrac{b}{c}$

**tangent** of $\angle A = \dfrac{\text{the length of the opposite side}}{\text{the length of the adjacent side}}$    $\tan A = \dfrac{a}{b}$

## Example 1    For $\triangle ABC$, find each ratio to three decimal places.

Find $\sin A$, $\cos A$, and $\tan A$.

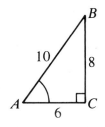

$\sin A = \dfrac{8}{10} = \dfrac{4}{5}$, or 0.800  ◄ $\sin A = \dfrac{opp.}{hyp.}$

$\cos A = \dfrac{6}{10} = \dfrac{3}{5}$, or 0.600  ◄ $\cos A = \dfrac{adj.}{hyp.}$

$\tan A = \dfrac{8}{6} = \dfrac{4}{3}$, or 1.333  ◄ $\tan A = \dfrac{opp.}{adj.}$

Find $\sin B$, $\cos B$, and $\tan B$.

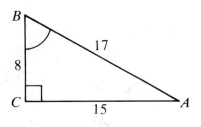

$\sin B = \dfrac{15}{17} = 0.882$

$\cos B = \dfrac{8}{17} = 0.471$

$\tan B = \dfrac{15}{8} = 1.875$

## Example 2    Find $\cos B$ to three decimal places.

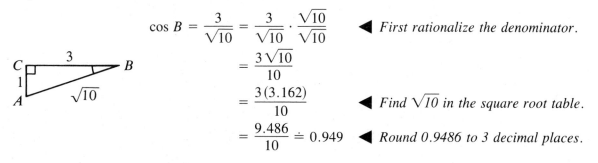

$\cos B = \dfrac{3}{\sqrt{10}} = \dfrac{3}{\sqrt{10}} \cdot \dfrac{\sqrt{10}}{\sqrt{10}}$  ◄ *First rationalize the denominator.*

$= \dfrac{3\sqrt{10}}{10}$

$= \dfrac{3(3.162)}{10}$  ◄ *Find $\sqrt{10}$ in the square root table.*

$= \dfrac{9.486}{10} \doteq 0.949$  ◄ *Round 0.9486 to 3 decimal places.*

Introduction to Trigonometric Ratios

## Oral Exercises

**Refer to the figure at the right.**
1. Name the leg adjacent to ∠ S. **r**
2. Name the leg opposite ∠ R. **r**
3. Name the hypotenuse. **t**
4. Name the leg opposite ∠ S. **s**
5. Name the leg adjacent to ∠ R. **s**

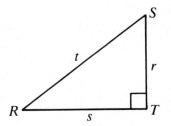

**Name each as a ratio of a pair of sides.**
6. sin R $\frac{r}{t}$
7. cos R $\frac{s}{t}$
8. tan S $\frac{s}{r}$
9. tan R $\frac{r}{s}$

## Written Exercises

1. 0.600, 0.800, 0.750, 0.800, 0.600, 1.333   2. 0.385, 0.923, 0.417, 0.923, 0.385, 2.400
**Find sin A, cos A, tan A, sin B, cos B, and tan B to three decimal places.**
3. 0.707, 0.707, 1.000, 0.707, 0.707, 1.000   4. 0.280, 0.960, 0.292, 0.960, 0.280, 3.429

**1.**
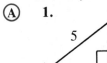

**2.**
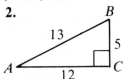

**3.**

**4.**
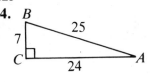

**5.**

**6.**

**7.**

**8.**

5. 0.500, 0.866, 0.577, 0.866, 0.500, 1.732   6. 0.661, 0.750, 0.882, 0.750, 0.661, 1.134
7. 0.447, 0.894, 0.500, 0.894, 0.447, 2.000   8. 0.707, 0.707, 1.000, 0.707, 0.707, 1.000

*Example*   **Find cos B to three decimal places.**

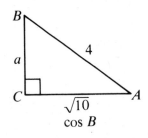

First use the Pythagorean property to find $a$.
$a^2 + (\sqrt{10})^2 = 4^2$   ◀ $a^2 + b^2 = c^2$
$a^2 + \quad 10 \quad = 16$   ◀ $(\sqrt{10})^2 = \sqrt{10} \cdot \sqrt{10} = 10$
$\quad\quad a^2 = 6$
$\quad\quad a = \sqrt{6}$

$\cos B = \dfrac{\sqrt{6}}{4} = \dfrac{2.449}{4}$, or 0.612

**Find the value of the ratio.**

**9.**
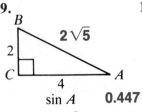
sin A   **0.447**

**10.**
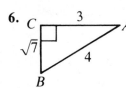
cos B   **0.447**

**11.**
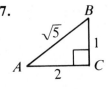
tan B   **2.828**

**12.**
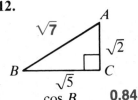
cos B   **0.845**

# USING TRIGONOMETRIC TABLES                    16.4

*Objectives*    **To find the values of the trigonometric ratios for a given angle measure using a table**
**To find the measure of an angle, given the value of a trigonometric ratio for that angle, using a table**

The table on page 533 gives decimal approximations of sine, cosine, and tangent for angles from 0° to 90°. Example 1 illustrates how to find the value of a trigonometric ratio for a given angle.

*Example 1*    **Use the table to find the value of cos 14° and tan 61°.**

*Find 14° in the angle column.*          *Find 61° in the angle column.*
*Read across to the cos column.*          *Read across to the tan column.*

| Angle Measure | Sin | Cos | Tan | Angle Measure | Sin | Cos | Tan |
|---|---|---|---|---|---|---|---|
| 13° | 0.2250 | 0.9744 | 0.2309 | 59° | 0.8572 | 0.5150 | 1.664 |
| 14° | 0.2419 | 0.9703 | 0.2493 | 60° | 0.8660 | 0.5000 | 1.732 |
| 15° | 0.2588 | 0.9659 | 0.2679 | 61° | 0.8746 | 0.4848 | 1.804 |
| 16° | 0.2756 | 0.9613 | 0.2867 | 62° | 0.8829 | 0.4695 | 1.881 |

cos 14° ≐ 0.9703                          tan 61° ≐ 1.804

For ordinary computation, you can write cos 14° = 0.9703.

Example 2 illustrates how to find the measure of an angle when the value of the trigonometric ratio is given.

*Example 2*    **Find m ∠A if cos A = 0.6293.**

*Find 0.6293 in the cos column.*
*Read across to the angle column.*

| Angle Measure | Sin | Cos | Tan | Angle Measure | Sin | Cos | Tan |
|---|---|---|---|---|---|---|---|
| 4° | 0.0698 | 0.9976 | 0.0699 | 50° | 0.7660 | 0.6428 | 1.192 |
| 5° | 0.0872 | 0.9962 | 0.0875 | 51° | 0.7771 | 0.6293 | 1.235 |
| 6° | 0.1045 | 0.9945 | 0.1051 | 52° | 0.7880 | 0.6157 | 1.280 |
| 7° | 0.1219 | 0.9925 | 0.1228 | 53° | 0.7986 | 0.6018 | 1.327 |

m∠A = 51°

Sin $B = 0.0860$ is a number that does not appear in the table. In this case, find the closest value to 0.0860 in the sin column.

*Example 3*   **Find m $\angle B$ to the nearest degree if sin $B = 0.0860$.**

*Read down the sin column. The closest value to 0.0860 is 0.0872. Read across to the angle column to 5°.*

| Angle Measure | Sin | Cos | Tan | Angle Measure | Sin | Cos | Tan |
|---|---|---|---|---|---|---|---|
| 3° | 0.0523 | 0.9986 | 0.0524 | 49° | 0.7547 | 0.6561 | 1.150 |
| 4° | 0.0698 | 0.9976 | 0.0699 | 50° | 0.7660 | 0.6428 | 1.192 |
| 5° | 0.0872 | 0.9962 | 0.0875 | 51° | 0.7771 | 0.6293 | 1.235 |
| 6° | 0.1045 | 0.9945 | 0.1051 | 52° | 0.7880 | 0.6157 | 1.280 |

**So, $m \angle B = 5$ to the nearest degree.**

# Written Exercises

**Find the value of the ratio. Use the table on page 533.**

(A)  **1.** sin 73° **.9563**  **2.** cos 58° **.5299**  **3.** tan 33° **.6494**  **4.** cos 60° **.5000**  **5.** sin 87° **.9986**
**6.** tan 55° **1.428**  **7.** cos 38° **.7880**  **8.** sin 25° **.4226**  **9.** cos 49° **.6561**  **10.** tan 24° **.4452**

**Find m $\angle B$ to the nearest degree. Use the table on page 533.**
**11.** tan $B = 0.1405$ **8°**  **12.** cos $B = 0.3420$ **70°**  **13.** sin $B = 0.6293$ **39°**
**14.** cos $B = 0.7453$ **42°**  **15.** sin $B = 0.3917$ **23°**  **16.** tan $B = 0.1399$ **8°**
**17.** cos $B = 0.9277$ **22°**  **18.** tan $B = 2.603$ **69°**  **19.** sin $B = 0.4069$ **24°**
**20.** tan $B = 1.799$ **61°**  **21.** cos $B = 0.7655$ **40°**  **22.** tan $B = 0.1080$ **6°**

**Simplify. Use the table on page 533.**
(B)  **23.** sin 50° − cos 40° **0**                          **24.** $(\sin 60°)^2 + (\cos 60°)^2$ **.999956**
**25.** $(\sin 45°)^2 − (\sin 45°)(\cos 45°)$ **0**            **26.** $(\sin 20°)(\cos 70°) − (\sin 20°)^2$ **0**

**True or false? (Ex. 27–28)**
**27.** If m $\angle A >$ m $\angle B$, then sin $A >$ sin $B$. **True.**
**28.** If m $\angle A <$ m $\angle B$, then tan $A <$ tan $B$. **True.**

**29. They must be complementary angles.**
(C)  **29.** What relationship must exist between m $\angle A$ and m $\angle B$ if sin $A =$ cos $B$?
**30.** For which trigonometric ratio does the value decrease as the degree measures of the angles increase? **Cosine**

# CUMULATIVE REVIEW

**1.** Solve $x^2 − 4x + 3 = 0$.        **2.** Solve $3 − (4 − x) = 2 + x$. ∅  **3.** Solve $5 = \dfrac{40}{x}$. $x = 8$
$x = 3$  $x = 1$

# RIGHT TRIANGLE SOLUTIONS    16.5

*Objectives*    **To find lengths of sides to the nearest tenth of a unit, using trigonometric ratios**
**To find the measures of angles to the nearest degree, using trigonometric ratios**

The trigonometric ratios can be used to form simple equations to find measures in a right triangle.

*Example 1*    **If m ∠A = 57 and c = 47, find a to the nearest tenth.**

Since leg $a$ is opposite $\angle A$ and 47 is the length of the hypotenuse, use $\sin A = \dfrac{opp.}{hyp.}$ to write an equation.

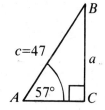

$\sin 57° = \dfrac{a}{47}$    ◀ $sin\ A = \dfrac{opp.}{hyp.}$

$0.8387 \doteq \dfrac{a}{47}$    ◀ *Use the table to find sin 57°.*

$(0.8387)47 \doteq \dfrac{a}{47} \cdot 47$    ◀ *Multiply each side by 47.*

$39.4189 \doteq a$

**So,** $a$ = 39.4.    ◀ *Round 39.4189 to the nearest tenth.*

*Example 2*    **If m ∠B = 43 and a = 12, find c to the nearest tenth.**

Since leg $a$ is adjacent to the 43° angle and $c$ is the hypotenuse, use $\cos B = \dfrac{adj.}{hyp.}$ to write an equation.

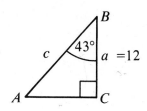

$\cos 43° = \dfrac{12}{c}$

$0.7314 = \dfrac{12}{c}$    ◀ *Use the table to find cos 43°.*

$(0.7314)c = \dfrac{12}{c} \cdot c$    ◀ *Multiply each side by c.*

$0.7314c = 12$

$c = \dfrac{12}{0.7314}$    ◀ *Divide each side by 0.7314.*

**So,** $c \doteq 16.4$.    ◀ *12 ÷ 0.7314 = 16.40689, which is 16.4 to the nearest tenth.*

Sometimes, you may be given the lengths of two sides of a right triangle and be asked to find the measure of an acute angle. You will still use a trigonometric ratio to write an equation. However, the value of this may not be listed in the table. You will then have to approximate by using the closest value appearing in the table. This is illustrated in Example 3.

## Example 3     If $a = 11$ and $b = 13$, find m $\angle A$ to the nearest degree.

Since leg $a$ is opposite $\angle A$ and leg $b$ is adjacent to $\angle A$, use $\tan A = \dfrac{\text{opp.}}{\text{adj.}}$ to write an equation.

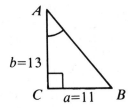

$\tan A = \dfrac{11}{13}$     ◀ $\tan A = \dfrac{opp.}{adj.}$

$\tan A = 0.84615$ ◀ *Carry the division to five significant digits.*

$\tan A = 0.8462$ ◀ *Round to four decimal places.*

m $\angle A \doteq 40$ ◀ *Find the closest value, 0.8391, in the tan column.*

**So,** m $\angle A$ is 40 to the nearest degree.

## Example 4     If $a = 3$ and $c = 7$, find m $\angle B$ to the nearest degree.

Since leg $a$ is adjacent to $\angle B$ (the angle you are asked to find) and $c$ is the hypotenuse, use $\cos B = \dfrac{\text{adj.}}{\text{hyp.}}$ to write an equation.

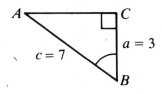

$\cos B = \dfrac{3}{7}$

$\cos B = 0.4286$ ◀ *Round to four decimal places.*

m $\angle B \doteq 65$ ◀ *Find the closest value, 0.4226, in the cos column.*

**So,** m $\angle B$ is 65 to the nearest degree.

# Oral Exercises

**Which trigonometric ratio could you use to find the indicated measure?**

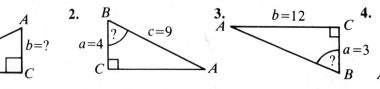

**1.**

$\sin 25° = \dfrac{b}{5}$

**2.**

$\cos \angle B = \dfrac{4}{9}$

**3.**

$\tan \angle B = \dfrac{12}{3}$

**4.**

$\tan 51° = \dfrac{15}{b}$

# Written Exercises

**Find the indicated measure (side to the nearest tenth or angle to the nearest degree).**

(A) **1.**

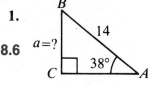

**8.6**

**2.   9.3**

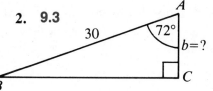

**3.**                            **4.5**

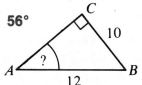

**4.      78°**

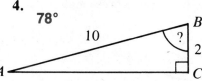

**5.**

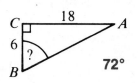

**72°**

**6.      56°**

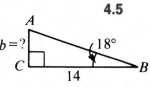

**Find the indicated measure (side to the nearest tenth or angle to the nearest degree). Use the drawing at the right.**

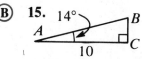

**7.** If $c = 9$ and $m\angle A = 25$, find $a$. **3.8**

**8.** If $a = 9$ and $c = 14$, find $m\angle B$. **50°**

**9.** If $b = 22$ and $c = 44$, find $m\angle A$. **60°**

**10.** If $b = 24$ and $m\angle A = 40$, find $c$. **31.3°**

**11.** If $b = 25$ and $m\angle A = 17$, find $c$. **26.1**

**12.** If $b = 15$ and $c = 30$, find $m\angle A$. **60°**

**13.** If $a = 27$ and $b = 54$, find $m\angle B$. **63°**

**14.** If $a = 19$ and $m\angle B = 22$, find $c$. **20.5**

**Find all the missing measures (sides to the nearest tenth and angles to the nearest degree.)**

(B) **15.**

**16.**
$m\angle B = 20°$
$AB = 40.9$
$BC = 38.5$

**16.**

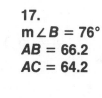

**17.**
$m\angle B = 76°$
$AB = 66.2$
$AC = 64.2$

**17.**

**14°**

$16$

**15.**
$m\angle B = 76°$
$AB = 10.3$
$BC = 2.5$

**18.**

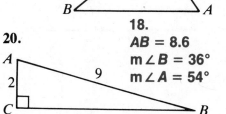

**18.**
$AB = 8.6$
$m\angle B = 36°$
$m\angle A = 54°$

**20.**

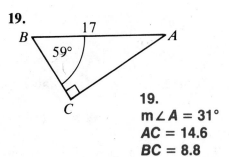

**20.**
$BC = 8.8$
$m\angle B = 13°$
$m\angle A = 77°$

**19.**

$59°$

**19.**
$m\angle A = 31°$
$AC = 14.6$
$BC = 8.8$

Right Triangle Solutions                                                        **469**

# APPLYING TRIGONOMETRIC RATIOS     16.6

*Objective*     **To solve problems involving trigonometric ratios**

Trigonometric ratios can be used to solve a variety of applied problems. Examples 1 and 2 illustrate two applications of trigonometric ratios to solve problems.

*Example 1*     **Find the distance across the lake from $A$ to $B$ to the nearest meter.**

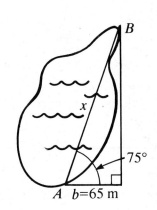

Let $x = AB$ in meters.
Since $x$ is the length of the hypotenuse of a right triangle and leg $b$ is adjacent to $\angle A$, use

$$\cos A = \frac{\text{adj.}}{\text{hyp.}}.$$

$$\cos 75° = \frac{65}{x}$$

$0.2588 = \frac{65}{x}$     ◄ *Use cos column.*

$0.2588x = 65$     ◄ *Multiply each side by x.*

$$x = \frac{65}{0.2588}$$

$x = 251$     ◄ *The answer is rounded to the nearest whole number.*

**So,** the distance from $A$ to $B$ is 251 m to the nearest meter.

*Example 2*     **Find the height of the cliff to the nearest meter.**

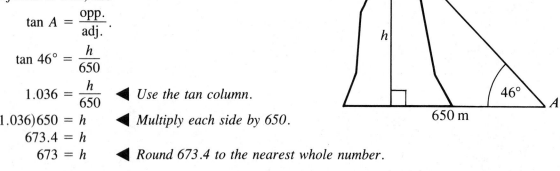

Let $h =$ height in meters.
Since $h$ is the length of the leg opposite $\angle A$ and 650 is the length of the leg adjacent to $\angle A$, use

$$\tan A = \frac{\text{opp.}}{\text{adj.}}.$$

$$\tan 46° = \frac{h}{650}$$

$1.036 = \frac{h}{650}$     ◄ *Use the tan column.*

$(1.036)650 = h$     ◄ *Multiply each side by 650.*

$673.4 = h$

$673 = h$     ◄ *Round 673.4 to the nearest whole number.*

**So,** the height of the cliff is 673 m to the nearest meter.

Some applications of trigonometric ratios involve the concepts of angle of elevation or angle of depression. They are illustrated in the diagram below.

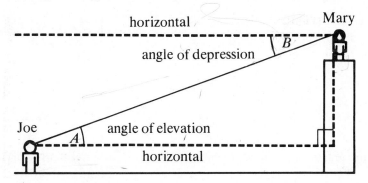

∠A is an **angle of elevation;** it is the angle by which Joe must raise, or elevate, his head with respect to the horizontal to see Mary. ∠B is an **angle of depression;** it is the angle by which Mary must lower, or depress, her head with respect to the horizontal to see Joe. Notice that m∠A, the measure of the angle of elevation, equals m∠B, the measure of the angle of depression.

*Example 3*  **A tree casts a 65-m shadow when the angle of elevation of the sun measures 55°. How tall is the tree to the nearest meter?**

Let $h$ = height of tree in meters.

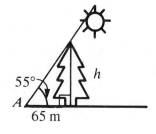

$\tan 55° = \dfrac{h}{65}$  ◀ $\tan A = \dfrac{opp.}{adj.}$

$1.428 = \dfrac{h}{65}$  ◀ *Use tan column.*

$(1.428)65 = h$

$92.82 = h$

**So,** the height of the tree is 93 m to the nearest meter.

*Example 4*  **A plane is flying at an altitude of 15,000 m. From the pilot, the angle of depression of an object on the ground is 32°. How far, to the nearest meter, is the object from the plane?**

Let $x$ = the distance from the plane to the object.

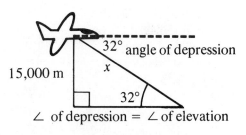

$\sin 32° = \dfrac{15{,}000}{x}$  ◀ $\sin A = \dfrac{opp.}{hyp.}$

$0.5299 = \dfrac{15{,}000}{x}$  ◀ *Use sin column.*

$0.5299x = 15{,}000$

$x = \dfrac{15{,}000}{0.5299}$   $x = 28{,}307.228$

**Thus,** the distance from the plane to the object is 28,307 m to the nearest meter.

Applying Trigonometric Ratios

**471**

# Written Exercises

**1.** Find the height of the building to the nearest meter. **75 m**

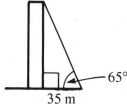

35 m
65°

**2.** Find the distance across the lake from $C$ to $D$ to the nearest kilometer. **77 km**

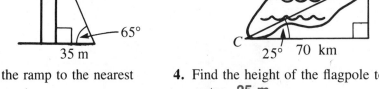

$D$
$C$
25°  70 km

**3.** Find the length of the ramp to the nearest meter. **17 m**

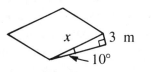

$x$  3 m
10°

**4.** Find the height of the flagpole to the nearest meter. **25 m**

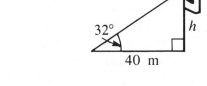

32°
$h$
40 m

**Find $x$ to the nearest meter.**

**5.** **12780 m**     **6.** **68 m**     **7.** **425 m**

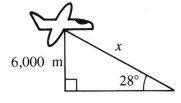

6,000 m
$x$
28°

$x$
42°
75 m

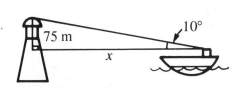

75 m
$x$
10°

**Give answers to the nearest unit.**

**8.** A tree casts a 40-m shadow when the angle of elevation of the sun is 49°. How tall is the tree? **46 m**

**9.** A ramp is 355 m long. It rises a vertical distance of 24 m. Find the measure of the angle of elevation. **4°**

**10.** A lighthouse is 85 m high. The angle of depression from the top of the lighthouse to a ship out at sea is 81°. How far from the base of the lighthouse is the ship? **13 m**

**11.** A kite is flying at the end of a 260-m string (straight). How high above the ground is the kite if the string makes an angle of 61° with the ground? **227 m**

**12.** A power-line tower casts a 190-m shadow when the angle of elevation of the sun is 30°. How high is the tower? **110 m**

**13.** The angle of elevation from a boat to the top of a 70-m high coastal hotel is 10°. How far from the coast is the boat? **397 m**

**14.** As it leans against a building, a 25-m ladder makes an angle of 62° with the ground. At what height does it touch the building? **22 m**

**15.** A plane is flying at an altitude of 14,000 m. The angle of elevation from a building on the ground to the plane measures 36°. How far is the building from the plane? **23818 m**

**16.** Each step of a stairway rises 14 cm for a tread width of 28 cm. What size angle does the stairway make with the floor? **27°**

28
14
floor

# INVESTMENT AND LOAN PROBLEMS     16.7

*Objective*     **To solve problems about investments and loans**

Interest is the amount of money paid to use money, either invested or borrowed. The formula for calculating simple interest is $i = prt$.

| $i$ | $=$ | $p$ | $\cdot$ | $r$ | $\cdot$ | $t$ |
|---|---|---|---|---|---|---|
| interest | | principal | | rate | | time |

$\begin{pmatrix} \text{money paid to use} \\ \text{money invested or} \\ \text{borrowed} \end{pmatrix}$ $\begin{pmatrix} \text{amount of} \\ \text{money invested} \\ \text{or borrowed} \end{pmatrix}$ $\begin{pmatrix} \text{percent, written} \\ \text{as a decimal, paid} \\ \text{to use the money} \end{pmatrix}$ $\begin{pmatrix} \text{number of years} \\ \text{the money is} \\ \text{used or borrowed} \end{pmatrix}$

If you deposit a principal of \$1,000 in a bank savings account at an interest rate of 7%, the amount of interest earned after 1 year is found by using the basic interest formula. The interest rate, 7%, is written as a decimal, 0.07.

$$i = p \cdot r \cdot t$$
$$i = \$1,000 \cdot 0.07 \cdot 1$$
$$i = \$70.00$$

$$\begin{array}{r} \$1,000 \\ 0.07 \\ \hline \$70.00 \end{array}$$

**Thus,** the amount of interest is \$70.00.

You can also use the interest formula to solve investment and loan problems.

*Example 1*     **The Wangs invested \$2,000 for 1 year. They earned \$160 in interest. What was the rate or percent of return?**

Let $x$ = rate of interest

$$i = p \cdot r \cdot t$$     ◀ *Interest formula*
$$\$160 = \$2,000 \cdot x \cdot 1$$
$$160 = 2,000x$$
$$\frac{160}{2,000} = x$$     ◀ *Divide each side by 2,000.*
$$0.08 = x$$     ◀ $\dfrac{160}{2,000} = 2{,}000\overline{)160.00}^{\,.08}$

**Thus,** the rate of interest is 8%.     ◀ *Write 0.08 as a percent, 8%.*

Sometimes a person will invest two different amounts of money at different rates. In the next example you will have to represent algebraically the two amounts invested.

Investment and Loan Problems     **473**

**Example 2**    **Ellen invested some money at 8% and twice as much at 9%. The total interest after 1 year was $260. How much did she invest at each rate?**

READ ▶    You are given the rates of interest, total interest, and the relationship between the investments.
You are asked to find the amount of each investment.

PLAN ▶    Represent algebraically the amounts being invested.
Let $x$ = the amount (in dollars) invested at 8%
   $2x$ = the amount (in dollars) invested at 9%

SOLVE ▶    Use $i = prt$ to write an equation for total interest.

$$\left(\begin{array}{c}\text{interest on}\\ \$x \text{ at 8\% for 1 yr}\end{array}\right) + \left(\begin{array}{c}\text{interest on}\\ \$2x \text{ at 9\% for 1 yr}\end{array}\right) = \text{total interest}$$

$$
\begin{array}{rcl}
x \cdot 0.08 \cdot 1 \quad + \quad 2x \cdot 0.09 \cdot 1 & = & 260 \\
0.08x \quad + \quad 0.18x & = & 260 \\
8x + 18x & = & 26{,}000 \\
26x & = & 26{,}000 \\
x & = & 1{,}000
\end{array}
$$

Amount invested at 8%, $x$, is $1,000.
Amount invested at 9%, $2x$, is $2 \cdot \$1{,}000$, or $2,000.

INTERPRET ▶    Amount at 9%, $2,000, is twice amount at 8%, $1,000.

$$\left(\begin{array}{c}\text{interest on } \$1{,}000\\ \text{at 8\% for 1 yr}\end{array}\right) + \left(\begin{array}{c}\text{interest on } \$2{,}000\\ \text{at 9\% for 1 yr}\end{array}\right) \bigg| \quad \text{total interest}$$

$$
\begin{array}{ccc}
\$1{,}000 \cdot 0.08 \cdot 1 \quad + \quad \$2{,}000 \cdot 0.09 \cdot 1 & \big| & \$260 \\
\$80 \quad + \quad \$180 & & \\
\$260 & &
\end{array}
$$

$$\$260 = \$260$$
True

**Thus,** Ellen invested $1,000 at 8% and $2,000 at 9%.

Sometimes it is convenient to solve investment or loan problems using a system of equations. In the next example one equation describes the relationship between the amounts borrowed while the other equation relates the amounts of interest charged. Also if the rate involves a fraction, you should rewrite it as a decimal. For example, $9\frac{3}{4}\% = 9.75\% = 0.0975$. These concepts are illustrated in Example 3 on the next page.

*Example 3*  **Irving borrowed $9,000, part from a bank at $9\frac{3}{4}\%$ a year and the rest from his father at $8\frac{1}{2}\%$ a year. At the end of 2 years he owed $1,630 in interest. How much did he borrow from each?**

Let $x$ = the amount borrowed at $9\frac{3}{4}\%$

$\quad y$ = the amount borrowed at $8\frac{1}{2}\%$

Total amount borrowed is $9,000

$$x + y = \$9,000 \qquad \blacktriangleleft \textit{ Write an equation relating the amounts borrowed.}$$

Use the interest formula to write an equation for the total amount of interest.

$$\left(\begin{matrix}\text{interest on} \\ \$x \text{ at } 9\frac{3}{4}\% \text{ for 2 yr}\end{matrix}\right) + \left(\begin{matrix}\text{interest on} \\ \$y \text{ at } 8\frac{1}{2}\% \text{ for 2 yr}\end{matrix}\right) = \text{total interest}$$

$$
\begin{aligned}
x \cdot 0.0975 \cdot 2 &+ y \cdot 0.085 \cdot 2 &= 1,630 \\
0.195x &+ 0.17y &= 1,630 \\
195x &+ 170y &= 1,630,000 \qquad \blacktriangleleft \textit{ Multiply each side by 1,000.}
\end{aligned}
$$

Solve the system of equations by the addition-multiplication method.

$$
\begin{aligned}
x + y &= 9,000 \qquad (1) \\
195x + 170y &= 1,630,000 \quad (2)
\end{aligned}
$$

$$
\begin{aligned}
-170(x + y) &= -170 \cdot 9,000 \qquad \blacktriangleleft \textit{ Multiply each side of equation (1) by } -170. \\
-170x - 170y &= -1,530,000 \quad (3)
\end{aligned}
$$

$$
\begin{aligned}
195x + 170y &= 1,630,000 \quad (2) \\
\underline{-170x - 170y} &= \underline{-1,530,000} \quad (3) \qquad \blacktriangleleft \textit{ Add equations (2) and (3).} \\
25x &= 100,000 \\
x &= 4,000
\end{aligned}
$$

To find $y$, let $x = 4,000$ in equation (1).

$$
\begin{aligned}
x + y &= 9,000 \quad (1) \\
4,000 + y &= 9,000 \\
y &= 5,000
\end{aligned}
$$

**Thus,** he borrowed $4,000 from the bank and $5,000 from his father.

## Oral Exercises

**Give each percent as a decimal.**
**1.** $6\frac{1}{2}\%$ **0.065**  **2.** $8\frac{3}{4}\%$ **0.0875**  **3.** $5\frac{1}{4}\%$ **0.0525**  **4.** $12\frac{1}{2}\%$ **0.125**  **5.** $11\frac{3}{4}\%$ **0.1175**  **6.** $15\frac{1}{4}\%$ **0.1525**

**Tell how to find the interest on each investment.** Multiply 10,000 by 0.1475 and multiply the answer
**7.** $3,000 invested at 6% for 1 yr **Multiply 3000**   **8.** $10,000 invested at $14\frac{3}{4}\%$ for 3 yr  by 3.
                                     by .06.

# Written Exercises

**Solve each problem.**

(A) **1.** Ben borrowed $4,000 at 9% interest for 1 year. How much interest did he pay at the end of 1 year? **$360**

**2.** Ms. Ruiz invested $6,000 for 1 year. She earned $420 interest. What was the rate of return? **7%**

**3.** Josh invested some money at 7% and three times as much at 8%. The total interest after 1 year was $930. How much did he invest at each rate? **3000 @ 7%; 9000 @ 8%**

**4.** Mike borrowed some money from friends at 6%. He borrowed twice as much from the bank at 9%. If the total interest after 1 year was $960, how much did he borrow from each? **$4000, $8000**

**5.** Ilene invested one sum of money at $5\frac{1}{2}\%$ and another sum at 6%. She invested $300 more at the 6% rate than at the $5\frac{1}{2}\%$ rate. If her total interest for 1 year was $133, find the amount she invested at each rate. **$1000, $1300**

**6.** The Yoshidas invested $10,000, part at 6% and the rest at $6\frac{1}{2}\%$. At the end of 4 years their total return was $2,490. How much did they invest at each rate? **$5500, $4500**

**7.** Mona had $10,000. She invested part of it at $5\frac{3}{4}\%$ and the rest at 5%. At the end of 6 years her total return was $3,360. How much did she invest at each rate? **$8000, $2000**

**8.** Donna borrowed some money at $8\frac{1}{2}\%$ and another amount at $6\frac{3}{4}\%$. The amount borrowed at $6\frac{3}{4}\%$ was $1,200 more than the amount at $8\frac{1}{2}\%$. If the total interest after 2 years was $711, find the amount of each loan. **$1800, $3000**

(B) **9.** Bluecloud borrowed $1,800, part from a bank at $6\frac{3}{4}\%$ and the rest from his father at $5\frac{1}{2}\%$. At the end of 6 months he owed $60.75 interest. How much did he borrow from each? [Hint: 6 months = $\frac{1}{2}$ year or 0.5 year] **$1800, 0**

**10.** Ann invested one-half of her money at $5\frac{1}{2}\%$ and one fourth at 6%. At the end of 18 months she had earned $637.50 in interest. What was the original sum of money? **$10,000**

**11.** Rhoda invested one-third of her money at $7\frac{3}{4}\%$ and three-fifths at $9\frac{1}{2}\%$. At the end of 3 years and 9 months she had earned $7,455 in interest. What was the original sum of money? **$24,000**

**12.** Alvin Jackson loaned $7,800, part at 6% and the rest at 8%. If each rate of interest had been interchanged, his interest for 1 year would have been $84 more. Find the amount loaned at each rate. **$6000 @ 6% $1800 @ 8%**

**13.** Mr. Johnson has twice as much money invested at 6% as he has at 8%. His annual income at the 6% rate is $32 more than his earning on the 8% investment. Find the amount invested at each rate. **6%: $1,600; 8%: $800**

**14.** Mrs. Dinero invests some money at 8%. She invests $400 more than this at 9%. The annual interest at 9% is $1 more than twice the annual interest from the other investment. Find the amount invested at each rate. **8%: $500; 9%: $900**

# COMPUTER ACTIVITIES

## Computing Trigonometric Ratios

All right triangles with an acute angle of the same measure are similar. For all similar triangles ABC with equal acute angles A, $\frac{a}{b}$, $\frac{a}{c}$ and $\frac{b}{c}$ are the same regardless of the lengths of the three sides of the triangle.

You can use computers for printing out tables. Programs can be written to put a heading on a table, set up columns, and then compute the values for the table. In presenting information about right triangle ratios, it is important to keep the values in the proper columns on a table.

Below is part of a program that will print out a table of ratios given the lengths of the three sides of a right triangle.

```
100   PRINT "THIS PROGRAM COMPUTES TRIGONOMETRIC RATI
      OS"
110   PRINT "FOR RIGHT TRIANGLE DEF."
120   PRINT "D  E  F  TAN D    SIN D    COS D"
130   READ D,E,F
      (HINT: Insert instructions for computing tan D,
             sin D and cos D given the sides d, e and f.
         Let D be the variable for side d;
             E be the variable for side e;
             F be the variable for side f;
             T be the variable for tan D;
             S be the variable for sin D;
             C be the variable for cos D.)
180   PRINT D TAB( 4)E TAB( 7)F TAB( 10)T TAB( 20)S TAB(
      30)C
190   GOTO 130
200   DATA  8, 6, 10, 5, 12, 13, 3, 4, 5
210   END
```

See the Computer Section beginning on page 488 for more information.

## Exercises (For Ex. 1–2 see TE Answer Section.)

1. Enter the completed program above and RUN it for the values in DATA statement 200.

2. Alter the program to make the table for an unlimited number of triangles.

## Vocabulary

acute [16.1]         similar [16.2]
adjacent [16.3]      sine [16.3]
complementary [16.1] supplementary [16.1]
cosine [16.3]        tangent [16.3]
opposite leg [16.3]  trigonometry [16.3]

Solve. [16.1]

1. An angle measures 40° more than its complement. Find the measure of the angle and its complement. **65°, 25°**

2. The measures of two supplementary angles are in the ratio 5:4. Find the measure of each angle. **100°, 80°**

3. The first angle of a triangle measures 10° less than the measure of the second angle. The third angle measures 20° more than the sum of the measures of the first two angles. Find the measure of each angle. **35°, 45°, 100°**

In Exercises 4–5, $\triangle ABC \sim \triangle DEF$. Find the indicated measures. [16.2]

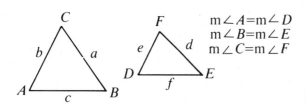

$m\angle A = m\angle D$
$m\angle B = m\angle E$
$m\angle C = m\angle F$

4. $a = 6$, $b = 5$, $e = 15$, $f = 10$. Find $c$ and $d$.                      $\frac{10}{3}$, **18**

5. $f = 12$, $d = 16$, $e = 10$, $b = 35$. Find $a$ and $c$. **56, 42**

6. Harvey was lying 22 m from the base of a tree 15-m tall. He could see the top of a 450-m building just beyond the top of the tree. What was his distance from the base of the building? **660 m**

Find sin $A$, cos $A$, tan $A$, sin $B$, cos $B$, and tan $B$ to three decimal places. [16.3]

7. **0.600, 0.800, 0.750, 0.800, 0.600, 1.333**

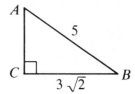

8. **0.667, 0.745, 0.894, 0.745, 0.667, 1.118**

9. Find cos $A$ to three decimal places. **0.529**

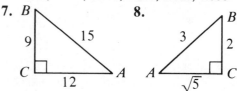

Use the table on page 531 to find each. [16.4]

10. sin 57°     11. cos 46°     12. tan 85°
**0.8387**       **0.6947**       **11.43**

Find m$\angle B$ to the nearest degree. Use the table on page 531.

13. tan $B = 0.1399$     14. cos $B = 0.1185$
**8°**                   **83°**

Use the table on page 531 to simplify each expression.

15. (cos 70°)(tan 70°) − sin 70° **0**
16. sin 80° − cos 80° **0.8112**

Find the indicated measure (side to the nearest tenth or angle to the nearest degree). [16.5]

17. If $c = 8$ and m$\angle A = 35$, find $a$. **4.6**
18. If m$\angle A = 12$ and $a = 18$, find $b$ and $c$. **84.7, 86.6**

Give the answer to the nearest whole number.

19. A tree casts a 30-m shadow when the angle of elevation of the sun measures 43°. How tall is the tree? [16.6] **28 m**

★ 20. Use the diagram for Exercise 8 to show that $(\sin A)^2 + (\cos A)^2 = 1$. [16.3]

$$\left(\frac{2}{3}\right)^2 + \left(\frac{\sqrt{5}}{3}\right)^2 = \frac{4}{9} + \frac{5}{9} = \frac{9}{9} = 1$$

Solve.

**1.** Two angles of a triangle have the same measure. The measure of the third angle is 60° more than the measure of the first or second. Find the measure of all three angles. **40°, 40°, 100°**

**2.** The first angle of a triangle measures 6° more than the measure of the second. The third angle measures 4° less than 3 times the sum of the measures of the first two angles. Find the measures of the three angles. **20°, 26°, 134°**

**3.** The measures of two complementary angles are in the ratio 7:2. Find the measure of each angle. **70°, 20°**

In Exercises 4–5, $\triangle ABC \sim \triangle DEF$. Find the indicated measures.

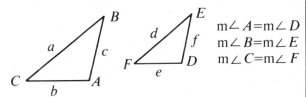

$m\angle A = m\angle D$
$m\angle B = m\angle E$
$m\angle C = m\angle F$

**4.** $d = 10, e = 8, f = 4, c = 6$. Find $a$ and $b$.

**5.** $a = 4, b = 8, e = 20, f = 15$. Find $c$ and $d$. **6, 10**　　　**4. 15, 12**

**6.** John walked 16 m up a ramp and was 6 m above the ground. If he had walked 24 m further up the ramp, how far above the ground would he be? **15 m**

**7. 0.882, 0.471, 1.875, 0.471, 0.882, 0.533**

Find $\sin A$, $\cos A$, $\tan A$, $\sin B$, $\cos B$, and $\tan B$ to three decimal places.

**7.**

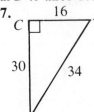

**8.**

**9.** Find $\cos B$ to three decimal places. **0.447**

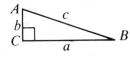

Use the table on page 531 to answer Ex. 10–20. Find each value.

**10.** $\sin 28°$　　**11.** $\cos 75°$　　**12.** $\tan 44°$
　**0.4695**　　　　**0.2588**　　　　**0.9657**

Find $m\angle B$ to the nearest degree.

**13.** $\cos B = 0.5394$　　**14.** $\tan B = 1.429$
　**57°**　　　　　　　　　**55°**

**15.** Simplify $(\sin 25°)^2 + (\cos 25°)^2$　　**1**

Find the indicated measure (side to the nearest tenth or angle to the nearest degree).

**16.** If $a = 12$ and $b = 14$, find $m\angle B$. **49°**

**17.** If $b = 4$ and $m\angle A = 38$, find $c$. **5.1**

**18.** If $c = 8$ and $b = 4$, find $a$ and $m\angle B$.

**19.** If $m\angle B = 24$ and $a = 16$, find $b$ and $c$.

**18. $4\sqrt{3}$, 30°　19. 7.1, 17.5**

Solve.

**20.** The angle of elevation from a boat to the top of a 90-m high beacon tower on the coast is 20°. How far from the coast is the boat? **247.3 m**

**21.** A 21-m ladder makes an angle of 57° with the ground as it leans against a wall. At what height does it reach the wall? **17.6 m**

★ **22.** Use the diagram to show that $\dfrac{1}{\sin A} \cdot \tan A = \dfrac{1}{\cos A}$.

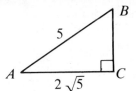

$$\frac{1}{\frac{a}{c}} \cdot \frac{a}{b} = \frac{c}{a} \cdot \frac{a}{b}$$

$$= \frac{c}{b} = \frac{1}{\frac{b}{c}} = \frac{1}{\cos A}$$

**8. 0.661, 0.750, .882, 0.750, 0.661, 1.134**

# CUMULATIVE REVIEW

Each exercise has five choices of answers.
Choose the one best answer. (Ex. 1–9)

**1.** Find the slope of $\overrightarrow{AB}$ for $A(3, 4)$, $B(6, 5)$.
(A) $\frac{1}{3}$  (B) $\frac{-1}{1}$  (C) $\frac{3}{1}$  (D) $-\frac{1}{3}$
(E) None of these  **A**

**2.** Solve $-2x + 4 < 12$. **C**
(A) $x < -4$  (B) $x > -8$  (C) $x > -4$
(D) $x > 4$  (E) None of these

**3.** Simplify $\dfrac{5}{x^2 - 4x + 3} + \dfrac{2}{x - 3}$. **B**

(A) $\dfrac{7}{x^2 - 4x + 3}$  (B) $\dfrac{2x + 3}{(x - 3)(x - 1)}$

(C) $\dfrac{2x + 5}{(x - 3)(x - 1)}$  (D) $\dfrac{2x + 2}{(x - 3)(x - 1)}$
(E) None of these

**4.** Simplify $(-2x^2)^3$. **D**
(A) $-2x^6$  (B) $-12x^6$  (C) $-8x^5$
(D) $-8x^6$  (E) None of these

**5.** Solve $3 - (4 - x) = 3x - 5$. **A**
(A) 2  (B) $-1$  (C) $-\frac{7}{2}$  (D) $-2$
(E) None of these

**6.** Solve $x^2 - 7x + 12 = 0$. **C**
(A) $(-4, 3)$  (B) $(6, 2)$  (C) $(3, 4)$
(D) $(4, -3)$  (E) None of these

**7.** Find the slope of the line $3y - 2x = 9$. **E**
(A) $\frac{3}{2}$  (B) $\frac{-2}{3}$  (C) 3  (D) 2
(E) None of these

**8.** Subtract $-x^2 - 2x + 1$ from **C**
$x^2 - 2x + 4$.
(A) $-2x^2 - 3$  (B) $2x^2 - 4x + 3$
(C) $2x^2 + 3$  (D) 4  (E) None of these

**9.** Solve: $x + y = 8$  **D**
$\quad\quad\quad\;\; x - y = 2$.
(A) $(5, -3)$  (B) $(3, 5)$  (C) $(-5, -3)$
(D) $(5, 3)$  (E) None of these

**10.** Solve for $x$, $ax - b = c$.  $\dfrac{b + c}{a}$

**11.** $y$ varies directly as $x$. $y = 20$ when
$x = -5$. Find $y$ when $x = 3$. $-12$

**12.** $y$ varies inversely as $x$. $y = 16$ when
$x = -4$. Find $y$ when $x = -8$. $8$

List the ordered pairs of each relation below.
Give the domain and range. Is the relation a
function?

**13.**  **14.**

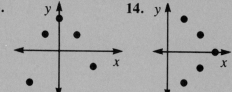

13. $(-2, -2), (-1, 1), (0, 2), (1, 1), (2, -1)$;
$D = \{-2, -1, 0, 1, 2\}$; $R = \{-2, -1, 1, 2\}$; function

**15.** $D$ is the domain of $f(x) = -x^2 + 5x - 2$.
$D = \{-1, 2, 3\}$ Determine the range.
$R = \{-8, 4\}$

Graph each relation. Which relations are
functions? Which functions are linear
functions? Which functions are linear and
constant? 16. const. lin. funct.  17. not a funct.
**16.** $y = -2$  **17.** $3x - 8 = 16$
**18.** $2x + 2y = 16$  **19.** $3x - 5 = 7$
18. lin. funct.  19. not a funct.
**20.** Find the fraction corresponding to $0.\overline{8}$. $\frac{8}{9}$

Indicate whether each number is rational or
irrational.
**21.** 0.323223222 . . .  **22.** $0.\overline{45}$ rat'l
irrat'l
**23.** Use the table on page 530 to find $\sqrt{32}$
to the nearest tenth. 5.7

14. $(1, 2), (1, -2), (2, 1), (2, -1), (3, 0)$ $D = \{1, 2, 3\}$
$R = \{-2, -1, 0, 1, 2\}$; not a function

**24.** Find the **9.375** height h. $A = 150$ m² 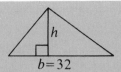 $b=32$

**25.** Use the divide and average method to approximate $\sqrt{87}$ to the nearest tenth. **9.3**

Simplify. **28. $2x^2y^3\sqrt{7x}$   29. $-6x^4y^4\sqrt{10x}$**
**26.** $\sqrt{45}$  **3$\sqrt{5}$**   **27.** $\sqrt{16x^4y^{12}}$  **4$x^2y^6$**
**28.** $\sqrt{28x^5y^6}$   **29.** $-3xy^2\sqrt{40x^7y^4}$
**30.** $5\sqrt{12} - 4\sqrt{3} + 2\sqrt{27}$ **12$\sqrt{3}$**
**31.** $4a^2\sqrt{8a^5b^4} + a^4b^2\sqrt{18a}$ **11$a^4b^2\sqrt{2a}$**
**32.** $\sqrt{3}(\sqrt{8} + 2\sqrt{3})$ **2$\sqrt{6}$ + 6**
**33.** $(2\sqrt{3} - \sqrt{5})(2\sqrt{3} + \sqrt{5})$ **7**
**34.** $\dfrac{1}{\sqrt{2}}$  $\dfrac{\sqrt{2}}{2}$ **35.** $\dfrac{12}{\sqrt{3}}$   **36.** $\dfrac{10ab}{\sqrt{20a^4b}}$  $\dfrac{\sqrt{5b}}{a}$
**35.** **4$\sqrt{3}$**

**37.** What value of $x$ will make $\sqrt{x^2 - 9x + 20}$ a real number?
**$x \geq 5$ or $x \leq 4$**

**38.** Find $a$ in **2$\sqrt{2}$** simplest radical form.
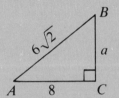

**39.** Tell whether the triangle with sides of length 7, $\sqrt{53}$, 2 is a right triangle.
**Yes, it is a rt △.**

Solve.
**40.** $\sqrt{5t} = 2$ **$\frac{4}{5}$**   **41.** $\sqrt{3p - 2} + 4 = 6$ **2**
**42.** $3a^2 = 60$   **43.** $(x + 3)^2 = 44$
**42.** **±2$\sqrt{5}$**   **43.** **−3 ± 2$\sqrt{11}$**

Solve by completing the square.
**44.** $b^2 + 4b = 32$   **45.** $x^2 + 6x = 2$
**44. 4, −8**   **45. −3 ± $\sqrt{11}$**

Solve by the quadratic formula. Leave irrational roots in simplest radical form.
**46.** $x^2 + 5x = -4$   **47.** $-\frac{1}{2}x - 1 = -\frac{1}{4}x^2$
**46. (−4, −1)**   **47. 1 ± $\sqrt{5}$**

**48.** Graph the parabola, showing the turning point $y = -x^2 + 4x - 2$. **T.P.: (2, 2)**

**49.** Find the minimum value of $y$ if $y = x^2 - 8x + 2$. **−14**

Find each value.
**50.** sin 32°     **51.** cos 49° **0.6561**
**50. 0.5299**

Find m∠$A$ to the nearest degree.
**52.** tan $A$ = 3.281     **53.** cos $A$ = 0.1400
**52. 73°   53. 82°**

**54.** Find sin $B$ to **0.816** three decimal places.
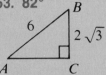

Find the indicated measure (side to the nearest tenth or angle to the nearest degree).
**55.** Find m∠$B$, **49** if $a = 6$ and $b = 7$.
**56.** Find $a$ and m∠$B$, if $c = 10$ and $b = 5$.
**$a$ = 8.7; m∠$B$ = 30**

**57.** $\triangle ABC \sim \triangle DEF$
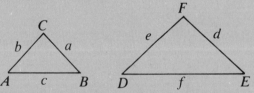

m∠$A$ = m∠$D$, m∠$B$ = m∠$E$, m∠$C$ = m∠$F$, $b = 2$, $c = 6$, $e = 3$, $d = 6$. Find $a$ and $f$. **4, 9**

Solve.
**58.** The current through an electric circuit varies inversely as the resistance. When the current is 45 amps, the resistance is 2 ohms. Find the current when the resistance is 15 ohms. **6 amps.**

**59.** The length of a rectangle is 6 m more than twice the width. The area is 32 m². Find the length and width. **11.6 m, 2.8 m**

**60.** If 5 is added to a number, the square of the result is 64. Find the number. **3, −13**

# COLLEGE PREP TEST

For each item you are asked to compare the quantity in Column 1 with the quantity in Column 2. Write the letter of the correct answer from these choices:

A. The quantity in Column 1 is greater than the quantity in Column 2.
B. The quantity in Column 2 is greater than the quantity in Column 1.
C. The quantity in Column 1 is equal to the quantity in Column 2.
D. The relationship cannot be determined from the information given.

Notes: The information centered over both columns refers to one or both of the quantities to be compared. A symbol that appears in both columns has the same meaning in each column.

| | Column 1 | Column 2 | |
|---|---|---|---|
| **1.** | $\dfrac{6 + \dfrac{2}{5}}{1 - \dfrac{1}{5}}$ | $2^3$ | **C** |
| **2.** | $\sqrt{\dfrac{1}{0.36}}$ | $2$ | **B** |

**3.**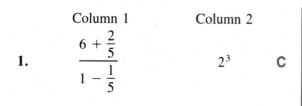

| | | | |
|---|---|---|---|
| | $x + y$ | $60°$ | **A** |

**4.**

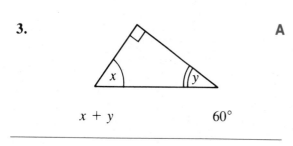

| | | | |
|---|---|---|---|
| | $m\angle A$ | $15°$ | **A** |

| | Column 1 | Column 2 | |
|---|---|---|---|
| **5.** | $a > 0, b < 0$ | | **C** |
| | $a - b$ | $a + |b|$ | |

| | Column 1 | Column 2 | |
|---|---|---|---|
| **6.** | $-5 < z < -2$ | | **B** |
| | $\dfrac{1}{z^3}$ | $\dfrac{1}{z^2}$ | |
| **7.** | $a \neq b, a \neq -b$ | | **D** |
| | $\dfrac{a^2 - b^2}{a - b}$ | $\dfrac{a^2 + 2ab + b^2}{2a + 2b}$ | |

**8.** 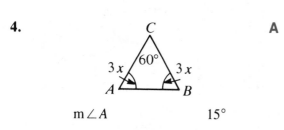 **D**

| | | |
|---|---|---|
| $x$ | $y$ | |

| | Column 1 | Column 2 | |
|---|---|---|---|
| **9.** | $x - y = 7$ | | **C** |
| | $2x = 8 - y$ | | |
| | $3x$ | $15$ | |
| **10.** | $x^2$ | $x^4$ | **D** |
| **11.** | $(0.6)^2$ | $\sqrt{0.16}$ | **B** |

**482**

# Probability and Statistics

## INTRODUCTION TO PROBABILITY: A SIMPLE EVENT

*Objectives*    **To determine the probability of a simple event**
**To determine the odds of a simple event**

The diagram at the right represents a disk with a spinner. Each sector is the same size. When the spinner is spun, it is *equally likely* that it will land on any one of the sectors. The possible outcomes are 1, 2, 3, 4, 5, 6. Since there are six sectors and only one of them has a 5, the chance of the spinner landing on 5 is one in six.

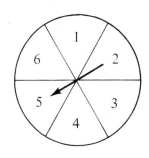

The probability of landing on any one of the sectors is $\frac{1}{6}$.

The probability of a simple event occurring is determined by comparing the number of favorable outcomes to the number of possible outcomes.

### PROBABILITY OF SIMPLE EVENT

**The probability of a simple event is given by the ratio**
$$P(\text{simple event}) = \frac{\text{number of favorable outcomes}}{\text{number of possible outcomes}}.$$

*Example 1*    A bag contains seven green marbles and two red marbles. One marble is drawn from the bag. Find the probability of drawing a red marble.

$$P(\text{red}) = \frac{\text{number of favorable outcomes}}{\text{number of possible outcomes}} = \frac{2}{7 + 2} = \frac{2}{9}$$

*Example 2*    A box contains ten quarters. One coin is removed. What is the probability that the coin is a quarter? What is the probability that the coin is a dime?

$$P(\text{quarter}) = \frac{\text{favorable outcomes}}{\text{possible outcomes}} = \frac{10}{10} = 1$$

Since all the coins are quarters, it is certain that a quarter will be picked.

$$P(\text{dime}) = \frac{\text{favorable outcomes}}{\text{possible outcomes}} = \frac{0}{10} = 0$$

Since there are no dimes, it is impossible to pick a dime. The probability of an event certain to occur and impossible events never changes.

# PROBABILITY OF 1 AND 0

**The probability of an event certain to occur is 1.**
**The probability of an impossible event is 0.**

The probability of a simple event $E$ ranges from 0 to 1:
$$0 \le P(E) \le 1$$
Probability can be used to predict how many times an event will occur in repeated happenings.

**Example 3**

A die (plural: dice) is rolled 100 times. About how many times should a number less than three come up?

Find $P$(number less than three).
$$P(\text{number less than three}) = \frac{2}{6} = \frac{1}{3}$$
Multiply the probability of the event by the number of times the die will be thrown.
$$\frac{1}{3} \times 100 = \frac{100}{3} = 33\frac{1}{3}$$
Since there cannot be a fractional number of rolls, round to the nearest whole number, 33.
So, a number less than three should come up about 33 times in 100 rolls.

If the die in Example 3 is thrown, the probability that a two will come up is $\frac{1}{6}$. The probability that a two will not come up is $\frac{5}{6}$. The sum of these probabilities is $\frac{1}{6} + \frac{5}{6} = \frac{6}{6}$ or 1. It is a certain event that either a two will come up or a two will not come up. If the probability of an event is $P(E)$, then the probability of that event not occurring is $1 - P(E)$ or $P(\overline{E})$. $P(E)$ and $P(\overline{E})$ are used to find the odds of an event.

# DEFINITION: ODDS

**The odds that a simple event will occur are**

$$\frac{P(E)}{1 - P(E)} \text{ or } \frac{P(E)}{P(\overline{E})}.$$

**Example 4**

A card is drawn from a standard deck of 52 cards. What is the probability that a club is not drawn? What are the odds of drawing a club?

$$P(\text{drawing a club}) = \frac{13}{52} = \frac{1}{4}$$

$$P(\text{not drawing a club}) = 1 - P(\text{drawing a club}) = 1 - \frac{1}{4} = \frac{3}{4}$$

$$\text{Odds of drawing a club} = \frac{P(\text{drawing a club})}{P(\text{not drawing a club})} = \frac{\frac{1}{4}}{\frac{3}{4}} = \frac{1}{3} \text{ or one to three}$$

A shortcut for finding odds is $\dfrac{\text{number of favorable outcomes}}{\text{number of non-favorable outcomes}} = \dfrac{13}{39} = \dfrac{1}{3}$.

Remember that the two parts of the odds ratio must add to the total number of possible outcomes: $13 + 39 = 52$.

# Exercises

**Find the probability of each event. Use the spinner shown at the right.**

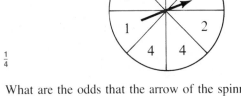

**1.** $P(2)$ $\frac{1}{4}$     **2.** $P(4)$ $\frac{3}{8}$     **3.** $P(1)$ $\frac{3}{8}$

**4.** $P(\text{odd number})$ $\frac{3}{8}$     **5.** $P(\text{prime number})$ $\frac{1}{4}$

**6.** $P(\text{not a 2})$ $\frac{3}{4}$     **7.** $P(7)$ **0**

**8.** If the spinner is spun 200 times, about how many times should it land on a two? on an even number? **50, 125**

**9.** What are the odds that the arrow of the spinner will land on a prime number? on a composite number? $\frac{1}{3}, \frac{3}{5}$

**10.** What is the probability that if a letter is picked from the alphabet, it will be a vowel? $\frac{5}{26}$

**11.** What is the probability that in a throw of a die, a five will appear? that a multiple of two will appear? $\frac{1}{6}, \frac{1}{2}$

**12.** A deck of cards contains 52 cards, 13 cards of each suit. One card is drawn. What is the probability that the card is an ace? that it is the seven of clubs? $\frac{1}{13}, \frac{1}{52}$

**13.** A bag contains 20 colored balls. There are 12 red balls and three blue balls. What is the probability of picking a ball that is neither red nor blue? $\frac{1}{4}$

**14.** What is the probability that if a month of the year is chosen, it will begin with an *M*? that it will have 31 days? $\frac{1}{6}, \frac{7}{12}$

**15.** What is the probability that if a letter in the word *mathematics* is chosen, it will be a consonant? that it will be a *t*? $\frac{7}{11}, \frac{2}{11}$

**16.** A die is rolled 200 times. About how many times should a three not appear? **167**

**17.** The probability of a machine putting a label on incorrectly is $\dfrac{1}{600}$. If 42,000 labels are put on in a day, how many will be put on incorrectly? **70**

**18.** If the probability that it will rain on a certain day is $\dfrac{3}{5}$, what is the probability that it will not rain? What are the odds that it will rain? $\frac{2}{5}, \frac{3}{2}$

**19.** A deck of cards contains 52 cards, 13 of each suit. One card is drawn. What is the probability of not drawing a red card? of drawing a card that is a ten? What are the odds of drawing a heart? a red jack? $\frac{1}{2}, \frac{1}{13}, \frac{1}{3}, \frac{1}{25}$

**20.** An envelope contains six ten-dollar bills, four five-dollar bills, and two one-dollar bills. A bill is picked from the envelope. What is the probability of not picking a ten-dollar bill? What are the odds of not picking a ten-dollar bill? of picking a two-dollar bill? $\frac{1}{2}, \frac{1}{1}$, **0**

# INTERPRETING STATISTICAL DATA

*Objective*   **To find the mean, mode, median, and range of a set of data**

The data at the right represents temperatures for a week in June. There are several ways of looking at the temperatures to pick one that seems to best represent all the temperatures. The three that you will study are the *mean*, the *mode*, and the *median*.

| | |
|---|---|
| Sunday | 73 |
| Monday | 90 |
| Tuesday | 70 |
| Wednesday | 60 |
| Thursday | 90 |
| Friday | 90 |
| Saturday | 80 |

## DEFINITION: MEAN

**The mean of a set of data is the sum of the data divided by the number of data items.**

*Example 1*   Find the mean temperature for a week when temperatures were 73, 90, 70, 60, 90, 90, 80.

$$\text{Mean} = \frac{73 + 90 + 70 + 60 + 90 + 90 + 80}{7} = \frac{553}{7} = 79$$

So, the mean is 79.

However, the mean does not always give a true picture of a set of data. Consider the following temperatures: 90, 90, 50, and 90. The mean is $\frac{90 + 90 + 50 + 90}{4} = \frac{320}{4} = 80$, which is not really representative of the data.

The most frequently occurring temperature was 90, the *mode* of the data.

## DEFINITION: MODE

**The mode of a set of data is the item that occurs most frequently.**

There can be more than one mode. To find the mode, list each item as many times as it occurs. Choose the one(s) that occurs most often.

*Example 2*   Find the mode(s) for the temperatures 60, 80, 60, 80, 90, 70, 80, 60.

60, 60, 60,      70,      80, 80, 80,      90
Both 60 and 80 occur three times.
So, the modes are 60 and 80. It is a *bimodal* set of data.
If each data item appears an equal number of times, there is *no* mode.
A third way to pick a representative data item is the *median*.

## DEFINITION: MEDIAN

**The median of a set of data is the middle data item or the average of the two middle data items when the items are arranged in numerical order.**

*Example 3*    Find the median temperature for a week when the temperatures were 73, 90, 70, 60, 90, 90, 80.

Arrange the temperatures in order and choose the middle one: 60, 70, 73, 80, 90, 90, 90.
So, the median is 80.
If the data contains an even number of items, you must average the two middle ones to find the median.

*Example 4*    Find the median temperature for the temperatures 60, 80, 60, 80, 90, 70, 80, 60.

Arrange the temperatures in order: 60, 60, 60, 70, 80, 80, 80, 90.
Average the two middle temperatures: $\frac{70 + 80}{2} = \frac{150}{2} = 75$.

So, the median is 75.
Since the mean, the mode, and the median each gives an idea of where the temperatures tended to *centralize*, they are called *measures of central tendency*. It is clear from this data set that the measures of central tendency,

| | |
|---|---|
| mean | 79 |
| mode | 90 |
| median | 80 |

are not necessarily the same. It might also be helpful to know how the data items *vary*. A simple *measure of variability* is the *range*.

## DEFINITION: RANGE

**The range in a set of data is the difference between the largest and the smallest data items in the set.**

*Example 5*    Find the range in temperatures for a week when the temperatures were 73, 90, 70, 60, 90, 90, 80.

Find the difference between the highest and the lowest temperatures.
$90 - 60 = 30$
So, the range is 30.
A more complex measure of variability is a standard deviation.

## Exercises

**Find the mean, mode, median, and range for each of the following sets of data.**

1. 6, 8, 2, 1, 14, 18, 12  $8\frac{5}{7}$, none, 8, 17
2. 18, 16, 14, 18, 16, 18, 14, 19, 20   **17, 18, 18, 6**
3. 80, 80, 70, 60, 60  **70, 60 and 80, 70, 20**
4. 4, 6, 2, 8, 6, 3, 13  **6, 6, 6, 11**
5. 48, 52, 64, 12, 40, 32, 19, 23
6. 31, 30, 29, 31, 30, 28, 27
   $36\frac{1}{4}$, none, 36, 52    $29\frac{3}{7}$, 30 and 31, 30, 4

# THE INPUT STATEMENT

**Objective**

**To use INPUT statements in programs that evaluate formulas**

The INPUT statement allows you to enter different data into a program each time you RUN it. The following program contains a simple INPUT statement.

```
10    PRINT "ENTER A NUMBER. "
20    INPUT X
30    PRINT 2 * X
40    END
```

Line 10 PRINTS the message within quotes.
Line 20 displays a ? and a blinking cursor, the signal for requesting data.
Line 30 computes the answer and displays it on the screen when the program is RUN.

To display a LISTing of your program, enter the command LIST. To debug any errors, enter the line number, correct the program line, and re-enter it. Now RUN the program. In response to the ? and blinking cursor, enter a number, say 6. The computer then stores 6 in the memory place X. The computer multiplies 6 by 2 and displays the answer, 12.

Another version of the INPUT statement prints a message in addition to requesting data. The format for this form of INPUT is:

10 INPUT "TYPE A VALUE FOR W. "; W

①—is the message to be printed. It is placed within " ".
②—is a space before the closing quotation mark. When the program is RUN, it separates the message from the blinking cursor.
③—is a semicolon that is necessary punctuation in this form of the INPUT statement.

**Problem 1**

Write a program that evaluates $3x - 2y$ for any values of $x$ and $y$.
RUN the program to find the value of $3x - 2y$ for $x = -3$ and $y = -2$.

**Program 1**

```
10    INPUT "ENTER THE VALUE OF X. ";X
20    INPUT "ENTER THE VALUE OF Y. ";Y
30    LET V = 3 * X - 2 * Y
40    PRINT "THE VALUE OF V IS "V"."
50    END
```

Lines 10 and 20 request values for X and Y.
Line 30 assigns the expression $3 * X - 2 * Y$ to V.
Line 40 PRINTS the result.

```
JRUN
ENTER THE VALUE OF X.  -3
ENTER THE VALUE OF Y.  -2
THE VALUE OF V IS -5.
```

In response to the two INPUT messages, you enter −3 for X and −2 for Y. Line 40 PRINTS the phrase in quotes, the *value* of V, and the period.

You will often find it necessary to translate familiar mathematical formulas into their BASIC forms. For example, the following BASIC formulas compute the sum and product of any two numbers, *a* and *b*.

LET S = A + B computes the *sum* of A and B and stores the result in S.
LET P = A * B computes the *product* of A and B and stores the result in P.

It is possible to enter two numbers, *a* and *b,* into a single INPUT statement by using the following form of INPUT:
INPUT "ENTER 1ST NUMBER, A COMMA, AND 2ND NUMBER. ";A,B

PROBLEM 2 uses the formulas and INPUT statement shown above.

**Problem 2**

Write a program that finds the sum and product of any two numbers. Then RUN the program to find the sum and product of −8 and −9.

**Program 2**

```
10   INPUT "ENTER 1ST NUMBER, A COMMA, AND 2ND NUMBER
     . ";A,B                    Line 10 requests values for A and B.
20   LET S = A + B              Lines 20 and 30 contain the formulas for
30   LET P = A * B              the sum and product of A and B.
40   PRINT "THE SUM IS "S"."    Lines 40 and 50 PRINT the results.
50   PRINT "THE PRODUCT IS "P"."
60   END
```

```
]RUN
ENTER 1ST NUMBER, A COMMA, AND 2ND NUMBER. −8, −9
THE SUM IS −17.          In response to the INPUT message, you type in −8 and −9,
THE PRODUCT IS 72.       which the computer stores in A and in B.
```

**Exercises**

Use algebra to predict the output of the program. Then enter and RUN the program to check your answers.

```
1. 10   INPUT "WHAT IS X? ";X
   20   INPUT "WHAT IS Y? ";Y
   30   LET A = X * Y − Y
   40   PRINT "A IS "A"."
   50   END
```

RUN for X = −5, Y = −8. **48**
*No first quotation mark, and*

2. Find the error(s) in the following program line: *no space before second*
10 INPUT TYPE IN A VALUE FOR M." ;M *quotation mark.*

3. If the numerical value of D is the desired output, find the error(s) in this program line: *No number for the line, and no quotation marks*
PRINT "THE VALUE OF D IS D." *before and after second D.*

Evaluate each expression for the values given. Then write, enter, and RUN a program to check each of your answers.
*For Ex. 5–8, see TE Answer Section.*

4. −8x − 4; x = 6 **−52**

5. 2r − $\frac{s}{4}$; r = −6, s = 12 **−15**

6. Write a program that finds the area of any rectangle, given its length and width. RUN for b = 6, w = 8. **48**

7. Write a program that finds the perimeter of any rectangle, given its length and width. RUN for l = 22, w = 36. **116**

# ORDER OF OPERATIONS IN BASIC

COMPUTER SECTION

*Objectives*

**To write programs that use BASIC algebraic expressions**
**To use the BASIC exponentiation symbol, $\wedge$**
**To use the order of operations in writing and evaluating algebraic expressions in computer programs**

When writing algebraic expressions in BASIC, you will often use grouping symbols. Consider the expression $\frac{8 + 6}{2}$. In algebra, $\frac{8 + 6}{2} = \frac{14}{2} = 7$. However, in BASIC, without grouping symbols, the expression becomes $8 + 6/2 = 8 + 3 = 11$, which is *not* 7. Therefore, to write $\frac{8 + 6}{2}$ in BASIC, you must use parentheses as follows: $(8 + 6)/2 = (14)/2 = 7$.

BASIC follows the rules of algebra for the order of operations. Compute within the parentheses first, then follow the rules for order of operations.

*Problem 1*

Write the following expressions in BASIC:

$$\frac{a}{2a - b} \qquad\qquad \frac{6 + m}{2m - 4}$$

BASIC form: A / (2 * A − B)          (6 + M) / (2 * M − 4)

To translate algebraic expressions containing exponents into BASIC, use the BASIC symbol for exponentiation, $\wedge$, as follows:

| Standard Notation | BASIC Notation |
|---|---|
| $5^4$ | $5 \wedge 4$ |
| $a^2 + b$ | $A \wedge 2 + B$ |

The rules for order of operations can now be extended to include exponents as follows:

First, compute within parentheses.
Second, do exponentiation: $\wedge$.
Third, do multiplication and division: * and /.
Fourth, do addition and subtraction: + and −.

*Problem 2*

Simplify $5 * 2 \wedge 3 − 6/3$.

| | |
|---|---|
| $5 * 2 \wedge 3 − 6/3$ | |
| $5 * 8 \qquad − 6/3$ | $2 \wedge 3 = 2^3 = 2 \cdot 2 \cdot 2 = 8$. |
| $40 \qquad − 2$ | Multiply and divide. |
| $38$ | Subtract. |

To write a program that evaluates an expression like $3a^2 + b$, you must first translate the expression into BASIC. Then you must write a BASIC formula that assigns the expression's value to a single variable.

*Examples*

Formula in Standard Notation

$$y = 3a^2 + b$$

$$f = \frac{2d^2 - 1}{h + 2^3}$$

Formula in BASIC Notation

LET Y = 3 * A ∧ 2 + B
LET F = (2 * D ∧ 2 − 1)/(H + 2 ∧ 3)

In the second example, note that parentheses are necessary to translate the algebraic formula into BASIC.

*Problem 3*

Write a program to evaluate $3a^2 + b$. RUN the program for $a = 5$ and $b = 4$.

*Program 3*

```
10   INPUT "TYPE IN A VALUE FOR A. ";A
20   INPUT "TYPE IN A VALUE FOR B. ";B
30   LET Y = 3 * A ^ 2 + B
40   PRINT "THE VALUE OF Y IS "Y"."
50   END
```

Line 30 contains the BASIC formula for $3a^2 + b$ and assigns its value to Y.

```
JRUN
TYPE IN A VALUE FOR A. 5
TYPE IN A VALUE FOR B. 4
THE VALUE OF Y IS 79.
```

You INPUT 5 for A and 4 for B.

*Problem 4*

Write a program to evaluate $\dfrac{x^2 - 9}{x + 3}$. RUN the program to evaluate the expression for $x = 7$.

*Program 4*

```
10   INPUT "WHAT IS THE VALUE OF X? ";X
20   LET Q = (X ^ 2 − 9) / (X + 3)
30   PRINT "THE VALUE OF Q IS "Q"."
40   END
```

```
JRUN
WHAT IS THE VALUE OF X? 7
THE VALUE OF Q IS 4.
```

*Exercises*

Simplify each of the following expressions. Check your answers on the computer.

**1.** $4(3^2) - 27 \div 9$ **33**　　**2.** $17 - 5^2 + 32 \div 8$ **−4**　**3.** $8^2 - 45 \div 5 + 7(3)$ **76**
**PRINT 4 * 3 ∧ 2 − 27/9**　**PRINT 17 − 5 ∧ 2 + 32/8**　**PRINT 8 ∧ 2 − 45/5 + 7 * 3**

Translate each of the following algebraic expressions into BASIC.

**4.** $\dfrac{x^2 - 49}{x + 7}$　　　　**5.** $\dfrac{a + b}{2c - d}$　　　　**6.** $\dfrac{x^2 - 8x + 12}{x - 6}$
**(X ∧ 2 − 49)/(X + 7)**　　**(A + B)/(2 * C − D)**　　**(X ∧ 2 − 8 * X + 12)/(X − 6)**

**7.** $x + \dfrac{y}{2}$ **X + (Y/2)**　　**8.** $\dfrac{x + y}{2}$ **(X + Y)/2**　　**9.** $\dfrac{a}{b^2 + c}$ **A/(B ∧ 2 + C)**

Write and RUN programs to evaluate each of the following expressions.

**10.** $4x^3$; $x = -7$ **−1,372**　　　　　　**11.** $5x^2 - 8x + 2$; $x = -19$ **1,959**

**12.** $\dfrac{a^2 - 7a + 12}{a - 3}$; $a = 49$ **45**　　　**13.** $\dfrac{x}{x - y}$; $x = 20, y = 15$ **4**

**For Ex. 10–13, see TE Answer Section.**

# FOR. . .STEP. . .NEXT LOOPS

*Objective*

**To write programs that use FOR. . .STEP. . .NEXT loops to perform repetitive calculations**

In the last lesson, you wrote programs to evaluate algebraic expressions for specific values of the variable. For example, one program evaluated $3a^2 + b$, for $a = 5$ and $b = 4$. To evaluate it for additional values of $a$ and $b$, however, you would have to enter RUN for each new set of values. An alternative to that process is the technique called *looping*. Programmers use looping to perform a variety of repetitive calculations and procedures. A simple way to put a loop in a computer program is by using the FOR and NEXT commands. A FOR. . .NEXT loop is formatted as follows:

```
10 FOR A = 1 TO 10
.
.                                        Final value of index
.             Index              Initial value of index
60 NEXT A
```

These two statements control the loop. All statements between the FOR and NEXT statements are inside the loop and are executed each time the computer passes through the loop. The index acts as a counter to keep track of how many times the loop is executed. The NEXT statement instructs the computer to branch back to the beginning of the loop, so the variable in the NEXT statement must be the same as the index. When the final value of the index is reached, the control passes to the statement immediately following the NEXT statement.

The use of a FOR. . .NEXT loop is illustrated below.

*Problem 1*

Write a program that finds the total cost of a purchase on which there is a 6% sales tax. Set up the program to handle three separate selling prices. Then RUN the program for $7.00, $27.50, and $105.

Recall the formula for finding the total cost of a purchase.

$$\text{Total Cost} = \text{Selling Price} + \text{Tax}$$
$$t = s \qquad\qquad + .06\,(s)$$

The tax is 6% of the selling price, or $.06\,(s)$.

This formula in BASIC is:
$$\text{LET } T = S + .06 * S$$

*Program 1*

```
10   FOR I = 1 TO 3
20   INPUT "ENTER SELLING PRICE. $";S
30   LET T = S + 0.06 * S
40   PRINT "THE TOTAL COST IS $"T"."
50   PRINT
60   NEXT I
70   END
```

Line 10 sets up the program to loop three times.
Line 30 computes T each time the loop is executed.
Line 60 makes the computer loop back to line 20 to compute T for each of the values of S.

```
RUN
ENTER SELLING PRICE. $7.00          You input $7.00 for the first selling price.
THE TOTAL COST IS $7.42.

ENTER SELLING PRICE. $27.50         You input values for the second and third
THE TOTAL COST IS $29.15.           selling prices.

ENTER SELLING PRICE. $105.00
THE TOTAL COST IS $111.3.
```

In this program, lines 20 through 50 are executed three times. Line 10 sets the index, I, equal to 1 and instructs the computer to execute the loop until $I = 3$. Each time line 60 is executed, I is incremented by 1. When the loop has been executed three times, $I = 3$. The computer then executes line 70, which ends the program.

An extension of the FOR. . .NEXT loop is the FOR. . .STEP. . .NEXT loop. STEP is added to a FOR statement to allow you to count by other than ones. The format for the FOR. . .STEP statement is as follows:

FOR I = 10 TO 45 STEP 5

Step value of the index

The step value indicates the increment or decrement of the index. A NEXT statement is used with a FOR. . .STEP statement to create a loop as before. If no STEP part is written, the STEP is understood to be an increment of 1. The use of STEP in a program is illustrated in Problem 2.

*Problem 2*

Write a program to find the squares of the multiples of 5 from 10 to 45. Provide headings (NUMBER, SQUARE OF NUMBER) for the output formatted in two columns.

*Program 2*

```
10   PRINT "NUMBER","SQUARE OF NUMBER"
20   FOR I = 10 TO 45 STEP 5
30   LET Y = I ^ 2
40   PRINT I,Y
50   NEXT I
60   END
```

Line 10 prints the two column headings. The comma makes the computer space over one column width before printing the second heading.
Line 20 specifies that values from 10 to 45 STEPped by 5 are to be placed in I.
Line 40 prints the number in I in the first column, and $I^2$ in the second.
Line 50 increments the number in I by 5 and continues to loop back to line 30 until $I = 45$.

```
]RUN
NUMBER          SQUARE OF NUMBER
10              100
15              225
20              400
25              625
30              900
35              1225
40              1600
45              2025
```

*Exercises*

**1.** Complete this program, which finds the value of *a* for each of 10 different values of *b,* when $a = 3b^2 - 7b$.

```
10 FOR X = 1 TO 10
20 INPUT "WHAT IS B? ":B
30 LET A = 3 * B ∧ 2 − 7 * B
40 PRINT _____
50 NEXT X
60 END
40 "3 * "B" ∧ 2 − 7 * "B" = " A
```

**For Ex. 3–9, see TE Answer Section.**

**2.** Complete this program, which finds the area of a parallelogram for three sets of values for the base and height.

```
10 FOR J = 1 TO 3
20 INPUT "WHAT IS THE
      HEIGHT? "; H
30 INPUT "WHAT IS THE BASE? "; B
40 LET A = B * H
50 PRINT "AREA = " A
60 NEXT J
70 END
```

**3.** Write a program that evaluates $y = 3x^2 - 5x - 7$ for three different values of *x*. Then RUN the program to find *y* for these values of *x:* −4, −6, 7. **61, 131, 105**

**4.** Write a program that evaluates $a = \dfrac{r + s}{2}$ for two different sets of values for *r* and *s*. Then RUN the program to find *a* for *r* = 78, *s* = 96, and for *r* = 100, *s* = 86. **87, 93**

**5.** The formula for the height *(h)* of a rocket at the end of *t* seconds is $h = 128t - 16t^2$. Write a program that finds the height *(h)* of the rocket for any three values of *t*. Then RUN the program to find *h* for *t* = 1, *t* = 3, and *t* = 4. **112, 240, 256**

Use FOR. . .STEP. . .NEXT to write and RUN programs that find solutions for each of the following exercises. Provide headings and two-column format for the output, as in Problem 2 above.

**6.** The squares of the numbers 7, 14, 21, 28, 35, 42, 49, 56.

**7.** The value of $x^2 - 25$ for *x* = 8, 10, 12, 14, . . ., 40, 42.

**8.** The value of $\dfrac{4x^2 + 2}{2x}$ for *x* = 10, 13, 16, 19, 22, 25, 28.

**9.** The squares of the numbers 40, 38, 36, 34, 32, 30, 28, 26.

# READ. . .DATA STATEMENTS

*Objective*
## To write programs that perform calculations on values entered in READ. . .DATA statements

You have learned to use the INPUT statement for entering data into a program. This allows you to enter new data into a program each time you RUN the program. If you know all the data to be used in a program at the time you are writing it, it might be easier to enter information into the program by using READ and DATA statements.

The READ statement reserves a memory place to store the values contained in the DATA statement. The DATA statement can be placed anywhere in a program prior to the END statement. The format for DATA statements is as follows:

DATA 4, −3, 7, 2

Punctuation (comma) is used *only* to separate DATA items.

Problem 1 illustrates the use of READ. . .DATA in a program.

*Problem 1*
Write a program using READ. . .DATA statements to evaluate $y = x^2 - 5x + 2$ for $x = 4, -3, 7, 2$. Provide headings X, VALUE OF Y for the output and format it in two columns. Enter and RUN the program.

*Program 1*

```
10    PRINT "   X","VALUE OF Y"
20    FOR K = 1 TO 4
30    READ X
40    LET Y = X ^ 2 - 5 * X + 2
50    DATA  4, -3, 7, 2
60    PRINT "   "X,"      "Y
70    NEXT K
80    END

]RUN
    X                VALUE OF Y
    4                   -2
   -3                   26
    7                   16
    2                   -4
```

Line 10 prints two column headings. Recall that the comma makes the output appear in columns.
Line 30 READs a value in the DATA statement each time the loop is executed. That value is stored in X.
Line 50 contains the values of X to be used in calculating Y.
Line 60 PRINTs X and Y under their proper headings. The spaces in quotes center the output.

Note that with each execution of the loop, line 30 READs a new value of X from the DATA line. The value of Y is then computed using this value.

READ statements can be used to store more than one DATA item at a time. For example, to evaluate an expression like $\frac{x+y}{2}$, you can use a single READ statement that stores *pairs* of values from a DATA statement. Similarly, you can write READ statements that store three or more DATA items at a time. It is also possible to store DATA items in more than one DATA statement. When two or more DATA statements are used, the computer READs the first DATA statement completely before READing any item in the following DATA statement(s). Program 2 illustrates the use of two DATA statements and a READ statement that stores pairs of DATA items at one time.

## Problem 2

Four students each received two test grades. Write and RUN a program that outputs each student's pair of grades and the average of those two grades. The students' grades were: 90, 80; 100, 90; 70, 80; 85, 75.

## Program 2

```
10   PRINT "GR.1","GR.2","AVE."
20   FOR I = 1 TO 4
30   READ X,Y
40   LET A = (X + Y) / 2
50   DATA  90,80,100,90,70,80
60   PRINT X,Y,A
70   DATA  85, 75
80   NEXT I
90   END

]RUN
GR.1          GR.2          AVE.
90            80            85
100           90            95
70            80            75
85            75            80
```

Line 10 prints three column headings.
Line 30 READs the values in the DATA statement two at a time, and stores those values in X and Y.
Line 50 stores three pairs of grades to be used in this program.
Line 60 prints each pair of grades and their average in separate columns.
Line 70 stores the last pair of grades to be used in this program.

Notice that with each execution of the loop, line 30 READs a new pair of grades from the DATA lines. The average is then computed using these grades.

## Exercises

In Exercises 1–7, evaluate each expression for all values given. Then write and RUN a program that uses READ. . .DATA statements to check each of your answers.
**For Ex. 1–9, see TE Answer Section.**

1. $x^2 - 7x + 4$; $x = 2, -5, 7$ **−6, 64, 4**   2. $-4x^3 - 8x$; $x = -2, -5, 1$ **48, 540, −12**

3. $5 - 4a^2$; $a = 8, -7, -1, 3$ **−251, −191, 1, −31**   4. $(2x - 5)^2$; $x = 1, -7, -4, 0$ **9, 361, 169, 25**

5. $\dfrac{x^2 - 5x + 4}{x - 4}$; $x = 9, -8, -2$ **8, −9, −3**   6. $\dfrac{x}{x - 8}$; $x = 8, 12, -4$ **−3, 3, $\frac{1}{3}$**

7. $-5a + 9b$; $a = 6, b = 8$; $a = -5$; $b = -5$; $a = -2, b = -6$ **42, −20, −44**

8. Write and RUN a program that finds the average of each of the following sets of four grades: 90, 80, 100, 90; 100, 80, 80, 75. **90, 83.75**

9. Write and RUN a program that finds the products of these pairs of numbers: −8, 9; −5, −4; −7, −5; −4, −2; 178, 987. **−72, 20, 35, 8, 175, 686**

# GOTO AND IF. . .THEN STATEMENTS

*Objective*   **To use GOTO and IF. . .THEN statements in writing programs**

In the programs you have seen so far, the program has been executed from the lowest line number to the highest line number. Sometimes it is necessary to branch from one part of a program to another. There are two types of branches, an unconditional branch, GOTO, and a conditional branch, IF. . .THEN.

The GOTO statement causes the computer to go to the line indicated and continue the program from there.

*Problem 1*   Write a program using a GOTO statement that adds pairs of numbers entered by READ and DATA statements.

*Program 1*
```
10   READ A,B
20   LET C = A + B
30   PRINT "C = "C
40   GOTO 10
50   DATA  2, 4, 6, 8, 10, 12
60   END

]RUN
C = 6
C = 14
C = 22

?OUT OF DATA ERROR IN 10
```

Line 30 prints each successive sum.
Line 40 loops back to line 10.

The GOTO statement in line 40 continues to branch to line 10 until all the DATA in line 50 is used. An OUT OF DATA message will appear when this happens.

The IF. . .THEN statement is a conditional command that enables the computer to make decisions. If the condition is true, the computer branches to the given line number or executes the given command. If the condition is false, the computer executes the next line.

The general form for the IF. . .THEN statement is as follows:

IF (relation) THEN (line number or command).

The following chart lists BASIC relation symbols used in stating conditions. Their algebraic equivalents and meanings are also listed.

| BASIC Symbol | Algebraic Symbol | Meaning |
| --- | --- | --- |
| = | = | is equal to |
| > | > | is greater than |
| < | < | is less than |
| > = | $\geq$ | is greater than or equal to |
| < = | $\leq$ | is less than or equal to |
| < > | $\neq$ | is not equal to |

Suppose you want to evaluate the expression $x^2 - 4$ for any three values of $x$. However, you want the computer to PRINT the result only if it is a positive number or zero. If the result is negative, you want the computer to PRINT the following message: "THE RESULT IS NEGATIVE." In this situation, you must write the program so that the computer can make the following decision:

1. PRINT the result if it is a positive number or zero.
   *or*
2. PRINT a message if the result is negative.

Program 2 illustrates the use of an IF. . .THEN statement and a GOTO statement to accomplish this task.

**Problem 2**

Write a program that evaluates $x^2 - 4$ for any three values of $x$. The computer should PRINT the result only if it is positive or zero.
Otherwise, have the computer display the following message: "THE RESULT IS NEGATIVE." Run the program for $x = 6, -1, 2$.

**Program 2**

```
10   FOR I = 1 TO 3
20   INPUT "WHAT IS X? ";X
30   LET Y = X ^ 2 - 4
40   IF Y < 0 THEN 80
50   PRINT "Y = "Y
60   PRINT
70   GOTO 100
80   PRINT "THE RESULT IS NEGATIVE."
90   PRINT
100  NEXT I
110  END
```

Line 40 checks to see if Y is negative. If it is, the computer branches to line 80, which PRINTs a message. If Y is not negative, line 50 PRINTs the result. In line 70, the GOTO statement causes the computer to branch to line 100.

```
JRUN
WHAT IS X? 6
Y = 32
```

The INPUT of 6 yields a positive result, 32. Line 50 is executed.

```
WHAT IS X? -1
THE RESULT IS NEGATIVE.
```

The INPUT of $-1$ yields a negative result, $-3$. Line 80 is executed.

```
WHAT IS X? 2
Y = 0
```

The INPUT of 2 yields zero. Line 50 is executed.

Suppose you also wanted the computer to print a message when a result is greater than 30. You can do this by entering the following two lines:

54 IF Y < = 30 THEN 100
56 PRINT "Y > 30."

Type LIST and press the RETURN key. The new lines are inserted between lines 50 and 60. Run the revised program for $x = 8, 0, -6$.

*Exercises*

Write IF. . .THEN statements for the following specifications.

**IF B < = −3 THEN 80**

**1.** If the value of B is less than or equal to −3, then skip to line 80.

**2.** If the value of B is positive, then skip to line 100. **IF B > 0 THEN 100**

Use the following program for answering Exercises 3–6.

```
10   FOR K = 1 TO 6
20   READ A,B
30   DATA  3,6,7,8,-4,-5,-9,-6,2,7,9,0
40   LET Y = A * B
50   IF Y > 45 THEN 80
60   PRINT "Y = "Y"."
70   GOTO 100
80   LET Y = A + B
90   GOTO 60
100   NEXT K
110   END
```

**3.** Explain what line 10 does. **Line 10 sets up a FOR. . .NEXT loop.**

**4.** Before you RUN the program, predict the value of Y for each set of DATA.

**18, 15, 20, −15, 14, 0**

**5.** During which program loops will line 80 be executed? **2nd and 4th**

**6.** Explain what line 90 does. **Line 90 branches to line 60.**

**7.** Complete the following program that computes the area of a triangle or a rectangle.

```
10  PRINT "ENTER 1 FOR TRIANGLE OR 2 FOR RECTANGLE."
20  INPUT X
30  IF X = 1 THEN 60
40  IF X = 2          THEN 90
50  GOTO 10
60  INPUT "ENTER BASE, A COMMA, AND HEIGHT. ";B,H
70  LET A = .5 * B * H
80  GOTO 110
90  INPUT "ENTER LENGTH, A COMMA, AND WIDTH. "; L, W
100 LET A = L * W
110 PRINT "AREA IS " A
120 END
```

RUN the completed program for a triangle with a base of 7 cm and a height of 20 cm, and a rectangle with a length of 9 in. and width of 14 in. Debug if necessary. **70 cm², 126 in.²**

**8.** Write and RUN a program that evaluates $\dfrac{6}{x-4}$ for $x = -4, -3, -2, -1, 0, 1, 2,$ 3, 4. Recall that you cannot divide by zero. The computer will display an error message if your program attempts to do so. Use an IF. . .THEN statement to test the denominator in the expression above for possible division by zero, and to output "YOU CANNOT DIVIDE BY ZERO." if a zero appears in the denominator.

*For Ex. 8, see TE Answer Section.*

# ACCUMULATORS

**Objective**

**To write programs that use accumulators in computing sums and averages**

In many computer programs, it is desirable to designate a memory storage place as a *counter* or an *accumulator*. Program 1 below illustrates the use of a counter in memory place H and an accumulator in memory place S.

**Problem 1**

Write and RUN a program that prints the integers 1, 2, 3, 4, and their sum.

**Program 1**

```
10   PRINT "INTEGER","RUNNING SUM"
20   FOR H = 1 TO 4
30   LET S = S + H
40   PRINT "    "H,"      "S
50   NEXT H
60   PRINT "THE SUM IS "S"."
70   END
```

Line 20 designates storage place H as a counter. It will count the integers 1, 2, 3, 4.
Line 30 designates storage place S as an accumulator. S will accumulate the sum of the integers 1 through 4.
Line 40 prints the contents of the counter H and the accumulator S.
Line 60 prints the sum of 1, 2, 3, 4.
In the first loop, line 30 replaces 0 in S with 0 + 1.

```
]RUN
INTEGER            RUNNING SUM
   1                    1
   2                    3
   3                    6
   4                    10
THE SUM IS 10.
```

In the second loop, 1 + 2 replaces 1 in S.
In the third loop, 3 + 3 replaces 3 in S.
In the fourth loop, 6 + 4 replaces 6 in S.
In each loop, line 40 prints the contents of H under the INTEGER heading.
Line 60 outputs the last number stored in S, the accumulated sum of 1, 2, 3, 4.

Accumulators are very useful for computing averages. Program 2 illustrates the use of an accumulator in computing averages.

**Problem 2**

Write a program that finds the average of any 10 bowling scores. RUN the program to find the average of these scores: 160, 180, 220, 195, 245, 290, 300, 180, 210, 200.

**Program 2**

```
10   FOR I = 1 TO 10
20   INPUT "ENTER BOWLING SCORE. ";B
30   LET S = S + B
40   NEXT I
50   LET A = S / 10
60   PRINT "AVERAGE = "A"."
70   END
```

Line 30 accumulates the sum of bowling scores in storage place S.
Line 50 computes the average score by dividing the sum of 10 scores by 10.

RUN the program on your own.

Suppose you wanted to find the averages of several lists of scores like the list in Program 2. You can alter the program for Program 2 so that it will compute any number of averages. You do this by inserting several lines into the program. Without re-entering the entire program, enter the following lines:

```
 5   LET S = O
62   PRINT "MORE AVERAGES? (YES OR NO)"
64   INPUT X$
66   IF X$ = "YES" THEN 5
```

Now list the program. Notice that the lines you added are listed in the program in their correct numerical order.

```
 5   LET S = O
10   FOR I = 1 TO 10
20   INPUT "ENTER BOWLING SCORE. ";B
30   LET S = S + B
40   NEXT I
50   LET A = S / 10
60   PRINT "AVERAGE = "A"."
62   PRINT "MORE AVERAGES? (YES OR NO)"
64   INPUT X$
66   IF X$ = "YES" THEN 5
70   END
```

Line 5 initializes the accumulator to zero.

Line 62 asks if there are more averages to compute. Line 64 accepts your YES or NO answer. Line 66 branches back to line 5 if you entered YES in line 64.

Line 5 sets the accumulator S to zero so that any previously stored sums in S will not be added to the new data. This process is called *initializing* the accumulator. The RUN command initializes all variables to zero automatically. Line 5 is necessary when more than one average is to be computed in the same RUN. Line 64 uses the variable X$ (called ''X-string'') to accept alphabetic INPUT. Single letters followed by $ designate memory storage places that will store alphabetic, numerical, and most other character data.

*Exercises*

1. RUN the revised program above. Make up your own scores for three sets of ten bowling scores and write the averages that are output by the computer.

**Averages will vary.**

Write and RUN programs utilizing accumulators to find solutions for each of the following exercises. **For Ex. 2–5, see TE Answer Section.**

2. PRINT the integers 1, 2, 3, 4, 5, 6, 7, 8, and find their sum. **36**

3. PRINT the even integers 2, 4, 6, 8, 10, 12, 14, and find their sum. **56**

4. Find the sum of the squares of any five numbers. RUN the program for 5, 9, 3, 2, 19. **480**

5. Find the average of any three test grades. RUN the programs for these grades: 90, 100, 80. **90**

# THE INT AND RND FUNCTIONS

**To use the INT and RND functions to simulate chance events**

The BASIC function INT finds the greatest integer contained in a decimal number. For example, the INT of 8.675924 is 8. For positive numbers, INT gives the whole number to the left of the decimal point. To display the INT of a number, use this statement form:

    PRINT INT(8.675924).

## Examples

Find each of the following:

    1) INT(4.232459)          2) INT(4.997659)          3) INT(4)

Solutions:

    1) INT(4.232459) = 4      2) INT(4.997659) = 4      3) INT(4) = 4

Note that the INT of a number is *not* the same as rounding a number. INT(6.897654) = 6, the whole number part of the decimal. 6.897654 rounded to the nearest whole number is 7, *not* 6.

The arrow of the spinner is equally likely to land on any of the numbers from 1 to 8. A computer can be programmed to simulate this by using the RND function to generate random numbers.

RND stands for RaNDom. The statement RND(1) will generate random 9-place *decimals* between 0 and 1. So, the program line 10 PRINT RND(1) will display numbers like .887234654, .520136748, etc. Combining the INT function with RND will produce random *whole* numbers. Using the two decimals generated by RND(1) above, you can see that the statement INT(7 * RND(1) + 1) produces random whole numbers from 1 through 7. **For TRS-80, replace RND(1) with RND(0).**

| | |
|---|---|
| INT(7 * RND(1) + 1) | INT(7 * RND(1) + 1) |
| = INT(7 * .887234654 + 1) | = INT(7 * .520136748 + 1) |
| = INT(6.21064258 + 1) | = INT(3.64095724 + 1) |
| = INT(7.21064258) | = INT(4.64095724) |
| =    7 | =    4 |

So, INT(7 * RND(1) + 1) generates random whole numbers from 1 through 7.

To generate random whole numbers from
1 through 8, use:                 5 through 35, use:
INT(8 * RND(1) + 1)            INT(31 * RND(1) + 5)

In general, to generate random whole numbers from A through B, use:
INT((B − A + 1) * RND(1) + A)

You can now use the computer to simulate the action of the spinner illustrated on the previous page. The probability of the arrow landing on 5 is 1 out of 8, or $P(5) = \frac{1}{8}$. So, if you were to spin the arrow 160 times, you would expect the arrow to land on 5 a total of 20 times, since $\frac{1}{8}(160) = 20$. This result is a mathematical expectation. Physically spinning the arrow 160 times does not mean the arrow will land on 5 exactly 20 times.

The program below simulates spinning the arrow 160 times. A counter (similar to an accumulator) is used to record the number of times a 5 is generated. That number should be close to the predicted value of 20.

**Problem 1**

Write a program that generates 160 random whole numbers from 1 through 8. Use a counter to count the number of times a 5 occurs.

**Program 1**

```
10   LET C = 0
20   FOR I = 1 TO 160
30   LET R =   INT (8 *   RND (1) + 1)
40   PRINT "SPIN "I,R" COMES UP."
50   IF R = 5 THEN 70
60   GOTO 80
70   LET C = C + 1
80   NEXT I
90   PRINT "5 OCCURS "C" TIMES."
100  END
```

Line 10 initializes the counter.
Line 30 generates random whole numbers from 1 through 8.
Line 40 prints the number of the spin and the result.
In line 50, if a 5 occurs, the computer branches to line 70, which keeps count of the number of times a 5 occurs.
Line 90 prints the number of times a 5 occurs in 160 spins.

**Exercises**

Find each of the following:

1. INT(3.453287) **3**
2. INT(3.786432) **3**
3. INT(0.4562363) **0**

Write BASIC expressions that will generate random whole numbers for the following range of values:

4. 1 through 12
   **INT(12 * RND(1) + 1)**
5. 1 through 50
   **INT(50 * RND(1) + 1)**
6. 5 through 15
   **INT(11 * RND(1) + 5)**

7. Enter Program 1 and RUN it four times. Are the results close to the prediction that a 5 will occur 20 times out of 160 spins? **Yes, in general.**

8. Modify Program 1 so that it will simulate 400 spins. RUN the program six times. By increasing the total number of spins, does the ratio of the number of 5s to the total number of spins come closer to the *mathematical* expectation of P(5)?
   **Change line 20: 20 FOR I = 1 TO 400. Yes, in general.**

9. If a die is tossed 300 times, how many times would you expect a 3 to come up, based on the probability, P(3)? **$P(3)(300) = \frac{1}{6}(300) = 50$**

10. Write a program that simulates tossing a die 300 times, and that counts the number of times a 3 occurs. RUN the program five times. How close do the results come to the probability, P(3), you found in Exercise 9? **See TE Answer Section. Answers will vary.**

# Selected Extra Practice

## Practice on Order of Operations and Evaluation for Chapter 1

**Compute.** *(page 3)* [1.1]

1. $7 \cdot 3 + 4$ **25**
2. $13 - 4 \cdot 3$ **1**
3. $8 \cdot 4 + 6$ **38**
4. $7 - 2 \cdot 2$ **3**
5. $9 \cdot 5 + 7$ **52**
6. $18 - 7 \cdot 2$ **4**
7. $8 \cdot 5 - 6$ **34**
8. $17 - 6 \cdot 2$ **5**
9. $6 + 7 \cdot 3 + 9$ **36**
10. $23 - 8 \cdot 2 + 8$ **15**
11. $8 + 4 \cdot 3 + 5$ **25**
12. $17 - 2 \cdot 3 + 6$ **17**
13. $28 \div 4 + 3$ **10**
14. $4 - 32 \div 16$ **2**
15. $(6 - 5)4 + 3$ **7**
16. $9 + 3(5 + 6)$ **42**
17. $4 + 18 \div 3 - 1$ **9**
18. $10 - 45 \div 9 + 9$ **14**
19. $5(4 + 3) - 36 \div 9$ **31**
20. $2(8 - 3) + 28 \div 14$ **12**
21. $\dfrac{20 - 2}{6}$ **3**
22. $\dfrac{41 + 4}{12 + 3}$ **3**
23. $\dfrac{35 - 7}{14}$ **2**
24. $\dfrac{37 + 8}{9 + 6}$ **3**
25. $\dfrac{5 \cdot 6 + 4}{21 - 4}$ **2**
26. $\dfrac{4 \cdot 3 + 9}{11 - 4}$ **3**
27. $\dfrac{8 \cdot 9 - 12}{15 - 5 \cdot 2}$ **12**
28. $\dfrac{6 + 8 \cdot 7 + 3}{13 - 4 \cdot 2}$ **13**
29. $\dfrac{4 \cdot 3 + 44 \div 11}{7 - 2 - 3}$ **8**
30. $\dfrac{7 \cdot 3 + 2 \cdot 13 - 2}{2 \cdot 6 + 3}$ **3**
★ 31. $12 \div [12 - 2(2 + 3)] + 6$ **12**
★ 32. $113 - 6[28 - 2(35 \div 7)]$ **5**
★ 33. $\dfrac{18 + 4[7 + 3(9 \div 3) + 2]}{12 - 54 \div [5 + 2(8 \div 4)]}$ **15**
★ 34. $\dfrac{8 \cdot 3 + 4[2 \cdot 21 - 4(6 \div 3) + 6]}{448 - 8[5 + 4(1 + 3 \cdot 2)]}$ **1**

**Evaluate for the given value(s) of the variable(s).** *(page 6)* [1.2]

35. $5x + 2$ for $x = 3$ **17**
36. $12 - 3y$ for $y = 2$ **6**
37. $12m - 7$ for $m = \frac{2}{3}$ **1**
38. $9 + 15x$ for $x = \frac{4}{5}$ **21**
39. $14 + 7y$ for $y = 3$ **35**
40. $8m - 6$ for $m = 3$ **18**
41. $18x - 4$ for $x = \frac{1}{3}$ **2**
42. $8 - 4a$ for $a = \frac{1}{2}$ **6**
43. $5a - 2 + 4b$ for $a = 6$ and $b = 3$ **40**
44. $6c + 3d - 4$ for $c = 7$ and $d = 2$ **44**
45. $\dfrac{7a + 4}{10 - 2b}$ for $a = 2$ and $b = 4$ **9**
46. $\dfrac{4x + 6y}{30y - 3x}$ for $x = 4$ and $y = \frac{1}{2}$ **$\frac{19}{3}$**
47. $\dfrac{6a + 2b}{b + a}$ for $a = 6$ and $b = 2$ **5**
48. $\dfrac{5x + 2y}{4y - 3x + 8}$ for $x = 6$ and $y = 5$ **4**
49. $5a + 3b + 4c$ for $a = 2$, $b = 3$, and $c = 4$ **35**
50. $8.3y + 0.2z - 0.03x$ for $x = 2$, $y = 8$, and $z = 4$ **67.14**
51. $5pq - 3r$ for $p = 4$, $q = 5$, and $r = 3$ **91**
52. $\dfrac{2xy - 4z}{x + y - 2z}$ for $x = 8$, $y = 4$, and $z = 2$ **7**
★ 53. $xy[xz + y(x + z)]$ for $x = 2$, $y = 3$, and $z = 4$ **156**
★ 54. $(2 + xy + x)/(x + yz)$ for $x = 3$, $y = 2$, and $z = 4$ **1**

**Simplify.**

**55.** $8(2m)$ **16m**      **56.** $9 \times (3)$ **27**      **57.** $(5k)8$ **40k**      **58.** $9(8b)$ **72b**

**59.** $4(3l)$ **12l**      **60.** $8t(6)$ **48t**      **61.** $7(9g)$ **63g**      **62.** $(10k)6$ **60k**

**Prove each of the following. Justify each step.**      *(page 9)*   [1.3]

★ **63.** $(x + y) + z = (z + x) + y$   $(x + y) + z = z + (x + y)$ comm.; $= (z + x) + y$ assoc.

★ **64.** $t + (m + h) = m + (t + h)$   $t + (m + h) = (t + m) + h$ assoc.; $= (m + t) + h$ comm.;
                                                          $= m + (t + h)$ assoc.

**Rewrite by using the distributive property. Then simplify.**      *(page 12)*   [1.4]

**65.** $7(3x + 5)$ **21x + 35**      **66.** $4(6m + 5)$ **24m + 20**      **67.** $(3k + 2)9$ **27k + 18**

**68.** $8(5 + 6m)$ **40 + 48m**      **69.** $(5 + 4k)7$ **35 + 28k**      **70.** $3(5 + 9t)$ **15 + 27t**

**71.** $\frac{1}{3}(6x + 12)$ **2x + 4**      **72.** $(8a + 16)\frac{1}{4}$ **2a + 4**      **73.** $\frac{1}{4}(20 + 12r)$ **5 + 3r**

**74.** $7 \cdot x + 9 \cdot x$ **(7 + 9)x; 16x**      **75.** $7 \cdot b + 5 \cdot b$ **(7 + 5)b; 12b**      **76.** $4 \cdot k + k \cdot 3 + 8 \cdot k$

**77.** $4 \cdot t + 3 \cdot t$ **(4 + 3)t; 7t**      **78.** $r \cdot 7 + 9 \cdot r$ **(7 + 9)r; 16r**      **79.** $5 \cdot b + b \cdot 8 + 7 \cdot b$

                                    **76.** **(4 + 3 + 8)k; 15k**    **79.** **b(5 + 8 + 7); 20b**

**Simplify.**      *(page 15)*   [1.5]

**80.** $7x + 5x + 4x$ **16x**      **81.** $7y + 4 + 3y$ **10y + 4**      **82.** $3k + 5b + 7k$ **10k + 5b**

**83.** $5a + 4b + 7a + 9b$      **84.** $3k + 5m + 7k + 4m$      **85.** $3c + 4a + 5c + 100a + 1$

**86.** $7 + 2(5a + 6) + 3a$ **13a + 19**          **87.** $2(3k + 5) + 4(5k + 2)$ **26k + 18**

**88.** $4m + \frac{1}{3}(6m + 15) + m$ **7m + 5**          **89.** $9b + 2 + (24 + 12b)\frac{2}{3} + 3$ **17b + 21**

**90.** $9 + \frac{2}{3}(6b + 9g) + 5b + \frac{3}{5}(10g)$ **9 + 9b + 12g**    **91.** $3(2n + 4) + 7n + 5(n + 3)$ **18n + 27**

**92.** $9 + (6 + 3n)2 + 3n + 2(3n + 4)$ **15n + 29**    ★ **93.** $9x + 2[5 + 4(3x + 1)]$ **33x + 18**

★ **94.** $3b + 5[(2b + 3)5 + 7b]$ **88b + 75**      ★ **95.** $4[3m + \frac{3}{4}(8m + 24) + 6] + 2(3m + 5)$

**83.** **12a + 13b**    **84.** **10k + 9m**    **85.** **8c + 104a + 1**             **42m + 106**

**Evaluate for the given value(s) of the variable(s).**      *(page 17)*   [1.6]

**96.** $x^3$ for $x = 6$ **216**      **97.** $m^4$ for $m = 6$ **1,296**      **98.** $k^5$ for $k = 5$ **3,125**

**99.** $5m^2$ for $m = 7$ **245**      **100.** $4a^6$ for $a = \frac{1}{2}$ **$\frac{1}{16}$**      **101.** $3k^3$ for $k = 0.07$ **0.001029**

**102.** $(3x)^2$ for $x = 2$ **36**      **103.** $(6m)^3$ for $m = \frac{2}{3}$ **64**      **104.** $(4a)^3$ for $a = 0.2$ **0.512**

**105.** $5x^3y^2$ for $x = 2$ and $y = 4$ **640**      **106.** $a^3b^2$ for $a = \frac{2}{3}$ and $b = \frac{3}{4}$ **$\frac{1}{6}$**

**Use the given formula to find each.**      *(page 19)*   [1.7]

**107.**

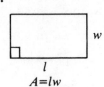

$A = lw$

**108.**

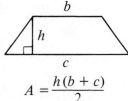

$A = \dfrac{h(b + c)}{2}$

**109.**

$A = \dfrac{hb}{2}$

**Find the area if $l = 6$ m, $w = 4$ m**   **24 m²**

**Find the area if $h = 4$ ft, $b = 10$ ft, $c = 12$ ft.**   **44 ft²**

**Find the area if $h = 8$ cm, $b = 4$ cm**   **16 cm²**

# Practice with Real Number Computation for Chapter 2

## Add.

*(page 36)* [2.3]

**1.** $-8 + 5$ **-3**    **2.** $-2 - 9$ **-11**    **3.** $6 - 7$ **-1**    **4.** $-8 + 4$ **-4**

**5.** $-9 - 3$ **-12**    **6.** $-9 + 9$ **0**    **7.** $-10 + 5$ **-5**    **8.** $-4 - 8$ **-12**

**9.** $-5 + 18$ **13**    **10.** $15 - 13$ **2**    **11.** $14 - 20$ **-6**    **12.** $-18 - 12$ **-30**

**13.** $-4.3 + 7.2$ **2.9**    **14.** $-5\frac{7}{12} + (-3\frac{1}{2})$ **$-9\frac{1}{12}$**    **15.** $5.3 + (-8)$ **-2.7**    **16.** $8\frac{1}{2} + (-5\frac{1}{4})$ **$3\frac{1}{4}$**

**17.** $-8 + 4\frac{2}{3}$ **$-3\frac{1}{3}$**    **18.** $9\frac{1}{4} + (-5\frac{7}{8})$ **$3\frac{3}{8}$**    **19.** $-6.2 + 8.1$ **1.9**    **20.** $-4\frac{2}{3} + (-1\frac{5}{9})$ **$-6\frac{2}{9}$**

**21.** $6\frac{1}{2} - 8\frac{1}{10}$ **$-1\frac{3}{5}$**    **22.** $-9 + 6\frac{3}{4}$ **$-2\frac{1}{4}$**    **23.** $-5\frac{1}{8} + 4\frac{3}{4}$ **$-\frac{3}{8}$**    **24.** $-8\frac{5}{7} + 12$ **$3\frac{2}{7}$**

**25.** $4 - 9 - 3 + 1$ **-7**    **26.** $-7 - 5 + 4 - 3$ **-11**    **27.** $7 - 4 - 3 + 2 - 1$ **1**

**28.** $-8 - 2 + 1 - 3 + 4$ **-8**    ★ **29.** $0.02 - 1.009 + 2.1$ **1.111**

★ **30.** $-106.43 - 17.2 - 0.009$ **-123.639**    ★ **31.** $3.018 - 9.004 + 11.2$ **5.214**

★ **32.** $-14.2 + 9.08 - 11.06$ **-16.18**

## Multiply.

*(page 43)* [2.6]

**33.** $4 \cdot -9$ **-36**    **34.** $-3 \cdot 18$ **-54**    **35.** $-9 \cdot -8$ **72**    **36.** $-5 \cdot 16$ **-80**

**37.** $-6 \cdot -8$ **48**    **38.** $9 \cdot -19$ **-171**    **39.** $-16 \cdot 8$ **-128**    **40.** $-8 \cdot -13$ **104**

**41.** $-\frac{3}{4} \cdot (-8)$ **6**    **42.** $2\frac{1}{3} \cdot (-\frac{3}{14})$ **$-\frac{1}{2}$**    **43.** $-\frac{9}{20} \cdot (-3\frac{1}{3})$ **$\frac{3}{2}$**    **44.** $-32 \cdot (-\frac{3}{4})$ **24**

**45.** $9 \cdot 8 \cdot -2$ **-144**    **46.** $-6 \cdot -6 \cdot 5$ **180**    **47.** $-9 \cdot -9 \cdot 2$ **162**    **48.** $5 \cdot -8 \cdot -6$ **240**

**49.** $5 \cdot -11 \cdot 2$ **-110**    **50.** $-8 \cdot 16 \cdot 4$ **-512**    **51.** $(-4)^2 \cdot (-2)^3$ **-128**

**52.** $(-2)^5 \cdot 3^2$ **-288**    **53.** $(-1)^7 \cdot (-2)^6$ **-64**    **54.** $(-10)^3 \cdot (-4)^3$ **64,000**

## Simplify.

★ **55.** $(-5 + 3 - 4) \cdot (-7 - 2 + 1)$ **48**    ★ **56.** $(-5 + 9) \cdot (-9 + 5 - 8 + 2)$ **-40**

★ **57.** $(-8 + 5)^3 \cdot (-7 + 2 - 4)$ **243**    ★ **58.** $(4 - 6)^5 \cdot (-8 + 2 - 7 + 12)^9$ **32**

## Divide.

*(page 46)* [2.7]

**59.** $\frac{-14}{7}$ **-2**    **60.** $\frac{-32}{-4}$ **8**    **61.** $\frac{0}{-9}$ **0**    **62.** $\frac{-19}{0}$ **undefined**

**63.** $-92 \div 2$ **-46**    **64.** $-84 \div -21$ **4**    **65.** $\frac{3}{5} \div -\frac{9}{10}$ **$-\frac{2}{3}$**    **66.** $\frac{9}{10} \div -6$ **$-\frac{3}{20}$**

**67.** $-4.2 \div 0.03$ **-140**    **68.** $-4.8 \div -0.024$ **200**    **69.** $(-48 \div 12) \div (18 \div 9)$ **-2**

**70.** $(-60 \div -3) \div (76 \div -38)$ **-10**    ★ **71.** $(-8 + 12 - 20) \div (-4 + 2)$ **8**

★ **72.** $(-9 + 7 - 15 - 11) \div (21 - 28)$ **4**    ★ **73.** $-8 \cdot 4 \div -2 + 8 + 4 \div -2$ **22**

## Simplify (mixed practice).

**74.** $-8 + 4$ **-4**    **75.** $-8 \cdot 4$ **-32**    **76.** $-8 \div 4$ **-2**    **77.** $-10 \div -2$ **5**

**78.** $10 - 5$ **5**    **79.** $10 \cdot -5$ **-50**    **80.** $10 \div -5$ **-2**    **81.** $-10 - 5$ **-15**

**82.** $-12(-3\frac{1}{2})$ **42**    **83.** $-4\frac{3}{4} + 11$ **$6\frac{1}{4}$**    **84.** $-0.04 \div 0.002$ **-20**    **85.** $-5.2 + 10$ **4.8**

**86.** $-6 - 9$ **-15**    **87.** $-6 \cdot -9$ **54**    **88.** $-18 \div 3$ **-6**    **89.** $3 - 18$ **-15**

**90.** $-4\frac{1}{2} \div -3$ **$\frac{3}{2}$**    **91.** $-17 + 8\frac{4}{5}$ **$-8\frac{1}{5}$**    **92.** $-8\frac{5}{6} + 1\frac{2}{3}$ **$-7\frac{1}{6}$**    **93.** $-2\frac{2}{3} \cdot \frac{9}{16}$ **$-\frac{3}{2}$**

**94.** $-7 \div -4\frac{2}{3}$ **$\frac{3}{2}$**    **95.** $3\frac{1}{3} \div (-20)$ **$-\frac{1}{6}$**    **96.** $-4\frac{1}{2} + 6\frac{3}{4} - \frac{1}{8}$ **$2\frac{1}{8}$**    **97.** $7 - 0.06 - 12$ **-5.06**

SELECTED EXTRA PRACTICE

**Evaluate for the given value(s) of the variable(s).** *(page 50)*  [2.9]

98. $6 - 3x$ for $x = 5$ **−9**
99. $7 - 4y$ for $y = -8$ **39**
100. $-4x - 8$ for $x = 2$ **−16**
101. $6 - 8p$ for $p = -2\frac{1}{2}$ **26**
102. $-5x - 4y$ for $x = -3$ and $y = -2$ **23**
103. $-2b + 3c$ for $b = -6$ and $c = -4$ **0**
104. $6r + 8s$ for $r = -2$ and $s = -\frac{3}{4}$ **−18**
105. $c - 5d$ for $c = -6$ and $d = 1.1$ **−11.5**
106. $-2y^3$ for $y = -3$ **54**
107. $4x^5$ for $x = -3$ **−972**
108. $-3b^4$ for $b = -2$ **−48**
109. $-5b^3$ for $b = -2$ **40**

**Simplify.** *(page 52)*  [2.10]

110. $-8 + 3b - 5 - 5b$ **−2b − 13**
111. $-8x - 7 + 6x - 2$ **−2x − 9**
112. $-5x - 4 - 8x + 2$ **−13x − 2**
113. $7m - 8 - 4m + 9$ **3m + 1**
114. $-7 - 2m + 8 + 1\frac{2}{3}m$ **−15 − $\frac{1}{3}$m**
115. $-4 + 3\frac{1}{4}k - 8 - 7\frac{1}{8}k$ **−12 − 3$\frac{7}{8}$k**
116. $-4 - 5.2t + 3.4t + 8$ **4 − 1.8t**
117. $-9 + 0.04y - 2 - 8.1y$ **−11 − 8.06y**
118. $-4y + 2x - 7y - 8x$ **−11y − 6x**
119. $-4t - 5w + 3w - 18t$ **−22t − 2w**
120. $-x + 5x$ **4x**
121. $2m - 3m + 4b - b$ **−m + 3b**
122. $-x - y - 2x + 4y$ **−3x + 3y**
123. $5m - b + 2b - 6m$ **−m + b**
124. $-2 - 5x + 2y + 4x - y - 8$ **−x + y −10**
125. $-t + 3 - t - 4w + 8 + 3w - 5$ **−2t − w + 6**

**Simplify.** *(page 57)*  [2.12]

126. $3x - 2(4x - 5)$ **−5x + 10**
127. $-5 - 2(4 - 3x) - 7$ **−20 + 6x**
128. $7 - (5 - x) - 2x + 3$ **−x + 5**
129. $3a - 2(1 - a) - 4a + 3$ **a + 1**
130. $-4d - (7 - 2d) + 1 - d$ **−3d − 6**
131. $-k - 1 - 3(k - 4) - 5 + k$ **−3k + 6**
132. $3(2m - 4) - \frac{1}{2}(8 - 2m)$ **7m − 16**
133. $\frac{2}{3}(6 - 9a) - (8 - a)$ **−4 − 5a**
134. $5(4 - b) - (2 - 3b)$ **−2b + 18**
135. $7 - (5 - x) - 2(x - 8) + x$ **18**
★ 136. $-x - [4 - (2 - x)]$ **−2x − 2**
★ 137. $-2a - [4 - (3 - a) - 2(5 + 3a)]$ **3a + 9**
★ 138. $7 - 2[4 - (6 - 2y)] - (8 - y)$ **−3y + 3**
★ 139. $-x - 3 - [7 - (3 - x)] + 4 - (2 - x)$ **−x − 5**

**Simplify. Then evaluate for the given value of the variable.**

140. $3(2y - 4) - 7y$ for $y = -3$ **−y − 12; −9**
141. $2(5x - 4) - 3(-7 - 3x)$ for $x = -4$ **19x + 13; −63**
142. $-3(4 - m) - (7 - m) + 2$ for $m = 0.08$ **4m − 17; −16.68**
143. $-(-c + 5) - 2(5 - c) - 4c - 8$ for $c = -3$ **−c − 23; −20**

**Subtract.** *(page 59)*  [2.13]

144. Subtract $-5x + 2$ from $-4x + 3$. **x + 1**
145. From $-4y - 3$, subtract $-2y + 8$. **−2y − 11**
146. Subtract $-7x + 3$ from $-8x - 3$. **−x − 6**
147. Subtract $-8m + 6$ from $-8m + 5\frac{3}{4}$. **−$\frac{1}{4}$**
148. From $3a - 6$, subtract $2\frac{1}{2}a + 9$. **$\frac{1}{2}$a − 15**
149. From $-3y - 6$, subtract $-2\frac{5}{7}y + 9$. **−$\frac{2}{7}$y − 15**
★ 150. Subtract the opposite of $3 - 4 - (2 - x)$ from $-5 - x$. **−8**
★ 151. From $8 - 26 - (3 - x)$, subtract the opposite of $7 - (3 - x)$. **2x − 17**
★ 152. Subtract $-5 - (8 - 3x) - 4 - 2(3 - x)$ from $3x - (8 - x) - 4x + 8$. **−5x + 23**

## Practice on Solving Equations for Chapter 3

**Solve.**

1. $a + 9 = -3$  **−12**
3. $-13 + p = 19$  **32**
5. $7 + x = -3$  **−10**
7. $-4 + x = 11$  **15**
9. $14 + x = -17$  **−31**
11. $14 = x - 7$  **21**
13. $x - 8 = 19$  **27**
15. $m + 9 = -5$  **−14**
17. $\frac{x}{3} - 6 = 1$  **21**
19. $x + 19 = -48$  **−67**
21. $t + 49 = -15$  **−64**
★ 23. $b - 7 = |9 - 14|$  **12**

★ 25. $-53 + 17 = x - 19$  **−17**

27. $42 = -2a$  **−21**
29. $-14 = \frac{t}{6}$  **−84**

31. $-9 = -y$  **9**
33. $0.004c = -8.8$  **−2,200**
35. $40k = -240$  **−6**
37. $-165 = -55k$  **3**
39. $-145 = 29x$  **−5**
41. $5t - 15 = 10$  **5**
43. $-8 = 12 - 2b$  **10**
45. $8 - 2.4x = -40$  **20**
47. $17 - 4y = -11$  **7**
49. $62 - 18r = -10$  **4**
51. $-14 = 8 - 11m$  **2**
53. $2x - 8.6 = -6.8$  **0.9**
55. $7 = 8 - 2m + m$  **1**
57. $-18 = -7k + 5k - 6$  **6**
59. $7 = 21 - 0.002x$  **7,000**

61. $c + \frac{5}{7} = 8$  **$7\frac{2}{7}$**

63. $0.0042b = 8.4$  **2,000**

65. $x - 7\frac{3}{4} = -9$  **$-1\frac{1}{4}$**

67. $3x + x + 9 = 14$  **$\frac{5}{4}$**

69. $4 - m - 7m = 13$  **$-\frac{9}{8}$**

71. $-8b - 4 - 2b = 8 - 15$  **$\frac{3}{10}$**
★ 73. $3x - 4 - 5x = 2|-8 + 2 - 3|$  **−11**
★ 75. $|-9 + 5 - 8| = 4x - |5 - 6|$  **$\frac{13}{4}$**

2. $5 = x + 3$  **2**
4. $x + 6 = -4$  **−10**
6. $m + \frac{2}{5} = 8$  **$7\frac{3}{5}$**
8. $e - 10 = 15$  **25**
10. $5 = -7 + h$  **12**
12. $-3 + x = -7$  **−4**
14. $4.3 + y = 9$  **4.7**
16. $7 = -15 + x$  **22**
18. $9 = \frac{x}{3} - 7$  **48**
20. $23 + a = -48$  **−71**
★ 22. $x + 8 = |-12|$  **4**
★ 24. $|-8 - (24 \div -4)| = x + 10$  **−8**

★ 26. $\dfrac{18 - 2 \cdot -3}{28 \div (5 - 19)} = x + |-4 + 3|$  **−13**

28. $-2b = -16$  **8**
30. $\frac{x}{5} = 9$  **45**

32. $7t = -3$  **$-\frac{3}{7}$**
34. $125 = -5x$  **−25**
36. $-b = -19$  **19**
38. $15b = -300$  **−20**
40. $2a + 5 = 9$  **2**
42. $7 - m = 8$  **−1**
44. $4x - 8 = -12$  **−1**
46. $-k - 8 = -19$  **11**
48. $8 - t = -32$  **40**
50. $-20 = -13k + 6$  **2**
52. $-3x + 5x + 19 = -3$  **−11**
54. $-8 + -12 - k = k$  **−10**
56. $3p - 7 - 5 = -48$  **−12**
58. $14 - 0.21g = -28$  **200**
60. $7 = 9 - 7m$  **$\frac{2}{7}$**

62. $7 = \frac{y}{11}$  **77**

64. $-4.2 = 2x - 7$  **1.4**

66. $5\frac{1}{6} = x + 10$  **$-4\frac{5}{6}$**

68. $14 - p - p = 17$  **$-\frac{3}{2}$**

70. $-4 - x - 2x = 9 - 5$  **$-\frac{8}{3}$**

72. $5 - 4x - 13x = 8 + 12$  **$-\frac{15}{17}$**
★ 74. $3x - 0.06 = |3.15 - 7.05|$  **1.32**
★ 76. $|0.04 - 6| = 4 - 4g$  **−0.49**

**SELECTED EXTRA PRACTICE**

## Problem Solving for Chapter 3

**Write each in mathematical terms.**

*(page 75)* [3.5]

**1.** 12 less than 6 times $x$  **$6x - 12$**

**2.** 8 more than $y$ divided by 3  **$\frac{y}{3} + 8$**

**3.** 5 increased by 3 times a number  **$3n + 5$**

**4.** 6 decreased by 5 times a number  **$6 - 5n$**

**5.** 10 less than 9 times a number  **$9n - 10$**

**6.** 3 more than twice a number  **$2n + 3$**

**7.** 36 decreased by 6 times a number  **$36 - 6n$**

**8.** 13 more than 8 times a number  **$8n + 13$**

**9.** $x$ divided by 5, increased by 3  **$\frac{x}{5} + 3$**

**10.** 7 less than a number divided by 5  **$\frac{x}{5} - 7$**

**11.** 17 increased by three times a number  **$3x + 17$**

**12.** 6 times $n$, increased by 22  **$6n + 22$**

**13.** $m$ decreased by $w$  **$m - w$**

**14.** $t$ less than $u$  **$u - t$**

**15.** 8 times y, increased by twice $x$  **$8y + 2x$**

**16.** $p$ more than 4 times $n$  **$4n + p$**

**17.** 13 more than $x$ times $w$  **$xw + 13$**

**18.** $q$ less than twice $a$  **$2a - q$**

★ **19.** 4 times a number, increased by two more than the same number  **$4n + (n + 2)$**

★ **20.** 11 less than 4 times a number, increased by 1 more than twice the number
$$4n - 11 + (2n + 1)$$

**Solve each word problem.**

*(page 79)* [3.6]

**21.** 12 more than a number is 18. Find the number.  **6**

**22.** 6 more than a number divided by 8 is 4. Find the number.  **−16**

**23.** $3\frac{1}{2}$ more than a number is 8. Find the number.  **$4\frac{1}{2}$**

**24.** A number increased by $4\frac{2}{3}$ is 7. Find the number.  **$2\frac{1}{3}$**

**25.** 100 more than 3 times Jake's bowling score is 700. Find his score.  **200**

**26.** If 6 is increased by twice a number the result is 20. Find the number.  **7**

**27.** The $480 selling price of a TV is $30 more than 3 times the cost. Find the cost.  **$150**

**28.** 9 less than a number divided by 4 is 2. Find the number.  **44**

**29.** If 5 times Tim's age is decreased by 30, the result is 45. How old is he?  **15 y**

**30.** The $120 selling price of a watch is the cost increased by twice the cost. Find the cost.  **$40**

**31.** Three kilograms more than twice Hank's mass is 83 kg. What is Hank's mass?  **40 kg**

**32.** Seven times Wanda's age increased by twice her age is 72. How old is Wanda?  **8 y**

**33.** 75 less than twice Tina's bowling score is 225. Find her score.  **150**

**34.** Six times a number, decreased by twice that number is 68. What is the number?  **17**

**35.** If 4 times Al's age is decreased by 6 more than twice his age, the result is 44. How old is he?  **25 y**

**36.** If 3 times a number is subtracted from 9 more than 6 times the number, the result is 45. Find the number.  **12**

**37.** Ev budgeted $167 for a rental car. How far can he travel if the rate is $95 plus $0.24/mi?  **300 mi**

**38.** Barbara budgeted $180 for car rental. How far can she travel if the rate is $85 plus $0.19/mi?  **500 mi**

**39.** Martin rents a car for $55 plus $0.24/mi. How far can he travel, to the nearest mile, on a budget of $340?  **1,187 mi**

**40.** Julia rents a car for $85 plus $0.22/mi. How far can she travel, to the nearest mile, on a budget of $225?  **636 mi**

## Practice with Equations Involving Several Steps for Chapter 4

**Solve each equation.**

*(page 88)* [4.1]

**1.** $6a - 7 = 4a + 9$  **8**

**2.** $5b + 11 = 8b - 16$  **9**

**3.** $4 + 7x = 8x - 9$  **13**

**4.** $-14 - n = -3n + 6$  **10**

**5.** $7 - a = 4 + 2a$  **1**

**6.** $7d - 4 = 9d$  **−2**

**7.** $-x = 7 - 4x$  $\frac{7}{3}$

**8.** $a + 13 = -3a + 18$  $\frac{5}{4}$

**9.** $10g - 18 = 8g + 20$  **19**

**10.** $7x = 5 - 2x - 3x$  $\frac{5}{12}$

**11.** $2 - 3x = 7x - 8 - 8x$  **5**

**12.** $-6t - 8 - t = 5 + t$  $-\frac{13}{8}$

**13.** $4 + 2k = -k + 8 - k$  **1**

**14.** $6 - x - 2x = 5x - 4 - x$  $\frac{10}{7}$

**15.** $-2b - 4 = 5 - b - 6 - 2b$  **3**

**16.** $3k - 5 - 4k = 7 + k - 8$  **−2**

**17.** $3 + 5t - 11t = 14 - t - t$  $-\frac{11}{4}$

**18.** $7c + 4 - c = 2c - 3$  $-\frac{7}{4}$

*(page 91)* [4.2]

**19.** $4(x - 2) = 20$  **7**

**20.** $8 - 4(y - 1) = -36$  **12**

**21.** $2(2a + 3) = -18$  **−6**

**22.** $y - (8 - y) = 32$  **20**

**23.** $7 - (3 - m) = 2m$  **4**

**24.** $3x - 5 = 8 - (1 - x)$  **6**

**25.** $2(3 - z) = 16 - 2(3 + 2z)$  **2**

**26.** $8y - 3(4 - 2y) = 6(y + 1) - 2$  **2**

**27.** $3 + 7(x + 1) = 6 - (5 + 2x)$  **−1**

**28.** $7(2x - 2) - 5x = 4x + 2$  $\frac{16}{5}$

**29.** $4z - 3 - 7(z + 1) = 6z$  $-\frac{10}{9}$

**30.** $-5y - 2(y + 4) = 6y - 9$  $\frac{1}{13}$

**31.** $8y + \frac{2}{3}(9 - 6y) = 2y - 6$  **−6**

**32.** $6b + 8 = 10 - \frac{4}{5}(10 - 5b)$  **−3**

**33.** $-3(6 - 2x) + 4x = -(2x - 8)$  $\frac{13}{6}$

**34.** $3 - (4 - x) + 2x + 7 = 5 - (x - 6)$  $\frac{5}{4}$

★ **35.** $6[5 - 3(x - 4)] = 4x + 14$  **4**

★ **36.** $-7x - [2(3x + 1) + 4] = 3x$  $-\frac{3}{8}$

★ **37.** $6x - [4 - (3 - x)] = 2x - (4 - 2x)$  **−3**

★ **38.** $7m - \frac{3}{4}[8 - (4m - 4)] = -(6 - m) - 5m$  $\frac{3}{14}$

*(page 99)* [4.4]

**39.** $\frac{2}{3}m = 12$  **18**

**40.** $12 = \frac{3}{2}a$  **8**

**41.** $\frac{7}{3}p = 14$  **6**

**42.** $\frac{3}{5}a = 9$  **15**

**43.** $\frac{2}{3}a = 10$  **15**

**44.** $6 = \frac{3}{4}x$  **8**

**45.** $\frac{2}{3}x - 3 = \frac{1}{2}x - 1$  **12**

**46.** $\frac{3}{10}x - \frac{2}{5} = \frac{1}{2}x - 6$  **28**

**47.** $\frac{2}{7}r - \frac{3}{14} = 3 + \frac{5}{28}r$  **30**

**48.** $\frac{3}{4}x - 5 = \frac{1}{2}x - 4$  **4**

**49.** $\frac{2}{3}a - \frac{5}{6} = \frac{1}{2}a - 4$  **−19**

**50.** $\frac{3}{4}p - 2 = \frac{1}{2}p$  **8**

**51.** $\frac{2}{5}x - 3 = \frac{1}{10}x + 3$  **20**

**52.** $\frac{1}{7}m - 2 = \frac{3}{14}m + 1$  **−42**

**53.** $\frac{3}{10}a - 2 = \frac{1}{5}a + 1$  **30**

**54.** $\frac{1}{8} - \frac{3}{4}x = \frac{1}{2} + \frac{1}{8}x$  $-\frac{3}{7}$

**55.** $\frac{2}{15} - \frac{1}{5}x = \frac{7}{15} + \frac{1}{3}x$  $-\frac{5}{8}$

**56.** $\frac{5}{9}m - \frac{1}{3} = 3 - \frac{1}{9}m$  **5**

★ **57.** $\frac{2}{3} - (4 - x) = \frac{1}{2}(4 - x)$  $\frac{32}{9}$

★ **58.** $5 - \frac{2}{3}(6 - 9x) = \frac{1}{2}x + 4$  $\frac{6}{11}$

★ **59.** $\frac{1}{9}x - \frac{2}{3}(x - 8) = \frac{1}{3}x + \frac{5}{9}$  $\frac{43}{8}$

★ **60.** $\frac{3}{10}m - \left|-\frac{1}{2} + \frac{1}{5}\right| = \frac{2}{5}m$  **−3**

*(page 102)* [4.5]

**61.** $0.03x = 0.2$  $\frac{20}{3}$

**62.** $0.016 = 0.32x$  **0.05**

**63.** $0.004t - 7.1 = 0.12$  **1,805**

**64.** $0.1m - 2.4 = 1.17$  **35.7**

**65.** $0.007 = 0.7b - 0.21$  **0.31**

**66.** $2.1g = 0.72 + 1.8g$  **2.4**

**67.** $0.18y - 24 = 0.1y + 0.6$  $\frac{615}{2}$

**68.** $5 - 0.03w = 0.7w - 0.11$  **7**

**69.** $0.7t - 0.2 = 0.13t - 0.08$  $\frac{12}{57}$

**70.** $0.12x - 4 = 0.112x + 1$  **625**

**71.** $0.7x - 1 = 0.6x + 0.002$  **10.02**

**72.** $0.5 - 0.08x = 0.004x + 0.2$  $\frac{25}{7}$

**73.** $0.03(4 - 0.2x) = 0.17x - 1.2$  $\frac{15}{2}$

**74.** $0.01(5 - 0.2x) = 0.75 + 0.198x$  **−3.5**

SELECTED EXTRA PRACTICE

# Problem Solving for Chapter 4

**Number Problems**                                  *(page 95)*   [4.3]

1. Tina's salary is twice Juan's salary. Together they earn $750 per week. Find the salary of each. **Tina: $500; Juan $250**

2. In a game, Ron's score was four times Jim's score. Together they scored 120 points. Find their scores. **Ron: 96   Jim 24**

3. The second of two numbers is 6 less than three times the first. Their sum is 26. Find the two numbers. **8, 18**

4. A 26 cm board is separated into two pieces. The longer is 5 cm longer than twice the shorter one. Find the length of each piece.   **19 cm, 7 cm**

5. The second of three numbers is five times the first. The third is 3 more than the second. If the second is decreased by twice the third, the result is −21. Find the three numbers. **3, 15, 18**

**Percent Problems**                                  *(page 105)*   [4.6]

6. 34% of 65 is what number? **22.1**

7. 7% of what number is 21? **300**

8. 40 is what % of 120? **$33\frac{1}{3}$%**

9. 26% of what number is 52? **200**

10. The selling price of a bicycle is $168. The profit is 40% of the cost. Find the cost. **$120**

11. A team won 7 games and lost 3. The losses were what % of the games played? **30%**

12. Last year the weekly cost of food for a family of five was $200. Today the cost is $250. The increase is what percent of last year's weekly cost? **25%**

**Perimeter Problems**                                  *(page 108)*   [4.7]

13. The length of a rectangle is 4 m more than twice the width. The perimeter is 56 m. Find the length and width. **20 m, 8 m**

14. The length of a rectangle is twice the width. The perimeter is 102 m. Find the length and the width. **34 m, 17 m**

15. The first side of a triangle is 4 cm longer than the second side. The third side is 3 cm shorter than twice the second side. The perimeter is 25 cm. How long is each side? **10 cm, 6 cm, 9 cm**

16. The length of each of the legs of an isosceles triangle is 3 m less than twice the length of the base. The perimeter of the triangle is 34 m. Find the length of each side. **13 m, 13 m, 8 m**

**Mixed Practice**                                  *(page 110)*

17. John's age is 5 more than twice Bill's. The sum of their ages is 50. Find the age of each. **Bill 15 y; John 35 y**

18. The length of a rectangle is 3 ft less than four times the width. The perimeter is 64 ft. Find the length and the width. **7 ft; 25 ft**

19. Separate 48 into two parts so that the larger is 3 times the smaller. Find both parts. **12, 36**

20. A team wins 8 games and loses 2. The wins are what percent of the games played? **80%**

21. John rents a car for $85 a week plus $0.26/mi. How far can he travel on a budget of $215? **500 mi**

22. Tim is paid $175 a week plus a commission of 6% on all sales. Find the total sales needed to give him a salary of $295. **$2,000**

## Practice with Equations and Inequalities for Chapter 5

**Find the solution set of each equation. Then graph the solution set.** *(page 120)* [5.1]

**1.** $5y - 8 = 20 + y$ **{7}**     **2.** $3x - 5 = 7 + 3x$ **φ**     **3.** $9z + 17 = 11z + 3$ **{7}**

**4.** $-16 - 5x = 2x + 5$ **{−3}**     **5.** $9 + 11z = 99 - 7z$ **{5}**     **6.** $2d + 18 = 23 + 2d$ **φ**

**7.** $2x + 8 + 3x = 5x + 8$ **{all nos.}**     **8.** $27 - 6y = -8y + 5y$ **{9}**     **9.** $3(x + 7) = 3x + 14$ **φ**

**10.** $4(t + 5) + 1 = 21 + 4t$ **{all nos.}**     **11.** $8 - (4 - x) = 5 + x$ **φ**     **12.** $9 - (3 - x) = 4x + 15$ **{−3}**

**13.** $8p - (7 - p) = 2p - (6 - 7p) - 1$ **{all numbers}**     **14.** $4(y + 3) - 5y = 8 - (5 - y) - 2y$ **φ**

**15.** $3m - (5 - m) = 6m + 2(m - 4) - 1$ **{1}**     **16.** $5y - 2(3 - y) = 4(1 + y) + 3y - 10$ **{all nos.}**

★ **17.** $10 + (3 - x)2 = 5 - [4 - (x + 3)]$ **{4}**     ★ **18.** $6x - [4 - (8 - x)] = 7 + 5x$ **φ**

**Solve and graph the solution set of each inequality.** *(page 130)* [5.4]

**19.** $2 + 5x > -8$ **x > −2**     **20.** $10x + 14 \geq -36$ **x ≥ −5**     **21.** $7 < -3x - 5$ **x < −4**

**22.** $7d - 6 < -24 + d$ **d < −3**     **23.** $7y + 3 > 10y - 12$ **y < 5**     **24.** $-3r + 13 \leq 2r - 7$ **r ≥ 4**

**25.** $25 - 4m \geq 4m + 1$ **m ≤ 3**     **26.** $8 - 7d \leq d + 8$ **d ≥ 0**     **27.** $-x + 12 > 3x - 4$ **x < 4**

**28.** $\frac{1}{2}x + 5 \leq \frac{1}{6}x + \frac{1}{3}$ **x ≤ −14**     **29.** $\frac{3}{5}x - \frac{1}{10} > \frac{1}{2}x + \frac{1}{5}$ **x > 3**     **30.** $\frac{1}{4}x + \frac{1}{2} < \frac{1}{8}x + \frac{3}{4}$ **x < 2**

**31.** $4(y + 1) < 7y - 2(3 - y)$ **y > 2**     **32.** $5a + 3(a - 7) > -(a + 8)$ **a > $\frac{13}{9}$**

★ **33.** $6 - 4x \geq -[8 - 10(1 - 2x)]$ **x ≥ −$\frac{1}{4}$**     ★ **34.** $7 - [2 - (5 - y)] > 3y + 2(3 + y)$ **y < $\frac{2}{3}$**

**Solve and graph each solution set.** *(page 135)* [5.6]

**35.** $x + 1 > 4 \wedge 3x - 2 < 2x + 8$ **3 < x < 10**     **36.** $2x + 5 \geq 7 \wedge 3x - 4 < 11$ **1 ≤ x < 5**

**37.** $2x + 5 < 9 \vee x + 1 > 3$ **all numbers ≠ 2**     **38.** $3x - 4 > 11 \vee 3x - 2 < 3x + 1$ **all nos.**

**39.** $x - 4 < 2 \vee 2x \geq 10$ **all numbers**     **40.** $3x - 2 \geq 7 \wedge 2x - 1 < 11$ **3 ≤ x < 6**

**41.** $-2 < 4x + 10 \wedge 5x - 4 < 16$ **−3 < x < 4**     **42.** $2x + 2 < 8 \wedge -3x - 6 \geq 9$ **x ≤ −5**

**43.** $5x - 2 > 3x + 4 \wedge -2x + 5 < 3 - x$ **x > 3**     **44.** $4x + 8 > 12 \wedge 2x + 5 > 3x + 7$ **φ**

**45.** $-7 < 3x + 2 < 20$ **−3 < x < 6**     **46.** $6 \geq -x + 4 > 4$ **−2 ≤ x < 0**

★ **47.** $2y - 4 < 2y + 6 < 2y + 10$ **all numbers**     ★ **48.** $3x + 5 < 3x + 1 < 3x - 8$ **φ**

**Solve each equation, if possible.** *(page 137)* [5.7]

**49.** $|t - 4| = 6$ **10, −2**    **50.** $|4 - m| = 8$ **−4, 12**    **51.** $|a| = -6$ **φ**     **52.** $|2 - y| = 7$ **−5, 9**

**53.** $|2x + 6| = 10$ **2, −8**   **54.** $|7 - 3k| = -1$ **φ**    **55.** $|7 - k| = 8$ **−1, 15**   **56.** $|y - 7| = 1$ **8, 6**

**57.** $3|x + 4| + 2 = 11$ **−1, −7**    **58.** $4 - 2|2p - 6| = -6$ **$\frac{11}{2}, \frac{1}{2}$**    **59.** $8 - 2|2 - a| = -2$ **−3, 7**

**60.** $-5 - |4 - 3b| = 7$ **φ**     **61.** $3 - 2|10 - 5x| = -7$ **1, 3**    **62.** $3|4 - t| + 8 = 20$ **0, 8**

★ **63.** $|2y - 6| = |y - 3|$ **3**     ★ **64.** $|2x - 6| = |4x + 8|$     ★ **65.** $4 - |8 - 2m| = m - 6$ **−2, 6**
                                                                 **−7, −$\frac{1}{3}$**

**Solve each inequality. Graph the solution set.**    **70.** $m < -1 \vee m > 3$     *(page 140)* [5.8]

**66.** $|2a - 4| < 8$ **−2 < a < 6**     **67.** $|3x - 6| \geq 9$ **x ≥ 5 ∨ x ≤ −1** **68.** $|5m - 20| < 10$ **2 < m < 6**

**69.** $|3x - 9| \leq 15$ **−2 ≤ x ≤ 8**     **70.** $|4 - 4m| > 8$     **71.** $2|8 - x| < 4$ **6 < x < 10**

**72.** $|y + 3| + 6 < 7$ **−4 < y < −2**     **73.** $|x - 8| - 4 \geq 5$ **x ≥ 17 ∨ x ≤ −1**

**74.** $7 + 3|2x - 4| < 13$ **1 < x < 3**     **75.** $4|2m - 6| + 8 > 12$ **m > 3$\frac{1}{2}$ or m < 2$\frac{1}{2}$**

★ **76.** $|p - 6| < p - 8$ **φ**     ★ **77.** $|x - 8| > x - 4$ **x < 6**

★ **78.** $|2x - 2| < 4$ and $|x| > 1$ **1 < x < 3**     ★ **79.** $|x| < 2$ or $|x| > 2$ **all nos. except −2 and 2**

**SELECTED EXTRA PRACTICE**

# Problem Solving for Chapter 5

## Inequalities in Word Problems

*(page 142)* [5.9]

1. A store makes a profit of at least $15 on each watch sold. How many watches must be sold to make a profit of at least $180? **at least 12**

2. The average cost for a used-car dealer for minor repairs under a 90-day warranty is $40. How many can occur in one month if he is to lose at most $360? **at most 9**

3. Jacques is paid $180 a week plus a commission of $15 on each stereo he sells. How many must he sell to make at least $330? **at least 10**

4. Rosa is paid $75 a week plus a commission of $12 on each suit she sells. How many must she sell to earn at least $315 a week? **at least 20**

5. Lois rents a car for $85 a week plus $0.15/km. How far can she travel if she wants to spend at most $160? **at most 500 km**

6. Tina rents a car for $74 a week plus $0.14/km. How far can she drive if she wishes to spend at most $130? **at most 400 km**

7. Rhonda earns $11,000 a year in salary and 6% commission on her sales. How much were her sales if her annual income was more than $13,400 and at most $14,000? **more than $40,000 and at most $50,000**

## Mixed Practice

8. 7 less than a number divided by 3 is 9. Find the number. **48**

9. Separate 32 into two parts so that the larger part is three times the smaller part. Find both parts. **8, 24**

10. The selling price of a suit is $40 more than twice the cost. Find the cost if the selling price is $320. **$140**

11. Four less than 6 times Mary's age is the same as 36 more than twice her age. How old is she? **10 y**

12. Sergio is paid $60 a week plus a 5% commission on all sales. Find the sales needed to give him a total salary of $260 for the week. **$4,000**

13. The length of a rectangle is 7 km more than 6 times the width. The perimeter is 126 km. Find the length and the width. **55 km, 8 km**

14. The sum of the ages of three sisters is 38. Jill's age is 4 less than Paula's age. Jane's age is 10 more than Jill's age. Find the age of each. **Paula 12 y, Jill 8 y, Jane 18 y**

15. Four times the sum of a number plus 6 is the same as 44 more than twice the number. Find the number. **10**

16. Ten more than three times a number is at most 34. Find the numbers. $n \le 8$

17. Six more than five times a number is at least 41. Find the numbers. $n \ge 7$

18. Fred's weight is 40 lbs more than 3 times Bob's weight. Together their weights total 240 lb. Find the weight of each. **Fred 190 lb, Bob 50 lb**

19. Donna rents a car for a week at the rate of $95 plus $0.15/km. How far can she go on a car rental budget of $230? **900 km**

20. A team won 12 games and lost 3 games. The number of games won is what percent of the games played? **80%**

21. The selling price of a stereo is $270. The profit is 35% of the cost. Find the cost. **$200**

22. The first side of a triangle is 4 cm longer than twice the second. The third side is 2 cm shorter than the second side. The perimeter is 30 cm. Find the length of each side. **18 cm, 7 cm, 5 cm**

23. The second of three numbers is one more than twice the first. The third is 10 less than the second. The sum of the three numbers is 27. Find each. **7, 15, 5**

# Practice on Simplifying and Factoring Polynomials for Chapter 6

**Simplify.**  (page 149)  [6.1]
1. $m^5 \cdot m^7$  $m^{12}$
2. $b^8 \cdot b^3$  $b^{11}$
3. $x^{11} \cdot y^3$  $x^{11}y^3$
4. $p \cdot p^8$  $p^9$
5. $(7a^3)(6a^5)$  $42a^8$
6. $(5n^3)(4n^2)$  $20n^5$
7. $(-3a)(7a^3)$  $-21a^4$
8. $(m^{10})(3m^2)$  $3m^{12}$
9. $(5a^4b^2)(3ab)$  $15a^5b^3$
10. $(6x^3y^5)(-4xy^2)$  $-24x^4y^7$
11. $(-4m^3n)(7mn^2)$  $-28m^4n^3$
12. $(-xy^3)(-2x^2y)$  $2x^3y^4$
13. $(-b^2m)(-2bm^4)$  $2b^3m^5$
14. $(ab)(-ab)$  $-a^2b^2$
15. $(-p^3)(5p)(-2p^2)$  $10p^6$
16. $(4r^2)(-r)(-2r^3)$  $8r^6$
17. $(5a^2)(-3a)(2a)$  $-30a^4$

**Find the value of $a$ that makes each sentence true.**
★ 18. $x^{3a-1} \cdot x^{2a+4} = x^{18}$  $a = 3$
★ 19. $p^{2a-3} \cdot p^{a+1} \cdot p^{3a} = p^{16}$  $a = 3$

**Simplify.**  (page 151)  [6.2]
20. $(x^4)^3$  $x^{12}$
21. $(m^2)^8$  $m^{16}$
22. $(p^3)^3$  $p^9$
23. $(x^4)^8$  $x^{32}$
24. $(g^2)^7$  $g^{14}$
25. $(3x^3)^2$  $9x^6$
26. $(-2a^7)^3$  $-8a^{21}$
27. $(-5n)^3$  $-125n^3$
28. $(-b^8)^5$  $-b^{40}$
29. $(8r^2)^3$  $512r^6$
30. $(-4a^2b^3c)^3$  $-64a^6b^9c^3$
31. $(-2x^5yz^2)^5$  $-32x^{25}y^5z^{10}$
32. $(5a^2b^4c)^2$  $25a^4b^8c^2$
33. $m(4m^2)^3$  $64m^7$
34. $-3m(2m)^2$  $-12m^3$
35. $8ab(-2ab^2)^3$  $-64a^4b^7$

**Find the value of $a$ that will make each sentence true.**
★ 36. $(x^{3a+1})^2 = x^{14}$  2
★ 37. $(y^{2a+1})^2 \cdot (y^a)^5 = y^{6a+11}$  3
39. $5y^3 - 4y^2 - y$  41. $-2m^5 - 3m^3 - m^2 + 5$  43. $-a^2b + 2ab^2 - 5ab$

**Simplify. Write the result in descending order of exponents of $x$, $y$, $a$, or $m$.**  (page 154)  [6.3]
38. $6x^2 - 5x + 7x^2 + 4x - 2$  $13x^2 - x - 2$
39. $-5y + 4y^3 + 4y - 3y^2 + y^3 - y^2$
40. $3a - 4a^4 + a^3 - 4a + 3a^4 - a^3$  $-a^4 - a$
41. $7 - m^5 - 4m^3 - m^5 + m^3 - 2 - m^2$
42. $4x^3y^2 - xy - x^3y^2 + 2xy$  $3x^3y^2 + xy$
43. $a^2b - ab^2 - 2a^2b + 3ab^2 - 5ab$
44. $x^2 + 2x^4 - 3x^3 - 5x^4 - x^3 + x^2 - x$
45. $-m^2 - 5mn^2 - m^2 + m^3 - mn^2 - 3mn$
44. $-3x^4 - 4x^3 + 2x^2 - x$
45. $m^3 - 2m^2 - 3mn - 6mn^2$

**Simplify.**  51. $7m^4 - 28m^3 - 14m^2 + 35m$  53. $-3g^4 + 4g^3 + 8g^2 - 5g$  (page 158)  [6.5]
46. $a^2 + 6a - 3 - 2(a^2 - 3a + 1)$  $-a^2 + 12a - 5$
47. $y^2 - y - 4 - (-y^2 - 2y + 8)$  $2y^2 + y - 12$
48. $p^2 - 3p - 4 - 2(p^2 - 7p + 1)$  $-p^2 + 11p - 6$
49. $y^2 - 8y - 8 - 3(y^2 - 2y + 1)$  $-2y^2 - 2y - 11$
50. $-2a(3a^3 - 4a^2 + 5a)$  $-6a^4 + 8a^3 - 10a^2$
51. $7m(m^3 - 4m^2 - 2m + 5)$
52. $x(-x^3 - x^2 + 7x - 1)$  $-x^4 - x^3 + 7x^2 - x$
53. $-g(3g^3 - 4g^2 - 8g + 5)$
54. $-2ab^3(-a^3 - 4a^2b + 5ab^2 - b^3)$
55. $-xy(x^3 - x^2y - xy^2 + xy^3)$
★ 56. $m^2n^3(4m^2 + 3mn - 5n^2) - m^2n^3(m^2 - mn + 6n^2)$  $3m^4n^3 + 4m^3n^4 - 11m^2n^5$
54. $+2a^4b^3 + 8a^3b^4 - 10a^2b^5 + 2ab^6$  55. $-x^4y + x^3y^2 + x^2y^3 - x^2y^4$  57. $3(a^2 + 7a + 5)$

**Factor out the GCF, if any, from each polynomial.**  64. $m(m^2 - m - 7)$  (page 163)  [6.7]
57. $3a^2 + 21a + 15$
58. $4a^2 - 8a + 16$  $4(a^2 - 2a + 4)$
59. $7x^2 - 35x + 42$  $7(x^2 - 5x + 6)$
60. $6t^2 - 30t - 18$  $6(t^2 - 5t - 3)$
61. $5x^2 - 10x + 2$  **none**
62. $8y^2 + 2y - 4$  $2(4y^2 + y - 2)$
63. $t^3 - 5t^2 + 4t$  $t(t^2 - 5t + 4)$
64. $m^3 - m^2 - 7m$
65. $a^5 - 4a^4 - 8a^3$  $a^3(a^2 - 4a - 8)$
66. $5b^3 - 25b^2 - 20b$
67. $3x^3 - 12x^2 - 30x$
68. $8a^3 - 12a^2 - 24a$
66. $5b(b^2 - 5b - 4)$  67. $3x(x^2 - 4x - 10)$  68. $4a(2a^2 - 3a - 6)$  70. $5m^2(m^2 - 7m - 4)$
69. $3b^3 - 9b^2$  $3b^2(b - 3)$
70. $5m^4 - 35m^3 - 20m^2$
71. $2x^3 - x^2$  $x^2(2x - 1)$
72. $7a^4 - 35a^3 - 21a^2 + 49a$
73. $3x^4 - 39x^3 - 33x^2 - 12x$
74. $6x^4 - 54x^3 + 150x^2$  $6x^2(x^2 - 9x + 25)$
75. $45x^3 - 15x^2 - 105x$  $15x(3x^2 - x - 7)$
★ 76. $x^{4a+3} + x^{2a+5}$
★ 77. $4y^{3m+2} + 28y^{6m+7}$
★ 78. $b + b^{a+6}$  $b(1 + b^{a+5})$
76. $x^{2a+3}(x^{2a} + x^2)$
77. $4y^{3m+2}(1 + 7y^{3m+5})$
72. $7a(a^3 - 5a^2 - 3a + 7)$  73. $3x(x^3 - 13x^2 - 11x - 4)$

SELECTED EXTRA PRACTICE

## Practice on Factoring Trinomials for Chapter 7

1. $6x^2 + 35x + 49$  2. $8a^2 + 10a - 7$  5. $14a^2 + 19a - 3$  6. $4y^2 + 16my - 9m^2$

**Multiply.**
8. $25a^2 + 5a - 12$
*(page 173)* [7.1]

1. $(3x + 7)(2x + 7)$
2. $(2a - 1)(4a + 7)$
3. $(2r - 3)(2r + 3)$  $4r^2 - 9$
4. $(3x - 1)(x + 2)$  $3x^2 + 5x - 2$
5. $(7a - 1)(2a + 3)$
6. $(2y - m)(2y + 9m)$
7. $(7x - 2)(x + 3)$  $7x^2 + 19x - 6$
8. $(5a - 3)(5a + 4)$
9. $(2b + 5)(b - 1)$  $2b^2 + 3b - 5$
10. $(2x + 7)(4x^2 - 3x + 2)$
    $8x^3 + 22x^2 - 17x + 14$
11. $(3x + 2)(x^3 - 7x^2 + 3x)$
    $3x^4 - 19x^3 - 5x^2 + 6x$
12. $(2x + 3)(3x^2 + 2x - 5)$
    $6x^3 + 13x^2 - 4x - 15$

**Multiply.**
*(page 174)* [7.2]

13. $(3y - 1)(3y + 1)$
    $9y^2 - 1$
14. $(5u - 6)(5u + 6)$
    $25u^2 - 36$
15. $(6a + 7b)(6a - 7b)$
    $36a^2 - 49b^2$

**Factor.**

16. $p^2 - 144$  $(p - 12)(p + 12)$
17. $36 - m^2$  $(6 - m)(6 + m)$
18. $49t^2 - 64p^2$  $(7t - 8p)(7t + 8p)$

19. $121b^2 - 4y^2$
    $(11b + 2y)(11b - 2y)$
20. $\frac{4}{9}x^2 - \frac{1}{25}$
    $\left(\frac{2}{3}x - \frac{1}{5}\right)\left(\frac{2}{3}x + \frac{1}{5}\right)$
★ 21. $(3x - 2)^2 - (2x + 5)^2$
    $(x - 7)(5x + 3)$

**Simplify.**
*(page 176)* [7.3]

22. $(5a - 1)^2$
    $25a^2 - 10a + 1$
23. $(b - 4c)^2$
    $b^2 - 8bc + 16c^2$
24. $(3k + 2p)^2$
    $9k^2 + 12kp + 4p^2$

**Determine if each is a perfect square. If so, factor the trinomial square.**

25. $a^2 - 14a + 49$  $(a - 7)^2$
26. $9m^2 + 16m + 1$  **no**
27. $4a^2 - 12a + 9$  $(2a - 3)^2$
28. $25x^2 - 40xy + 4y^2$  **no**
29. $49a^2 + 42ab + 9b^2$
    $(7a + 3b)^2$
30. $64a^2 - 48ab + 9b^2$
    $(8a - 3b)^2$

**Factor.**
*(page 179)* [7.4]

31. $g^2 + 7g + 12$  $(g + 4)(g + 3)$
32. $x^2 - 12x + 35$  $(x - 7)(x - 5)$
33. $x^2 + 2x - 24$  $(x + 6)(x - 4)$
34. $b^2 - 20b - 44$
    $(b - 22)(b + 2)$
★ 35. $x^{2m} - 24x^m + 128$
    $(x^m - 8)(x^m - 16)$
★ 36. $x^{4b} - 23x^{2b} - 140$
    $(x^{2b} - 28)(x^{2b} + 5)$

*(page 182)* [7.5]

37. $2x^2 + 7x + 5$
38. $2a^2 - 7a - 4$
39. $2a^2 + 7a + 6$  $(2a + 3)(a + 2)$
40. $3y^2 - 10y - 25$
    $(3y + 5)(y - 5)$
41. $3b^2 + 23b + 30$
    $(3b + 5)(b + 6)$
42. $6a^2 + a - 1$
    $(3a - 1)(2a + 1)$

*(page 183)* [7.5]

43. $2a^2 - 4a - 96$
44. $2b^2 - 6b - 80$
45. $3k^2 + 3k - 18$
46. $2r^3 - 10r^2 - 28r$
    $2r(r - 7)(r + 2)$
47. $4m^3 - 36m$
    $4m(m - 3)(m + 3)$
48. $6k^3 - 2k^2 - 20k$

*(page 185)* [7.6]

49. $a^2 + 4ab + 3b^2$
50. $6w^2 - 26wy - 20y^2$
51. $16a^2 - 25y^2$
52. $21m^2 + 29mn - 10n^2$
53. $2r^3 - 2r^2t - 84rt^2$
54. $4a^3 + 6a^2b - 40ab^2$
55. $6x^3y^3 - 150xy^5$
    $6xy^3(x - 5y)(x + 5y)$
56. $2a^4 - 68a^2b^2 + 450b^4$
    $2(a - 5b)(a + 5b)(a - 3b)(a + 3b)$
★ 57. $2p^5 - 244p^3q^2 + 242q^4p$

*(page 187)* [7.7]

58. $by + 5b + y^2 + 5y$
59. $tw + uw + ty + uy$
60. $ab + 3b + ac + 3c$
61. $xt - 9t + xu - 9u$
    $(x - 9)(t + u)$
62. $mh - 7h + mg - 7g$
    $(m - 7)(h + g)$
63. $x^2a - 9a - x^2b + 9b$
    $(a - b)(x - 3)(x + 3)$

**Solve each equation.**
*(page 190)* [7.8]

64. $x^2 - 14x + 48 = 0$  **8, 6**
65. $2a^2 + 9a - 5 = 0$  $\frac{1}{2}, -5$
66. $2p^2 - 5p - 3 = 0$  $-\frac{1}{2}, 3$
67. $2x^2 - 9x - 56 = 0$  $-\frac{7}{2}, 8$
68. $x^3 - 13x^2 + 22x = 0$
    **0, 11, 2**
★ 69. $x^5 - 34x^3 + 225x = 0$

*(page 194)* [7.9]

70. $5z = -z^2$  **0, −5**
71. $5r = r^2 - 50$  **10, −5**
72. $2p^2 = -3p + 20$  $\frac{5}{2}, -4$
73. $40 - 11x = 2x^2$  $\frac{5}{2}, -8$
74. $7a^2 - 20 = -23a$  $\frac{5}{7}, -4$
75. $5x - 50 = -6x^2$  $-\frac{10}{3}, \frac{5}{2}$
★ 76. $(3x - 4)(x + 7) = 2x^2 + 24x - 40$  **4, 3**
★ 77. $3y^2 + 32y - 25 = (4y - 5)(y + 7)$  **10, −1**

**For Ex. 37, 38, 43–45, 48–54, 57–60, 69, see TE Answer Section.**

# Problem Solving for Chapter 7

### Word Problems and Quadratic Equations

*(page 194)* [7.9]

1. The square of a number is 30 less than 11 times the number. Find the number. **5, 6**

2. The square of a number is the same as 5 times the number. Find the number. **5, 0**

3. If twice a number is subtracted from the square of a number, the result is 48. Find the number. **8, −6**

4. If the square of a number is increased by 42, the result is the same as 13 times the number. Find the number. **6, 7**

★ 5. The square of 2 more than a number is the same as 10 more than 9 times the number. Find the number. **6, −1**

### Consecutive Integer Problems

*(page 198)* [7.10]

6. Find two consecutive integers whose sum is 37. **18, 19**

7. Find three consecutive odd integers whose sum is 63. **19, 21, 23**

8. Find two consecutive odd integers such that the sum of their squares is 130. **7, 9 or −7, −9**

9. Find three consecutive integers such that the square of the first is 18 more than the last. **5, 6, 7 or −4, −3, −2**

★ 10. Find three consecutive integers such that the product of the second and the third is 2 less than 8 times the first. **4, 5, 6 or 1, 2, 3**

### Mixed Practice

11. Find three consecutive integers whose sum is 27. **8, 9, 10**

12. A number is the same as its square. Find the number. **0,1**

13. Each leg of an isosceles triangle is 2 m longer than the base. The perimeter is 19 m. Find the length of the legs and the base. **7, 7, 5**

14. The square of a number is 12 more than the number. Find the number. **4, −3**

15. Eight less than three times a number is at most 37. Find the number. **x ≤ 15**

16. José rents a car for $90 plus $0.25/km. How far can he go on a budget of $140? **200 km**

17. Separate 40 into two parts so that the larger part is 7 times the smaller part. Find each part. **5, 35**

18. Find two consecutive odd integers whose product is 143. **11, 13 or −11, −13**

19. Jim is paid $160 a week plus 4% commission on all sales. Find the sales needed to give him a total salary of $200. **$1,000**

20. The length of a rectangle is 18 m more than 3 times the width. The perimeter is 92 m. Find the length and the width. **39 m, 7 m**

21. A graduating class of 170 this year is only 85% of the size of last year's class. How many were graduated last year? **200**

22. Find two consecutive even integers such that the sum of their squares is 100. **6 and 8 or −6 and −8**

**SELECTED EXTRA PRACTICE**

# Practice on Simplifying Rational Expressions for Chapter 8

**Simplify.**
1. $\dfrac{a-5}{2a-1}$
2. $\dfrac{m+2}{2m+1}$
3. $\dfrac{b+5}{b-2}$
4. $\dfrac{x+4}{4}$
10. $\dfrac{(x-3)(x+1)}{1}$  *(page 211)*  [8.2]

1. $\dfrac{a^2-8a+15}{2a^2-7a+3}$
2. $\dfrac{m^2-3m-10}{2m^2-9m-5}$
3. $\dfrac{b^2-25}{b^2-7b+10}$
4. $\dfrac{2x^2+3x-20}{8x-20}$

5. $\dfrac{3x^2+16x-35}{2x^2+17x+21}$  $\dfrac{3x-5}{2x+3}$
6. $\dfrac{5b^2-19b+12}{10b^2+27b-28}$  $\dfrac{b-3}{2b+7}$
7. $\dfrac{2b^4-b^3-21b^2}{3b^4+4b^3-15b^2}$  $\dfrac{2b-7}{3b-5}$

8. $\dfrac{x^2y+x^2-9y-9}{x^2-9}$  $y+1$
9. $\dfrac{bc-5c+bd-5d}{b^2-7b+10}$  $\dfrac{c+d}{b-2}$
★10. $\dfrac{5x^5-50x^3+45x}{5x^3+10x^2-15x}$

12. $-\dfrac{2}{b-1}$  13. $-(p+5)$  14. $\dfrac{x+7}{x+3}$  15. $-\dfrac{t-9}{t-8}$  16. $-\dfrac{x}{x-8}$  *(page 214)*  [8.3]

11. $\dfrac{x-4}{16-x^2}$  $-\dfrac{1}{x+4}$
12. $\dfrac{8-2b}{b^2-5b+4}$
13. $\dfrac{p^2+p-20}{4-p}$
14. $\dfrac{49-x^2}{-x^2+4x+21}$

15. $\dfrac{-t^2+4t+45}{t^2-3t-40}$
16. $\dfrac{2x^2+12x}{96+4x-2x^2}$
17. $\dfrac{x^2+2x-35}{-x^2-x+42}$
18. $\dfrac{-x^2+y^2}{2x^2+3xy-5y^2}$

19. $\dfrac{-16a^2+36a+10}{4a^2-11a-3}$
★20. $\dfrac{-x^4+13x^2-36}{x^2+x-6}$
★21. $\dfrac{a^2b-49b+a^2c-49c}{-ba-ac+7b+7c}$

17. $-\dfrac{x-5}{x-6}$  18. $-\dfrac{x+y}{2x+5y}$  19. $\dfrac{-2(2a-5)}{a-3}$  20. $-1(x-3)(x+2)$  *(page 217)*  [8.4]

22. $\dfrac{a^8}{a^7}$  $a$
23. $\dfrac{r^{10}}{r^{11}}$  $\dfrac{1}{r}$
24. $\dfrac{a^4b^6}{a^3b^9}$  $\dfrac{a}{b^3}$
25. $\dfrac{6x^2y^5}{20xy^7}$  $\dfrac{3x}{10y^2}$
26. $\dfrac{8t^{10}w^7}{20t^9w^4}$  $\dfrac{2tw^3}{5}$

27. $\dfrac{m^5(a^2-5a+6)}{m^9(4a-12)}$  $\dfrac{a-2}{4m^4}$
28. $\dfrac{x^2y^6(10x^2+13x-3)}{x^3y^5(9-4x^2)}$
29. $\dfrac{a^4b^2(64-x^2)}{a^3b^5(ax+8a)}$  $-\dfrac{x-8}{b^3}$

30. $\dfrac{b^3y^2-8b^3y-48b^3}{b^2y^2-11b^2y-12b^2}$  $\dfrac{b(y+4)}{y+1}$
★31. $\dfrac{cx-4cy+dx-4dy}{x^3y-x^2y^2-12xy^3}$  $\dfrac{c+d}{xy(x+3y)}$

21. $\dfrac{a+7}{-1}$  28. $\dfrac{y(5x-1)}{-x(2x-3)}$  32. $2(a+2)$  *(page 220)*  [8.5]

32. $\dfrac{a^2+7a+10}{x+9} \cdot \dfrac{4x+36}{2a+10}$
33. $\dfrac{2b+14}{4b^2+26b-14} \cdot 2b-1$  $1$
34. $\dfrac{5a^4b^3}{a^2-7a} \cdot \dfrac{7-a}{15a^3b^4}$  $-\dfrac{1}{3b}$

35. $\dfrac{5-x}{x^2+7x+12} \cdot \dfrac{x^2+8x+16}{x^2-x-20}$  $-\dfrac{1}{x+3}$
36. $\dfrac{4m^2-n^2}{2m^2+3m-35} \cdot \dfrac{m+5}{n-2m}$  $-\dfrac{2m+n}{2m-7}$

37. $\dfrac{x-2}{3(x+6)}$  38. $\dfrac{2(x-1)}{5x^6}$  *(page 223)*  [8.6]

37. $\dfrac{x^2-6x+8}{x^2-36} \div \dfrac{3x-12}{x-6}$
38. $\dfrac{x^2-5x+4}{5x^8} \div \dfrac{3x^2-12x}{6x^3}$
39. $\dfrac{3-m}{m-6} \div \dfrac{m^2+2m-15}{m^2-m-30}$  $-1$

40. $\dfrac{a^2-5a+6}{a^2-3a+2} \cdot \dfrac{a^2-1}{2a^2-a-3} \div \dfrac{a^2+a-12}{2a^2+5a-12}$  $1$

**Solve each proportion.**  *(page 227)*  [8.7]

41. $\dfrac{3}{7}=\dfrac{x}{4}$  $\dfrac{12}{7}$
42. $\dfrac{x-2}{5}=\dfrac{x+6}{2}-\dfrac{34}{3}$
43. $\dfrac{7}{x}=\dfrac{4}{2x-3}$  $\dfrac{21}{10}$
44. $\dfrac{x}{7}=\dfrac{4}{x+12}$  $-14, 2$

45. $\dfrac{y}{6}=\dfrac{4}{y-10}$
46. $\dfrac{10}{x}=\dfrac{x+16}{8}$
47. $\dfrac{x-5}{6}=\dfrac{3}{x-8}$
48. $\dfrac{7}{x+6}=\dfrac{x-12}{-8}$

$12, -2$  $-20, 4$  $11, 2$  $8, -2$

# Problem Solving for Chapter 8

## Ratio and Proportion

*(page 227)*  [8.7]

1. In Centerville 3 out of 5 people belong to a union. How many union members are there if the population is 70,000? **42,000**

2. Pam's batting average is 0.385 (385:1,000). How many hits should Pam get in 6,000 times at bat? **2,310**

3. One out of four people earn less than $5,000 per year. How many people in a city of 40,000 earn less than $5,000 per year? **10,000**

4. Four out of 5 freshmen study algebra. How many study algebra in a freshman class of 400? **320**

5. Two out of 5 people shave with Dull Blades. How many use this brand in a town of 60,000? **24,000**

6. Five out of 7 people use Nodor Deodorant. How many people in a city of 42,000 do NOT use this brand? **12,000**

## Applications: Ratio and Proportion in Travel

*(page 228)*

7. A sales representative's car averages 40 mi/gal. Find the cost of gas for a 240-mile trip if gas sells at $2.25/gal. **$13.50**

8. Tanya's motor bike averages 50 mi/gal. Find the cost of the gas for a 6,000 mi cross-country trip if gas costs $2.15/gal. **$258**

9. Ali budgets $65 for gas for a vacation. His car averages 45 mi/gal. If gas costs $2.29/gal, is a 1,400 mile trip within his gas budget? **no**

10. An insurance agent travels 20,000 mi in a year. Her car averages 49 mi/gal. Find the agent's yearly gas expense if gas sells at $2.23/gal. **$910.20**

11. Carl commutes 475 miles a week to work. His car averages 39 mi/gal. If gas sells at $2.17/gal, how much does he spend on gas in 4 weeks? **$105.72**

## Mixed Practice

12. A store manager lists the selling price of a TV at $540. The profit is 35% of the cost. Find the cost of the TV. **$400**

13. Louella is paid $145 a week plus a commission of 5% on all sales. Find her sales for the week if she is to earn a total of $270. **$2,500**

14. A basketball team won 14 games and lost 6 games. What percent of the games played did the team win? **70%**

15. Doreen budgeted $155 for renting a car. How far can she travel if the rental rate is $85 plus $0.14/km? **500 km**

16. If 2 out of 7 people buy Frizz Hair Tonic, how many people use this product in a city with a population of 35,000? **10,000**

17. A car that costs the dealer $8,000 is sold for a 9% profit. What is the selling price? **$8,720**

18. The length of a rectangle is 3 more than twice the width. The perimeter is 54. Find the length and width of the rectangle. **19,8**

19. The second of two consecutive odd integers is 16 less than three times the first. The sum of the three integers is 33. Find the three integers. **9, 11, 13**

20. A gain of 6 lb would put a welterweight boxer out of his class. To be in the welterweight class, one's weight must be more than 135 lb but at most 145 lb. How much could the boxer weigh? **no more than 139 lb**

SELECTED EXTRA PRACTICE

# Practice on Combining Rational Expressions for Chapter 9

**Add. Simplify, if possible.**

*(page 241)* [9.2]

1. $\dfrac{5x}{8} + \dfrac{3x}{2} + \dfrac{x}{4}$ **$\dfrac{19x}{8}$**

2. $\dfrac{3m}{7} + \dfrac{m}{14} + \dfrac{5m}{2}$ **$3m$**

3. $\dfrac{3x-5}{6} + \dfrac{2x+1}{4}$ **$\dfrac{12x-7}{12}$**

4. $\dfrac{2x+1}{3} + \dfrac{x}{15} + \dfrac{3x-1}{5}$ **$\dfrac{20x+2}{15}$**

5. $\dfrac{3x+5}{12} + \dfrac{5x-1}{4} + \dfrac{5x}{3}$ **$\dfrac{19x+1}{6}$**

6. $\dfrac{4y-1}{5} + \dfrac{y}{10} + \dfrac{3y-2}{6}$ **$\dfrac{21y-8}{15}$**

★7. $\left(\dfrac{2a}{3} + \dfrac{1}{6}\right) \div \dfrac{16a^2-1}{12}$ **$\dfrac{2}{4a-1}$**

*(page 244)* [9.3]

8. $\dfrac{5}{a-2} + \dfrac{12}{a^2-4} + \dfrac{3}{a+2}$ **$\dfrac{8}{a-2}$**

9. $\dfrac{5}{m-7} + \dfrac{-35}{m^2-7m}$ **$\dfrac{5}{m}$**

10. $\dfrac{k^2-7k}{k^2-9} + \dfrac{2}{k-3} + \dfrac{1}{k+3}$ **$\dfrac{k-1}{k+3}$**

11. $\dfrac{-6x-38}{x^2-4x-21} + \dfrac{x+1}{x-7}$ **$\dfrac{x+5}{x+3}$**

12. $\dfrac{3x-2}{x^2-5x} + \dfrac{4}{x^2-7x+10}$

13. $\dfrac{6k-22}{k^2-5k+6} + \dfrac{k+1}{k-3}$ **$\dfrac{k+8}{k-2}$**

*(page 247)* [9.4]

14. $\dfrac{5}{a} + 6$ **$\dfrac{6a+5}{a}$**

15. $9 + \dfrac{4}{m}$ **$\dfrac{9m+4}{m}$**

16. $\dfrac{7}{a-3} + \dfrac{4}{a+6}$

17. $\dfrac{3}{b-4} + \dfrac{2}{b+5}$

18. $x - 5 + \dfrac{6}{x-2}$

19. $\dfrac{5}{3x} + \dfrac{4+3x}{15x^3} + \dfrac{4-x}{5x^2}$

20. $\dfrac{x+5}{x-4} + 2x - 1$

21. $\dfrac{7}{2a} + \dfrac{2-a}{3a^2} + \dfrac{3+5a}{6a^3}$

22. $x - 5 + \dfrac{7}{3x+2}$

★23. $\dfrac{4}{a-2} + \dfrac{3}{a-1} + \dfrac{2}{a+1}$

*(page 250)* [9.5]

24. $\dfrac{6}{x-5} - \dfrac{x-3}{x^2-25}$

25. $\dfrac{5}{3b-2} - \dfrac{4}{2b+5}$

26. $\dfrac{7}{12y^3} - \dfrac{3}{8y^2}$ **$\dfrac{14-9y}{24y^3}$**

27. $\dfrac{2}{9m^2} - \dfrac{5}{3m}$ **$\dfrac{2-15m}{9m^2}$**

28. $\dfrac{5}{k^2-5k-14} - \dfrac{2}{k-8}$

29. $\dfrac{5}{k^2-5k-14} - \dfrac{k-1}{k-7}$

★30. $\dfrac{4}{xy-3x+ay-3a} - \dfrac{a+1}{y^2+y-12}$

★31. $\dfrac{3}{xy-3xb+2y^2-6yb} - \dfrac{7}{y^2-9b^2}$

*(page 252)* [9.6]

32. $\dfrac{4a}{a^2-9} + \dfrac{2}{3-a}$ **$\dfrac{2}{a+3}$**

33. $\dfrac{5y+2}{y^2-4y-12} + \dfrac{4}{6-y}$ **$\dfrac{1}{y+2}$**

34. $\dfrac{-17b+18}{b^2+2b-48} - \dfrac{b}{6-b}$ **$\dfrac{b-3}{b+8}$**

35. $\dfrac{4}{m-8} + \dfrac{3}{4-m}$

36. $\dfrac{3}{t} - \dfrac{7-t}{2t-t^2}$ **$\dfrac{2t+1}{t(t-2)}$**

37. $\dfrac{9y-22}{y^2-2y} - \dfrac{y}{2-y}$ **$\dfrac{y+11}{y}$**

38. $\dfrac{5y-36}{y^2-4y} - \dfrac{y}{4-y}$

39. $\dfrac{8}{6-m} + \dfrac{2}{m-2}$

40. $\dfrac{5x+15}{x^2-25} + \dfrac{4}{5-x}$

**Divide.**

*(page 254)* [9.7]

41. $a^{13} \div a^{10}$ **$a^3$**

42. $15x^3 \div 5x$ **$3x^2$**

43. $35a^4 \div 7a^2$ **$5a^2$**

44. $(6a^3 - 4a^2 + 8a) \div 2a$ **$3a^2 - 2a + 4$**

45. $(10n^3 - 15n^2 - 35n) \div 5n$ **$2n^2 - 3n - 7$**

46. $(6a^5 - 8a^4 - 2a^3 + 10a^2 - 4a) \div 2a$
  **$3a^4 - 4a^3 - a^2 + 5a - 2$**

★47. $(x^{5a} - x^{3a} + x^a) \div x^a$ **$x^{4a} - x^{2a} + 1$**

*(page 257)* [9.8]

48. $(x^2 - 14x + 45) \div (x - 9)$ **$x - 5$**

49. $(20k^2 + 27k - 14) \div (5k - 2)$ **$4k + 7$**

50. $(2p^3 + 7p^2 - 20p + 8) \div (2p + 1)$ **$p^2 + 4p - 8$**

51. $(2x^3 - 5x^2 + x - 4) \div (x - 3)$
  **$2x^2 + x + 4$ R8**

**For Ex. 12, 16–25, 28–31, 35, 38–40, see TE Answer Section.**

SELECTED EXTRA PRACTICE

# Practice with Rational Expressions for Chapter 10

**Solve.**

*(page 267)* [10.1]

**1.** $\dfrac{2x + 12}{x^2 - 10x + 16} = \dfrac{3}{x - 8} + \dfrac{4}{x - 2}$  **10**

**2.** $\dfrac{8}{m^2 - 6m} - \dfrac{3}{m - 6} = \dfrac{2}{m}$  **4**

**3.** $\dfrac{3}{x^2 - 4} + \dfrac{5}{x - 2} = \dfrac{7}{x + 2}$  $\dfrac{27}{2}$

**4.** $\dfrac{8}{y^2 - 7y + 12} = \dfrac{5}{y - 3} + \dfrac{2}{y - 4}$  $\dfrac{34}{7}$

**5.** $\dfrac{2}{b^2 - 5b - 14} = \dfrac{3}{b - 7} - \dfrac{4}{b + 2}$  **32**

**6.** $\dfrac{4}{a + 3} + \dfrac{2}{3 - a} = \dfrac{4}{a^2 - 9}$  **11**

**Solve. Check for extraneous solutions.**

**7.** $\dfrac{4}{p^2 - 8p + 12} = \dfrac{p}{p - 2} + \dfrac{1}{p - 6}$  **−1**

**8.** $\dfrac{a + 8}{16} = \dfrac{-2}{a - 10}$  **8, −6**

**9.** $\dfrac{x^2 + 7x}{x - 2} = 4 + \dfrac{36}{2x - 4}$  **−5**

**10.** $\dfrac{y}{y - 2} - 1 = \dfrac{4}{y + 3}$  **7**

**Simplify.**

*(page 275)* [10.3]

**11.** $\dfrac{3a + \dfrac{1}{5}}{\dfrac{a}{2} + \dfrac{7}{10}}$  $\dfrac{30a + 2}{5a + 7}$

**12.** $\dfrac{\dfrac{7}{y^2} - \dfrac{2}{y}}{\dfrac{13}{y} + \dfrac{1}{y^2}}$  $\dfrac{7 - 2y}{13y + 1}$

**13.** $\dfrac{1 - \dfrac{11}{a} + \dfrac{28}{a^2}}{\dfrac{1}{a} - \dfrac{4}{a^2}}$  $a - 7$

**14.** $\dfrac{1 - \dfrac{41}{m^2} + \dfrac{400}{m^4}}{\dfrac{20}{m^4} + \dfrac{1}{m^3} - \dfrac{1}{m^2}}$

*(page 278)* [10.4]

**15.** $\dfrac{\dfrac{3}{a^2 - 7a + 10} + \dfrac{2}{a - 5}}{\dfrac{4}{a - 5} + \dfrac{2}{a - 2}}$

**16.** $\dfrac{\dfrac{7}{x^2 - 7x + 12} + \dfrac{3}{x - 4}}{\dfrac{2}{x - 4} + \dfrac{7}{x - 3}}$

**17.** $\dfrac{m + 2 + \dfrac{3}{m - 5}}{\dfrac{4}{m - 5} + 1}$

**18.** $\dfrac{\dfrac{4}{m + 5} + \dfrac{-20}{m^2 + 5m}}{\dfrac{2}{m + 5} - \dfrac{1}{m}}$  **4**

**19.** $\dfrac{\dfrac{x + 1}{x} - \dfrac{5}{x + 2}}{\dfrac{x + 1}{x^2 + 2x} + \dfrac{3}{x + 2}}$

**20.** $\dfrac{\dfrac{x + 1}{x + 2} + \dfrac{x + 7}{x - 5}}{\dfrac{5}{x^2 - 3x - 10}}$  $\dfrac{2x^2 + 5x + 9}{5}$

**21.** $\dfrac{\dfrac{a}{b} + 2 + \dfrac{b}{a}}{\dfrac{a^2 - b^2}{ab}}$  $\dfrac{a + b}{a - b}$

★ **22.** $\dfrac{\dfrac{1}{cx^2 - 4c + dx^2 - 4d} + \dfrac{2}{cx + 2c + xd + 2d}}{\dfrac{3}{x^2 - 4} - \dfrac{4}{xc + xd - 2c - 2d}}$  $\dfrac{2x - 3}{3c + 3d - 4x - 8}$

**Solve for x.**

*(page 282)* [10.5]

**23.** $tx = y$  $\dfrac{y}{t}$

**24.** $2c + x = b$  **b − 2c**

**25.** $mx = g + px$  $\dfrac{g}{m - p}$

**26.** $5ax - 3ax + t$  $\dfrac{t}{2a}$

**27.** $ax = 13 - bx$  $\dfrac{13}{a + b}$

**28.** $5x - 8a = 2x + 7a$  **5a**

**29.** $\dfrac{x}{4} = \dfrac{a}{2} + \dfrac{b}{8}$  $\dfrac{4a + b}{2}$

**30.** $m = \dfrac{x - 7}{w}$  **mw + 7**

**31.** $\dfrac{1}{c} - \dfrac{1}{d} = \dfrac{2}{x}$  $\dfrac{2cd}{d - c}$

**32.** $a^2 - ax + 44 = 15a - 11x$  ★ **33.** $p = \pi m(m + x)$  $\dfrac{p - \pi m^2}{\pi m}$   ★ **34.** $xa + ya - xb - by = a^2 - b^2$

**32.** **a − 4**   **34.** **a + b − y**

**14.** $\dfrac{(m + 5)(m - 4)}{-1}$  **15.** $\dfrac{2a - 1}{6a - 18}$  **16.** $\dfrac{3x - 2}{9x - 34}$  **17.** $\dfrac{m^2 - 3m - 7}{m - 1}$  **19.** $\dfrac{x^2 - 2x + 2}{4x + 1}$

**520**

SELECTED EXTRA PRACTICE

# Problem Solving for Chapter 10

## Work Problems

*(page 272)*  [10.2]

1. Working together, Janice and Lola can clean a house in 6 hours. It takes Lola 3 times longer than Janice working alone. How long would it take each girl alone? **Janice 8 h, Lola 24 h**

2. Fay can prepare surgical equipment in 3 hours. Another nurse, Carlo, can do it in 4 hours. How long will it take if they work together? **$1\frac{5}{7}$ h**

3. To do a job alone it would take Rose 4 hours, Bill 3 hours, and Marc 5 hours. How long would it take if they all work together? **$1\frac{13}{47}$ h**

4. Bill can do a job in 8 hours. Henry takes 3 times as long as Clara. How long would it take Clara if all three could finish it together in 4 hours? **$10\frac{2}{3}$ h**

★ 5. A pipe can fill a pool in 20 h, while an outlet pipe can empty the pool in 30 hours. How long would it take to fill the empty pool if both pipes are operating at the same time? **60 h**

## Distance Problems

*(page 286)*  [10.6]

6. Two cars started toward each other at the same time from towns 500 km apart. One car's rate was 65 km/h, the other's 60 km/h. After how many hours did they meet? **4 h**

7. Two bicyclists traveled in opposite directions, one at a rate of 8 km/h, and the other at a rate of 7 km/h. In how many hours were they 45 km apart? **3 h**

8. The O'Rileys drove to a resort at 50 km/h. They returned at 40 km/h. The return trip took 1 hour longer than the trip to the resort. How long did it take them to get home? **5 h**

9. Keith and Kevin drove to a service station at 60 km/h. They bicycled home at 15 km/h. The entire trip took 4 hours. How many km did they bicycle? **48 km**

10. Herb drove 600 km to visit his parents. He returned driving 10 km/h faster. The total time for the round trip was 22 h. Find his rate for each part of the trip. **50 km/h going** **60 km/h returning**

## Mixed Practice

*(page 287)*

**Marsha $10\frac{2}{3}$ days; Bill 32 days**

11. Working together, it takes Bill and Marsha 8 days to paint a house. It takes Bill 3 times as long as Marsha to do it alone. How long would it take each to do the job alone?

12. A train left a station traveling at 110 km/h. Two hours later, a train left the station traveling in the opposite direction at 90 km/h. How many hours later were they 420 km apart? **1 h**

13. Find two consecutive odd integers such that the square of the first is 40 more than the second. **7, 9**

14. The length of a rectangle is 12 m more than 3 times the width. The perimeter is 88 m. Find the length and the width. **36 m; 8 m**

15. Brenda is paid $165 a week plus a commission of 6% on all sales. Find her sales if she earns a total of $285. **$2,000**

16. In a high school with 1,500 students, 4 out of 5 students go out for sports. How many students are in the sports program? **1,200**

# Practice on Graphing for Chapter 11

Find the slope of $\overleftrightarrow{AB}$ for the given points. **(page 301)** [11.3]

**1.** $A(5, 5)$, $B(7, 6)$ $\frac{1}{2}$  **2.** $A(-3, -4)$, $B(7, -2)$ $\frac{1}{5}$  **3.** $A(-9, 1)$, $B(0, -4)$ $-\frac{5}{9}$

**4.** $A(5b, 2k)$, $B(8b, 11k)$ $\frac{3k}{b}$  **5.** $A(3a - 5b, 7a - 3b)$, $B(2a - b, 5a + 2b)$ $\frac{(-2a + 5b)}{(-a + 4b)}$

Determine the value(s) of $a$ so that $\overleftrightarrow{AB}$ has the given slope.

★ **6.** $A(4, 5)$, $B(a, 9)$; slope $\frac{4}{5}$ **9**  ★ **7.** $A(1, 5a)$, $B(3, a^2)$; slope 3 **6, −1**

Find the slope of $\overleftrightarrow{PQ}$ for the given points. Describe the slant of $\overleftrightarrow{PQ}$ left to right. **(page 304)** [11.4]

**8.** $P(-2, 8)$, $Q(5, 3)$ $\frac{-5}{7}$ slants down  **9.** $P(-2, -4)$, $Q(7, -4)$ 0 horiz. **10.** $P(-8, -1)$, $Q(2, -4)$ $\frac{-3}{10}$ slants down

**11.** $P(-1, 8)$, $Q(2, 11)$  **12.** $P(1, -2)$, $Q(-7, 3)$  **13.** $P(3, 4)$, $Q(-2, -1)$

   **1 slants up**     $\frac{-5}{8}$ **slants down**    **1 slants up**

Write an equation for $\overleftrightarrow{PQ}$ with the given coordinates. **(page 307)** [11.5]

**14.** $P(2, 3)$, $Q(4, 7)$ $y = 2x - 1$  **15.** $P(0, 4)$, $Q(-2, 10)$  **16.** $P(1, -4)$, $Q(-2, 8)$ $y = -4x$

**17.** $P(2, 7)$, $Q(-1, -11)$  **18.** $P(0, -1)$, $Q(1, 4)$ $y = 5x - 1$ **19.** $P(-6, 2)$, $Q(3, -7)$

   $y = 6x - 5$     **15.** $y = -3x + 4$    $y = -x - 4$

Write an equation of the line passing through the given point and having the given slope.

**20.** $P(-5, 2)$; slope 4 $y = 4x + 22$  **21.** $P\left(\frac{2}{5}, 0\right)$; slope 10 k $y = 10\,kx - 4\,k$

Write an equation for $\overleftrightarrow{PQ}$ using the given points. **(page 309)** [11.6]

**22.** $P(0, 2)$, $Q(4, 5)$ $y = \frac{3}{4}x + 2$  **23.** $P(-2, 0)$, $Q(0, 5)$ $y = \frac{5}{2}x + 5$ **24.** $P(0, 8)$, $Q(3, 6)$ $y = -\frac{2}{3}x + 8$

**25.** $P(5, 4)$, $Q(10, 2)$  **26.** $P(-8, -4)$, $Q(4, -1)$  **27.** $P(3, 3)$, $Q(-2, 5)$

   $y = -\frac{2}{5}x + 6$     $y = \frac{1}{4}x - 2$    $y = -\frac{2}{5}x + \frac{21}{5}$

Write an equation for $\overleftrightarrow{AB}$. Then use the equation to find the $y$-coordinate of point $C$ on $\overleftrightarrow{AB}$.

**28.** $A(0, 2)$, $B(-4, -1)$, $C(4, y)$ $y = \frac{3}{4}x + 2$; 5  **29.** $A(-3, -5)$, $B(3, -1)$, $C(6, y)$ $y = \frac{2}{3}x - 3$; 1

Graph the line described by the given equation. **33.** slope $-\frac{1}{2}$; int $-6$ **(page 313)** [11.7]

**30.** $y = \frac{2}{3}x - 4$  **31.** $y = -2x + 2$  **32.** $y = 4x$  **33.** $y = -\frac{1}{2}x - 6$

   **30.** slope $\frac{2}{3}$; int $-4$  **31.** slope $-\frac{2}{1}$; int 2  **32.** slope $\frac{4}{1}$; int 0

**34.** $y = -x$  **35.** $y = -2x + 1$  **36.** $y = \frac{3}{5}x - 2$  **37.** $y = -\frac{1}{3}x + 8$

   **34.** slope $-1$; int 0   **35.** slope $-\frac{2}{1}$; int 1   **36.** slope $\frac{3}{5}$; int $-2$   **37.** slope $-\frac{1}{3}$; int 8

★ **38.** Graph the line passing through the point $(2, -3)$ and parallel to the line with equation $y = -\frac{3}{4}x + 10$. through $(2, -3)$; slope $-\frac{3}{4}$

   **39.** $y = \frac{3}{2}x - 2$  **40.** $y = -\frac{3}{4}x + 2$  **41.** $y = -\frac{2}{5}x - 2$  **42.** $y = 4$; horiz.

Graph. **(page 316)** [11.8]

**39.** $3x - 2y = 4$  **40.** $4y + 3x = 8$  **41.** $5y + 2x = -10$  **42.** $-6 - y = -10$

**43.** $x - y = 8$ $y = \frac{1}{1}x - 8$  **44.** $3y - 12 = 3$ $y = 5$; horiz.  **45.** $8 - (4 - 2y) = 6x$ $y = \frac{3}{1}x - 2$

   **(page 318)** [11.9]

**46.** $x = -7$  **47.** $12 + 3x = 0$  **48.** $3x - (4 - x) = 0$ ★ **49.** $|6 - 2x| = 8$

   vert.; $x = -7$   $x = -4$ vert.   $x = 1$ vert.   two verticals $x = -1$, $x = 7$

Graph. **(page 321)** [11.10]

**50.** $y < \frac{1}{2}x - 1$  **51.** $y \geq 2x - 4$  **52.** $x \leq -8$  **53.** $-2y \geq -6$

**50.** below line with slope $\frac{1}{2}$, $y$-int $-1$   **51.** incl. & above line with slope 2, $y$-int $-4$

**52.** left of and incl. vert. through $(-8, 0)$   **53.** below & incl. horiz. through $(0, 3)$

SELECTED EXTRA PRACTICE

## Practice with Systems of Equations for Chapter 12

**Solve each system by graphing.** *(page 332)* [12.1]

**1.** $y = -2x + 1$ **(2, −3)**
$y = \frac{1}{2}x - 4$

**2.** $y = \frac{1}{2}x + 6$ **(2, 7)**
$y = -2x + 11$

**3.** $y = -2x + 4$ **(0, 4)**
$y = 3x + 4$

**4.** $y = 3x - 6$ **(3, 3)**
$y = -\frac{2}{3}x + 5$

**5.** $y = 2x + 1$ **(2, 5)**
$y = -\frac{3}{2}x + 8$

**6.** $3x - 2y = 6$ **(2, 0)**
$x - y = 2$

**7.** $y = 2x$ **(3, 6)**
$x + y = 9$

**8.** $x = 3y + 6$ **(6, 0)**
$x + 3y = 6$

**9. left of line $x = -2$ and below and inc. line $y = 4$**

**Solve each system by graphing.** **10. right of line $x = 1$ and below line $y = -6$** *(page 334)* [12.2]

**9.** $x < -2$
$y \le 4$

**10.** $x > 1$
$y < -6$

**11.** $y < x + 3$
$y > -x + 2$

**12.** $y \le 3x - 1$
$y > \frac{1}{2}x - 2$

**13.** $3x - 2y \le 4$
$x - y > 1$

**14.** $2x - 3y < 6$
$4y > -2x - 8$

**15.** $y - 2x > -2$
$x - 2y < 8$

**16.** $y - x < 0$
$5y \ge 3x - 10$

★**17.** $x + y \le 4$
$x \ge 1$
$y \ge 2$

★**18.** $3x - y \le 9$
$x \ge 2$
$y > -3$

**11. below line with slope 1 and $y$-int 3 and above line with slope −1, $y$-int 2**

**Solve each system by substitution. For Ex. 12–18, see TE Answer Section.** *(page 337)* [12.3]

**19.** $y = 2x$ **(4, 8)**
$x + y = 12$

**20.** $x + y = 10$ **(6, 4)**
$y = x - 2$

**21.** $y = 2x - 5$ **(3, 1)**
$x - y = 2$

**22.** $x + y = 7$ **(3, 4)**
$y = x + 1$

**23.** $3x - 2y = 9$ **(3, 0)**
$x + 2y = 3$

**24.** $2x - y = 3$ **(2, 1)**
$5x + 3y = 13$

**25.** $2x - 3y = 8$ **(4, 0)**
$x + y = 4$

**26.** $x - 3y = 2$ **(2, 0)**
$3x - 2y = 6$

**27.** $x - 2y = 6$ **(4, −1)**
$x + 2y = 2$

**28.** $3x + y = 8$ **(2, 2)**
$x - 2y = -2$

**29.** $2x + y = 7$ **(3, 1)**
$4x + 3y = 15$

**30.** $3x - y = 18$ **(6, 0)**
$4x + 2y = 24$

**31.** $2x + 3y = 22$ **(5, 4)**
$3x + y = 19$

**32.** $4x + y = 29$ **(7, 1)**
$2x - 3y = 11$

**33.** $2x + 5y = 21$ **(3, 3)**
$4x - y = 9$

**34.** $-5x - 2y = -43$
$x - y = 3$ **(7, 4)**

★**35.** $2x + 4y = 12$ **(2, 2)**
$5x - 2y = 6$

★**36.** $2x - 5y = -1$ **(7, 3)**
$\dfrac{x + y}{2} = 5$

★**37.** $x + y = 14$ **(6, 8)**
$\dfrac{x}{3} + \dfrac{y}{2} = 6$

**Solve each system by addition.** **47. (−5, −46)** *(page 346)* [12.6]

**38.** $3x + 2y = 8$ **(2, 1)**
$2x + 5y = 9$

**39.** $5x - 2y = 3$ **(1, 1)**
$2x + 7y = 9$

**40.** $2x + 2y = 8$ **(2, 2)**
$5x - 3y = 4$

**41.** $5x - 3y = 2$ **(1, 1)**
$4x + 2y = 6$

**42.** $5x + 4y = 29$ **(5, 1)**
$3x - 2y = 13$

**43.** $3x - 5y = 1$ **(2, 1)**
$2x + 10y = 14$

**44.** $7x + 2y = 9$ **(1, 1)**
$3x + 8y = 11$

**45.** $4x - 7y = 5$ **(3, 1)**
$2x + 8y = 14$

**46.** $8x - 6y = 10$ **(2, 1)**
$4x - 5y = 3$

**47.** $-12x + y = 14$
$-7x + y = -11$

**48.** $11x - 3y = 16$ **(2, 2)**
$4x + 2y = 12$

**49.** $7x + 5y = 31$ **(3, 2)**
$-14x + 6y = -30$

**50.** $x + y = 4$ **(−2, 6)**
$2x + y = 2$

**51.** $x - 7y = 4$
$x + 7y = 16$ $\left(10, \dfrac{6}{7}\right)$

**52.** $6x - 2y = 4$ **(2, 4)**
$4x + 4y = 24$

**53.** $5x - y = 13$
$x - y = -1$ $\left(\dfrac{7}{2}, \dfrac{9}{2}\right)$

## Problem Solving for Chapter 12

### Systems in Problem Solving

*(page 340)* [12.4]

1. Bill drove 4 times as far as Jan did. The difference between their distances was 255 km. How far did each drive? **Bill 340, Jan 85**

2. Tina's age is 5 times Joyce's age. The sum of their ages is 42. Find their ages. **35, 7**

3. One number is 4 more than another. If twice the smaller is added to the larger, the result is 28. Find the numbers. **12, 8**

4. One number is 2 less than another. If twice the larger is decreased by 3 times the smaller, the result is $-10$. Find the numbers. **14, 16**

5. The difference of Ron's and Joan's ages is 8 years. The sum of twice Ron's age and 3 times Joan's age is 66. How old is each? **Ron: 18 y; Joan: 10 y**

6. The perimeter of a rectangle is 70 m. The difference between 3 times the length and 5 times the width is 25 m. Find the length and the width. **length: 25 m   width: 10 m**

### Coin and Mixture Problems

*(page 350)* [12.7]

7. Jim has 11 coins in dimes and quarters. Their value is $1.70. How many of each does he have? **7 d   4 q**

8. There are 5 less dimes than nickels. The total value is $4.00. How many coins are there of each type? **25 d   30 n**

9. For a school play, 738 tickets valued at $856 were sold. Some cost $1 and some cost $1.50. How many $1 tickets were sold? **502**

10. Mary has $3.20. She has 2 more dimes than half-dollars. How many coins are there of each kind? **5 h. d.   7 d**

11. Cashews cost $12.15/kg. Pecans cost $13.00/kg. The number of kg of cashews is 2 less than the number of kg of pecans. How many kg of each should be mixed to make a box that will sell for $76.30? **cashews: 2 kg; pecans 4 kg**

12. A $7.65 assortment of pads contains $0.55 pads and $0.60 pads. If the number of $0.55 pads is 2 less than $\frac{1}{2}$ the number of $0.60 pads, find the number of each type of pads in the mixture. **10 $0.60 pads; 3 $0.55 pads**

### Digit Problems

*(page 354)* [12.8]

13. The sum of the digits of a two-digit number is 15. If the digits are reversed, the new number is is 27 less than the original number. Find the original number. **96**

14. The tens digit of a two-digit number is twice the units digit. If the digits are reversed, the new number is 36 less than the original number. Find the original number. **84**

15. The units digit of a two-digit number is 1 more than 3 times the tens digit. If the digits are reversed, the new number is 9 less than 3 times the original number. Find the new number. **72**

16. Six times the tens digit of a two-digit number is 4 more than the units digit. If the digits are reversed, the sum of the new number and twice the original one is 138. Find the original number. **28**

17. The tens digit of a two digit number is 3 less than 4 times the units digit. If the digits are reversed, the new number is 8 more than one third of the original number. Find the original two digit number. **93**

SELECTED EXTRA PRACTICE

# Practice with Relations and Functions for Chapter 13

**1.** *D*: {0, 1, 2, 5} *R*: {4, 3, 7, 8} **yes**     **2.** *D*: {1, 2, −4, 7} *R*: {5, 6, 3} **yes**

Give the domain and range of each relation. Is it a function? *(page 367)* [13.1]

**1.** {(0, 4), (1, 3), (2, 7), (5, 8)}                     **2.** {(1, 5), (2, 5), (−4, 6), (7, 3)}

**3.** {(−1, 2), (3, 2), (5, −1), (3, −4)} *D*: {−1, 3, 5}     **4.** {(−4, 5), (−3, 5), (−1, 5), (4, 5)}

                           *R*: {2, −1, −4}  **no**                *D*: {−4, −3, −1, 4}  *R*: {5}  **yes**

For what values of *k* will the relation not be a function?

**5.** {(3*k* + 1, 4), (*k* + 11, −2)} **5**     **6.** {(−10, 1), ($k^2$ − 8*k* + 2, 4)}     ★ **7.** {(|*k* + 2| + 4, 6), (8, 3)}

                       **6. 6, 2**                     **7. 2, −6**

Graph. Which relations are functions? Which functions are linear functions?
Which functions are constant linear functions? *(page 370)* [13.2]

**8.** *y* = 3*x* **linear funct.**   **9.** 3*x* − 2 = 13     **10.** *y* − 4 = 6     **11.** 16 = −4*x* **not funct.**

**12.** 3*x* − *y* = 4     **13.** 3*y* − 4 = 17     **14.** 8 − *x* = 4 **not funct. 15.** *y* − (4 − *x*) = 2

**9. not funct.   10. linear cons. f.   12. linear funct.   13. linear const. funct.   15. linear funct.**

For *g*(*x*) = 2*x* − 5, find the indicated value. *(page 372)* [13.3]

**16.** *g*(−3) **−11**          **17.** *g*(0) **−5**          **18.** *g*(1) **−3**          **19.** *g*(−4) **−13**

Find the range of each function for the given domain.  **21.** {0, 1, −15}

**20.** *f*(*x*) = $x^2$ − *x* + 2 *D* = {−2, −1, 3} **{8, 4}**     **21.** *f*(*x*) = −$x^2$ − 2*x* *D* = {−2, −1, 3}

**22.** *g*(*x*) = $(x − 3)^2$ *D* = {−4, 0, $\frac{1}{2}$} **{49, 9, $\frac{25}{4}$}**     **23.** *g*(*x*) = *x* − |*x*| *D* = {−4, 1, 3} **{−8, 0}**

For *f*(*x*) = 2*x* − 5 and *g*(*x*) = −$x^2$ + 2, find the indicated value.

★ **24.** *f*(−3) + *g*(−3) **−18**          ★ **25.** *f*[*g*(−$\frac{1}{4}$) − 1] **−3$\frac{1}{8}$**

Determine if *y* varies directly as *x*. If so, find the constant of variation. *(page 375)* [13.4]

**26.**

| *x* | *y* |
|---|---|
| 7 | 14 |
| −2 | −4 |
| 30 | 60 |

**yes; 2**

**27.**

| *x* | *y* |
|---|---|
| −3 | −18 |
| 8 | 48 |
| 2 | −12 |

**no**

**28.**

| *x* | *y* |
|---|---|
| 9 | 27 |
| 3 | 30 |
| 7 | 28 |

**no**

**29.**

| *x* | *y* |
|---|---|
| −3 | 12 |
| 15 | −60 |
| −8 | 32 |

**yes; −4**

In each of the following, *y* varies directly as *x*.

**30.** *y* is 64 when *x* is 16. Find *y* when *x* is −7. **−28**   ★ **31.** *y* is $\frac{1}{3}$ when *x* is $\frac{5}{6}$. Find *y* when *x* is 4. **$\frac{8}{5}$**

Determine if *y* varies inversely as *x*. If so, find the constant of variation. *(page 379)* [13.5]

**32.**

| *x* | *y* |
|---|---|
| 8 | −6 |
| −24 | 2 |
| −16 | 3 |

**yes; −48**

**33.**

| *x* | *y* |
|---|---|
| −14 | 4 |
| −8 | 7 |
| 2 | −28 |

**yes; −56**

**34.**

| *x* | *y* |
|---|---|
| 8 | −9 |
| −24 | 3 |
| 6 | 12 |

**no**

**35.**

| *x* | *y* |
|---|---|
| −24 | −4 |
| 16 | 6 |
| −32 | −3 |

**yes; 96**

In each of the following, *y* varies inversely as *x*.

**36.** *y* is 27 when *x* is 4.     **37.** *y* is 2 when *x* is −60.     **38.** *y* is −7 when *x* is 9.
Find *y* when *x* is 2. **54**     Find *y* when *x* is 4. **−30**     Find *y* when *x* is 21. **−3**

Determine if *y* varies directly or inversely as *x*. If so, find the constant of variation.

**39.**

| *x* | *y* |
|---|---|
| 6 | 8 |
| 24 | 2 |
| −12 | −4 |

**Inversely**
**48**

**40.**

| *x* | *y* |
|---|---|
| 100 | 20 |
| 25 | 5 |
| −15 | −3 |

**Directly**
**$\frac{1}{5}$**

**41.**

| *x* | *y* |
|---|---|
| 8 | 3 |
| 12 | 2 |
| −6 | 4 |

**Neither**

**42.**

| *x* | *y* |
|---|---|
| 32 | −2 |
| −8 | 8 |
| 16 | −4 |

**Inversely**
**−64**

# Problem Solving for Chapter 13

## Direct Variation

(*page 376*) [13.4]

1. On a scale drawing, 3 cm represents 15 m. How many meters are represented by 7 cm? **35 m**
2. The cost of gold varies directly as its mass. If 5 g of gold cost $375, find the cost of 9 g of gold.
3. The bending of a beam varies directly as its mass. A beam is bent 30 mm by a mass of 50 kg. How much will the beam bend with a mass of 200 kg? **120 mm**
4. A car's gas consumption is directly proportional to the distance traveled. A car uses 25 L of gas to travel 200 km. How much gas is needed for a 1,000 km trip? **125 L**

★ 5. The distance that a free-falling body falls varies directly as the square of the time it falls. A stone falls 48 ft in 6 s. How far will it fall in 3 s? **12 ft**

     **2. $675**

## Inverse Variation

(*page 379*) [13.5]

6. The length of a rectangle with a constant area varies inversely as the width. When the length is 24 the width is 8. Find the length when the width is 6. **32**
7. The volume of a gas varies inversely as the pressure. If the volume is 60 $m^3$ under 6 kg of pressure, find the volume under 30 kg of pressure. **12 $m^3$**

★ 8. The brightness of the illumination of an object varies inversely as the square of the distance of the object from a source of illumination. If a light meter reads 25 luxes at a distance of 2 m from a light source, find the reading at a distance of 5 m from the light source. **4 luxes**

## Applying Functions and Variation

(*page 381*)

9. The amount of money a family spends on food varies directly as their income. A family making $32,000 a year will spend $8,000 on food. How much would a family earning $52,000 spend on food? **$13,000**
10. The current in an electrical circuit varies inversely as the resistance. The current is 15 amps when the resistance is 40 ohms. Find the current when the resistance is 30 ohms. **20 amps**

11. $B(s) = 0.004\ s^2$ represents the approximate distance in meters traveled by a car from the time the driver applies the brakes until the car actually stops; $s$ is the speed in km/h. Find $B(55)$. What is the significance of your answer? **Car travels about 12 m before it comes to a stop.**

## Age Problems

(*page 383*) [13.6]

12. Bert is 4 times as old as Harry. Three years ago he was 7 times as old as Harry. How old is each now? **Bert 24; Harry 6**
13. Sue is 12 years older than Maria. Four years ago she was 4 times as old as Maria. How old is each now? **Sue 20; Maria 8**
14. Rose is 20 years older than Lisa. In 8 years she will be 3 times as old as Lisa. How old is each now? **Rose 22; Lisa 2**
15. Earl is 2 years younger than Tony. Eight years ago the sum of their ages was 14. How old is each now? **Earl 14; Tony 16**

SELECTED EXTRA PRACTICE

# Practice with Radicals for Chapter 14

**Simplify.** *(page 398)* [14.4]

1. $\sqrt{18}$  $3\sqrt{2}$
2. $\sqrt{28}$  $2\sqrt{7}$
3. $\sqrt{72}$  $6\sqrt{2}$
4. $\sqrt{54}$  $3\sqrt{6}$
5. $\sqrt{32}$  $4\sqrt{2}$
6. $4\sqrt{8}$  $8\sqrt{2}$
7. $5\sqrt{12}$  $10\sqrt{3}$
8. $4\sqrt{27}$  $12\sqrt{3}$
★ 9. $\sqrt{405}$  $9\sqrt{5}$
★ 10. $\sqrt{1152}$  $24\sqrt{2}$

*(page 401)* [14.5]

11. $\sqrt{a^6}$  $a^3$
12. $\sqrt{b^{18}}$  $b^9$
13. $\sqrt{m^{20}}$  $m^{10}$
14. $\sqrt{25a^2b^4}$  $5ab^2$
15. $\sqrt{49x^4y^2z^6}$  $7x^2yz^3$
16. $\sqrt{121a^6b^4c^{12}}$  $11a^3b^2c^6$
17. $\sqrt{28x^6y^{10}}$  $2x^3y^5\sqrt{7}$
18. $\sqrt{32m^{10}n^4z^6}$  $4m^5n^2z^3\sqrt{2}$
★ 19. $\sqrt{\sqrt{81a^{12}}}$  $3a^3$

23. $2mn^2\sqrt{6mn}$   24. $5x^3y\sqrt{2y}$   25. $2a^3b^5\sqrt{10ab}$   26. $2a^2c^2\sqrt{11c}$   *(page 403)* [14.6]

20. $\sqrt{x^9}$  $x^4\sqrt{x}$
21. $\sqrt{18a^7}$  $3a^3\sqrt{2a}$
22. $\sqrt{a^4b^5}$  $a^2b^2\sqrt{b}$
23. $\sqrt{24m^3n^5}$
24. $\sqrt{50x^6y^3}$
25. $\sqrt{40a^7b^{11}}$
26. $\sqrt{44a^4c^5}$
27. $\sqrt{98x^{11}y}$  $7x^5\sqrt{2xy}$
28. $-3xy^2\sqrt{80x^5y^6}$  $-12x^3y^5\sqrt{5x}$
29. $3ab^2c^3\sqrt{75a^4bc^7}$  $15a^3b^2c^6\sqrt{3bc}$
★ 30. $\sqrt{a^{6k}b^{4k+1}}$  $a^{3k}b^{2k}\sqrt{b}$

*(page 405)* [14.7]

31. $5\sqrt{3} + 4\sqrt{3}$  $9\sqrt{3}$
32. $6\sqrt{27} - 3\sqrt{48}$  $6\sqrt{3}$
33. $9\sqrt{2} - 3\sqrt{8}$  $3\sqrt{2}$
34. $-\sqrt{32} + 5\sqrt{18} - 7\sqrt{98}$  $-38\sqrt{2}$
35. $2\sqrt{10x} + \sqrt{40x} - 5\sqrt{90x}$  $-11\sqrt{10x}$
36. $7\sqrt{8a^2b^3} - 4b\sqrt{50a^2b}$  $-6ab\sqrt{2b}$
37. $2x\sqrt{45xy^2} + 3\sqrt{20x^3y^2} - \sqrt{80x^3y^2}$  $8xy\sqrt{5x}$
★ 38. $\sqrt{by^2 - 10by + 25b} - \sqrt{y^2b} + 4\sqrt{b}$  $-\sqrt{b}$

41. $-27\sqrt{2} + 24\sqrt{3}$   *(page 408)* [14.8]

39. $5\sqrt{6} \cdot 4\sqrt{2}$  $40\sqrt{3}$
40. $6\sqrt{2} \cdot 8\sqrt{2}$  $96$
41. $-3\sqrt{6}(3\sqrt{3} - 4\sqrt{2})$
42. $(2\sqrt{3} - \sqrt{2})(3\sqrt{3} + \sqrt{2})$  $16 - \sqrt{6}$
43. $(\sqrt{5} - \sqrt{3})(2\sqrt{5} + 4\sqrt{3})$  $-2 + 2\sqrt{15}$
44. $(2\sqrt{7} - 3\sqrt{5})(2\sqrt{7} + 3\sqrt{5})$  $-17$
45. $(3\sqrt{6} + 5\sqrt{3})(3\sqrt{6} - 5\sqrt{3})$  $-21$
46. $(3\sqrt{6} + 2\sqrt{2})^2$  $62 + 24\sqrt{3}$
47. $(4\sqrt{5} - 2\sqrt{10})^2$  $120 - 80\sqrt{2}$
★ 48. $(\sqrt{5} + \sqrt{3})(\sqrt{5} + \sqrt{8} + \sqrt{3})$
48. $8 + 2\sqrt{15} + 2\sqrt{10} + 2\sqrt{6}$
★ 49. $(\sqrt{8} + \sqrt{3})^3$  $34\sqrt{2} + 27\sqrt{3}$

*(page 411)* [14.9]

50. $\dfrac{18}{\sqrt{2}}$  $9\sqrt{2}$
51. $\dfrac{6}{\sqrt{3}}$  $2\sqrt{3}$
52. $\dfrac{20}{\sqrt{5}}$  $4\sqrt{5}$
53. $\dfrac{8}{\sqrt{20}}$  $\dfrac{4\sqrt{5}}{5}$
54. $\dfrac{16}{\sqrt{12}}$  $\dfrac{8}{3}\sqrt{3}$

55. $\dfrac{24ab}{\sqrt{a^4b}}$  $\dfrac{24\sqrt{b}}{a}$
56. $\dfrac{6x^3y^2}{\sqrt{8xy^4}}$  $\dfrac{3x^2\sqrt{2x}}{2}$
57. $\dfrac{12x^4y^4}{\sqrt{6xy}}$
58. $\dfrac{9x^4y}{\sqrt{18x^3y^2}}$
59. $\dfrac{4xy}{\sqrt{12x^5y^3}}$

★ 60. $\dfrac{15}{4 - \sqrt{3}}$  $\dfrac{60 + 15\sqrt{3}}{13}$

57. $2x^3y^3\sqrt{6xy}$   58. $\dfrac{3x^2\sqrt{2x}}{2}$

★ 61. $\dfrac{4 - \sqrt{2}}{4 + \sqrt{2}}$  $\dfrac{9 - 4\sqrt{2}}{7}$

59. $\dfrac{2\sqrt{3xy}}{3x^2y}$

*(page 414)* [14.10]

62. If the lengths of the legs of a right triangle are 6 cm and 4 cm, find the length of the hypotenuse in simplest radical form.   $2\sqrt{13}$ cm

63. The length of the hypotenuse of a right triangle is 8 m. The length of one leg is 4 m. Find the length of the other leg in simplest radical form.   $4\sqrt{3}$ m

**For each right triangle with legs *a* and *b* and hypotenuse *c*, find the missing length in simplest radical form.**

64. $a = 6$, $b = 6$  $c = 6\sqrt{2}$
65. $b = 4$, $c = 6$  $a = 2\sqrt{5}$
66. $a = 10$, $b = 2$  $c = 2\sqrt{26}$
67. $c = 4\sqrt{2}$, $a = \sqrt{3}$  $b = \sqrt{29}$
68. $a = 4\sqrt{2}$, $b = 2\sqrt{6}$  $c = 2\sqrt{14}$
69. $a = 2\sqrt{5}$, $c = 4\sqrt{5}$  $b = 2\sqrt{15}$

# Practice with Quadratic Functions for Chapter 15

Solve. *(page 428)*  [15.1]

**1.** $\sqrt{y} = 8$ **64** **2.** $\sqrt{3x} = 9$ **27** **3.** $\sqrt{x} + 3 = 8$ **25** **4.** $7 = \sqrt{z} + 3$ **16**

**5.** $\sqrt{2x - 3} - 2 = 3$ **14** **6.** $11 = 9 + \sqrt{y - 2}$ **6** **7.** $\sqrt{3x - 2} + 5 = 9$ **6** **8.** $7 - \sqrt{4x + 8} = 1$ **7**

**9.** $\sqrt{2x - 1} = \sqrt{x + 3}$ **4** **10.** $\sqrt{x - 3} = \sqrt{2x + 7}$ **11.** $2\sqrt{x} = \sqrt{48}$ **12**

**12.** $x - 3 = \sqrt{13 - 3x}$ **4** **13.** $\sqrt{4p + 5} = 10 - p$ **5** ★**14.** $\sqrt{x} + 3 = \sqrt{5x - 9}$ **9**

**10. no solution** *(page 431)*  [15.2]

**15.** $x^2 = 40$ **$\pm 2\sqrt{10}$** **16.** $y^2 = 48$ **$\pm 4\sqrt{3}$** **17.** $m^2 = 75$ **$\pm 5\sqrt{3}$**

**18.** $(y - 6)^2 = 64$ **14, −2** **19.** $(2y - 15)^2 = 25$ **10, 5** **20.** $(4a + 2)^2 = 16$ **$\frac{1}{2}, -\frac{3}{2}$**

**21.** $6x^2 = 24$ **$\pm 2$** **22.** $4y^2 = 128$ **$\pm 4\sqrt{2}$** **23.** $3y^2 = 120$ **$\pm 2\sqrt{10}$**

**24.** $(m - 6)^2 = 32$ **$6 \pm 4\sqrt{2}$** **25.** $(b + 4)^2 = 44$ **$-4 \pm 2\sqrt{11}$** **26.** $(a - 7)^2 = 63$ **$7 \pm 3\sqrt{7}$**

**27.** $(l - 2)^2 = 45$ **$2 \pm 3\sqrt{5}$** ★**28.** $4(2x - 5)^2 + 6 = 54$ **$\frac{5 \pm 2\sqrt{3}}{2}$** ★**29.** $2(4x - 2)^2 - 4 = 96$ **$\frac{2 \pm 5\sqrt{2}}{4}$**

Solve by completing the square. **40.** $-2 \pm \sqrt{14}$ **41.** $-4 \pm \sqrt{10}$ *(page 434)*  [15.3]

**30.** $x^2 - 10x = -16$ **8, 2** **31.** $x^2 + 2x = 3$ **−3, 1** **32.** $x^2 + 16x = -15$ **−15, −1**

**33.** $x^2 + 2x = 8$ **−4, 2** **34.** $y^2 - 2y - 48 = 0$ **8, −6** **35.** $a^2 - 16a + 28 = 0$ **2, 14**

**36.** $l^2 - 18l + 32 = 0$ **16, 2** **37.** $m^2 - 24m + 128 = 0$ **16, 8** **38.** $k^2 - 30k + 200 = 0$ **20, 10**

**39.** $a^2 - 2a - 6 = 0$ **$1 \pm \sqrt{7}$** **40.** $p^2 + 4p - 10 = 0$ **41.** $y^2 + 8y + 6 = 0$

**42.** $x^2 - 5x - 6 = 0$ **6, −1** **43.** $p^2 - 3p - 28 = 0$ **7, −4** **44.** $k^2 - 7k - 18 = 0$ **9, −2**

**45.** $2a^2 - 5a = 12$ **$-\frac{3}{2}, 4$** **46.** $3m^2 + 2m - 1 = 0$ **$\frac{1}{3}, -1$** **47.** $2y^2 - y - 3 = 0$ **$\frac{3}{2}, -1$**

Solve for $x$ in terms of $b$ or $c$.

★**48.** $x^2 - 8x + c = 0$ **$4 \pm \sqrt{16 - c}$** ★**49.** $x^2 - 6bx = 16b^2$ **$8b, -2b$**

Solve by using the quadratic formula. Leave the answer in simplest
radical form. *(page 437)*  [15.4]

**50.** $x^2 + 6x + 8 = 0$ **−4, −2** **51.** $2x^2 + 7x - 4 = 0$ **$\frac{1}{2}, -4$** **52.** $x^2 - 5x - 14 = 0$ **7, −2**

**53.** $x^2 - 5x + 1 = 0$ **$\frac{5 \pm \sqrt{21}}{2}$** **54.** $y^2 + y - 4 = 0$ **$\frac{-1 \pm \sqrt{17}}{2}$** **55.** $t^2 - 3t - 5 = 0$ **$\frac{3 \pm \sqrt{29}}{2}$**

**56.** $x^2 - 6x + 1 = 0$ **$3 \pm 2\sqrt{2}$** **57.** $x^2 - 8x = 4$ **$4 \pm 2\sqrt{5}$** **58.** $-3x - 6 = -x^2$ **$\frac{3 \pm \sqrt{33}}{2}$**

**59.** $x^2 + 4x = 2$ **$-2 \pm \sqrt{6}$** **60.** $-10x + 2 = -x^2$ **$5 \pm \sqrt{23}$** **61.** $y^2 - 6 = -2y$ **$-1 \pm \sqrt{7}$**

**62.** $\frac{1}{x} = \frac{x - 8}{1}$ **$4 \pm \sqrt{17}$** **63.** $\frac{x^2}{2} - x + \frac{1}{4} = 0$ **$\frac{2 \pm \sqrt{2}}{2}$** **64.** $\frac{x^2}{2} - \frac{1}{8} = \frac{1}{4}x$ **$\frac{1 \pm \sqrt{5}}{4}$**

★**65.** $\sqrt{3}\, x^2 - 2x - 2\sqrt{3} = 0$ **$\frac{\sqrt{3} \pm \sqrt{21}}{3}$** ★**66.** $x^2 + 4 = +4\sqrt{2}x$ **$2\sqrt{2} \pm 2$**

**67.** (4, −15) **68.** (5, 1) **69.** (2, −16) **70.** (3, −17) **71.** (2, −12) **72.** (−1, 3) **73.** (2, −13)
**74.** (1, −4)

Find the coordinates of the turning point of each parabola.
Graph the parabola. *(page 445)*  [15.7]

**67.** $y = x^2 - 8x + 1$ **68.** $y = -x^2 + 10x - 24$ **69.** $y = 5x^2 - 20x + 4$ **70.** $y = x^2 - 6x - 8$

**71.** $y = 3x^2 - 12x$ **72.** $y = -2x^2 - 4x + 1$ **73.** $y = 3x^2 - 12x - 1$ **74.** $y = 4x^2 - 8x$

**75.** Find the maximum value of $y$ if $y = -x^2 - 8x + 6$. **22**

**76.** Find the minimum value of $y$ if $y = 4x^2 + 16x - 8$. **−24**

**77.** Find the maximum value of $y$ if $y = -2x^2 - 16x$. **32**

**78.** Find the minimum value of $y$ if $y = 3x^2 - 3x$. **$-\frac{3}{4}$**

# Problem Solving for Chapter 15

## Area Problems

*(page 440)*  [15.5]

1. The length of a rectangle is 3 m more than the width. The area is 40 m². Find the length and the width.  **8 m, 5 m**

2. The length of a rectangle is 3 cm more than twice the width. The area is 44 cm². Find the length and the width.  **4 cm, 11 cm**

3. The length of a rectangle is 4 cm less than twice the width. The area is 30 cm². Find the length and the width.  **6 cm, 5 cm**

4. The base of a triangle is 4 km more than twice the height. The area is 48 km². Find the height and base.  **6 km, 16 km**

5. The height of a triangle is 2 cm more than twice the base. The area is 12 cm². Find the base and height.  **3 cm, 8 cm**

6. The length of a rectangle is twice the width. The area is 200 m². Find the length and the width.
**6. 20 m, 10 m**

★ 7. The perimeter of a rectangle is 24 cm. The area is 35 cm². Find the length and the width.  **7 cm, 5 cm**

## Wet Mixture Problems

*(page 443)*  [15.6]

8. How many liters of water must be added to 9 liters of a 40% solution of hydrochloric acid to make it a 30% acid solution?  **3 L**

9. Dr. Rivers has 3 liters of a 60% salt solution. How many liters of water must be added to obtain a 20% solution?  **6 L**

10. How many liters of water must be added to 26 liters of a 15% antifreeze solution to dilute it to a 12% solution?  **6.5 L**

11. How many cubic centimeters of a solution that is 65% alcohol must be added to 50 cm³ of a 25% alcohol solution to make a 40% solution?

12. Dr. Green has one solution that is 40% salt and another solution that is 65% salt. **11. 30 cm³**
    How much of the 65% solution must be added to 5 deciliters of the original solution to obtain a solution that is 55% salt?  **7.5 dL**

## Mixed Practice

*(page 444)*

13. Monica has 7 more nickels than dimes. Their total value is $1.55. How many coins of each type does she have?  **8 dimes, 15 nickels**

14. The selling price of a camera is $195. The profit is 30% of the cost. Find the cost.  **$150**

15. The cost of a certain metal varies directly as its mass. If 9 kg cost $27, find the cost of 45 kg.  **$135**

16. Five out of 12 people use a certain product. How many people use this product in a town of 48,000?  **20,000**

17. Find three consecutive odd integers such that the square of the first is 16 more than the last.  **5, 7, 9**

18. The length of a rectangle is 2 m less than 5 times the width. The area is 39 m². Find the length and the width.  **13 m, 3 m**

19. The tens digit of a two-digit number is 1 more than 4 times the units digit. The sum of the number and the number with its digits reversed is 121. Find the number.  **92**

20. Working together, Jake and Felicia can clean a house in 3 hours. It takes Jake 4 times longer than Felicia to do it alone. How long would it take each alone?  **Jake 15 h, Felicia $3\frac{3}{4}$ h**

# Practice with Indirect Measurement for Chapter 16

**Find sin A, cos A, tan A, sin B, cos B, and tan B to three decimal places.** *(page 464)* [16.3]

**1.**
0.8
0.6
1.333
0.6
0.8
0.75

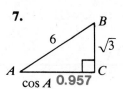

**2.**
0.471
0.882
0.533
0.882
0.471
1.875

**3.**
0.6
0.8
0.75
0.8
0.6
1.333

**4.**
0.28
0.96
0.292
0.96
0.28
3.429

**5.**
0.667
0.745
0.894
0.745
0.667
1.118

**6.**
0.866
0.5
1.732
0.5
0.866
0.577

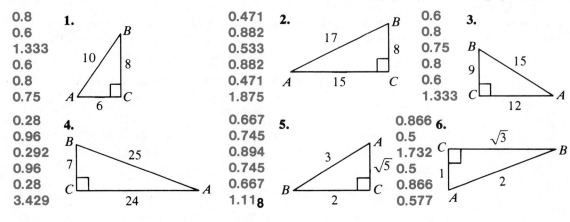

**Find the indicated value to three decimal places.**

**7.**
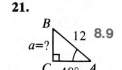
cos A **0.957**

**8.**
cos A sin A **0.791**

**9.**
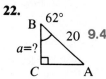
tan B **3.873**

★ **10.** The cosine of an angle is always between what two numbers? **0 and 1**

**Find the value of the ratio. Use the trigonometry table on page 533.** *(page 466)* [16.4]

**11.** sin 74° **0.9613**    **12.** cos 18° **0.9511**    **13.** tan 89° **57.29**    **14.** tan 65° **2.145**

**Find m ∠B to the nearest degree. Use the trigonometry table on page 533.**

**15.** tan B = 0.1398 **8°**    **16.** cos B = 0.9281 **22°**    **17.** sin B = 0.7583 **49°**

**Simplify. Use the trigonometry table on page 533.**

**18.** sin 40° − cos 50° **0**    **19.** (sin 30°)² + (cos 30°)² **1**    **20.** 2 · sin 45° · cos 45° **1**

**Find the indicated measure (side to the nearest tenth or angle to the nearest degree).** *(page 469)* [16.5]

**21.**
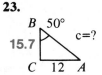
a=? 12 **8.9**
C 48° A

**22.**
B 62°
a=? 20 **9.4**
C A

**23.**
B 50°
15.7 c=?
C 12 A

**24.**
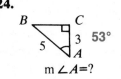
B C
5 3 **53°**
A
m ∠A=?

★ **25.** Find the area of the rectangle *ABCD* to the nearest square unit.

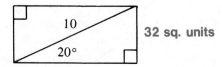
10
20°
**32 sq. units**

★ **26.** Find the area of the triangle to the nearest square unit.

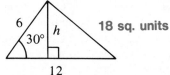
6
30° h
12
**18 sq. units**

SELECTED EXTRA PRACTICE

# Problem Solving for Chapter 16

**Angle Properties** *(page 458)* [16.1]

1. An angle measures 60° less than its complement. Find the measure of the angle and its complement. **15°, 75°**

2. An angle measures 60° more than twice its supplement. Find the angle and its supplement. **140°, 40°**

3. The measures of two supplementary angles are in the ratio 11:9. Find the measure of each angle. **99°, 81°**

4. The measures of the three angles of a triangle are in the ratio 2:3:10. Find the measure of each angle. **24°, 36°, 120°**

5. The first angle of a triangle measures 40° more than the second angle. The third angle measures 20° less than the sum of the measures of the first two angles. Find the measure of each angle. **70°, 30°, 80°**

**Applying Trigonometric Ratios** *(page 472)* [16.6]

**Give the answers to the nearest unit.**

6. The angle of elevation from a ship to the top of a 50-m lighthouse on the coast measures 13°. How far from the coast is the ship? **217 m**

7. A tree casts a 60-m shadow when the angle of elevation of the sun measures 58°. How tall is the tree? **96 m**

8. A plane is flying at an altitude of 10,000 m. The angle of elevation from a building on the ground to the plane measures 28°. How far is the building from the plane? **18,808 m**

9. A 30-m ladder makes an angle of 55° with the ground as it leans against a building. At what height does it touch the building? **25 m**

10. A 12-m diagonal of a rectangle makes an angle of 56° with a side of the rectangle. Find the dimensions of the rectangle. **7 m, 10 m**

**Investment and Loan Problems** *(page 476)* [16.7]

11. Bill borrowed $5,000 at $7\frac{1}{2}$% interest for 1 year. How much interest did he pay at the end of 1 year? **$375**

12. Ms. Diaz invested $5,000 for 1 year. She earned $300 interest. What was the rate of return? **6%**

13. Liz invested $1,800, part at 5% and the rest at 6%. If her total interest at the end of 1 year was $96, find the amount she invested at each rate. **$1,200 at 5%; $600 at 6%**

14. Ann invested one-half of her money at $5\frac{1}{2}$% and one-fourth at 6%. At the end of two years she had earned $595 in interest. Find the original sum of money. **$7,000**

15. Bob invested one-third of his money at 6% and three-fifths at $6\frac{1}{2}$%. At the end of 3 years he had earned $531 in interest. Find the original sum of money. **$3,000**

16. Maria borrowed money from her brother at 6% and twice as much money from the bank at 15%. The total interest at the end of 1 year was $1080. How much money did she borrow from the bank? **$6,000**

# Table of Roots and Powers

| No. | Sq. | Sq. Root | Cube | Cu. Root | No. | Sq. | Sq. Root | Cube | Cu. Root |
|---|---|---|---|---|---|---|---|---|---|
| 1 | 1 | 1.000 | 1 | 1.000 | 51 | 2,601 | 7.141 | 132,651 | 3.708 |
| 2 | 4 | 1.414 | 8 | 1.260 | 52 | 2,704 | 7.211 | 140,608 | 3.733 |
| 3 | 9 | 1.732 | 27 | 1.442 | 53 | 2,809 | 7.280 | 148,877 | 3.756 |
| 4 | 16 | 2.000 | 64 | 1.587 | 54 | 2,916 | 7.348 | 157,564 | 3.780 |
| 5 | 25 | 2.236 | 125 | 1.710 | 55 | 3,025 | 7.416 | 166,375 | 3.803 |
| 6 | 36 | 2.449 | 216 | 1.817 | 56 | 3,136 | 7.483 | 175,616 | 3.826 |
| 7 | 49 | 2.646 | 343 | 1.913 | 57 | 3,249 | 7.550 | 185,193 | 3.849 |
| 8 | 64 | 2.828 | 512 | 2.000 | 58 | 3,364 | 7.616 | 195,112 | 3.871 |
| 9 | 81 | 3.000 | 729 | 2.080 | 59 | 3,481 | 7.681 | 205,379 | 3.893 |
| 10 | 100 | 3.162 | 1,000 | 2.154 | 60 | 3,600 | 7.746 | 216,000 | 3.915 |
| 11 | 121 | 3.317 | 1,331 | 2.224 | 61 | 3,721 | 7.810 | 226,981 | 3.936 |
| 12 | 144 | 3.464 | 1,728 | 2.289 | 62 | 3,844 | 7.874 | 238,328 | 3.958 |
| 13 | 169 | 3.606 | 2,197 | 2.351 | 63 | 3,969 | 7.937 | 250,047 | 3.979 |
| 14 | 196 | 3.742 | 2,744 | 2.410 | 64 | 4,096 | 8.000 | 262,144 | 4.000 |
| 15 | 225 | 3.873 | 3,375 | 2.466 | 65 | 4,225 | 8.062 | 274,625 | 4.021 |
| 16 | 256 | 4.000 | 4,096 | 2.520 | 66 | 4,356 | 8.124 | 287,496 | 4.041 |
| 17 | 289 | 4.123 | 4,913 | 2.571 | 67 | 4,489 | 8.185 | 300,763 | 4.062 |
| 18 | 324 | 4.243 | 5,832 | 2.621 | 68 | 4,624 | 8.246 | 314,432 | 4.082 |
| 19 | 361 | 4.359 | 6,859 | 2.668 | 69 | 4,761 | 8.307 | 328,509 | 4.102 |
| 20 | 400 | 4.472 | 8,000 | 2.714 | 70 | 4,900 | 8.357 | 343,000 | 4.121 |
| 21 | 441 | 4.583 | 9,261 | 2.759 | 71 | 5,041 | 8.426 | 357,911 | 4.141 |
| 22 | 484 | 4.690 | 10,648 | 2.802 | 72 | 5,184 | 8.485 | 373,248 | 4.160 |
| 23 | 529 | 4.796 | 12,167 | 2.844 | 73 | 5,329 | 8.544 | 389,017 | 4.179 |
| 24 | 576 | 4.899 | 13,824 | 2.884 | 74 | 5,476 | 8.602 | 405,224 | 4.198 |
| 25 | 625 | 5.000 | 15,625 | 2.924 | 75 | 5,625 | 8.660 | 421,875 | 4.217 |
| 26 | 676 | 5.099 | 17,576 | 2.962 | 76 | 5,776 | 8.718 | 438,976 | 4.236 |
| 27 | 729 | 5.196 | 19,683 | 3.000 | 77 | 5,929 | 8.775 | 456,533 | 4.254 |
| 28 | 784 | 5.292 | 21,952 | 3.037 | 78 | 6,084 | 8.832 | 474,552 | 4.273 |
| 29 | 841 | 5.385 | 24,389 | 3.072 | 79 | 6,241 | 8.888 | 493,039 | 4.291 |
| 30 | 900 | 5.477 | 27,000 | 3.107 | 80 | 6,400 | 8.944 | 512,000 | 4.309 |
| 31 | 961 | 5.568 | 29,791 | 3.141 | 81 | 6,561 | 9.000 | 531,441 | 4.327 |
| 32 | 1,024 | 5.657 | 32,768 | 3.175 | 82 | 6,724 | 9.055 | 551,368 | 4.344 |
| 33 | 1,089 | 5.745 | 35,937 | 3.208 | 83 | 6,889 | 9.110 | 571,787 | 4.362 |
| 34 | 1,156 | 5.831 | 39,304 | 3.240 | 84 | 7,056 | 9.165 | 592,704 | 4.380 |
| 35 | 1,225 | 5.916 | 42,875 | 3.271 | 85 | 7,225 | 9.220 | 614,125 | 4.397 |
| 36 | 1,296 | 6.000 | 46,656 | 3.302 | 86 | 7,396 | 9.274 | 636,056 | 4.414 |
| 37 | 1,369 | 6.083 | 50,653 | 3.332 | 87 | 7,569 | 9.327 | 658,503 | 4.431 |
| 38 | 1,444 | 6.164 | 54,872 | 3.362 | 88 | 7,744 | 9.381 | 681,472 | 4.448 |
| 39 | 1,521 | 6.245 | 59,319 | 3.391 | 89 | 7,921 | 9.434 | 704,969 | 4.465 |
| 40 | 1,600 | 6.325 | 64,000 | 3.420 | 90 | 8,100 | 9.487 | 729,000 | 4.481 |
| 41 | 1,681 | 6.403 | 68,921 | 3.448 | 91 | 8,281 | 9.539 | 753,571 | 4.498 |
| 42 | 1,764 | 6.481 | 74,088 | 3.476 | 92 | 8,464 | 9.592 | 778,688 | 4.514 |
| 43 | 1,849 | 6.557 | 79,507 | 3.503 | 93 | 8,649 | 9.644 | 804,357 | 4.531 |
| 44 | 1,936 | 6.633 | 85,184 | 3.530 | 94 | 8,836 | 9.695 | 830,584 | 4.547 |
| 45 | 2,025 | 6.708 | 91,125 | 3.557 | 95 | 9,025 | 9.747 | 857,375 | 4.563 |
| 46 | 2,116 | 6.782 | 97,336 | 3.583 | 96 | 9,216 | 9.798 | 884,736 | 4.579 |
| 47 | 2,209 | 6.856 | 103,823 | 3.609 | 97 | 9,409 | 9.849 | 912,673 | 4.595 |
| 48 | 2,304 | 6.928 | 110,592 | 3.634 | 98 | 9,604 | 9.899 | 941,192 | 4.610 |
| 49 | 2,401 | 7.000 | 117,649 | 3.659 | 99 | 9,801 | 9.950 | 970,299 | 4.626 |
| 50 | 2,500 | 7.071 | 125,000 | 3.684 | 100 | 10,000 | 10.000 | 1,000,000 | 4.642 |

# Trigonometric Ratios

| Angle Measure | Sin | Cos | Tan | Angle Measure | Sin | Cos | Tan |
|---|---|---|---|---|---|---|---|
| 0° | 0.000 | 1.000 | 0.000 | 46° | .7193 | .6947 | 1.036 |
| 1° | .0175 | .9998 | .0175 | 47° | .7314 | .6820 | 1.072 |
| 2° | .0349 | .9994 | .0349 | 48° | .7431 | .6691 | 1.111 |
| 3° | .0523 | .9986 | .0524 | 49° | .7547 | .6561 | 1.150 |
| 4° | .0698 | .9976 | .0699 | 50° | .7660 | .6428 | 1.192 |
| 5° | .0872 | .9962 | .0875 | 51° | .7771 | .6293 | 1.235 |
| 6° | .1045 | .9945 | .1051 | 52° | .7880 | .6157 | 1.280 |
| 7° | .1219 | .9925 | .1228 | 53° | .7986 | .6018 | 1.327 |
| 8° | .1392 | .9903 | .1405 | 54° | .8090 | .5878 | 1.376 |
| 9° | .1564 | .9877 | .1584 | 55° | .8192 | .5736 | 1.428 |
| 10° | .1736 | .9848 | .1763 | 56° | .8290 | .5592 | 1.483 |
| 11° | .1908 | .9816 | .1944 | 57° | .8387 | .5446 | 1.540 |
| 12° | .2079 | .9781 | .2126 | 58° | .8480 | .5299 | 1.600 |
| 13° | .2250 | .9744 | .2309 | 59° | .8572 | .5150 | 1.664 |
| 14° | .2419 | .9703 | .2493 | 60° | .8660 | .5000 | 1.732 |
| 15° | .2588 | .9659 | .2679 | 61° | .8746 | .4848 | 1.804 |
| 16° | .2756 | .9613 | .2867 | 62° | .8829 | .4695 | 1.881 |
| 17° | .2924 | .9563 | .3057 | 63° | .8910 | .4540 | 1.963 |
| 18° | .3090 | .9511 | .3249 | 64° | .8988 | .4384 | 2.050 |
| 19° | .3256 | .9455 | .3443 | 65° | .9063 | .4226 | 2.145 |
| 20° | .3420 | .9397 | .3640 | 66° | .9135 | .4067 | 2.246 |
| 21° | .3584 | .9336 | .3839 | 67° | .9205 | .3907 | 2.356 |
| 22° | .3746 | .9272 | .4040 | 68° | .9272 | .3746 | 2.475 |
| 23° | .3907 | .9205 | .4245 | 69° | .9336 | .3584 | 2.605 |
| 24° | .4067 | .9135 | .4452 | 70° | .9397 | .3420 | 2.747 |
| 25° | .4226 | .9063 | .4663 | 71° | .9455 | .3256 | 2.904 |
| 26° | .4384 | .8988 | .4877 | 72° | .9511 | .3090 | 3.077 |
| 27° | .4540 | .8910 | .5095 | 73° | .9563 | .2924 | 3.270 |
| 28° | .4695 | .8829 | .5317 | 74° | .9613 | .2756 | 3.487 |
| 29° | .4848 | .8746 | .5543 | 75° | .9659 | .2588 | 3.732 |
| 30° | .5000 | .8660 | .5774 | 76° | .9703 | .2419 | 4.010 |
| 31° | .5150 | .8572 | .6009 | 77° | .9744 | .2250 | 4.331 |
| 32° | .5299 | .8480 | .6249 | 78° | .9781 | .2079 | 4.704 |
| 33° | .5446 | .8387 | .6494 | 79° | .9816 | .1908 | 5.145 |
| 34° | .5592 | .8290 | .6745 | 80° | .9848 | .1736 | 5.671 |
| 35° | .5736 | .8192 | .7002 | 81° | .9877 | .1564 | 6.314 |
| 36° | .5878 | .8090 | .7265 | 82° | .9903 | .1392 | 7.115 |
| 37° | .6018 | .7986 | .7536 | 83° | .9925 | .1219 | 8.144 |
| 38° | .6157 | .7880 | .7813 | 84° | .9945 | .1045 | 9.514 |
| 39° | .6293 | .7771 | .8098 | 85° | .9962 | .0872 | 11.43 |
| 40° | .6428 | .7660 | .8391 | 86° | .9976 | .0698 | 14.30 |
| 41° | .6561 | .7547 | .8693 | 87° | .9986 | .0523 | 19.08 |
| 42° | .6691 | .7431 | .9004 | 88° | .9994 | .0349 | 28.64 |
| 43° | .6820 | .7314 | .9325 | 89° | .9998 | .0175 | 57.29 |
| 44° | .6947 | .7193 | .9657 | 90° | 1.000 | 0.000 | |
| 45° | .7071 | .7071 | 1.000 | | | | |

# Glossary

The explanations given in this glossary are intended to be brief descriptions of the terms listed. They are not necessarily definitions.

**abscissa** (p. 294)   The abscissa of an ordered pair is the $x$-coordinate, the first number in the pair.

**absolute value** (p. 30)   The absolute value of a positive number or zero is the number itself. The absolute value of a negative number is the opposite of the number. We read $|x|$ as the absolute value of $x$. *Examples:* $|-3| = 3$, $|2| = 2$, $|0| = 0$

**acute** (p. 457)   An angle is an acute angle if its measure is less than 90°. *Example:* $m\angle A = 72$; $\angle A$ is an acute angle.

**addition property for equations** (p. 64)   We can add the same number to each side of an equation. If $a = b$ is true, then $a + c = b + c$ is also true, for all numbers $a$, $b$, and $c$.

**addition property for inequalities** (p. 124)   We can add the same number to each side of a true inequality, and the result is another true inequality of the same order. If $a < b$, then $a + c < b + c$.

**additive identity** (p. 37)   Zero is the additive identity since adding zero to a number gives the same number.

**additive inverse** (p. 38)   The additive inverse of a number is the opposite of the number. $-6$ is the additive inverse of 6. 8 is the additive inverse of $-8$. 0 is its own additive inverse.

**adjacent side** (p. 462)   In a triangle, a side is adjacent to an angle if it is contained in the angle.

**angle** (p. 455)   An angle is a figure formed by two rays with a common endpoint.

**area** (p. 18)   The area of a geometric figure is the number of unit squares it contains. The area of a rectangle is given by the formula $A = l \cdot w$, where $l$ is the length and $w$ is the width.

**base** (p. 18)   In $3^4$, the 3 is the base. $3^4 = 3 \cdot 3 \cdot 3 \cdot 3$. The base is used 4 times as a factor.

**binomial** (p. 152)   A binomial is a polynomial with two terms.

**combine** (p. 13)   To combine like terms in an expression such as $5y - 9y$, we use the distributive property.   $5y - 9y = (5 - 9)y$
$$= -4y$$

**complementary angles** (p. 454)   Two angles are complementary if the sum of their measures is 90°.

**completing the square** (p. 432)   Completing the square is a method for finding the solution set of a quadratic equation.

**conjunction** (p. 131)   A conjunction is a compound sentence with the connective *and*. The symbol for *and* is $\wedge$.

**constant function** (p. 370)   A constant function is a function whose graph is a horizontal line or a subset of a horizontal line.

**convenient form** (p. 212)   A polynomial is in convenient form if its terms are in descending order of exponents, and the coefficient of the first term is positive. $6x - 2x^2 + 3$ is in convenient form when it is expressed as $-1(2x^2 - 6x - 3)$.

**converse** (p. 413)   The converse of an if–then statement is formed by reversing the if and the then parts.

**coordinate(s) of a point** (p. 28)   On a number line, the coordinate of a point is the number that corresponds to the point. In a coordinate plane, the coordinates of a point make up the ordered pair that corresponds to the point.

**corresponding angles** (p. 459)   The corresponding angles of two similar figures are the pairs of angles that have the same measure.

**corresponding sides** (p. 459)   The pairs of corresponding sides of two similar figures are the sides that lie opposite the pairs of corresponding angles.

**cosine of an angle** (p. 463)  The cosine of an acute angle of a right triangle is the ratio of the length of the side adjacent to the angle to the length of the hypotenuse.

**degree** (p. 152)  The degree of a monomial is the sum of the exponents of all its variables. $3x^2y^5$ has degree 7. The degree of a polynomial is the highest degree of any of its terms. The polynomial $7x^3 - 4x + 1$ has degree 3.

**denominator** (p. 2)  In the fraction $\frac{7}{8}$, 8 is the denominator.

**difference** (p. 40)  The difference is the result of a subtraction. In $17 - 8 = 9$, 9 is the difference.

**difference of two squares** (p. 173)  $a^2 - b^2$ represents the difference of two squares. $a^2 - b^2$ may be factored as $(a + b)(a - b)$.

**direct variation** (p. 373)  A direct variation is a function in which the ratio $y$ to $x$ is always the same. $\frac{y}{x} = k$, or $y = kx$ is a direct variation. $y$ varies directly as $x$, or $y$ is directly proportional to $x$.

**disjunction** (p. 131)  A disjunction is a compound sentence with the connective *or*, symbol $\vee$.

**distributive property of multiplication over addition** (p. 10)  Multiplication is distributive over addition. $a(b + c) = a \cdot b + a \cdot c$ and $(b + c)a = b \cdot a + c \cdot a$, for all numbers $a$, $b$, and $c$.

**division property for equations** (p. 67)  We can divide each side of an equation by the same nonzero number. If $a = b$ is true, then $\frac{a}{c} = \frac{b}{c}$ is also true, for all numbers $a$, $b$, and $c$, $[c \neq 0]$.

**division property for inequalities** (p. 126)  We can divide each side of a true inequality by the same positive number and the result is another true inequality of the same order. If $a < b$ and $c > 0$, then $\frac{a}{c} < \frac{b}{c}$. We can divide each side of a true inequality by the same negative number, and the result is another true inequality of the reverse order. If $a < b$ and $c < 0$, then $\frac{a}{c} > \frac{b}{c}$.

**domain of a relation** (p. 365)  The domain of a relation is the set of all first elements of the ordered pairs in the relation. For the relation $\{(0, 1), (2, -5), (4, 3)\}$, the domain is $\{0, 2, 4\}$.

**elements** (p. 118)  The objects that belong to a set are the elements, or members, of the set.

**empty set** (p. 119)  The empty set is the set containing no elements. The symbol $\varnothing$ means the empty set.

**equation** (p. 64)  A sentence with $=$ is an equation.

**equation of a line** (p. 311)  $y = mx + b$ is an equation of a line in a coordinate plane. $m$ is the slope of the line and $b$ is the $y$-intercept.

**even integer** (p. 195)  An even integer has a factor of 2. $\ldots, -4, -2, 0, 2, 4, 6, \ldots$ are even integers.

**exponent** (p. 16)  In $3^4$, the 4 is an exponent. $3^4 = 3 \cdot 3 \cdot 3 \cdot 3$. A positive integer exponent tells how many times the base (3) is used as factor.

**extraneous solution** (p. 266)  An extraneous solution of an equation is an apparent solution that does not check.

**extremes** (p. 225)  Is the proportion $\frac{a}{b} = \frac{c}{d}$, $a$ and $d$ are the extremes.

**factor a trinomial** (p. 174)  To factor a trinomial means to express it as the product of two binomials. $x^2 - 5x + 6$ can be factored as $(x - 2)(x - 3)$.

**factors** (p. 16)  Factors are numbers that are multiplied. In $5 \cdot 8 = 40$, 5 and 8 are factors.

**finite set** (p. 118)  A set is finite if it contains a definite number of elements. $\{2, 4, 6\}$ is a finite set.

**fraction** (p. 2)  A fraction is an indicated quotient of two numbers.

**function** (p. 366)  A function is a relation in which no two ordered pairs have the same first element.

**graph of an equation (inequality)** (p. 314, p. 319) In a coordinate plane, the graph of an equation (inequality) is the set of all points and only those points whose coordinates satisfy the equation (inequality).

**graph of an ordered pair** (p. 293) In a coordinate plane, the graph of an ordered pair is the point that corresponds to the ordered pair.

**greatest common factor (GCF)** (p. 161) The greatest common factor of two numbers is the greatest number that is a factor of both numbers. 8 is the GCF of 16 and 24. The greatest common variable factor of $a^3$ and $2a^2$ is $a^2$.

**horizontal line** (p. 293) A horizontal line in a coordinate plane is parallel to the $x$-axis.

**hypotenuse** (p. 412) The hypotenuse of a right triangle is the side opposite the right angle.

**increase** (p. 74) 3 increased by 7 means $3 + 7$, or 10.

**inequality** (p. 74) An inequality is a sentence with $\neq$, $<$, $>$, $\leq$, or $\geq$. $x + 3 < 7$ is an inequality.

**infinite set** (p. 118) A set is infinite if it contains no definite number of elements. $\{1, 2, 3, 4, \ldots\}$ is an infinite set.

**integer** (p. 28) The numbers $\ldots, -4, -3, -2, -1, 0, 1, 2, 3, 4, \ldots$ are integers.

**irrational number** (p. 33, p. 390) An irrational number cannot be expressed in the form $\frac{a}{b}$, where $a$ and $b$ are integers and $b \neq 0$. Nonrepeating decimals are irrational numbers.

**least common multiple (LCM)** (p. 97, p. 238) The least common multiple of two or more numbers is the smallest number that has each as a factor. The LCM of 2, 3, and 4 is 12.

**leg** (p. 412, p. 462) The legs of a right triangle are the two sides that form the right angle.

**less than** (p. 74, p. 121) 7 less than 9 means $9 - 7$, or 2.

**like terms** (p. 13) Like terms contain the same variable or variables. $5xy$ and $-3xy$ are like terms. $-8b$ and $11b^2$ are unlike terms.

**linear function** (p. 369) A linear function is a function whose graph is a line or a subset of a line that is neither vertical nor horizontal.

**means** (p. 225) In the proportion $\frac{a}{b} = \frac{c}{d}$, $b$ and $c$ are the means.

**mode** (p. 486) The mode of a set of data is the member(s) that occurs most frequently.

**monomial** (p. 161) A monomial is a polynomial with one term.

**more than** (p. 74) 8 more than 5 means $5 + 8$, or 13.

**multiplication property for equations** (p. 67, p. 96) We can multiply each side of an equation by the same number. If $a = b$ is true, then $a(c) = b(c)$ is also true, for all numbers $a$, $b$, and $c$.

**multiplication property for inequalities** (p. 125) We can multiply each side of a true inequality by the same *positive* number, and the result is another true inequality of the *same order*. If $a < b$ and $c > 0$, then $a \cdot c < b \cdot c$.
We can multiply each side of a true inequality by the same *negative* number, and the result is another true inequality of the *reverse order*. If $a < b$ and $c < 0$, then $a \cdot c > b \cdot c$.

**multiplicative inverse** (p. 221) Two numbers are multiplicative inverses (reciprocals) if their product is 1.

**negative number** (p. 28) A number is negative if it lies to the left of zero on a number line. $-5$, $-1$, $-\frac{1}{2}$, and $-\frac{4}{3}$ are negative numbers.

**numerator** (p. 2) In the fraction $\frac{7}{8}$, 7 is the numerator.

**numerical coefficient** (p. 5) A numerical coefficient is the number by which a variable is multiplied. In $6a - 3b + 8$, 6 is the numerical coefficient of $a$, and $-3$ is the numerical coefficient of $b$.

**odd integer** (p. 195)   An odd integer does not have a factor of 2.  . . . , $-5, -3, -1, 0, 1, 3, 5, 7, \ldots$ are odd integers.

**open sentence** (p. 21)   An open sentence is a sentence that contains a variable. $7y - 3 = 5$ is an open sentence because it contains the variable $y$.

**opposite of a number** (p. 38)   *See* additive inverse. The symbol $-x$ means the opposite of $x$.

**opposite side** (p. 462)   In a triangle, a side is opposite an angle if it is not contained in the angle.

**order of operations** (p. 1)   When both multiplications and additions occur, we agree to multiply first and then add.
*Example:* $5 + 7 \cdot 3 = 5 + 21$
$$= 26$$

**ordered pair** (p. 293)   $(-4, 1)$ is an ordered pair of numbers. Each ordered pair of numbers corresponds to a point in a coordinate plane, and vice versa.

**ordinate** (p. 294)   The ordinate of an ordered pair is the $y$-coordinate, the second number in the pair.

**origin** (p. 28)   The origin is the point for zero on a number line. The origin is the point for $(0, 0)$ in a coordinate plane.

**parallel** (p. 304)   Lines are parallel if they lie in the same plane and never meet.

**perimeter** (p. 18)   The perimeter of a figure is the distance around the figure. If a triangle has sides measuring 6 inches, 7 inches, and 10 inches, then its perimeter is $6 + 7 + 10$, or 23 inches.

**perpendicular lines** (p. 293)   Two lines are perpendicular if they form right angles (90°). $\perp$ means *is perpendicular to*.

**plane** (p. 293)   A plane is a flat surface.

**polynomial** (p. 152)   A polynomial contains one or more terms.
*Examples:* $2x^2, 5x + 3, 7a^2 - 5a - 2$

**positive number** (p. 28)   A number is positive if it lies to the right of zero on a number line.
$8, 2, \frac{1}{3}$, and $\frac{16}{5}$ are positive numbers.

**power** (p. 16)   $3^4$ means the fourth power of 3. $3^4 = 3 \cdot 3 \cdot 3 \cdot 3$.  3 is used 4 times as a factor.

**prime factorization** (p. 159)   The prime factorization of 20 is $2 \cdot 2 \cdot 5$, since $2 \cdot 2 \cdot 5 = 20$, and 2 and 5 are prime numbers.

**prime number** (p. 159)   A prime number is a whole number greater than 1 whose only factors are itself and 1. 2, 3, 5, 7, 11, . . . are prime numbers.

**product** (p. 16)   The product is the result of a multiplication. In $5 \cdot 6 = 30$, 30 is the product.

**property of additive identity** (p. 37)   Adding zero to a number gives the same number. $a + 0 = a$ and $0 + a = a$, for each number $a$.

**property of additive inverse** (p. 38)   Adding two opposite numbers (additive inverses) gives zero. $a + (-a) = 0$, for each number $a$.

**property of multiplicative identity** (p. 13)   Multiplying any number by 1 gives the same number. $(1)(a) = a$ and $(a)(1) = a$, for each number $a$.

**property of zero for multiplication** (p. 41)   Multiplying any number by zero gives zero. $a \cdot 0 = 0$ and $0 \cdot a = 0$, for each number $a$.

**proportion** (p. 225)   A proportion is an equation that states that two ratios are equal.
$\frac{a}{b} = \frac{c}{d}$ is a proportion.

**Pythagorean property** (p. 415)   In any right triangle, the square of the length of the hypotenuse equals the sum of the squares of the lengths of the other two sides. If $\triangle ABC$ is a right triangle, then $a^2 + b^2 = c^2$.

**quadrant** (p. 294)   The $x$- and $y$-axes divide the coordinate plane into four quadrants.

**quadratic equation** (p. 188)   In a quadratic equation, one variable is raised to the second power, but no higher. $3x^2 - 5x + 4 = 0$ and $6x^2 = 2$ are quadratic equations.

**quotient** (p. 41)   The quotient is the result of a division. In $54 \div 9 = 6$, 6 is the quotient.

**radical equation** (p. 426)   In a radical equation, the variable is the radicand. $\sqrt{2x} = 6$ is a radical equation.

**range of a relation** (p. 365)   The range of a relation is the set of all second elements of the ordered pairs in the relation. For the relation $\{(0, 1), (2, -5), (4, 3)\}$, the range is $\{1, -5, 3\}$.

**ratio** (p. 225)   The ratio of $a$ to $b$ is the quotient $\frac{a}{b}$, or $a : b$. A ratio is a comparison of two numbers by division.

**rational expression** (p. 205)   Expressions that contain polynomials in the numerator, or in the denominator, or in both, are called *rational expressions*. $\frac{4}{x}$ and $\frac{x-4}{x^2+2}$ are rational expressions.

**rational number** (p. 32, p. 389)   A rational number is a number that can be written in the form $\frac{a}{b}$, where $a$ and $b$ are integers and $b \neq 0$. $\frac{3}{5}, \frac{-24}{7}$, 8, and .63 are rational numbers.

**real number** (p. 33)   The set of real numbers includes all the rational numbers and all the irrational numbers.

**reciprocal** (p. 45, p. 222)   Two numbers are reciprocals (multiplicative inverses) if their product is 1.   5 and $\frac{1}{5}$ are reciprocals, since $5\left(\frac{1}{5}\right) = 1$.

**relation** (p. 365)   A relation is a set of ordered pairs. $\{(0, 1), (2, -5), (-4, 2)\}$ is a relation.

**relatively prime** (p. 209)   Two numbers are relatively prime if their only common factor is 1.

**repeating decimal** (p. 32, p. 389)   A repeating decimal has a digit or a group of digits that repeats forever. The decimal .5858585858. . . , or .58̄5̄8̄, is a repeating decimal. The bar indicates that the digits repeat forever.

**right triangle** (p. 42)   A right triangle is a triangle with one right angle.

**root** (p. 189)   A root of an equation is a solution of the equation.

**sentence** (p. 21)   A mathematical sentence contains either $=$, $\neq$, $>$, $<$, $\geq$, or $\leq$. $4 + 5 = 9$ and $7 \leq x - 3$ are mathematical sentences.

**set** (p. 118)   A set is a collection of objects. We use braces { } to show a set.

**similar** (p. 459)   Two figures are similar if the corresponding angles are congruent and the lengths of the corresponding sides have the same ratio.

**simplest radical form** (p. 399)   An expression is in simplest radical form if it contains no factor that is a perfect square.

**simplify** (p. 13)   To simplify an expression means to replace it with the least complicated equivalent expression.

**sine of an angle** (p. 462)   The sine of an acute angle of a right triangle is the ratio of the length of the side opposite the angle to the length of the hypotenuse.

**slope of a line** (p. 302)   The slope of a line in a coordinate plane is the slope of any segment on the line. The slope of a horizontal line is zero. The slope of a vertical line is undefined.

**slope of a segment** (p. 299)   The slope of a segment in a coordinate plane is
$$\frac{\text{difference of } y\text{-coordinates}}{\text{difference of } x\text{-coordinates}}$$

**solution** (p. 21)   A solution of an open sentence is a replacement of the variable that makes the sentence true. In $5x - 3 = 32$, 7 is a solution since $5(7) - 3 = 35 - 3 = 32$.

**solution set** (p. 118)   The solution set of an open sentence is the set of all members of the replacement set that are solutions of the sentence.

**solve** (p. 64)   To solve an open sentence means to find all of its solutions.

**square of a number** (p. 391)   The square of a number is the product of the number and itself.

**square root** (p. 391)   $x$ is a square root of $n$ if $x \cdot x = n$. 6 is a square root of 36 since $6 \cdot 6 = 36$. Also, $-6$ is a square root of 36 since $(-6)(-6) = 36$.

**subtract** (p. 40)   To subtract $b$ from $a$ means to add the opposite of $b$ to $a$. $a - b = a + (-b)$, for all numbers $a$ and $b$.

**sum** (p. 31)   The sum is the result of an addition. In $7 + 8 = 15$, 15 is the sum.

**system of equations** (p. 330)   Two equations in two variables form a system of equations.
$$2x + 3y = 6$$
$$x - 4y = -3$$
is a system.

**tangent of an angle** (p. 463)   The tangent of an acute angle of a right triangle is the ratio of the length of the side opposite the angle to the length of the side adjacent to the angle.

GLOSSARY

**terminating decimal** (p. 32, p. 389)   A terminating decimal has a finite number of digits.

**terms** (p. 5)   In $7x - 3y + 8$, the terms are $7x$, $-3y$, and 8. Terms are added.

**trinomial** (p. 152)   A trinomial is a polynomial with three terms.

**twice** (p. 75)   Twice a number means two times the number. Twice $x$ means $2x$.

**undefined rational expression** (p. 205)   A rational expression is undefined if its denominator is zero.

**value of a function** (p. 371)   If $(x, y)$ is an ordered pair in function $f$, then the value of $f$ at $x$ is $y$. We read $f(x) = y$ as the value of $f$ at $x$ is $y$, or as $f$ at $x$ is $y$.

**variable** (p. 4)   A variable takes the place of a number. In $5x - 3 = 7$, $x$ is a variable.

**vertical line** (p. 293)   A vertical line in a coordinate plane is parallel to the $y$-axis.

**vertical line test** (p. 368)   A relation is a function if no vertical line crosses its graph in more than one point.

**whole number** (p. 28)   The numbers 0, 1, 2, 3, . . . are whole numbers.

**$x$-axis** (p. 293)   The $x$-axis is the horizontal number line in a coordinate plane.

**$x$-coordinate** (p. 293)   The $x$-coordinate of an ordered pair of numbers is the first number in the pair. For $(5, -2)$, 5 is the $x$-coordinate.

**$y$-axis** (p. 293)   The $y$-axis is the vertical number line in a coordinate plane.

**$y$-coordinate** (p. 293)   The $y$-coordinate of an ordered pair of numbers is the second number in the pair. For $(5, -2)$, $-2$ is the $y$-coordinate.

**$y$-intercept** (p. 310)   The $y$-intercept of a line in a coordinate plane is the $y$-coordinate of the point of intersection of the line with the $y$-axis. In the equation of a line, $y = mx + b$, $b$ is the $y$-intercept.

**zero** (p. 28)   Zero (0) lies between the positive and the negative numbers on a number line. Zero is neither positive nor negative.

# Annotated Answers

This section contains those answers that do not appear on the TE text pages.

## Page 9

**56.** $(a + b) + c = a + (b + c)$, associative; $a + (c + b)$, commutative   **57.** $(x \cdot y) \cdot z = x \cdot (y \cdot z)$, associative; $x \cdot (z \cdot y)$, commutative; $(z \cdot y) \cdot x$, commutative   **58.** $m \cdot (g \cdot v) = (g \cdot v) \cdot m; g \cdot (v \cdot m)$, associative   **59.** $r + (t + w) = r + (w + t)$, commutative; $(r + w) + t$, associative; $(w + r) + t$, commutative

## Page 23

**1.**
```
10  LET R = 32 - 45 / 15 + 8 * 9
20  PRINT R
30  END
```

**2.**
```
10  LET B = 42 / 2.1 + 13 * 6 - 4
20  PRINT B
30  END
```

**3.**
```
10  LET A = 42 * 12 / 2
20  PRINT A
30  END
```

## Page 39

**32.** $(-9 + 4) + (-8) = (-8) + (-9 + 4)$, commutative; $(-8) + (4 + -9)$, commutative; $(-8 + 4) + (-9)$, associative
**33.** $(-8 + (-3)) + 8 = 8 + ((-8) + (-3))$, commutative; $(8 + (-8)) + (-3)$, associative; $0 + (-3)$, additive inverse; $-3$, additive identity
**34.** $(-x + y) + x = x + (-x + y)$, commutative; $(x + -x) + y$, associative; $0 + y$, additive inverse; $y$, additive identity   **35.** $-m + (-r + m) = -m + (m + (-r))$, commutative; $(-m + m) + (-r)$, associative; $0 + (-r)$, additive inverse; $-r$, additive identity

## Page 54

**37.** $-x \cdot -y = (-1)x \cdot (-1)y$, multiplication by $-1$; $(-1)(-1) \cdot xy$, commutative; $((-1)(-1)) \cdot xy$, associative; $1 \cdot xy$, multiplication fact; $xy$, multiplication identity   **38.** $a \cdot -b = a \cdot (-1)b$, multiplication by $-1$; $(-1) \cdot a \cdot b$, commutative; $(-1)ab$, associative; $-ab$, multiplication by $-1$

## Page 57

**39.** $a - b = a + -b$, definition of subtraction; $1 \cdot a + (-b)$, multiplication by 1; $(-1)(-1) \cdot a + -b$, multiplication fact; $(-1) \cdot (-1)a + (-1)b$, multiplication by $-1$; $(-1)(-a + b)$, distributive; $(-1)(b + (-a))$, commutative; $(-1)(b - a)$, definition of subtraction; $-(b - a)$, multiplication by $-1$

**40.** $-(x + y) = (-1)(x + y)$, multiplication by $-1$; $(-1)x + (-1)y$, distributive; $-x + (-y)$, multiplication by $-1$; $-y + -x$, commutative; $-y - x$, definition of subtraction

## Page 60

**47.** $-b \cdot a = (-1)b \cdot a$; multiplication by $-1$; $(-1) \cdot a \cdot b$, commutative; $(-1)ab$, associative; $-ab$, multiplication by $-1$   **48.** $-(a - b) = -(a + (-b))$, definition of subtraction; $(-1)(a + (-b))$, multiplication by $-1$; $(-1)(a + (-1)b)$, multiplication by $-1$; $(-1)a + (-1)(-1)b$, distributive; $(-1)a + b$, multiplication fact; $-a + b$, multiplication by $-1$

## Page 61

**45.**

| Expression | Reason |
|---|---|
| $-(-x - y)$ | Given |
| $-1(-x - y)$ | Mult. Prop. $-1$ |
| $1x + 1y$ | Distributive |
| $x + y$ | Mult. Prop. 1 |

## Page 111

**3.**
```
10  FOR I = 1 TO 2
20  INPUT "WHAT IS A? ";A
30  INPUT "WHAT IS C? ";C
40  LET X = C / A
50  PRINT "THE SOLUTION OF "A"X = "C" IS "X"."
60  NEXT I
70  END
```

**6.**
```
10  FOR I = 1 TO 2
20  INPUT "WHAT IS A? ";A
25  INPUT "WHAT IS B? ";B
30  INPUT "WHAT IS C? ";C
40  LET X = (C - B) / A
50  PRINT "THE SOLUTION OF "A"X + "B" = "C" IS "
    X"."
60  NEXT I
70  END
```

## Page 135

For the solution shade:   **1.** between open circles at 5 and 7   **2.** left of open circle at 4 and right of open circle at 8   **3.** left of open circle at $-4$ and right of open circle at 2   **4.** between open circles at $-6$ and $-3$   **5.** between closed circles at $-9$ and 1   **6.** left of closed circle at 3 and right of closed circle at 12   **7.** between closed circle at $-8$ and open circle at 7   **8.** left of open circle at $-5$ and right of closed circle at 5   **9.** left of closed circle at $-5$ and right of closed circle at 5   **10.** between open circles at 3 and 6   **11.** between closed circle at $-1$ and open circle at 5   **12.** between open circle at $-8$ and closed circle at 2

ANNOTATED ANSWERS

## Page 143

```
2.  10  PRINT "THIS PROGRAM DETERMINES IF THE"
    20  PRINT "ENTERED NUMBER IS A SOLUTION OF"
    30  PRINT "-7 <= 2X-5 < 3."
    40  FOR I = 1 TO 7
    50  INPUT "TRY A VALUE FOR X. ";X
    60  IF - 7 < = 2 * X - 5 AND 2 * X - 5 < 3 THEN
          90
    70  PRINT X" IS NOT A SOLUTION."
    80  GOTO 100
    90  PRINT X" IS A SOLUTION."
    100  PRINT : PRINT
    110  NEXT I
    120  END
```

```
3.  10  PRINT "THIS PROGRAM DETERMINES IF THE"
    20  PRINT "ENTERED NUMBER IS A SOLUTION OF"
    30  PRINT "THE ABSOLUTE VALUE OF (3A + 6) > 18."
    40  FOR I = 1 TO 5
    50  INPUT "TRY A VALUE FOR A. ";A
    60  IF 3 * A + 6 > 18 OR 3 * A + 6 < - 18 THEN
          90
    70  PRINT A" IS NOT A SOLUTION."
    80  GOTO 100
    90  PRINT A" IS A SOLUTION."
    100  PRINT : PRINT
    110  NEXT I
    120  END
```

**4.** Change lines 30 and 60 of the program in Ex. 3 to read:
30  PRINT ''THE ABSOLUTE VALUE OF (2A − 4) < 6.''
60  IF 2 * A − 4 < 6 AND 2 * A − 4 > −6 THEN 90.

## Page 147

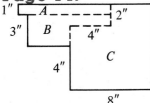
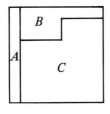

## Page 172

**7.** $3m^2 + 14m - 49$
**8.** $4y^2 + 26y - 14$
**9.** $6k^2 + 7k - 20$
**10.** $10x^2 - 29x + 10$
**11.** $8a^2 + 10a - 3$
**12.** $2h^2 - 10h - 48$
**13.** $12a^2 + ab - 6b^2$
**14.** $4y^2 + 16ym - 9m^2$
**15.** $6x^2 - 7xy - 3y^2$
**16.** $3p^2 + p - 2$
**17.** $2a^2 - 5a + 3$
**18.** $40m^2 - 20m + 25$
**19.** $10x^2 - 7xy + y^2$
**20.** $x^2 + 2xy + y^2$

**21.** $10a^2 - 3ab - b^2$
**22.** $2m^2 + 6m - 8$
**23.** $3x^2 - xy - 4y^2$
**24.** $3d^2 + 7d - 6$
**25.** $6x^3 + 10x^2 + 5x - 3$
**26.** $8a^3 + 22a^2 - 17a + 14$
**27.** $3x^3 + 20x^2 - 2x + 35$
**28.** $t^3 - 12t^2 + 36t - 5$
**29.** $3p^3 - 19p^2 - 5p + 6$
**30.** $2g^3 - 13g^2 - 32g - 12$
**31.** $3k^4 - 20k^3 + 27k^2 - 10k$
**32.** $2x^4 + x^3 - 8x^2 - x - 15$
**33.** $28 - 29a + 10a^2 - 3a^3$
**34.** $3x^3 + 8x^2 - 68x - 48$
**35.** $6x^3 - 7x^2 - 14x + 8$

## Page 174

**21.** $(2b - 5)(2b + 5)$
**22.** $(3a - 8)(3a + 8)$
**23.** $(4m + 5)(4m - 5)$
**24.** $(7a + 10)(7a - 10)$
**25.** $(1 + 8c)(1 - 8c)$
**26.** $(1 + 3x)(1 - 3x)$
**27.** $(3a + 5)(3a - 5)$
**28.** $(12 + 7a)(12 - 7a)$
**29.** $(12a - 13)(12a + 13)$
**30.** $(10x - 11)(10x + 11)$
**31.** $(11m - 12)(11m + 12)$
**32.** $(6m - 13)(6m + 13)$
**33.** $(7 - 13a)(7 + 13a)$
**34.** $(x - 14)(x + 14)$
**35.** $(13p - 6)(13p + 6)$
**36.** $(11m - 15)(11m + 15)$
**37.** $\left(\frac{2}{5}x + \frac{3}{7}\right)\left(\frac{2}{5}x - \frac{3}{7}\right)$
**38.** $\left(\frac{4}{9} + \frac{5}{12}a\right)\left(\frac{4}{9} - \frac{5}{12}a\right)$
**39.** $\left(\frac{11}{3}r + \frac{1}{2}\right)\left(\frac{11}{3}r - \frac{1}{2}\right)$
**40.** $\left(\frac{7}{6} + \frac{4}{7}t\right)\left(\frac{7}{6} - \frac{4}{7}t\right)$
**41.** $(4x + 3)(2x - 7)$
**42.** $(11x - 4)(-x + 10)$
**43.** $(3k - 14)(-k - 2)$
**44.** $(4a - b - 6c)(-2a + 3b + 2c)$

## Page 176

**13.** no
**14.** $(a + 4)^2$
**15.** $(y - 6)^2$
**16.** $(t + 2)^2$
**17.** no
**18.** $(u + 8)^2$
**19.** no
**20.** $(3a - 1)^2$
**21.** $(3a - 2)^2$
**22.** $(2m + 3)^2$
**23.** no
**24.** $(4a - 3)^2$
**25.** $(3x + 5y)^2$
**26.** $(5a - 2b)^2$
**27.** no
**28.** $(7x - 4y)^2$
**29.** $(5x - 6y)^2$
**30.** $(8x + 3y)^2$
**31.** $(11m + 2n)^2$
**32.** $(5k + 9j)^2$
**33.** $(5a + 7b)^2$

## Page 181

**7.** $6m$ and $m$ or $3m$ and $2m$; $-5$, $-1$
**8.** $10y$ and $y$ or $5y$ and $2y$; $-7$, $-1$
**9.** $8b$ and $b$ or $4b$ and $2b$; $-3$, $-1$

## Page 182

**1.** $(2x + 3)(x + 2)$
**2.** $(3y + 5)(y + 2)$
**3.** $(3x - 2)(x - 2)$
**4.** $(2a - 7)(a + 2)$
**5.** $(3x - 4)(x + 1)$
**6.** $(2b + 1)(b - 4)$
**7.** $(3m - 1)(2m - 5)$
**8.** $(5y - 7)(2y - 1)$
**9.** $(4b - 3)(2b - 1)$
**10.** $(6a - 1)(a + 1)$
**11.** $(4n + 5)(n + 1)$
**12.** $(3a - 8)(a + 1)$
**13.** $(4x - 5)(x + 1)$
**14.** $(2a - 3)(a + 1)$
**15.** $(2y - 3)(y - 2)$
**16.** $(2x - 3)(x + 7)$
**17.** $(2x - 5)(x + 4)$
**18.** $(2y - 5)(y + 2)$
**19.** $(2g + 3)(g - 5)$
**20.** $(2a + 3)(a + 5)$

ANNOTATED ANSWERS

**21.** $(3b + 2)(b + 3)$  **22.** $(3x - 5)(x + 5)$
**23.** $(6k + 1)(6k + 1)$  **24.** $(2p - 5)(p - 5)$
**25.** $(3a + 1)(2a - 1)$  **26.** $(3y + 4)(y - 6)$
**27.** $(2m - 5)(m - 1)$  **28.** $(2b + 5)(b + 6)$
**29.** $(2x + 1)(x + 2)$  **30.** $(4y + 1)(3y + 1)$
**31.** $(7y - 1)(4y - 2)$  **32.** $(2a - 5)(2a - 5)$
**33.** $(3x + 5)(x - 6)$  **34.** $(2x + 5)(x - 8)$
**35.** $(2a + 5)(a + 9)$  **36.** $(3b - 8)(b + 6)$
**37.** $(2a - 5)(a - 10)$  **38.** $(3b - 7)(b + 5)$
**39.** $(2m - 5)(m - 12)$  **40.** $(3m + 5)(m + 8)$
**41.** $(5x + 3)(x - 9)$  **42.** $(3a - 7)(a - 8)$
**43.** $(6a + 5)(a - 3)$  **44.** $(4r - 5)(2r - 9)$
**45.** $(3p - 7)(2p + 5)$  **46.** $(5q - 3)(3q - 2)$
**47.** $(5y + 9)(3y - 2)$  **48.** $(7b + 3)(2b - 3)$
**49.** $(8y - 5)(2y + 3)$  **50.** $(6a + 7)(3a - 5)$
**51.** $(7p - 2)(2p + 5)$  **52.** $(5 - z)(2 + z)$
**53.** $(9 + m)(2 - m)$  **54.** $(5 + a)(3 - a)$
**55.** $(3 - 4a)(1 + 2a)$  **56.** $(5 + 2a)(1 - 2a)$
**57.** $(1 + x)(8 - 5x)$

## Page 185

**16.** $2t(t - 3)(t - 4)$  **17.** $2g(3g + 1)(g - 5)$
**18.** $3r(2r - 5)(r + 2)$  **19.** $5(a - 5)(a - 2)$
**20.** $2a(a + 3)(a + 5)$  **21.** $3(4w - 1)(w + 3)$
**25.** $2(y + 9z)(y + z)(y + z)$  **26.** $2(m - 5n)(m + 4n)$
**27.** $3(x - 5y)(x + 2y)$  **28.** $3(x + 5y)(x - 5y)$
**29.** $4(a + b)(a - b)$  **30.** $3(2q + 5k)(2q - 5k)$
**34.** $3(2x - 7)(x + 3)$  **35.** $a(11 - 7a)(11 + 7a)$
**36.** $p^2(3p + 7)(p - 4)$  **37.** $2e(5e + 7)(3e - 2)$
**38.** $3r(5r - 4)(2r + 1)$  **39.** $2b(5b - 3)(b + 4)$
**40.** $b(ab + 5)(ab - 5)$  **41.** $r(r - 4t)(r - t)$
**42.** $xy(x + y)(x - y)$  **43.** $x(x - 2y)(x - y)$
**44.** $ab(3a^2 + 13ab - 3b^2)$
**45.** $3my(2m - 5y)(2m + 5y)$
**49.** $2(x - 4)(x + 4)(x^2 + 16)$
**50.** $3(3 - y)(3 + y)(9 + y^2)$
**51.** $(5m - 2)(5m + 2)(25m^2 + 4)$
**52.** $x(x - 4)(x + 4)(x - 4)(x + 4)$
**53.** $2(a - 3)(a + 3)(a - 2)(a + 2)$
**54.** $(r - 1)(r + 1)(2r - 5)(2r + 5)$
**55.** $3(a - 4)(a + 4)(a - 3)(a + 3)$
**56.** $3t(t - 4)(t + 4)(t - 2)(t + 2)$
**57.** $2y^2(2y - 3)(2y + 3)(y - 2)(y + 2)$
**58.** $(x + 3y)(x - 3y)(x + 2y)(x - 2y)$
**59.** $(2a - 5b)(2a + 5b)(4a^2 + 25b^2)$
**60.** $2xy(x - 3y)(x + 3y)(x - y)(x + y)$

## Page 202

**5.**
```
10   PRINT "THIS PROGRAM FINDS THE SUM OF 4 CONSE
     CUTIVE ODD OR EVEN INTEGERS."
20   PRINT "TYPE IN THE FIRST NUMBER."
30   INPUT X
40   LET S = X + (X + 2) + (X + 4) + (X + 6)
50   PRINT "THE FOUR INTEGERS ARE "X", "X + 2", "
     X + 4", "X + 6"."
60   PRINT "THE SUM IS "S"."
70   END
```

**6.**
```
10   PRINT "THIS PROGRAM FINDS THE SUM OF 3 CONSECU
     TIVE ODD OR EVEN INTEGERS."
20   PRINT "TYPE IN THE FIRST NUMBER."
30   INPUT X
40   LET S = X + (X + 2) + (X + 4)
50   PRINT "THE THREE INTEGERS ARE "X", "X + 2", "X
     + 4"."
60   PRINT "THE SUM IS "S"."
70   END
```

**7.**
```
10   PRINT "THIS PROGRAM FINDS 3 CONSECUTIVE INTE
     GERS"
20   PRINT "FOR A GIVEN SUM."
30   PRINT "ENTER THE SUM."
40   INPUT S
50   LET X = (S - 3) / 3
60   IF INT (X) < > X THEN 100
70   PRINT "THE 3 CONSECUTIVE INTEGERS "
80   PRINT "ARE: "X", "X + 1", "X + 2"."
90   GOTO 120
100  PRINT S" DOES NOT EQUAL THE SUM OF"
110  PRINT "THREE CONSECUTIVE INTEGERS."
120  END
```

**8.**
```
10   PRINT "THIS PROGRAM FINDS 3 CONSECUTIVE EVEN
     OR ODD INTEGERS"
20   PRINT "FOR A GIVEN SUM."
30   PRINT "ENTER THE SUM."
40   INPUT S
50   LET X = (S - 6) / 3
60   IF INT (X) < > X THEN 100
70   PRINT "THE 3 CONSECUTIVE EVEN OR ODD INTEGER
     S "
80   PRINT "ARE: "X", "X + 2", "X + 4"."
90   GOTO 120
100  PRINT S" DOES NOT EQUAL THE SUM OF THREE"
110  PRINT "CONSECUTIVE EVEN OR ODD INTEGERS."
120  END
```

## Page 208

**15.** $\dfrac{-2x^2 + x + 3}{2x^2 + 7x + 6}$  **16.** $\dfrac{3a^2 + 11a - 4}{3a^2 + 4a + 1}$

**17.** $\dfrac{-2b^2 - b + 15}{2b^2 + 17b + 35}$  **18.** $\dfrac{6p^2 + 7p - 3}{p^2 + 2p - 24}$

**19.** $\dfrac{7d^2 + 26d - 8}{d^2 - 5d}$  **20.** $\dfrac{3r^2 - 7r - 20}{r^2 + 4r}$

**24.** $\dfrac{2m^2 - 9m - 5}{m - 6}$  **25.** $\dfrac{-2a^2 - 3a + 2}{a + 4}$

**26.** $\dfrac{2r^2 + 9r - 18}{3r - 4}$  **27.** $\dfrac{a^2 - b^2}{x^2 + 2xy - 3y^2}$

**28.** $\dfrac{6m^2 + mn - n^2}{3m^2 + 8mn + 4n^2}$  **29.** $\dfrac{20r^2 - 9rt + t^2}{r^2 + 5rt + 6t^2}$

**30.** $\dfrac{x^2 - y^2}{x + 7y}$ **31.** $\dfrac{a^2 - 9b^2}{x^2 - 4y^2}$ **32.** $\dfrac{9x^2 + 12xy + 4y^2}{m^2 + 2mn + n^2}$

**33.** $\dfrac{4a^2 + 4ab + b^2}{9a^2 + 6ab + b^2}$ **34.** $\dfrac{a^2 - 25b}{3a - 4b}$

**35.** $\dfrac{-4a^2 - 7ab + 2b^2}{2a^2 - 5ab - 3b^2}$

## Page 211

**5.** $\dfrac{a - 4}{2a - 3}$ **6.** $\dfrac{m - 9}{2m - 1}$ **7.** $\dfrac{b + 6}{b - 2}$ **8.** $-\dfrac{x - 3}{x + 4}$

**13.** $-\dfrac{x - 1}{2}$ **14.** $\dfrac{5}{b + 4}$ **15.** $\dfrac{2a + 3}{4a + 3}$ **16.** $\dfrac{a - 2}{3}$

**17.** $\dfrac{m - 8}{m}$ **18.** $\dfrac{y + 2}{y}$ **19.** $\dfrac{3(x - 4)}{2(x - 2)}$

**20.** $\dfrac{2(x + 5)}{3}$ **21.** $\dfrac{2n - 1}{n - 2}$ **22.** $\dfrac{3a - 7}{2a - 1}$ **23.** $\dfrac{b - 7}{3}$

**24.** $\dfrac{3k - 2}{k}$ **25.** $\dfrac{2k - 5}{3k - 7}$ **26.** $\dfrac{5a - 1}{3a - 1}$ **27.** $\dfrac{x - 4}{x - 1}$

**28.** $\dfrac{3y + 2}{y - 6}$ **29.** $\dfrac{2a - 5}{a - 6}$ **30.** $\dfrac{b - 6}{3b - 1}$ **31.** $\dfrac{4m - 3}{2m + 1}$

**32.** $\dfrac{k + 4}{2k + 1}$ **33.** $\dfrac{3x + 1}{x + 8}$ **34.** $\dfrac{(x + 2)(y + 1)}{x - 8}$

**35.** $\dfrac{y - r}{b + 5c}$ **36.** $\dfrac{(x^2 - 9)(a + 1)}{(x - a)}$

**37.** $\dfrac{(x - 3)(x + 2)}{x}$ **38.** $(3p - 1)(p + 4)$

**39.** $\dfrac{y^2(5y + 2)(y - 1)}{3}$ **40.** $\dfrac{(x - 5y)(x + 3y)}{7}$

**41.** $\dfrac{1}{3(b + 4)(b - 1)}$ **42.** $\dfrac{2}{a^4(a + 4)(a + 2)}$

## Page 220

**17.** $-4(a + 2)$ **31.** $\dfrac{b + 5y}{5b - y}$ **32.** $\dfrac{5c - d}{2(c - 5d)}$

**33.** $-\dfrac{(a - 1)(a + 1)}{(a - 3)(2a - 3)}$ **34.** $\dfrac{5(2b + 3)}{2b(b + 9)}$ **36.** $1$

**37.** $\dfrac{2a + 7}{(a - 2)(a + 3)}$ **38.** $\dfrac{x + 7}{s - r}$

**39.** $-(x + y)(a + 5)$

## Page 224

**26.** $-\dfrac{x(4x + 5)(2x + 5)}{4(x - 5)(x + 1)}$ **27.** $\dfrac{-(x - 4)(x - 2)}{x(x - 6)}$

**28.** $\dfrac{y(y - 2d)(y - 5d)(y + d)}{(y + 3d)(y - d)(y + 5d)}$ **29.** $-3$ **30.** $-\dfrac{1}{2}$

**31.** $-\dfrac{1}{b}$ **32.** $-\dfrac{x - 8}{2}$ **33.** $-3x(x - 5)$

**34.** $\dfrac{2c - b}{a - d}$ **35.** $\dfrac{b - 2c}{a + 4d}$ **36.** $\dfrac{x - 4y}{a + 4d}$

## Page 227

**1.** $\dfrac{7}{4} = \dfrac{a}{3}$; means 4, $a$; ext 7, 3; $7 \cdot 3 = 4 \cdot a$; $\dfrac{21}{4}$

**2.** $\dfrac{2}{5} = \dfrac{x}{4}$; means 5, $x$; ext 2, 4; $2 \cdot 4 = 5 \cdot x$; $8 = 5x$;

$\dfrac{8}{5}$ **3.** $\dfrac{3}{7} = \dfrac{m}{5}$; means 7, $m$; ext 3, 5; $3 \cdot 5 = 7 \cdot m$;

$15 = 7m$; $\dfrac{15}{7}$ **4.** $\dfrac{5}{b} = \dfrac{3}{11}$; means 3, $b$; ext 5, 11;

$5 \cdot 11 = 3 \cdot b$; $55 = 3b$; $\dfrac{55}{3}$ **5.** $\dfrac{x - 5}{7} = \dfrac{3}{5}$;

means 7, 3; ext $x - 5$, 5; $5(x - 5) = 7 \cdot 3$;

$5x - 25 = 21$; $5x = 46$; $\dfrac{46}{5}$ **6.** $\dfrac{2}{x} = \dfrac{3}{x + 6}$;

means $x$, 3; ext 2, $x + 6$; $2(x + 6) = 3x$; $2x + 12 = 3x$;

$12 = x$; 12 **7.** $\dfrac{m - 2}{4} = \dfrac{m + 3}{5}$; means 4, $m + 3$;

ext $m - 2$, 5; $5(m - 2) = 4(m + 3)$;

$5m - 10 = 4m + 12$; $m - 10 = 12$; 22

**8.** $\dfrac{x - 1}{4} = \dfrac{3 + x}{2}$; means 4, $3 + x$; ext $x - 1$, 2;

$2(x - 1) = 4(3 + x)$; $2x - 2 = 12 + 4x$;

$-2 = 12 + 2x$; $-14 = 2x$; $-7$

**9.** $\dfrac{a + 3}{5} = \dfrac{14}{10}$; means 5, 14; ext $a + 3$, 10;

$10(a + 3) = 5 \cdot 14$; $10a + 30 = 70$; $10a = 40$; 4

**10.** $\dfrac{x - 1}{5} = \dfrac{2x + 7}{3}$; means 5, $2x + 7$; ext $x - 1$, 3;

$3(x - 1) = 5(2x + 7)$; $3x - 3 = 10x + 35$;

$-3 = 7x + 35$; $-38 = 7x$; $-\dfrac{38}{7}$

**11.** $\dfrac{5}{2m + 5} = \dfrac{2}{4m - 1}$; means $2m + 5$, 2; ext 5,

$4m - 1$; $5(4m - 1) = 2(2m + 5)$; $20m - 5 = 4m + 10$;

$16m - 5 = 10$; $16m = 15$; $\dfrac{15}{16}$ **12.** $\dfrac{x - 6}{2} = \dfrac{x + 4}{3}$;

means 2, $x + 4$; ext $x - 6$, 3; $3(x - 6) = 2(x + 4)$;

$3x - 18 = 2x + 8$; $x - 18 = 8$; 26

## Page 229

**4.**
```
10   FOR I = 1 TO 5
20   INPUT "TYPE IN A VALUE FOR X.";X
30   LET A = 2 * X - 3
40   LET B = X - 7
50   LET C = X + 5
60   LET D = X + 4
70   IF B = 0 OR D = 0 THEN 110
80   LET Q = (A / B) * (C / D)
90   PRINT "THE VALUE OF Q IS "Q"."
100  GOTO 120
110  PRINT "THERE IS NO DIVISION BY 0."
120  LET N = 2 * X ^ 2 + 7 * X - 15
130  LET F = X ^ 2 - 3 * X - 28
140  IF F = 0 THEN 180
150  LET V = N / F
160  PRINT "THE VALUE OF V IS "V"."
170  GOTO 190
180  PRINT "THERE IS NO DIVISION BY 0."
190  NEXT I
200  END
```

ANNOTATED ANSWERS

## Page 230

**19.** $\frac{x}{4} = \frac{3}{2}$; means 4, 3; ext $x$, 2; $2x = 12$; 6

**20.** $\frac{a + 2}{4} = \frac{a - 5}{3}$; means 4, $a - 5$; ext $a + 2$, 3;
$3(a + 2) = 4(a - 5)$; $3a + 6 = 4a - 20$; $6 = a - 20$;
$26 = a$   **21.** $\frac{x + 5}{4} = \frac{-1}{x}$; means 4, $-1$; ext $x + 5$, $x$;
$x(x + 5) = 4(-1)$; $x^2 + 5x = -4$; $x^2 + 5x + 4 = 0$;
$(x + 4)(x + 1) = 0$; $x + 4 = 0$ or $x + 1 = 0$; $-4$, $-1$

**22.** $\frac{a - 7}{4} = \frac{-9}{a + 6}$; means 4, $-9$; ext $a - 7$, $a + 6$;
$(a - 7)(a + 6) = 4 \cdot -9$; $a^2 - a - 42 = -36$;
$a^2 - a - 6 = 0$; $(a - 3)(a + 2) = 0$; $a - 3 = 0$ or
$a + 2 = 0$; 3, $-2$

## Page 259

**5.**
```
10   PRINT "PLEASE TYPE IN THE DENOMINATORS OF TH
     E THREE NUMBERS "
20   PRINT "FOR WHICH YOU WANT TO FIND THE LCD. T
     YPE IN THE "
30   PRINT "SMALLEST NUMBER FIRST, A COMMA, THEN
     THE SECOND"
40   PRINT "NUMBER, THEN THE THIRD NUMBER."
50   INPUT A,B,C
60   IF A > B THEN 10
70   REM  LINE 60 CHECKS IF A IS LESS THAN OR EQU
     AL TO B.
80   FOR X = 1 TO B * C
90   LET M = X * A
100  IF  INT (M / B) < > M / B THEN 120
110  IF  INT (M / C) = M / C THEN 130
120  NEXT X
130  PRINT M" IS THE LCD OF "A", "B", "C"."
140  END
```

**6.**
```
10   PRINT "THIS PROGRAM FINDS THE GREATEST"
20   PRINT "COMMON FACTOR OF TWO NUMBERS."
30   PRINT "ENTER TWO NUMBERS."
40   INPUT N1,N2
50   REM  THE LOOP FINDS FACTORS COMMON TO EACH N
     UMBER.
60   FOR N = 1 TO N1
70   LET A =  INT (N1 / N)
80   LET B = N1 - N * A
90   LET C =  INT (N2 / N)
100  LET D = N2 - N * C
110  IF B < > 0 THEN 140
120  IF D < > 0 THEN 140
130  LET F = N
140  NEXT N
150  PRINT "GCF = "F"."
160  END
```

## Page 290

**4.**
```
10   PRINT "DO YOU WANT TO SOLVE FOR K "
20   PRINT "IN THE FORMULA K = .5 * F * D " 2?"
30   PRINT "YES = 1, NO = 2"
40   INPUT A
50   IF A = 1 THEN 140
60   PRINT "DO YOU WANT TO SOLVE FOR F?"
70   INPUT A
80   IF A = 1 THEN 190
90   PRINT "TO SOLVE FOR D, TYPE IN VALUES FOR K,
     F."
100  INPUT K,F
110  LET D =  SQR (K / (.5 * F))
120  PRINT "D = "D", THE FORMULA IS D = SQR(K /
     .5 * F)."
130  GOTO 230
140  PRINT "TYPE IN VALUES FOR F, D."
150  INPUT F,D
160  LET K = .5 * F * D ^ 2
170  PRINT "K = "K", THE FORMULA IS K = .5 * F
     * D ^ 2."
180  GOTO 230
190  PRINT "TYPE IN VALUES FOR K, D."
200  INPUT K,D
210  LET F = K / (.5 * D ^ 2)
220  PRINT "F = "F", THE FORMULA IS F = K / (.5
     * D ^ 2)."
230  END
```

## Page 298

**1–23.**

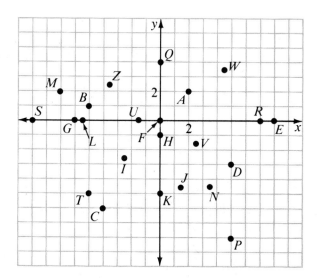

# Page 313

**10.**

**11.**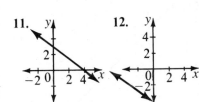

**12.**

**13.**

**14.**

**15.**

**16.**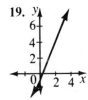

**17.**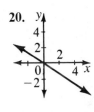

**18.**

**19.**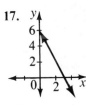

**20.**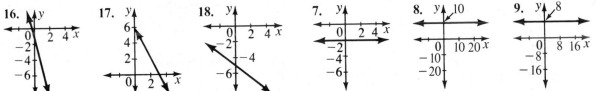

**21.**

**22.**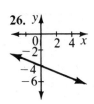

**24.**

**25.**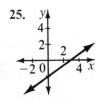

**26.**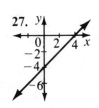

**27.**

**28.**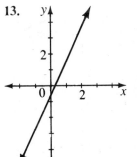

# Page 316

**1.**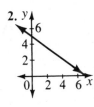

**2.**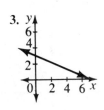

**3.**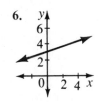

**4.**

**5.**

**6.**

**7.**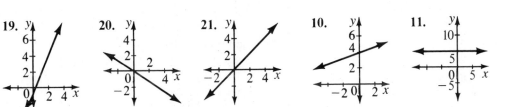

**8.**

**9.**

**10.**

**11.**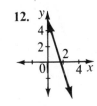

**12.**

**13.**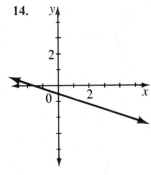

**14.**

ANNOTATED ANSWERS

**15.**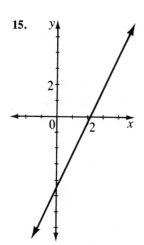

**16.**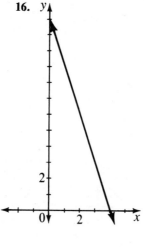

**23.**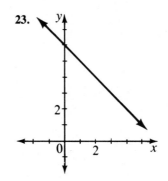

## Page 318

**1.–18.** are vertical lines, with the following x-intercepts:
**1.** −3  **2.** −1  **3.** 3  **4.** −6  **5.** −5  **6.** 6
**7.** 0  **8.** −7  **9.** −2  **10.** 9  **11.** 4  **12.** −4
**13.** 3  **14.** −6  **15.** 3  **16.** −9  **17.** 1  **18.** 2
**19.** vertical lines through $(4, 0)$ and $(-4, 0)$
**20.** vertical lines through $(-1, 0)$ and $(5, 0)$
**21.** vertical lines through $(-4, 0)$ and $(16, 0)$

## Pages 321–322

**1.**

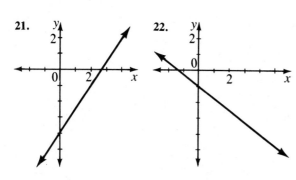

**2.**

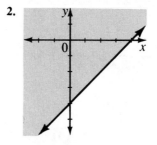

**3.**

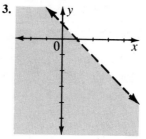

**17.**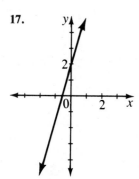

**18.**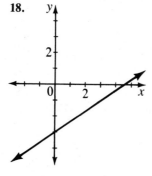

**19.** **20.**

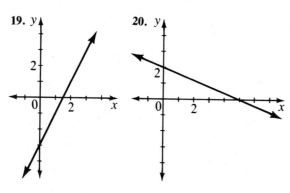

**21.** **22.**

ANNOTATED ANSWERS

**4.**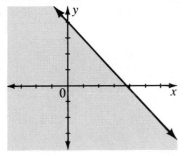

**5.** 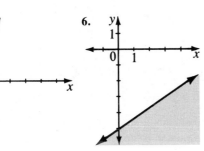 **6.**

**7.**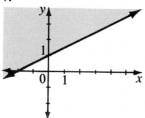

**8.**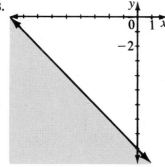

**9.** 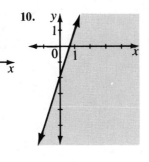 **10.**

**11.**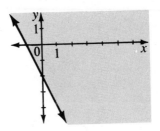

**12.**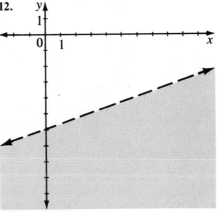

**13.** 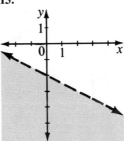 **14.**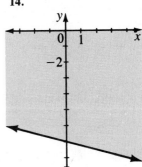

**15.** **16.**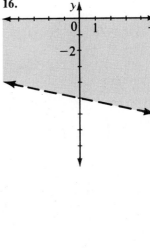

**16.**

**17.**
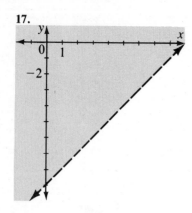

**18.**
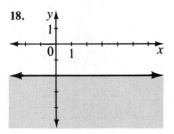

**19.** **20.**
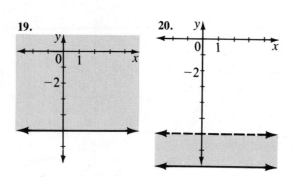

**21.** **22.**
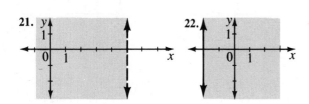

**23.** **24.**
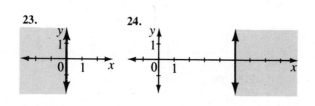

**25.** **26.**
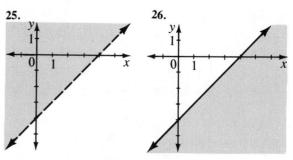

**27.**
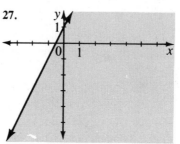

**28.**
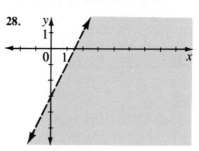

**45.** **46.**
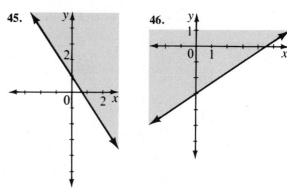

**47.** **48.**
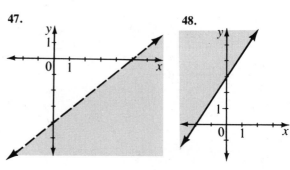

ANNOTATED ANSWERS

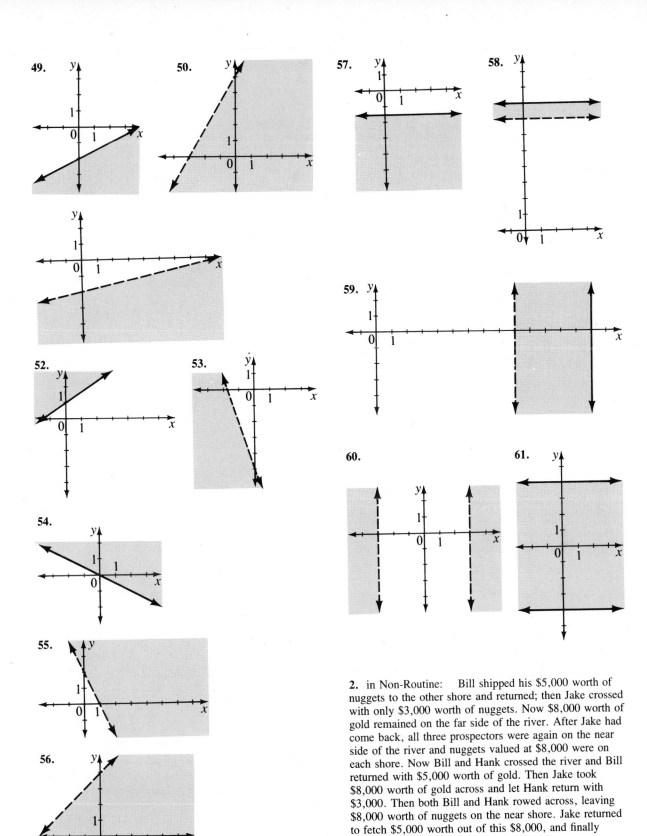

**2.** in Non-Routine: Bill shipped his $5,000 worth of nuggets to the other shore and returned; then Jake crossed with only $3,000 worth of nuggets. Now $8,000 worth of gold remained on the far side of the river. After Jake had come back, all three prospectors were again on the near side of the river and nuggets valued at $8,000 were on each shore. Now Bill and Hank crossed the river and Bill returned with $5,000 worth of gold. Then Jake took $8,000 worth of gold across and let Hank return with $3,000. Then both Bill and Hank rowed across, leaving $8,000 worth of nuggets on the near shore. Jake returned to fetch $5,000 worth out of this $8,000, and finally Hank came back for the remaining $3,000.

# Page 326

**10–13.**

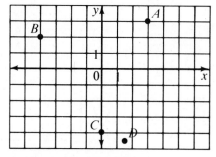

**30.** line through $(0, 6)$ and $(3, 4)$
**31.** line through $(0, 0)$ and $(1, 4)$
**32.** line through $(0, -5)$ and $(4, -4)$
**33.** line through $(0, -3)$ and $(4, 0)$
**34.** line through $(0, 11)$ and $(4, 5)$

**35–36.**    **38.**

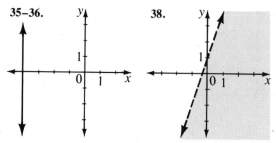

**39.**    **40.**

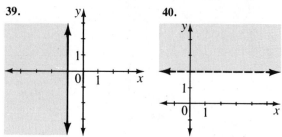

**41.**    **42.**

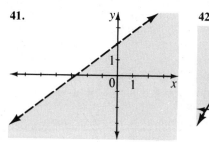 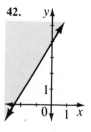

# Page 327

**4–6.**    **22.**

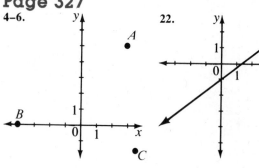

**23.** line through $(0, -3)$ and $(3, 1)$
**24.** horizontal line through $(0, 1)$
**25.** vertical line through $(-3, 0)$
**26.** line through $(0, 11)$ and $(3, 8)$

**30.**    **31.**

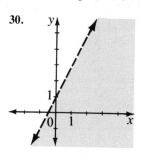 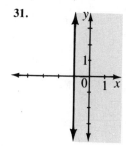

**32.**    **33.**

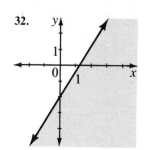 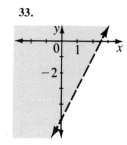

**35.** line through $(0, -2)$ and $(-3, -6)$

# Page 332

**1.–8.** are intersecting lines; the point of intersection and one other point on each line are given.   **1.** $(2, -3)$, $(0, 5)$, $(3, 0)$   **2.** $(-2, 4)$, $(0, 6)$, $(1, -2)$   **3.** $(2, 1)$, $(0, 7)$, $(0, -3)$   **4.** $(2, 5)$, $(0, -3)$, $(0, 9)$   **5.** $(2, 1)$, $(0, -2)$, $(3, -1)$   **6.** $(2, 0)$, $(0, -3)$, $(4, 2)$
**7.** $(6, 0), (0, 2), (3, -1)$   **8.** $\left(4\frac{1}{2}, -\frac{1}{2}\right), (0, 4), (5, 0)$
**9.** line through $(0, 4)$ and $(2, 1)$; line through $(-9, 0)$ and $(-5, 1)$   **10.** both lines are through $(6, 0)$ and $(0, 6)$   **11.** line through $(0, 4)$ and $(-4, 0)$ and parallel line through $(0, 5)$ and $(-5, 0)$   **12.** line through $(0, -8)$ and $(6, -5)$; line through $(0, 4)$ and $(4, 0)$

ANNOTATED ANSWERS

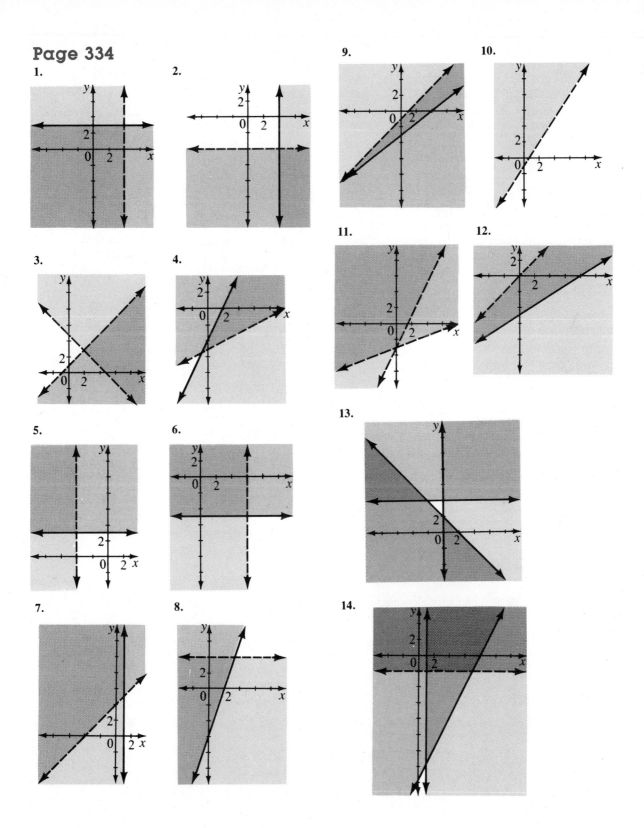

1.

2.

3.

4.

5.

6.

7.

8.

9.

10.

11.

12.

13.

14.

**15.**

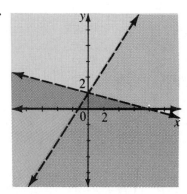

## Page 337
**2.** line through $(0, -2)$ and $(4, 1)$

## Page 346
**2.** line through $(0, -6)$ and $(4, 0)$

## Page 354
**1.** line through $(0, -2)$ and $(4, 1)$

## Page 358
**1.** lines through $(0, 0)$ and $(6, 6)$ intersecting at $(4, 8)$
**2.** lines through $(1, 3)$ and $(1, 0)$ intersecting at $(0, 1)$
**3.** lines through $(0, 1)$ and $(0, 0)$ intersecting at $(4, 4)$
**4.** line through $(0, -2)$ and $(2, 1)$; line through $(.5, .5)$ and $(0, -.25)$; inconsistent  **5.** both lines through $(0, -5)$ and $(2.5, 0)$; dependent and consistent  **6.** line through $(4, 0)$, $(0, -4)$; line through $(0, 6)$, $(3, 3)$; independent and consistent

**7.**

**8.**

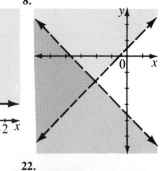

**9.**

**22.**

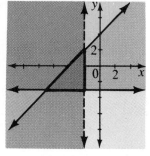

**5.**

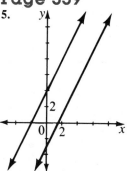

inconsistent

**6.**

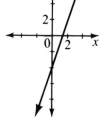

dependent
consistent

**9.**

**10.**

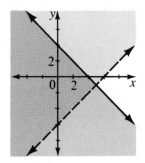

## Page 360
**1.** For $y = 2x - 1$  For $y = 6x - 1$

| X | Y | X | Y |
|---|---|---|---|
| 0 | −1 | 0 | −1 |
| .5 | 0 | .5 | 2 |
| 1 | 1 | 1 | 5 |
| 1.5 | 2 | 1.5 | 8 |
| 2 | 3 | 2 | 11 |
| 2.5 | 4 | 2.5 | 14 |
| 3 | 5 | 3 | 17 |
| 3.5 | 6 | 3.5 | 20 |
| 4 | 7 | 4 | 23 |
| 4.5 | 8 | 4.5 | 26 |
| 5 | 9 | 5 | 29 |

**2.** For $y = 2x - 1$  For $y = 6x - 1$

| X | Y | X | Y |
|---|---|---|---|
| −5 | −11 | −5 | −31 |
| −4.5 | −10 | −4.5 | −28 |
| −4 | −9 | −4 | −25 |
| −3.5 | −8 | −3.5 | −22 |
| −3 | −7 | −3 | −19 |
| −2.5 | −6 | −2.5 | −16 |
| −2 | −5 | −2 | −13 |
| −1.5 | −4 | −1.5 | −10 |
| −1 | −3 | −1 | −7 |
| −.5 | −2 | −.5 | −4 |
| 0 | −1 | 0 | −1 |

ANNOTATED ANSWERS

**3.** The printout is:

## Page 362

**32.** line through $(0, -4)$ and $\left(2\frac{2}{3}, 0\right)$

**33.** vertical line through $(6, 0)$

**34.** vertical line through $(0, -2)$

**35.**

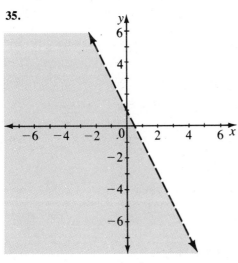

**44.**

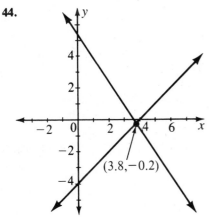

**45.**

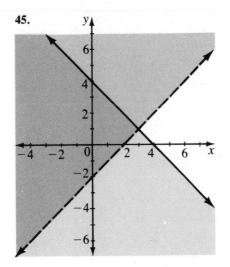

**55.**

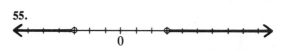

**56.**

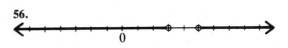

**57.**

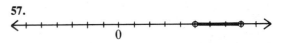

## Page 367

**1.** $\{(-3, -3), (-1, -3), (2, 3), (3, 0), (4, -3)\}$; $D = \{-3, -1, 2, 3, 4\}$; $R = \{-3, 0, 3\}$; function

**2.** $\{(-3, 1), (-1, -3), (-1, 3), (3, -1), (4, 2)\}$; $D = \{-3, -1, 3, 4\}$; $R = \{-3, -1, 1, 2, 3\}$; not a function

**3.** $\{(-3, 2), (-2, 1), (-1, 0), (1, -2), (2, -3)\}$; $D = \{-3, -2, -1, 1, 2\}$; $R = \{-3, -2, 0, 1, 2\}$; function

## Page 370

**5.–13.** are lines; two points on each are given.

**5.** $(0, 0), (2, 4)$ **6.** $(0, 0), (2, 2)$ **7.** $(0, -4), (4, 0)$

**8.** $(0, -5), (3, -3)$ **9.** $(0, 4), (2, 4)$ **10.** $(-4, 0), (-4, 3)$ **11.** $(0, -1), (2, -1)$ **12.** $(0, 1), (2, 1)$

**13.** $(0, -4), (-3, -2)$

**14.** vertical line through $(6, 0)$

**15.** horizontal line through $(0, -4)$

**16.** vertical line through $(7, 0)$

**17.** line through $(0, 4)$ and $(4, 0)$

**18.** horizontal line through $(0, 4)$

**19.** vertical line through $(-2, 0)$

**20.** horizontal line through $(0, 4)$

ANNOTATED ANSWERS

21. line through $(0, -2)$ and $(2, 1)$

22. line through $(0, 2)$ and $(-2\frac{2}{3}, 0)$

23. line through $(0, -4)$ and $(-4, 0)$

24. vertical line through $(-2, 0)$

25. horizontal line through $(0, 10)$

26. vertical line through $(2\frac{1}{9}, 0)$

27. line through $(0, 4)$ and $(4, 0)$

28. line through $(0, 0)$ and $(-4, -1)$

29. line through $(0, 0)$ and $(1, 2)$

30. vertical line through $(7, 0)$

## Page 384

**1.** $(-2, -1), (-1, 1), (0, 2), (1, -1), (2, 3)$;
$D = \{-2, -1, 0, 1, 2\}$; $R = \{-1, 1, 2, 3\}$; function
**2.** $(-3, 2), (-2, 1), (-1, 0), (-1, -2), (0, -1)$;
$D = \{-3, -2, -1, 0\}$; $R = \{-2, -1, 0, 1, 2\}$; not a
function   **3.** not a function because $(-2, -2)$ and
$(-2, 3)$ have the same first coordinate   **6.–11.** are lines;
two points on each are given.   **6.** $(0, 0), (1, 3)$
**7.** $(7, 0), (7, 3)$   **8.** $(0, 7), (2, 7)$   **9.** $(0, 6), (6, 0)$
**10.** $(0, 3), (2, 3)$   **11.** $(4, 0), (2, -4)$

## Page 385

**1.–5.** are lines; two points on each are given.
**1.** $(0, 5), (2, 5)$   **2.** $(22, 0), (22, 4)$   **3.** $(12, 0), (0, 12)$
**4.** $(2, 0), (2, 2)$   **5.** $(0, 11), (3, 11)$   **9.** $(-2, 0)$,
$(-1, 1), (-1, -1), (1, 2), (1, -2)$; $D = \{-2, -1, 1\}$;
$R = \{-2, -1, 0, 1, 2\}$   **10.** $(-2, 2), (-1, 1), (0, -2)$,
$(1, 0), (2, 2)$; $D = \{-2, -1, 0, 1, 2\}$ $R = \{-2, 0, 1, 2\}$
**11.** a function, since no two pairs have the same first
coordinate

## Page 386

Program for Ex. **7.-9.**
```
10  REM  THIS PROGRAM GIVES VALUES FOR THE DOMAI
    N AND RANGE.
20  REM  VALUES FOR A, B, AND C ARE INPUT.
30  FOR X = 1 TO 3
40  PRINT "ENTER THE COEFFICIENTS A, B, C."
45  INPUT A,B,C
50  LET F = (A * X ^ 2) + (B * X) + C
60  PRINT X,F
70  NEXT X
80  PRINT "FIRST VALUES ARE VALUES FOR THE DOMAI
    N."
90  PRINT "SECOND VALUES ARE VALUES FOR THE RANG
    E."
100 END
```
**10.**
```
10  PRINT "THIS PROGRAM DETERMINES WHETHER"
20  PRINT "OR NOT A FUNCTION VARIES DIRECTLY."
30  DIM R(10)
```

```
40  PRINT "ENTER THE NUMBER OF SETS OF VALUES YO
    U HAVE TO INPUT."
50  INPUT N
60  FOR I = 1 TO N
70  PRINT "ENTER DOMAIN, RANGE."
80  INPUT X,Y
90  LET R(I) = Y / X
100 NEXT I
110 FOR I = 1 TO N - 1
120 IF R(I) < > R(I + 1) THEN 160
130 NEXT I
140 PRINT "THE FUNCTION VARIES DIRECTLY."
150 GOTO 170
160 PRINT "THE FUNCTION DOES NOT VARY DIRECTLY.
    "
170 END
```

## Page 415

**37.** $(a + b)^2 = 4\left(\frac{1}{2} ab\right) + c^2$; $a^2 + 2ab + b^2 =$
$2ab + c^2$; $a^2 + b^2 = c^2$

## Page 421

**7.**
```
10  PRINT "THIS PROGRAM FINDS THE LENGTH"
20  PRINT "OF A SIDE OF A RIGHT TRIANGLE."
30  PRINT "ENTER HYPOTENUSE, SIDE."
40  INPUT C,B
50  LET A = SQR (C ^ 2 - B ^ 2)
60  PRINT "SIDE A = "A","
70  PRINT "DO YOU HAVE MORE VALUES ?"
80  PRINT "ENTER 1 FOR YES, 2 FOR NO."
90  INPUT A
100 IF A = 1 THEN 30
110 END
```
**8.**
```
10  PRINT "THIS PROGRAM DETERMINES WHETHER OR NO
    T"
20  PRINT "A TRIANGLE IS A RIGHT TRIANGLE."
30  PRINT "ENTER THE 3 SIDES OF THE TRIANGLE."
40  INPUT X,Y,Z
50  IF X * X = Y * Y + Z * Z THEN 100
60  IF Z * Z = X * X + Y * Y THEN 100
70  IF Y * Y = X * X + Z * Z THEN 100
80  PRINT "THE TRIANGLE IS NOT A RIGHT TRIANGLE.
    "
90  GOTO 110
100 PRINT " THE TRIANGLE IS A RIGHT TRIANGLE."
110 END
```

## Page 448

**7.–12.** are parabolas; the turning point and two other
points on each are given.   **7.** $(2, 2), (0, 6), (4, 6)$

**8.** $(2, 9), (0, 1), (4, 1)$   **9.** $(2, -10), (1, -6), (3, -6)$
**10.** $(1, 8), (0, 5), (2, 5)$   **11.** $(2, -14), (0, -6), (4, -6)$
**12.** $(-1, -7), (1, -11), (-3, -11)$

## Page 450

**33.** parabola through $(3, 10), (0, 1), (6, 1)$
**34.** parabola through $(2, -12), (0, 0), (4, 0)$

## Page 451

**20.** parabola through $(-3, 10), (-6, 1), (0, 1)$

## Page 452

**1.**
```
10   PRINT "THIS PROGRAM SOLVES QUADRATIC EQUATIO
     NS."
20   PRINT "ENTER COEFFICIENTS A, B, C."
30   INPUT A,B,C
40   LET S1 = B ^ 2 - (4 * A * C)
50   LET S2 = SQR (S1)
60   LET X1 = ( - B + S2) / (2 * A)
70   LET X2 = ( - B - S2) / (2 * A)
80   PRINT "THE SOLUTIONS ARE "X1" AND "X2
90   PRINT "FOR THE EQUATION "A" X   2 + "B" X +
     "C" = 0."
100  PRINT "DO YOU HAVE MORE COEFFICIENTS?"
110  PRINT "ENTER 1 FOR YES, 0 FOR NO."
120  INPUT A
130  IF A = 1 THEN 20
140  END
```

**4.**
```
10   PRINT "THIS PROGRAM SOLVES QUADRATIC EQUATIO
     NS AND PRINTS THE NUMBER OF SOLUTIONS."
20   PRINT "ENTER THE COEFFICIENTS A, B, C."
30   INPUT A,B,C
40   LET S1 = B ^ 2 - (4 * A * C)
50   IF S1 = 0 THEN 120
60   IF S1 < 0 THEN 150
70   LET S2 = SQR (S1)
80   LET X1 = ( - B + S2) / (2 * A)
90   LET X2 = ( - B - S2) / (2 * A)
100  PRINT "THE SOLUTIONS ARE "X1" AND "X2"."
110  GOTO 160
120  LET X1 = - B / (2 * A)
130  PRINT "THE SOLUTION IS "X1"."
140  GOTO 160
150  PRINT "THERE ARE NO REAL SOLUTIONS."
160  PRINT "DO YOU HAVE MORE COEFFICIENTS?"
170  PRINT "ENTER 1 FOR YES, 0 FOR NO."
180  INPUT A
190  IF A = 1 THEN 20
200  END
```

**5.**
```
10   PRINT "THIS PROGRAM PRINTS QUADRATIC"
20   PRINT "EQUATIONS IN STANDARD FORM."
30   PRINT "ENTER COEFFICIENTS A, B, C."
40   INPUT A,B,C
50   PRINT A" X ^ 2 + "B" X + "C" = 0"
60   PRINT "DO YOU HAVE MORE COEFFICIENTS?"
70   PRINT "ENTER 1 FOR YES, 0 FOR NO."
80   INPUT A
90   IF A = 1 THEN 30
100  END
```

## Page 477

**1.**
```
100  PRINT "THIS PROGRAM COMPUTES TRIGONOMETRIC
     RATIOS"
110  PRINT "FOR RIGHT TRIANGLE DEF."
120  PRINT "D E F  TAN D    SIN D    COS D"
130  READ D,E,F
140  LET T = D / E
150  LET S = D / F
160  LET C = E / F
180  PRINT D TAB( 4)E TAB( 7)F TAB( 10)T TAB( 20
     )S TAB( 30)C
```

**2.**
```
100  PRINT "THIS PROGRAM COMPUTES TRIGONOMETRIC"
110  PRINT "RATIOS FOR RIGHT TRIANGLES ABC."
120  DIM A(20),B(20),C(20),T(20),S(20),I(20)
130  PRINT "ENTER THE NUMBER OF TRIANGLES YOU HA
     VE."
140  INPUT N
150  FOR K = 1 TO N
160  PRINT "ENTER THE SIDE, SIDE, HYPOTENUSE OF
     THE RIGHT TRIANGLE."
170  INPUT A(K),B(K),C(K)
180  LET T(K) = A(K) / B(K)
190  LET S(K) = A(K) / C(K)
200  LET I(K) = B(K) / C(K)
210  NEXT K
220  PRINT "A B C TAN A    SIN A    COS A"
230  FOR K = 1 TO N
240  PRINT A(K) TAB( 4)B(K) TAB( 7)C(K) TAB( 10)
     T(K) TAB( 20)S(K) TAB( 30)I(K)
250  NEXT K
260  END
```

ANNOTATED ANSWERS

## Page 489

4.  ```
    10  INPUT "TYPE VALUE OF X ":X
    20  LET Y = 8 * X - 4
    30  PRINT "Y = "Y
    40  END
    ```

5.  ```
    10  INPUT "TYPE VALUE OF R ":R
    20  INPUT "WHAT IS S ? ":S
    30  LET Y = 2 * R - S / 4
    40  PRINT "Y = "Y
    50  END
    ```

6.  ```
    10  INPUT "WHAT IS THE LENGTH? ":L
    20  INPUT "WHAT IS THE WIDTH? ":W
    30  LET A = L * W
    40  PRINT " AREA IS "A
    50  END
    ```

7.  ```
    10  INPUT "WHAT IS THE LENGTH? ":L
    20  INPUT "WHAT IS THE WIDTH? ":W
    30  LET P = 2 * L + 2 * W
    40  PRINT "THE PERIMETER IS "P
    50  END
    ```

## Page 491

10.  ```
     10  INPUT "WHAT IS X? ":X
     20  LET Y = 4 * X ^ 3
     30  PRINT "Y = "Y"."
     40  END
     ```

11.  ```
     10  INPUT "ENTER THE VALUE OF X. ":X
     20  LET A = 5 * X ^ 2 - 8 * X + 2
     30  PRINT "A = "A"."
     40  END
     ```

12.  ```
     10  INPUT "WHAT IS VALUE OF A? ":A
     20  LET Q = (A ^ 2 - 7 * A + 12) / (A - 3)
     30  PRINT "THE VALUE OF Q IS "Q"."
     40  END
     ```

13.  ```
     10  INPUT "WHAT IS X? ":X
     20  INPUT "WHAT IS Y? ":Y
     30  LET R = X / (X - Y)
     40  PRINT "R = "R"."
     50  END
     ```

## Page 494

3.  ```
    10  FOR K = 1 TO 3
    20  INPUT "WHAT IS X? ":X
    30  LET Y = 3 * X ^ 2 - 5 * X - 7
    40  PRINT "Y IS "Y"."
    50  NEXT K
    60  END
    ```

4.  ```
    10  FOR K = 1 TO 2
    20  INPUT "WHAT IS R? ":R
    30  INPUT "WHAT IS S? ":S
    40  LET A = (R + S) / 2
    50  PRINT "A = "A
    60  NEXT K
    70  END
    ```

5.  ```
    10  FOR I = 1 TO 3
    20  INPUT "WHAT IS THE TIME T? ":T
    30  LET H = 128 * T - 16 * T ^ 2
    40  PRINT "THE HEIGHT OF THE ROCKET IS "H"."
    50  NEXT I
    60  END
    ```

6.  ```
    10  PRINT "NUMBER","SQUARE OF NUMBER"
    20  FOR I = 7 TO 56 STEP 7
    30  LET S = I ^ 2
    40  PRINT I,S
    50  NEXT I
    60  END
    ```

7.  ```
    10  PRINT "NUMBER","VALUE OF EXPRESSION"
    20  FOR X = 8 TO 42 STEP 2
    30  LET Y = X ^ 2 - 25
    40  PRINT X,Y
    50  NEXT X
    60  END
    ```

8.  ```
    10  PRINT "X","(4 * X ^ 2 + 2) / (2 * X)"
    20  FOR X = 10 TO 28 STEP 3
    30  LET Q = (4 * X ^ 2 + 2) / (2 * X)
    40  PRINT X,Q
    50  NEXT X
    60  END
    ```

9.  ```
    10  PRINT "NUMBER","SQUARE OF NUMBER"
    20  FOR I = 40 TO 26 STEP - 2
    30  LET S = I ^ 2
    40  PRINT I,S
    50  NEXT I
    60  END
    ```

## Page 496

1.  ```
    10  PRINT "X","Y"
    20  FOR I = 1 TO 3
    30  READ X
    40  LET Y = X ^ 2 - 7 * X + 4
    50  DATA 2,-5,7
    60  PRINT X,Y
    70  NEXT I
    80  END
    ```

2.
```
10  PRINT "X","Y"
20  FOR K = 1 TO 3
30  READ X
40  LET Y = - 4 * X ^ 3 - 8 * X
50  DATA -2,-5,1
60  PRINT X,Y
70  NEXT K
80  END
```

3.
```
10  PRINT "A","Y"
20  FOR I = 1 TO 4
30  READ A
40  LET Y = 5 - 4 * A ^ 2
50  DATA 8,-7,-1,3
60  PRINT A,Y
70  NEXT I
80  END
```

4.
```
10  PRINT "X","A"
20  FOR J = 1 TO 4
30  READ X
40  LET A = (2 * X - 5) ^ 2
50  DATA 1,-7,-4,0
60  PRINT X,A
70  NEXT J
80  END
```

5.
```
10  PRINT "X","Q"
20  FOR I = 1 TO 3
30  READ X
40  LET Q = (X ^ 2 - 5 * X + 4) / (X - 4)
50  DATA 9, -8, -2
60  PRINT X,Q
70  NEXT I
80  END
```

6.
```
10  PRINT "X","Q"
20  FOR K = 1 TO 3
30  READ X
40  LET Q = X / (X - 8)
50  DATA 6,12,-4
60  PRINT X,Q
70  NEXT K
80  END
```

7.
```
10  PRINT "A","B","M"
20  FOR I = 1 TO 3
30  READ A,B
40  LET M = - 5 * A + 9 * B
50  DATA 6,8,-5,-5,-2,-6
60  PRINT A,B,M
70  NEXT I
80  END
```

8.
```
10  PRINT "G1   G2   G3   G4   AVG"
20  FOR I = 1 TO 2
30  READ W,X,Y,Z
40  LET A = (W + X + Y + Z) / 4
50  DATA 90,80,100,90,100,80,80,75
60  PRINT W"    "X"    "Y"    "Z"    "A
70  NEXT I
80  END
```

9.
```
10  PRINT "IST NUM","2ND NUM","PROD"
20  FOR I = 1 TO 5
30  READ X,Y
40  LET P = X * Y
50  DATA  -8,9,-5,-4,-7,-5,-4,-2,178,987
60  PRINT X,Y,P
70  NEXT I
80  END
```

# Page 499

8.
```
10  FOR X = - 4 TO 4 STEP 1
20  IF X - 4 = 0 THEN 60
30  LET Q = 6 / (X - 4)
40  PRINT X,Q
50  GOTO 70
60  PRINT "YOU CANNOT DIVIDE BY ZERO."
70  NEXT X
80  END
```

# Page 501

2.
```
10  PRINT "NUMBER","RUNNING SUM"
20  FOR H = 1 TO 8
30  LET C = C + H
40  PRINT H,C
50  NEXT H
60  PRINT "THE SUM OF 1 TO 8 IS "C"."
70  END
```

3.
```
10  PRINT "NUMBER","RUNNING SUM"
20  FOR H = 2 TO 14 STEP 2
30  LET C = C + H
40  PRINT H,C
50  NEXT H
60  PRINT "THE SUM OF EVEN NUMBERS FROM 2 TO 14
       IS "C"."
70  END
```

4.
```
10  FOR H = 1 TO 5
20  INPUT "ENTER A NUMBER. ";X
30  LET Y = X * X
```

Annotated Answers

**T-557**

ANNOTATED ANSWERS

```
40   LET C = C + Y
50   NEXT H
60   PRINT "THE SUM OF THE SQUARES OF THE ENTERED
        FIVE NUMBERS IS "C"."
70   END
```

5.
```
10   FOR H = 1 TO 3
20   INPUT "ENTER TEST GRADE. ";X
30   LET C = C + X
40   NEXT H
50   LET A = C / 3
60   PRINT "THE AVERAGE IS "A"."
70   END
```

## Page 503

10.
```
10   LET C = 0
20   FOR I = 1 TO 400
30   LET R = INT (8 * RND (1) + 1)
40   IF R = 5 THEN 70
50   PRINT "SPIN "I,R" COMES UP"
60   GOTO 80
70   LET C = C + 1
80   NEXT I
90   PRINT "5 OCCURS "C" TIMES."
100  END
```

## Page 515

**37.** $(2x + 5)(x + 1)$    **38.** $(2a + 13)(a - 4)$
**43.** $2(a - 8)(a + 6)$    **44.** $2(b - 8)(b + 5)$
**45.** $3(k - 2)(k + 3)$    **48.** $2k(3k + 5)(k - 2)$
**49.** $(a + b)(a + 3b)$    **50.** $2(3w + 2y)(w - 5y)$
**51.** $(4a - 5y)(4a + 5y)$    **52.** $(3m + 5n)(7m - 2n)$
**53.** $2r(r - 7t)(r + 6t)$    **54.** $2a(2a - 5b)(a + 4b)$
**57.** $2p(p - 11q)(p + 11q)(p - q)(p + q)$
**58.** $(b + y)(y + 5)$    **59.** $(t + u)(w + y)$
**60.** $(a + 3)(b + c)$    **69.** $0, 5, -5, 3, -3$

## Page 519

**12.** $\dfrac{3x^2 - 4x + 4}{x(x - 5)(x - 2)}$    **16.** $\dfrac{11a + 30}{(a - 3)(a + 6)}$

**17.** $\dfrac{5b + 7}{(b - 4)(b + 5)}$    **18.** $\dfrac{x^2 - 7x + 16}{x - 2}$

**19.** $\dfrac{22x^2 + 15x + 4}{15x^3}$    **20.** $\dfrac{2x^2 - 8x + 9}{x - 4}$

**21.** $\dfrac{19a^2 + 9a + 3}{6a^3}$    **22.** $\dfrac{3x^2 - 13x - 3}{3x + 2}$

**23.** $\dfrac{9a^2 - 9a - 6}{(a - 2)(a - 1)(a + 1)}$    **24.** $\dfrac{5x + 33}{(x - 5)(x + 5)}$

**25.** $\dfrac{-2b + 33}{(3b - 2)(2b + 5)}$    **28.** $\dfrac{-2k - 1}{(k - 8)(k + 3)}$

**29.** $\dfrac{-k^2 - k + 7}{(k - 7)(k + 2)}$    **30.** $\dfrac{4y + 16 - xa - x - a^2 - a}{(x + a)(y - 3)(y + 4)}$

**31.** $\dfrac{-11y + 9b - 7x}{(x + 2y)(y - 3b)(y + 3b)}$    **35.** $\dfrac{m + 8}{(m - 8)(m - 4)}$

**38.** $\dfrac{y + 9}{y}$    **39.** $\dfrac{-6m + 4}{(m - 6)(m - 2)}$    **40.** $\dfrac{1}{x + 5}$

## Page 523

**12.** below and inc. line with slope 3, $y$-int $-1$ and above line with slope $\dfrac{1}{2}$, $y$-int $-2$

**13.** above and inc. line with slope $\dfrac{3}{2}$, $y$-int $-2$ and below line with slope 1, $y$-int $-1$

**14.** above line with slope $\dfrac{2}{3}$, $y$-int $-2$ and above line with slope $\dfrac{1}{2}$, $y$-int $-4$

**15.** above line with slope 2, $y$-int $-2$ and above line with slope $\dfrac{1}{2}$, $y$-int $-4$

**16.** below line with slope 1 through $(0, 0)$ and above and inc. line with slope $\dfrac{3}{5}$, $y$-int $-2$

**17.** below and inc. line with slope $-1$, $y$-int 4, right of and inc. line $x = 1$, above and inc. $y = 2$
**18.** above and inc. line with slope 3, $y$-int $-9$, right of and inc. line $x = 2$, above line $y = -3$

# Index

**Index**

**Index**

Opposite numbers, 38
Order of numbers, 29
Order of operations, 1–3, 23 (computer)
Ordered pair, 293–294
Ordinate, 294
Origin
    on a number line, 28
    on a plane, 293

**Parabola, 445–448**
Parallel lines, 304, 325
Parentheses, 2
    algebraic expressions with, 55–56
    equations with, 89–90
Percent, 103–105
Perfect square(s), 392
Perfect square trinomial, 175–176, 432–434
Perimeter, 18, 106–108, 339, 355
Perpendicular line, 293
Pi($\pi$), 33, 91, 164–165
Plane
    coordinates of points in, 293–294
    origin on, 293
    plotting points in, 296–297
    quadrants in, 293–294
    $x$-axis in, 293–294
    $y$-axis in, 293–294
Point(s)
    coordinates of, 28–29, 293–294
    corresponding to a number, 33, 28–29
    determining if on line, 302–306
    locating in a plane, 293–294
    plotting in a plane, 296–297
Polynomial(s)
    addition of, 155
    classifying, 152–153
    as denominators, 242–243
    determining degree of, 152–153
    division of by binomial, 255–256
    division of by monomial, 253–254
    factoring, 159–163, 177–187
    multiplication of, 157, 171–172
    products of, 171–173
    simplifying, 152–153, 157
    subtraction of, 155–156
Positive numbers, 28
Power(s)
    of a number, 16–17
    of a power, 150–151
    of a product, 150–151
    product of, 148–149
    of variables, 402–403
Prime factorization, 159–160
Prime number, 159–160

Problems
    age, 87, 93, 338, 382–383
    angle, 455–458
    area, 438–439
    coin, 347–348
    commission, 109
    compound interest, 168
    consecutive integer, 195–197, 202
    cost formulas, 20
    digit, 351–353
    direct variation, 373–375
    distance, 283–285
    fulcrum, 379–380
    inverse variation, 377–378
    investment and loan, 473–475
    mixed types, 110, 199, 287, 355, 444
    mixture, 348–349
    motion, 283–285
    Non-Routine Problems, x, 12, 79, 99, 130, 147, 170,
    174, 214, 250, 263, 278, 298, 322, 346, 364, 380,
    398, 431, 461
    number, 92–94
    perimeter, 106–107
    similar triangles, 459–460
    trigonometry, 465–466
    wet mixtures, 441–443
    work, 269–271
Problem solving
    formulating the problem, 85, 117, 235, 329, 454
    logo for, 63, 76–78, 85, 92–94, 104, 106, 117, 193,
    199, 235, 329, 425, 454
    steps in, 63, 76–78
    using systems of equations, 338–339
Product(s)
    of fractions, 4, 14, 96–98
    of means and extremes, 226–227
    of polynomials, 171–172
    of powers, 148–149
    powers of a, 150–151
    property of zero, 188
    of radicals, 406–407
    of rational expressions, 206
    of signed numbers, 41–42
    simplest form of, 218–219
    of square roots, 396–397
Property(ies)
    addition for equations, 64
    addition for inequalities, 124
    addition for rational expressions, 236–237
    additive identity, 37, 80
    additive inverse, 38
    of angles, 455–458
    associative of addition, 7–8, 37, 47, 80
    associative of multiplication, 7–8, 41, 80

Index